INTRODUCTION TO GENOMICS

INTRODUCTION TO GENOMICS

FOURTH EDITION

ARTHUR M. LESK
The Pennsylvania State University

Immured the whole of Life
Within a magic Prison
– Emily Dickinson

OXFORD
UNIVERSITY PRESS

Great Clarendon Street, Oxford, OX2 6DP,
United Kingdom

Oxford University Press is a department of the University of Oxford.
It furthers the University's objective of excellence in research, scholarship,
and education by publishing worldwide. Oxford is a registered trade mark of
Oxford University Press in the UK and in certain other countries

First edition 2007
Second edition 2012
Third edition 2017

Published in the United States of America by Oxford University Press
198 Madison Avenue, New York, NY 10016, United States of America

British Library Cataloguing in Publication Data
Data available

Library of Congress Control Number: 2024946526

ISBN 978–0–19–886689–3

Printed and bound by
CPI Group (UK) Ltd, Croydon, CR0 4YY

For Victor, Valerie, Abigail, Sarah, and Theodore

PREFACE

Of all the claims on our curiosity, we want most to understand ourselves. What are we? How have we come to be what we are? What lies in our future?

Many features of our lives depend on accidents of history. The time and place of our birth largely determine what language we first learn to speak and whether we are likely to be well-fed and well-educated and receive adequate medical care. Many aspects of our future depend on events outside ourselves and beyond our control. Yet, within us, there are constraints on our lives that brook relatively little argument.

In many respects, we are at the mercy of our genomes. Under normal circumstances, our basic anatomy and physiology, including eye colour, height, intelligence, and basic personality traits, are ingrained in our DNA sequences. This is not to say that our genomes dictate our lives. Some constraints are tight—for instance, eye colour—but our genetic endowment also confers on us a remarkable robustness. This robustness also is a product of evolution. Within the last century, lifestyles have changed with a rapidity hitherto unknown (except for the instants of asteroid impacts). We can meet and survive brutal stresses. Our talents have many opportunities to nurture themselves and to develop in novel ways. We are beginning to understand how our lives shape themselves; under the influence of our genes, epigenetic modifications, and our surroundings and life histories. This is the gift of our genomes: Genes control the response of an organism to its environment.

The human genome is only one of the many complete genome sequences known. Taken together, genome sequences from organisms distributed widely among the branches of the tree of life give us a sense, only hinted at before, of the very great unity in detail of all life on Earth. This recognition has changed our perceptions, much as the first pictures of the Earth from space engendered a unified view of our planet. Superimposed on this underlying unity is great variety. We ask: What is special about us? What do we share with our parents and siblings and how do we differ from them? What do we share with all other human beings and what makes us different from the other members of our species? What are the sources of our differences from our closest extant non-human relatives, the chimpanzees and bonobos? What do we have in common and how have we diverged from other species of primates? Of mammals? Of vertebrates? Of eukaryotes? Of all other living things? The complete sequences of human and other genomes reveal the underlying text of this story.

We are also beginning to be able to intervene. Genetic engineering of microorganisms is an established technique. Genetically modified plants and animals exist and are the subjects of lively debate. To override the genes for hair colour is trivial. Changes in lifestyle or behaviour can—to some extent—postpone or even avert the development of diseases to which we are genetically at risk. Gene therapy has cured some inborn defects. Novel gene editing techniques based on the CRISPR-Cas system offer revolutionary power in reshaping individuals, whole species, and ecosystems.

Fast, inexpensive sequencing has transformed genomics. It is very likely that within the lifetime of many readers, human genome sequencing will be nearly universal. Already, hundreds of thousands of human individuals have had their full genomes sequenced, and the pace is increasing. This explosive growth of the data produced—in quality, quantity, and type—has altered the landscape of genomics itself, and its influence has spread to surrounding fields. No area of biology has been left unscathed. Study of human disease through sequencing continues to be a major effort. Applications to improve human health were the major motivation for support of the Human Genome Project and continue to be the major elicitor of largesse for pursuing its consequences.

Understanding the relationships between genes and disease will allow more precise diagnosis and warnings of increased risk of disease in individuals and their offspring. For many important diseases, a patient can expect more precise diagnosis and prognosis, and more precise recommendations for treatment. Design of treatment based on a patient's genome, called pharmacogenomics, is already part of clinical practice.

Identification of genes responsible for particular diseases permits testing, genetic counselling, and risk assessment and avoidance. For example, mutations in *BRCA1, BRCA2,* and *PALB2* suggest an increased likelihood of developing breast or ovarian cancer. Many people have had these regions sequenced to provide information about their personal risk. Cancer genomics—the comparison of sequences from normal and tumour cells from single patients—has become a heavy industry.

Genomes of other organisms also have implications for human health, especially those of pathogenic organisms that have developed, or are threatening to develop, antibiotic resistance. We study the biology of viruses and bacteria to take advantage of their vulnerabilities, and the biology of humans to ward off the consequences of our own.

Other applications of genomics include improvement of crops and domesticated animals, enhancing food production, support of conservation efforts dedicated to preserving endangered species, and environmental remediation. Development of alternative energy sources is a challenge to both physics and biology.

Beyond these applications, genomics offers us a profound understanding of fundamental principles of biology, and of the pageant represented by the history of life.

In this book, I have tried to present a balanced view of the background of the subject, the technical developments that have so greatly increased the data flow, the current state of our knowledge and understanding of the available information, and applications to medicine and other fields. With power derived from knowledge goes commitment to act wisely. We have responsibilities, to ourselves, to other people, to other species, and to ecosystems; ranging up to the entire biosphere.

Ethical, legal, and social issues have been a prominent component of the Human Genome Project. Most technical questions in genomics, as in other scientific subjects, have objectively correct answers. We do not know all the answers, but they are out there for us to discover. Ethical, legal, and social issues are different. Many choices are possible. Their selection is not the privilege of scientists in individual laboratories, but of society as a whole. Scientists do have a responsibility to contribute to the informed public discussion that is essential for wise decisions.

Genomics is an interdisciplinary subject. The phenomena we want to explain are biological. But many fields contribute to the methods and the intellectual approaches that we bring to bear on the data. Physicists, mathematicians, computer scientists, engineers, chemists, clinical practitioners and researchers have all joined in the enterprise. This book will appeal, at least in part, to all of them. However, the central point of view here remains focused on the biology. More specifically, the focus is on human biology. In fact, on the biology of humans who are curious about other species, albeit primarily for what the other species tell us about ourselves. This choice naturally reflects the potential readership of this book. (If bacteria or fruit flies could read, genomics textbooks would look very different.)

In the new edition, extended coverage is given both to clinical applications to humans and to the applications of genome sequences to working out of evolutionary relationships in microorganisms, plants, and animals. Non-clinical applications to human biology and history are sufficient to justify a chapter of their own. The genomics of plant and animal domestication not only sheds light on human—as well, obviously, as plant and animal—history, but also emphasizes the reciprocal interactions between humans and the rest of the biosphere.

One aspect of the earlier editions of this book that I liked was their concision. Unfortunately, this has been quashed under the stampeding progress of the field. A major problem encountered in writing about genomics is the need to pick and choose from an *embarras de richesses*. The list of topics that cannot be left out is too long and threatens to reduce the treatment of each to superficiality. There is also a serious organizational challenge: many phenomena must be approached from several different points of view. A reader may be relieved to conclude that a topic has been beaten thoroughly into submission in one chapter, only to encounter it again, alive and kicking, in a later section, in a different context.

The speed at which the field is moving causes other problems. An author is often pleased with a draft of a section only to find the carefully described conclusions overturned in next week's journals, or even newspapers. Yet, there is a great pleasure in seeing nature's secrets emerging before one's eyes.

Another casualty of rapid progress is the dismissal of attention to history and biography. We are fantastically interested in the development of the sea urchin and the fruit fly, but not in the development of molecular genomics. This is too bad: those who do not learn from the successes of history will find it harder to emulate them. Intellectual struggles that occupied entire careers leave behind only the terse conclusions, often without any appreciation of the experiments that established the facts, much less of the alternative hypotheses tested and rejected. We remember the breakthroughs but not the frustrations that their achievement required. The force of the scientists' personalities, and their foibles, are forgotten. Describing results isn't the same thing as discovering them: 'To imitate the *Iliad* is not to imitate Homer.'

Progress is happening so fast as to make unavoidable a feeling of frustration in aiming at a moving target. The hope is that this fourth edition has erected for the reader a sound framework, both intellectual and factual, that will make it possible, when encountering subsequent developments, to appreciate where and how they fit in.

This book assumes that the reader already has some acquaintance with modern molecular biology. It builds on and develops this background, as a self-contained presentation. It is suitable as a textbook for undergraduates or starting postgraduate students.

Exercises and problems appearing at ends of chapters test and consolidate understanding, and provide opportunities to practise skills and explore additional subjects. Exercises are short and straightforward applications of material in the text. Problems also do not make use of information not contained in the text, but require lengthier answers or in some cases calculations.

A third category of challenges, weblems, requires access to the World Wide Web. Weblems are designed to give readers practice with the skills and tools required for further study and research in the field. Some of them are suitable for use as practical or laboratory assignments, or even as class projects.

PLAN OF THE FOURTH EDITION

Chapter 1, Introduction and Background, sets the scene, and introduces all of the major players: DNA and protein sequences and structures, genomes, transcriptomes, and proteomes, databases and information retrieval, and bioinformatics and the World Wide Web. Chapter 1 sketches the framework of how the pieces fit together and sets genomics in its context among the biomedical, physical, and computational sciences. It aims for breadth in its introduction to the subject; in consequence it is rather lengthy. Subsequent chapters develop topics in depth.

The message is clear: Genomics is the hub of biology. The 'central dogma' and the genetic code underlie the implementation of the genome, in terms of the synthesis of RNAs and proteins. Absent from Crick's original statement of the central dogma is the crucial role of regulation in making cells and organisms stable and robust, two characteristics essential for survival.

Whereas genome sequences are determined from individuals, to 'see life steadily and see it whole' requires extending our point of view spatially, to populations and interacting populations; and temporally, to consider life as a phenomenon with a history. We can study the characteristics of life in the present, we can determine what came before, and we can—at least to some extent— extrapolate to the future.

Chapter 2, The Human Genome Project: Achievements and Applications, focuses on the human genome. It reports the current state of the data, although that gives an inadequate sense of their explosive growth. Clinical applications are maturing quickly, from hype to serious promise to clinical practice. Putting our species in context involves, most narrowly, comparing our genomes with those of our closest relatives, including Neandertals and Denisovans; and our nearest extant relatives, the chimpanzee and bonobo. More general applications to anthropology appear in Chapter 9.

Applications of genome sequences in personal identification are well known. These include determinations of paternity, and, less frequently, maternity. Crime-scene investigation has established the guilt or innocence of many suspects.

Although much genome sequence investigation is carried out by clinical or forensic organizations, sequencing has gone public with 'pop' applications such as personal genealogy companies. There are even 'dating sites' that use a reported correlation between MHC haplotype and mate selection to help identify mutually attractive individuals.

Many of these applications involve ethical, legal, and social issues. Different jurisdictions have established different guidelines or regulations.

Chapter 3, Bioinformatics, speaks to the essential role of computing in genomics. Computers are integral to the measurements that produce the data. Chapter 3 also introduces and surveys the databanks that archive, curate, and distribute the data, and some of the information-retrieval tools that make the data accessible to scientific enquiry and clinical application.

We use programs to analyse the information that databases contain, to answer scientific questions, and to make decisions and predictions in research and in clinical practice. Readers who have ambitions to create novel software to solve challenging unsolved problems need advanced training in computer science and programming. The requirements for the molecular biologist at the lab bench are more modest. What is necessary is expertise in using tools available on the web, and in the intelligent interpretation of their results.

Chapter 4, Classical Genetics, Mapping, and Sequencing, describes how genomics has emerged from classical genetics and molecular biology. The first nucleic acid sequencing, by groups led by F. Sanger and W. Gilbert in the 1970s, was a breakthrough comparable to the discovery of the double helix of DNA. The challenges of sequencing stimulated spectacular improvements in technology. First came automation of the Sanger method. The original sequences of the human genome were accomplished by batteries of automated Sanger sequencers. Subsequently a series of 'new generations' of novel approaches have achieved and surpassed the landmark goal, the US$1000 human genome. Sequencing power is very widely distributed. There are major specialized institutions, such as the Beijing Genomics

Institute (now in Shenzhen). The DNBSeq-T20x2 sequencer can generate 22 terabytes of raw data per day. (You do the math: that's the equivalent of nearly five human genomes per minute!) Smaller installations are common in universities, hospitals, and companies.

Chapter 5, Evolution and Genomic Change, treats relationships. Life has been shaped by evolution, acting primarily through natural selection. T. Dobzhansky famously said, 'Nothing in biology makes sense except in the light of evolution.'

Genomics allows us to trace many aspects of this process, if not always to make sense of them. Of course most of evolution took place in the past. We cannot observe it directly. However, we can observe and draw inferences from its products. These products are for the most part the genomes—and phenotypes—of extant organisms, plus sporadic data from recently extinct species. In addition, many past evolutionary events have left their traces in contemporary genomes.

Chapter 5 explores some of the tools that scientists have developed to analyse sequence and structure data for what it can reveal about evolution. Prominent among these are methods for alignment, and calibration of the results, and the computation of phylogenetic trees.

We are entering an era when evolution, even in natural populations, may be under the direct control of molecular biologists wielding tools based on CRISPR/Cas and gene drive.

The goal of *Chapter 6, Comparative Genomics,* is to prepare for the surveys of viral, prokaryotic, and eukaryotic genomes presented in Chapters 7 and 8. We begin by comparing the different modes of genome organization with which living things have experimented. If the great variety of living things has arisen through evolution by natural selection—the change in allele frequency in a population through differential reproductive success among different variants—how does the variation arise? We must consider the nature and extent of the variability in the genomes within individual populations, and the mechanisms that generate this variability.

Chapter 7, Genomes of Prokaryotes and Viruses, surveys the genomes of viruses, bacteria, and archaea in more depth. Taxonomy and phylogeny present problems because of extensive horizontal gene transfer, especially in prokaryotes. Indeed, horizontal gene transfer challenges the whole idea of the Linnaean hierarchy of biological classification. In the past, many bacteria have been cloned and studied in isolation, especially those responsible for disease. In contrast, a *metagenome* is a collection of genomes from many individuals within an environmental sample. The new field, metagenomics, deals with the entire complement of living things in a sample, allowing us to address questions about interspecies interaction in the 'real world'. Sources include ocean water, soil, and the human body. Clinical applications include measurements and remediation of the microbiomes on and within a human body.

Chapter 8 surveys *Genomes of Eukaryotes.* It starts with yeast, which is about as simple as a eukaryote can get. Selected plant, invertebrate, and chordate genomes illuminate the many profound common features of eukaryotic genomes; and the very great variety of structures, biochemistry, and lifestyles that are compatible with the underlying similarities. There are examples of recovery and sequencing of DNA from extinct organisms.

Chapter 9, Genomics and Anthropology: Human Evolution, Migration, and Domestication of Plants and Animals, develops additional applications to the study of our own species. Although clinical applications are undoubtedly the most important, genomics has important contributions to make to human palaeontology, anthropology, and the law. The ability to extract DNA from extinct species, including Neandertals and Denisovans, sheds light on our early evolution. Events in our history, including migrations, and domestication of crops and animals, have left their traces in DNA sequences.

Chapter 10, The Impact of Genomics on Human Health and Disease, describes clinical applications of genome sequencing. Sometimes, when particular genes are known to correlate with disease or risk of disease, only specific regions in a patient's genome are sequenced. More and more, though, we see complete genome sequencing in clinical contexts.

Applications include improved diagnosis and prognosis of the causes underlying presenting syndromes, genetic counselling of parents with family histories of a dangerous genetic condition, and 'pharmacogenomics', the tailoring of treatment to the individual patient, based on DNA sequence information. 'Gene therapy', the replacement of defective genes with correct ones, has already had some successes and, with the development of CRISPR/Cas technology of genome editing, will undoubtedly grow in applicability. Several individuals have been cured of sickle-cell disease by editing their stem cells—precursors of blood cells. However, a set of guidelines that emerged from the March 2023 Third International Summit on Human Genome Editing suggests that *hereditable* human genome editing remains unacceptable at this time.

With *Chapter 11, Transcriptomics,* we move on from the (relatively) static genome to the selectivity and dynamics of expression patterns. Following the Central Dogma, there are two stages: Transcription, in which DNA makes RNA (Chapter 11); and translation, in which RNA makes protein (Chapter 12). Measurements of cellular RNAs describe the transcription patterns of regions of the genome. These patterns vary in response to changes in the environment: The Lac operon of *E. coli* is a classic example. In higher organisms, expression patterns vary among different tissues, different physiological states, and different developmental stages. For example, the human embryo and neonate synthesize foetal haemoglobin, switching to expression of adult haemoglobin at about six months post-birth age.

The inventory of mRNAs in the cell—a component of the transcriptome—is naturally of interest for what it can tell us about the distribution of cellular proteins. Certainly protein synthesis does involve many RNAs, including messenger RNAs, the RNA of the ribosome, transfer RNAs, and the RNAs of the spliceosome, which removes introns from pre-messenger RNAs. However, other RNAs have a variety of roles, including catalytic activity. Catalytically active RNAs are called ribozymes. (In both the ribosome and the spliceosome, the catalytic activity is in the RNA, not the protein component.) Other RNAs, such as microRNAs (miRNAs), short interfering RNAs, and silencing RNAs (siRNAs), control gene expression by interacting with messenger RNAs. In bacteria, the RNA transcripts of CRISPR sequences, bound to CRISPR-associated nucleases, identify viral invaders.

Chapter 12, Proteomics, describes the principles of protein structure and the high-throughput data streams that provide information about sets of proteins in cells, including methods for predicting protein structures from amino-acid sequences. Proteomics is an essential complement to genomics; to a large extent, proteins are the 'executive branch' of molecular biology. (A colleague once entitled a keynote lecture: 'Genes are from Venus; proteins are from Mars.') In fact, one of the major messages of Chapters 11 and 12 is how much the genome *doesn't* tell us about cellular proteins. Missing information includes protein oligomer formation, variable splicing, and post-translational modification. Nevertheless, the interactions and relationships between the genome and proteome are intimate, both during cellular activity and in the longer term in evolution.

Chapter 13, Metabolomics, deals with the networks of reactions that manipulate matter and energy. Cells require a source of energy, from nutrients or from light, and use it to drive metabolites through series of metabolic pathways. The flows through the paths in this network must be kept under control. To achieve this, another network, a logical one, keeps cellular activities organized. Thus, cells contain parallel sets of networks, based on physical and logical interactions among molecules. Each network also has static and dynamic aspects.

Chapter 14 introduces *Systems Biology,* a relatively new field that presents a description of how biological organization and activity are controlled. Components present in a cell at any instant are subject to direct controls; for instance, feedback inhibition by an end product to shut down a metabolic pathway. Transcription is subject to very extensive oversight: turning genes on and off. Systems biology deals with the structures of biological networks that regulate transcription and metabolism, and analyses and models of the mechanisms that control the traffic patterns.

A reasonable definition of life would include the criterion that a living thing executes controlled manipulations of matter, energy, and information. The genius of classical biochemistry was to take cells apart and show that the components could function in isolation. Our job now is to put things back together.

ADDITIONAL READING

Where else might the interested reader turn? This book is designed as a companion volume to three others: *Introduction to Protein Architecture: the Structural Biology of Proteins*; *Introduction to Protein Science: Architecture, Function, and Genomics;* and *Introduction to Bioinformatics* (all published by Oxford University Press). Of course there are many fine books and articles by many authors, some of which are listed as recommended reading at the ends of the chapters. The goal is that each reader will come to recognize his or her own interests and be equipped to follow them up.

For the recommendations for additional reading, I have limited myself to books, and to articles in the scientific literature. However, if one wants an introduction to a topic, there are many quite good lectures on the web, which might well be useful. Animations are particularly perspicuous presentations of dynamic processes, superior to prose descriptions. Or, if one wants up-to-date details about a topic, there are blogs. Indeed, blog entries are often more useful than a full scientific paper. With a full paper there is the problem of extracting a take-home message from a mass of detail, whereas the blog entries often contain one or two terse and to-the-point paragraphs.

It is undeniable that the web contains many useful sources of information outside of the printed scientific literature. Nevertheless, I am reluctant to include these in the recommendations. This is partly out of a commitment to the refereed scientific literature. But also: (1) There is no control over whether a recommended website will remain available. The scientific literature at least has a permanent presence. (2) Blogs can contain a mixture of some contributions that are very useful and others that are not; and it may be difficult to distinguish.

Many applications of genomics to health care are discussed in the book. However, nothing here should be taken as offering medical advice to anyone about any condition.

ACKNOWLEDGEMENTS

I thank S. Ades, G.F. Anderson, E. Axelsson, M.M. Babu, S.L. Baldauf, P. Berman, P. Bork, B. de Bono, D.A. Bryant, C. Cirelli, A. Cornish-Bowden, C. dePamphilis, R. Diskin, R.B. Eckhardt, D. Fairbanks, N.V. Fedoroff, J.G. Ferry, R. Flegg, J.R. Fresco, S. Girirajan, N. Goldman, J. Gough, D. Grove, D. Halitsky, R. Hardison, E. Holmes, H. Klein, E. Koc, A.S. Konagurthu, T. Kouzarides, H.A. Lawson, E.L. Lesk, M.E. Lesk, V.E. Lesk, V.I. Lesk, D.A. Lomas, B. Luisi, P. Maas, K. Makova, W.B. Miller, C. Mitchell, J. Moult, E. Nacheva, A. Nekrutenko, C.A. Orengo, G. Otto, A. Pastore, D. Perry, C. Praul, K. Reed, G.D. Rose, J. Rossjohn, S. Schuster, R.L. Serrano, B. Shapiro, J. Shi, J. Tamames, A. Tramontano, A.A. Travers, A. Valencia, G. Vriend, L. Waits, R. Wayne, J.C. Whisstock, A.S. Wilkins, G.P. Winter, V.E. Womble, and E.B. Ziff for helpful advice. I thank several colleagues who read drafts of the book for very helpful suggestions.

I thank the staff of Oxford University Press for their skills and patience in producing this book.

TABLE OF CONTENTS

LIST OF ABBREVIATIONS

ADP	adenosine diphosphate
AHL	*N*-acyl homoserine lactone
AHPC	alkyl hydroperoxidase reductase subunit C
AIM	ancestry-informative marker
AMP	adenosine monophosphate
ATP	adenosine triphosphate
BAC	bacterial artificial chromosome
BLAST	basic local alignment sequence tool
bp	base pair
BPG	2,3-bisphosphoglycerate
BSE	bovine spongiform encephalopathy
cAMP	cyclic adenosine monophosphate
CAP	catabolite-gene activator protein
CASP	critical assessment of structure prediction
cDNA	complementary DNA
CJD	Creutzfeldt–Jakob disease
CoA	coenzyme A
CODATA	Committee on Data of the International Council of Scientific Unions
CODIS	Combined DNA Index System
COI	cytochrome oxidase subunit I
CPI	Combined Paternity Index
cryoEM	cryo electron microscopy
DFR	dihydroflavonol reductase
DNA	deoxyribonucleic acid
EBI	European Bioinformatics Institute
EC	Enzyme Commission
ENCODE	Encyclopedia of DNA Elements
EST	expressed sequence tag
ESU	evolutionary significant unit
EU	European Union
FISH	fluorescent *in situ* hybridization
FOXP2	forkhead box protein P2
GBIF	Global Biodiversity Information Facility
GDP	guanosine diphosphate
GLUT4	glucose transporter type 4
GMP	guanosine monophosphate
GO	Gene Ontology
GPCR	G protein-coupled receptor
GTP	guanosine triphosphate
GWAS	genome-wide association studies
HER2	human epidermal growth factor receptor 2
HGPRT	hypoxanthine-guanine phosphoribosyltransferase
HIV	human immunodeficiency virus
HMDB	Human Metabolome Database
Hz	Hertz (unit of frequency)

INSDC	International Nucleotide Sequence Database Collaboration
IPTG	isopropylthiogalactopyranoside
ISI	iron-stress-inducible
ITIS	Interagency Taxonomic Information System
IUCN	International Union for the Conservation of Nature and Natural Resources
KEGG	Kyoto Encyclopedia of Genes and Genomes
K_a	number of non-synonymous substitutions per site
K_M	Michaelis constant
K_s	number of synonymous substitutions per site
LAMP	loop-mediated isothermal amplification
LINE	long interspersed nuclear element
LSST	Large Synoptic Survey Telescope
LTP	long-term potentiation
LUCA	last universal common ancestor
MALDI-TOF	matrix-assisted laser desorption ionization–time of flight
MAOA	monoamine oxidase A
MePV	medial amygdala posteroventral
MHC	major histocompatibility complex
MIC	minimum inhibitory concentration
MIPS	Munich Institute for Protein Sequences
mRNA	messenger RNA
MSMS	tandem mass spectrometry
mtDNA	mitochondrial DNA
NAD^+	nicotinamide adenine dinucleotide
$NADP^+$	nicotinamide adenine dinucleotide phosphate
NADPH	reduced nicotinamide adenine dinucleotide phosphate
NBS-LRR	nucleotide-binding site/leucine-rich repeat
NCBI	National Center for Biotechnology Information
NDB	Nucleic Acid Structure Data Bank
NDIS	National DNA Index System
NDNAD	National DNA Database
NMDA	*N*-methyl-D-aspartate
NMR	nuclear magnetic resonance
OMIA	online Mendelian inheritance in animals
OMIM	online Mendelian inheritance in man
ORF	open reading frame
PAGE	polyacrylamide gel electrophoresis
PAH	phenylalanine hydroxylase
PCR	polymerase chain reaction
PDB	Protein Data Bank
PDBe	Protein Data Bank in Europe
PDBj	Protein Data Bank in Japan
PDH	pyruvate dehydrogenase
PI	paternity index
PIR	Protein Information Resource
PKU	phenylketonuria
ppm	parts per million

PrP^C	prion protein–cellular
PrP^{Sc}	prion protein–scrapie
Pyl	pyrrolysine
R	gas constant
RCSB	Research Collaboratory for Structural Bioinformatics
RFLP	restriction fragment length polymorphism
r.m.s.d.	root-mean-square deviation
RNA	ribonucleic acid
RNAi	interfering RNA
rRNA	ribosomal RNA
RT-PCR	reverse transcription PCR
RUBISCO	ribulose-1,5-bisphosphate carboxylase/oxygenase
S	Svedberg unit (ultracentrifugation)
S	shear number
SCOP	Structural Classification of Proteins
SDS	sodium dodecyl sulphate
SDS–PAGE	sodium dodecyl sulphate–polyacrylamide gel electrophoresis
SINE	short interspersed nuclear element
siRNA	small interfering RNA
SNP	single-nucleotide polymorphism
STOL	sequenced/species Tree of Life
STR	short tandem repeat
STRP	short tandem repeat polymorphism
STS	sequence-tagged site
TAP	tandem affinity purification
TH	tyrosine hydroxylase
TIGR	The Institute for Genome Research
TIM	triose phosphate isomerase
TLR	Toll-like receptor
TrEMBL	translations to amino acid sequences of DNA sequences in EMBL databank
tRNA	transfer RNA
UPGMA	unweighted pair group method with arithmetic mean
vCJD	variant Creutzfeldt–Jakob disease
VISA	intermediate stage of vancomycin resistance
V_{max}	maximum initial velocity
VNTR	variable number tandem repeat
VRSA	vancomycin-resistant *Staphylococcus aureus*
wwPDB	Worldwide Protein Data Bank

Collect a sample of your cells by rinsing out your mouth with dilute salt water. Add some detergent to dissolve the cell and nuclear membranes, releasing the contents. Precipitate out the proteins by adding alcohol; stir, and let settle.

In a few minutes, fibres will rise out of the murk to the top of the vessel.

These fibres are your DNA (plus that of some bacteria that were living in your mouth). The sequence of your DNA is your lifelong endowment. It determined that you are human, that you are male or female, that you are brown- or blue-eyed, that you are right- or left-handed. But think of these fragile cables not as a leash that constrains you but as the cords that raise the curtains on the drama of your life.

— From *The Slurry with the Fringe on the Top*

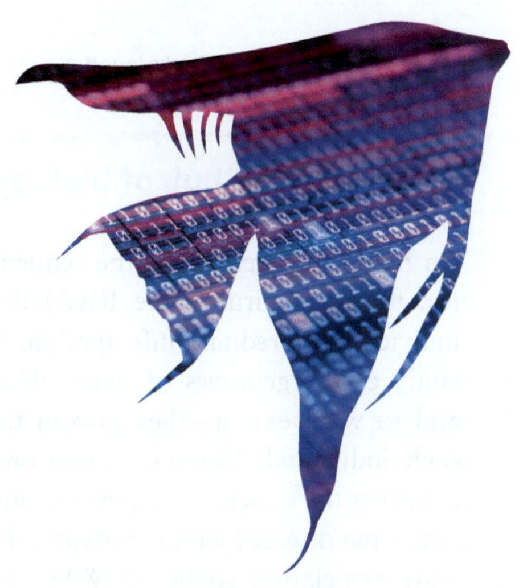

CHAPTER 1

Introduction and Background

LEARNING GOALS

- *Describe the basic central dogma,* that DNA is transcribed to RNA, which is translated to protein.
- *Explain the contributions to any person's phenotype* from his or her genome sequence, life history, and epigenetic signals within the fertilized egg.
- *Represent the diversity of genome organization* in viruses, organelles, prokaryotes, and eukaryotes.
- *Be able to sketch the general structure of a eukaryotic protein-coding gene.*
- *Know why even a complete genome sequence is insufficient information to specify the repertoire of expressed proteins.*
- *Be aware, with appropriate scepticism, that the goal of recovering extinct organisms* is being pursued.
- *Be able to distinguish the several types of potential clinical applications of genome sequence data,* including ways in which genome sequences of large numbers of people can support clinical research, and ways in which sequences of individuals can help treat their diseases more effectively, or—even better—prevent them.
- *Recognize that diseases have variable genetic components.* Some are traceable to a specific mutation; for instance, sickle-cell disease. Others depend primarily on external assaults: These include diseases arising from bacterial and viral infections. Genomic changes play a prominent role in the development and progression of cancers. For maintenance of health in general, the role of genomics is to determine the person's risk factors and capacity for defence, to inform diagnosis and prognosis, and to suggest the most appropriate therapy.
- *Understand the importance of comparative genome sequencing projects,* to reveal processes of evolution, and to help interpret the function of regions in human genomes.
- *Appreciate the large number of active genome projects treating different species,* widely distributed among life forms, and including metagenomics, which produces very large amounts of data from samples from the environment or from locations within the human body.
- *Understand how different types of human DNA sequencing projects are organized,* including those carried out by large international organizations; collection of data by law enforcement organizations; specific clinical tests; and direct-to-the-public sequencing usually motivated by questions of genealogy, including parentage testing and/or disease risk.
- *Recognize the importance of computer science and bioinformatics* in producing the raw sequence data, in creating databases, archiving and careful curation of the data, distributing them via the web, and creating information-retrieval tools to allow effective data mining for research and applications.

Genomics: the hub of biology

An organism's genome is the sequence of the DNA (or, for some viruses, the RNA) that contains the individual's hereditary information. **Genomics** is the study of the genomes of living things, and how—and to what extent—they govern the properties of each individual. Genomic variation within species describes the structures of populations and can correlate some diseases with genotypes, through genome-wide association studies (GWAS). Comparisons of genomes among different species illuminate pathways of evolution.

Just as a genome is central to the life of an individual, genomics is central to the science of biology. New techniques have created many high-throughput data streams: These prominently include but are not limited to genome sequences. Protein sequences and structures; expression patterns in different cells and tissues, and in different states; and metabolic pathways, are others.

We appeal to the science of genomics to make sense of this information. The effects are pervasive: From the computational molecular biologist sitting at a laptop to the clinician treating a patient, to what Ernst Mayr—one of the twentieth century's leading scientists of evolution—has called the 'dirty-fingernails' field biologist, genomics has transformed the life sciences.

Computing is an essential component of the enterprise (see Box 1.1).

The invention that unlocked the data and applications of genomes has been the technology to determine DNA sequences.

Do not underestimate the impact of this revolution. Recent discoveries call into question core concepts such as species and even organism and individual. Examples of inheritance of acquired characteristics are real. It is not only the data that are exploding, but also the conceptual framework.

BOX 1.1 Genomics and Computing

The immense amount of available data for modern genomics, and their complexity, makes computing essential for storage, retrieval, and analysis of the information. The new discipline of **bioinformatics**—a marriage of biology and computer science—has amply demonstrated its fecundity.

The products of different data streams populate databases. For example, the European Nucleotide Archive (ENA) based at EMBL-EBI in Hinxton, UK, collects and presents DNA and RNA sequences. Software allows analysis of the data; for instance, working out a phylogenetic tree from a set of sequences, or predicting the three-dimensional structure of a protein from its amino-acid sequence derived by translating a gene.

Databases do not just collect and curate data, they impose a structure on them, allowing selective retrieval of information. An example would be a search for nucleotide or amino-acid sequences similar to a probe sequence. Databases also provide links from any particular collection to the entire information ecosystem; for instance, starting from the nucleotide sequence of a gene, the three-dimensional structure and function of the protein it encodes is a single 'click' away.

This state of affairs has created roles for participation in genomics research by individuals expert in biology or in computer science—in many cases in both. Many genome institutes have as many software specialists on their payrolls as 'wet-lab' scientists.

There are a very large number of databases for molecular biology. A particular type of database linked most intimately with genomics is a **genome browser**—genome browsers integrate genomic sequence and annotation data from different sources and provide platforms to search, browse, retrieve, and analyse genomic data. Comprehensive genome browsers include the Ensembl Genome Browser, based at the Sanger Centre in Hinxton, UK; the UCSC Genome Browser (UCSC = University of California at Santa Cruz), in the US; and the NCBI Genome Data Viewer, in Maryland, US. Other genome browsers deal with individual species or sets of species.

We shall discuss bioinformatics in Chapter 3.

At the epicentre are human genomes. No other data sets command so strongly our curiosity and imagination, and no others offer the same hope of improving human health. Human genomes contains ~3.2×10^9 base pairs, distributed among the 46 chromosomes, plus mitochondrial DNA. There are about 23 000 regular protein-coding genes.[1] In addition, populations of bacteria occupy microenvironments on and within our bodies: our microbiome. And that is only the static picture. The genome equips us with a complex network of regulatory interactions, to integrate the activities of the individual components, including different expression patterns in different tissues or even individual cells, and developmental programmes.

To infer biological phenomena from the genome is the fundamental challenge, requiring attention to integrative and even historical aspects. No one underestimates the difficulty. From a computing frame of reference, Robbins has expressed the situation unimprovably (see Box).

Exploration of the topics mentioned so briefly in this section is the goal of this book.

'... Consider the 3.2 gigabytes of a human genome as equivalent to 3.2 gigabytes of files on the mass-storage device of some computer system of unknown design. Obtaining the sequence is equivalent to obtaining an image of the contents of that mass-storage device. Understanding the sequence is equivalent to reverse engineering that unknown computer system (both the hardware and the 3.2 gigabytes of software) all the way back to a full set of design and maintenance specifications.

'Reverse engineering the sequence is complicated by the fact that the resulting image of the mass-storage device will not be a file-by-file copy, but rather a streaming dump of the bytes in the order they were entered into the device. Furthermore, the files are known to be fragmented. In addition, some of the device contains erased files or other garbage. Once the garbage has been recognized and discarded and the fragmented files reassembled, the reverse engineering of the codes can be undertaken with only a partial, and sometimes incorrect, understanding of the CPU on which the codes run. In fact, deducing the structure and function of the CPU is part of the project, since some of the 3.2 gigabytes are the binary specifications for the computer-assisted manufacturing process that fabricates the CPU. In addition, one must also consider that the huge database also contains code generated from the result of literally millions of maintenance revisions performed by the worst possible set of kludge-using, spaghetti-coding, opportunistic hackers who delight in clever tricks like writing self-modifying code and relying upon undocumented system quirks.'

—Robbins, R.J. (1992). Challenges in the human genome project. IEEE Eng Med. Biol. **11**: 25–34 (©1992 IEEE).

What's in a genome?

A walk through a typical eukaryotic genome is like a tour of a continent. One encounters centres of bustling activity, rich in genes and their regulatory elements. These are like villages and even cities. One passes also through large tracts of emptiness, or regions with unrelieved monotony of repeated elements.

An inventory of a typical eukaryotic genome includes:

The most prominent and familiar aspects of the genome, the regions that code for RNAs and proteins. Protein-coding genes are transcribed into messenger RNA (mRNA). After processing, ribosomes translate mature mRNA to polypeptide chains. (Only about 1% of human genomes codes for proteins.)

In addition to proteins, genomes encode large numbers of RNA molecules other than messenger RNAs, 'non-coding RNAs'. Some of these are involved in metabolism of nucleic acids, including DNA replication and RNA splicing, as well as providing the catalytic activity of the ribosome itself in protein synthesis. Many others are active in regulatory processes.

Francis Crick encapsulated this scheme in the original Central Dogma of Molecular Biology: DNA makes RNA makes Protein (see Box 1.2).

Crick promulgated this principle in 1957, to emphasize the *one-way direction of information flow:* DNA sequences inform RNA sequences, and RNA sequences inform amino-acid sequences, but the information does not travel back directly from protein to RNA to DNA. In Crick's words: 'Once

[1] This number is dwarfed by 10^{18} antibodies.

BOX 1.2 The Central Dogma

Francis Crick's statement of the Central Dogma, in 1957, described the pathway of information transfer in cells (see Figure 1.1).

'DNA makes RNA makes protein'

The codons shown here are those appearing in DNA rather than RNA; that is, the codons contain t (thymine) rather than u (uracil). Both three-letter and one-letter abbreviations for the amino acids appear. (For detailed information about the 20 canonical amino acids, including the chemical structures, see Chapter 12.) In most but not all cases, the one-letter code is the initial letter of the amino acid name.

Note that the genetic code is redundant: Except for Met and Trp, multiple codons specify the same amino acid. A mutation that changes a codon to another codon for the same amino acid is called a synonymous or 'silent' mutation. Three triplets are reserved as STOP signals, terminating translation.

Variations from this standard code occur naturally in mitochondria and chloroplasts, and sporadically in individual species.

Molecular biologists have engineered genomically recoded strains of *Escherichia coli,* with altered genetic codes. Such strains should be blocked from horizontal gene transfer, so that any genes they carry could not be released into the environment. They should be immune from phage infection. However, there is some evidence that these barriers may not be absolute.

The standard genetic code

ttt	Phe	F	tct	Ser	S	tat	Tyr	Y	tgt	Cys	C
ttc	Phe	F	tcc	Ser	S	tac	Tyr	Y	tgc	Cys	C
tta	Leu	L	tca	Ser	S	taa	STOP		tga	STOP	
ttg	Leu	L	tcg	Ser	S	tag	STOP		tgg	Trp	W
ctt	Leu	L	cct	Pro	P	cat	His	H	cgt	Arg	R
ctc	Leu	L	ccc	Pro	P	cac	His	H	cgc	Arg	R
cta	Leu	L	cca	Pro	P	caa	Gln	Q	cga	Arg	R
ctg	Leu	L	ccg	Pro	P	cag	Gln	Q	cgg	Arg	R
att	Ile	I	act	Thr	T	aat	Asn	N	agt	Ser	S
atc	Ile	I	acc	Thr	T	aac	Asn	N	agc	Ser	S
ata	Ile	I	aca	Thr	T	aaa	Lys	K	aga	Arg	R
atg	Met	M	acg	Thr	T	aag	Lys	K	agg	Arg	R
gtt	Val	V	gct	Ala	A	gat	Asp	D	ggt	Gly	G
gtc	Val	V	gcc	Ala	A	gac	Asp	D	ggc	Gly	G
gta	Val	V	gca	Ala	A	gaa	Glu	E	gga	Gly	G
gtg	Val	V	gcg	Ala	A	gag	Glu	E	ggg	Gly	G

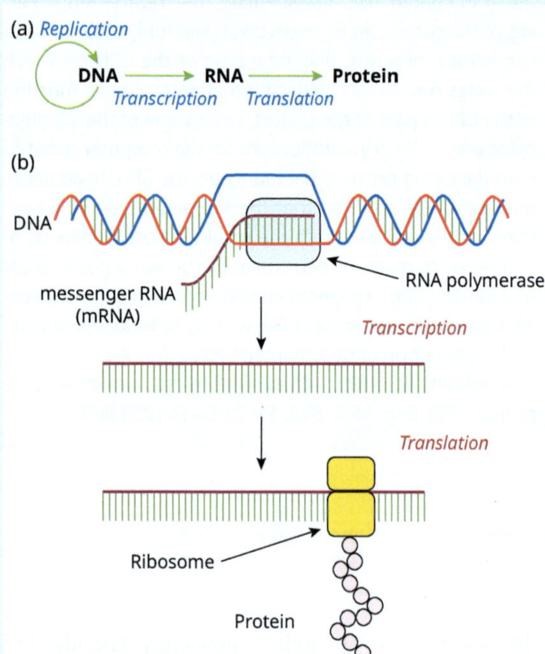

Figure 1.1 (a) Replication of DNA preserves and transmits genetic information. The pathway of information then follows The 'Central Dogma' of molecular biology: ***DNA makes RNA makes protein***. (b) To implement the Central Dogma: Protein-coding regions of a genome are transcribed into messenger RNA. The messenger RNAs, after processing, are translated by ribosomes into the amino-acid sequences of proteins. The amino-acid sequences of proteins are specified by the base sequence of messenger RNA, with successions of three bases—called codons—corresponding to successive amino acids, and in three cases to STOP signals.

information has got into a protein it can't get out again.' (Since then we have learned that reverse transcription can copy RNA into DNA, allowing certain viruses to integrate versions of their RNA genomes into the host's DNA; and bacteria to 'remember' fragments of viral genomes after infection.)

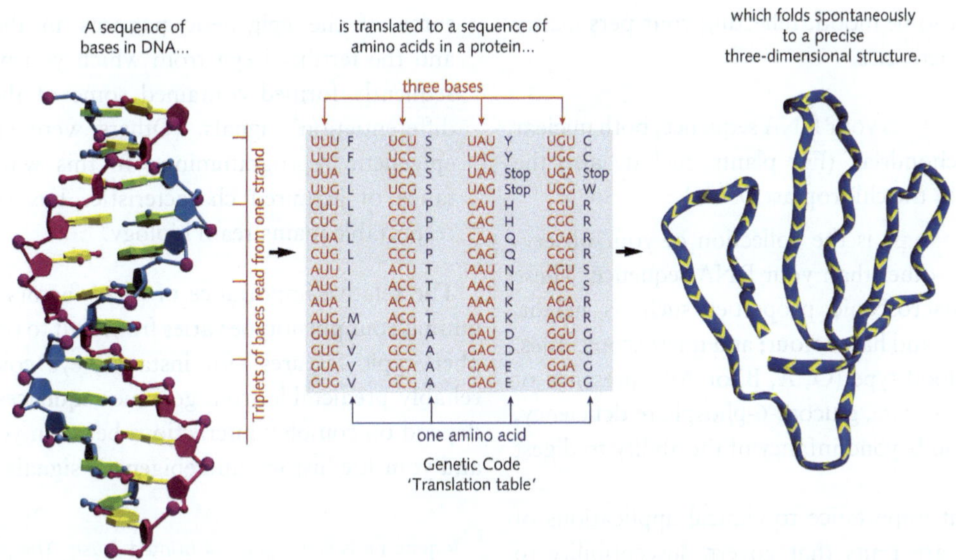

Figure 1.2 From one dimension to three: The spontaneous folding of proteins to form their native states is the point at which nature makes the giant leap from the one-dimensional world of gene and protein sequences to the three-dimensional world that we inhabit.

Also, even though information cannot generally flow *directly* from protein to RNA to DNA, changes in proteins arising from mutations in DNA can influence the DNA through selection. An extreme case: if a mutation destroys the activity of an essential protein, the cell will not survive, and the mutant DNA will be eliminated from the population.

The Central Dogma takes us as far as amino-acid sequences. But to carry out their functions, most proteins must adopt an exquisitely precise three-dimensional structure. There is 'a most ingenious paradox': Ironically, the translation of RNA into protein is *logically* very simple: The genetic code specifies the translation of triplets of nucleotides (called *codons*) into amino acids; but the *implementation* of this directed polymer synthesis requires immensely complicated machinery, including the ribosome. Conversely, and also ironically, the way the amino-acid sequence specifies the three-dimensional structure is logically extremely complicated—but it happens spontaneously (under suitable physicochemical conditions) (see Figure 1.2).

KEY POINT

Both DNA and proteins are polymers with variable sequences. In DNA, the sequences adopt the same three-dimensional structure—the double helix—but in proteins different sequences give rise to different three-dimensional structures.

Phenotype ← genotype + environment + life history + epigenetics

Each reader of this book is an individual, with physical, biochemical, and psychological characteristics. (Do not be surprised if the distinctions among these categories become more and more nebulous!) Each of you has a general form and metabolism that is common to all humans. At the molecular level, you have much in common with other species as well. But there is also substantial variation within our species, to give you your individual appearance and character. You are in a state of health somewhere along the spectrum between robust good health and morbid disease. You are currently in some psychological state, and in some mood, reflecting your personality and current activities.

- Your **genotype** is your DNA sequence, both nuclear and mitochondrial. (For plants, include also the sequence of the chloroplast DNA.)
- Your **phenotype** is the collection of your observable traits, other than your DNA sequence. These include macroscopic properties such as height, weight, eye and hair colour; and microscopic ones, such as blood type (O, A, B, or AB), presence of **sickle-cell disease**, glucose-6-phosphate deficiency, or retention beyond infancy of the ability to digest lactose.

  Of great importance to clinical applications of genomics are traits that govern susceptibility to disease and risk factors; and those that determine the effectiveness of different drugs in different individuals. These allow for personalized prevention and treatment of disease based on DNA sequences.
- Your **life history** includes the integrated total of your experiences, and the physical and psychological environment in which you developed. Your nutritional history has influenced your physical development. A nurturing environment and educational opportunities have influenced your cognitive and psychological development. Less obvious than most aspects of your life history is the growing recognition of the importance of your prenatal *in utero* environment in determining your developmental trajectory and your adult characteristics.
- At the interface between the genome and life experience are **epigenetic** factors. It is largely true that all cells of your body, except sperm or egg cells (haploid), erythrocytes, and cells of the immune system, have almost the same DNA sequence (subject to usually modest amounts of accumulated mutations). Yet your tissues are differentiated, with different sets of genes expressed or silenced in liver, brain, etc., under control of epigenetic regulators. Some of these regulatory signals survive cell division. (When a liver cell divides, it divides into two liver cells.)

  Some epigenetic factors implement species-wide developmental programmes. Others are personal: your parents' own life histories might have altered some of the epigenetic patterns in their cells, and the fertilized egg from which you were subsequently formed contained some of these 'pre-differentiation' signals. (Others were erased by epigenetic reprogramming.) In this way, inheritance of acquired characteristics has re-entered respectable mainstream biology.

The relative importance of these factors in determining your phenotype varies from trait to trait. Some phenotypic features—for instance, eye colour—are reliably predictable from genome sequences. Others depend on complex interactions between your genes and your life history, and epigenetic signals.

> A genome is like a page of printed music. The page is a fixed physical object, but the printed notes are consistent with realizations, in time and space, in a variety of ways—a *limited* variety of ways. The often-heard metaphor of a genome as a 'blueprint', specifying a precise implementation, is extremely misleading and should be deprecated.

Some environmental effects have specific windows of responsiveness to give results that are subsequently irreversible. Growth hormone treatments and the endocrinological consequences of surgery to produce the famous castrati as opera singers are examples. In turtles and other reptiles, incubation temperature affects whether eggs develop into males or females. This effect depends on the temperature dependence of the expression of the enzyme aromatase, which converts androgens to oestrogens. (The corresponding enzyme in humans is the target of inhibitors used in the treatment of cancers that are stimulated to grow by oestrogen, to reduce the exposure of the tumour to oestrogen.)

Thus a genome constrains but does not dictate the features of an organism. Varied surroundings and experiences lead organisms to explore different states consistent with their genomes. Even in microorganisms, expression of many genes is conditional. Synthesis of enzymes for lactose utilization in *E. coli* depends on the composition of the medium. The shift of yeast between anaerobic and aerobic metabolism is a temporary and reversible change in physiological state in response to changes in environmental conditions.

Most people would not regard simple and transient alterations in appearance that have no genetic component at all as true changes in phenotype. Trimming one's nails, getting a haircut, and wearing makeup are common examples. Conversely, many environmental effects, thought of as components of lifestyles, have long-term effects on development. Exercise enhances musculature. Education enhances intellectual development. A phenylalanine-controlled diet can prevent the harmful effects of the metabolic disease phenylketonuria. Malnutrition, disease, injury, and cruel treatment can be physically and mentally debilitating. Viral infection during pregnancy can lead to birth defects.

Science has provided means of interposition between genotype and phenotype. These are widely applied and sometimes abused. Examples include simple changing of hair colour, tanning, cosmetic surgery, and 'cosmetic endocrinology' (including both the treatment of children deficient in human growth hormone and the use of performance-enhancing drugs by athletes), and even—at least arguably—psychoanalysis. The CRISPR/Cas system is a powerful technique for genome editing. Its potential for modifying human cells is well-established but as yet largely untapped.

Given the variability of the contributions of heredity, environment, life history, and epigenetics in determining phenotype, how can we measure their relative importance for any particular trait? The classic method of distinguishing effects of genetics from those of surroundings and experience—often called 'nature' *v.* 'nurture'—is controlled experiments with genetically identical organisms. For human beings, this means monozygotic twins. (Monozygotic human triplets are extremely rare.) Many studies have compared the similarities of identical twins reared apart. Such individuals have the same genes but different environments. Less well-controlled are comparisons of identical with fraternal twins, or of non-twin siblings with adopted children reared in the same family. There have been suggestions that fraternal twins share greater similarities than non-twin siblings as a result of the shared uterine experience.

Human twin and sibling studies to distinguish hereditary components of traits have a long and contentious history, as they have formed the basis for important social decisions. Intelligence quotient, or IQ, is the ratio of mental age to chronological age. When properly measured, it remains constant in early childhood. Objective and quantitative measurement of intelligence in adults is extremely difficult. Attempts to measure inheritance of intelligence—or even of score on intelligence quotient tests; by no means the same thing if the tests are culturally biased—have produced controversial and unsatisfying results. It has proved very difficult to devise tests free of socioeconomic bias.[2]

KEY POINT

Phenotypic traits—both macroscopic and molecular—depend on a combination of influences from genome sequences themselves, the individual's life history, and epigenetic signals in the fertilized egg.

Varieties of genome organization

Chromosomes, organelles, and plasmids

The biosphere as we know it includes living things based on cells, and viruses. The most general classification of cells, according to both their structure and molecular biology, divides prokaryotes, simple cells without a nucleus, from eukaryotes, cells with nuclei (Table 1.1). From the point of view of genomics, the most relevant difference between prokaryotic and eukaryotic cells is the form and organization of the genetic material.

In structuring their DNA, cells have two requirements. The first is packaging. The DNA of *E. coli* is 1.6 mm long but must fit into a cell 2 μm long and 0.8 μm wide (1 μm = 0.001 mm). The discrepancy is even higher in eukaryotic cells: the nucleus of a typical human cell has a diameter of 6 μm, but the total length of the DNA is 1 m. How can the DNA

[2] See Gould, S.J. (1996). *The Mismeasure of Man*. New York: W.W. Norton.

Table 1.1 Differences between prokaryotic and eukaryotic cells

Feature	Typical feature of	
	Prokaryotic cell	**Eukaryotic cell**
Size	10 μm	100 μm
State of major component of genetic material	Circular loop, few proteins permanently attached	Complexed with histones to form chromosomes
Internal differentiation	No organized subcellular structure; in particular, no nucleus	Nuclei, mitochondria, chloroplasts, cytoskeleton, endoplasmic reticulum, Golgi apparatus
Cell division	Binary fission	Mitosis (or meiosis)

be deployed in cells in a form that is compact, but accessible to proteins active in replication and transcription? The second requirement is how, during cell division, after DNA replication, to ensure that each daughter cell gets one copy of the DNA. Different types of cells, and viruses, satisfy these requirements in different ways.

In the typical prokaryotic cell, most of the DNA typically has the form of a single closed, or circular, molecule. The DNA is complexed with proteins to form a structure called a nucleoid. The DNA is attached to the inside of the plasma membrane but is accessible to molecules in the cytoplasm. (Some bacteria have multiple circular DNA molecules; others have linear DNA.) In addition, prokaryotic cells can contain plasmids, small pieces of circular DNA, neither complexed permanently with protein nor attached to the membrane. The development and spread of antibiotic resistance in pathogenic bacteria, an increasingly serious public health problem, is often associated with exchange of plasmids among strains—an example of 'horizontal gene transfer'. In the laboratory, bacterial plasmids are also used as vectors for genetic engineering.

In a eukaryotic cell, most of the DNA is sequestered in the nucleus; mitochondria and chloroplasts contain additional small amounts of DNA. The nucleus is the site of DNA replication and RNA synthesis in gene transcription. Nuclear DNA is complexed with histones and other proteins to form chromosomes, large nucleoprotein complexes. Each chromosome contains a single linear molecule of DNA. The nuclei of different species contain different numbers of chromosomes, and in each species the chromosomes vary in length. Humans contain 46 chromosomes—22 pairs, plus two X chromosomes in females or one X and one Y chromosome in males. Deviations from the normal complement of chromosomes have clinical consequences; for example, the presence of three copies of chromosome 21 (trisomy 21) is associated with Down's syndrome, characterized by a distinctive physical appearance, a variable degree of intellectual disability, and enhanced risk of early-onset Alzheimer's disease. Other disorders arise from an autosomal chromosome present in only one copy (monosomy), or partial chromosomal deletions or translocations.

The state—notably the accessibility to transcriptional machinery—of different regions of DNA in eukaryotic chromosomes is modulated by the local structure of the chromosome, notably the interaction with histones (see Figure 1.3).

Subcellular organelles, including mitochondria and chloroplasts, are believed to have originated as intracellular parasites (see Box 1.3). These organelles contain DNA in the form of single closed or circular molecules; uncomplexed with histones, like the DNA of prokaryotes. Mitochondria and chloroplasts also contain their own protein-synthesizing machinery, using a slightly different dialect of the nearly universal genetic code.

Eukaryotic cells may also contain plasmids. Yeast artificial chromosomes (YACs)—plasmids in yeast cells—were of great utility in the early human genome sequencing projects.

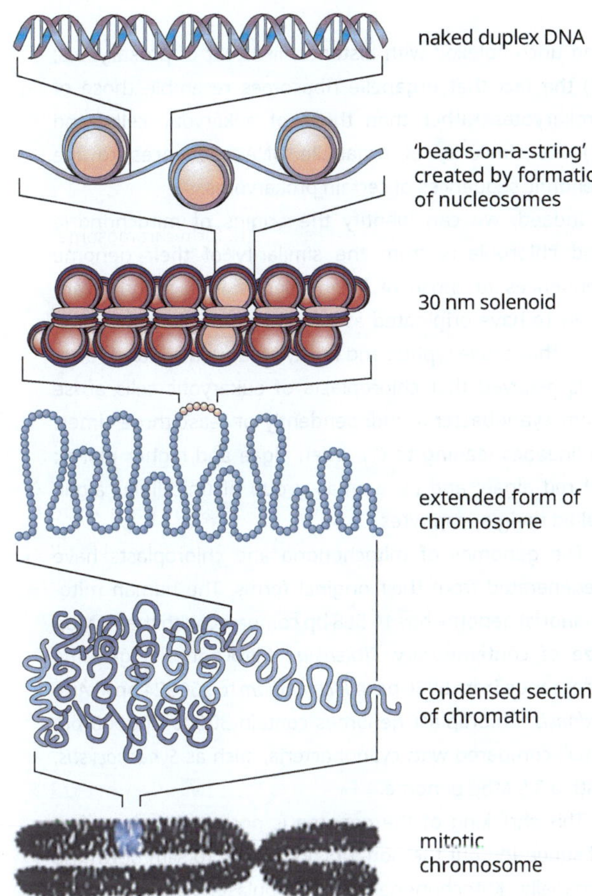

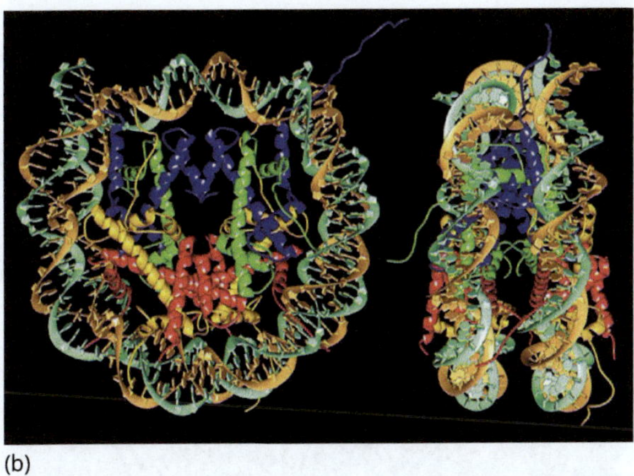

Figure 1.3 (a) Hierarchical organization of structure of DNA in eukaryotic chromosomes. From top: DNA double helix; if entirely straight, it would be much longer than dimensions of cell. DNA in cells has the form of chromatin, containing nucleosomes formed of a protein core of histones, around which nearly two turns of DNA (about 150 bp) are wrapped. The nucleosomes condense to form a 30 nm helical fibre called the solenoid. Further levels of compaction produce the bottom image, corresponding to the typical appearance of a chromosome in a metaphase karyotyping spread. (b) In the left frame, the DNA appears on the outside, recognizable by the depiction of base pairs, and the histones (proteins) appear within the DNA. In the right frame, the DNA double helices appear, vertical, in the front of the picture, and the histones behind the DNA. (a) Used with permission of Springer Nature BV, from Felsenfeld, G., & Groudine, M. (2003). Controlling the double helix. Nature 421: 448–453. https://doi.org/10.1038/nature01411. (b) Used with permission from Springer Nature BV, from Luger, K., Mäder, A., Richmond, R., *et al.* (1997). Crystal structure of the nucleosome core particle at 2.8 Å resolution. Nature 389: 251–260. https://doi.org/10.1038/38444. Permission for both conveyed through Copyright Clearance Centre, Inc.

BOX 1.3

Origin of intracellular organelles: the endosymbiont hypothesis

Mitochondria and chloroplasts are subcellular particles involved in energy transduction (see Figure 1.4). Mitochondria carry out oxidative phosphorylation, the conversion into ATP of reducing power derived from metabolizing food. Chloroplasts carry out photosynthesis, the capture of light energy in the form of NADPH and ATP, used to synthesize carbohydrate from CO_2 and H_2O.

Both types of particles lead a quasi-independent life within the cell. They are surrounded by membranes, they have their own genetic material and protein-synthesizing machinery, and they reproduce themselves within cells independently of cell division.

To perform their functions, it is essential for mitochondria and chloroplast thylakoids to be enclosed. Electrochemical and photochemical energy conversion require establishment of a pH gradient across a membrane. The passage of protons through the membrane is coupled to generation, first of mechanical energy, and then of

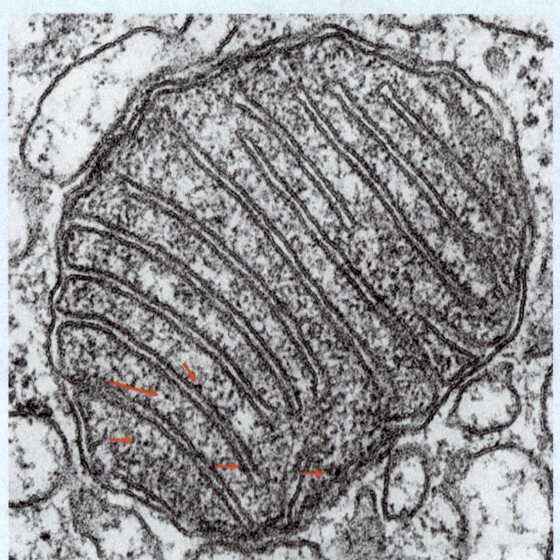

Figure 1.4 Structure of a mitochondrion. The convoluted inner membrane is easily visible. Arrows indicate ribosomes. (C023/1020—Don W. Fawcett/SCIENCE PHOTO LIBRARY)

chemical bond energy through the action of the molecular motor ATP synthase:

chemiosmotic energy stored in pH gradient →
mechanical energy in ATP synthase →
chemical bond energy of ATP.

What might provide a source of a small, self-enclosed, and largely self-sufficient object to serve as the basis of an organelle? Why, another cell! There is consensus that mitochondria and chloroplasts originated as prokaryotic endosymbionts that originally lived independently but took up residence inside a eukaryotic cell. Evidence for this includes: (1) the state of the DNA: organelle DNA is circular and uncomplexed with histones, like that of prokaryotes; (2) the fact that organelle ribosomes resemble those of prokaryotes rather than those of eukaryotic cells; and (3) the similarity of organelle DNA sequences to the genomic sequences of certain prokaryotes.

Indeed, we can identify the origins of mitochondria and chloroplasts from the similarity of their genome sequences to those of living species. Mitochondria are likely to have originated as a relative of Rickettsiae, parasites that cause typhus and Rocky Mountain spotted fever. It is believed that chloroplasts of eukaryotic cells arose from cyanobacteria, independently at least three times, in lineages leading to (1) green algae and higher plants; (2) red algae; and (3) a small group of unicellular algae called the glaucophytes.

The genomes of mitochondria and chloroplasts have degenerated from their original forms. The human mitochondrial genome has 16 568 bp compared with the >1 Mbp size of contemporary *Rickettsia*.[3] Mitochondrial genome sizes range from 5966 bp in *Plasmodium* to 366 924 bp in *Arabidopsis*. Chloroplast genomes contain 30 000–200 000 bp, small compared with cyanobacteria, such as *Synechocystis*, with a 3.5 Mbp genome.

This shrinking of the genome is not entirely the effect of squeezing out the non-functional DNA to slim down the organelle. Mitochondria and chloroplasts synthesize relatively few proteins, and certainly not all that they need. There has been substantial transfer of genes from the mitochondria and chloroplast genomes to the nucleus. Mitochondrial and chloroplast proteins encoded in nuclear DNA and synthesized in the cytoplasm contain an N-terminal targeting sequence specifying their translocation into the appropriate organelle.

[3] bp = base pair. 1 Mbp = 1 megabase pair = 10^6 bp.

The scope and applications of genome sequencing projects

The static contents of human genomes, and its dynamic aspects, are similar in general features to what other genomes contain. As genome sequencing techniques become easier, the field is progressing in two directions: (a) to determine more and more human genome sequences, especially those that may prove useful for research into disease anticipation, diagnosis, prevention, and treatment; and (b) more and more different species have now had their genomes sequenced, for at least one individual (Table 1.2). The Genomes OnLine Database contains comprehensive information about genome and metagenome sequencing projects.

There are many reasons for sequencing non-human genomes. They reveal and illuminate the processes of evolution. They help us to understand the functions of different regions of the human genome. An important principle is that if evolution conserves something, it is essential. If evolution does not conserve

Table 1.2 Genome sequencing projects. Figures refer to numbers of species and strains, not numbers of individuals[a]

Organism type	Genomes
Viruses and viroids	18 325
Archaea	4 978
Bacteria	406 317
Eukaryotes	46 550

[a] Source: https://gold.jgi.doe.gov/

something, it is not essential. In trying to understand the function of the regions of human genomes that do not code for proteins or non-coding RNAs, we gain important clues by comparing human genomes with other mammalian genomes. Regions that are conserved must be conserved for a reason. We can focus on these regions to try to determine their function. Regions that are not conserved can be set aside, if only for the moment.

> He who does not know foreign languages does not know anything about his own.
>
> —Goethe, Kunst und Alterthum

Sequences of genomes of other species also have direct application to human welfare. The genomes of pathogens that have developed antibiotic resistance, or are threatening to, give clues that we can use to try to keep ahead of them. Other practical applications include improving agriculture and domesticated animals, and biotechnological applications such as deriving fuel from crops. Genomics can also support conservation efforts aimed at preserving endangered species.

Many genome projects target individual species. In others, a major component of the DNA sequence data comes from metagenomic data. These are sequences determined from environmental samples, without isolating individual organisms. Sources include ocean water, soil samples, and the human gut.

> Some readers will be surprised to learn that within the volume of their bodies, there are as many prokaryotic cells as human ones.

Nevertheless, the focus of genome sequencing efforts has been human subjects. Clinical applications have elicited a very large amount of sequence data for individual human genes. Many people have undergone genetic testing; for example, many prospective parents determine whether they are carriers of cystic fibrosis. Many women test for potentially dangerous mutations in the genes *BRCA1* and *BRCA2* and *PALB*, which can predispose an individual to cancer. Formerly, these tests have been carried out when a family history suggests increased risk. Reduced sequencing costs have led to more widespread testing for these and other risk-alerting genes. Some people make use of commercial 'over-the-counter' genotyping services, outside medical supervision, to explore disease risk, and also to trace genealogy, including but not limited to parentage. Law enforcement agencies determine DNA sequences from samples from crime scenes.

A separate area of comparative genomics involves comparing the genome sequences of normal cells and cancer cells from individual patients. These can differ in essential ways, which assist in precise diagnosis and guide treatment.

Comparisons of genome sequences

A genome sequence belongs to an individual organism. But, to 'see life clearly and to see it whole' it is essential to relate genomes to one another. Genome comparisons are necessary across both space and time. We analyse differences in genomes among species that are currently extant, or that are accessible from preserved specimens of extinct species. Within species, we study genetic variation within and between populations. Sequence variation in humans has applications in medicine; in anthropology, to trace migration patterns; in genealogy; and in personal identification to prove parentage or for crime investigation. Sequence variability in other species gives clues to the history of those species, including but not limited to understanding the history of domestication of animals and crop plants. The data also guide efforts to enhance desirable traits.

Any two people—except for identical siblings—have genomic sequences that differ at approximately 0.1% of the positions. Measurements of multiple human genomes permit distinction between random components of this variation, and those that systematically characterize different populations.

KEY POINT

Human genomic variation reveals the history of our origin and dispersal, and how our species has adapted to different environments and lifestyles. For example, adult lactose tolerance is a concomitant of the dietary change associated with cattle domestication. The original trait, loss of the ability to digest lactose after infancy—at the time of weaning—persists in many populations. For these populations, alternative sources of calcium include fermented dairy products such as cheese or yoghurt—bacteria hydrolyse the lactose—or soya bean. (It is said that 'The soybean is the cow of Asia'.)

Comparing genomes between species can reveal their relationships and how they achieve their similarities and differences. We share 99% of our genomes with chimpanzees and 98% with bonobos, our nearest relatives, and many proteins are identical. This small amount of genomic change must account for the differences between humans, and chimps and bonobos. Indeed, genomics is essential to understanding and drawing species boundaries. But, as useful a concept as species may be, there is consensus that its definition is a very tricky problem.

Mutations and disease

Many mutations, even if they are not synonymous mutations, are consistent with a healthy life, and undiminished lifespan. (A synonymous, or 'silent', mutation is a base change that does not change the encoded amino-acid sequence.) Such mutations contribute to our population and individual variety. Loss of some proteins is surprisingly innocuous. Mice lacking myoglobin thrive, and even show athletic prowess comparable to normal mice.

In some cases, species-wide loss of biosynthetic enzymes is not generally considered a disease, but contributes to a list of essential nutrients. ('Not a bug but a feature.') For instance, whereas most animals can synthesize vitamin C, we must provide it in our diet. If not, the result is the disease scurvy.

Nevertheless, evolution has largely optimized proteins for their roles in healthy organisms. Therefore, most mutations causing amino-acid sequence changes are deleterious, impairing protein function and threatening to produce disease. Mutations associated with human disease are collected in Online Mendelian Inheritance in Man (OMIM)™ (http://www.ncbi.nlm.nih.gov/omim). A corresponding site, Online Mendelian Inheritance in Animals (OMIA) (http://www.ncbi.nlm.nih.gov/omia) collects mutations associated with diseases in animals, other than human and mouse.

By what mechanisms can mutations affect human health?

Some mutations destroy the activity of a protein. For instance the common forms of haemophilia arise from mutation of a protein in the blood coagulation cascade, either Factor VIII or IX. The most common form of phenylketonuria arises from mutation of phenylalanine hydroxylase, the enzyme that converts phenylalanine to tyrosine.

Other mutations cause disease only in combination with unusual features of the environment or specific triggering events. For instance, α_1-antitrypsin is an inhibitor of elastase in the lining of the alveoli of our lungs. The Z-mutation of α_1-antitrypsin, Glu342→Lys, an impediment to normal function, creates a predisposition to emphysema and chronic obstructive pulmonary disease, from lung damage by elastase unrestrained by the normal inhibitor. Smoking makes the development of emphysema all but certain. Heavy smokers homozygous for Z-antitrypsin generally die from respiratory disease by the age of 50. The disease is brought on by a combination of genetic and environmental factors.

> Genetics loads the gun; environment pulls the trigger.
>
> —J. Stern

Because loss-of-function mutations are often recessive—one good copy of the gene suffices—homozygosity for the mutant allele typically has more severe consequences than heterozygosity. Each of us is heterozygous for some deleterious mutations that, if homozygous, would be lethal.

Some mutations that produce defective proteins show complex interactions with other traits, with the result that some deleterious mutations have not been eliminated from populations by selection,

BOX 1.4 Glucose-6-phosphate deficiency, food taboos, folk medicine, pharmacogenomics, and mosquito breeding seasons

Glucose-6-phosphate dehydrogenase (G6PDH) is the most common enzyme deficiency, affecting over 400 million people worldwide. It is a recessive X-linked genetic defect, appearing in up to 10% of individuals in populations in which mutations are common.

Glucose-6-phosphate dehydrogenase (G6PDH) catalyses the reaction:

$$\text{glucose-6-phosphate} + \text{NADP} \rightarrow \text{6-phosphogluconate} + \text{NADPH},$$

the first step in the pentose phosphate shunt. This pathway produces reduced glutathione needed to dispose of the toxin hydrogen peroxide (H_2O_2). It is particularly important in red blood cells, which, lacking nuclei and mitochondria, are metabolically impoverished and have no alternative mechanism for detoxifying H_2O_2. Without active G6PDH, build-up of H_2O_2 will oxidize and denature haemoglobin, leading to the destruction of red blood cells, producing the condition haemolytic anaemia.

Eating fava beans, especially if uncooked, can induce anaemic episodes in people deficient in G6PDH. The danger of eating fava beans has been recognized since antiquity and has been associated with food taboos, and preparation techniques designed to reduce toxicity. For example, Pythagoras banned eating fava beans in his school. We now know that fava beans contain the compounds vicine and convicine, metabolized in the intestine to isouramil and divicine, which react with oxygen to produce hydrogen peroxide, subjecting cells to oxidative stress.

Other chemicals, including certain drugs, present the same danger to G6PDH-deficient people. During the First World War, some patients were observed to suffer dangerous side effects from the antimalarial drug primaquine. Many drugs, including sulphonamides, are now contraindicated for G6PDH-deficient patients, as is the taking of large doses of vitamin C. (The observation of variations in effectiveness and toxicity of different drugs in different people has developed into the new field of pharmacogenomics: the tailoring of drug treatments to the genotype of the individual patient.)

Why have dysfunctional G6PDH genes remained at such a high level in the population? Why does primaquine produce haemolytic anaemia in G6PDH-deficient patients and does this have a relationship to its antimalarial activity? And why have fava beans continued to be grown if non-toxic alternatives are available?

The malarial parasite invades the red blood cell of its host and competes metabolically with normal activity. Primaquine and related drugs, such as chloroquine, subject the red blood cells to oxidative stress. Cells stressed by *both* parasite and drug are the most vulnerable, and if they die they take the parasite down with them. Because consumption of fava beans subjects cells to oxidative stress, they also provide an antimalarial effect, recognized in folk medicine. Indeed, fava beans have some effect against malaria even for people with normal G6PDH activity; those with abnormal G6PDH have an even greater advantage, until the maturing *Plasmodium* produces its own G6PDH.

The link with malaria is the likely explanation for the persistence of the gene in the population and the fava bean in agriculture. A final clue appears in the calendar: there is a good overlap between the fava bean harvest period and the peak breeding season of the *Anopheles* mosquito, the vector of malaria.

because they carry some compensating advantages. For example, the genes for sickle-cell anaemia and for glucose-6-phosphate deficiency confer resistance to malaria (see Box 1.4).

Dysfunction of a regulatory protein or receptor can disorganize the operation of a pathway even if all components of the pathway regulated are normal. Some abnormal regulatory proteins cannot be activated at all, whereas others are constitutively activated and cannot be shut off. The effects include:

Physiological defects: A number of diseases are associated with mutations in G protein-coupled receptors, a large and diverse group of cell-surface receptors in eukaryotes. For example, some mutations in opsins are associated with colour blindness. Certain mutations in the common G protein target of olfactory receptors lead to loss of sense of smell.

Developmental defects: Several types are traceable to mutations in hormone receptors. For instance, Laron syndrome, a phenotype including diminished stature, arises from a mutation in the human growth hormone receptor. Administration of exogenous growth hormone does not restore normal growth.

There are even genome sequence variations within a single individual. Healthy cells accumulate mutations at a modest rate. But cancer cells that have lost checks on accuracy of DNA replication accumulate mutations copiously. Cancer genome sequencing is of extremely intense interest, with implications for understanding the origins of cancer, and for more precise diagnosis, prognosis, and choice of therapy.

Many diseases are associated with the formation of insoluble aggregates, usually of misfolded proteins. These include classical amyloidoses, Alzheimer's and Huntington's disease, aggregates of misfolded α_1–antitrypsin, and prion diseases. Polymerization of insulin creates problems in production, storage, and delivery for diabetes therapy. Mutations that destabilize proteins can increase the proportion of misfolded proteins, which show a greater tendency to form aggregates.

KEY POINT

Many diseases are caused directly by mutations; many others have a genetic component and arise in the context of an interaction between genetics and lifestyle.

Haplotypes

Mutations in the same DNA molecule in diploid chromosomes will become unlinked by recombination events that occur between their loci. The greater the distance between two sites, the greater the frequency of recombination. However, recombination rates vary widely along the genome, by several orders of magnitude. SNPs on opposite sides of recombinational 'hot spots' are more likely to be separated in any generation. (SNP = single-nucleotide polymorphism; that is, an isolated base change.) SNPs lying within recombination-poor ('cold') regions will tend to stay together.

In humans, many 100 kb regions tend to remain intact. They show the expected number of SNPs, but relatively few of the possible combinations. An average SNP density of 0.1%, or 1 SNP/kb, suggests ~100 SNPs per 100 kb. The genome of any individual may possess, or may lack, each of them, giving a very large number (2^{100}) of possible combinations. However, many 100 kb regions show fewer than five combinations of SNPs. These discrete combinations of SNPs in recombination-poor regions define an individual's haplotype.

Haplotypes provide a very economical characterization of entire genomes. They simplify the search for genes responsible for diseases, or any other phenotype–genotype correlations. For field biologists, including anthropologists, haplotypes permit detection of migratory and interbreeding patterns in populations.

In looking for genes responsible for diseases, or other phenotypic traits, haplotypes provide a magnifying glass. The goal is to correlate phenotype with genome sequence. The target may be to identify one base out of 3.2×10^9. By correlating phenotype with haplotype, only enough sequence must be collected to localize the site to within the typical length of a haplotype block, perhaps ~100 kb, containing only a few genes.

KEY POINT

Haplotypes are local combinations of genetic polymorphisms that tend to be co-inherited. Think of boundaries between haplotype blocks as being like the grooves in a bar of chocolate that permit it to be broken easily into bite-size fragments.

The International HapMap Project collects and curates haplotype distributions from several human populations. SNPs are its raw material, from which it identifies the correlations among them. Phase I of the project, published in October 2005, had the goal of measuring the distributions of at least one SNP every 5 kb across the whole human genome. Blood samples were provided by 269 individuals from four continents (see Box 1.5). Even this is a very sparse representation of the populations of those geographical areas. Over 1 million SNPs of significant frequency (>5%) were documented. In addition, ten selected 500 kb regions were fully sequenced from 48 of the samples. Phases II and III have extended the analysis of the samples to determine an additional 4.6 million SNPs from the same individuals.

The work of the International HapMap Consortium, together with other studies, shows that:

- Most of the variations appear in all populations sampled. Some of the inter-population differences reflect different relative amounts of the same SNPs.
- A very few SNPs are unique to particular populations. For example, out of over 1 million SNPs, only 11 are consistently different—in the sample studied—

BOX 1.5

Origin of samples for the International HapMap Project, Phase I

Population origin	Location	Number of individuals	Relationships
Yoruba	Ibadan, Nigeria	90	30 parent–offspring trios
Northern and western European descent	Utah, US	90	30 parent–offspring trios
Han Chinese	Beijing, China	45	—
Japanese	Tokyo, Japan	44	—

Why the choice of parent–offspring combinations? A difficulty in determining haplotypes in heterozygous regions of diploid chromosomes is how to determine which SNPs lie in the same DNA molecule. Comparison of parental and child sequences can sort the observed SNPs into haploid contributions.

Source: The International HapMap Consortium. (2005). A haplotype map of the human genome. Nature 437: 1299–1320.

ETHICAL ISSUES FACED IN THE HAPMAP PROJECT

The International HapMap Consortium paid due attention to ethical, legal, and social issues. Informed consent of the donors preceded collection of samples. The procedure for informed consent involved not only individual agreement, but also community engagement, including interactive public explanation of the project.

Samples were labelled anonymously. In fact, more samples were collected than used (similar in some ways to the principle of issuing blank cartridges to some members of a firing squad). Nevertheless, characteristics of a population constitute personal information, the release of which may affect all individuals in the population, including those who were never asked to contribute a sample, and even those who refused. For this reason, the HapMap Consortium did not collect potentially useful medical information about the sample contributors, even under the protection of consent and anonymity.

between all individuals of European origin and all individuals of Chinese or Japanese origin.

- The genomes of individuals from Japan and China are very similar, suggesting more recent common ancestry than other population pairs in the study.
- The X chromosome varies more between different populations than other chromosomes. This may arise from the fact that males contain only one X chromosome, the genes on which are, therefore, more subject to selective pressure. Recombinations of X chromosomes can occur, but only in females.
- Lengths of haplotype blocks vary among the different sources of samples. They tend to be shorter among populations from Africa, consistent with the idea of an African origin of the human species. The idea is that the older the population, the larger the number of generations, and the greater the chance of recombination.

A clinically important haplotype: the major histocompatibility complex

In the human genome, proteins of the major histocompatibility complex (MHC; known in humans as the human leucocyte antigen (HLA) system) are encoded in a ~4 Mb region on chromosome 6. In vertebrate species, each individual expresses a set of MHC proteins selected from a diverse genetic repertoire of the species. The system is highly polymorphic, with 50–150 alleles per locus, higher sequence variability than found for most proteins. The set of MHC proteins expressed defines a partial haplotype of an individual. Compared with other haplotype blocks, the MHC region shows unusually wide individual variation.

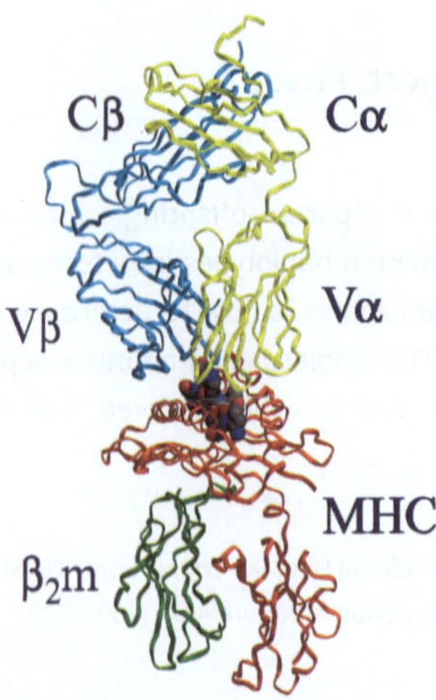

Figure 1.5 The immunological distinction between 'self' and 'non-self' resides in the proteins of the major histocompatibility complex (MHC) and their interaction with T-cell receptors. This picture shows a human T-cell receptor in complex with a class I MHC protein and a viral peptide. MHC proteins have broad specificity, each binding many peptides including those of self and non-self origins. Cell surfaces contain large numbers of MHC–peptide complexes, among which those binding foreign peptides are a small minority. T-cell receptors, in contrast, have narrow specificity, and pick out the complexes containing foreign peptides, like a professional antiques dealer spotting a valuable item in a car boot sale. (T-cell receptor domains: α and β constant domains, Cα and Cβ, α and β variable domains, Vα and Vβ; MHC domains include β_2-microglobulin.) Independent of their scientific significance, many people have found attractive the purely calligraphic features of these diagrams. They appear on stationery, mugs, clothing, and even tattoos.

The MHC region contains over 120 expressed genes, coding for proteins that:

- provide the mechanism by which the immune system distinguishes 'self' molecules—those to be tolerated—from 'non-self' molecules—those recognized as foreign invaders that must be repelled (see Figure 1.5);
- determine individual profiles of competence for resistance to diseases;
- are useful markers for determining relationships among populations of humans and animals, and for tracing large-scale migrations and population interactions;
- help determine suitable donor–recipient pairs for organ transplants.

MHC proteins mediate 'self–non-self' recognition in the immune response: MHC proteins bind peptides and present them on cell surfaces. The triggering event in alerting the immune system to the presence of a foreign protein is the recognition by a T-cell receptor of a complex between an MHC protein and a peptide derived from denaturation and cleavage of the foreign protein (see Figure 1.5). In addition to triggering immune responses in mature individuals, MHC–peptide complexes are also involved in the removal of self-complementary T cells in the thymus during development, at the stage when the distinction between self and non-self is 'learned'.

MHC haplotype influences autoimmune diseases—breakdowns in self–non-self-distinguishability that result in a person's immune system attacking his or her own tissues. Examples of autoimmune diseases include rheumatoid arthritis, multiple sclerosis, type I diabetes, and systemic lupus erythematosus.

MHC haplotypes determine patterns of disease resistance: Different MHC molecules have different binding specificities and can present different sets of peptides. People whose MHC molecules do not effectively present epitopes from a particular pathogen will be more susceptible to infection. For instance, MHC haplotype is a predictor of survival horizon in people infected by human immunodeficiency virus.

MHC haplotypes control donor–recipient compatibility in transplants: Surgical patients, if not immunosuppressed by drugs, will reject transplanted organs—unless the donor is an identical sibling—because the transplant is recognized as foreign.

KEY POINT

A person's MHC haplotype is the set of alleles for over 100 highly polymorphic sites. MHC haplotype is an explicit signature of individuality, governing transplant rejection, resistance to infection, and, reportedly, even mate selection.

Genome projects and our current library of genome information

The first genome sequenced was that of the single-stranded DNA virus, bacteriophage ϕX-174. F. Sanger and co-workers published this result, 5386 bases, in 1977. Recognition of the importance of sequencing stimulated intensive efforts to improve and automate the techniques. A major breakthrough was the replacement of the autoradiography of gels, each nucleotide occupying a separate lane, with four fluorescent dyes, permitting a 'one-pot' reaction. Leroy Hood and colleagues automated this technique, developing a machine that supported a generation of sequencing projects. Recently, a number of novel approaches, or 'next-generation' sequencing techniques, have been developed. We shall discuss these in detail in Chapter 4.

High-throughput sequencing

The past few years have seen truly astounding progress in the development of high-throughput sequencing techniques (see Box 1.6). Initial determination of a draft of a human genome took ten years, at an estimated cost of US$ 3×10^9. At the time of writing, instruments exist that can produce 22 terabytes of sequence data per day. You do the math: this amounts to five human genomes per minute! The estimated cost of a human genome is ~US$100.

Moreover, there is no reason to think that the technical progress will not continue to accelerate. Undoubtedly, by the time this book is published, these specs will be outdated (see Weblem 1.1).

There are two aspects of a large-scale sequencing project. One is the generation of the raw data. Many methods sequence long DNA molecules by fragmenting them, and partially sequencing the pieces. To determine the first genome from a species, these short

BOX 1.6 Vocabulary for high-throughput DNA sequencing

Fragment: a small piece of genomic DNA—typically several hundred bp in length—subject to an individual partial sequence determination, or 'read'.

Single-end read: technique in which sequence is reported from only one end of a fragment (see Figure 1.6).

Paired-end read: technique in which sequence is reported from both ends of a fragment (with a number of undetermined bases between the reads that is known only approximately).

Read length: the number of bases reported from a single experiment on a single fragment.

Assembly: the inference of the complete sequence of a region, by piecing together data on sequences of individual fragments from the region.

Contig: a partial assembly of data from overlapping fragments into a contiguous region of sequence.

Single-end read

Paired-end read

Figure 1.6 Sequence data reported from fragments using single-end sequencing and paired-end sequencing techniques. Data reported arise from red and green regions. The lengths of the black regions are known only approximately, from the fragment length. Typically, fragments used are about 200 bp long, varying in length by about 10%.

De novo *sequencing:* determination of a full-genome sequence without using a known reference sequence from an individual of the species to avoid the assembly step.

Resequencing: determination of the sequence of an individual of a species for which a reference genome sequence is known. The assembly process is replaced by mapping the fragments onto the reference genome.

Exome sequencing: targeted sequencing of regions in DNA that code for parts of expressed proteins (exons). There are approximately 180 000 exons in the human genome.

Variability, single-nucleotide polymorphisms (SNPs): variability is the differences among genome sequences from different individuals of the same species. Much observed variability has the form of isolated substitutions at individual positions, or SNPs.

RNAseq: sequencing the contents and composition of the RNAs in the cell, the transcriptome, by conversion of RNA to complementary DNA and sequencing the result, or directly sequencing the RNA.

ChIP-seq: sequencing of the fragments of DNA to which particular proteins are known to bind. (ChIP = Chromatin ImmunoPrecipitation.) ChIP-seq permits the identification of the targets, in the genome, of DNA-binding proteins.

Methylation-pattern determination: comparison of sequences of native DNA, and DNA after treatment with bisulphate to convert unmethylated C to U. Methylation of C residues in CpG dinucleotides at the 5′ position is the most common signal in animals for repression of transcription.

sequences must be assembled into the whole sequence, using overlaps between the individual fragments.

The typical length of the individual short sequences reported is called the read length of the method. The goals of contemporary technical development are to increase not only the number of bases sequenced per unit time and per unit cost, and the accuracy of the sequence reported, but also the read length. This is particularly important in sequencing regions containing repetitions. The very high throughput sequencing capacity of new instruments allows addressing several types of biological questions.

Both generation of raw data and assembly depend crucially on effective and efficient computer programs.

De novo sequencing

The most ambitious type of sequencing project is the determination of the complete sequence of the first genome from a species. In older sequencing methods, the total DNA must be broken into fragments, typically about 200 bp long. High-throughput sequencing can produce partial sequence information for such fragments, either from one end only, or from paired ends (see Figure 1.6). Paired-end sequencing reports partial sequences from both ends of the same fragment. In either case, the number of bases reported is the read length. The number of unknown bases between the paired ends is limited (by the estimate of the overall fragment length) but the exact value is unknown.

Next, it is necessary to assemble the genome, from the sequences of overlapping fragments. A contig is a partial assembly of fragments, into a contiguous stretch of sequence. As the assembly proceeds, the contigs grow in length, like the completed portion of a jigsaw puzzle. Assembly requires a sufficient number of fragments to cover the entire genome, with enough replicates to be able to detect errors. The ratio of the total number of bases sequenced to the genome length is the coverage of the data set. There is now consensus that to achieve complete and accurate assembly of a novel genome requires collection of data with a coverage of 30 or 50 (30X or 50X). That is, each region in the genome aligns with 30 or 50 individual reads.

For prokaryotic genomes, the process outlined, which corresponds to 'shotgun' sequencing of fragments, allows accurate assembly. For a large eukaryotic genome, achieving the highest quality result requires a genetic map, breaking the assembly problem into smaller pieces. Eukaryotic genome assembly is a very computer-intensive problem. Development of algorithms and computer programs for effective assembly of fragmentary sequence data is a thriving field of research. The resources of many genome centres are such that assembly is the rate-limiting step in the process. New methods with very long read lengths greatly simplify assembly, or even avoid it completely.

Resequencing

Once a reference genome for an individual of a species is available—for example, a published human genome—the sequences of genomes from other individuals of the species are considerably easier to determine. It is not necessary to assemble fragment sequences *de novo,* but merely to map them onto the reference genome. Except for highly repetitive regions, this is fairly straightforward. Coverage must be adequate so that the error rate of sequence determination is less than the frequency of natural variation.

Human genome sequencing

Large numbers of individuals have had their genome sequences determined.

The original academic human DNA sequencing project used DNA primarily from a donor from Buffalo, New York. His identity has not been revealed. Others include J. Craig Venter, the subject of the original Celera Corporation genome sequencing project, James D. Watson, Bishop Desmond Tutu (perhaps analysis might reveal a genetic locus associated with the Nobel Prize), and Stanford University professor Stephen Quake. Actress Glenn Close had her genome sequenced, motivated by a family history that included several individuals who suffered from mental illness.

Venter has also sequenced the genome of his poodle, Shadow. They are the first human–pet combination with both sequences known. (The ultimate 'one-man-and-a-dog' project.)

Several companies now offer personal genome sequencing. Many provide sequencing of mitochondrial DNA or individual loci in nuclear DNA, for tracing of ancestry, or reporting of health-risk factors. The application of DNA sequencing to demonstration of legal relationships—most commonly, paternity testing—is well established.

Until recently, the cost of a full-genome sequence has been prohibitive for most people if the motivation is casual curiosity. However, it is already true that the cost of sequencing a person's DNA is comparable or even less than the cost of a night's stay in hospital in the US. As costs fall, and equipment becomes smaller, and we can derive more and more clinically useful information from sequence data, the conclusion seems inescapable that anyone entering a hospital will have a DNA sequence determination along with taking his or her pulse rate, temperature, and blood pressure (see Problem 1.7).

It took 30 years for DNA sequencing to make the transition from Nobel Prize breakthrough research to secondary school science projects:

- The teacher in a secondary school in New Jersey, US, organized a project to analyse DNA samples of the students in her class. This was not full-genome sequencing but produced limited, genealogy-oriented data. The students compared the results with their own cultural backgrounds.
- The following example is not human DNA sequencing, but usefully hints at the potential variety of applications, and the relative ease with which they can be carried out. For a secondary-school science project, two teenage students sequenced samples of fish from New York City sushi bars. They discovered that, of the samples they could identify, 25% were mislabelled, in fact originating in less-expensive species than advertised.

Human genomes and medicine

The development and delivery of health care challenge biomedical science, engineering, industry, and government—both separately and in their coordination. Not all components of these challenges are within the scope of this book. What is relevant is the increasing scientific sophistication of medical treatment and the shrinking of the distance and time between laboratory discovery and clinical practice: walls between 'pure' and 'applied' science have come tumblin' down.

Of course, immense improvements in human health could be achieved by quite low-tech measures. Obvious examples include educating people about the long-term dangers of obesity, smoking, and alcohol abuse; use of seatbelts in automobiles and aeroplanes; and the provision, to the many people who lack them, of basic medical care and hygienic living conditions—for instance, clean water. In all communities, preserving a mutually nurturing relationship with nature will improve people's mental and spiritual health, which—although in ways not entirely demonstrable yet in the laboratory—has important effects on physical well-being.

Nevertheless, genomics and proteomics have played a central role in the recent transformation of medicine and surgery, making contributions to prevention of disease, to detection and precise diagnosis, and to effective treatment.

Prevention of disease

- Vaccinations are pre-emptive strikes against infectious diseases. They prime the immune system to recognize pathogens. Some vaccines are (almost) completely effective. Vaccines have eliminated smallpox entirely, and polio from much of the world. Even a partially effective vaccine can tip the balance between a devastating pandemic—for example, of influenza—and a controllable set of localized outbreaks. Speedy and innovative development of vaccines provided major weapons against the current COVID-19 pandemic.
- Understanding individual genetic predispositions to disease can, for some conditions, help to preserve health through suggesting changes in lifestyle, and/or medical treatment. An example of a risk factor detectable at the genetic level involves the genes *BRCA1*, *BRCA2*, and *PALB*, mutations in which predispose individuals to a higher rise of breast and ovarian cancers.
- In other cases, such as Huntington's disease, detection of genetic abnormalities will not prevent a disease but can dispel fear of the unknown (see Box 1.7).
- Genetic counselling is a potential preventative approach to avoiding abnormalities or diseases in children that arise from dangerous combinations of parental genes.

BOX 1.7

Huntington's disease

Huntington's disease is an inherited neurodegenerative disorder affecting a total of approximately 0.003–0.007% of people with European ancestry. Its symptoms are quite severe, including uncontrollable dance-like (choreatic) movements, mental disturbance, personality changes, and intellectual impairment. Death usually follows within 10–15 years of the onset of symptoms. The gene arrived in New England, US, during the colonial period in the seventeenth century. It may have been responsible for some accusations of witchcraft. The gene has not been eliminated from the population, because the age of onset—30–50 years—is after the typical reproductive period.

Formerly, members of affected families had no alternative but to face the uncertainty and fear, during youth and early adulthood, of not knowing whether they had inherited the disease. The discovery of the gene related to Huntington's disease in 1993 made it possible to identify affected individuals. The gene contains expanded repeats of the trinucleotide CAG, corresponding to polyglutamine blocks in the corresponding protein, huntingtin. (Huntington's disease is one of a family of neurodegenerative conditions resulting from trinucleotide repeats.) The larger the block of CAGs, the earlier the onset and more severe the symptoms. The normal gene contains 11–28 CAG repeats. People with 29–34 repeats are unlikely to develop the disease and those with 35–41 repeats may develop only relatively mild symptoms. However, people with >41 repeats are almost certain to suffer full Huntington's disease.

Nancy Wexler, the molecular biologist who led the team that discovered the gene responsible for Huntington's disease, came from a family containing many relatives who had the disease. Despite her discovery, ironically she did not test her own genome. It is sad to report that she eventually did develop the disease.

Although formerly there was no treatment for Huntington's disease, drugs are now available that at least mitigate the symptoms, although they do not prolong the patient's lifespan. Tetrabenazine (Xenazine) acts by depleting brain levels of dopamine, serotonin, and norepinephrine. It is applicable to Huntington's disease and other hyperkinetic disorders such as Tourette's syndrome. (Samuel Johnson was the most famous sufferer of Tourette's.) An antisense RNA targeting the huntingtin gene is now in clinical trials.

Detection and precise diagnosis

Early detection of many diseases permits simpler and more successful treatment. With a range of therapeutic strategies often available, precise classification may allow better prediction of the probable course of a disease, and dictate optimal treatment. For instance, oncologists classify leukaemias into seven subtypes. Determination of the subtype from gene expression patterns permits better prognosis and treatment.

Genetic counselling—carrier status

The inheritance of many diseases is simple Mendelian autosomal recessive (albeit in some cases with different degrees of penetrance or expressivity). Sickle-cell disease, cystic fibrosis, and Tay-Sachs disease are examples. Prospective parents may themselves be free of such a disease, but family history may alert them to the possibility of being carriers. Similar considerations apply, to the prospective mother only, for sex-linked diseases such as haemophilia. Companies offering personal DNA test kits report carrier status for many common conditions. However, many people feel that genetic counselling requires consultation with a trained professional, who can arrange for appropriate diagnostic tests, and explain the implications of the results, and discuss choices available.

For instance, phenylketonuria (PKU) testing of neonates is still done by measuring blood phenylalanine concentrations. (Phenylketonuria is an inherited defect in phenylalanine metabolism, which if untreated leads to accumulation, to toxic levels, of phenylalanine and its metabolic by-products.) Most PKU arises from a mutation in the gene for the enzyme phenylalanine hydroxylase, which converts phenylalanine to tyrosine. However, a related condition, hyperphenylalaninemia, can arise from a defect in the metabolism of tetrahydrobiopterin, a cofactor of phenylalanine hydroxylase, by dihydropteridine reductase. DNA sequencing could detect a mutant phenylalanine hydroxylase, but not a low expression level of dihydropteridine reductase. Gene sequencing is used for genetic counselling of prospective parents who might be carriers of a mutant phenylalanine hydroxylase gene, and for prenatal testing.

Discovery and implementation of effective treatment

Advances in basic science provide the background of understanding, the basis for applications. We study the biology of viruses and bacteria to take advantage of their vulnerabilities, and the biology of humans to ward off the consequences of ours.

A large component of the progress in effective treatments for diseases involves the development of drugs. It is a sobering experience to ask a classroom full of students how many would be alive today without at least one course of drug therapy during a serious illness (ignoring diseases escaped through vaccination) or to ask the students how many of their surviving grandparents would be leading lives of greatly reduced quality without regular treatment with drugs. The answers are eloquent. They engender fear of antibiotic-resistant strains of infectious microorganisms.

The traditional drug development process involved identifying a target—usually a protein—either from host or pathogen, the behaviour of which it is desired to affect. A drug is a molecule that intervenes in a living process by interacting with the target. Recent scientific advances have accelerated the drug development process. Identifying metabolic features unique to a pathogen helps to identify targets for antibacterial and antiviral agents. Human proteins provide other drug targets to deal with molecular dysfunction or to adjust regulatory controls. Knowing the structure of a target permits computer-assisted drug design by molecular modelling. Artificial intelligence has proved an effective tool.

An attractive alternative to drugs that interfere with the action of human-encoded proteins is the use of small RNA molecules to suppress expression.

Health care delivery

Some countries in which advanced treatments are available face severe economic impediments to the equitable delivery of medical and surgical care to their citizens. Research creates novel treatments, but many of them are extremely costly. In the US, treatment for serious diseases is already too expensive to be paid for out of most people's earnings, health insurance has not been universal and 'safety nets' to protect the poor and older adults have been cut back. Baby-boomers, born in the late 1940s, are now approaching older adult status. All of these factors will put further pressure on the system. The Health Care and Education Reconciliation Act, which became US law in March 2010, has ameliorated the situation to some extent, provided it is not rescinded.

Considerations of social and economic policy are outside the scope of this book. However, science can make a contribution by improving the efficiency of use of the resources available. For example, many drugs vary in their effectiveness in different patients. A drug may be effective for some patients and useless for others. Some patients may tolerate a treatment easily; others may suffer side effects ranging from discomfort through disability to death.

Analysis of patients' genes and proteins permits selection of drugs and dosages optimal for individual patients, a field called pharmacogenomics (see Figure 1.7). Physicians can thereby avoid experimenting with different therapies, a procedure that is dangerous in terms of side effects—sometimes even fatal—and in any case is wasteful and expensive. Treatment of patients for adverse reactions to prescribed drugs consumes billions of dollars in health care costs. Conversely, being able to predict individual patients' responses can make it possible to rescue drugs that are safe and effective in a minority of patients, but which have been rejected before or during clinical trials because of inefficacy or severe side effects in the majority of patients.

Some specific examples:

- Acute lymphoblastic leukaemia is a childhood cancer treated by thiopurines. In the patient, the enzyme thiopurine methyltransferase breaks down the drug. A genetic variant producing an inactive enzyme threatens build-up of toxic levels of the drug in patients. Screening for the deficiency allows monitoring to determine appropriate dosages.
- Abacavir is a drug used in treatment of AIDS. Four to eight per cent of patients show a serious, potentially fatal hypersensitivity reaction. This is correlated with MHC allele HLA-B*5701. Genomic screening can thereby detect potential hypersensitivity, and guide treatment.

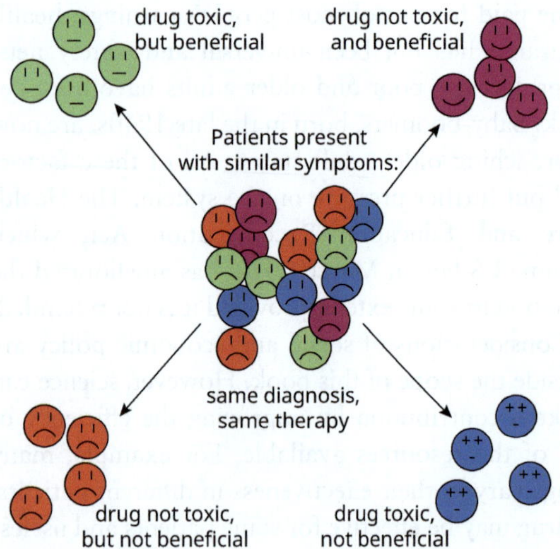

Figure 1.7 The ability to rationalize and anticipate individual differences in responses to drugs can improve the rate of clinical success of treatment. In improving the application of resources by avoiding treatments that are useless or even harmful, science can have a direct effect on the quality and economics of health care delivery.

In current medical practice, a physician may encounter many patients presenting with the same symptoms. One of several possible drugs is prescribed. The result may be (a) upper right: the drug is not toxic and beneficial. Satisfactory! The problem is that it is a matter of hit or miss whether this result will be achieved. (b) Lower left: the drug is not toxic, but not beneficial. The patient returns to the physician with the same complaint, and another drug is tried. In the meantime, the patient has been taking useless drugs, for which someone has paid, and, without effective treatment, the symptoms may have got worse. (c) Upper left: the drug is beneficial but toxic. The patient must cope with the side effects, which may be serious. Would a different drug have been beneficial but without the side effects? (d) Lower right: worst case, the drug is toxic but does not do any good. The patient is worse off than before. The idea of pharmacogenomics is that by use of the patient's genome sequence, the physician can make an informed choice of the best drug, and appropriate dosage, to prescribe for that patient with that condition. A greater proportion of those treated will take the route to the upper right.

- Cytochromes P450 are a family of enzymes in the liver responsible for metabolizing a wide variety of drugs. Sequence variations affect the activities of these enzymes, to the point where lowered activity, or loss of activity, can cause drug toxicity. Genetic tests for variations in cytochrome P450 genes warn of potential overdose dangers. When J.D. Watson's genome sequence was determined, it emerged that he is homozygous for an unusual allele of the drug-metabolizing cytochrome gene CYP2D6. Individuals with this genotype metabolize some drugs relatively slowly. Watson had been taking β–blockers to reduce his blood pressure; however, the treatment made him unacceptably sleepy. Based on information from his genome he is now taking a lower dose.

KEY POINT

Applications of genome sequencing to medicine already include: detection of diseases irrevocably implied by gene sequences, such as Huntington's disease; detection of diseases that an individual has an enhanced risk of developing but for which the risks can be lessened by changes in lifestyle or prophylactic surgery; genetic counselling of prospective parents; and the identification of optimal therapy depending on detailed diagnosis of the variety of the disease, and on the expected reaction of the patient to different drugs.

This century will see a revolution in health care development and delivery. It is possible that a reader of this book will discover a cure for a disease that would otherwise kill him or her. It is extremely likely that Szent-Györgi's quip, 'Cancer supports more people than it kills', will come true. One hopes that this will happen because the research establishment succeeds in developing therapeutic or preventative measures against tumours rather than merely by imitating their uncontrolled growth.

Genes

As they come off the sequencing machines, genomes are long strings of As, Ts, Gs, and Cs, without punctuation marks, captions, or signposts. The challenge is to identify the functional regions and figure out what they do.

- *Protein-coding regions.* Proteins are polymers of amino acids. There are 20 canonical amino acids from which proteins are constructed. The genetic code specifies the relationship between the sequence of bases in DNA and the sequence of amino acids in a protein (see Box 1.2). Because three consecutive bases correspond to one amino acid, any nucleic-acid sequence can be translated into an amino-acid sequence in six ways: Beginning at nucleotide 1, 2, or 3 gives three possible phases of translation, and translating the reverse complement—the other strand—gives another three possible phases. A protein-coding region will contain open reading frames (ORFs) in at least one of the six reading phases. An ORF is a region of DNA sequence, of reasonable length, that begins with an initiation codon (ATG) and ends with a stop codon. An ORF is a potential protein-coding region.

In prokaryotes, a gene is a contiguous region of DNA. In eukaryotes, the coding regions of genes—called exons—may be interrupted by non-coding regions called introns, which must be removed to form the mature messenger RNA that specifies the amino-acid sequence of the protein.

Approximately 1% of human genomes codes for protein.

- *Some regions are expressed as non-protein-coding RNA.* They will show regions of local self-complementarity corresponding to hairpin loops. For instance, genes for transfer RNA (tRNA) will contain the signature cloverleaf pattern (see Figure 1.8).
- *Other regions are targets of regulatory interactions.*

Gene identification is easier in prokaryotes than in eukaryotes. Prokaryotic genomes are smaller and contain fewer genes. Genes in bacteria are contiguous—they lack the introns characteristic of eukaryotic genomes. The intergene spaces are small. In *E. coli*, for instance, approximately 90% of the sequence is coding. Features such as ribosome-binding sites are conserved. If identification of genes in mammalian genomes is like hunting for needles in a haystack, identification of genes in prokaryotes is no worse than hunting for needles in a sewing kit.

It is estimated that human genomes contain about 23 000 protein-coding genes (exclusive of antibody genes—we can produce 10^{18} antibodies, each a distinct protein!). Some regions of the genome are relatively poor in protein-coding genes. These include the subtelomeric regions, on all chromosomes, and chromosomes 18 and X. In contrast, chromosomes 19 and 22 are relatively rich in protein-coding genes.

Most human protein-coding genes contain exons (expressed regions) interrupted by introns (regions spliced out of mRNA and not translated to protein) (see Figure 1.9). The average exon size is about 200 bp. It is primarily the variability in intron size that causes the large size differences among protein-coding genes: the gene for insulin is 1.7 kb long, encoding

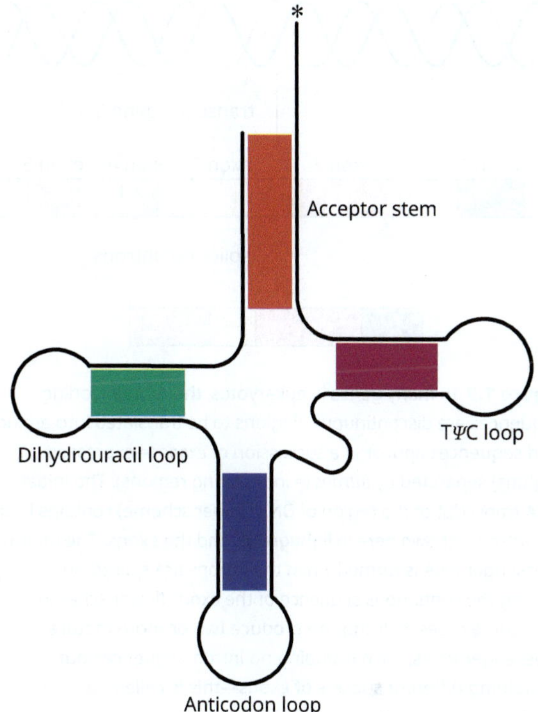

Figure 1.8 The framework of the structure of transfer RNA (tRNA). The * marks the 3′ end, the site of attachment of the amino acid during protein synthesis. tRNA contains several regions of double-helical structure, indicated by coloured patches. These regions have almost perfect complementarity of base pairs. They therefore induce a pattern observable in the DNA sequence, by which it is possible to recognize genes for tRNAs in genomes.

a proinsulin polypeptide of 86 amino acids; and the dystrophin gene is 2400 kb, encoding a protein of 3685 amino acids.

Most multi-exon genes in human genomes show variable splicing affecting the function of the encoded proteins. Approximately 15% of human genetic diseases and cancers involve alternative splicing. For example, Alzheimer's disease (AD) is the most common cause of progressive dementia in older adults in the UK. It is caused by accumulation of protein aggregates in brain cells, notably β-amyloid and tau proteins. Variable splicing in several proteins affects development of Alzheimer's disease:

Protein	Effect of variable splicing
APBB2	Increase in β-amyloid release
RBFox	Increase in β-amyloid production
Presenilin 1, 2	Improper cleavage of amyloid precursor protein increasing β-amyloid deposition
Tau	Different relative amounts of splice variants promote tau protein aggregation

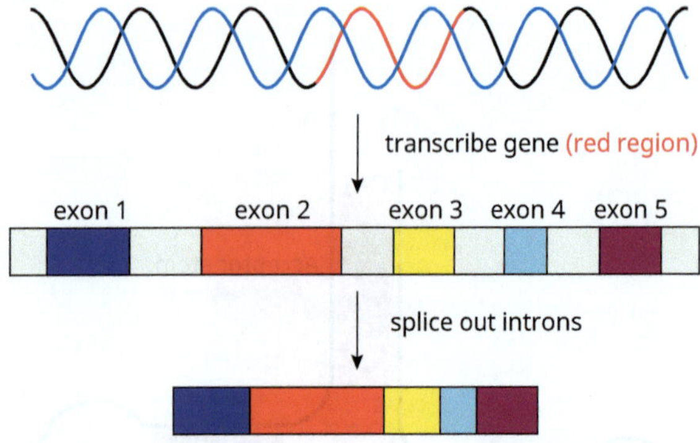

Figure 1.9 In many genes in eukaryotes, the protein-coding sequences are discontinuous. Regions to be translated into amino-acid sequences appear in a succession of *exons* (= expressed regions) separated by *introns* (= intervening regions). The initial RNA transcript of the region of DNA (upper scheme) contains both the introns (shown here in light green) and the exons. The mature messenger RNA is formed when the introns are spliced out, leaving the continuous sequence of the exons (lower scheme).

In some cases, splicing can produce two or more mature messenger RNAs, each including no intron sequences but containing different subsets of exons—this is called **variable splicing**.

In addition, the e4 allele of apolipoprotein E increases the risk and severity of Alzheimer's disease, but this is a difference in DNA sequence, not variable splicing.

Despite their importance, protein-coding genes occupy a small fraction of human genomes—no more than about 1% of the overall sequence. They are distributed across the different chromosomes, but not evenly. Many protein-coding genes appear in multiple copies, either identical or diverged into families. For instance, humans have over 900 related olfactory-receptor genes, and some animals have many more.

In cells, protein-coding genes appear on either strand of DNA. In many cases, unrelated genes are fairly well separated. However, there are examples of genes that partially overlap; and cases of an entire gene appearing, on the complementary strand, within an intron of another gene.

A typical eukaryotic protein-coding gene locus contains the exons and introns, with splice-signal sites at the intron–exon junctions. Transcription of the gene may be under the control of cis-regulatory elements near the gene, either upstream or down. Other regulatory elements may appear elsewhere in the genome, even on different chromosomes.

Often a neighbourhood of a gene contains a set of closely linked related genes. This is because a common mechanism of evolution is gene duplication followed by divergence. It is often possible to follow evolution through a set of successive duplications, producing *tandem* duplicates. However, in rare cases, multiple copies of a human gene appear on different chromosomes. For instance, three copies of the gene for a small nucleolar RNA appear on chromosomes 5, 6, and 9. And of course, polyploids contain multiple copies of genes.

Following identification of a genome sequence, ideally one would hope to be able to infer the corresponding proteome—that is, the amino acid sequences of the proteins expressed. Indeed, amino acid sequences in protein databases are now derived by translating gene sequences. However, there is more variety in the genome–proteome relationship than meets the eye (see Box 1.8).

Some regions of the genome encode non-protein-coding RNA molecules. RNAs, exclusive of messenger RNAs, include but are not limited to transfer RNAs, the RNA components of ribosomes, and microRNAs and small interfering RNAs that regulate transcription (miRNAs, siRNAs, piwiRNAs).

There are about 3000 genes coding for RNAs, exclusive of the mRNAs translated to proteins. It is becoming clear that the RNA-ome is much richer than had been suspected. Except for RNAs involved in the machinery of protein synthesis, such as transfer RNAs and the ribosome itself, most non-coding RNAs are involved in control of gene expression.

Some regions of the genome contain pseudogenes. Pseudogenes are degenerate genes that have mutated so far from their original sequences that the polypeptide sequence they encode will not be functional. In some cases, processed pseudogenes have been picked up by viruses from mRNA, and reverse transcribed. This is recognizable from the fact that the introns have been lost. Because they lack promoters, processed pseudogenes are not expressed as the original proteins. However, sometimes they are transcribed and play a regulatory role by competing with miRNAs for binding to messenger.

BOX 1.8 Gene sequences determine protein structures and functions—but not entirely

Several mechanisms complicate the relationship between genes and proteins.

- In eukaryotes, a mechanism of generating variety at the protein level from a single gene sequence is variable splicing. Variable splicing involves forming a mature messenger RNA from different selections of exons from a gene, retaining the order in which they appear in the genome. It is estimated that 95% of multi-exon protein-coding genes in human genomes produce splice variants. Splice variants enrich both the protein repertoire and the network of protein isoform regulation. Regulatory interactions allow adjustments in response to environmental conditions or developmental programmes. These features support increase in organism complexity.

    Abnormal splice variants contribute to a large fraction of human genetic disorders, and in particular to development of cancer.
- In some cases, multiple promoters lead to transcription of parts of the same region into different proteins. If the reading frames of the different transcripts are not in phase, there will be no relationship between the protein sequences.
- In both prokaryotes and eukaryotes, RNA editing can produce one or more proteins for which the amino-acid sequence may differ from that predicted from the genome sequence. For instance, in the wine grape (*Vinus vinifera*), the mRNAs arising from every mitochondrial protein-coding gene are subject to multiple C→U editing events, most of which alter the encoded amino acid. In humans, nuclear protein-coding genes are subject to editing that changes adenine to inosine (inosine has the coding properties of guanine). The editing, and hence the final amino-acid sequence, can be tissue-specific.

Variable splicing and RNA editing affect the relationship between genome sequences and proteins potentially encoded in them.

But even if we could infer from the genome the amino-acid sequences of all the expressed proteins, our knowledge of the proteome would still be incomplete. A large proportion of cellular activity is dedicated to the regulation of gene expression—the selection of which potentially encoded proteins are expressed, and in what amounts. Moreover, proteins are subject to post-translational modifications, including binding of prosthetic groups (sometimes but not always covalently). Phosphorylation of specific sidechains is a common mechanism for regulating protein activity.

Many proteins are complexes of two or more separately encoded polypeptide chains. None of this information is deducible from a genome sequence. For instance, there is no direct way to tell from a human genome sequence that haemoglobin is a tetramer of two α chains and two β chains, and contains four haem groups.

The immune system stands outside the general assertions about the protein-coding regions of the human genome. The number of antibodies that we produce dwarfs all the other proteins—it is estimated that each of us synthesizes 10^{18} antibodies. The generation of such high diversity arises by special combinatorial splicing at the DNA, not the RNA, level.

KEY POINT

The human genome, and other eukaryotic genomes, encodes genes for proteins and non-coding RNAs (that is, other than messenger RNAs), control regions, pseudogenes (usually non-functional sequences derived from genes by degeneration), and a wide variety of repetitive sequences. Genes that encode proteins in eukaryotes contain exons—regions that can be translated—and introns—regions that are spliced out before translation. The possibility of selecting different subsets of the exons, called variable splicing, adds complexity to the relationship between base sequences of genes and amino-acid sequences of proteins. The expression pattern of different splice variants of the same gene may vary among different tissues or different developmental stages. In many cases, RNA editing creates additional differences between the DNA sequences in the genome and the amino-acid sequences of the proteins.

Other regions of the genome contain binding sites for ligands responsible for regulation of transcription. In assessing the total amount of the genome dedicated to control, one would need to include both the regulatory sites themselves, and all the proteins and RNAs encoded that have regulatory functions.

Repetitive elements of unknown function account for surprisingly large fractions of our genomes. Long and Short Interspersed Nuclear Elements (LINEs and SINEs) account for 21% and 13% of the human genome. Even more highly repeated sequences—minisatellites and microsatellites—may appear as tens or even hundreds of thousands of copies, in aggregate amounting to 15% of the genome.

Protein evolution: divergence of sequences and structures within and between species

Mechanisms of protein evolution

Protein evolution involves generation of variants that differ in sequence, structure and function, expression level, or lifetime. Natural selection then acts to enhance the frequency of alleles with favourable properties. (In some cases, particularly in small populations, genetic drift can alter allele frequencies even in the absence of selection.)

There are two major sources of variants:

1. A protein can explore other proteins in the neighbourhood of its amino-acid sequence, by mutations. These may be nucleotide substitutions, or insertions or deletions, in the genome. Some substitutions do not change the amino acid; some change the amino acid conservatively (for instance, leucine to isoleucine); others produce amino acid substitutions that more severely change the physicochemical character of the amino acid (glycine to arginine substitution is an extreme case). In respect to insertions and deletions, a protein is more likely to survive an insertion or deletion of a multiple of three bases in the genome than an insertion or deletion of 1, 2, 4, 5 . . . bases, that would change the reading frame.

   In many cases, mutations conserve many of the general topological features of protein folding patterns without retaining the exact structure—think of one letter printed in different type fonts. In evolution with retention of function, typically the active site is the best-preserved region of the structure.

2. Proteins can 'mix-and-match' domains. This is a relatively safe way to generate diversity, for recombining domains is likely to produce a protein that at least forms a native structure. The domains may retain their function in the new context, but do not always do so.

What if a protein plays an essential role in the cell? How can it evolve without depriving the cell of the essential function? One possibility, suggested by S. Ohno, is gene duplication. With the formation of two copies of the gene, one copy may retain function, leaving the other free to explore other possibilities. The globins are an example of protein evolution by gene duplication and divergence (see Box 1.9). A second possibility is for a single protein to show multiple functions. Then, depending on the cell's requirements, evolution in different species or strains might optimize the protein for one or another function, or if necessary retain both.

To acquire proteins with novel functions, prokaryotes frequently do not modify proteins already present, but import plasmids containing proteins with the required properties. This is particularly true of the spread of antibiotic resistance among bacteria. It is as if there were a World Wide Web of plasmids from which bacteria can download resistance genes.

Let us survey the subject from increasingly broad fields of view: genes, species, populations, and the entire biosphere.

BOX 1.9 Evolution of human globins

The human genome encodes several proteins related to haemoglobin. The two chains that form the haemoglobin tetramer appear in α and β clusters on chromosomes 16 and 11, respectively (see Figure 1.10). Different haemoglobin genes from these clusters are expressed at different developmental stages.

Differences in gene sequences, differences in the corresponding amino-acid sequences, and differences in three-dimensional structure reflect gene duplication and evolutionary divergence. The globins within either of the α and β clusters are more closely related to one another than members of the α cluster are to members of the β cluster. Other globins in the human genome—myoglobin, neuroglobin, and cytoglobin—are more distant relatives.

Corresponding globins in related species, such as human and horse haemoglobin, have also diverged. In general, the extent of divergence at the molecular level runs parallel to the divergence of the species according to classical taxonomic methods. But the power of comparative genomics

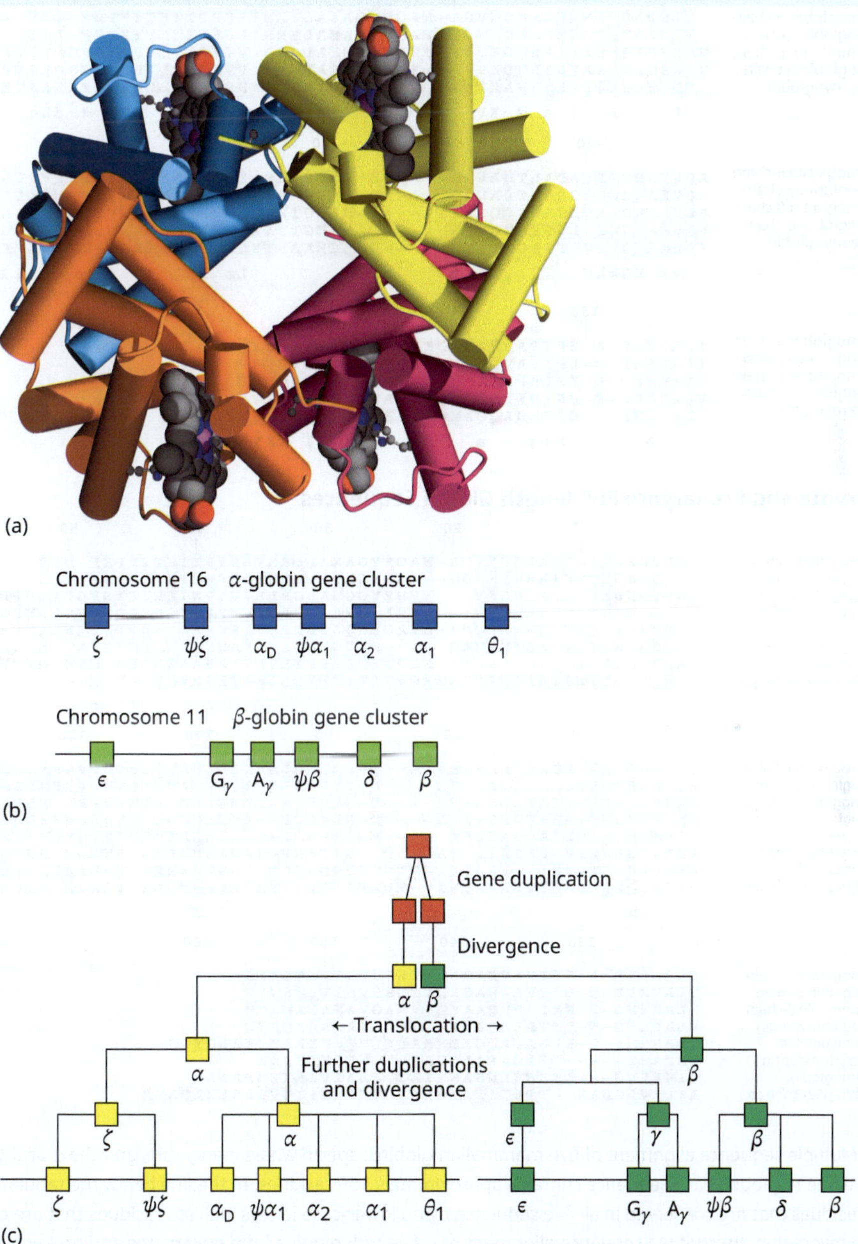

Figure 1.10 (a) The structure of the human haemoglobin tetramer, two α chains and two β chains. Cylinders represent helical regions. (b) Distribution of protein-coding genes and pseudogenes in the α-globin cluster on chromosome 16, and the β-globin cluster on chromosome 11. (c) Haemoglobin genes and pseudogenes are distributed on their chromosomes in a way that appears to reflect their evolution via duplication and divergence. That is, adjacent genes are similar in sequence. The evolutionary tree can be drawn in juxtaposition with the region of the genome, without any intersecting lines.

and proteomics in tracing precise relationships both within and among species is immense.

The basic tool for investigating *sequence* divergence is the multiple sequence alignment (see Figure 1.11). The amino acids are colour-coded by physicochemical type. (Colour coding: green, amino acids with medium-sized and large hydrophobic sidechains; yellow, small sidechains; magenta, polar sidechains; blue, positively charged

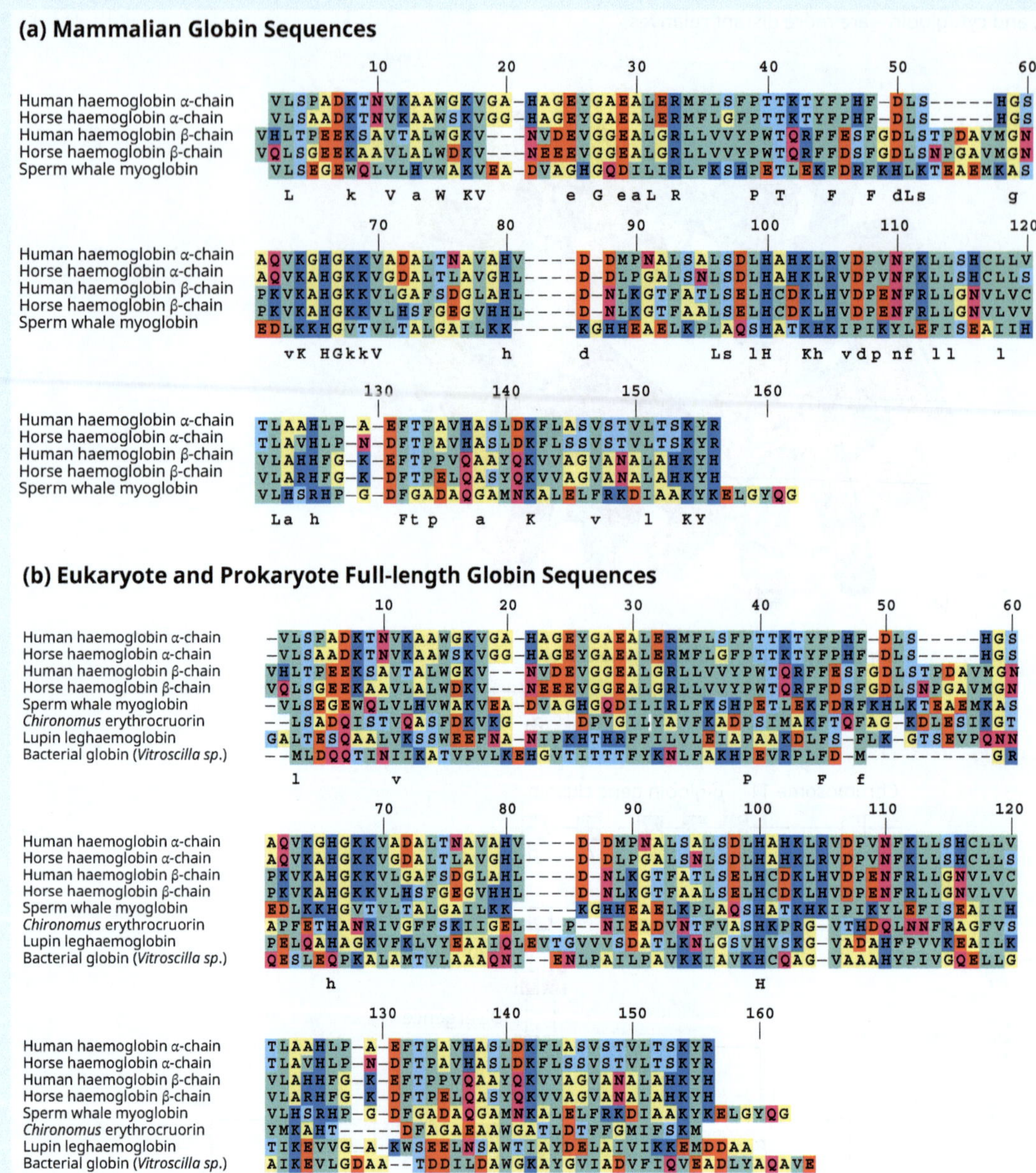

Figure 1.11 (a) Multiple sequence alignment of five mammalian globins: sperm whale myoglobin, and the α and β chains of human and horse haemoglobin. Each sequence contains approximately 150 residues. In the line below the tabulation, upper-case letters indicate residues that are conserved in all five sequences, and lower-case letters indicate residues that are conserved in all but sperm whale myoglobin. (b) Multiple sequence alignment of full-length globins from eukaryotes and prokaryotes. Many fewer positions are conserved than in the mammals-only case. In the line below this tabulation, upper-case letters indicate residues that are conserved in all eight sequences, and lower-case letters indicate residues that are conserved in all but the bacterial globin.

Colours: yellow, small non-polar; light blue, small polar; green, large hydrophobic; purple, large polar; red, negatively charged; blue, positively charged.

sidechains; red, negatively charged sidechains. Upper-case letters in black on the line below the sequences indicate amino acids conserved in all sequences. Lower-case letters in the line below the sequences indicate amino acids conserved in all but one of the sequences.) Therefore, a mutation that leaves the colour unchanged is likely to be a conservative mutation, usually making only minor changes in structure and function. Even among mammalian globins (Figure 1.11a), the patterns of residue conservation and change suggest a hierarchical classification. Position 44, for example, contains tyrosine in the human and horse α-chains, phenylalanine in the human and horse β-chains, and lysine in myoglobin. The reader can easily identify other such positions. The α-chains are more similar to each other than to the β-chains or to myoglobin, the β-chains are more similar to each other than to the α-chains or to myoglobin, but the α- and β-chains are more similar to each other than either is to myoglobin. Given another mammalian globin sequence, it would be easy to identify it as a haemoglobin α-chain, a haemoglobin β-chain, or a myoglobin. This is consistent with the evolutionary tree shown in Figure 1.12. Note that this tree represents the evolutionary divergence of the molecules, not the species.

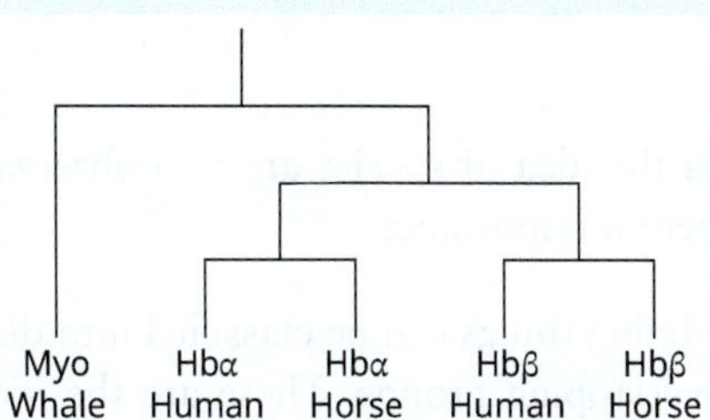

Figure 1.12 Evolutionary tree of five mammalian globins. This tree shows only the *topology* of the relationships. By varying the lengths of the links it would be possible to suggest the relative evolutionary distances also. The split between myoglobin and haemoglobin probably occurred about 800 million years ago. The split between horse and human probably occurred about 70 million years ago.

The basic tool for investigating *structure* divergence is superposition (see Figure 1.13).

Applications of these techniques have become heavy industries in molecular biology. We shall make abundant use of them in the following chapters.

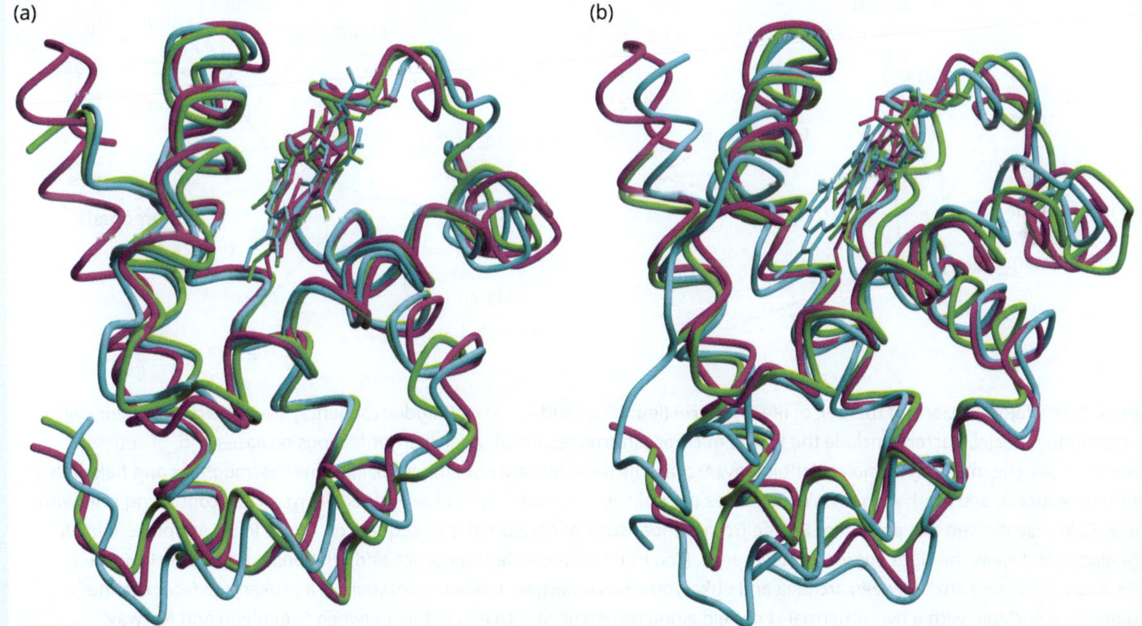

Figure 1.13 (a) Superposition of three closely related mammalian globins: sperm whale myoglobin (magenta), human haemoglobin, α-chain (green) and β-chain (cyan). (b) Superposition of three more-distantly-related mammalian globins: myoglobin (sperm whale) (magenta), insect (*Chironomus*) (green), and plant (yellow lupin) (cyan).

Observe that the overall topology of the structure is conserved. More distantly related species show greater structural distortion.

Species

Underlying the idea of species are two observations of fundamental importance:

(1) Many living things can be classified into discrete non-overlapping groups. These are the individual species.

(2) A catalogue of these groups can be organized into a hierarchy, or tree structure, based on their similarities (see Box 1.10).

Biology has had a long-standing and profound commitment to the idea of species. The species concept is an organizing principle of research and exposition. The word appears in the titles of classic books by Linnaeus and Darwin; that alone would make the concept virtually unchallengeable.

And yet a satisfactory definition of the term species has proved elusive (see Box 1.11). Now, it may be conceded that biology is intrinsically messy. It

BOX 1.10 The 'Tree of Life'

Observing the similarities and differences among living things, species can be organized into a hierarchical scheme, with successively ramifying branches, like a tree. Linnaeus In 1735 published such a systematic classification of species. He introduced the levels of the hierarchy still used today: kingdom, phylum, class, order, family, genus, and

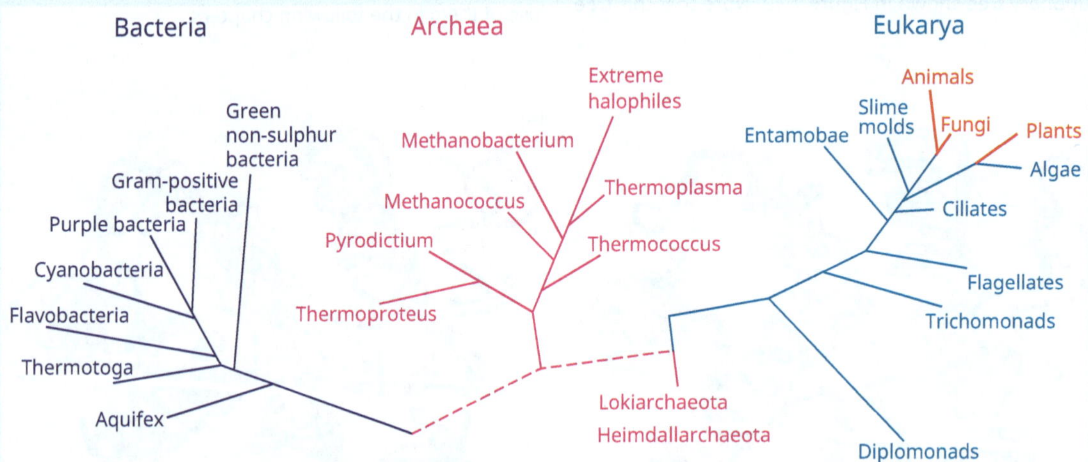

Figure 1.14 Major divisions of the tree of life. **Bacteria** (left; blue) and **archaea** (middle; magenta) are prokaryotes; their cells do not contain nuclei. Bacteria include the typical microorganisms responsible for many infectious diseases and, of course, *Escherichia coli*, the mainstay of molecular biology. Archaea include, but are not limited to, extreme thermophiles and halophiles, sulphate reducers, and methanogens. We ourselves are **eukaryotes** (right; light blue and red)—organisms containing cells with nuclei. C. Woese derived this phylogenetic tree from comparisons of ribosomal RNA sequences. These RNAs are present in all organisms, and show the right degree of divergence. (Too much or too little divergence and relationships become invisible.)

Recently, a 'missing link' between archaea and eukaryotes has emerged. Lokiarchaeotaeum is a prokaryote from marine sediments associated with a hydrothermal vent field along the Arctic Mid-Ocean Ridge, between Greenland and Norway. Lokiarchaeum is archaeal, but is more complex than any other prokaryote. (Lokiarchaea was named for the Castle of Loki, the Norse god of trickery; the nomenclature maintained this convention.)

The Lokiarchaeum genome encodes ~175 proteins heretofore known only in eukaryotes. These include proteins involved in cell membrane remodelling, cytoskeletal activity, and intracellular trafficking and transport processes.

Lokiarchaeotaeum was the first discovered intermediate and is perhaps the best known. But Heimdallarchaeota may be the known member closest to the common ancestor of archaea and eukaryotes. Lokiarchaea and Heimdallarchaeota are members of an archaeal superphylum Asgard.

species. Species are identified by their genus and species; for instance, *Homo sapiens* (common name human), *Bos taurus* (common name cow).[4]

Linnaeus produced a taxonomy, or classification scheme. A century later, the theory of evolution suggested that this is actually a phylogenetic tree—a set of ancestor–descendant relationships. Early classification systems were based on the classical methods of comparative anatomy and embryology. Modern classifications are based on comparisons of genome sequences, a more precise tool.

The most general tree of life reflects the discovery of C. Woese that life can be divided into three domains (see Figure 1.14).

The hierarchical nature of phylogenetic classification depends on the underlying idea that relationships among organisms are 'ancestor–descendant'. This is true for multicellular species, and their relationships do for the most part fit into a hierarchy or tree. But relationships among prokaryotes are more complex, as the result of horizontal gene transfer. (A famous example of horizontal gene transfer was the transformation experiments of Griffith and of Avery, MacLeod and MacCarty on *Streptococcus pneumoniae*, which first showed that DNA was the carrier of genetic information.) Horizontal gene transfer gives relationships that are not simple ancestor–descendant, and the results do not conform to a simple tree (see Figure 1.15).

[4] Classically, the discoverer of a new species has the opportunity to choose the name, and the responsibility to deposit a 'type specimen' in a museum collection, with open access to scholars. (An exception is *Homo sapiens:* the type specimen is Linnaeus himself; his body is buried in the Uppsala cathedral.)

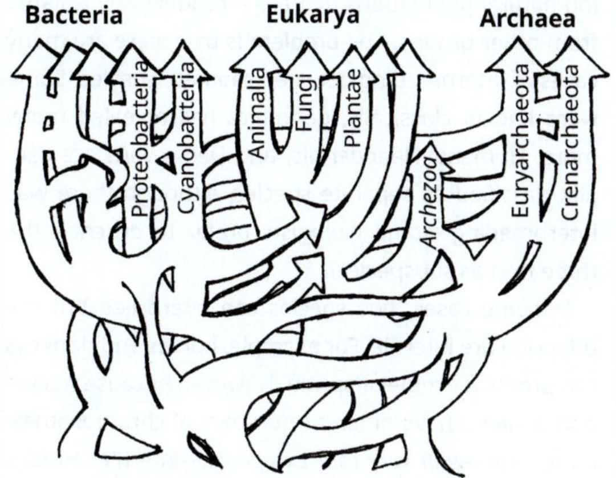

Figure 1.15 W.F. Doolittle produced a somewhat 'tongue-in-cheek' idea of what the 'tree' of life actually looks like, below the Linnaean level, taking into account the non-ancestor–descendant relationships implied by horizontal gene transfer. Used with permission of American Association for the Advancement of Science, from Doolittle, W.F. (1999). Phylogenetic classification and the universal tree. Science (New York, N.Y.) 284, 5423: 2124–2129. https://doi.org/10.1126/science.284.5423.2124. Permission conveyed through Copyright Clearance Centre, Inc.

is unreasonable to demand definitions as 'hard-edged' as in, say, engineering.[5] And although many biologists—some not much less authoritative than Linnaeus and Darwin—have worked hard at it, non-trivial exceptions challenge all the definitions that have been proposed.

The crucial question is: Why have many groups of higher plants and animals coalesced into a hierarchy of discrete groups, with differentiating sets of features? The basic answer, of course, is evolution. More specifically, part of the answer is the discreteness of the evolutionary niches for which different populations compete. Another part is the emergence of reproductive isolating mechanisms, including both geographic barriers and genomic ones.

Evolution requires a mechanism to generate variety in a population of reproducing individuals. All living things are subject to mutations. For diploid organisms, independent assortment of chromosomes, and recombination, provides additional sources of variation. These are not available to haploid organisms. For higher organisms, sexual reproduction is the rule and horizontal gene transfer the exception. We are dealing with ancestors and descendants. Given a differentiation into discrete groups, the hierarchical organization is a consequence of the hierarchy of the 'family tree'. In contrast, prokaryotes must use gene transfer to propagate useful variants. They do not fit the hierarchical classification schemes of classical taxonomy.

Genome sequencing has raised new opportunities for, and new challenges to, the idea of species. One would expect genomic sequencing to clarify the situation:

[5] One of my colleagues begins her molecular biology courses with the following admonition: 'This is biology. If I tell you something is true, that means it happens 90 to 95% of the time. If I tell you that something is impossible, that means it happens no more than 5% of the time. If you don't like it, go major in physics.'

BOX 1.11 Attempts to define the term species include

1. **The Biological Species Concept:** Ernst Mayr defined species as groups of actually or potentially interbreeding natural populations that are reproductively isolated from other groups. The problem is that there are many cases of interbreeding between separate species. Some were known classically; genomics has revealed many more. Humans, Neandertals, and Denisovans are usually classified as separate species, although there was interbreeding (some biologists prefer to describe the three taxa as subspecies).

   In some cases, two species can interbreed but the offspring are infertile. For example, horses and donkeys can produce a mule offspring. However, because horses and donkeys have different numbers of chromosomes, mules are—*with very rare exceptions*—infertile. Horses and donkeys are different species: *Equus caballus* and *Equus asinus.*

2. **The Evolutionary Species Concept:** George Gaylord Simpson defined a species as a lineage—that is, an ancestor-descendant sequence of populations—with its own separate and individual evolutionary history. Reproductive isolation is not a requirement.

3. **The Phylogenetic Species Concept (PSC):** Willi Hennig defined a species as a maximal group, all members of which are descended from a common ancestor and share a common set of fixed heritable differences. There is no explicit reference to the evolutionary history of the group.

   The defining features can include genomic sequence characters. Indeed, even when two species have diverged recently, and not yet developed morphological differences, it is possible to detect *genetic* differences. This is both a strength and an alleged weakness of the phylogenetic species approach. The ease of finding features that differ between populations has led to multiplication of species. Many biologists reject it for this reason: There have been accusations of 'taxonomic inflation'. Moreover, there is at least tacit consensus to limit the consideration to natural populations. Even very closely related laboratory plants and animals, such as transgenic mice, would under the phylogenetic species definition be considered as different species.

   The splitting of traditional species has import beyond academic systematics. Promoting a subspecies might make it eligible for protection under laws protecting endangered species.

DNA sequences of members of different species should differ more than the variation among individuals of the same species. This is true of many species of higher organisms. However, the extents of sequence diversity that distinguish species are quite variable. Genomic distance does not provide a definition of species. In microbiology, a standard in widespread use is to define a bacterial species as a set of strains that maintain ≥97% sequence identity in 16S ribosomal RNA (rRNA). This is both arbitrary and unsatisfactory.

Darwin himself, in *The Origin of Species*, adopted a pragmatic attitude, of course then against the background of the belief that species were immutable. He expressed the hope that, with acceptance of the theory of evolution:

> '. . . we shall at least be freed from the vain search for the undiscovered and undiscoverable essence of the term species.'

'Fraid not.

Populations

A population is an interacting and interbreeding group of individuals of the same species inhabiting the same geographical area. Study of how genomes vary in a population reveals the population structure and its history (see Box 1.12). A population that has developed in isolation, from a small 'founder' group, or that has passed through a bottleneck, will show very narrow variation. All cheetahs, for example, are as closely related as human siblings. A population with relatively high

BOX 1.12 Application of haplotypes to infer relationships between populations: the Barbary macaques

Comparison of genomes between populations can reveal histories of migrations. Here we consider the Barbary macaques; in Chapter 9, we shall treat human history.

The macaques (*Macaca sylvanus*) on the island of Gibraltar are the only wild primates on the continent of Europe—other than football fans. The island has been host to these animals for several hundreds of years. Other populations of the same species exist across the Mediterranean, in northwest Africa.

Although there is fossil evidence for the ancient presence of Barbary apes in Europe, Gibraltar is not a refuge for survivors of that population. Instead, almost all the population arrived in Gibraltar during the Second World War, deliberately imported to enhance morale during those dark days. Politically, Gibraltar is the last remnant of the British Empire in continental Europe: It is said that Gibraltar will remain British as long as the macaque population survives. With just four individuals left on the island in 1943, Winston Churchill ordered that the population be restocked. He minuted the Colonial Secretary: 'The establishment of the apes should be 24. Action should be taken to bring them up to this number at once and maintain it thereafter.'

Where did they come from? Modolo, Salzburger, and Martin analysed a 428 bp segment of mitochondrial DNA from individuals from the Gibraltar colony and from seven natural populations in Algeria and Morocco. Among the individuals sampled, this region contained 24 different haplotypes. The haplotypes differed by between 1 and 26 mutations, all but one mutation being an SNP.

Figure 1.16 shows the clustering of the sequences. Each haplotype is represented by a circle with a radius proportional to the number of individuals bearing the haplotype.

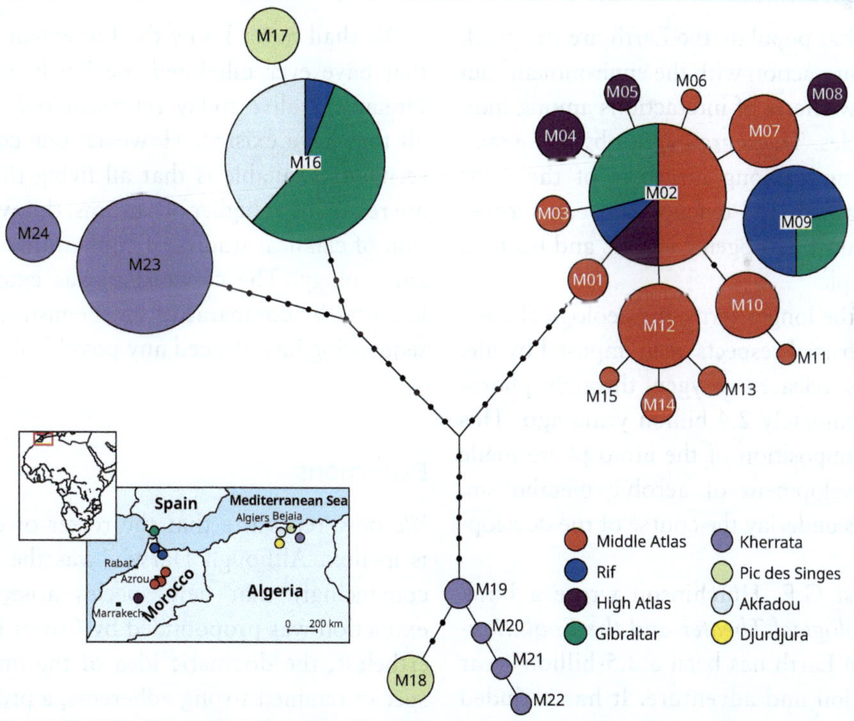

Figure 1.16 Haplotype network from 428 bp segments of mitochondrial DNA collected from free-living Barbary macaques (*Macaca sylvanus*) inhabiting Gibraltar, Algeria, and Morocco. The size of each coloured circle is proportional to the number of individuals bearing each haplotype. Each line segment represents a single mutation. Thus, the sequences represented by M18 and M19, just to the right of the inset map, differ by three mutations. The colours of the circles in the graph indicate locations of sample collection (see inset map). Modolo, L., Salzburger, W., & Martin, R.D. (2005). Phylogeography of Barbary macaques (Macaca sylvanus) and the origin of the Gibraltar colony. Proc. Natl. Acad. Sci. USA 102: 7392–7397. Copyright (2005) National Academy of Sciences, U.S.A.

Colour coding indicates the location of the sample collection. For instance, individuals of haplotype M16 are found mostly in Gibraltar, sometimes in Algeria, and infrequently in Morocco.

The topology of the relationship has several interesting implications.

- There are three major clusters. One cluster (upper right) contains most of the individuals from Morocco. The individuals from Algeria show two populations, from different collection areas, a major one (upper left) and a minor one (bottom centre). One location, *Pic des Singes*, provided individuals with haplotypes linked to each of the well-separated Algerian clusters.
- Dating of the divergence suggests that the Moroccan and Algerian populations separated over 1.2 million years ago.
- It is likely that the female line in the current Barbary macaques originated from both the Moroccan and Algerian populations.

variation is likely to have a longer evolutionary history. After all, most genomic change arises from random drift rather than selection. It is such observations, together with a richer and more ancient fossil record, that suggest that humans originated in Africa.

The biosphere

The organisms that populate the Earth are the product not only of interaction with the environment, but very largely show effects of interactions among individuals and species. There are many obvious examples of competition among members of the same species, and of attack and defence between species. Our struggle against pathogenic viruses and bacteria is a salient example.

Moreover, in the longer term, the geological environment has in crucial respects been imposed by life. Early organisms released oxygen through photosynthesis, approximately 2.4 billion years ago. This change in the composition of the atmosphere made possible the development of aerobic metabolism. Such interactions underlay the course of the development of life.

Palaeontologist G.E. Hutchinson wrote a book entitled *The Ecological Theater and the Evolutionary Play*. Life on Earth has been a 3.5-billion[6] year epic of exploration and adventure. It has included comedy and tragedy. Hutchinson emphasized the interdependence of environment and living things, which reciprocally affect each other. The photosynthetic origin of atmospheric oxygen is a good example.

We shall never know the full extent of the species that have ever inhabited the Earth (see Box 1.13). Organisms alive today represent only a fraction of all that have existed. However, one conclusion that seems unassailable is that all living things on Earth are related. For higher organisms, this was an implication of classical studies of comparative anatomy and embryology. The conclusion was extended to prokaryotes by comparative biochemistry, and genomic sequencing has silenced any possible dissent.

Extinctions

We now recognize that the roster of extant species is in flux. Although Darwin was the first to argue convincingly that new species arise, the idea of extinction was propounded by Cuvier in 1796. Nevertheless, the dogmatic idea of the immutability of species retained strong adherents, a problem Darwin faced later. When US President Thomas Jefferson sent Meriwether Lewis and William Clark to explore the North American continent in 1804, he urged them to be on the lookout for mammoths. Anyone could have hoped that mammoths, extinct in Eurasia,

[6] Throughout this book 1 billion = 10^9.

BOX 1.13 A census of species?

We simply do not know the number of species that have ever existed. Even for the species alive today, only a fraction have been formally described. About 1.5 million species are known to science (see below), but probably at least an order of magnitude more than that exist.

Most palaeontologists agree that living species amount to less than 1% of the number that have ever existed.

Group of organisms	Number of species described
Insects	751 000
Other animals	281 000
Higher plants	248 400
Fungi	69 000
Protozoa	30 800
Algae	26 900
Prokaryotes	4 800
Viruses	1 000

Over half of the described species are insects. Almost 40% of the insects are beetles (order Coleoptera). When J.B.S. Haldane was asked what he had learned about God from his study of biology, he replied that God must have '. . . an inordinate fondness for beetles.'

Given the large numbers of fairly closely related species, one might be tempted to think that the species that lack scientific description are only minor variations on familiar themes.

This is not true.

The Burgess Shale, in the Canadian Rockies near Banff, British Columbia, Canada, contains some of the world's best-preserved fossils from the Cambrian era, ~530 million years ago. The locality contained a rich, predominantly invertebrate, community that left fossils in an unusually high-quality state of preservation. Some of the forms are similar to known animals, both extant and extinct, but others show profound structural differences, including completely different body architectures. The loss of these species has impoverished the natural world in general and the science of biology in particular.

These fossils strongly suggest that the living forms we see are, to a large extent, the result of historical accident and that the course of evolution might easily have been very different.

might have survived in the New World; the idea of a fixed roster of species shaded hope into confidence.

Cuvier believed that extinction of species was the result of major natural catastrophes. Indeed, sometimes, natural catastrophes do cause large-scale simultaneous extinctions (see Figure 1.17 and Table 1.3). An example is the asteroid that landed in what is now the Yucatan peninsula of Mexico, approximately 65 million years ago, at the boundary between the Cretaceous and Tertiary ages. However, in other cases, species reach a more peaceful end of their existence—'dying peacefully in their beds', so to speak—as a result of gradual environmental change.

The origin of new species and the extinction of others are features of the history of life. In some cases, extinctions arise from external events such as an asteroid impact. In others, extinction is the result of action by humans—sailors killing the dodos of Mauritius, or twentieth-century scientists exterminating the smallpox virus in the wild. Preceding these recent events are ancient extinctions of mammoths, and even of Neandertals and Denisovans. Indeed, there is a correlation between expansion of human habitats and extinction. Today, habitat destruction threatens many other species, often an inadvertent result of human activity. To some threatened extinctions we can plead not guilty: Examples include the mollicutes infection ('elm yellows') which is devastating the elm trees of the eastern US and southern Ontario (see Figure 1.18), the fungal 'white-nose disease' threatening bats in eastern North America, and devil facial tumour disease, a transmissible cancer in Tasmanian devils (*Sarcophilus harrisii*), believed to have originated as a single mutation in a single individual.

Climate change, if unchecked, will seriously aggravate threats of extinction for many species.

A major mass extinction is going on right now and we are largely to blame. Human activity is responsible for the disappearance of species through excessive

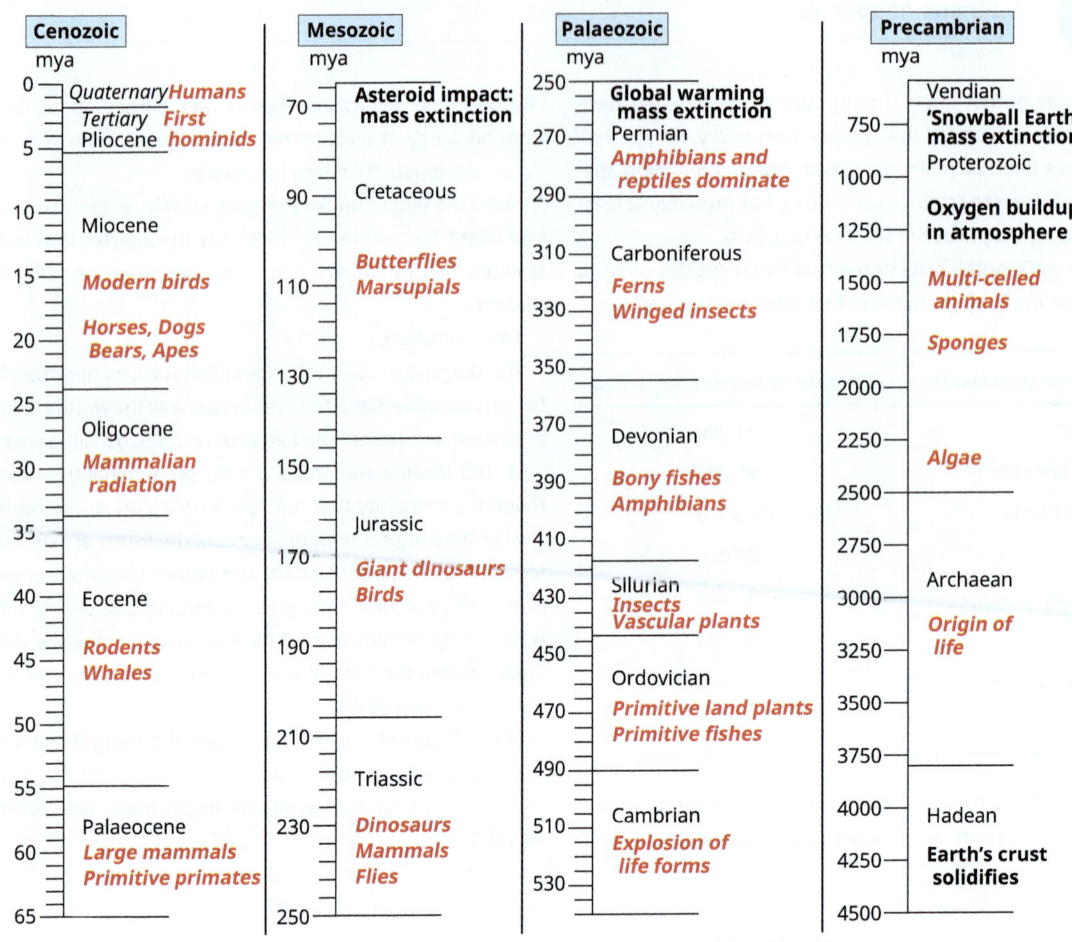

Figure 1.17 Geological ages (e.g., Cenozoic), epochs (e.g., Quaternary), and cataclysmic events (e.g., asteroid impact: mass extinction) in black. First appearance of, or prevalence of, different life forms in red italics. (mya = millions of years ago)

Table 1.3 Ancient extinctions

Geological age	Date (mya)	Cause	% species extinct	% genera extinct
Ordovician–Silurian	443	climate change	86	57
Late Devonian	359–380	climate change	75	35
Permian–Triassic	251	volcanic activity in Siberia	96	56
Triassic–Jurassic	201	volcanic activity	80	47
Cretaceous–Palaeogene	65.5	asteroid impact	76	40

Source: https://education.nationalgeographic.org/resource/mass-extinctions/

hunting, habitat destruction—including but not limited to the effects of climate change—transport of species that replace native ones, and the introduction of toxic chemicals into the environment. Moreover, extinction of one species may cause the loss of others. For instance, some plants are dependent on a particular insect for pollination. Loss of the insect is fatal to the plant. Such networks of interdependence may be quite far-ranging.

North American bees are being killed in great numbers, by the Varroa mite, a parasite that has developed resistance to the pesticides that formerly

Figure 1.18 Elm tree infected with elm yellows. Image UGA5021038, Pennsylvania Department of Conservation and Natural Resources—Forestry Archive, Bugwood.org made available under a Creative Commons Attribution 3.0 License.

controlled it, and by the mysterious colony collapse disorder, which may be at least partly viral in origin. Many Californian almond growers must now import honeybees to ensure the pollination of their trees. The problem is not restricted to almonds: about one-third of our food depends on honeybee pollination.

Extinction of the thylacine (Thylacinus cynocephalus)

It is a sad event when the last individual of a species dies. The last thylacine, or Tasmanian wolf, died in the Hobart Zoo on 7 September 1936 (see Figure 1.19).[7] His name was Benjamin.

Conversely, it is a happy event when a threatened species is rescued.

Survival of Père David's deer (Elaphurus davidianus)

In 1865, Père Armand David (1826–1900), a French Basque priest and biologist, learned of an unusual animal living in the Imperial Hunting Park, near Beijing, China, within precincts reserved to the Emperor. He was able to observe the animals and to take an antler and two hides back to Europe. Recognized as a novel species, the deer were named for the missionary (see Figure 1.20). (Père David also sent the first pelt of a giant panda to Europe.) The animal formerly had a range extending over a large area of western and northern China, but it became extinct in the wild. (A picture of this deer illustrates a seventeenth-century map of Xensi (Shanxi) province, about 500 km southwest of Beijing.) A herd in the hunting park survived. The Emperor gave about a dozen deer as gifts to zoos in Europe and Japan. Herbrand Arthur Russell, 11th Duke of Bedford, installed a small herd on his Woburn Abbey estate.

It was well that he did. The Beijing deer staggered under several blows. There were severe floods

Figure 1.19 The Tasmanian wolf, or thylacine (*Thylacinus cynocephalus*), was a marsupial showing striking convergent evolution in overall body form to the placental wolf/dog family. It had stripes on its back and tail. It lived on 'mainland' Australia and New Guinea several thousand years ago, but its recent distribution was confined to southwestern Tasmania. (A thylacine appears in the state coat of arms of Tasmania, and on the label of a well-known Tasmanian beer.) The thylacine was challenged by the introduction of dogs and hunted to extinction in the early twentieth century. Despite sporadic claims of sightings since the 1930s, no convincing evidence for survival has emerged.

This photograph shows two thylacines that lived in the US National Zoological Park in Washington DC from 1902 to 1905. Photograph by F.W. Goding.

Figure 1.20 Père David's Deer. (Photograph by Tony Pearse, taken at Woburn Abbey in 1993.)

[7] On the same day, Boulder Dam, now Hoover Dam, started operation. Straddling the Colorado River, it provides water and electrical power to the southwestern US, including the rich agricultural lands of Southern California and the city of Los Angeles. The opening of Hoover Dam and the extinction of the thylacine bear simultaneous witness to human control over nature.

in 1894–1895, reducing the Chinese herd to 20–30 animals. The survivors were lost in 1900 during the military campaign to lift the siege of European legations during the Boxer Rebellion. The remaining Père David's deer were collected from European zoos and pooled in the Woburn Abbey herd. This contained 18 individuals, of which 11 were fertile. Some—probably not all—of these were the founder individuals of the entire current population. The deer bred, with the size of the Woburn Abbey herd reaching 250 in 1945.

In 1956, some deer were sent to Beijing. In 1957, a calf was born in China for the first time in over 50 years. They are now maintained in dedicated nature reserves and have also been reintroduced into the wild. The site of the Emperor's hunting garden is now a park devoted to Père David's deer. The deer have come home!

Is extinction reversible?

Will it be possible to recreate living mammoths, thylacines, passenger pigeons, and dodos? There are two basic approaches. You could try:

(1) Selective breeding—to recover a mammoth, start with an elephant, select individuals with long hair, long tusks, hump behind skull, etc. Continue over many generations. If you are impatient, or worried that your grant will run out, choose a species with a shorter generation time; perhaps the passenger pigeon. (Asian elephants take 14 years to reach sexual maturity and have a gestation period of 645 days; but pigeons reproduce every year.) Even were it possible to recreate an animal with a suitable phenotype, there is no guarantee that it would have the genotype of the extinct species. The microbiome also might well not be authentic. If your goal is to restore the ecology of some region, or to populate a theme park, reproducing the phenotype might suffice.

(2) Use the CRISPR/Cas system to modify the genome of a living relative to approach that of the extinct species. The power of this technique is so high that success cannot be ruled out. This approach assumes that you know the genome sequence of the target species, and that there is an extant related species available to act as host. Sadly, the extraordinary fauna from the Burgess Shale fail on both accounts.

There is an active project to reconstruct the extinct thylacine, using CRISPR. The thylacine genome has been determined, from preserved specimens. An extant species, the fat-tailed dunnart (*Sminthopsis crassicaudata*) is a close relative of the thylacine, that might provide a starting point for genetic modification. However, the last common ancestor of the thylacine and dunnart lived 40 million years ago. The divergence is much greater than between mammoths and elephants—it is roughly the same divergence time as between New World monkeys and humans.

Other efforts to recover extinct animals include the quagga (an extinct relative of the zebra), the aurochs, a large bovid widespread in Eurasia (it went extinct in the early seventeenth century), the Pyrenean ibex, the passenger pigeon once superabundant in North America (in 1866, a flock in southern Ontario was described as 1.5 km wide and 500 km long, took 14 hours to pass, and held in excess of 3.5 billion birds), the woolly mammoth, and the dodo.

The future?

We can study the present as thoroughly as we wish, and the past as extensively as we can. What of the future? One thing of which we can be sure is that molecular biology will give us greater control over living things, including but not limited to ourselves. Molecular biologists used to be like astronomers, in that we could observe our subjects but not affect them. Now, methods of genetic engineering are straightforward for bacteria and have had spectacular successes in eukaryotes, including humans. Genetically modified plants and animals are in the field; not without opposition.

It is already possible to replace some dysfunctional proteins through direct administration; for example, insulin therapy for diabetes.

Early approaches to human genetic modification for clinical purposes primarily involved viruses to deliver modified DNA. There have been sporadic successes. Conditions successfully treated include adenosine deaminase deficiency, carcinoma, HIV, metachromatic leukodystrophy, and haemophilia. One of the problems with viral-mediated gene therapy is that one does not have control over where the exogenous DNA will be inserted. There would be danger if it were inserted within (and knocked out) an essential tumour-suppressor gene, for example.

The genome-editing technique, the CRISPR/Cas system, is a real 'game-changer'. It is simple, precise as to target, and can be applied to individuals and to whole ecosystems. However, there are valid fears of (a) 'collateral damage'—genome changes other than those intended—and (b) the ethical, legal, and social consequences of applying these methods.

On 6–8 March 2023, the UK Royal Society, UK Academy of Medical Sciences, US National Academies of Sciences and Medicine, and The World Academy of Sciences organized The Third International Summit on Human Genome Editing, at the Francis Crick Institute, in London. The goals were to discuss progress and challenges and to propose guidelines for the regulation of applications and research.

The participants distinguished:

- *Heritable human genome editing:* editing of human embryos or gametes to be used for human reproduction.
- *Germline human genome editing in research settings:* editing of human embryos or gametes *not* for human reproduction.
- *Somatic human genome editing:* editing of non-reproductive cells, that affect only the person whose cells are edited, and are not hereditable.

It was concluded that *heritable human genome editing* should not be acceptable now. Public discussions and policy debates will be required to enunciate governance frameworks and ethical principles, including standards of safety and efficacy. Governments must delineate what is allowed. (At present, laws vary in different jurisdictions.)

In 2018, a Chinese scientist, Jiankui He, used CRISPR/Cas to edit embryos to remove the CCR5 gene, which encodes the attachment site for HIV-1. Individuals with defective CCR5 have natural immunity to AIDS. The embryos were implanted, and twin baby girls were born, who survived. Dr He was sentenced to three years in prison, and fined the equivalent of UK£300 000. In France, his actions could have led to a maximum of 20 years imprisonment.

Human germline genome editing can proceed in a research context, but human embryos whose genomes have been edited should not be used to create a pregnancy. (In the UK, this is explicitly forbidden by the Human Fertilisation and Embryology Authority, a body set up by The UK Department of Health under the Human Fertilisation and Embryology Act, to regulate human embryo treatment and research.)

Somatic (non-hereditable) human genome editing is acceptable. CRISPR/Cas techniques have cured patients suffering from sickle-cell disease and transthyretin amyloidosis and have been used in the treatment of cancer. Applications to many other diseases are being studied in animal models and human tissue culture. Clinical applications of CRISPR have been approved in the UK, Europe, and the US.

There was considerable discussion at the London meeting about how to make gene-editing therapies accessible to patients in non-wealthy countries. A large percentage of patients with sickle-cell disease live in communities without advanced health care infrastructure. Participants in the Summit expressed concern about how the global medical community can provide patients, worldwide, with cost-effective, affordable therapies, specifically including therapies based on genome editing.

The European Parliament is considering regulations governing the cultivation of plants produced by New Genome Techniques (NGTs, including CRISPR/Cas). In January 2024, a large number of scientists published an open letter, addressed to the Members of the European Parliament, urging the acceptance of NGTs in agriculture.

CRISPRs (clustered regularly interspaced short palindromic repeats) and their applications to genetic engineering

CRISPRs are regions in prokaryotic genomes that provide a defence to viral infection. They are *very* loosely analogous to vertebrate immune systems, in that they can be created in response to a viral

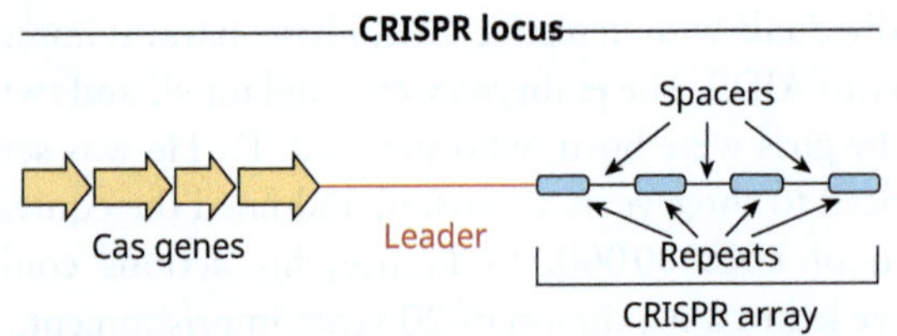

Figure 1.21 Structure of a region containing the CRISPR locus, and genes for CRISPR-associated (Cas) proteins. The CRISPR locus itself contains repeats equivalent to segments extracted from previously encountered-but-survived viral infections. Cas genes encode CRISPR-associated proteins, nucleases that cleave DNA encountered in subsequent infections by viruses containing the sequences stored in the repeats.

challenge. In addition, they survive cell division. (This provides a clear example of inheritance of acquired characteristics.)

CRISPR loci contain copies of the repeat sequences, separated by spacers. They are linked to genes encoding Cas (CRISPR-associated) proteins, which function as nucleases (see Figure 1.21). The repeat sequences are copies of segments of viral sequences. They store a memory of previous infection events. CRISPRs have been likened to 'Wanted Posters'. Upon subsequent reinfection by the virus, the cell detects DNA matching a CRISPR sequence, triggering the Cas nuclease to cleave it.

Large mimiviruses also contain CRISPR-like systems. This is a defence mechanism against 'virophages' that infect the viruses.

When a phage infects a bacterium, assuming that the cell survives, regions of phage DNA are clipped out, replicated, and integrated into a new CRISPR locus, separated by spacers. Transcription of the regions produces CRISPR RNAs (crRNAs) that bind to Cas proteins. The Cas proteins use the bound RNA as a kind of probe. When a match is detected to a region of DNA from a new invading virus, the Cas protein clips the viral DNA. This degradation of the viral DNA is a defence mechanism of the cell (see Figure 1.22). The 'memory' and triggering of a response to infection is reminiscent of the vertebrate immune system, although the mechanism is completely different.

Several groups have brought the CRISPR/Cas system into the laboratory. The mechanism works in eukaryotic cells. The method is applicable to mammalian cells for medical research and clinical intervention, to plants and animals for application

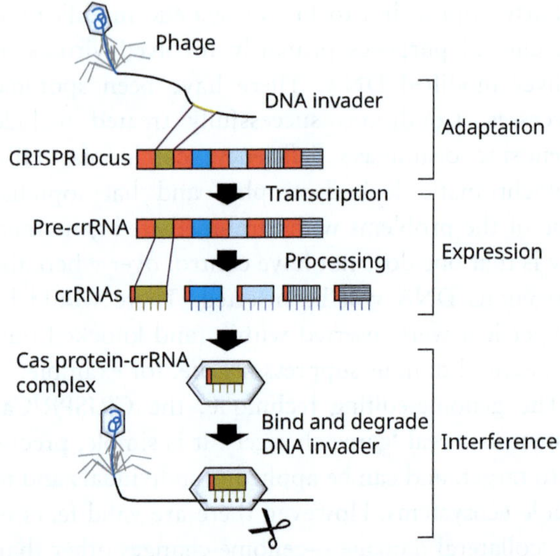

Figure 1.22 Mechanism of bacterial defence using the CRISPR/Cas mechanism. Infection by a virus, the genome of which contains the segment stored in the CRISPR locus, triggers expression of genes to produce the complex between the Cas nuclease and the transcribed guide RNA from the CRISPR repeats. This complex binds to the invading viral DNA, recognized by its complementarity to the CRISPR repeats. The complex cleaves the viral DNA. Exploration of Origin of Acquired Immunity (Press release)—SPRIng-8 web site. (n.d.). http://www.spring8.or.jp/en/news_publications/press_release/2013/131011/. Reproduced with permission. Copyright © RIKEN/JASRI.

in agriculture, and to insects in a possible approach to controlling malaria and other insect-borne pathogens such as Zika virus. For instance, insertion of Cas nuclease, together with CRISPRs derived from HIV-1 LTR promoter U3 sequences, created a line of human cells in tissue culture that were immune to AIDS. This is a direct mimicking of the bacterial usage.

By providing 'guide RNAs', CRISPR-associated nucleases can target endogenous genes. This provides a clean and simple knockout procedure. It is possible to clip out many genes at once. Alternatively, the system can 'paste' a new gene into the space where it 'cut', using eukaryotic repair mechanisms to close the gap. This provides the capability for easy genome editing (see Figure 1.23). Potential clinical applications include repair of mutant genes that cause disease, and knocking out drug resistance genes.

Scientists are also applying CRISPR techniques to the goal of growing, in pigs, human pancreata for transplant. The idea is to use CRISPR to knock out regions of the pig genome that would normally

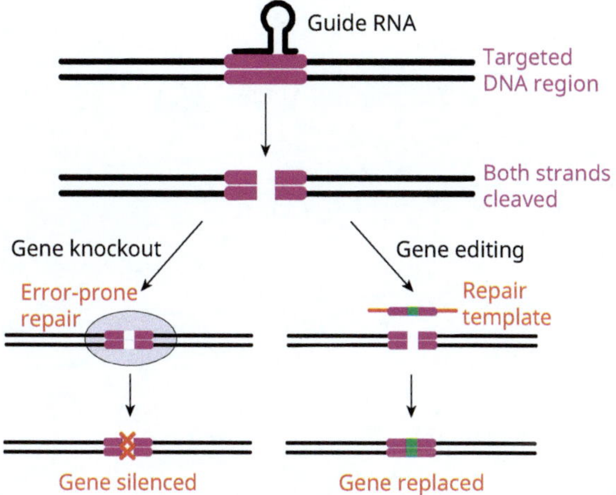

Figure 1.23 Application of CRISPR/Cas techniques to knock out or edit genomes. Adaptation of the CRISPR/Cas system to genetic engineering involves providing a synthetic guide RNA targeting a genomic region of interest. In gene-knockout mode (left), cleavage of both strands leaves the region at the mercy of error-prone repair systems. Their product, albeit an intact double helix, does not replicate the original sequence to give a functional gene product (in the case of a protein-coding region). In gene-editing mode, the repair system is provided with a template, containing the new sequence desired to be inserted. Although originally prokaryotic in origin, the CRISPR/Cas system has been extended to eukaryotic cells.

lead to development of a pig pancreas, and to insert human stem cells that would develop, in the pigs, into a human pancreas.

CRISPR technology is a revolutionary technique. One can look forward to a very large set of important developments.

Gene drive

Gene drive applies CRISPR/Cas technology to accelerate the dispersal of a chosen gene throughout a population. It does not depend on selective pressure. Imperial College biologist Adrian Burt first proposed the idea. K.M. Esvelt, A.L. Smidler, F. Catteruccia, and G.M. Church of Harvard University showed the applicability of the CRISPR/Cas system to implement it.

Normally a novel gene spreads through a population through selective breeding. This can be a slow process, unless the selective advantage is very, very strong and the population quite small, as the process will take many generations.

The following scenario illustrates CRISPR/Cas-mediated gene drive: Suppose you insert into a mosquito a gene that prevents transmission of the malarial parasite, plus a CRISPR locus that cuts the original but not the antimalarial copy of the gene. Then the offspring of a mating between a genetically modified individual and a 'wild-type' mosquito would result in a zygote that was initially heterozygotic. However, the CRISPR locus would cleave the normal gene and the repair process would replace it with a copy of the modified gene. Instant homozygosity for the changed gene. The same would happen in all subsequent generations. The gene would spread rapidly through the population, even if it had no selective advantage (to the mosquito!) (see Figure 1.24).

Gene drive has been discussed as a weapon against the mosquito *Aedes aegypti* that carries the Zika virus that is causing birth defects in Central and South America.[8] The same mosquito transmits dengue fever and yellow fever.

Gene drive can affect not only individual cells or organisms, but entire species—and hence entire ecosystems.

KEY POINT

An enzyme from *Natronobacterium gregoryi*, Argonaute, is a DNA-guided endonuclease capable of gene knockout or editing in mammalian cells. It has some advantages over CRISPR/Cas, including fewer restrictions on target sequences, and a much lower tolerance for base-pair mismatches. The Argonaute and other systems are likely to be developed and applied. At the moment, however, the CRISPR/Cas system is the leader in the clubhouse.

[8] Other biotechnological methods for controlling mosquitoes include releasing:

- mosquitoes infected with bacteria of the genus *Wolbachia*, which reduces the mosquito population and prevents transmission of the virus. A number of different mechanisms contribute to this effect.
- mosquitoes engineered by Oxfordshire-based British biotech company Oxitec to contain a 'self-destruct' gene that causes the next generation of mosquitoes to die in the larval stage. Tetracycline suppresses expression of this gene, permitting creating populations of mosquitoes that carry the gene, and live long enough after release to breed. Oxitec has released its mosquitoes in Brazil, with favourable results.

Resistance by governments to the release of genetically modified mosquitoes was the subject of an article on the editorial page of *The New York Times* on 6 April 2016.

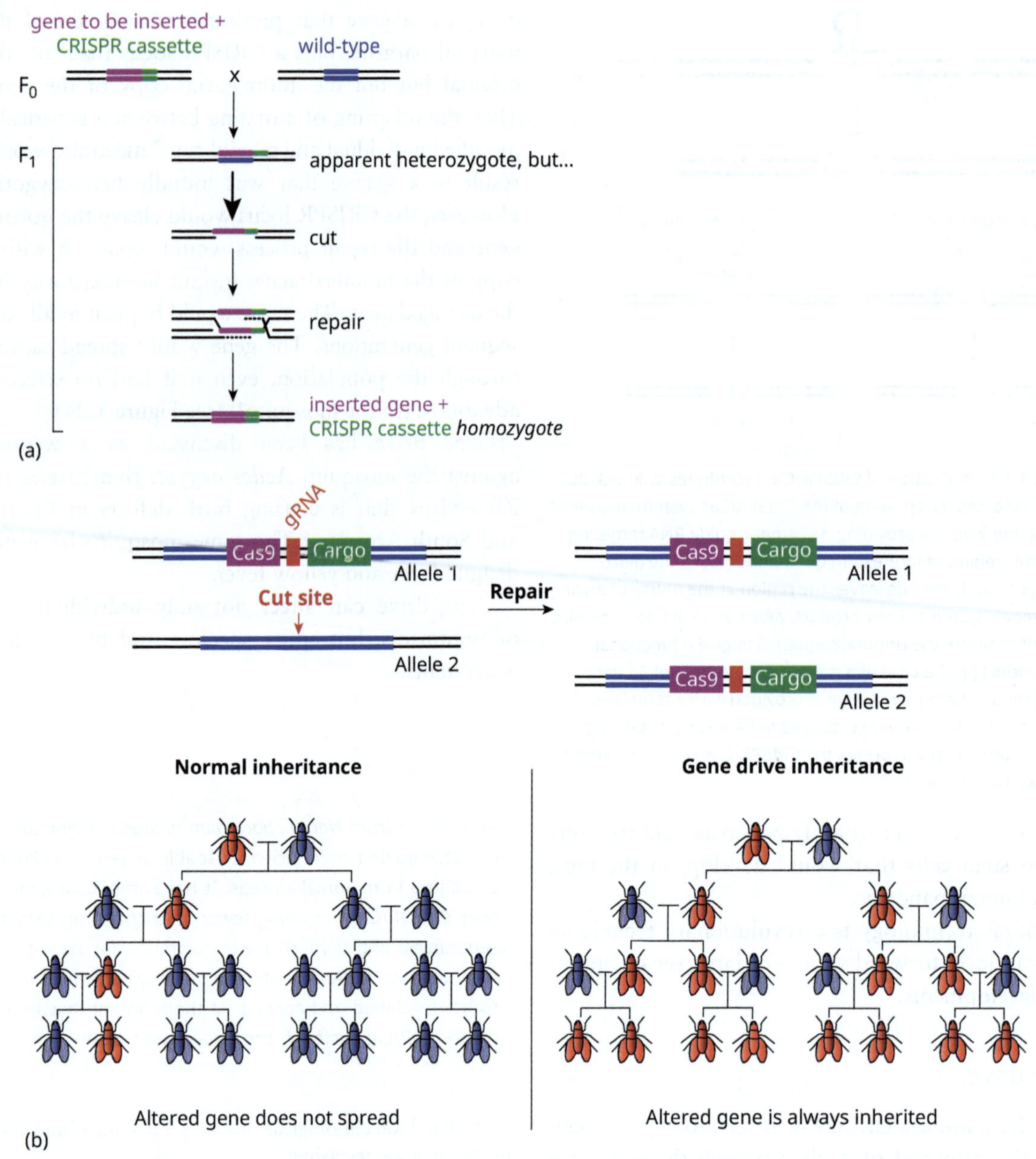

Figure 1.24 Gene drive. (a) A hybrid combining a wild-type allele (blue) with a gene to be inserted (purple) equipped with a CRISPR cassette (green) will produce an apparent heterozygote, converted by CRISPR to a homozygote containing two copies of the inserted gene (purple). (b) Left: unmodified by CRISPR, the altered gene will not spread through the population. Right: inheritance of the altered gene with the CRISPR cassette will convert all descendants to homozygotes of the altered gene. [(b) Picture by Dr Marius Walter.]

Dynamic components of genomes

Transposable elements are skittish segments of DNA, found in all organisms, that move around the genome. They were discovered by B. McClintock in the 1940s in studies of maize (or corn). Transposable elements in Indian maize create a genetic mosaic, giving the ears a mottled appearance (see Figure 1.25). In this case, transposition is fast enough to affect an individual organism. Other transposable elements move more slowly, on evolutionary timescales.

Figure 1.25 The ears of Indian maize are mosaics. The dark pigments are anthocyanins. Yellow sectors arise when a jumping element or transposon interferes with expression or function of the genes for biosynthesis of anthocyanins, during development of individual kernels. (Photograph courtesy of L.D. Graham, Rockingham, VA, US.)

The focus of this section is transposable elements within nuclear genes of eukaryotes. We shall discuss the traffic of genes between organelle (mitochondria and chloroplasts) and nuclear genomes in Chapter 6.

Biological effects of transposable elements include:

- *Sequence broadcasting:* Multiple copies of elements of a sequence may be distributed to various locations in the genome.
- *Altering properties of genes:* The arrival of a fragment of sequence within or in the vicinity of a gene may, if inserted into a coding region, render the gene product non-functional, creating a 'knockout' effect. An inserted segment near a gene may affect its regulation or alter its splicing pattern. (Approximately 20% of human genes have transposable elements in flanking non-coding sequences.) Even an insertion within an intron can affect rates of transcription by slowing down the RNA polymerase as it passes through.
- *Transposable elements are an important engine of evolution:* They provide a mechanism for gene evolution by gene fusion or exon shuffling. Transposable element insertion can cause species-specific alternative splicing patterns. This can produce new protein isoforms. Variant splicing can lead to disease; for example, the cause of a case of ornithine aminotransferase deficiency was a single base change that activated a cryptic 5' splice site in an Alu repeat. This introduced an in-frame stop codon, producing a truncated protein.
- *Causing chromosomal rearrangements:* This can include inversions, translocations, transpositions, and duplications, perhaps through mispairing of chromosomes during cell division. The deletions leading to Prader–Willi and Angelman syndromes (see Chapter 4) are associated with a mutation in the sequence of a nearby transposable element.
- *Leakage of epigenetic modification:* From the landlord's point of view, transposable elements are squatters. They make up 70% of the maize genome. In one sense, it is bad enough that they clutter up the DNA, but, even worse, eukaryotes must defend themselves against the expression of transposable elements. The tactics of defence are to methylate transposable elements or to use small interfering RNAs (siRNAs). Some cancers and other diseases that lead to hypomethylation of DNA can cause transcriptional reactivation of some transposable elements. Methylation also cuts down on the mobility of those transposable elements that require transcription for mobility. However, mechanisms of silencing transposable elements can also affect neighbouring genes.

Genomics and developmental biology

As they emerge from sequencing machines and assemblers, genomes are static data sets. Their implementation in cells is dynamic: genomes contain developmental programmes governing expression patterns of genes in different life stages, differentiated tissues, and reactions and responses to environmental stimuli.

Just as different species show very different body plans despite high similarities in their genes and proteins, different taxa can also show high similarities in their developmental toolkits—that is, sets of genes active in guiding development. For example, homeotic genes are responsible for organization of anterior–posterior (head-to-tail) patterning in the body plans of flies and humans, and even *Caenorhabditis elegans*. The human paired box gene *PAX6* is required for proper eye development, but if expressed in *Drosophila* can transform embryonic wing tissue to an ectopic (= out-of-place) eye. The implication is that despite the very great differences in gross anatomy of the eyes in vertebrates, insects, and octopus, the visual systems arose from a common ancestor. Both the molecular structures of the initial light receptors—the rhodopsins—and the architecture of the neural pathways confirm this conclusion.

Full-genome sequences allow the tracking of similarities and differences in developmental processes during evolution (known colloquially as 'evo-devo'). In particular, it is possible to identify homologues of genes from the developmental toolkit across different phyla. The results both make use of, and illuminate, phylogenetic relationships. This can be thought of as an extrapolation, to the molecular level, of the classical application of embryology to taxonomy. We shall discuss 'evo-devo' in lungfish in Chapter 8.

- Correct taxonomic assignment of species with unusual features can reveal not only phylogenetic relationships, but developmental ones also. The phylum cnidaria contains almost 10 000 species of aquatic animals. Sea anemones and jellyfish are typical examples (Figure 1.26a,b). The worm *Buddenbrockia plumatellae* (Figure 1.26c) was until recently an enigma. Its body plan did not strikingly resemble the familiar radially symmetric cnidaria such as sea anemones or jellyfish; it might easily be thought to be related to nematodes. However, Holland and co-workers have shown, on the basis of sequence alignments of 129 proteins, that *B. plumatellae* is a cnidarian. The significance of this observation for developmental biology is that it extends the body plan observed for cnidaria, requiring investigation of unsuspected aspects of the developmental pathways in this species.

(a)

(b)

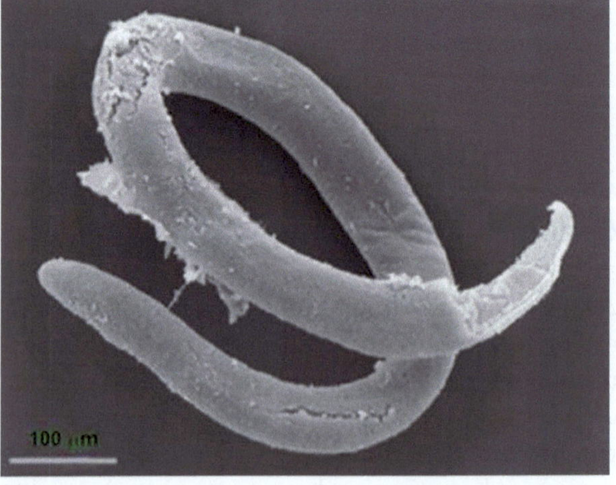

(c)

Figure 1.26 (a) Sea anemone, *Anthopleura sola*, showing the typical radially symmetric cnidarian body plan. Photograph by Jerry Kirkhart made available under a Creative Commons Attribution 2.0 license. (b) Jellyfish—photographed by David Burdick in the Mariana Islands (from NOAA Photo Library, image ref1133, National Oceanic and Atmospheric Administration/US Department of Commerce). (c) Scanning electron microscopy image of a *Buddenbrockia plumatellae*. From: Jiménez-Guri, E., Philippe, H., Okamura, B., *et al.* (2007). Buddenbrockia is a cnidarian worm. Science (New York, N.Y.) 317, 5834: 116–118. https://doi.org/10.1126/science.1142024

- HOX genes are a classic example of how genomes illuminate developmental biology, both within a species and among species. Organisms with bilateral symmetry, including insects and vertebrates, contain HOX genes, which encode a family of DNA-binding proteins (see Figure 1.27). The expression of these genes varies along the anterior–posterior (head-to-tail) body axis and controls the setting out of the body plan. HOX genes have overlapping domains of expression within the body. Different regions of the embryo develop into anatomically distinct regions of the adult, based on the subset of HOX genes expressed.
- Indeed, there is a fascinating mapping between (1) the order of the genes on the chromosome, (2) the relative times during development of the

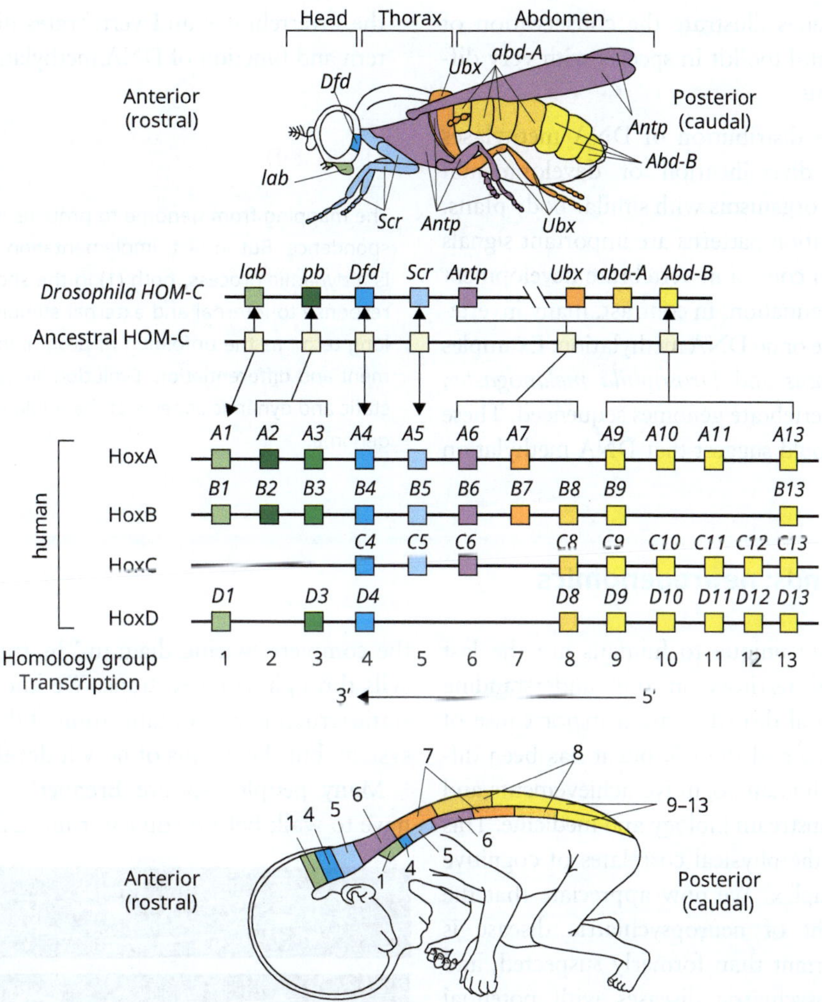

Figure 1.27 Homeotic genes control the development of body plans. In *Drosophila*, a series of genes appearing in tandem along the genome are differentially expressed in different body segments of the fly. The anterior–posterior body axis is collinear with the appearance of the expressed genes in the DNA. Abbreviations over the genes indicate the phenotypic effect of mutations; for instance, Antp = antennapedia, mutations which convert antennae to ectopic legs.

Homologous genes in the HOX clusters of vertebrates show an analogous pattern of expression, also collinear with order of appearance in the genome. By comparing *Drosophila* homeotic genes with human HOX clusters it is possible to infer the outlines of a common ancestor.

In this picture, the distribution of expression patterns in both fly and human is simplified to suppress overlapping domains of expression: Each colour shows the anterior limit of expression of a given gene. Mark, M., Rijli, F., & Chambon, P. (1997). Homeobox genes in embryogenesis and pathogenesis. Pediatr Res 42: 421–429. https://doi.org/10.1203/00006450-199710000-00001, Springer Nature.

onset of their activity, and (3) the order of their action along the body. (The β–globin locus shares the first two of these but not the third.)

HOX genes reveal the duplications that have occurred during vertebrate evolution. Insects and amphioxus have a single HOX cluster. Humans have four HOX clusters. Zebrafish have seven HOX clusters, interpretable as a series of duplications: 1→2→4→8 followed by loss of one to reduce 8→7.

The HOX genes illustrate the conservation of the developmental toolkit in species with very different body plans.

- Conversely, the distribution of DNA methylases illustrates the diversification of developmental toolkits even in organisms with similar body plans.

DNA methylation patterns are important signals for transcription control in vertebrate development and tissue differentiation. In contrast, many invertebrates show little or no DNA methylation. Examples include *C. elegans* and *Drosophila melanogaster*, the first two invertebrate genomes sequenced. These observations would suggest that DNA methylation arose in the vertebrate lineage. However, although *C. elegans* lacks genes for DNA methylases, and *D. melanogaster* has but an incomplete complement, the genome sequences of related nematodes and insects show that the genes were present in invertebrates, and some or all of them have been lost in the specific lineages leading to *C. elegans* and *D. melanogaster*. For instance, the honeybee contains a full, functional complement of DNA methylase genes, and its DNA is methylated. However, it appears that the invertebrates and vertebrates differ in the pattern and function of DNA methylation.

KEY POINT

The mapping from genome to proteins is a static correspondence. But in fact, implementation of the genome is a dynamic process, both (1) in the short term, as the response to internal and external stimuli, and (2) in the long term, as the unfolding of programmes of development and differentiation. Evolution illuminates both the static and dynamic aspects of the implementation of the genome.

Genes and minds: neurogenomics

Cognitive processes unique to humans are the last major unexplored territory in our understanding of life. Psychological disorders are a major cause of disability in Europe and the US, but it has been difficult to integrate human cognitive achievements and problems into mainstream biology and medicine. This is partly because the physical correlates of cognitive events are so complex. We now appreciate that the genetic component of neuropsychiatric disease is much more important than formerly suspected, and that even neuropsychiatric diseases with potential environmental contributions to their aetiology have biochemical correlates—for instance, hyperactive dopamine transmission in schizophrenia.

One approach to cognitive processes in humans is to work our way up from simpler nervous systems. This was Sydney Brenner's motive for choosing *C. elegans* as a model organism. The adult hermaphrodite form has exactly 302 nerve cells and 7000 synapses (see Figure 1.28). J. White worked out the complete 'wiring diagram' by tracing individual cells through serial sections. We know not only the static structure and organization of the adult nervous system, but the details of how it develops.

Many people espouse Brenner's approach: 'You have to walk before you can run.' Others retort that

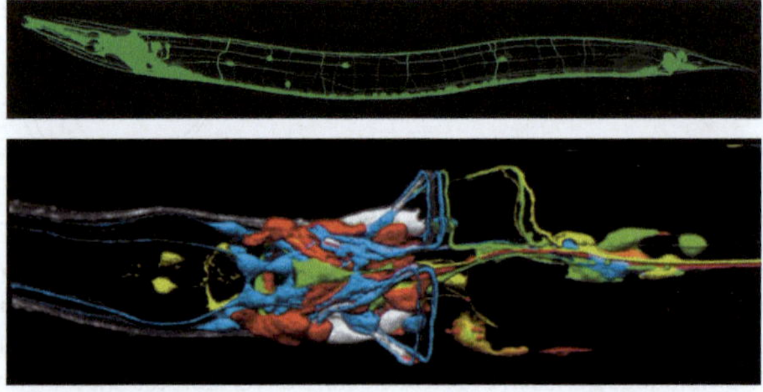

Figure 1.28 Top: the nervous system of *C. elegans*, head at left, labelled with green fluorescent protein. Bottom: the brain of *C. elegans*, with different groups of neurons labelled with different-coloured fluorescent proteins. Reproduced by permission of Prof. H. Hutter.

human cognitive achievements go far beyond those of *C. elegans*. It is undeniable that understanding the minds of worms can have only limited applicability to humans. Nevertheless, many surprising similarities have appeared, even in such distantly related organisms. What analogues of human cognitive phenomena do model organisms show? The basic mechanisms of sensation, such as vision and olfaction, are similar in many animals. Learning and memory are widespread throughout the animal kingdom. Flies learn; even worms learn. Language is one uniquely human attribute. Parts of its genetic component have been traced through clinical disorders. One might assume that 'higher' emotions and sophisticated talents—love, despair, the ability to play chess, or talent for art or music or science—would also be absent from simple model organisms, but in many cases tantalizing analogies do exist. Fruit flies show courtship behaviour under the control of known genes. The objection that fruit flies are merely displaying unsophisticated instinctive behaviour raises the question of how rational and sophisticated is most human activity, even courtship (*especially* courtship).

Genomics provides the bridge between the minds of the different species. Two ways to use model organisms are the study of homologues to illuminate functions of human proteins and the insertion of human genes into the model organisms to study the effects of their expression.

Models for neurological disease, in natural or transgenic *Drosophila* and *C. elegans*, include Parkinson's and Huntington's disease, Friedrich's ataxia (a degenerative disease of the nervous system), and early-onset dystonia (neuromuscular dysfunction producing sustained involuntary and repetitive muscle contractions or abnormal postures). Species more closely related to humans, including zebrafish and mice, provide models for most neuropsychological and neuropsychiatric diseases.

Traditionally, there was a fairly rigid distinction between neurological/physiological/biochemical diseases that affect cognitive performance, and psychiatric diseases believed to have emotional causes and effects. However, we now recognize that even if an illness is caused by an unhealthy emotional environment (and leaving aside for the moment the genetic component of susceptibility to disease in response to such an environment), organic changes—down to the molecular level—underlie the psychological manifestations.

There is now good evidence for a substantial genetic component in numerous neuropsychological and neuropsychiatric disorders, including autism, schizophrenia, attention-deficit hyperactivity disorder, and others. A number of conditions appear in both familial forms, showing relatively simple patterns of inheritance, and sporadic forms, with more complex genetic components and greater influence of environment. Familial forms often show early onset. For instance, familial Alzheimer's disease is a relatively early-onset form of the condition, with early onset defined in this case as appearing before the age of 65. It affects about 10% of those with Alzheimer's disease. The familial form is associated with mutations in genes on chromosomes 1, 14, and 21. In contrast, a mutation in the gene for apolipoprotein E on chromosome 19 causes increased risk of late-onset Alzheimer's disease.

Genetics of behaviour

- *Different strains of* C. elegans *show different social behaviour in dining. C. elegans* can be grown on an agar lawn in a Petri dish. The wild type collected from Australia will congregate in groups to feed. The wild type collected from Britain will eat separately (see Figure 1.29). The difference has been traced to a single amino acid change in a seven-transmembrane helix protein, NPR-1.

> The unity of life is a theme of this book, but perhaps it would be wrong to read too much into this example.

- *The Lesch–Nyhan syndrome was the first correlation discovered between a specific human genetic defect and behavioural anomalies.* It is an X-linked deficiency of a single protein, the enzyme hypoxanthine–guanine phosphoribosyltransferase. The consequent inability to metabolize uric acid properly leads to physical symptoms including gout and kidney stones. But patients also show poor muscle control, intellectual disability, facial grimacing, writhing and repetitive limb movements, and uncontrollable lip and finger biting to the point of severe self-mutilation.

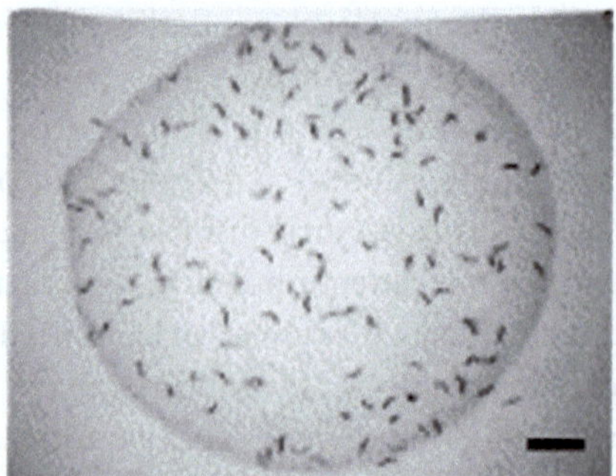

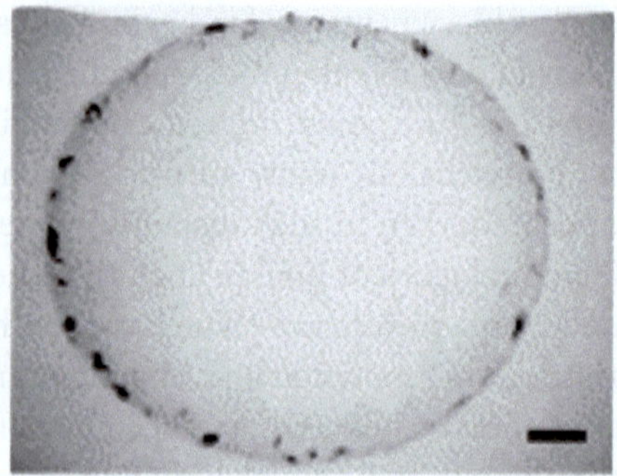

Figure 1.29 Alternative social behaviours of *C. elegans*, feeding on a lawn of agar in a Petri dish. Left: solitary feeding. Right: clumping. The agar is slightly thicker at the edges of the dish, which is why the worms congregate there. From: de Bono, M., & Bargmann, C.I. (1998). Natural variation in a neuropeptide Y receptor homolog modifies social behavior and food response in C. elegans. Cell 94, 5: 679–689. https://doi.org/10.1016/s0092-8674(00)81609-8

- *In fruit flies, T. Tully and co-workers have identified a gene associated with memory.* CREB (cyclic AMP response element-binding protein) encodes a transcription factor, part of a large family of paralogues in mammals and distributed widely in eukaryotes and prokaryotes. Some engineered changes in CREB produced flies that could learn but not store memories; other changes produced flies that learned substantially faster than normal. Alterations in mouse CREB have produced memory impairment.
- *Seasonal affective disorder, a mood change in response to prolonged darkness, is related to the regulation of circadian rhythms.* Jet lag is a related condition. The system of genes involved in circadian rhythms was originally worked out in fruit flies and has homologues in many animals, including humans and *C. elegans*, and in plants.
- *Mutations in an X-linked human gene, DLG3, cause severe learning disability.* Knockout of the mouse homologue PSD95 produces learning-impaired mice. The protein encoded by PSD95 binds to the NMDA receptor, a protein involved in synaptic plasticity. Overexpression of NMDA receptor gives mice superior learning and memory abilities.
- *Several genes are implicated in schizophrenia.* A common effect, hypersensitivity to the neurotransmitter dopamine in the brain, plays a role in schizophrenia. Some antipsychotic drugs block dopamine receptors on neuronal surfaces; conversely, amphetamines, which release dopamine, aggravate the symptoms.
- *Allelic variation in the serotonin transporter gene and neuropsychiatric disorders.* The development of debilitating neuropsychiatric disease in response to stressful events in childhood depends on alleles in the promoter region of the gene for the serotonin transporter (5-HTT). Individuals with one or two copies of the shorter allele of this gene are more likely to exhibit depression and suicidal tendencies than individuals homozygous for the longer allele.
- *A genetic component of antisocial behaviour.* The likelihood of development of antisocial behaviour in response to childhood maltreatment depends on the allele for monoamine oxidase A (MAOA). Maltreated children with a genotype producing high expression levels of MAOA are less likely to become antisocial or violent offenders. A. Caspi and colleagues found that, although only 12% of a sample of people had the combination of low-activity MAOA genotype and childhood maltreatment, these individuals accounted for 44% of subsequent convictions for violent offences. (The gene did not show any large effect on people who had not been maltreated as children.)

KEY POINT

Cognitive abilities are our species' proudest achievement. Neuropsychiatric illness is a correspondingly difficult problem to understand and treat. Both have genetic components that are beginning to be understood, in part through study of analogues in other species.

Proteomics

Proteins are, largely, the executive branch of the cell. Some proteins are structural, such as the keratins that form our hair and the outer horny layer of our skin. Some are catalytic: the enzymes that catalyse metabolic reactions. Others are involved in regulation, including but not limited to the proteins that bind DNA to control expression. Some are involved in signal transduction, receiving signals at a cell surface and transmitting the signal to intracellular regulatory proteins.

To carry out such a wide variety of functions, most proteins show a great diversity of three-dimensional conformations. How do these arise? For most natural amino-acid sequences, there is a unique stable **native state** that under proper conditions is taken up spontaneously. The evidence for this is from the **reversible denaturation** of proteins: A native protein that is heated, or otherwise brought to conditions far from its normal physiological environment, will **denature** to a disordered, biologically inactive state. When normal conditions are restored, proteins **renature**, readopting the native structure, indistinguishable in structure and function from the original state. No information is available to the protein, to direct its renaturation, other than the amino-acid sequence.

There are complications, however, including ligands, post-translational modifications, and formation of homo- or hetero-oligomeric proteins. Prediction of protein function, even knowing detailed protein structure, is an even more difficult problem.

Moreover, the experiments showing that proteins fold to native states based on information contained in the amino-acid sequence were carried out on isolated proteins in dilute solution. Cells are much more crowded, and this makes a difference. For instance, proteins are in danger of aggregation, with fatal consequences for the cell. If you boil an egg and then cool it down, the proteins do not renature. What you have is an aggregate of denatured proteins. It is the aggregation, not the renaturation, that is irreversible. Intracellular protein aggregates are features of many diseases, including Alzheimer's and the prion diseases. Cells have mechanisms to protect themselves: chaperone proteins that inhibit protein aggregation.

Nevertheless, the general conclusion that amino-acid sequence determines protein structure suggests that it should be possible to write computer programs to predict protein structure from amino acid sequence. There has recently been a major breakthrough in successful structure prediction by a program AlphaFold2 from Google subsidiary DeepMind. There is consensus that this is a real 'game-changer'. Two scientists associated with the AlphaFold2 project, John Jumper and Sir Demis Hassabis, were awarded the 2024 Nobel Prize in Chemistry.

Given the disparity between the very large amount of available sequence data, and the relative paucity of experimentally determined structures, this success should unlock the structural information inherent in the sequence databases. We shall discuss this in detail in Chapter 12.

Organization and regulation

The triumph of classical biochemistry was to show that purified components of cells could retain their individual activities after isolation. But within the cell these functions must be integrated and regulated. There is a network of metabolic pathways that governs the possible interconversions of metabolites. A parallel network of regulatory interactions controls the traffic through these pathways. For instance, feedback inhibition may act to shut off a pathway, if the end product is present in adequate concentration. Each network—metabolic or regulatory—has both static and dynamic features.

There is a clear logical distinction between these two networks, but not so clear a physical distinction, as signalling molecules may themselves be metabolites. In feedback inhibition (see Figure 1.30),

A → B → C → D → E

Figure 1.30 A generic metabolic pathway, with the end product E acting as an inhibitor of the enzyme that catalyses the first step, the conversion of A to B. The 'T' symbol conventionally signifies repression. The effect is to shut the pathway down if the end product E is present in adequate amounts.

a molecule participates in both metabolic and control networks.

KEY POINT

Think of intermediary metabolism, catalysed by enzymes, as the 'smokestack industries' of the cell, and the regulatory systems, including signal transduction and expression control, as the 'silicon valley'.

Cells are extremely rich in control mechanisms. The Central Dogma of DNA→RNA→protein suggests several possible leverage points for regulation of protein expression and activity (Figure 1.31). It is approximately true that mechanisms that apply at the levels of expressed proteins have faster effects than those that control gene expression. (When Jacques Monod investigated diauxy in yeast in the 1940s he noted in several cases a 'lag phase' as the cells converted between alternative active metabolic pathways. The cells needed time to 'retool' by changing patterns of gene expression.)

In prokaryotes, a specific focus of transcriptional regulation is at or near the binding site of RNA polymerases to DNA, just upstream of (5' to) the beginning of the gene (see Chapter 11). Repressors can turn off transcription by occluding the binding site, blocking polymerase activity. In contrast, promoters can actively recruit polymerases through cooperative binding, along with polymerase, to a site on the DNA.

Gene regulation in eukaryotes is more complex. Transcription regulators bind to DNA at positions proximal to the gene as in prokaryotes, but also at remote sites. The control of human β–globin expression illustrates such a scheme (see Box 1.14). Regulatory interactions also govern the expression of other transcription factors. Eukaryotic control networks show far greater complexity, in both their logic and their dynamics, than those of viruses or prokaryotes.

The gift of complexity is robustness. Eukaryotic control networks show an ability to reprogramme themselves, to respond to stimuli by changing cell state. The source of robustness appears to be redundancy. Yeast (*Saccharomyces cerevisiae*) has about 6000 genes. Under 'normal'—that is, non-stress—conditions, about 80% of them are being expressed. It is also true that yeast can survive approximately

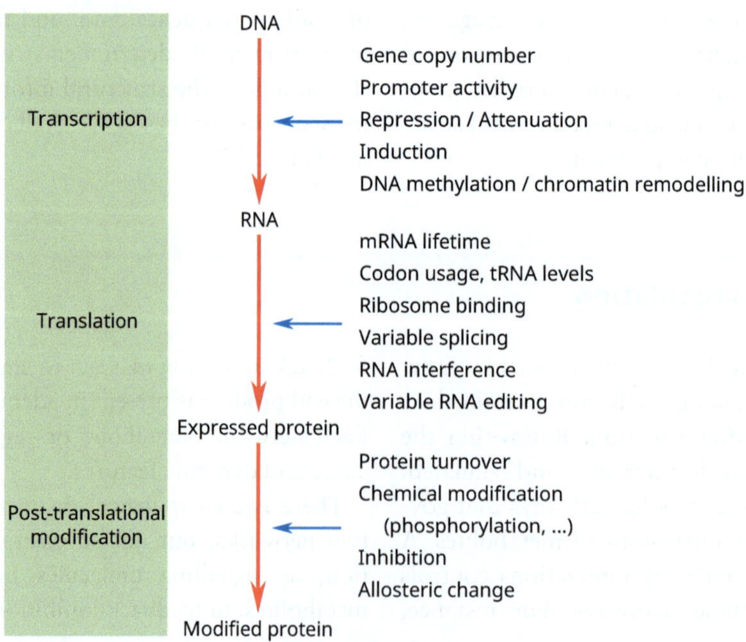

Figure 1.31 Regulation of protein activity involves many mechanisms, active at different stages of the Central Dogma.

BOX 1.14 Control of β–globin gene expression

The protein-coding genes of the globin loci interact with many control regions. The β–globin region includes promoters proximal to individual genes; a locus control region occupying a region between 6 and 22 kb upstream of the most 5′ gene (ε); a 250 bp pyrimidine-rich region 5′ to the δ gene (YR). Enhancer regions may appear on the same chromosome, in some cases near to and in other cases distant from the gene they control, or on entirely different chromosomes.

Control of globin gene expression is asserted for both tissue specificity and developmental progression. In humans, transcription of the β–globin region switches from the embryonic ε–globin in the yolk sac to the foetal γ–globins in the liver (G_γ and A_γ) and finally to the adult β–globin produced in cells derived from bone marrow.

An essential mechanism of control is **modulation of chromatin conformation**. Eukaryotic chromosomes contain complexes of DNA with proteins called histones (see Figure 1.3). Chromatin remodelling is an important mechanism of transcriptional control. Reversible chemical modification of histones, by a variety of reactions including deacetylation, methylation, decarboxylation, phosphorylation, ubiquitinylation, and sumoylation, leads to alterations of the DNA–histone interactions that render transcription-initiation sites more or less accessible. Chromatin conformational changes can also be induced by binding of specific chromatin-remodelling proteins.

Differential sensitivity of sites to DNAse I digestion measures differences in exposure. Some regions near actively transcribed genes are hypersensitive to DNAse I digestion (see Figure 1.32a). The **locus control region** upstream of the β–globin locus consists of five hypersensitive regions. The degree of their exposure is correlated with transcriptional activity.

Regulation of β–globin expression involves interaction of the locus control region with proximal promoters associated with specific genes. The interactions are mediated by a large complex of proteins recruited to the site. The alternative interactions of the locus control region, with foetal- and with adult-expressed genes, suggest that expression is determined by a competition between these interactions (see Figure 1.32b,c).

A well-known enhancer of globin expression is erythropoietin, a glycoprotein hormone encoded on chromosome 7. Erythropoietin does not interact with sequences in the vicinity of the globin locus but works indirectly by binding to a receptor to activate intracellular signalling pathways. Erythropoietin expression is sensitive to oxygen tension. Hypoxia increases erythropoietin production. This occurs naturally at high altitudes; as a result, people who live at

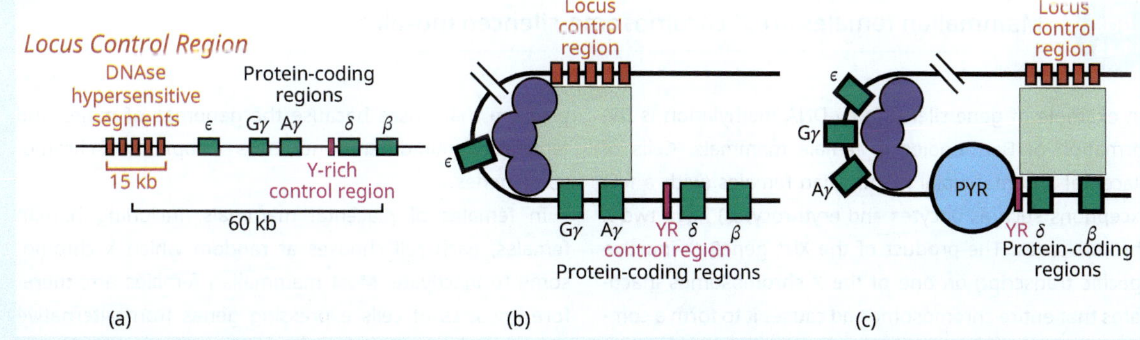

Figure 1.32 (a) β–Globin region showing protein-coding genes and locus control region, consisting of five DNAse hypersensitive segments. Pseudogene $\phi\beta$ not shown. (b,c) Schematic structural model for the control of globin gene expression. Part (b) shows the foetal structure, with interaction between the locus control region and the Gγ and Aγ genes, mediated by proteins (green). Purple circles indicate chromatin-remodelling complexes. In this configuration, the Gγ and Aγ genes will be expressed. In part (c), the cyan circle indicates the PYR complex of proteins, which binds to the pyrimidine-rich (YR; Y stands for pyrimidine) region just 5′ to the δ gene. This binding reconfigures the system: PYR blocks the foetal mode of interaction of the locus control region with the γ region (b). Instead, the locus control region interacts with, and promotes the expression of, the β gene. Used with permission of American Society for Clinical Investigation, from Bank A. (2005). Understanding globin regulation in beta-thalassemia: It's as simple as alpha, beta, gamma, delta. The Journal of Clinical Investigation 115, 6: 1470–1473. https://doi.org/10.1172/JCI25398. Permission conveyed through Copyright Clearance Centre, Inc.

2500 m above sea level have about 12% more haemoglobin than people who live at sea level. Athletes take advantage of this. About three months of adaptation time is necessary to build up this differential.

A surprising player in the game of globin expression regulation is acetylcholinesterase, most famous for its physiological role in neural synapses and neuromuscular junctions. Acetylcholinesterase also regulates globin synthesis.

80% of single-gene knockouts. (It would be interesting to know the overlap of these sets!) Many expressed genes must be redundant, and redundancy provides robustness. Gene regulatory networks are the subject of Chapter 14.

Other mechanisms of transcription regulation in eukaryotes involve changes in patterns of methylation of DNA associated with changes in the structure of chromatin. In differentiation, DNA methylation is a regulatory mechanism that survives cell division (see Box 1.15).

Some mechanisms of regulation act at the level of transcription

Antisense RNA will form a double helix with mRNA, and block transcription. (Antisense RNA is single-stranded RNA complementary to mRNA.)

Introduction of genes for antisense RNA can silence genes. For example, ethylene is a plant hormone that stimulates ripening. Genetic modification—a controversial activity in the context of agriculture—has produced a tomato with longer shelf life. An artificial gene in the 'Flavr Savr' tomato is transcribed to an antisense RNA that greatly reduces translation of a gene involved in ethylene synthesis, delaying ripening.

In RNA interference, a short stretch of double-stranded RNA (~20 bp) elicits degradation, by a ribonucleoprotein complex, of mRNA complementary to either of the strands. RNA interference may have a natural function in defence against viruses. It has been applied in the laboratory to achieve effective gene knockouts in studies aimed at deducing gene functions.

Another mechanism of translation control is the attachment of ligands to the Shine–Dalgarno sequence, blocking the ribosomal binding site in prokaryotic messenger RNA. In *E. coli*, vitamins B1 (thiamin) and B12 (adenosylcobalamin) bind to an mRNA containing transcripts of genes encoding

Mammalian females are X-chromosome-silenced mosaics

An example of gene silencing by DNA methylation is the formation of Barr bodies in female mammals. Cells of placental and marsupial mammalian females (with a few exceptions such as oocytes and erythrocytes) have two X chromosomes. The product of the Xist gene9 (X-inactive specific transcript) on one of the X chromosomes inactivates that entire chromosome and causes it to form a compact, transcriptionally inert object, called a Barr body. In each cell, the Xist gene product remains associated with, and inactivates only, the X chromosome on which it is expressed. Expression of the Xist gene on the other X chromosome is suppressed by cytosine methylation of its promoter, leaving that chromosome normal in structure and activity. As a result, cells of both males and females have one active X chromosome. This is the placental and marsupial mammalian 'solution' of the dosage compensation problem that arises because the genomes of males and females contain different numbers of copies of X chromosome genes.

In females of placental mammals, including human females, each cell chooses at random which X chromosome to inactivate. Most mammalian females are, therefore, mosaics of cells expressing genes from alternative X chromosomes. A visible example of this mosaicity is a calico cat, necessarily a female[9] (see Figure 1.33b). A calico cat has an orange-coat allele on the X chromosome inherited from one parent and a black-coat allele on the other X chromosome. (In the white patches of the coat, neither

[9] Not absolutely true: There are very rare cases of male calico cats with the analogue of Klinefelter's syndrome, karyotype XXY (see Chapter 2).

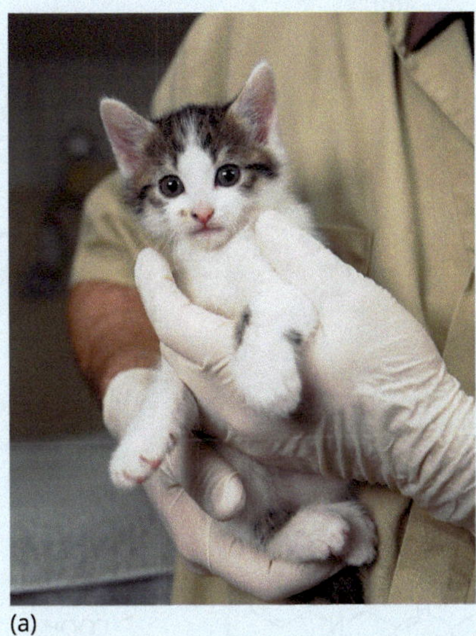

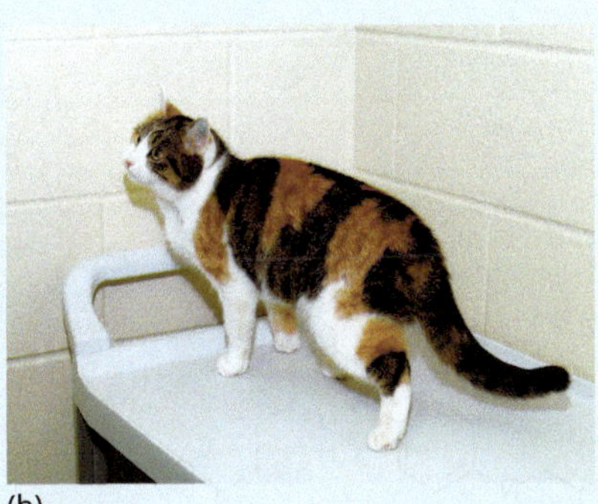

(a) (b)

Figure 1.33 (a) Cc—or Copycat—a cloned cat. (b) Rainbow—a calico cat, the source organism for Cc. The varied-colour appearance is reminiscent of Indian maize (Figure 1.26), and indeed there are some similarities to the mechanisms that created them. Photos reproduced courtesy of The College of Veterinary Medicine & Biomedical Sciences, Texas A&M University.

allele is expressed.) The size of the coloured patches on the coat reveals when the genes were inactivated. (In contrast, in female marsupials, all cells inactivate the paternally derived X chromosome. This is an example of the general phenomenon of genetic **imprinting**, the dependence of phenotype on the parental origin of a gene.)

A difficulty in cloning of higher animals is how to restore pluripotency to the single cell from which the animal will develop. Figure 1.33 shows (a) a cloned cat, named Cc (for Copycat), and (b) its source organism (very loosely, its mother), Rainbow, a calico cat. Although the two cats are genetically identical as far as the nucleotide sequences of their DNA are concerned, Rainbow is a mosaic and Cc is not. The cell that produced Cc had an inactivated X chromosome bearing the orange-coat allele and this inactivation was replicated in all of Cc's cells.

proteins involved in their biosynthesis. This is a kind of feedback inhibition—adequate amounts of a product inhibit its synthesis. However, unlike the more familiar product inhibition of an enzyme, this control is applied at the level of RNA, rather than protein.

Different modes of transcriptional control vary in their dependence on external conditions and many are reversible—some more readily than others. The lactose operon control in *E. coli* is a 'toggle' switch that can respond to both the appearance of lactose and its subsequent withdrawal. Other control processes are cyclic, such as cell-cycle control and diurnal rhythms. Cellular differentiation in mammals is usually irreversible, except in certain forms of cancer. In *Drosophila,* certain stages of metamorphosis involve dedifferentiation.

Some mechanisms of regulation act at the level of translation

Translation of eukaryotic genes may produce several different splice variants. Regulation of translation in eukaryotes may affect the distribution of splice variants produced. Mechanisms include (1) degradation of specific mRNA variants by microRNAs (miRNAs) and siRNAs, which may repress translation of specific splice variants; and (2) splicing factors, RNA-binding proteins that interact with the transcripts of specific exons (or even introns) and interact with the splicing machinery to direct maturation of the mRNA. Back at the transcriptional level, chromatin remodelling may render certain exons inaccessible and affect the splice variant expressed.

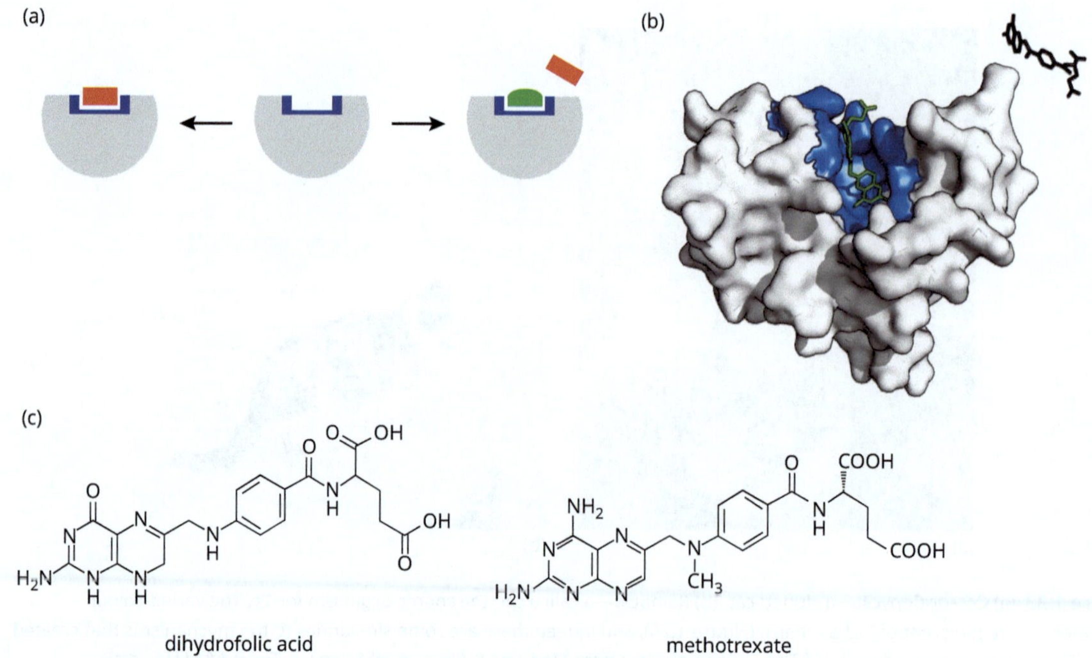

Figure 1.34 Dihydrofolate reductase (DHFR) is an enzyme that catalyses the reaction:

7,8-dihydrofolate + NADPH → 5,6,7,8-tetrahydrofolate + $NADP^+$.

NADPH → $NADP^+$

dihydrofolic acid → tetrahydrofolic acid

Tetrahydrofolate is an essential metabolite in amino acid and nucleic acid metabolism. Its production, by DHFR, is essential for cell survival.

Methotrexate is an inhibitor of dihydrofolate reductase. It is used in the treatment of a variety of cancers.

(a) Schematic diagram of enzyme with empty active site (centre), active site occupied by substrate (red, left), or active site occupied by inhibitor (green, right), blocking access by substrate.

(b) Structure of dihydrofolate reductase binding inhibitor methotrexate (green), also showing substrate dihydrofolate displaced from active site, not able to bind and undergo the reaction.

[(a) and (b) by Thomas Shafee made available under a Creative Commons Attribution 4.0 License.]

(c) Structures of dihydrofolate and methotrexate.

Some regulatory mechanisms affect protein activity

After expression of proteins, cells can modulate the activities of proteins by post-translational modifications, some of which are reversible. Binding of ligands can affect protein activities: examples include inhibition and allosteric changes. Reversible phosphorylation of tyrosine residues is a common mechanism for turning on and off proteins in signalling networks.

A competitive inhibitor is a molecule that interacts with the active site of an enzyme, to block binding of the normal substrate (see Figure 1.34).[10] Such an inhibitor shares structural features with the substrate, so that it can also bind to the active site. But it cannot undergo the reaction to form product and be released.

[10] A colleague uses the following analogy to explain competitive inhibition to her biochemistry class: 'Imagine a pizza restaurant with a single server. A local cinema has just completed its show, and there is a long queue of people waiting to buy a slice of pizza. Then one of the customers requests an order of sushi. The server explains that they don't do sushi—only pizza. The customer insists, and an argument ensues, bringing the progress of the queue to an abrupt and prolonged halt. The customers wanting pizza are substrates; the one who wants sushi is the inhibitor.'

Summary

Thus, cells regulate the activities of their proteins by mechanisms applied at the levels of transcription, translation, post-translational modification of proteins, and response of proteins to ligation. Although these processes are biochemically distinct, cells apply them in coordinated ways. Chemical modifications of transcription factors can regulate amounts of proteins in cells quantitatively, rather than simply switching transcription on and off. Different control processes are effective over different timescales: cells must sometimes react quickly, to threat or stress; at other times rhythmically, over cell cycles; and sometimes in programmes unfolding over years or even decades during development of an organism. It is not possible to sort the different control mechanisms into fast- and slow-acting categories. There are layers of complexity here that we are only beginning to understand.

KEY POINT

There are many mechanisms, and types of targets, for regulation. These include control over expression patterns, and control over the activities of proteins and non-coding RNAs in the cell. For instance, allosteric changes are ligand-induced conformational changes in proteins that modify activity.

LOOKING FORWARD

The goal of this chapter has been to introduce the reader to the subject, setting out broadly the various component topics. It is also intended to make sure the reader is equipped with the basic background required for the rest of the book. Just as genomics is central to all of biology, all of biology is central to genomics; that is why this introduction has had to be so long.

Crucial to the background is the basic idea of a genome, and the nature and contents of genomes in prokaryotes and eukaryotes. The Central Dogma (DNA makes RNA makes protein) motivates introduction of RNAs, including but not limited to messenger RNA, and protein sequences and structures. Computing, so essential to contemporary genomics, will enter the discussion in Chapter 3.

Subsequent chapters will expand the material presented here (following Chekov's famous advice, 'If there is a shotgun hanging over the fireplace in Act I, it must be fired in Act III.'). Chapter 2 will focus on the human genome, some of its applications, and the associated ethical, legal, and social issues; that chapter is also to some extent introductory.

RECOMMENDED READING

The first three items are books that are sources for the history of the development of molecular biology after the Second World War. Rosenfield, Ziff, & Van Loon's is a less formal but not less serious account of the founding people and events.

de Chadarevian, S. (2002). *Designs for Life/Molecular Biology after World War II*. Cambridge: Cambridge University Press.

Judson, H.F. (1980). *The Eighth Day of Creation: Makers of the Revolution in Biology*. London: Jonathan Cape.

Rosenfield, I., Ziff, E.B., & Van Loon, B. (1983). *DNA for Beginners*. London: Writers & Readers Publishing.

Choudhuri, S. (2003). The path from nuclein to human genome: A brief history of DNA with a note on human genome sequencing and its impact on future research in biology. Bull. Sci. Technol. Soc. 23: 360–367.

Cobb, M. (2017). 60 years ago, Francis Crick changed the logic of biology. PLoS Biol. 18: e2003243.

The next publications deal with genomics and the Human Genome Project, placing it in broad context. These are selections from a very large literature.

National Human Genome Research Institute (2022). Human Genome Project Fact Sheet. https://www.genome.gov/about-genomics/educational-resources/fact-sheets/human-genome-project.

Moskowitz, C., & Krzywinski, M. (2022). See how scientists put together the complete human genome. Scientific American 327, 2: 92.

Smetana, J., & Brož, P. (2022). National genome initiatives in Europe and the United Kingdom in the era of whole-genome sequencing: A comprehensive review. Genes 13: 556.

Formenti, G., Theissinger, K., Fernandes, C., *et al*. (2022). The era of reference genomes in conservation genomics. Trends Ecol. Evol. 37: 197–202.

van Heyningen, V. (2022). A journey through genetics to biology. Ann. Rev. Genomics and Human Genetics 23: 1–27.

Birney, E. (2021). The International Human Genome Project. Hum. Mol. Genet. 30: R161–R163.

Ridley, M. (1999). *Genome: The Autobiography of a Species in 23 Chapters*. New York: HarperCollins Publishers.

Wolfsberg, T.G., Wetterstrand, K.A., Guyer, M.S., *et al*. (2002). A user's guide to the human genome. Nat. Genet. 32 (Suppl.): 1–79.

Vandiedonck, C., & Knight, J.C. (2009). The human Major Histocompatibility Complex as a paradigm in genomics research. Brief. Funct. Genom. Proteom. 8: 379–394.

Sequencing projects for other species

Supple, M.A., & Shapiro, B. (2018). Conservation of biodiversity in the genomics era. Genome Biol. 19: 131.

Kozlov, M. (2023). Huge cache of mammal genomes offers fresh insights on human evolution. Nature 616.

Vignieri, S. (2023). Zoonomia. Science 380: 356–357.

Blaxter, M., Archibald, J.M., Childers, A.K., *et al*. (2022). Why sequence all eukaryotes? Proc. Natl. Acad. Sci. USA 119: e2115636118.

Hotaling, S., Kelley, J.L., & Frandsen, P.B. (2021). Toward a genome sequence for every animal: Where are we now? Proc. Natl. Acad. Sci. USA 118: e2109019118.

Schweizer, R.M., Saarman, N., Ramstad, K.M., *et al*. (2021). Big data in conservation genomics: Boosting skills, hedging bets, and staying current in the field. J. Hered. 112: 313–327.

Pharmacogenomics

Cecchin, E., & Stocco, G. (2020). Pharmacogenomics and personalized medicine. Genes 11: 679.

Pirmohamed, M. (2023). Pharmacogenomics: Current status and future perspectives. Nat. Rev. Genet. 24: 350–362.

Extinctions

Dirzo, R., Ceballos, G., & Erlich, P.R. (2022). Circling the drain: The extinction crisis and the future of humanity. Phil. Trans. Roy. Soc. B377: 20210378. (Part of a theme issue 'Ecological complexity and the biosphere: the next 30 years'.)

Kolbert, E. (2014). *The Sixth Extinction: An Unnatural History*. New York: Henry Holt & Co.

Ceballos, G., & Ehrlich, P.R. (2023). Mutilation of the tree of life via mass extinction of animal genera. Proc. Natl. Acad. Sci. USA 120: e2306987120.

Genomics and neurobiology

Benson, D.L. (2020). Of molecules and mechanisms. J. Neurosci. 40: 81–88.

Spiess, K., & Won, H. (2020). Regulatory landscape in brain development and disease. Curr. Opin. Genet. Dev. 65: 53–60.

Fitzgerald, J., Morris, D.W., & Donohoe, G. (2020). Cognitive genomics: Recent advances and current challenges. Curr. Psychiatry Rep. 22: 2.

Shapiro, B. (2021). *Life as We Made It: How 50,000 Years of Human Innovation Refined—and Redefined—Nature*. New York: Basic Books.

Genetic modification using the CRISPR/Cas system

Hille, F., Richter, H., Wong, S.P., *et al.* (2018). The biology of CRISPR-Cas: Backward and forward. Cell 172: 1239–1259.

Balderston, S., Clouse, G., Ripoll, J.J., *et al.* (2021). Diversification of the CRISPR toolbox: Applications of CRISPR-Cas systems beyond genome editing. CRISPR J. 4: 400–415.

Prasad, K., George, A., Ravi, N.S., *et al.* (2021). CRISPR/Cas based gene editing: Marking a new era in medical science. Mol. Biol. Rep. 48: 4879–4895.

Gonzalez-Avila, L.U., Vega-López, J.M., Pelcastre-Rodríguez, L.I., *et al.* (2021). The challenge of CRISPR-Cas toward bioethics. Front. Microbiol. 12: 657981.

Ishino, Y., Krupovic, M., & Forterre, P. (2018). History of CRISPR-Cas from encounter with a mysterious repeated sequence to genome editing technology. J Bacteriol. 200: e00580-17.

Zimmer, C. (2022). CRISPR, 10 years on: Learning to rewrite the code of life. *The New York Times* (27 June 2022) https://www.nytimes.com/2022/06/27/science/crispr-gene-editing-10-years.html

Kan, M.J., & Doudna, J.A. (2022). Treatment of genetic diseases with CRISPR genome editing. J. Amer. Med. Assoc. 328: 980–981.

Davies, K. (2021). *Editing Humanity: The CRISPR Revolution and the New Era of Genome Editing*. New York & London: Pegasus Books.

Doudna, J.A., & Sternberg, S.H. (2017). *A Crack in Creation: Gene Editing and the Unthinkable Power to Control Evolution Hardcover*. Boston & New York: Houghton Mifflin Harcourt.

Nidhi, S., Anand, U., Oleksak, P., *et al.* (2021). Novel CRISPR-Cas systems: An updated review of the current achievements, applications, and future research perspectives. Int. J. Mol. Sci. 22: 3327.

Bier, E. (2021). Gene drives gaining speed. Nat. Rev. Genet. 23: 5–22.

Annual Reviews. Special Article Collection Archive CRISPR: Editing the Genome, https://www.annualreviews.org/page/crispr.

Gostimskaya, I. (2022). CRISPR-Cas9: A history of its discovery and ethical considerations of its use in genome editing. Biochemistry (Mosc). 87: 777–788.

Boti, M.A., Athanasopoulou, K., Adamopoulos, P.G., *et al.* (2023). Recent advances in genome-engineering strategies. Genes (Basel). 14: 129.

Hardouin, G., Magrin, E., Corsia, A., *et al.* (2023). Sickle cell disease: From genetics to curative approaches. Ann. Rev. Gen. Hum. Genet. 24: 255–275.

Wang, J.Y., & Doudna, J.A. (2023). CRISPR technology: A decade of genome editing is only the beginning. Science 379: eadd8643.

Specter, M. (2015). The gene hackers. *The New Yorker* (15 November 2015), pp. 52–61.

Genome informatics: a selection of articles and books

Bromberg, Y., & Radivojac, P. (2022). Computational interpretation of human genetic variation. Hum. Genet. 141: 1545–1548. (Introduction to a special thematic collection from Human Genetics including reviews and original research about advances in computational interpretation of human genetic variation.)

Krishna, R., & Elisseev, V. (2021). User-centric genomics infrastructure: Trends and technologies. Genome 64: 467–475.

Publications dealing with ethical, legal, and social aspects of genomics and its applications

Labude, M.K., Xafis, V., Lai, P.S., *et al*. (2022). Vulnerability and the ethics of human germline genome editing. CRISPR J. 5: 358–363.

Croydon, S. (2022). In it together: Why writing our genetic future should not be left to the scientists alone. Voices in Bioethics: 8.

Gürsoy, G. (2022). Genome privacy and trust. Annu. Rev. Biomed. Data. Sci. 5: 163–181.

Gostimskaya, I. (2022). CRISPR-Cas9: A history of its discovery and ethical considerations of its use in genome editing. Biochemistry (Mosc). 87: 777–788.

Cohen, I.G., Sherkow, J.S., & Adashi, E.Y. (2022). Handle with Care: The WHO Report on human genome editing. Hastings Cent Rep. 52: 10–14.

Statement from the Organising Committee of the Third International Summit on Human Genome Editing, London (2023). https://royalsociety.org/news/2023/03/statement-third-international-summit-human-genome-editing/

Schmitz, D., & Henn, W. (2021). The fetus in the age of the genome. Hum. Genet. 141: 1017–1026.

Middleton, A. (2018). Society and personal genome data. Hum. Mol. Genet. 27: R8–R13.

Companion volumes to this one

Lesk, A.M. (2014). *Introduction to Bioinformatics*, 4th edn. Oxford: Oxford University Press.

Lesk, A.M. (2016). *Introduction to Protein Science: Architecture, Function, and Genomics*, 3rd edn. Oxford: Oxford University Press.

EXERCISES, PROBLEMS, AND WEBLEMS

Exercises

Exercise 1.1 Make a very rough estimate of the average density of protein-coding genes in the human genome. (Total genome size ~3×10^9 bp; total number of genes ~3×10^4 genes.)

Exercise 1.2 A randomly chosen pair of humans (not identical twins) will show an average nucleotide diversity that lies between 1 base pair in 1000 and 1 in 1500. Any human and any chimpanzee differ in approximately 1 base pair in 100. (Note that the chimpanzee genome is somewhat larger than the human.) (a) Estimate the number of differences in the total sequences between two randomly chosen humans. (b) Estimate roughly the number of differences in the total sequences between a human and a chimpanzee.

Exercise 1.3 Different humans, not in a close family relationship, have genomes that differ in approximately 0.1% of the sequence. (a) If all the differences were SNPs, and the sites were evenly distributed (which is not true), what would be the estimated density of SNPs/1000 base pairs? (b) If 1% of the human genome codes for proteins, then under the same (incorrect) assumptions as in part (a), what fraction of the sequence differences would be in protein-coding regions?

Exercise 1.4 Assume that most eukaryotes have approximately 25 000 protein-coding genes, and that the average eukaryotic protein has a length of 300 amino acids. Assume that other functional regions, including RNA-coding genes and control regions, do not require more base pairs than the protein-coding genes themselves (an assumption for which there exists not the slightest justification other than ignorance). Estimate the minimum size that a eukaryotic genome could potentially have.

Exercise 1.5 For the standard genetic code (Box 1.2), give an example of a pair of codons related by a synonymous single-site substitution (a) at the third position and (b) at the first position. (c) Give an example of a pair of codons related by a non-synonymous single-site substitution at the third position. (d) Can a change in the second position of a codon ever produce a synonymous mutation? (e) Is it possible to convert Phe to Tyr by a single base change? If so, what would be possible wild-type and mutant codons? (f) Is it possible to convert Ser to Arg by a single base change? If so, what would be possible wild-type and mutant codons? (g) What is the minimum number of base substitutions that would convert Cys to Glu? (h) In the evolution of an essential protein encoded by a single gene, a Trp is converted to a Gln by two successive single base changes. What is the intermediate codon? (i) To what amino acids could (1) phenylalanine and (2) serine be converted by a single nucleotide substitution?

Exercise 1.6 RNA editing in the mitochondria of higher plants often changes cytosine to uracil at the second position of codons. What amino acid changes could this effect? Higher plant mitochondria use the standard genetic code.

Exercise 1.7 A single base-pair deletion in an exon in a protein-coding gene would be very serious because it would throw off the reading frame. What would you expect to be the effect of a single base-pair deletion in a gene for a structural RNA molecule? (Consider transfer RNA, for example; see Figure 1.8.)

Exercise 1.8 (a) How many *potential* splice variants in the form of mature mRNAs might arise from the exons in Figure 1.9? (Note that: 1. Each splice variant must contain at least one exon. 2. The included introns must appear in the same order as they are in the original transcript.) (b) List all the potential splice variants that contain the exons appearing in red and cyan in Figure 1.9. Report them as a sequence of colours; thus, the mRNA containing all the exons—the one shown in the lower part of the figure—would be reported as: 'blue-red-yellow-cyan-magenta'.

Exercise 1.9 In Figure 1.16 showing the data on Barbary macaques, where did the animal corresponding to the isolated orange point come from?

Exercise 1.10 What is the reasoning behind the statement that: It is likely that the female line in the current Barbary macaques originated from both the Moroccan and Algerian populations?

Exercise 1.11 Why is it possible to have calico cats but not calico kangaroos (barring abnormal cell division at an early embryonic stage or a skin disease)?

Exercise 1.12 Most newborn mammals can digest lactose in infancy, consistent with their dependence on maternal milk for feeding. The enzyme lactase hydrolyses the disaccharide lactose, the major carbohydrate component in milk, to glucose + galactose. In most mammalian species, expression of lactase ceases at the time of weaning. Many human populations follow the mammalian paradigm: lactase expression is permanently turned off when a child is about four years old. Loss of lactase expression produces subsequent intolerance to dairy products.
A single-site mutation that causes lifetime lactase expression—or lactase persistence—was selected for in populations that domesticated cattle and depended on dairy products in their adult diet. In Europe, lactase persistence is more common in the north, where less exposure to sunlight makes northern Europeans more dependent on dairy products as sources of vitamin D precursors (as well as calcium), and the colder weather inhibits fermentation of milk. Ninety-six per cent of Swedes and Finns are lactose persistent.

(a) Would you expect the mutation to be found within the coding regions of the lactase gene itself?

(b) Heterozygotes produce approximately half the amount of lactase. (This is sufficient to avoid the symptoms of lactose intolerance.) From this observation, what additional inference can you make about the location of the mutation?

Exercise 1.13 Why might the prescription of monoamine oxidase inhibitors to a child presenting with symptoms of anxiety, in a case of suspected maltreatment, be contraindicated?

Exercise 1.14 Suggest a molecular mechanism of effecting the feedback loop in Figure 1.34, in which an adequate amount of a product turns off or at least slows down the pathway that produced it. (Hint: Consider Figure 1.30.) What structural relationship between A and E would be required for this to work?

Exercise 1.15 What are the structural differences between dihydrofolate and methotrexate? (Suggestion: Print a copy of Figure 1.34c and circle the regions of difference.)

Exercise 1.16 How might genome sequencing be applied to the conservation of endangered species?

Problems

Problem 1.1 Draw a rough sketch of the outline of a small eukaryotic plant cell (representing 10 000 nm diameter) to fill almost a whole page. Within this outline, draw in, at scale, the cellular components listed in the table. At the same scale, draw an *E. coli* cell (approximate diameter 2000 nm).

Structure	Diameter
Globular protein	4 nm
Cell membrane thickness	10 nm
Ribosome	11 nm
Large virus	100 nm
Mitochondrion	3000 nm
Length of chloroplast	5000 nm
Cell nucleus	6000 nm

Problem 1.2 From the following data, compute the number of genes per Mb for mitochondria, rickettsia, chloroplasts, and cyanobacteria. Compare the mitochondria with the rickettsia and the chloroplasts with the cyanobacteria. Is the smaller genome size of the organelles largely the result of eliminating non-functional DNA, or loss of genes, some of which were transferred to the nuclear genome?

Genome	Number of bp	Number of genes
Human mitochondrion	16 569	37
Rickettsia prowazekii	1 111 523	834
Arabidopsis thaliana chloroplast	154 478	128
Synechocystis sp. PCC 6803	3 573 470 (plus plastids)	3500

Problem 1.3 (a) Suppose that there are ten SNPs in a 50 kb region. If the region is on the Y chromosome, how many possible haplotypes are there? (b) If the region is on a diploid chromosome, how many possible haplotypes are there? (c) Assuming a typical single-nucleotide polymorphism (SNP) density of 1 SNP/5 kb in a human genome, and only two possible bases observed at the position of any SNP, how many sequences could you expect to find throughout a population, within a 100 kb region, if recombination were common at every position in the region? (d) If only three of the possible combinations of SNPs—that is, three haplotypes—are observed, what fraction of possible sequences does this represent?

Problem 1.4 Survival of an SNP. Suppose a person is heterozygous for a novel, selectively neutral mutation on a somatic chromosome (i.e., not X). Suppose the person has two children that survive to reproductive age. The probability of loss of the mutation in that one generation is 25%. If each descendant has two children that survive to reproductive age, what is the probability of complete disappearance of the mutation in 200 years? Assume 25 years per generation.

Problem 1.5 Replication of RNA viruses is error-prone. It is estimated that the replication of HIV-1 in an untreated patient introduces one mistake per replication of its ~10^5 bp genome. If the estimated generation time of HIV-1 in a human body is ~1 day and 10^{10} progeny viruses are produced per patient per day, and assuming that an AIDS patient is initially infected by viruses with identical genomes, estimate whether the patient will (a) generate a mutation at every possible site in the genome every day; and (b) generate mutations at every possible pair of sites in the genome every day.

Problem 1.6 Use the model of expression switching in the human β–globin region shown in Figure 1.32b,c. (a) What developmental progression of globin expression would you expect if the region containing the G_γ and A_γ genes were interchanged with the region containing the δ and β genes? (b) What developmental progression would you expect in patients with deletions in the δ- and β-regions, including the pyrimidine-rich control region YR?

Problem 1.7 (a) In the experiment that produced Cc (Copycat) as a clone of Rainbow (see Figure 1.33), the X-linked inactivation of the cell selected was not reversed. If a number of other cats were cloned by the same procedure, from other cells taken from Rainbow, what coat colours would the cats produced show? (b) Suppose a way were discovered to reverse the X-linked inactivation in the cell from Rainbow from which another cat were cloned. Would the cloned cat have a coat indistinguishable in appearance from Rainbow's? (c) Would monozygotic twins (= 'identical' twins, in genome sequence at least), natural daughters of Rainbow, show identical coat colour patterns? (monozygotic twins are rare in cats) (d) If your answer to (c) suggests that the twin daughters might not look the same, how might you be sure that two daughter cats are monozygotic twins, without sequencing their entire genomes?

Problem 1.8 It is estimated that the human immune system synthesizes 10^{10}–10^{18} antibodies. The portion of a typical antibody active in binding antigen consists of two variable domains each containing 100 amino acids. If every variable domain were separately encoded in the genome but could pair promiscuously, then the number of variable domains that needed to be encoded would be of order of magnitude the square root of the total number of antibodies. How many bases would be required to encode separately all the variable domains? Compare this with the size of the human genome.

Problem 1.9 Outline how you would search for tRNA genes in a genome sequence. Assume that the lengths of the double-helical regions and the lengths of the single-stranded regions between them vary within relatively tight limits and that there are only a limited number of deviations from perfect base pairing in the helical regions. How would your method have to be modified if tRNA genes contained introns?

Problem 1.10 The annual birth rate in the UK is approximately 700 000. Two 'back-of-the-envelope calculations': (a) Now that the goal of a US\$1000 (~UK£800) genome has been attained, what would be the total cost of sequencing 700 000 babies? The current annual budget of the UK National Health Service is approximately £182 billion. What percentage of the current annual NHS budget would be required to sequence the genomes of all babies born in the UK? (Whether the money would be available, and, if it were, whether sequencing would be the best use of it, are separate questions.) (b) If the sequence data were stored completely and independently, what fraction would they represent of the current holdings of the international Nucleotide Sequence Databanks?

Problem 1.11 Outline the arguments for and against the use of animals for testing in drug development. What restrictions might be imposed that would mitigate any of the arguments against the use of animals in testing? How and to what extent would the effectiveness of drug development suffer if these restrictions came into force?

Problem 1.12 A recent meeting in London recommended that editing of heritable genes in humans should not at this time be pursued. Consider two questions: (a) If some organization were to edit human genes, in contravention of these proposed guidelines, what would be some obvious candidates for editing? (b) What would be some genes in other organisms that one might consider editing? Would there be any reason to suggest not carrying out such experiments, and, if you can think of any possible 'side effects', would there be a way of testing for them and/or avoiding them? Consider, for example, applications to agriculture, and gene editing in mosquitoes to prevent carrying malaria, dengue, Zika, and other viruses.

Weblems

Weblem 1.1 At the time of preparing the manuscript of this book, the sequencing capacity of the largest dedicated genomics institute was estimated as 3.2×10^{12} bp per day. This corresponds to 1000 human genome equivalents. What is the corresponding current maximal sequencing capacity?

Weblem 1.2 For each of the following traits of any human being, estimate the contributions of (1) sequences of particular genes, (2) life history and environment, (3) epigenetic factors: (a) blood type; (b) adult height; (c) native language; (d) whether or not a person will develop Huntington's disease; (e) whether or not a person will develop emphysema; (f) whether or not a person will develop Angelman syndrome.

Weblem 1.3 (a) How many chromosomes do zebras of different species have? (b) Would you expect it to be possible for these species of zebra to have offspring if cross-bred with a horse? (c) If your answer to (b) is yes, would you expect these offspring to be generally fertile?

Weblem 1.4 The gene for insulin is 1.7 kb long; the LDL receptor gene is 5.45 kb; and the dystrophin gene is 2400 kb. (a) What are the lengths of the amino-acid sequences of these proteins? (b) What are the ratios of total exon length/total gene length and total intron length/total gene length for these three genes?

Weblem 1.5 On a photocopy of the amino-acid sequence of the human haemoglobin α-chain in Figure 1.11a, indicate the regions arising from the three exons in the gene.

Weblem 1.6 Using the Online Mendelian Inheritance in Man™ site (http://www.ncbi.nlm.nih.gov/omim), what are the clinical consequences of the mutations associated with (a) haemoglobin Adana and (b) haemoglobin Malhacen?

Weblem 1.7 What animal models are available for the human mental diseases schizophrenia, depression, and bipolar disorder?

Weblem 1.8 Female placental and marsupial mammals achieve dosage compensation by silencing one of the X chromosomes in each cell. In birds, however, males are homogametic (ZZ) and females are heterogametic (ZW), with the W chromosome being gene deficient like the mammalian Y chromosome. Do male birds achieve dosage compensation by silencing one of their Z chromosomes?

Weblem 1.9 Find a feature of a human MHC haplotype that is associated with a relatively slow progression to AIDS after HIV infection.

Weblem 1.10 Which MHC haplotypes give indications for effectiveness of use of the following drugs: (a) carbamazapine (epilepsy, bipolar disorder); (b) ximelagatran (an anticoagulant)?

Weblem 1.11 What are the most common mutations in *BRCA1* among Swedish women?

Weblem 1.12 The three most frequent mutations in *BRCA1* and *BRCA2* genes in Ashkenazi Jews are 185delAG and 5382insC in *BRCA1* and 6174delT in *BRCA2*. In what exons of these genes do these mutations appear?

Weblem 1.13 Comparison of haemoglobin and lactase expression: Haemoglobin expression in humans is characterized by a succession of different proteins expressed at different stages of life (embryonic, then foetal, then adult). These are different proteins encoded by different genes. Control mechanisms regulate their expression. In cases of sickle-cell anaemia affecting the adult form of the β–chain of haemoglobin, neonates still synthesizing foetal haemoglobin are asymptomatic until the switch, at about two years of age. Once an affected individual synthesizes the adult HbS β–chain, symptoms of sickle-cell disease appear.

In lactase persistence, neonates can digest lactose, which they ingest in milk; but fail to lose the ability to digest lactose at about four years of age, the ancestral condition. Consider the following questions: Is the lactase story analogous to that of haemoglobin? That is, are there neonatal and adult forms of lactase? Or is lactase impersistence purely an effect of control regions? (a) Consider the idea of a separate adult form of lactase, non-functional in lactase impersistence, but functional in lactase persistence. Why is this idea in principle unlikely? (b) How many lactase genes are there in the human genome? What is the relevance of this question to part (a)? (c) What are the common mutations that lead to lactase persistence? Does a single mutational event account for the lactase persistence trait in all people who show this phenotype? (d) In the condition congenital lactose intolerance, neonates cannot digest lactose and never develop lactose tolerance. Would there be any way, other than sequencing the relevant genes, to decide whether the cause of the condition were a mutation in the control region or in the gene for the lactose-hydrolysing enzyme itself? What mutations have been found to be associated with congenital lactose intolerance? Are any of these in the same region as those that lead to lactase persistence?

Weblem 1.14 Which organism has a more effective control over levels of expression of an enzyme that hydrolyses the disaccharide lactose to monosaccharides glucose + galactose, in response to changes in amounts of lactose available—humans or *E. coli?*

Weblem 1.15 Open a browser at the site "Open the Superfamily database website" [http://supfam.org/SUPERFAMILY/]. Select Phylogenetic Trees, under Tools. Select the following species: *Homo sapiens, Pan troglodytes, Otolemur garnettii, Tarsius syrichta, Ochotona princeps, Felis catus, Sarcophilus harrisi, Xenopus laevis, Ciona intestinalis, Octopus bimaculoides, Bombyx mori, Drosophila melanogaster,* and *Aedes aegypti.* Scroll down to the bottom of the list and select Submit. Go back to the top; select Plot Tree. You will see that the tree is interactive, in that you can pick a species with your (computer) mouse.

It is likely that you will recognize many of the animals in the list; the first four are primates: human, chimpanzee, lemur, and tarsier. What is the common name of *Ochotona princeps?* To find it from within the site, choose the corresponding node in the tree with your mouse, and in the page that appears select Taxonomy browser entry. Note that this will activate a link to a site at the US National Center for Biotechnology Information (NCBI). On a copy of the tree insert the common names of the other species. Using only one mouse selection, restrict the drawing to the subtree containing only insects. Save or print a copy. Using only one mouse selection, restrict the drawing to the subtree containing only mammals. Save or print a copy.

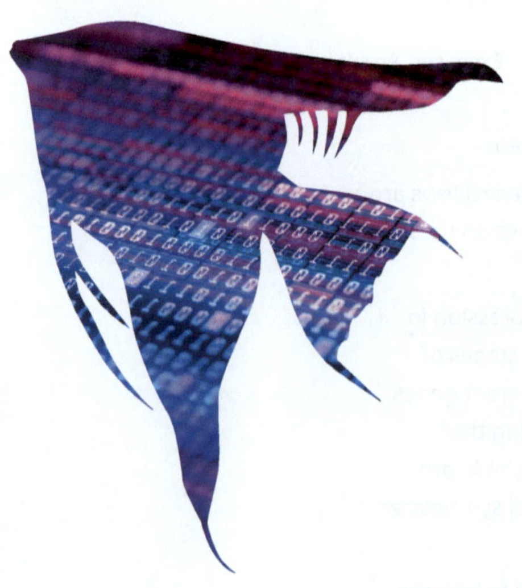

CHAPTER 2

The Human Genome Project: Achievements and Applications

LEARNING GOALS

- *To become acquainted with the history of the Human Genome Project*. To recognize that the application to improvement of health was the driving force in generating the support, and continues to be the major focus of attention.
- *Know that diseases differ in their genetic components*—from mild proclivity through serious risk to inevitability.
- *Understand that identifying the genetic substrate underlying diseases can lead to clinical treatments*, including (a) recommendation for changes in lifestyle to reduce risk, (b) genetic counselling for parents with a family history of genetic disease, and (c) more effective treatments, including optimal choice of available drugs, and development of new ones. Gene editing using the CRISPR/Cas technology (see Chapter 1) allows repair of defective genes.
- *Anticipate the continued growth of pharmacogenomics*—the tailoring of treatment to the individual patient on the basis of his or her genome sequence.
- *Appreciate the improvements in technology that have turned genome sequencing into a very high-throughput data-generation engine.*
- *Understand the basic function of restriction enzymes that cleave DNA at specific sequences.*
- *Know what variable number tandem repeats (VNTRs) and restriction fragment length polymorphisms (RFLP) are*, and their significance for personal identification.
- *Understand the basis of the application of genomics to personal identification*, particularly by law enforcement agencies. Know that a person's identity code in databases is a series of lengths of short tandem repeats, plus an indicator of gender.
- *Relate the application of genomics in personal identification to parentage determination*. What is meant by Paternity Index and Combined Paternity Index (PI and CPI)? How are they determined?
- *Understand the use of DNA sequence information to determine physical features*, such as eye colour, and its possible application to law enforcement.
- *Know how mitochondrial DNA sequences can also be used for identification*, and their applications to the initial examination of remains of the late Russian Royal family.
- *Be aware of commercial 'personal genomics' services that accept a sample and report on salient features of the genome sequence*. These include parentage testing, disease risks, and more general genealogical and ancestry information.
- *Understand what is meant by an ancestry-indicative marker and what these markers tell us about family history.*

- Recognize the very serious ethical, legal, and social issues raised by genome sequence databanks, how different jurisdictions vary in their procedures with respect to what types of data are collected and preserved, and how access to the data is controlled. Appreciate the conflicting demands of public safety and individual privacy. Understand what is meant by 'research goal creep' and how it threatens to erode informed consent.

'. . . the end of the beginning'

On 26 June 2000, the sciences of biology and medicine changed forever. Prime Minister of the UK Tony Blair and President of the US Bill Clinton held a joint press conference, linked via satellite, to announce the completion of the draft of the Human Genome.[1] *The New York Times* ran a banner headline: 'Genetic Code of Human Life is Cracked by Scientists'.[2] The sequence of over 3 billion bases was the culmination of over a decade of work, during which the goal was always clearly in sight and the only questions were how fast the technology could progress and how generously the funding would flow. Box 2.1 shows some of the landmarks along the way.

BOX 2.1 Landmarks in the Human Genome Project

1953 Watson–Crick structure of DNA announced.

1975 F. Sanger and colleagues, and independently A. Maxam and W. Gilbert, develop methods for sequencing DNA.

1977 Bacteriophage ϕX-174 sequenced: first 'complete genome'.

1980 US Supreme Court holds that genetically modified bacteria are patentable. This decision was the original basis for patenting of genes.

1981 Human mitochondrial DNA sequenced: 16 568 base pairs.

1984 Epstein–Barr virus genome sequenced: 172 281 base pairs.

1990 International Human Genome Project launched—target horizon 15 years.

1991 J. Craig Venter and colleagues identify active genes via Expressed Sequence Tags—sequences of initial portions of DNA complementary to messenger RNA.

1992 Complete low-resolution linkage map of the human genome.

1992 Beginning of the *Caenorhabditis elegans* sequencing project.

1992 Wellcome Trust and United Kingdom Medical Research Council establish The Sanger Centre for large-scale genomic sequencing, directed by John Sulston.

1992 J. Craig Venter forms The Institute for Genome Research (TIGR), associated with plans to exploit sequencing commercially through gene identification and drug discovery.

1995 First complete sequence of a bacterial genome, *Haemophilus influenzae*, by TIGR.

1996 High-resolution map of human genome—markers spaced by 600 000 base pairs.

1996 Completion of yeast genome, first eukaryote genome sequence.

May 1998 Celera claims to be able to finish human genome by 2001. Wellcome responds by increasing funding to Sanger Centre.

1998 *Caenorhabditis elegans* genome sequence published.

1 September 1999 *Drosophila melanogaster* genome sequence announced, by Celera Genomics; released Spring 2000.

1999 Human Genome Project states goal: working draft of human genome by 2001 (90% of genes sequenced to >95% accuracy).

1 December 1999 Sequence of first complete human chromosome published.

26 June 2000 Joint announcement of complete draft sequence of human genome.

2003 Fiftieth anniversary of discovery of the structure of DNA. Announcement of completion of human genome sequence.

[1] https://www.youtube.com/watch?v=C22JlzHlLJQ

[2] https://timesmachine.nytimes.com/timesmachine/2000/06/27/issue.html

Next to the politicians stood the scientists. John Sulston (now Sir John), Director of The Sanger Centre in the UK, had been a key player since the beginning of high-throughput sequencing methods. He had grown with the project from the earliest 'one man and a dog' stages to the current international consortium. In the US, appearing with President Clinton were Francis Collins, Director of the US National Human Genome Research Institute, representing the US publicly funded efforts, and J. Craig Venter, President and Chief Scientific Officer of Celera Genomics Corporation, representing the commercial sector. It is difficult to introduce these two without thinking, 'In this corner . . . and in this corner . . .' Although never actually coming to blows, there was certainly intense competition, in the later stages a race.

The race was more than an effort to finish first and receive credit for priority. It was a race after which the contestants would be tested not for whether they had taken drugs, but whether they and others could discover them. Clinical applications were a prime motive for support of the Human Genome Project. At the time, US courts had held that gene sequences were patentable (see Box 2.2)—with enormous potential payoffs for drugs based on them. The commercial sector rushed to submit patents on sets of sequences that they determined, and the academic groups rushed to place each bit of sequence that *they* determined into the public domain to *prevent* Celera—or anyone else—from applying for patents.

The academic groups lined up against Celera were a collaborating group of laboratories primarily but not exclusively in the UK and the US. These included The Sanger Centre in England; Washington University in St. Louis, Missouri; the Whitehead Institute at the Massachusetts Institute of Technology in Cambridge, Massachusetts; Baylor College of Medicine in Houston, Texas; the Joint Genome Institute at Lawrence Livermore National Laboratory in Livermore, California; and the RIKEN Genomic Sciences Center, now in Yokahama, Japan.

Both sides could dip into deep pockets. Celera had its original venture capitalists; its current parent company, PE Corporation; and, after going public, anyone who cared to take a flutter. The Sanger Centre was supported by the UK Medical Research Council and The Wellcome Trust. The US academic labs were supported by the US National Institutes of Health and Department of Energy.

On 26 June 2000, the contestants agreed to declare the race a tie, or at least a carefully out-of-focus photo finish.

The sequencing of the human genome sequence ranks with the Manhattan project that produced

BOX 2.2 Patenting of organisms and genes

Different jurisdictions apply different criteria and have reached different conclusions about whether genes can be patented.

- In the US, the Supreme Court (the highest court in the US) handed down the first ruling on this subject in 1980. They held that a life form was patentable. DNA sequencing was in its infancy. The only completed sequences were bacteriophage ϕX-174 (5836 nucleotides, single-stranded DNA) and simian virus 40 (5224 bp). It was before the invention of polymerase chain reaction (PCR). But this decision served later as the precedent for attempts to patent genes.

The 1980 case arose from Ananda Chakrabarty's development of a bacterium that could degrade hydrocarbons and organic toxins. It was applicable to oil spills and other environmental remediations. Attempts to patent the bacterium were initially rejected by the US Patent Office, but lower courts overturned this ruling. Sidney Diamond, Commissioner of Patents and Trademarks, appealed against the patent. The Supreme Court held that despite the fact that the bacterium was a living thing, it constituted a manufacture. (There is a long history of plant breeders' rights over new varieties; the rules overlap with patents but are not identical.)

In 1997 and 1998, Myriad Genetics, a biotech company based in Utah (in the Western US), together with the University of Utah and the National Institute of Environmental Health Sciences (a component of the US National Institutes of Health), patented the *BRCA1* and *BRCA2* genes. Mutations in these genes increase the risk of breast and ovarian cancer. The company was ruthless in enforcing its monopoly: They charged thousands of US$ per patient for a *BRCA1* genetic test. They asserted their rights under the patent to inhibit not only competitive genetic testing of patients, but also research projects involving the genes. The latter was an unusual proscription.

Dissatisfaction was widespread in the clinical and research communities. In 2009, they went to court. A coalition, including The Association of Molecular Pathologists and the American Civil Liberties Union, filed a lawsuit against Myriad Genetics, the US Patent and Trademark Office, and the University of Utah. The plaintiffs challenged the validity of the *BRCA1* and *BRCA2* gene patents, arguing that—in contrast to Chakrabarty's bacterium—the genes are a product of nature and not an invention, under the terms of US patent law. The dispute percolated up through the US court system, eventually reaching the US Supreme Court.

In their 2013 decision, the Court ruled unanimously that genes isolated but unmodified were products of nature, and therefore not eligible for patents. However, they stated that cDNAs did not occur in nature, that their creation *did* involve invention, and therefore that cDNAs could be patented.

This decision represents current US law.

In contrast, a novel *technique*, such as the extension to mammalian cells of the CRISPR/Cas system, is eminently patentable.

- In the UK, biotech company Human Genome Sciences patented the DNA and amino-acid sequence of neutrokine-α.[3] Pharmaceutical company Eli Lilly challenged the patent, on the grounds that the molecule failed the European legal test of being 'susceptible of industrial application': Despite the reasonable potential of such applications, none had yet been demonstrated. However, in 2009, the European Patent Office upheld the patent, ruling that the plausibility of potential for application had been adequately demonstrated. In 2011, the UK Supreme Court subsequently also ruled in favour of upholding the patent. One of the justices, Lord Walker of Gestingthorpe, said that one of the 'strong policy arguments' for accepting the patent was 'to align this country's [i.e., the UK's] interpretation of the European Patent Convention more closely with that of other [European] states'.

[3] Human Genome Sciences filed the application in 1996. The European Patent Office granted the patent in 2005. Readers surprised at the delay might have a look at Dickens's 1850 short story, 'A Poor Man's Tale of a Patent'.

Confirming the UK's subservience to Brussels at that time, in 2013 the EU countries—including the UK—agreed on a unified patent system. Some non-EU countries, including Norway and Switzerland, and, now, the UK, are members of the European Patent Organization.

Scotland, which differs from England and Wales in many legal respects, no longer awards its own patents. The Patent Law Amendment Act of 1852 created a common patent office for the entire UK.[4] Part of the impetus for this legislation was the Great Exhibition of 1851, held in the Crystal Palace. The feeling that visitors duly awed by British inventions would scurry home to copy them engendered a feeling that inventors deserved protection, at least within the country.

If the Scots had voted for independence in the 2014 referendum, or should they do so in the future, then they would be able to make independent decisions about patentability of genes.

- In Australia, a Queensland cancer patient, Yvonne D'Arcy, apparently acting as an individual, successfully challenged the Myriad Genetics patent of *BRCA1* in the Australian High Court. In this matter, Australian law is in alignment with that of the US but not that of Europe and the UK. Ironically, the Myriad *BRCA1* patent in Australia expired on 11 August 2015, making the High Court ruling handed down on 7 October 2015 moot. However, the decision set a precedent.

Conclusion: Genes can be patented in the UK and Europe. Genes cannot be patented in the US and Australia.

[4] For copyright, a different category of intellectual property from patentable inventions, the union of England and Scotland in 1707 led to resentment among Scottish booksellers of loss of rights, as they became subject to the London Stationers' Company. The result of this dispute was the 1707 Statute of Queen Anne, which applied to the entire Kingdom. In this law, for the first time, rights over literary creations were granted in the first instance to authors rather than publishers. Authors could choose their publishers (and, of course, then as now, *vice versa*).

There is no question that in the eighteenth century London was the centre of the British literary world. However, many of Walter Scott's novels were published in Scotland (some jointly in England). But many Scottish authors had their works published in London; notably Arthur Conan Doyle, and James Boswell, biographer of Dr. Johnson and famously apologetic about his nationality:

Boswell: 'Mr. Johnson, I do indeed come from Scotland, but I cannot help it.'

Johnson retorted: 'That, Sir, I find, is what a very great many of your countrymen cannot help.'

atomic weapons during the Second World War, and the space programme that sent people to the Moon, as one of the great bursts of technological achievement of the last century. These projects share a grounding in fundamental science, and large-scale and expensive engineering development and support. For biology, neither the attitudes nor the budgets will ever be the same.

The human genome is only one of the many complete genome sequences known. Taken together, genome sequences from organisms distributed widely among the branches of the tree of life give us a sense, only hinted at before, of the very great unity *in detail* of all life on Earth. The genome has changed our perceptions, much as the first pictures of the Earth from space engendered a unified view of our planet.

Human genome sequencing

Current estimates are that 40 million complete human genomes have been sequenced, by academic researchers and by companies. A few of the individuals are known (see Box 2.3). J. Craig Venter was himself the subject of the Celera Corporation genome sequencing project. Others include James D. Watson and Bishop Desmond Tutu. The academic human DNA sequencing project used DNA from several anonymous human donors. Samples were obtained after discussions with genetic counsellors and granting of informed consent.

Some prominent individuals whose genome sequences have been determined include the following (in alphabetical order). Unsurprisingly, many come from the genomics community.

George Church (molecular geneticist)

Glen Close (actress, advocate of mental health awareness; motivated by family history that included several individuals who suffered from mental illness)

Esther Dyson (journalist, entrepreneur, philanthropist; daughter of physicist Freeman Dyson)

Stephen Hawking (physicist)[5]

Inuk ('Ice Man' from Greenland, died ~4000 years ago)

Larry King (television interviewer)

Marjolein Kriek (Leiden University geneticist; first woman whose genome was sequenced)

Stephen Quake (bioengineer)

Desmond Tutu (Anglican Archbishop, winner of Nobel Prize for Peace)

J. Craig Venter (American molecular biologist and biotechnologist; leader of one of the groups originally sequencing the human genome; many other important scientific contributions include study of genetic diversity in marine microbial populations; founder of The Institute for Genomic Research (TIGR) and the Venter Institute)

James Watson (American molecular biologist, most famous for discovery with Francis Crick of the double-helix of DNA, for which he shared the 1962 Nobel Prize in Physiology or Medicine; later an important force in organizing the Human Genome Project)

Comparison of sequences from multiple individuals reveals human variation. From the many available sequences, the Pangenome Project juxtaposed full genomes of 47 men and women of African, Asian, Caribbean, American, and European ancestry (see Figure 2.1).

In addition:

- Genome sequences of historical figures include those of Ludwig van Beethoven and Gregor Mendel (see Box 2.3).
- Sequence information is available for over 20000 ancient human genomes. (https://reich.hms.harvard.edu/allen-ancient-dna-resource-aadr-downloadable-genotypes-present-day-and-ancient-dna-data)

[5] Hawking's genome has a claim to interest not only because of his intellectual gifts, but because he is the longest known survivor of amyotropic lateral sclerosis (ALS). There is evidence that enhanced expression of Ephrin receptor A4 (EphA4), a tyrosine kinase, is associated with duration of survival of patients with this disease.

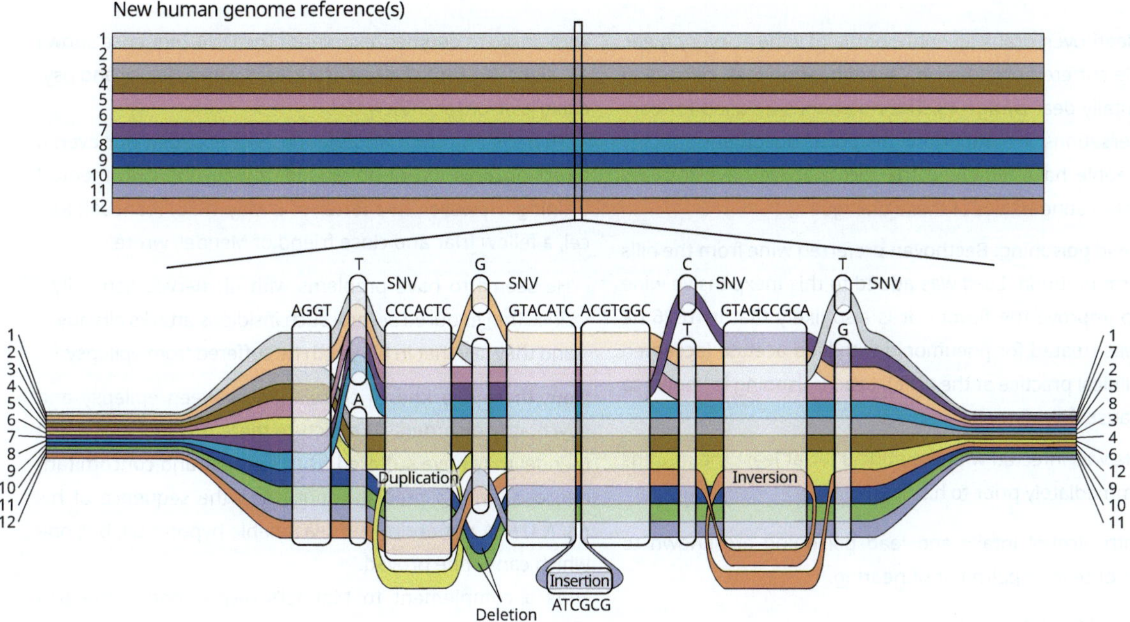

Figure 2.1 In addition to isolated single-nucleotide variations (SNV) the alignment of a few sequences shows many categories of diversity:

yellow	pink	green, blue	light blue
duplication	inversion	deletion	insertion

Image by Darryl Leja, NHGRI. Scientists release a new human 'pangenome' reference, Genome.gov. (20 April 2023), https://www.genome.gov/news/news-release/scientists-release-a-new-human-pangenome-reference

DNA sequences of historical figures

Ludwig van Beethoven

Beethoven died in 1827, age 58, of liver disease. Friends and admirers preserved locks of his hair as mementos.

In 2023, Begg *et al.* collected the available material and determined Beethoven's genome sequence.[6] Although the sequence does not illuminate Beethoven's achievements as a composer, it does help to understand an important medical problem: his cirrhosis of the liver, ultimately a fatal condition. Given a genetic predisposition, the disease was likely the result of high alcohol intake and lead poisoning. It is likely that these factors also contributed to his deafness.

Beethoven's genome sequence indicated elevated risk of liver cirrhosis, from variants in genes *PNPLA3* and *HFE*.

PNPLA3 encodes the protein adiponutrin. Its function is not well understood, but there is evidence that it helps regulate development of adipocytes and the production and breakdown of fats. The *PNPLA3* mutation rs738409 (Ile148→Met) observed in Beethoven's genome is, in conjunction with alcohol intake, strongly associated with liver disease.

The *HFE* gene encodes a homeostatic iron absorption regulator. Mutations in *HFE* also enhance the risk of cirrhosis in connection with heavy drinking. Beethoven's *HFE* gene showed the mutations rs1799945 (His63→Asp) and rs1800562 (Cys282→Tyr).

Given these enhanced risk factors, it appears that contributors to Beethoven's medical problems include:

1. Beethoven was a heavy drinker.

   Beethoven's father was a severe alcoholic. Beethoven's own wine consumption increased after the death of his mother when he was 17 years old. Tinnitus developed and grew more severe over the course of the following ten years. Then, Beethoven reportedly began consuming even more wine to help deal with gastrointestinal problems. According to friends and local tavern owners,

[6] Begg, T.J.A., *et al.* (2023). Genomic analyses of hair from Ludwig van Beethoven. Curr. Biol. 33: 1–17.

Beethoven drank an entire bottle of wine at every meal. He suffered progressive severe hearing loss, becoming totally deaf by age 45. The notebooks he used for conversations are important historical documents. (Many people have lamented the fact that we have the less interesting halves of these dialogues!)

2. Lead poisoning: Beethoven preferred wine from the hills around Buda. Lead was added to this inexpensive wine to improve the flavour. It is also likely that in 1826 he was treated for pneumonia with lead acetate (accepted clinical practice at the time). Lead poisoning is known to cause liver damage.
3. He was infected with hepatitis B for at least the months immediately prior to his death.

Both alcohol intake and lead poisoning are known to contribute to impairment of hearing.

Gregor Mendel

To commemorate the 200th anniversary of Gregor Mendel's birth, scientists in his city of Brno—then in Moravia, now in the Czech Republic—exhumed his body and sequenced his DNA.[7] A choice both appropriate and ironic: Mendel played a crucial role in the history of the subject, and now makes a contribution to developments that flowed from his work. With permission of the Augustinian hierarchy his grave was opened, and identification of the body was confirmed using a hair found in a book on astronomy in Mendel's personal library.

Does the sequence illuminate Mendel's medical history? Throughout his life he suffered from nervous breakdowns—in one case leading to his abandoning a series of exams for certification as a teacher, after passing a preliminary part with answers deemed 'excellent'. The DNA sequence shows an allelic variant in a gene associated with epilepsy and psychiatric disorders: *CACNA1H*.

There is no physician's diagnosis of epilepsy, nor even a direct observation of an episode by anyone with medical training. However, in a letter of 8 May 1856, František Klácel, a fellow friar and close friend of Mendel, wrote:

> 'He seems to have problems with his nerves generally since [he] endured several such insidious attacks already, and they say that in his youth he suffered from epilepsy.'[8]

Now, there is a known correlation between epilepsy and psychiatric disorders. The picture therefore suggests itself: Mendel may have suffered from epilepsy and concomitant neurosis, with a predisposition from the sequence of his *CACNA1H*. A not entirely unreasonable hypothesis, but one which cannot be proved.

As a complement to Mendel's own genome, the pea plant *Pisum sativum*, the subject of his experiments, has been sequenced. Readers of the Mendel literature will be aware that Mendel has been attacked for: (1) Publishing data that are 'better' statistically than might be expected from his sample sizes; (2) In dihybrid crosses, either failing to observe linkage or failing to report it. From the *P. sativum* genome sequence, it has been possible to evaluate these criticisms objectively.[9] The conclusions show the criticisms to be unjustified and support the integrity of Mendel's work.

[7] Drozdová, E., Fialová, D., Pardy, F., *et al.* (2022). *Ways to the Genome of the Founder of Genetics*. Brno: Masaryk University Press.

[8] Klein, J., & Klein, N. (2013). *Solitude of a Humble Genius—Gregor Johann Mendel* 363–364 Vol. 1. Formative Years. Berlin: Springer.

[9] Fairbanks, D.J., & Rytting, B. (2001). Mendelian controversies: A botanical and historical review. Amer. J. Bot. 88: 737–752; Ellis, T.H.N., Hofer, J.M.I., Swain, M.T., *et al.* (2019). Mendel's pea crosses: Varieties, traits and statistics. Hereditas 156: 33; Sussmilch, F.C., Ross, J.J., & Reid, J.B. (2022). Mendel: From genes to genome. Plant Physiol. 190: 2103–2114.

Large-scale sequencing projects

Sequencing is now done under a variety of auspices. A number of international organizations pursue ambitious targets.

- The International HapMap Project http://hapmap.ncbi.nlm.nih.gov/ focuses on variations in genome sequences among populations distributed around the world. They are collecting in particular an atlas of single-nucleotide polymorphisms (SNPs = substitutions of individual bases). Clusters of SNPs that appear to be inherited in tandem are called haplotypes (see Chapter 1).
- The 1000-genome project and the Human Genome Diversity Project are extensions of the HapMap Project towards complete genome data, with an emphasis on discovering the conditions required to ensure appropriate data quality

(http://www.1000genomes.org/). Their cup runneth over: Results already include over 4000 individuals. Specific goals include careful sequencing of family groups (including trios: mother + father + child), and detailed sequencing of 1000 protein-coding regions in 1000 individuals. Although one may wonder to what extent the stated goals will be overtaken by the growing 'background' of whole-genome sequencing, the importance of the commitment of the international organizations to data quality control, curation, and free distribution must not be underestimated.

- The UK Biobank has released complete genome sequences of all entries in its database, derived from 500 000 British volunteers.
- Genomics England has the goal of sequencing 100 000 genomes, including tumour cells and healthy ones from cancer patients.
- In the UK, the NHS sponsors the 100 000 Genomes Project, to sequence whole genomes from NHS patients and their families.
- The UK Newborn Genomes Programme has the goal of whole-genome sequencing of up to 200 000 newborn babies.
- The US National Institutes of Health organizes The All of Us Research Project, with the goal of sequencing one million human genomes. In March 2022, they released 100 000 sequences.
- The Mount Sinai Health System and the Icahn School of Medicine at Mount Sinai, together with biotechnology company Regeneron, seek to enrol one million patients for sequence determination.
- GenomeAsia 100K (GA100K), a consortium of Nanyang Technological University (NTU), Macrogen, and MedGenome, seek to sequence 100 000 Asian individuals.
- The Personal Genome Project has the goal of supporting developments in genome-informed and personalized medicine by publishing complete genomes and medical records of 100 000 volunteers.
- Sequence determination of particular genes is common, in clinical settings. A well-known example is the sequencing of breast cancer predisposition genes *BRCA1*, *BRCA2*, and *PALB2* (partner and localizer of *BRCA2*), in cases where family history suggests the possibility of elevated risk. Mutations in these genes are strongly correlated with early development of breast and ovarian cancer. Actress Angelina Jolie had lost eight family members to cancer, including her mother, who died of ovarian cancer at age 56. When sequencing turned up mutation in her *BRCA1* gene, she opted for prophylactic double mastectomy, followed by removal of her ovaries and fallopian tubes. Her announcement of her decisions attracted wide publicity, with the intent to encourage others not to avoid treatment out of reticence, embarrassment, or denial.
- Several companies now offer personal genome sequencing. Many provide sequencing of mitochondrial DNA or individual loci in nuclear DNA, for tracing of ancestry and reporting of disease risks. The application of DNA sequencing to demonstrate legal relationships—most commonly, paternity testing—is well established.

What makes us human?

Two questions compelling our curiosity are: How do humans differ from other species or subspecies? How do we differ among ourselves?

Two approaches help in addressing these questions.

- *Comparative genomics*. We can compare the human and chimpanzee genomes, for example, and ask how differences between these genomes might give rise to differences between the species.
- *Study of human mutations*. Many mutations alter phenotypes and give clues to the functions of the affected regions. Many mutations cause disease. The regions affected may encode structural proteins or enzymes or regulatory proteins or RNAs, or they may be DNA sequences that are targets of regulatory mechanisms. Understanding the effects of such mutations illuminates human biology and often has immediate clinical application.

On the other hand, it is futile to think that a study of the one-dimensional genome will provide complete answers. The implementation of genetic information in any organism involves a complex and dynamic pattern of interactions among genes and gene products, including responses to pressures from the environment. Genomes set in motion processes over which they cannot entirely retain control.

Comparative genomics

The human and chimpanzee genomes are about 99% identical. To understand what makes us human—or at least what makes us not chimpanzees—we can focus on the ~3 Mb of different sequences, rather than the full 3.2 billion. One might argue that it would be better to compare our genomes with those of closer relatives, such as the Neandertals, despite the fact that we have only limited access to Neandertal anatomy, development, and behaviour. However, humans and chimpanzees express very similar sets of proteins, and most of the homologous proteins of chimpanzees and humans are identical or very similar. The amino-acid sequences of chimpanzee and human proteins are even more similar than the nucleotide sequences of the corresponding genes. About 30% of homologous human and chimpanzee proteins show no differences at all. On average, there are only two amino acid differences.

How, then, do humans and chimpanzees develop differently? The ultimate answer *must* lie within the static sequence of the genome. However, a satisfactory answer will require an understanding of the dynamics of patterns of regulation of gene expression.

With enormous prescience, Mary-Claire King and Allan C. Wilson wrote in their 1975 paper *Evolution at two levels in humans and chimpanzee* that 'Their macromolecules are so alike . . . that regulatory mutations may account for the major biological differences between humans and chimpanzees'. (Science 188: 107–116)

There is a paradox here. On the one hand, living systems are fairly robust to perturbations. Yeast, for example, survives individual knockout of 80% of its genes. On the other hand, the 4% differences between chimpanzee and human genomes make profound differences in phenotype. This suggests a *chaotic system*, one in which tiny perturbations can lead to large changes in the subsequent trajectory. Superposed on the general overall robustness are specific changes that exert immense leverage.

There are two ways to identify these crucial sequences. One is to look closely at the differences between human and chimpanzee genomes and try to figure out what the changed loci are doing. (We shall do this in Chapter 6.) Another is to examine human mutations that affect phenotypic properties that chimpanzees do *not* share with humans, such as language.

Genomics and language

Of all the differences between modern humans and other species, surely the most significant, and scientifically the most fascinating, are our cognitive and linguistic abilities.

The relationships between genomics and language demand interest for several reasons. One is that the phenomenon of language acquisition and practice is still largely mysterious. There are innate—i.e., genome-dependent—components to language learning by children. No other comparably complex activity is learned so naturally. As Darwin wrote, 'No child has an instinctive tendency to bake, brew, or write.'

There is a childhood window of opportunity for easy language learning. Consider the ease of developing fluency in a cradle language. In contrast, learning a second language in adulthood requires mental and study habits comparable to learning calculus. Evidence from strokes in older adults shows that cradle languages are stored in different regions of the brain than languages acquired later. Although these phenomena are developmental, they must ultimately be encoded in the genome—as they are species-wide traits.

Language is also a pre-eminent social phenomenon. It underlies interpersonal communication (both face-to-face and worldwide), commerce, and

BOX 2.4 Systematic language impairment: the *FOXP2* gene

Many people suffer from diseases that interfere with production or comprehension of language, or both. Some of these are associated with trauma, or with complex genetics. However, one abnormality with simple Mendelian inheritance appears in a family in London. Members of the 'KE' family have a severe developmental disorder affecting both the facial motor control involved in producing speech and also the mental processing of language.

The mutation responsible for this condition is an SNP in a gene called *FOXP2*, which encodes a transcription factor. The major protein encoded by *FOXP2* is 715 amino acids long. It is quite a stable protein from the evolutionary point of view, with only one substitution between mouse and the identical chimpanzee, rhesus macaque, and gorilla sequences. However, the human protein has two mutations relative to the other extant primate sequences.

This example illustrates the power of the combination of studies of human phenotypes and comparative sequence analysis. And yet the observation that the phenotype shown by members of the KE family arises from two SNPs in a single gene may be deceptively simple. We cannot conclude that the expression of only one gene is involved in creating the phenotype, as is the case for sickle-cell disease, for example. The *FOXP2* gene product is a transcription factor. Its activity affects the expression of many genes. The effectiveness of their coordinated expression required co-evolution, i.e. sequence changes in other genes.

In other types of language impairment, we must also look beyond the genome sequence. Epigenetic factors are important in Rett syndrome, a progressive neurodevelopmental disorder characterized by a slowing of development, regression of spoken language, cognitive impairments, and autism. The mutation responsible appears in McCP2 (methyl CpG-binding protein 2), a protein associated with DNA-methylation-based gene regulation, and with control of alternative splicing.

literature. Many important social decisions, notably the design of educational systems, depend on appreciating the biological substrates of language. These are subtle and our knowledge of them is far from complete.

Language, as a biological phenomenon, links genomics with several other disciplines, including neurobiology, development, medicine, and anthropology. Clinically, it is interesting to understand the genetic component underlying specific language impairment syndromes (see Box 2.4). The question of whether Neandertals and Denisovans had language is unresolved, but the gene sequences contain clues.

Language is a feature of all human populations. Of the approximately 6500 languages currently spoken, a few major ones such as Chinese, English, Hindi, and Spanish can each claim about 400 million speakers or more (see Table 2.1). Others have much smaller communities: Europe retains niche languages such as Basque and Breton. There were estimated to be about 250 languages of indigenous Australians at the time of settlement by Europeans, of which about two-thirds survived. Over 800 distinct languages are spoken on the island of New Guinea.

Table 2.1 Major languages of the world

Native language	Number of speakers
Mandarin Chinese	939 000 000
Spanish	485 000 000
English	380 000 000
Hindi	345 000 000

Human spoken languages have many features in common with biological species. Both languages and species exhibit varying degrees of similarity and diversity. Languages and species can be classified into taxonomies that, at the highest levels at least, are hierarchical. In both languages and species, ancestor–descendant relationships can be observed. Languages diverge, as species do. The Romance languages, divergent descendants of Latin, are the Galapagos finches of linguistics. Latin itself sits within the larger class of Indo-European languages, of which Sanskrit, the classical language of India, is the extant language believed to be closest to their common ancestor.

Dialects of languages are analogous to genetic haplotypes. Some dialects distinguish social classes—still true in Britain today. This effect may curtail panmyxis (free gene mixing) in a population. More extreme examples are formal court languages, in stasis relative to freely evolving colloquial speech and writing. Examples appeared in the Ottoman Empire and in China, up until the early twentieth-century revolutions. Emperor Hirohito's speech bringing the Second World War to its end, broadcast on the wireless at noon on 15 August 1945, was expressed in classical Japanese very difficult for his hearers to understand.

Just as species have become extinct, so have many languages. Many languages are currently 'endangered', with only a few older adult speakers remaining. Fortunately, like species, languages can be rescued from extinction. Navajo, in decline until the 1940s, is now thriving. Hebrew, which had survived since ancient times as a liturgical language, re-emerged 75 years ago as the colloquial language of the new country of Israel. In some cases, it is possible to reconstruct an extinct common ancestor of surviving descendant languages. This can have interesting implications about the lifestyles of its speakers. For instance, the similarities in some words in Indo-European languages, and dissimilarities in others, suggest that the speakers of the ancestral language had domesticated dogs and horses but not cats and camels.

Linguists have developed quantitative methods to measure divergence of languages. Languages develop variation in pronunciation (or spelling, if the language is written) and in structure. An example of systematic variation in pronunciation would be the difference between the English word *brother* and the German cognate *Bruder*; another such pair is *leather/Leder*. An example of a structural feature of a language is word order: In sentences in English and many other languages, the typical word order is *subject–verb–object*, as in 'The mouse ate the cheese'. In some languages, such as Turkish, the typical word order is *subject–object–verb*. Phonological change is much more rapid than structural change—molecular biologists might find a useful analogy to the rapidity of sequence change relative to structural change in proteins!

There are obvious correlations between people's native language and physical features; for instance, blue or brown eyes correlate well with native speakers of Swedish or Mandarin, respectively. However, native language choice is clearly transmitted culturally rather than genetically: Any child can become a fluent speaker of any language to which he or she is exposed in the cradle.

Although there is no *direct* genetic link with native language, L.L. Cavalli-Sforza and co-workers found substantial similarities between clustering of genetic variation among human populations and language groups. For instance, just as the Basque language is an isolate, unrelated to any other language spoken in Europe, the Basque people are also genetically distinct. This link between languages and genetics is useful because we can estimate rates of genetic change and thereby infer when languages arose and diverged. One conclusion is that most of the world's language families arose, in a burst, between 6000 and 25 000 years ago. When such inferences from genetics can be compared with datable archaeological evidence or historical records, the results are sometimes, although not always, gratifyingly consistent.

Comparison of genetic and linguistic data can illuminate what happens when two populations collide. Some discrepancies between correlations of genetics and languages appear when conquerors impose a new language on an indigenous population.

- This happened in (what is now) Great Britain after the Anglo-Saxon invasions around the fifth century. The indigenous Celtic languages were pushed to far reaches of the British Isles (and Brittany; Breton is a Celtic language). Less effective, linguistically, was the Norman Conquest of 1066, after which English absorbed many French words but retained its own vocabulary alongside them.

> In *Ivanhoe*, Sir Walter Scott contrasted the French word veau for the meat, with calf for the animal, noting that '... he is Saxon when he requires tendance, and takes a Norman name when he becomes matter of enjoyment'.

- Hungary was converted to speaking a Uralic language upon invasion by the Magyars in 896 AD.
- In contrast, the fall of Rome in 476 AD was *not* followed by a replacement of Latin in Italy by a Germanic tongue.

Conversely, although native language is not determined by genes, language differences can retard genetic mixing among populations by creating cultural barriers to intermarriage.

Linguistics will continue to be an area in which the very subtle links between genomics and culture can be explored.

'Pop' applications of genome sequencing

Clinical and research contexts characterize most genome sequencing. However, several goals motivate use of the personal sequencing services offered by several companies. These include:

- identification of parentage, including but not limited to paternity testing, and determining birth parents of adopted children, and *vice versa*; and
- curiosity about genealogy, ethnic background, and health risks.

For these purposes it is not necessary to sequence the entire genome.

For exploring a person's ethnic background, it is sufficient to check a set of loci known as ancestry-informative markers (AIM). These are a set of polymorphic loci for which the distribution frequency of bases differs substantially between different geographic regions or populations. Most companies analyse 700 000 of these loci from a subject's DNA to estimate a percentage breakdown of ancestry admixture from different populations.

This can be motivated by more than curiosity. In some cases, membership in a particular group affords special privileges. For instance, individuals who have greater than 50% Hawaiian ancestry are entitled to homestead leases on very favourable terms (a 99-year lease for US$1 per year; in effect a 'peppercorn' rent). Formerly, paper documentation was required to prove ancestry, but the rules have been changed to allow DNA sequence evidence to be adduced. AIM markers are so far not capable of distinguishing descendants of indigenous Hawaiians from general Polynesian ethnic groups. Therefore, the use of DNA sequencing under consideration would involve proof of family relationship to individuals of proven Hawaiian ancestry.

In New Zealand, seven seats in Parliament are reserved for people of Maori descent, and only people of Maori descent can vote for candidates for those seats. In Canada, people of North American Indian descent are eligible for special tax exemptions and social services.

Elizabeth Warren was a member of the faculty of Harvard Law School, in Cambridge, US. In 1995, the university announced that she had part Native American ancestry, bolstering its record for diversity in hiring. In 2012, Warren ran for the US Senate in Massachusetts, US. Unlike the situation in some Western states, capture of the Native American vote was not of significant importance to her electoral prospects. Nevertheless, her opponent challenged her claim. A DNA sequence test showed that Warren's ancestry is primarily European but does contain segments of Native American origin.

Genomics in personal identification

Legal applications of DNA sequencing depend on several facts:

- The genomes of all individuals (*even* monozygotic; that is, 'identical' siblings) are unique. Like fingerprints, genomes provide a unique personal identification. A bloodstain at a crime scene, like a set of fingerprints, can identify or exclude specific individuals.
- The genome of every person combines chromosomes from his or her parents. Unlike fingerprints, therefore, genomes can indicate familial relationships; notably, identification of parentage.

- Each person's genome contains genes that influence, even if they do not inevitably determine, recognizable features such as eye colour. DNA left by an unknown individual at a crime scene can be analysed to suggest a physical description of the source individual.
- Thus, unlike fingerprints, genomes contain much more information about a person than simple identification. The treatment of this information by governmental authorities raises ethical and legal questions.

The use of molecular characteristics to identify people and relationships is over a century old. The earliest methods applied the classical blood groups—A, B, AB, and O. A suspect with blood type O must be innocent of a crime committed by a person of blood type A. A person of blood type O could not be the parent of a child of type AB. However, many people share the same blood type. Therefore, blood typing can prove innocence but not guilt. Tests based on a wider panel of serotypes, including but not limited to major histocompatibility complex (MHC) haplotypes, have far better discrimination. In contrast, DNA sequences can provide much more reliable indication of guilt or paternity.

DNA identification has provided evidence in several very high-profile cases:

- The trial of famous (American) football player and movie actor O.J. Simpson in 1994–1995 for the murder of his wife; he was acquitted despite the presentation of evidence by the prosecution that he was the source of fresh bloodstains found at the scene of the crime.
- A stain on a White House intern's dress provided evidence of inappropriate behaviour against US President William J. Clinton in 1998.
- Comparison of DNA from descendants of early nineteenth-century US President Thomas Jefferson and Sally Hemings, a slave on his Virginia plantation, found matches in the Y-chromosome sequences, indicating that a male member of the Jefferson family was the father of Hemings' children. Historical evidence—taken together with this result—suggests that the father was Thomas Jefferson himself.

Less sensational, but more important in everyday law enforcement, is the fact that DNA evidence is sufficiently definitive and widely accepted to *avoid* many trials, by not indicting innocent people.

DNA 'fingerprinting'

A.J. Jeffreys (now Sir Alec) at the University of Leicester discovered DNA fingerprinting in 1984, when he and his colleagues compared the sizes of the restriction fragments (see Box 2.5) of DNA samples including a human family group (father, mother, and child). Different individuals give different patterns—the gel provided a 'bar code' unique to each individual. (This bar code is different from the one used for species identification—see Chapter 4.) Moreover, the pattern from the child's DNA was a combination of those of the parents (see Figure 2.2).

Two restriction enzymes in common use in DNA profiling are *Hae*III and *Hin*fI. HaeIII produces 'blunt ends'. HinfI produces 'sticky ends'.

*Hae*III	5'...GG | CC...3'	*Hin*fI	5'... G | ANTC...3'
	3'...CC | GG...5'		3'... CTNA | G...5'

In specifying the sequence specificity, N = any nucleotide.

Because the sequences at which the enzyme acts are distributed through the DNA, cleavage produces a set of 'restriction fragments' of variable length (see Figure 2.4). Spreading the fragments out on a gel allows visualizing the pattern of lengths. Such patterns characterize and distinguish different DNA molecules.

Cutting the same DNA with other restriction enzymes, with different specificities, produces overlapping fragments. From the sizes of the fragments

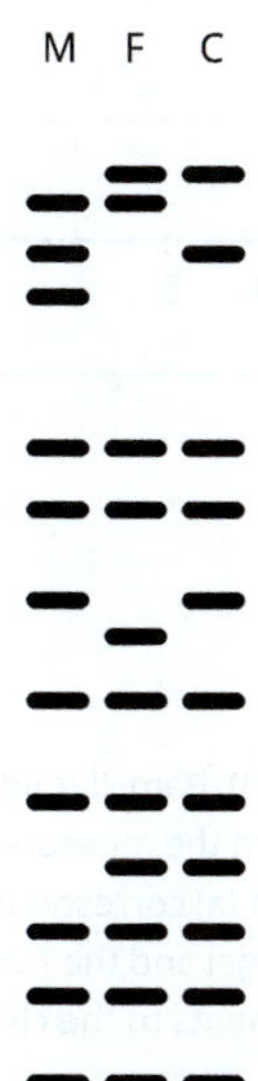

Figure 2.2 Pattern of a gel showing DNA fingerprints from a mother (M), father (F), and child (C). Every band in lane C matches one appearing in lane F or lane M or both. The bands on the gel correspond to restriction fragments from a complete digest by HinfI. The gel separates fragments according to size.

produced by individual enzymes and in combination, it is possible to construct a **restriction map**, stating the order and distance between the restriction enzyme cleavage sites (see Box 2.5 and Fig. 2.4).

KEY POINT

As an analogue of a restriction map: Walk the entire length of Broadway in New York City (see https://globetrottergirls.com/wp-content/uploads/2013/08/broadway-map-manhattan-13.3-miles.jpg). Mark on the map the location of every Starbucks coffee shop and calculate the number of blocks between successive sites; then mark the location of every CitiBank office and calculate and plot on a graph the number of blocks between successive sites; then calculate the number of blocks between every occurrence of *either* Starbucks *or* Citibank. Observe that a mutation in a cleavage site—tantamount to opening or closing a branch of Starbucks or Citibank—will change the sizes of the fragments, allowing the mutation to be located in the map.

(Thanks to Professor B. Misra, New York University)

BOX 2.5 Restriction maps

A **restriction endonuclease** cuts DNA at a specific sequence (see Figure 2.3). For example, the specificity of EcoRI is: (| indicates sites of cleavage)

```
G | A A T T C
C T T A A | G
```

Note that the enzyme cleaves both strands, in this case not at opposing positions, leaving dangling bits of single strand, called 'sticky ends'. Other restriction enzymes cleave at opposing positions, leaving 'blunt ends'.

*Eco*RI illustrates the nomenclature of restriction enzymes: *Eco* abbreviates the name of the species of origin (*Escherichia coli*) as the first letter of the genus name and the first two of the species name. The letter R specifies the strain. (Not all restriction enzyme names contain a strain identifier.) The Roman numeral, in this case I, distinguishes different restriction enzymes from the same strain of the same organism. Thus, EcoRI is the first restriction enzyme from *E. coli*, strain RY13. Other examples are HaeIII, from *Haemophilus aegyptius*, HinfI, from *Haemophilus influenzae*, and TaqI, from *Thermus aquaticus*.

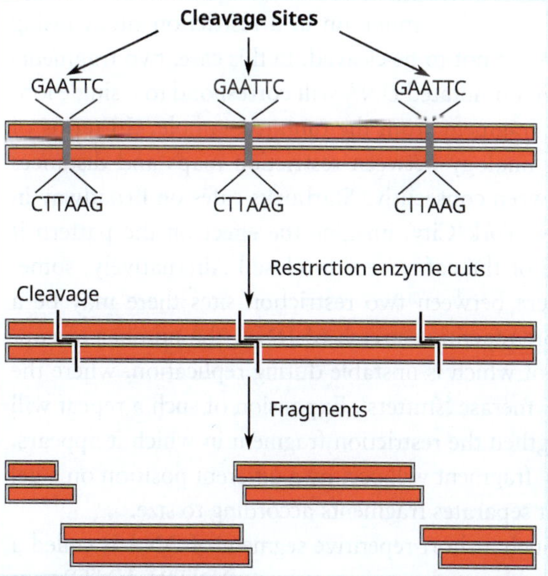

Figure 2.3 A restriction enzyme cuts at sites containing particular sequences, distributed throughout a DNA molecule. The distances between restriction sites vary, and therefore the lengths of the fragments vary. Spreading out the fragments on a gel gives a kind of 'bar code' characterizing the DNA sample (see Exercise 2.1).

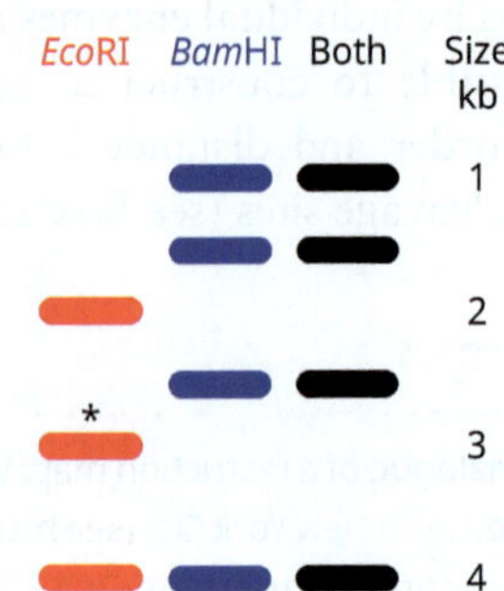

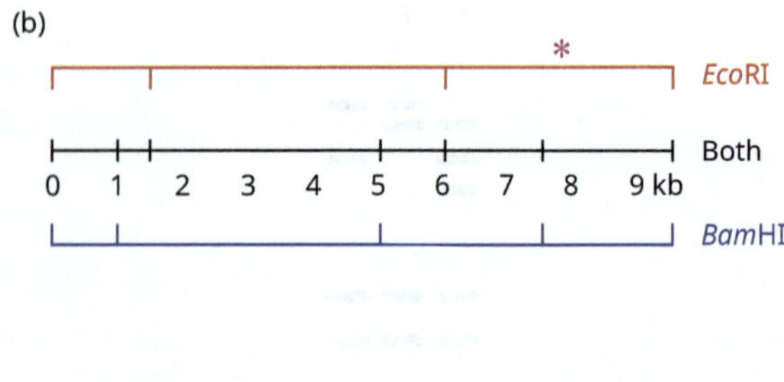

Figure 2.4 (a) Pattern of a gel showing sizes of restriction fragments produced by EcoRI (left; red), BamHI (middle; blue), or both (right; black). (Direction of migration: up the page.) (b) Positions of the cutting sites, deduced from the measurements in (a). Each lane of (a) corresponds to a strip of the same colour in (b) in the following way: each band on the gel in (a) corresponds to the length of a fragment between two successive cutting sites in (b). For instance, the * indicates a band on the gel and the fragment that gave rise to it. In contrast with Figure 2.3, in this case, we do *not* want to limit our observation of the fragments to the cleavage sites of a single enzyme.

In this case, it is relatively easy to determine the unique order of fragments that accounts for all the data. In more complicated cases, it is possible to simplify the problem by placing radioactive tags on either end of the segment mapped, or by carrying out partial digests to produce additional fragments.

Why do different individuals give different patterns of restriction fragment sizes? One possible cause of the difference is a mutation in a restriction site, causing that site not to be cleaved. In this case, two fragments from unmutated DNA will correspond to a single longer fragment from the mutated sample. (In terms of our analogy between restriction maps and distances between consecutive Starbucks cafés on Broadway in New York City, imagine the effect on the pattern if one of the cafés were to close.) Alternatively, somewhere between two restriction sites there may be a short repetitive stretch of DNA, the number of copies of which is unstable during replication, where the polymerase 'stutters'. Expansion of such a repeat will lengthen the restriction fragment in which it appears. The fragment will occupy a different position on a gel that separates fragments according to size.

Such a short repetitive segment of DNA is called a **variable number tandem repeat** (VNTR). VNTRs are generally flanked by recognition sites for the same restriction enzyme, which will neatly excise them, producing fragments of different lengths. It is these fragment lengths that vary between individuals, known as **restriction fragment length polymorphism** (RFLP). The fragments can be separated on a gel according to size and detected by Southern blotting.

KEY POINT

VNTRs are characteristics of genome sequences; RFLPs characterize artificial mixtures of short stretches of DNA created in the laboratory in order to identify VNTRs.

The patterns were easy to determine from a sample of DNA. Jeffreys and his co-workers quickly established that they were unique to individuals, providing a 'genetic fingerprint'.

The first legal application, in 1985, was to a case of disputed identification involving a family of UK citizens. A child in the family visited Ghana. When he returned to the UK, immigration authorities suspected him of being an impostor, not entitled to UK residency. None of the classical blood tests, including A, B, AB, O, and other blood groups, and even MHC haplotyping—which gives much higher discrimination—produced definitive results. Indeed, there was a possibility that the boy was related to the woman who claimed to be his mother, but perhaps he was her nephew rather than her son. Quite fine distinctions were therefore essential. Jeffreys' DNA fingerprints, comparing the patterns from the child's DNA with that of members of the UK family, proved his identity to the satisfaction of the Home Office. The family were reunited.

Jeffreys also applied his method to criminal identification. In the first case, DNA fingerprinting proved the innocence of a suspect who had actually confessed to two crimes. The true criminal was discovered after a survey of DNA samples from almost 4000 people living in the region. In this single case, DNA fingerprinting proved both the innocence of a man under arrest, in serious danger of conviction and punishment, and the guilt of the real criminal.

A substantial number of persons convicted and sent to jail before Jeffreys' discovery have subsequently been proved innocent by analysis of samples saved from the evidence presented at their trial.

Personal identification by amplification of specific regions has superseded the RFLP approach

Despite its successes, identification by gel separation of RFLPs has disadvantages in practice. It requires relatively large amounts of undegraded DNA (10–50 ng of material no shorter than 20 000–25 000 bp). Since the development of PCR, DNA-based identification methods have tested for the presence of selected regions known to vary in the population, using PCR to amplify those present. This greatly improves sensitivity. Subnanogram amounts suffice to identify 100 bp regions. It is possible to get a positive identification from a single hair (of a person, or in one case from the cat of a criminal's parents), or from the saliva on a licked envelope.

KEY POINT

Sensitivity: Among people with a condition, the proportion that gives a positive test result.

Specificity: Among people who do *not* have the condition, the proportion that gives a negative test result.

For a perfect test, sensitivity and specificity are both 100%: no false positives, no false negatives.

The method in common use now is to PCR amplify a short tandem repeat (STR) typically containing 2–5 bp, repeated between a few and a dozen times (see Box 2.6). Amplification produces fragments about 200–500 bp long. Loci in common use show 5–20 common alleles. Typically 11 or 13 autosomal loci are tested, plus the amelogenin test for gender.

The human amelogenin gene appears on both X and Y chromosomes. The X copy shows a 6 bp deletion in intron 1, relative to the Y version. In the conventional gender test, PCR using suitable primers amplifies fragments of 106 bp from the Y chromosome (if present) and 112 bp from the X chromosome. Two bands on a gel imply a male DNA source, and a single band, female. If mutations invalidate this test, other Y-chromosome markers are available.

In some cases, X-chromosome-specific STRs are useful. Consider a case of questioned paternity if two possible fathers are themselves father and son. Y-chromosome-specific STRs are also useful if a specimen contains a mixture of body fluids from different individuals, as in cases of male-aggressor/female-victim rape.

Mitochondrial DNA

Human mitochondrial DNA is a circular molecule 16 568 bp long, containing 37 genes densely packed in a coding region. There are 13 genes coding for polypeptides, 2 for ribosomal RNAs, and 22 for tRNAs. Most mitochondrial proteins are encoded in nuclear DNA and the proteins imported into

BOX 2.6 DNA profiling

Governmental DNA databases, used in criminal investigation and parentage testing, do not require full-genome sequencing. The lengths of selected STRs, plus gender, identify individuals.

In the population, each of the selected STRs shows variability in length. The number of alleles at each site is relatively small: About 5–20% of the population will have any particular STR length. It therefore requires a combination of many sites to identify a person's DNA uniquely.

The UK National DNA Database uses the SGM+ protocol, specifying the number of tandem repeats, on both strands, in ten STRs, plus an amelogenin test result with values XX or XY. That is, a person's DNA fingerprint contains 20 numbers and two letters. The UK Crown Prosecution Service estimates the probability of a false positive—that is, a match between unrelated individuals—at 1 in 10^9. (The current UK population is about 6.5×10^7.) The probability of a match between siblings is higher, approximately 1 in 1000. The corresponding US scheme involves 13 STRs, plus amelogenin reporting gender. The two systems have eight loci in common.

People with karyotype abnormalities, such as Klinefelter's syndrome (XXY), will appear in the databanks with the attribution of normal chromosome content consistent with their gender; male in the case of Klinefelter's. A small number of people are chimaeric, containing two genetically distinct populations of cells. This is believed to arise by exchange of cells *in utero* between twins. It has occasionally led to ambiguities in using DNA fingerprints for identification.

The UK government also maintains the Missing Persons DNA Database (MPDD), containing DNA profile records of missing persons, relatives of missing persons, unidentified bodies, and some crime-scene specimens that might be traceable to missing persons or unidentified bodies; for instance, from scenes of apparent murders with no body or fragment of a body. The US has similar databases: The National Missing and Unidentified Persons System (NamUS) and the UnClaimed Persons database (UCP) contain information about deceased persons with no identified next of kin.

The UK and the US military collect DNA sequence information from members of the armed forces to assist in identification of battlefield victims. The information is not merged with the general national DNA databases.

For readers thinking of writing murder mystery novels: Two obvious ploys (clichés by now) are: (1) let the criminal be an identical twin, or (2) let a bank robber, for instance, leave someone else's blood at the crime scene. Depending on the sophistication of the intended readership, the first one will not really work: Comparison of genomes between identical twins shows a few SNPs. In one case, there were five reliably determined SNPs (out of the 3.2×10^9!). These mutations occurred after the splitting of the fertilized egg. Clearly, this requires full-genome sequencing, not just a few STR lengths, and places a heavy burden on sequencing accuracy. (Defence barristers take note!) Easier to determine are differences in methylation patterns between the twins' DNA.

In the unlikely event that a molecular biologist takes to violent crime, he or she might attempt to rort the system by synthesizing DNA to match some entry in the national database, to leave at the crime scene. This is possible without access to the DNA of the individual from whom the entry was derived. In such a case, it would be necessary to check the crime-scene sample for epigenetic modifications, absent from synthetic DNA.

the mitochondrion. Mitochondrial DNA contains a single major non-coding region, 1122 bp long, which is hypervariable relative to the coding regions. Formerly, two segments from this region were sequenced for population and forensic studies. However, now it is easy enough, and better practice, to sequence the entire mtDNA. The mitochondrial DNA of unrelated people typically differs at eight positions. Mitochondrial DNA is very abundant and survives very well.

However, each person receives mitochondrial DNA exclusively from his or her mother. Mitochondrial DNA sequence comparisons would therefore be useless for paternity testing. However, they can identify an individual from a sequence match to a specimen, are useful in tracking migratory patterns (see Chapter 9), and can identify relationships among people linked by a maternal line of descent. For instance, mitochondrial DNA was used in the initial identification of the remains of the Russian royal family (see Box 2.7).

BOX 2.7 Identification of the remains of the family of Tsar Nicholas II

For most of us, all of our mitochondria are genetically identical, a condition called homoplasmy. However, in some individuals, different mitochondria contain different DNA sequences; this is called heteroplasmy. Such sequence variation in a disease gene in the mitochondrial genome can complicate the observed inheritance pattern of the disease.

The most famous case of heteroplasmy involved Tsar Nicholas II of Russia. After the revolution in 1917, the Tsar and his family were taken into exile in Yekaterinburg in Central Russia. During the night of 16–17 July 1918, the Tsar, Tsarina Alexandra, their five children, their physician, and three servants who had accompanied the family were killed and their bodies were buried in secret graves. When the remains were rediscovered, assembly of the bones and examination of the dental work suggested—and sequence analysis confirmed—that the remains included most of the expected family group. The identity of the remains of the Tsarina was proved by matching her mitochondrial DNA sequence with that of a maternal relative, the late Prince Philip, former Chancellor of the University of Cambridge, Duke of Edinburgh—and grandnephew of the Tsarina.

However, comparisons of mitochondrial DNA sequences of the putative remains of Nicholas II with those of two maternal relatives revealed a difference at base 16169: The Tsar had a C and the relatives a T. Extreme political and even religious sensitivities mandated that no doubts were tolerable. Further tests showed that the Tsar was heteroplasmic; T was a minor component of his mitochondrial DNA at position 16169. To confirm the identity beyond any reasonable question, the body of Grand Duke Georgij, brother of the Tsar, was exhumed and was shown to have the same rare heteroplasmy.

In 2007, bone fragments from two individuals were found at a site about 45 miles away from the grave of the Tsar, his family members, and their entourage.

This discovery prompted a complete reexamination of all the available material, including sequencing of nuclear DNA. DNA sequences from additional relatives were brought into the picture. These included specimens gathered by opening the tombs of Tsar Alexander III (reigned 1881–1894, father of Nicholas II) and Tsarina Alexandra's sister Grand Duchess Elizabeth Fyodorovna, buried in Jerusalem.

The results were conclusive: They confirmed the identification of Tsar Nicholas and showed that the 2007 remains were those of Tsarevich Alexei and the missing daughter, Grand Duchess Marie. Y-chromosome markers from the remains of the Tsar matched sequences from suitable male relatives; from the remains of Tsarevich Alexei (one set of the 2007 fragments); and from a specimen of blood known to be the Tsar's, from a shirt held in Hermitage Museum in St. Petersburg.[10] Mitochondrial DNA from the shirt showed the expected heteroplasmy.

Haemophilia is a genetic feature of several of Queen Victoria's descendants. That Tsarevich Alexei inherited the disease had an immense impact, not only on the life of the patient but on history. Haemophilia affected the royal families of Spain and Germany also. In those countries too—especially Spain—it had consequences not only on the patients and their relatives, but also on government and revolution. (Prince Philip's shared maternal line with Tsarina Alexandra, a carrier, means that in principle his chances of suffering from haemophilia were 12.5%.)

The most common form of haemophilia arises from Factor VIII deficiency; a less common type from Factor IX deficiency. (Factors VIII and IX are proteases in the blood-coagulation cascade.) A mutation in the gene for Factor IX appeared in the genomes of Tsarevich Alexei, his sister Grand Duchess Maria, and his mother Tsarina Alexandra. None of the other daughters was a carrier. The DNA sequencing revealed an SNP: an A→G substitution affecting the splice site upstream of exon 4 of the gene for Factor IX. This SNP creates a cryptic splice acceptor site, resulting in a frameshift (see Exercise 2.13).

[10] In 1891, while heir to the throne, Nicholas visited Japan, on the way to Vladivostok for ceremonies to mark the start of construction of the Trans-Siberian Railroad. In Osaka, he survived an assassination attempt by Tsuda Sanzō, a member of the police escort. Prince George of Greece and Denmark (Prince Philip's uncle) saved him by parrying the attacker's second sabre blow with his cane. This was 13 years before war broke out between the two countries: The battle of the Tsushima strait, on 27–28 May 1905, was a terrible defeat for the Russian Navy. (Admiral Heihachirō Tōgō, commander of the Japanese fleet, had studied naval science in England for six years, from 1871. He returned to England in 1911 to represent the Japanese government at the coronation of King George V.) The consequences of the Russo-Japanese war in terms of worsening of the Russian economy by increases in military spending, and the discrediting of the Tsar and his advisors, contributed to the 1917 revolution.

Identification of non-human DNA sequences

Applications of DNA identification techniques to animals include the proof of claims that Dolly the sheep was indeed a clone, testing of horses and dogs to confirm breeders' claims of pedigrees (confirmation of racehorse bloodlines is essential given the prices of thoroughbred horses: The most expensive horse in the world was sold for US$70 000 000 = £53 700 000), restoring lost pets to their owners, identifying the claimed species of origin of supermarket meats, and checking of commercial whale meat for endangered species. A number of localities have required submission of DNA samples as a condition for issuing dog licences, in order to be able to identify—and fine—owners who do not clean up after their pets. (Barking, in East London, is one of them. Really!) We have already mentioned the sequencing of samples from sushi bars to test the 'truth in labelling' of menus.

Identification of samples from animals has proved useful in crime investigation. Animals can cause harm, or be victims of crime, or provide links to human criminals. The association of hair from a suspect's pet with trace material found on a victim has contributed to many convictions.

- In 1995, the body of Shirley Duguay, a Canadian woman from Prince Edward Island, was found in a shallow grave. Near the body, Royal Canadian Mounted Police found a leather jacket stained with the victim's blood, and also over two dozen white hairs from a cat. DNA from the crime-scene hairs matched that of the white cat, Snowball, pet of Duguay's estranged husband. The husband was convicted of second-degree murder and sentenced to 15 years in prison.

This was the first use of non-human DNA in a criminal trial.

Parentage testing

Paternity testing is the most common. Every child receives VNTRs from their mother and father. For VNTRs on autosomes, the child inherits only one of each chromosome pair from each parent. Therefore, half of each parent's VNTRs appear in the child. Conversely, every VNTR in the child must come from one of the parents (see Figure 2.2). Assume that the mother is known, and that there is a candidate father. If the child has any VNTR that does not appear in the mother or in the candidate father, paternity is disproved. If all tested VNTRs in the child appear in both mother and candidate father, probability of paternity depends on how many loci are studied.

Typically, 15–20 autosomal loci are tested. The results are the two VNTR lengths at these loci for the child, mother, and candidate father. If father and mother share a VNTR length, then its appearance in the child is non-discriminatory. For each VNTR shared by child and candidate father—but absent in the mother—the strength of the inference of paternity varies with the frequency of appearance of that VNTR length in the population. The Paternity Index (PI), at one VNTR locus, is the likelihood of occurrence of the observed phenotypes if the candidate father is the biological father, relative to the likelihood of occurrence of the observed phenotype in a randomly selected man from the appropriate population. Thus, if the child and candidate father share a *rare* allele, the associated PI is high. The values are typically in the range 0–40. The **Combined Paternity Index (CPI)** is the product of the paternity indices at all loci. In many countries, a CPI over 100 is the legal criterion for presumed paternity. CPI is conventionally translated into Probability of Paternity, abbreviated (unsurprisingly) POP.

Test results from 'over-the-counter' paternity test kits are not accepted by courts, if only because the origin of the samples is not adequately controlled.

Reverse paternity testing

It is now standard to decide paternity if samples are available from the mother and child or foetus, and from a candidate for paternity. What if no candidate sample is available? The reader can think of many ways in which this situation might arise (see Exercise 2.8).

If there is an alleged father from whom a sample is unavailable, but sequences from other children or close relatives of his are available, it is possible to compare them with data from the mother and child. If not, it might still be possible to 'reconstruct' the genetic makeup of the father. This might make it possible to identify certain of his phenotypic features (see below), or to search for him in genetic sequence databases.

Maternity testing is, for obvious reasons, less common. However, the question does sometimes arise:

- Adopted children seeking their birth parents; or parents seeking a child whom they gave up for adoption.
- During the military dictatorship in Argentina (1976–1983) many children were abducted and placed in other homes. In many cases, the fathers were killed. The Mothers of the Plaza de Mayo is an Argentine organization with the goal of reuniting children with their natural mothers. In other cases, pregnant women survived in captivity only until after delivery of the child. A parallel group, the Grandmothers of the Plaza de Mayo, seeks to trace their grandchildren as well as the fate of their daughters.
- There are very rare cases of babies being swapped accidentally in hospitals.

Foetal DNA is present in maternal blood during pregnancy, as early as seven weeks post-conception. Taking a sample of maternal blood is a much safer procedure than amniocentesis. It is possible to extract and sequence foetal DNA, which allows for prenatal (a) paternity testing and (b) detection of disease-associated mutant genes. Also, although paternity might be in question, it might be thought obvious that prenatal *maternity* cannot be in doubt, with the foetus *in situ*; however, in cases of *in vitro* fertilization and surrogate pregnancy it is a genuine question.

Inference of physical features, and even family name

Suppose a sample containing DNA is collected at a crime scene, and there is reason to believe that a criminal deposited it. It is possible to use the sample for identification. But suppose the source individual is not represented in the forensic databanks, and is not one of the suspects—usual or unusual—rounded up. It is still technically feasible to make some inferences about the perpetrator.

It is possible to predict certain physical characteristics from analysis of DNA sequences. Gender, obviously, but also colour of hair, eyes (see Figure 2.5), and skin, and ethnic background. Use of these inferences for suspect profiling is controversial.

In some cases, it is possible to infer the source individual's family name! Oxford don Brian Sykes discovered that all males named Sykes in the UK are descendants of a single founder individual, and all carry specific diagnostic features of their Y-chromosome sequences. Police could deduce, from a sample left at a crime scene, whether or not the source individual was named Sykes (unless, of course, he changed his name).

The extent of DNA methylation correlates with age. Certain chemicals, for instance melatonin, vary in concentration in blood and saliva following regular circadian rhythms. Therefore, analysis of a blood sample left at a crime scene might provide an estimate of the time of day of its deposition. Such a test does not involve DNA.

It might not be too fanciful to imagine a police investigator asking:

> 'Now then, Grandfather Sykes, where were you at 11 pm last night?'

KEY POINT

In addition to matching a DNA sample with an individual, it is possible to analyse crime-scene samples to infer several characteristics, including eye and hair colour, complexion, and ethnicity. Use of these inferences in criminal investigation remains controversial, and there is substantial variation in what different jurisdictions permit.

Solutions of 'cold cases'

- The Golden State Killer

A series of burglaries, rapes, and murders took place during the 1970s and 1980s, in California in the Western US. The press named the perpetrator 'The Golden State Killer' (California is known as the 'Golden State' as a result of the gold rush that began in 1848). The crimes were not solved, and the investigation was shelved as a 'cold case'.

Hair & eye colour phenotype | **Prediction result**

Hair	Dark 0.839	Light 0.161	
Black 0.631	Brown 0.309	Red 0.001	Blond 0.059
Eye	Blue 0.058	Int. 0.183	Brown 0.759

(a) Black hair, brown eyes
Report: The most probable hair colour of the individual is black/dark brown with an accuracy of 87.5% for the black hair prediction category based on a >300 European test set.
The most probable eye colour is brown with an accuracy of 95% based on a European dataset of >3800 individuals.

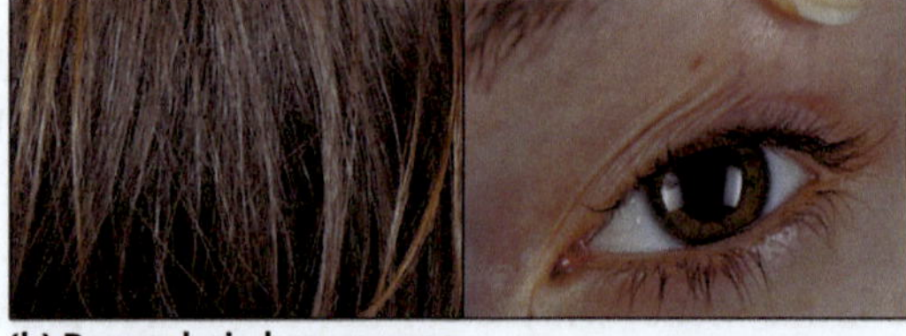

Hair	Dark 0.856	Light 0.144	
Black 0.393	Brown 0.511	Red 0.014	Blond 0.082
Eye	Blue 0.215	Int. 0.221	Brown 0.564

(b) Brown hair, brown eyes
Report: The most probable hair colour of the individual is brown with an accuracy of 78.5% for the brown hair prediction category based on a >300 European test set.
The most probable eye colour is brown with an accuracy of 91% based on a European dataset of >3800 individuals.

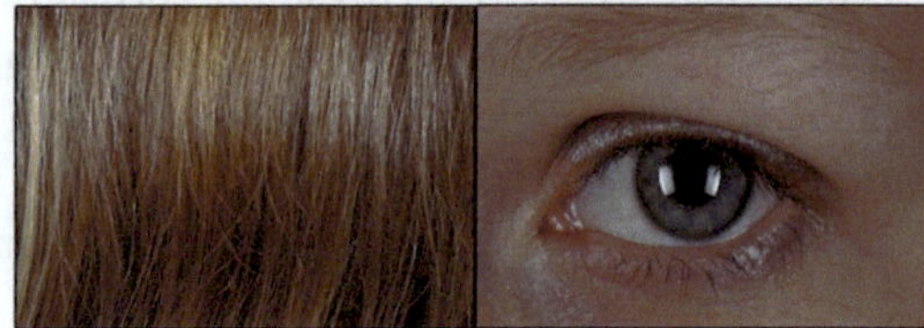

Hair	Dark 0.055	Light 0.945	
Black 0.631	Brown 0.152	Red 0.01	Blond 0.78
Eye	Blue 0.95	Int. 0.03	Brown 0.02

(c) Blond hair, blue eyes
Report: The most probable hair colour of the individual is light blond with an accuracy of 69.5% for the blond hair prediction category based on a >300 European test set.
The most probable eye colour is blue with an accuracy of 99% based on a European dataset of >3800 individuals.

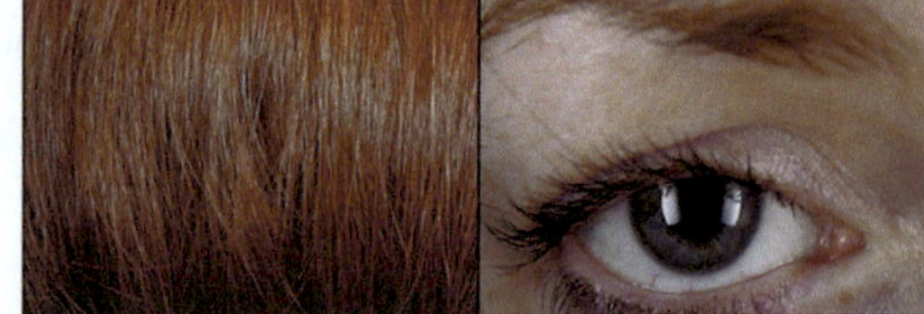

Hair	Dark 0.033	Light 0.967	
Black 0.631	Brown 0.077	Red 0.816	Blond 0.1
Eye	Blue 0.965	Int. 0.027	Brown 0.007

(d) Red hair, blue eyes
Report: The most probable hair colour of the individual is red with an accuracy of 80% for the red hair prediction category based on a >300 European test set.
The most probable eye colour is blue with an accuracy of 99% based on a European dataset of >3800 individuals.

Figure 2.5 The HIrisPlex system predicts both hair and eye colour from DNA sequences. It assays 24 eye and hair colour-predictive DNA variants, from a total of 11 genes, including sites within exons, within introns, and intergenic. The figure contains examples of four individuals, showing predicted and observed hair and eye colour.

Reprinted from Walsh, S., Liu, F., Wollstein, A., *et al.* (2013). The HIrisPlex system for simultaneous prediction of hair and eye colour from DNA. Forensic Science International: Genetics 7, 1: 98–115. https://doi.org/10.1016/j.fsigen.2012.07.005, with permission from Elsevier.

In 2018, police reopened the investigation. More powerful tools had become available to make use of samples from the old crime scenes. Barbara Rae-Venter, a genealogist turned detective, made important contributions to the analysis.

Checking the sequences against nearly 1 million entries in the public genealogy website GEDmatch identified distant relatives of the criminal, equivalent to third cousins. Based on these, it was possible to build a family tree, to identify additional relatives who had not contributed to GEDmatch. From the DNA samples, it was possible to infer physical characteristics of the criminal: Caucasian, green or hazel eyes, blond hair, and a likely tendency to early baldness. Combining these with descriptions from surviving witnesses allowed focus on a specific individual.

Surveillance of the suspect led to a collection of material from which his DNA could be sequenced: a discarded tissue and a swab taken from the handle of a car door. Matching DNA sequences from these samples with the evidence from the crime scenes confirmed the identification; an arrest followed. The suspect, John DeAngelo, confessed to the crimes and pleaded guilty at trial. He was sentenced to life without possibility of parole.

- On 9 August 1998, in the Netherlands, an 11-year-old boy named Nicky Verstappen disappeared during the night from the summer camp he was attending. His body was found two days later, in a pine grove 2 km away. DNA from a tissue and cigarette found close to the body allowed determination of the DNA profile of the killer. But the investigations proved unsuccessful in this case also.

In 2018, the Dutch police reopened the case. They decided to carry out comprehensive screening: Over 15 000 samples were collected from men living in the area. From the DNA collected the police focused on a 55-year-old man, Joseph Brech. Neither Brech nor his immediate family had contributed to the DNA collection, but police were led to Brech from samples from more distant relatives. Indeed, Brech was missing, but police were able to retrieve and characterize a DNA sample for Brech himself from personal items left behind. It matched the DNA from the material recovered from the crime scene.

Brech was found in Spain, arrested on 26 August, and extradited to the Netherlands on 6 September. On 28 January 2022, he was sentenced to 16 years in prison.

If the country, or region, had a universal DNA-sequence databank, it would have been possible simply to look up the crime-scene sample in the databank.

Ethical, legal, and social issues

Knowledge creates power. Power requires control. Control requires decisions.

Advances in genomics have created problems that individuals and societies must face. In setting up the Human Genome Project, the US Department of Energy and National Institutes of Health recognized the importance of ethical, legal, and social issues by allocating 3–5% of the funding to them.

DNA databases containing information about every citizen of a country are technically feasible. Routine testing of all newborns for specific genetic diseases is common. In 2015, a consortium of hospitals in the US embarked on a project to sequence whole genomes of newborns (subject to parental consent). It is even possible to sequence the genome of a foetus. Should this information be determined? If it is determined, who should have access to it?

The Faroe Islands, population 50 000, has offered full-genome sequencing to any citizen who wants it. Kuwait, population 4 million, requires all citizens and residents to submit DNA samples. Both Iceland and the UK have undertaken large-scale whole-genome sequencing projects. It is very likely that within the lifetimes of many readers full-genome sequencing will be universal in many countries.

No one doubts that DNA sequence information could benefit people—as individuals they could receive better health care, and society would benefit from the results of research based on the collected information. What is the downside?

Controversy arises over allowing the information to be collected into a widely accessible databank. Most people would have no problem accepting an effort that would assist in the capturing of criminals, especially those who are likely to repeat their offences. Most people would have no problem accepting an effort that would make it easier to identify victims of death on a battlefield or after a terrorist attack. Genome sequence data, correlated with clinical records, would be an extremely valuable resource for research. Questions have been raised, however, in particular over the use of genome sequence data by law enforcement agencies:

- *Privacy issues*: Should inclusion in a databank of genomic information about individuals require the individuals' consent?
- *What data should be retained?* Should the data be limited to the minimum required for standard identification procedures, or be more extensive? (For instance, should sufficient additional data be kept to identify physical features or ethnic characteristics?)
- *Access*: Who should have access to the information?

Databases containing human DNA sequence information

There are two major national repositories of human DNA sequence information in the UK (see Box 2.8).

- The National DNA Databank (NDNAD) primarily supports law enforcement agencies.
- The UK BioBank has the goal of improving prevention, diagnosis, and treatment of illness. It has amassed a large collection of clinical data and biological samples to provide sequence information. It has recently released complete genome sequences of its 500 000 entries.

Also based in the UK are the comprehensive nucleotide sequence databanks at the European Bioinformatics Institute and The Sanger Centre.

In England and Wales, police have been allowed to take samples, from any individual arrested on suspicion of all but the most minor crimes, and determine partial sequences. Consent is not required. Access to the database is not strictly limited to police, but has been used to support research projects without consent of the individuals represented in the database. Until 2012, the sequence data *and samples* were retained even if the person arrested was never charged, or had been tried but not convicted. In 2012, the law changed. Now, data and samples from people not charged or not convicted are to be destroyed (with some exceptions).

In the current political climate, there is intense pressure to tip the scales towards giving governments powers that can be used to protect their citizens. The dangers are that such powers, although they appear to be harmless to innocent people, may be compatible with abuse if safeguards are inadequate, or—in the worst case—deliberately violated. Even in the UK, a House of Commons committee in 2005 described as 'extremely regrettable' the fact 'that for most of the time that the NDNAD has been in existence there has been no formal ethical review of applications to use the database and the associated samples for research purposes'.[11] If the situation in the UK could have been described as 'extremely regrettable', one need not be a Colonel Blimp to shudder to think what things must be like elsewhere (see Box 2.8).

It has been suggested that the NDNAD be extended to the entire population. The current contents of the database show ethnic and gender biases, which a universal database would eliminate. A higher fraction of reported crimes might be solved, but there is no consensus about how significant this would be. There is intense debate over whether the loss of individual privacy would justify the benefits to society. Even if there is consensus that certain data *should* be kept confidential, computer security is simply not up to the task of ensuring that it will be.

In the US, the analogue of the UK NDNAD is the Combined DNA Index System and the National DNA Index System (CODIS/NDIS), maintained by the Federal Bureau of Investigation (FBI). In late 2015, NDIS contained over 21 million DNA profiles. This is about three times the size of the corresponding UK NDNAD, but represents a smaller percentage of the national population. Practices dealing with expungement of records in cases of no charge or no conviction are similar to the European standard.

Individual states of the US maintain their own DNA databases. In the absence of overriding federal legislation, laws of individual states govern DNA collection. The variety of guidelines for collection and retention of samples expressed in state laws have had a patchy career in the (state) courts. Some have been declared unconstitutional.

[11] House of Commons Science and Technology Committee, Forensic Science on Trial, 29 March 2005, http://www.publications.parliament.uk/pa/cm200405/cmselect/cmsctech/96/96i.pdf

BOX 2.8 DNA sequence databases, law enforcement, and the courts

The UK National DNA Database (NDNAD) is one of the largest forensic DNA collections in the world. It stores both a database of sequence-derived information and the biological samples from which the sequences were determined. In 2022, it contained profiles of ~7 million individuals. Of these, 80% are male, 19% female; a small number are unassigned. The NDNAD contains personal identification 'DNA fingerprints' and gender. The probability that sequences from a sample collected at a crime scene will match an entry in the database is over 50%!

Entries in the NDNAD are of two types:

(a) Samples from known individuals. Police can collect samples, without consent, from anyone suspected of a 'recordable' offence, not only before trial and conviction, but even before being charged with a crime. The list of recordable offences is steadily growing through incremental legislation. Roughly speaking, recordable offences include all but the most trivial antisocial actions. For example, under the Football (Offences) Act of 1991, 'unlawfully going onto the playing area' is a recordable offence. For purposes of elimination during active investigations, the NDNAD also contains samples from persons present at a crime scene, and from police officers. Under the new (2012) law, in most cases, samples taken from people not charged or not convicted will be destroyed when the investigation is complete.

(b) Samples from unidentified individuals collected at a crime scene. They might originate from the perpetrator of a crime, from a victim, or even from someone who had been at the scene at some time other than during the commission of the crime. In the event of a match, these become samples from known individuals. (In the case of a partial match, they may suggest that the crime scene sample came from a near relative of the individual matched.)

In the UK, the governing legislation has been in a state of flux. The Criminal Justice Act of 2003 authorized the widespread collection of samples for DNA profiles in England and Wales, without consent. (This was extended to Northern Ireland in the next year.) The act also allowed the indefinite retention of the samples and sequence data, even if the suspect were never charged with a crime. In 2008, the Counter-Terrorism Act extended the criteria that allow police to demand samples for DNA sequencing.

The law in Scotland, in this as in many other respects, varied from that of England and Wales. Scotland maintains a separate DNA database and shares results with England. One salient legal difference is that Scotland did not permit the automatic indefinite retention of samples and data from people not convicted of a crime.

In a 2008 decision, The European Court of Human Rights aligned itself more closely with the Scottish law.

A case against the UK government was brought to the European Court by two individuals who wanted their identifying information expunged from the NDNAD and their samples destroyed. One was tried for attempted robbery but acquitted, the other was never charged. The court held that there had been violation of Article 8 of the European Convention for the Protection of Human Rights and Fundamental Freedoms:

> 'In conclusion, the Court finds that the blanket and indiscriminate nature of the powers of retention of the fingerprints, cellular samples and DNA profiles of persons suspected but not convicted of offences, as applied in the case of the present applicants, fails to strike a fair balance between the competing public and private interests and that the respondent State has overstepped any acceptable margin of appreciation in this regard. Accordingly, the retention at issue constitutes a disproportionate interference with the applicants' right to respect for private life and cannot be regarded as necessary in a democratic society.'[12]

This case led to provisions in The Protection of Freedoms Act 2012, which completed its passage through Parliament and received Royal assent on 1 May 2012. This brought the law of England and Wales more into line with Scotland and the European Union.

[12] http://www.bailii.org/eu/cases/ECHR/2008/1581.html

The conclusion is that, in the UK and US at least, legislation is moving in the direction of greater protection of privacy of DNA sequence information. Belief in this protection, which may in some respects be illusory (see Box 2.9), may lead to increased genetic testing, both in regular medical practice and by private companies. However: (a) Like mailing lists, testing companies may have the right to sell

BOX 2.9 Databases of DNA sequences: Privacy issues

There are several categories of databases of human DNA sequences:

- Research databases, such as that of the '1000-genome' project. Participants are promised anonymity.
- Clinical databases. Privacy guarantees are in accord with other medical records.
- Databases associated with law enforcement organizations. Access limited to members of legitimate official organizations. International data sharing is increasing.
- Personal genealogy databases. Companies provide data to subjects who submit samples. The subject can decide whether to put his or her data on the web. Often these postings are not anonymous—this is the point of searching for unknown relatives.

The problem is that the public information contained in genealogy postings makes it possible to break the anonymity of databases such as the 1000-genome project.

genetic information to outside parties. (b) Given the increased degree of international sharing of identification information, individuals need to be concerned not with the countries with the most secure databases, but those with the least secure ones. (c) Experience has shown that much private information of any type in fact becomes disseminated, through either accident or malice.[13]

Use of DNA sequencing in research on human subjects

In principle, use of DNA samples in research should require consent of the individuals donating the material: consent not only for its collection, but for the specific uses to which the samples will be put. There have been cases of 'research goal creep' in which permission was granted for analysis of restricted scope, but the samples were subsequently used for other studies.

A US case testing the propriety of use of samples collected voluntarily from subjects was settled in April 2010.

A Native American group, the Havasupai, who inhabit an inaccessible area of the Grand Canyon, in the Western US, has among the highest known incidences of Type II diabetes. In 1991, 55% of Havasupai women and 38% of Havasupai men were affected. Scientists from Arizona State University collected samples from the group and carried out research projects, believed—by the subjects—to be focused on susceptibility to diabetes. However, using the same samples, the scientists also investigated genetic susceptibility to schizophrenia and evidence for migration patterns. The subjects objected. Researchers pointed to the wording in the consent form; representatives of the Havasupai alleged (among other things) that the wording was too vague to constitute true informed consent. The ensuing lawsuit was settled, with the Arizona state agency responsible for the university agreeing to pay the Havasupai US$700 000, and to return the samples.

KEY POINT

Genomic databases are useful in identifying individuals from samples collected at crime scenes. In many cases, these databases contain data about selected regions of the genome, and the biological samples from which they were derived. In drafting the laws granting law enforcement agencies authorization to maintain such databases, there is a tension between the desire to protect society against offenders and upholding individuals' rights to privacy.

[13] See, for instance: https://www.nytimes.com/2023/12/04/us/23andme-hack-data.html

LOOKING BACK AND FORWARD

Following the previous chapter, which presented an overview of the entire subject, this chapter focused on its cynosure, the Human Genome Project. We discussed the history of human genome sequencing, some of the applications to medicine and law, and some of the ethical, legal, and social issues involved with creation of databanks containing human genome sequences. Some of these databanks support law enforcement efforts; others support personal curiosity about genealogy, parentage testing, and disease risk.

In Chapter 4, we shall turn to the techniques of DNA sequencing, that are generating the data stream on which the subject is riding. Scientific, clinical, and commercial rewards motivate progress in the technology. The goal is to minimize speed and expense and maximize accuracy. The proposed goal of the (US)$1000 genome has been achieved, but there is consensus that the instrumentation and algorithms will continue to improve.

RECOMMENDED READING

Personal accounts by two of the major players

Ferry, G., & Sulston, (Sir) J. (2002). *The Common Thread: A Story of Science, Politics, Ethics and the Human Genome*. London: Bantam Press.

Jeffreys, A. (2009). The eureka moment: An interview with Sir Alec Jeffreys. Interview by Jane Gitschier. PLoS Genet. 5: e1000765.

A useful short summary, although, like anything else in print, somewhat out of date

Sundermann, U., Kushnir, S., & Schulz, F. (2010). The development of DNA sequencing: From the genome of a bacteriophage to that of a Neanderthal. Angew. Chem. Int. Ed. 49: 8795–8797.

The general consequences of mutation

Hill, W.G., & Loewe, L. (2010). The population genetics of mutations: Good, bad and indifferent. Philos. Trans. R. Soc. Lond. B Biol. Sci. 365: 1153–1167.

Lannoy, N., & Hermans, C. (2010). The 'royal disease'—haemophilia A or B? A haematological mystery is finally solved. Haemophilia 16: 843–847.

Sherman, R.M., & Salzberg, S.L. (2020). Pan-genomics in the human genome era. Nat. Rev. Genet. 21: 243–254.

Liverpool, L. (2023). First human 'pangenome' aims to catalogue genetic diversity. Nature 617: 444–445.

Identification of remains of Russian royal family

Rogaev, E.I., Grigorenko, A.P., Moliaka, Y.K., *et al.* (2009). Genomic identification in the historical case of the Nicholas II royal family. Proc. Natl. Acad. Sci. USA 106: 5258–5263.

Coble, M.D., Loreille, O.M., Wadhams, M.J., *et al.* (2009). Mystery solved: The identification of the two missing Romanov children using DNA analysis. PLoS One 4: e4838.

Childebayeva, A., & Zavala, E.I. (2023). Review: Computational analysis of human skeletal remains in ancient DNA and forensic genetics. iScience 26: 108066.

Interactions, and collisions, between scientific advances and social policy; including, but not limited to, issues of privacy

Levitt, M. (2007). Forensic databases: Benefits and ethical and social costs. Brit. Med. Bull. 83: 235–248.

Norrgard, K. (2008). Forensics, DNA fingerprinting, and CODIS. Nature Education 1: 35.

In March 2010, The Home Affairs Committee of the UK House of Commons published a report on The National DNA Database: http://www.publications.parliament.uk/pa/cm200910/cmselect/cmhaff/222/22202.htm.

Harmanci, A., & Gerstein, M. (2016). Quantification of private information leakage from phenotype-genotype data: Linking attacks. Nat. Methods 13: 251–256.

Lacaze, P., Pinese, M., Kaplan, W., *et al.* (2019). The Medical Genome Reference Bank: A whole-genome data resource of 4000 healthy elderly individuals. Rationale and cohort design. Eur. J. Hum. Genet. 27: 308–316.

Kling, D., Phillips, C., Kennett, D., *et al.* (2021). Investigative genetic genealogy: Current methods, knowledge and practice. Forensic Sci. Int. Genet. 52: 102474.

Lappalainen, T., Scott, A.J., Brandt, M., *et al.* (2019). Genomic analysis in the age of human genome sequencing. Cell 177: 70–84.

Pope, S., & Puch-Solis, R. (2021). Interpretation of DNA data within the context of UK forensic science—investigation. Emerg. Top. Life Sci. 5: 395–404.

Ram, N., Murphy, E.E., & Suter, S.M. (2021). Regulating forensic genetic genealogy. Science 373: 1444–1446.

Arshad, S., Arshad, J., Khan, M.M., *et al.* (2021). Analysis of security and privacy challenges for DNA-genomics applications and databases. J. Biomed. Inform. 119: 103815.

Mateen, R.M., Sabar, M.F., Hussain, S., *et al.* (2021). Familial DNA analysis and criminal investigation: Usage, downsides and privacy concerns. Forensic Sci. Int. 318: 110576.

Haddrill, P.R. (2021). Developments in forensic DNA analysis. Emerg. Top. Life Sci. 5: 381–393.

Lek, M., & Mardis, E.R. (2021). Envisioning the next human genome reference. Dis. Model Mech. 14: dmm049426.

Dabas, P., Jain, S., Khajuria, H., *et al.* (2022). Forensic DNA phenotyping: Inferring phenotypic traits from crime scene DNA. J. Forensic Leg. Med. 88: 102351.

Wickenheiser, R.A. (2022). Expanding DNA database effectiveness. Forensic Sci. Int. Synerg. 4: 100226.

Koplin, J.J., Skeggs, J., & Gyngell, C. (2022). Ethics of buying DNA. J. Bioeth. Inq. 19: 395–406.

Rafnar, T., Holm, H., Stefansson, H., *et al.* (2022). The sequences of 150119 genomes in the UK Biobank. Nature 607: 732–740.

Neves, C., & Zieger, M. (2023). 'Total Human DNA Sampling'—Forensic DNA profiles from large areas. Forensic Sci. Int. Genet. 67: 102939.

EXERCISES, PROBLEMS, AND WEBLEMS

Exercises

Exercise 2.1 The restriction fragments from Figure 2.3 are run on a gel. Alongside a photocopy of Figure 2.3, sketch the appearance of the gel. Indicate the direction of migration. Indicate, with labels or arrows, which band on your gel corresponds to which restriction fragment(s).

Exercise 2.2 On a photocopy of Figure 2.2, (a) Assuming that M is known to be the mother, sketch in and indicate by the letter X a band, the appearance of which would prove that F (F = the person whose DNA produced lane F) is *not* the father of child C. (b) Sketch in and indicate by the letter Y a band, the appearance of which would prove that M was *not* the mother of child C, but F could be the father. (c) Sketch in and indicate by the letter Z a single band, the appearance of which would prove that M was *not* the mother of child C *and* F was *not* the father.

Exercise 2.3 On a photocopy of Figure 2.2, relabel lane M as C, lane F as M, and lane C as F. Now M sues F for support of child C, alleging paternity. Can F prove from these data that he is not the father of C?

Exercise 2.4 A woman, M, brings a newborn baby to her solicitor. During the period during which the child was conceived, she was marooned for a month on a remote desert island with two sailors, S1 and S2. After they were rescued, M returned home and the sailors rejoined their ships. M wants to sue the father for support of the child.

VNTR analysis showed the following:

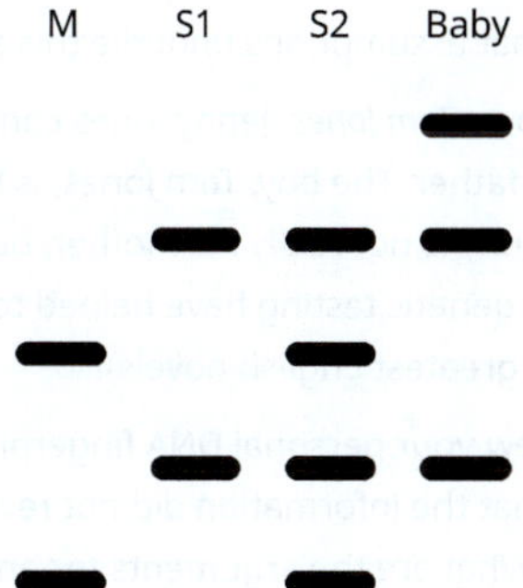

What can you conclude about the baby's parentage?

Exercise 2.5 Why could DNA analysis not decide whether or not a *woman* in the UK was named Sykes?

Exercise 2.6 An individual who left a bloodstain at a crime scene is male, but had a Y-chromosome mutation in the primer site for PCR amplification of the amelogenin locus. (a) What effect would this have on identification of the suspect? (b) How might the difficulty be overcome (provided it were suspected)?

Exercise 2.7 The mitochondrial DNA sequence determined from the remains of Richard III matched those of two living relatives who share an unbroken line of female descent. In a similar analysis of the Y-chromosome sequence, data from the remains did not match living people with a putatively unbroken line of male descent. What is the most likely explanation?

Exercise 2.8 A paternity case with limited access to relatives: The diagram shows a family tree, with squares representing males and circles representing females. Sequences from the two males with red lines through them are unavailable. This includes the actual father (D) and paternal grandfather (B) of the child (E). It is known that D is the son of A and that E is the daughter of C. The question is whether D could be the father of E. How could you answer this? If E had been male rather than female, would the method you proposed work?

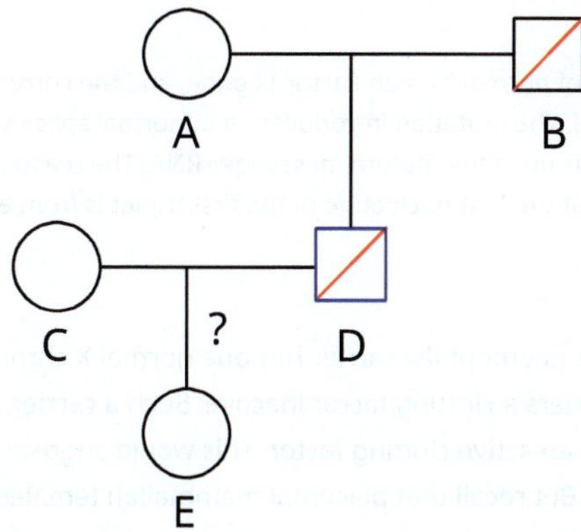

Exercise 2.9 Haemophilia A arises from a deficiency of Factor VII. Haemophilia B arises from a deficiency of Factor IX. These are two different proteins in the blood coagulation cascade. One of Queen Victoria's sons, Leopold, Duke of Albany, suffered from haemophilia B. He married Princess Helena of Waldeck and Pyrmont, who was not a carrier of haemophilia. They had a daughter and a son. Assuming no mutation in Leopold or Helena: (a) Was the daughter a carrier of haemophilia? (b) Could the son have suffered from haemophilia? (c) If Leopold had married a carrier of haemophilia B, what unusual feature might have been observed in their children? (d) If Leopold had married a carrier of haemophilia A, what would be the possible genotypes and phenotypes of their children?

Exercise 2.10 It is stated in the chapter that Prince Philip's chance of having haemophilia was 12.5%. How was this computed? What assumptions underlie this assertion?

Exercise 2.11 In Henry Fielding's novel *Tom Jones*, Jenny Jones confesses to being the mother of a foundling, but refuses to name the father. The boy, Tom Jones, is brought up in Squire Allworthy's household. It later turns out that Jenny is not really his mother, but was paid to conceal the baby's true parentage. How would today's genetic testing have helped to sort out the situation (aborting the plot development of one of the greatest English novels)?

Exercise 2.12 Suppose that you knew your personal DNA fingerprint in the format stored in national DNA databases. Assume that the information did not reveal that you were unusually susceptible to any known disease. What are the arguments for and against your voluntarily depositing these data in the national DNA database of the country of your residence? What conditions would you want to impose on the use of the data?

Exercise 2.13 The human gene for blood-clotting Factor IX has eight exons. The mutation in Tsarevich Alexei Nikolaevich Romanov, that accounted for his haemophilia, was an SNP just upstream of exon 4 (see Figure 2.6). The mutation introduced an abnormal splice site, just before the normal one. What would be the differences between normal Factor IX and the protein expressed from the mutant gene, assuming that the splicing machinery recognized exclusively the abnormal splice signal?

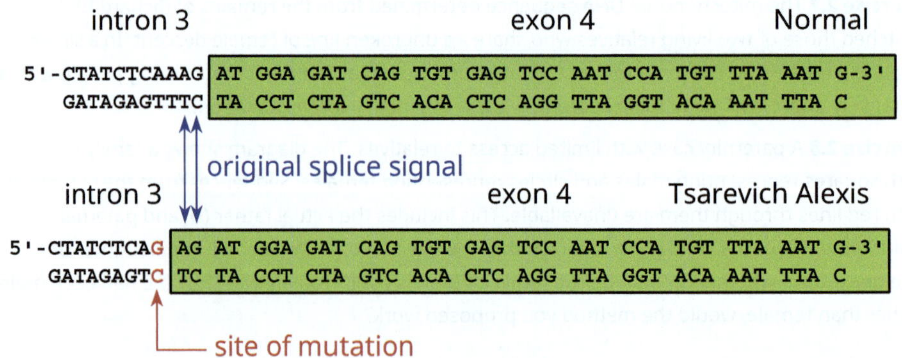

Figure 2.6 Partial sequences of normal human Factor IX gene, and the corresponding sequence from the genome of Tsarevich Alexei. The mutation introduces an abnormal splice site. In both cases, the material in the box is what ends up in the mature messenger RNA. The reason the division into triplets starts with a dinucleotide is that the first nucleotide of the first triplet is from exon 3.

Problems

Problem 2.1 A typical female haemophilia carrier has one normal X chromosome and one carrying a mutation that renders a clotting factor inactive. Such a carrier, just like a normal male, has one X chromosome with an active clotting factor. This would suggest that carriers should show normal blood clotting. But recall that placental mammalian females randomly inactivate one X chromosome in each cell (see Chapter 1). (a) What is the most likely fraction of active X chromosomes in a carrier that has an unmutated clotting factor gene? (b) Clinically, carriers with 30% or more active unmutated X chromosomes do not show clotting problems.[14] What fraction of carriers would be expected to show <30% active unmutated X chromosomes? These carriers would show at least mild symptoms of haemophilia. (Colour blindness is also an X-linked trait. There has been a report of monozygotic twins, one of whom has normal vision and the other is colour blind.) (c) Consider a female kangaroo that is a carrier of an X-linked recessive trait. What phenotypes might such an animal show? Would it correspond to the range of phenotypes shown by a human haemophilia carrier? On what would the phenotype depend?

Problem 2.2 Princess Alix of Hesse and by Rhine, a granddaughter of Queen Victoria and a carrier of haemophilia, was born in 1872. In 1894, she married Tsar Nicholas II of Russia.

Princess Alix had received an earlier proposal. This was from Prince Albert Victor, a grandson of Queen Victoria and Alix's first cousin. He was second in line of succession to the British throne; after his father, who was later to reign as King Edward VII.[15] Victoria supported the match.

Reader, she didn't marry him.

Very briefly suggest how history might have been different if she had. Consider the possibility that Nicholas might have married someone who was not a haemophilia carrier, and who bore him a healthy heir; and the effect of reintroducing the haemophilia gene on the British Royal family in particular, and on Britain in general.

Problem 2.3 In a study of brown bear (Figure 2.7) populations in northwest North America, samples of mitochondrial DNA were collected and sequenced from 317 free-ranging brown bears (*Ursus arctos*) from 22 localities. Forty-six variable sites corresponded to 29 haplotypes, which clustered into four major clades (see Table 2.2). Table 2.3 identifies the location(s) at which bears with the corresponding haplotypes were found.

Figure 2.7 North American brown bear cubs playing.
(Photograph by S. Hillebrand, from US Fish and Wildlife digital library.)

[14] Shoukat, H.M.H., Ghous, G., Tarar, Z.I., *et al.* (2020). Skewed inactivation of X chromosome: A cause of hemophilia manifestation in carrier females. Cureus 12: e11216. Another relatively common cause of clotting disorder, occurring in both males and females, and not limited to haemophilia carriers, is a patient's raising antibodies against clotting factors, most often Factor VIII. For these patients the clotting disorder is an **auto-immune disease**.

[15] Albert Victor predeceased his father, who was succeeded by his second son as George V.

Table 2.2 Mitochondrial sequence data from brown bears in northwest North America. These data do not represent continuous sequences, but only the variable positions. A dot indicates that the base at this position is the same as in the reference sequence in the first line.

	Sequence data	Location of collection
	CCCTCCCAACGTTAACATTACGTAATCGAACAGCGGGTTAGGGAAC	O
	..T.T.T.......................................	Q
Clade I	..T.TTT.......................................	PU
	....T...	Q
	....T.T.......................................	MOPRSV
	....TTT....................................G..	U
	....TTT.......................................	T
	......TGG......T.......G.........T.......AA...	HN
	......TGG.....GT.........................AA.G.	G
	......TGG......T............G..........G.AA.G.	LM
Clade II	......TGG......T............G............AA.G.	IJK
	......TGG......TG................T.......AA.G.	K
	......TGG.....GT...G.................C...AA.G.	H
	......TGG......T.................T...C...AA.G.	H
	......TGG......T.................T.......AA.G.	H
	......TGG.....GT.......G.............C...AA.G.	H
	TT..T.T..T..CG...................T......A.A..T	BE
	TT..T.T..T..CG........C..........T........A..T	ABCDE
	TT..T.T.....CG........C.........ATA.......A..T	B
Clade III	TT..T.T.....CG........C..........T...........T	BC
	TT..T.T..T..CG........C..........T...........T	A
	TT..T.T..T.CCG........C.........ATA.......A..T	B
	TT..T.T..T..CG........C..........T..A.....A..T	A
	TT..T.T..T..CG...............G...T..A.....A..T	C
	.T.C......A......C..TA.GGCT...T....A..C...A..T	F
Clade IV	.T.C....G.A......C.GTA.GGCT...T....A..C...A..T	F
	.T.C......A......C.GTA.GGCT...T....A..C...AG.T	F

(a) On a copy of a map showing Alaska, northwestern Canada, and the lower 48 states as far south as northern Wyoming, mark in different colours the sites of appearance of bears with mitochondrial DNA in the different classes. Use the data in Table 2.2. Describe the geographical distribution of the different classes. Do they overlap substantially?

(b) You are lost somewhere in northwest North America. You collect a tissue sample from a brown bear in the vicinity and determine the following mitochondrial DNA haplotype (relative to the reference sequence in Table 2.2):

.T.C......A......C.GTA.GGCT...T....A..C...A..T

Where are you?

Table 2.3 Sample collection locations of bears with sequences in Table 2.2. For instance, bears with the first haplotype of clade III were found at location B (latitude 48.0°N, longitude 113.0°W) and location E (latitude 51.0°N, longitude 118.1°W).

Location	Latitude, N	Longitude, W
A	44.4	110.3
B	48.0	113.0
C	48.5	116.5
D	51.0	114.2
E	51.0	118.1
F	57.2	134.6
G	58.7	133.5
H	60.8	139.5
I	67.8	115.3
J	69.2	124.0
K	69.4	129.0
L	69.0	138.0
M	69.2	143.8
N	60.1	142.5
O	63.0	145.5
P	60.0	154.8
Q	63.1	151.0
R	68.5	158.0
S	65.5	165.0
T	55.2	162.7
U	58.3	155.0
V	60.9	161.2

(c) Additional sequences were collected from four polar bears (*Ursus maritimus*) from zoos:

```
Reference sequence    CCCTCCCAACGTTAACATTACGTAATCGAACAGCGGGTTAGGGAAC
Polar bear 1          .T........A......CCGTA.GG.....T....A..C...A..T
Polar bear 2          .T........A......CCGTA.GG..A..T....A..C...A..T
Polar bear 3          .T........A......CCGTA.GG..A...GA..A..C...A..T
Polar bear 4          .T........A......CCGTA.GG..A....A..A......A..T
```

(The reference sequence, from a brown bear, is the same as the reference sequence in Table 2.2.) To which class of brown bears are these polar bears most closely related? Where are the brown bears most closely related to polar bears found?

(d) For the brown bear sequences, compute the average number of sequence differences between classes. Assuming a divergence rate of ~0.125 sequence changes per 10 000 years, estimate the times since divergence of the classes. Note that the total length of the region sequenced was 294 bases.

Brown bears immigrated to North America from Asia over a temporary land bridge over the Bering Strait. The first fossil evidence of brown bears in the New World is from 50 000 to 70 000 years ago. Is it likely that the classes diverged in North America or that they had already diverged in Asia?

(e) Assume that (1) the original North American brown bear population contained all of the currently observed haplotypes; and (2) there is now a continuous brown bear habitat covering all areas listed in Table 2.2. What then accounts for the current geographical distribution of the different haplotype classes? There are two questions to address:

- What accounted for their initial separation?
- What is continuing to keep them separated?

Consider the following possible scenarios:

- Climate changes in the past, associated with glaciation, fragmented the population and left pockets that reflect 'founder effects'.
- Climate changes in the past, associated with glaciation, fragmented the population, and mutations within each group of bears accumulated to form separate haplotype groups.
- Female bears tend to be more philopatric than males, i.e. male bears have larger home ranges and disperse greater distances than females. Daughters often set up home ranges within their mother's home range. (Recall that mitochondrial DNA sequences tell us *only* about patterns of maternal inheritance.)

To what extent do these scenarios account for the observations? What additional experiments would you suggest to illuminate the situation further?

(f) Ethical, legal, and social issues. The division of North American brown bears into subspecies affects conservation issues. The brown bear populations of the lower 48 states of the US are endangered. If these populations represent an **evolutionarily significant unit** (ESU), they could be protected as a **distinct population segment** under the US Endangered Species Act. To qualify as an ESU, a population must (1) be substantially reproductively isolated from other populations of that species; and (2) contain an important component in the evolutionary legacy of a species (69 Fed. Reg. at 31355). Outline a petition to the US Secretary of the Interior or the US Fish and Wildlife Service arguing that the US brown bear population of the lower 48 states should be declared an ESU and listed as a distinct population segment. What, if any, additional data would you recommend collecting that might strengthen the case? (In fact, such a petition has been successful, and the brown bears in the lower 48 states are currently protected under the Endangered Species Act.)

Weblems

Weblem 2.1 What are the specificity sequences for the restriction enzymes: (a) EcoRV, (b) BamHI, (c) BsiWI, and (d) XhoI? For each enzyme, does it produce sticky or blunt ends?

Weblem 2.2 A set of primers for the amelogenin gender test are:

Forward 5'-CCCTGGGCTCTGTAAAGAA-3'

Reverse 5'-ATCAGAGCTTAAAACTGGGAAGCTG-3'

Download the sequences of the human amelogenin genes from X and Y chromosomes, gene names *AMELX* and *AMELY*. Mark their intron–exon structures. What are the sequences of the amplicons arising from PCR of each of the two genes using the given primers? Where, relative to the intron/exon structure of the genes, do these amplicons appear (i.e., in which intron or exon)? What are the lengths of these amplicons?

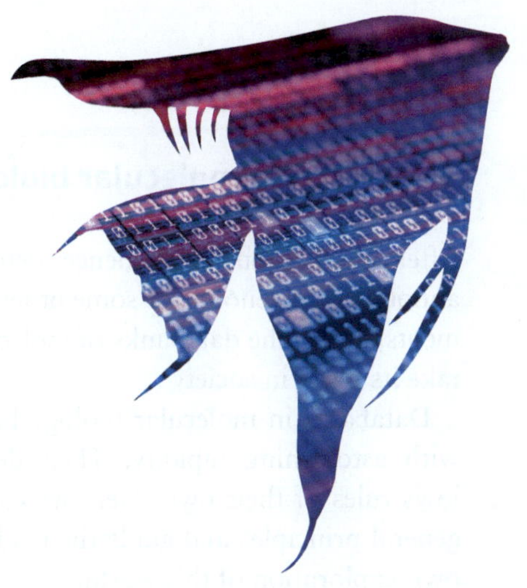

CHAPTER 3

Bioinformatics

LEARNING GOALS

- *Understand the idea of a database*—the requirements to accept, curate, organize (= impose a logical structure on), and distribute information, and to provide retrieval tools and links.
- *Know the major databases in molecular biology and genomics.*
- *Understand the general composition of a database as a collection of individual, discrete entries;* for instance, the amino-acid sequence of a particular protein, or the atomic coordinates of the three-dimensional structure of a protein or nucleic acid.
- *Recognize the factors that have led to the current organization of databases:* some are primary, comprehensive databases, many involved in international collaborations; others are specialized, secondary, or derived, 'boutique' databases, of more restricted scope.
- *Appreciate some of the general problems associated with curating and distributing information in a high-quality scientific database.*
- *Understand the distinction between data and annotation,* and the different kinds and sources of error to which they are subject.
- *Know what types of information would appear in annotations* of a gene sequence, of the amino-acid sequence of a protein, and of a protein structure.
- *Appreciate the necessity for quality control of databases,* and the difficulty of achieving it.
- *Understand the types of links integral to databases.*
- *Be aware of large-scale efforts at digitization of library materials* and how the results will be made available.
- *Come to terms with the explosion of scientific information,* and how to find personal modes of keeping track of what you need to know.
- *Be expert in identifying books, articles, and other presentations* relevant to subjects of interest to you.
- *Understand the rights and responsibilities associated with Open Access.*
- *Be knowledgeable and active in use of various avenues of dissemination of novel scientific results, including social media.*
- *Know the principles of markup languages,* and the idea of a schema in specifying the format of a database.
- *Be able to distinguish different types of computer languages,* to understand their relative strengths, and know how to decide which to use for different purposes.
- *Appreciate the power and limitations of natural language processing by computer,* and its applications in bioinformatics.

Databases in molecular biology

After completion of sequence determination and annotation, a genome, or some other set of measurements, enters the databanks of molecular biology 'to take its place in society'.

Databases in molecular biology have proliferated with astonishing rapidity. Their development follows rules of their own. Here we can only describe general principles and guide the reader to his or her own exploration of this world.

The original databases were small, specialized, and—by post-web standards—isolated. It has always been true that different types of information need to be curated by people with the appropriate expertise. Originally, specialists in different areas of biology organized the archiving of data related to their interests. For instance, crystallographers ran the database of protein structures—the Protein Data Bank. One problem was that even where data overlapped—for instance, amino-acid sequences are common to protein sequence and protein structure databases—there was relatively little effort to use controlled vocabularies and to make storage formats compatible (much less to reconcile discrepancies in the data). The International Scientific Unions and CODATA (the Committee on Data of the International Council of Scientific Unions) made important contributions, but problems remained.

With (1) the growing recognition of the importance of bioinformatics to research in biology, (2) the spectacular increases in the quantity of data, and—above all—(3) the emergence of the internet and World Wide Web, there came pressure towards growth and integration of databanks. Requirements for large-scale funding, and the need for combining biological and computational expertise, led to the creation of national and international institutions responsible for archiving and curating the data.

The term 'databank' suggests a metaphor that perhaps is outdated: a bank as a safe place for something valuable, from which you can make a withdrawal and then go off shopping up and down the high street, leaving the bank behind. This description emphasizes the archiving, curation, and distribution activities of the databanks. Of course, these activities remain essential. However, the databanks also provide facilities—or at least links—for the computational analysis of the information recovered. Databanks are not merely passive accumulators and distributors of data—libraries—but, by providing tools for information retrieval and analysis, are actively involved in research projects based on their contents.

Moreover, the many databases form interlocking networks. Release of a genome into any of the major archival projects is like casting a stone into a lake, sending ripples through the whole system. A genome itself is a nucleic-acid sequence, but the protein-coding genes it contains will, after translation into amino-acid sequences, contribute to protein-sequence databanks.

The realization that different data types were not intellectual islands, but that researchers needed coordinated access to them, led to the forging of links among the databases. The World Wide Web made this possible. The results were (1) systematization of formats and vocabularies, so that data in different collections became compatible; (2) pointers or links from each databank to others, facilitating access to related material; and (3) development of information retrieval software that would streamline access to different databanks and smooth the passage from retrieving data to analysis of the results.

Suppose, for example, that a researcher wishes to understand the differences between homologous proteins in humans and birds. This would involve first, assembling the data—the relevant sequences; and second, analysing relationships among them, for which the starting point would be sequence alignment, which might reveal systematic human-bird differences in this protein family. Gathering all sequences from birds, that are homologous to a human protein family, is a problem in information retrieval; forming a multiple sequence alignment of these sequences is computational analysis. Smooth passage means a simple pipeline from the sequences returned by the database search into a multiple sequence alignment program.

Together with the great growth and proliferation of databases there has—perhaps paradoxically—emerged a trend towards integration into larger and more comprehensive ones, to combine different categories of information that were formerly the provinces of individual projects. This has been driven by both academic and political forces.

In summary, the requirements of a major database project include the following:

1. Harvest the data plus annotations, curate them—that is, check both for accuracy and format—and distribute them.
2. Track and back up the data so that it does not get lost. Record provenance. Should any question arise, it must be possible to trace the data back to their origin and review all subsequent actions performed on them.
3. Provide links from the data to relevant items in other databanks, including bibliographical collections such as PubMed.
4. Provide information retrieval and analysis software to support research that includes both recovery of selected data and calculations with them.
5. Provide ample documentation and tutorial information, so that users can make effective use of the facilities.
6. Keep up with scientific advances in both biology and informatics. These may suggest improvements in the presentation and facilities.
7. Be responsive to users' needs.

The acronym FAIR (Findability, Accessibility, Interoperability, and Reuse) enunciates general principles for scientific data management and stewardship.

Databases are the silos containing the harvest from the scientific literature.

A database is an organized collection of information, in computer-readable form. Defining characteristics of databases include:

- the contents;
- the ontology: the list of valid terms and their definitions;
- the logical structure, or the expression of the interrelationships among the data, called the *schema*;
- the format of the data;
- the routes for selective retrieval of data, for presentation of the results, and for passing them on to programs for analysis;
- links to other information resources—other databases, references to original publication of data, tutorial background, etc.

Any database project must assemble all of these. In addition, many independent avenues of retrieval are possible: Anyone can write his or her own 'front end' to offer access to any distributed or web-available database. However, usually only the institution that maintains the archives takes responsibility for curation (that is, applying quality control, and standardizing format), providing annotation of the data, and exercising control of the contents.

Observables and data archives

A database includes (1) an archive of information, (2) a logical organization or 'structure' of that information, called a *schema*, and (3) tools to gain access to it. Databases in molecular biology include (but are by no means limited to):

- archival databases of biological information:
  - DNA and protein sequences, including annotation (see Box 3.1);
  - variations, such as catalogues of single-nucleotide polymorphisms (SNPs), compilations of haplotypes, or disease-associated mutations;
  - protein and nucleic acid structures, including annotation;
  - databases linked to species, including genome browsers;
  - databases of protein expression patterns;
  - databases of metabolic pathways;
  - databases of interactions and of regulatory networks;
- derived databases: these contain information collected from the archival databases, and inferred from analysis of their contents. For instance:
  - sequence motifs (characteristic 'signature patterns' of families of proteins);
  - classifications or relationships (connections between, and common features of, entries in archives). Examples include databases of protein sequence families, taxonomic databases containing phylogenetic trees, and hierarchical classifications of protein folding patterns;
- bibliographic databases: the scientific literature itself is data. Researchers 'datamine' PubMed as they do any other database;

BOX 3.1

Archives of nucleic acid and protein sequences

The archive of nucleic-acid sequences is maintained by a triple partnership, the International Nucleotide Sequence Database Collaboration, comprising the European Nucleotide Archive, ENA, based at the European Bioinformatics Institute (EBI), in Hinxton, UK; GenBank, based at the US National Center for Biotechnology Information, in Bethesda, Maryland; and the Center for Information Biology and DNA Data Bank of Japan, at the National Institute of Genetics in Mishima, Japan. The three sites exchange incoming submissions daily to ensure common coverage. However, the general format, annotation, and embedded links differ among the corresponding entries released by the different databases.

The archive of amino-acid sequences of proteins, now determined almost exclusively from translation of gene sequences, is maintained by the United Protein Database (UniProtKB), a merger of the databases SWISS-PROT, the Protein Identification Resource (PIR), and Translated EMBL (TrEMBL).

Associated with the archives are tools for selection and retrieval of sequences. The EBI has a number of search engines pointed at different components of its databases. The US National Center for Biotechnology Information offers ENTREZ. Both allow parallel searches in multiple data archives.

Many derived databases assemble families of proteins or subunits based on the similarities of their sequences. An 'umbrella' database, InterPro, integrates the contents, features, and annotation of several individual databases of protein families, domains, and functional sites; and contains links to others, including the Gene Ontology Consortium functional classification. InterPro assimilates other databases, including structural databases.

Genome browsers: Some institutions maintain sets of databases focused on individual species for which complete genomes have been determined, annotating the DNA sequence as far as possible, and integrating other types of relevant data. Notable are Ensembl (Wellcome Trust Sanger Institute, Hinxton, UK) and The University of California at Santa Cruz browsers for the human and other genomes, and FlyBase.

- databases of websites:
  - databases of databases containing biological information;
  - links between databases.

Annotation

A typical entry in a database in molecular biology might contain the sequence of a gene. However, the entry will contain more than the bare nucleotide sequence. It will also contain:

- *Reference information:* citations of the publications that served as the source of the entry, the history of the entry in the database, and accession, or cataloguing, information assigned by the database.
- *Interpretative information:* for example, the limits of exons within the sequence.
- *Links to other information:* perhaps to a protein sequence database containing information about the product encoded and the function attributed to that product, or to other entries in the same or other databases describing homologous genes.

When databases were more thematically focused and isolated, there was a comfortable and clear distinction between the primary data and the annotations. Annotations tended to be free-form comments, some expressed more casually than others. Recently, many database mergers have occurred, in response to the need to assemble a wide spectrum of information about gene sequences (and many other items). As a result of mergers, and of the importance of ontologies and computer-interpretable formats, entries in databases have taken more formal structures. It is growing more difficult to draw as sharp a distinction between data and annotation.

The information in entries varies in reliability. Nucleic-acid sequences, determined by modern techniques with generous coverage allowing confident assembly, and long reads, are quite accurate. On the other hand, assignment of function to gene products in the absence of direct experimental information is an important challenge in database annotation. It is a common practice to transfer functional annotation from a previously annotated homologous protein. This approach relies on the assumptions that (1) because homologous proteins have similar sequences and structures, they have similar functions, and (2) the annotation of the homologue is correct. Often, but certainly not always, these assumptions are valid. When they are not, annotation errors can result. For example, because of the phenomenon of 'recruitment', proteins very similar or even identical in sequence can adopt different functions. This can lead to misannotation.

Database quality control

Given that virtually all work in the biomedical field depends on databases, it is clear that quality of data directly affects the quality of research. If errors do enter databases—either in data or in annotations—they tend to propagate into other databases, and are very difficult to extirpate.

In principle, there are two approaches to improving database quality: keeping errors out in the first place, and removing them when they have been detected.

As part of the 'get it right the first time' approach, database curation and annotation has emerged as a new profession. Curators bring to their activities a constellation of skills and attitudes. The quality of their work translates directly into the quality of the databases.

Nevertheless, the high volume and diversity of subjects of scientific papers make it difficult for database staff to keep up adequately with the workload. An alternative approach is to involve the scientists who publish papers in the harvesting of database entries based on their results. For example, the Protein Data Bank accepts from authors a virtually complete entry, including annotations—the preparation guided by a website—corresponding to the structure deposited. Databank staff carry out validation procedures, but rarely add significant amounts of material.

Nevertheless, despite the professionalism of the curators, and the assiduity of their checking, errors will appear. The first problem is to identify them and the second is to remove them. One approach to identifying errors is to enlist experts as external curators to examine database entries in their own specialities. Often, database users call attention to errors.

Once identified, errors can be corrected in a 'master copy' of a database, particularly if the database management is in the hands of a single institution or a close-coupled partnership.

However, correction at source is not enough, because:

1. Many users create local versions of databases. These copies will contain the errors that were present at the time of downloading. The dissemination and implementation of any corrections are at the mercy of the frequency of updating the downloaded versions.
2. Many other databases assimilate, reintegrate, and redisseminate data, processes which may shield errors from correction, especially if items are not carefully tagged with their site and date of origin.

What is possible are programs that offer 'health checks' of versions of databases. An example is The PDBREPORT database,[1] which contains the results of validation software, WHAT_CHECK, applied to each entry in the Protein Data Bank. The program tests the validity and consistency of the format, and also analyses the structures, detecting outliers in stereochemical properties, such as bond lengths or angles, and looking for inconsistencies in hydrogen bonding patterns. It has been pointed out by crystallographers—very, very emphatically indeed—that outliers do not necessarily signal errors in the structure determination. (Of course, non-outliers also may or may not be errors.)

The dynamic quality of databases creates additional problems: The proliferation of divergent copies, of an object that is continually changing anyway, makes it difficult to reproduce published investigations.

Electronic publication

We are in an era of a transition to paper-free publishing. More and more publications are appearing on the web. A scientific journal may post only its table of contents, or a table of contents together with abstracts of articles, or complete articles. Many printed articles now contain references to links to websites containing supplementary material that never appears on paper. Many institutional publications—newsletters and technical reports—and

[1] See: http://swift.cmbi.ru.nl/gv/pdbreport/ and Hooft, R.W., Vriend, G., Sander, C., *et al.* (1996). Errors in protein structures. Nature 381: 272.

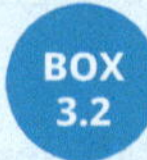

BOX 3.2 Open access

Open access redefines the author/publisher/reader relationship. Open access retains the peer-review process as the criterion for publication. Then:

- Accepted articles are placed on the web with free access, immediately upon publication. (Some commercial journals impose a delay—typically six months—between paper publication and electronic posting.)
- Payments for publication are transferred from readers to authors.
- Authors retain copyright, rather than assigning it to the publisher as most commercial journals require. However, not only are articles accessible, the content is freely usable by readers, subject to giving proper credit to authors. For example, it is permissible to reproduce and distribute articles in a set of teaching materials, or to import pictures into presentations. Such use of material not subject to open access would require consent of the author or publisher as copyright holder, or, for sufficiently old material, expiration of copyright protection.
- Guidelines for usage of material from PLoS journals appear in the Creative Commons Attribution License: https://journals.plos.org/plosone/s/licenses-and-copyright for summaries, https://plos.org/terms-of-use/, or http://creativecommons.org/licenses/by/2.5/legalcode for the full details.
- For a directory of open-access journals see: http://www.doaj.org

Open access has wide popular appeal in the scientific community. Supporters include both individual scientists in their dual roles of producers and consumers of the literature, and funding agencies. The US National Institutes of Health, and the several Research Councils and the Wellcome Trust in the UK, require open access to publications reporting work that they have supported. The European Commission has imposed analogous requirements on research it supports, and urges member states to follow its lead. Several universities have imposed requirements on faculty for open-access publication.

The US Congress (the analogue of Parliament) considered—first in 2006—but never passed a law that would require public availability, of *all* journal articles resulting from work supported by funding from US Government agencies involved in large-scale research subsidy. Recently, in August 2022, the White House Office of Science and Technology Policy issued a directive mandating that all federal departments and agencies make results of taxpayer-supported research publicly available at no cost to readers. The deadline for implementation is 31 December 2025.

many newspapers and magazines aimed at the general public are web-accessible.

Major forces in the conversion of paper to electronic libraries are the advent of electronic-format-only journals and Google's project to scan in the contents of a number of major libraries. Copyright restrictions do impede information flow. This has motivated movement towards *open-access* publication (see Box 3.2).

Today, any research group might simply post their results on the web. But scientists also depend on journals for another important feature: quality control. What journals provide, through the review process used in deciding whether or not to accept a submitted manuscript, is a (putative!) guarantee of trustworthiness and quality in the articles published. Despite a few well-publicized exceptions, and many more unpublicized ones, the system works fairly well. A European Commission report notes that: 'Scientific journals fulfil a double role of *certification* and *dissemination* of knowledge.'[2]

The peer-review process also acts as an (again, putative) guarantee and ranking of the quality of *scientists*. A major component of judging and comparing scientists, for career milestones of hiring and promotion, depends on the record of *peer-reviewed* publication. This too works well—not without exceptions—but it is a very expensive solution of

[2] Dewatriont, M., Ginsburgh, V., Legros, P., *et al.* (2006). Study on the economic and technical evolution of the scientific publication markets in Europe. European Commission, Directorate-General for Research, Brussels.

the problem, and there does not seem to be any logical link between the quality of a scientist's achievements and the medium in which they are reported. A cynic has quipped: 'The day Harvard or Stanford gives tenure to someone for web postings, 90% of scientific journals will disappear overnight.'

The World Wide Web

All readers will have used the World Wide Web for reference material, for news, for access to databases in molecular biology, for checking out personal information about individuals—friends or colleagues or celebrities—or just for browsing. The web is a means of interpersonal and intercomputer contact over networks. It provides a complete global village, containing the equivalent of library, post office, shops, and schools.

As a repository, the web can be thought of as a giant worldwide multimedia notice board. It contains text, images, videos, and sound. Virtually anything that can be stored on a computer can be made available and accessed via the web. An interesting example is a site treating Walt Whitman (http://www.whitmanarchive.org). The highest level page contains a table of contents. The site contains printed text of different poems and letters. You can compare different editions of the poems. You can access critical analysis of the poems. You can see versions of some poems in manuscript. There is even a link to an audio file, from which you can hear Whitman himself reading part of a poem.

Another interesting and comprehensive site focused on an individual, more closely related to the subject of this book, is: http://darwin-online.org.uk/.

Unlike books, websites are non-linear. Embedded links are wormholes, allowing direct transport between distant points. Links can be internal or external. Internal links take you to other portions of the text of a current document, or to associated images, cinema, or sounds. External links may allow you to move *down* to more specialized documents, *up* to more general ones (perhaps providing background to technical material), *sideways* to parallel documents (other papers on the same subject), or *over,* to directories that show what other relevant material is available.

Nor is the web solely a one-way street. Many web documents include forms in which you can enter information, and launch a program. Search engines are common examples. Many calculations in bioinformatics are now launched via such web servers. If the calculations are lengthy the results may not be returned within the session, but sent by e-mail.

The main thing to do, to get started using the web effectively, is to find useful entry points. Once a session is active, links will take you where you want to go. Among the most important sites are *search engines,* such as Google, that index the entire web and permit retrieval by keywords. You can enter one or more terms, such as 'phosphorylase', 'allosteric change', or 'crystal structure', and the search program will return a list of links to sites on the web that contain these terms.

Once you have completed a successful session, when you next log in the intersession memory facilities of the browsers allow you to pick up cleanly where you left off. During any session, should you find yourself viewing a document to which you will want to return, you can save the link in a file of *bookmarks* or *favourites.* In a subsequent session, you can return directly to any site on this list, not needing to retrace the trail of links that led you there in the first place.

A personal *home page* is a short autobiographical sketch (with links, of course). You and your colleagues will have your own home pages which typically include name, institutional affiliation, addresses for paper and electronic mail, telephone and fax numbers, a list of publications, and current research interests. It is common for home pages to include personal information, such as hobbies, pictures of the individual with his or her spouse and children, and even the family dog! (It is important, however, *not* to include information that would create vulnerability to identity theft.)

Links

The utility of a database depends on the quality of its links as well as on its contents. Internal links allow navigation around the database itself.

External links make connection to other databases, including literature databases containing references.

As an example, the UniProtKB entry for R-phycoerythrin, α chain, part of a light-harvesting antenna system in *Polysiphonia urceolata (syn. Conferva urceolata)*, extends over several pages in a browser window: http://www.uniprot.org/uniprot/P84861. Figure 3.1 shows a small portion of the entry.

Only four lines contain the amino-acid sequence of the protein! The web page contains many links, not shown in Figure 3.1. These include:

- Relevant reference information, some specific to the entry (for instance, bibliographical information about papers reporting the sequence and structure) or relevant but not specific to the entry (for instance, information about the taxonomic classification of the source organism).
- Links to other databases, including InterPro, Gene3D, Pfam, SUPFAM, PROSITE, and GO.
- A list of similar proteins with entries in UniProtKB, identified as other proteins with at least 50% sequence identity.
- The *feature table,* indicating annotations of structural roles of different residues, including the assignments of secondary structure: helices and strands of sheet, and the quaternary structure of the complex of which it forms a part.

Another important type of link that appears on web pages has the effect of launching a calculation, to analyse selected data.

Consider the retrieval of amino-acid sequences from a page at the European Institute of Bioinformatics. Searching for serpins in *Caenorhabditis elegans* returned nine entries (Figure 3.2a). It is possible to select any or all of them, by checking the boxes,

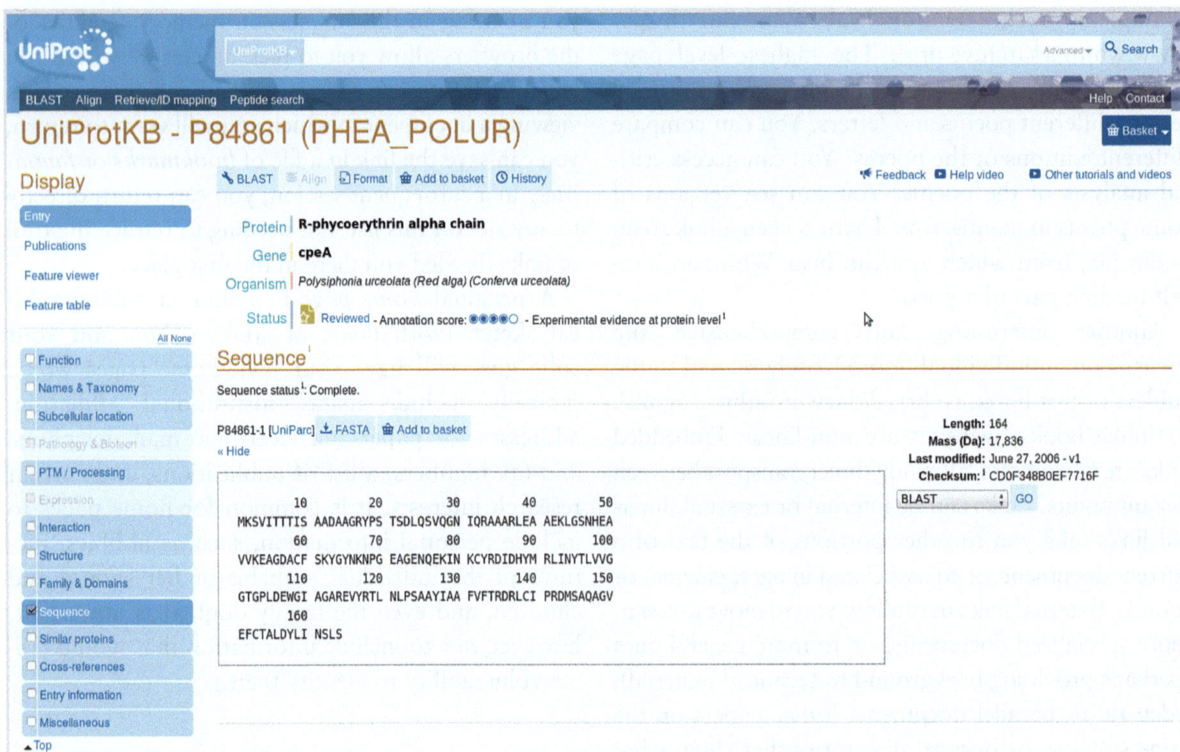

Figure 3.1 Small part of UniProtKB entry for R-phycoerythrin, α chain, from the red alga *Polysiphonia urceolata* (synonym of *Conferva urceolata*). The sequence is a relatively small portion of the entire entry. Readers should bring up this page in a browser and select additional components of this entry in the boxes at the left of the window. UniProt. (n.d.). https://www.uniprot.org/uniprotkb/P84861/entry. Made available under Creative Commons Attribution 4.0 International.

and pass them directly to the multiple sequence alignment program Clustal Omega. It is not necessary to save the sequences, or to cut-and-paste them into a different window. An initial portion of the result appears in Figure 3.2b. (The entire alignment extends to almost 400 residues.)

(a)

9 result(s) selected. (Clear selection)

Entry	Entry name	Protein names	Gene names	Organism	Length
O17365	O17365_CAEEL	SeRPin	**srp-2** C05E4.1, CELE_C05E4.1	Caenorhabditis elegans	359
G5EFH4	G5EFH4_CAEEL	SeRPin	**srp-6** C03G6.19, CELE_C03G6.19	Caenorhabditis elegans	375
G5EE40	G5EE40_CAEEL	SeRPin	**srp-1** C05E4.3, CELE_C05E4.3	Caenorhabditis elegans	366
A9Z1L1	A9Z1L1_CAEEL	SeRPin	**srp-7** CELE_F20D6.4, F20D6.4	Caenorhabditis elegans	373
G5ECY2	G5ECY2_CAEEL	SeRPin	**srp-3** CELE_Y32G9A.4, Y32G9A.4	Caenorhabditis elegans	362
Q19651	Q19651_CAEEL	SeRPin	**srp-8** CELE_F20D6.3, F20D6.3	Caenorhabditis elegans	372
Q6QUQ6	Q6QUQ6_CAEEL	SeRPin	**srp-7** CELE_F20D6.4, F20D6.4	Caenorhabditis elegans	366
Q6QUQ5	Q6QUQ5_CAEEL	SeRPin	**srp-7** CELE_F20D6.4, F20D6.4	Caenorhabditis elegans	365
G5EBT3	G5EBT3_CAEEL	SeRPin	**srp-7** CELE_F20D6.4, F20D6.4	Caenorhabditis elegans	366

(b)

```
CLUSTAL O(1.2.4) multiple sequence alignment


TR|G5EFH4|G5EFH4_CAEEL MSDSSSDEKMGLLLHSETDFGLSLLRQQN--LTESFVFSPLSIALALSLVHVAAKGETRD 58
TR|G5EE40|G5EE40_CAEEL -----MSFSLINSTFTEAQFGIKLLSDLTSDQLTPCVFSPVSILLSLALVHLGAKGHTRH 55
TR|G5ECY2|G5ECY2_CAEEL -----------MFDVAERRFALNFLNTLPV-HNESLVFSPLSIALVLSLVHTGVRGSSRD 48
TR|Q6QUQ6|Q6QUQ6_CAEEL ---------MALLLQSETDFGLGLLRQQN--ISESLAFSPLSIALALSLVHVAAKGETRD 49
TR|Q6QUQ5|Q6QUQ5_CAEEL ---------MALLLQSETDFGLGLLRQQN--ISESLAFSPLSIALALSLVHVAAKGETRD 49
TR|G5EBT3|G5EBT3_CAEEL ---------MALLLQSETDFGLGLLRQQN--ISESLAFSPLSIALALSLVHVAAKGETRD 49
TR|Q6QUR0|Q6QUR0_CAEEL ---------MSDNATSQTDFALKLLATLP--HSGSVVLSPLSISLGLALIHAGACGSTQK 49
TR|Q6QUQ4|Q6QUQ4_CAEEL ---------HEFLSKSEAEFALKLLHQQD--HSQSFVFSPIAISLALYSLYEAARGETRS 49
                                      ::  *.: :*           .:**::* * *  :: .. * ::


TR|G5EFH4|G5EFH4_CAEEL EIRKALLNGATDEELEQHFSNISAGLLVAEKGTEVNVANHIFSRKTFTIKKLYLNDVKKL 118
TR|G5EE40|G5EE40_CAEEL DIRNSVVNGSTDEQFIEHFSFINKLLNSSVNDVETLIANRLFVSPEQAIRKAFTDELREH 115
TR|G5ECY2|G5ECY2_CAEEL QIRNTLLSGATDEQLVEHFSFVSKEVKNGTKGVEVYLANKVYLKKGFTVNPTFLSTALKN 108
TR|Q6QUQ6|Q6QUQ6_CAEEL QIREALVKGSTDEQLEQHFANISAALLAAERGTEVKLANHVFTRAGFKIKQSYLDDVKKL 109
TR|Q6QUQ5|Q6QUQ5_CAEEL QIREALVKGSTDEQLEQHFANISAALLAAERGTEVKLANHVFTRAGFKIKQSYLDDVKKL 109
TR|G5EBT3|G5EBT3_CAEEL QIREALVKGSTDEQLEQHFANISAALLAAERGTEVKLANHVFTRAGFKIKQSYLDDVKKL 109
TR|Q6QUR0|Q6QUR0_CAEEL ELEDVLGGS----RIFEEFSGLMEAVGDTDNGVETKIVNRVFVNQAYTIHQDYLETVEKL 105
TR|Q6QUQ4|Q6QUQ4_CAEEL QIHDCLFKNATDGEMVTHFSNVSTALSAMKKDPQVKICNHIICRNGIKTESLCLDKVKLL 109
                       ::.. :  .    .:   *: :   :    .. :. : *::        .   .
```

Figure 3.2 (a) Search of UniProtKB for *C. elegans* serpins. Nine sequences were recovered. By selecting all and activating Align, the system transfers the sequences directly to Clustal Omega, with the initial blocks of the resulting alignment appearing in part (b). UniProt. (n.d.). https://www.uniprot.org/uniprotkb/P84861/entry. Made available under Creative Commons Attribution 4.0 International.

Information flow in bioinformatics

Data enter the bioinformatics ecosystem when a scientist deposits an experimental result in an archive, or a database records a result appearing in the literature. The archive curates and annotates the data to create an entry of proper contents and format. Quality checks are part of the curation process. The new entry then appears in the public release of the archive.

The division of the archive into entries is determined by the provenance of the data; that is, an entry corresponds to one coherent set of experimental measurements, often corresponding to one published article. In some cases, fragments of a complete sequence appear in several articles. A database can join the results to form an entry containing the complete biological entity. Currently, many nucleotide sequence data sets enter the databases as annotated genomes, or as unassembled metagenomic fragments.

Other information-retrieval projects, either associated with an archive or independent, may integrate newly released entries into their individual systems. Such secondary, or derived, databases may select or reorganize the data structure, and provide novel tools for analysis.

Reorganization of the data may involve:

- simply integrating the new entries into a general or specialized search engine. For example, typing 'aardvark myoglobin' into Google will return a link to the UniProtKB entry.
- extracting useful subsets of the data. Examples include (1) identification of genes in a connected DNA sequence, such as a bacterial genome or a eukaryotic chromosome, and (2) the extraction of a non-redundant set of protein sequences (that is, choosing only one representative of a set of very closely related sequences), both to shorten searches and to reduce statistical bias.
- deriving new types of information from the original data. A simple example: release of a protein-coding gene by a DNA sequence archive will trigger the appearance of its amino-acid sequence translation in databases of protein sequences. (A not-so-simple aspect: DNA sequences don't tell us about splice variants or about other important information related to the protein).
- recombining data in different ways. Many projects group sequences or structures of families of homologous proteins, or proteins that share function. The MEROPS protease database and the G-protein-coupled receptor database (GPCRdb) are examples. Some archives tend to keep related entries separate to preserve clarity of provenance.

  Some databases integrate data about a particular organism or sets of related organisms. FlyBase is a database containing *Drosophila* genes and genomes, plus material of interest to the community of scientists carrying out research on *D. melanogaster* and its near relatives. FlyBase includes a bulletin board showing schedules of meetings and courses, an atlas of pictures and movies, and links to other relevant sites.

  The International Nucleotide Sequence Database Collaboration and FlyBase contain overlapping material, presented from different points of view, and set within different contexts of additional material, facilities, and links.
- reannotating the data, including provision of different constellations of links. The integration may be horizontal or vertical. That is, links may indicate relationships to other entries of the same type (for instance, correspondences within a genome among homologous genes or among genes associated with the same metabolic pathway). Or, links may adduce a variety of information about a gene or protein (for instance, links between a gene and the clinical consequences of mutations).

Many sites serve as gateways between the archives and the computational tools available for data analysis. Information retrieval permits selection and extraction of data to provide the ingredients of a research project. Many bioinformatics resources not only offer information retrieval, but facilitate the 'downstream' processing of the entries selected. A typical example would be to retrieve the sequences of a set of homologous genes and then to align them (see Figure 3.2). The goal is to provide smooth integration of all the data-processing steps required for a research project, by intimate links among the tools for data storage, retrieval, and analysis.

The growing importance of simultaneous access to databases has led to research in database interactivity: how can databases 'talk to one another' without too great a sacrifice of the freedom of each one to structure its own data in ways appropriate to the individual features of the material it contains? Despite efforts to construct distributed query systems, the most effective solution is reflected in a strong trend towards merging and integration of data resources in bioinformatics. Some of the reasons for this are admittedly political: the 'empire-building' allele is present at fairly high frequency in the scientific population, and then the 'too big to fail' argument for continued and enhanced funding takes over. Nevertheless, scientifically, integrating databases allows for 'one-stop shopping', makes for ease of handling queries that require access to different categories of information, and facilitates cross-category consistency checks in curating the data. Moreover, large database organizations have the resources to provide tutorial guides: to usage of the site and to scientific background. A large organization can support a help desk. Frustrated users may retort that the integration of the data produces a site so complex as to *require* guidance. However, there are plenty of small, specialized databases that users also find confusing.

A reasonable extrapolation suggests the concept of specialized 'virtual databases' grounded in the archives but providing individual scope and function, tailored to the needs of individual research groups or even individual scientists.

Although there are good arguments for centralized, or no more than partnership, control over the contents of the archives, there is no need to limit the ways to access them: colloquially, the design of the 'front end' of the database. Specialized user communities may extract subsets of the data, or recombine data from different sources, and provide specialized avenues of access. Such 'boutique' databases depend on the primary archives as the source of the information they contain, but redesign the organization and presentation. Indeed, different derived databases can slice and dice the same information in different ways.

Curation, annotation, and quality control

The scientific and medical communities are dependent on the quality of databases. Indices of quality, even if they do not permit correction of mistakes, may help to avoid arriving at wrong conclusions.

Database entries comprise raw experimental results, and supplementary information or annotations. Each of these has its own sources of error.

The most important determinant of the quality of the data themselves is the state of the art of the experiments. Older data were limited by older techniques; for instance, amino-acid sequences of proteins were once determined by peptide sequencing, but are now translated from DNA sequences (except for partial sequencing by mass spectrometry). One consequence of the data explosion is that most data are new data, governed by current technology, which in most cases does quite a good job.

Annotations include information about the source of the data and the methods used to determine them. They identify the investigators responsible and cite relevant publications. They provide links to related information in other databases. In some sequence databases, the annotations include *feature tables*, which are lists of segments of the sequences that have biological significance; for instance, regions of a DNA sequence that code for proteins. These appear in computer-parsable formats, their contents restricted to a controlled vocabulary. Note that a statement by each database on a controlled vocabulary, and the definitions of the terms that appear in the vocabulary, is essential for information-retrieval operations. (This is like a 'convention card' at a bridge tournament.)

Formerly, a typical DNA sequence entry was produced by a single research group, investigating a particular gene and its products in a coherent way. Annotations were grounded in experiments, and written by specialists. In contrast, full-genome sequencing projects offer no experimental confirmation of the expression of most putative genes, nor characterization of their products. Curators at databases base much of their annotation on the analysis of the sequences by computer programs.

Annotation is the weakest component of the genomics enterprise. Automation of annotation is possible only to a limited extent; getting it right remains labour-intensive, and allocated resources are inadequate. But the importance of proper annotation cannot be underestimated. P. Bork pointed out that errors in gene assignments vitiate the high quality of the sequence data themselves.

Growth of genomic data will permit improvement in the quality of annotation as statistical methods increase in accuracy. This will allow improved *reannotation* of entries. Improvement of annotations will be a good thing. It implies, however, the disturbing concomitant that annotation will be in flux. We shall have to give up the safe idea of a stable database composed of entries that were correct when first distributed and which will stay fixed. The problem is aggravated by the proliferation of websites with increasingly dense networks of links. Networks of websites provide useful avenues for applications. But the web is also a vector of contagion, propagating errors in raw data, in immature data subsequently corrected but the corrections not passed on, and in variant annotations.

A database without effective modes of access is merely a data graveyard

Useful access to data requires a set of tools for answering questions, such as:

- 'Does the database contain the information I require?' (Example: can I retrieve the amino-acid sequence of human alcohol dehydrogenase?)
- 'How can I assemble selected information from the database in a useful form?' (Example: compile a list of globin sequences; or even better, a table of aligned globin sequences.)
- Indices of databases are useful in asking 'Where can I find some specific piece of information?' (Example: what databases contain the amino-acid sequence of porcupine trypsin?) Of course, if I know and can specify exactly what I want, the problem is relatively straightforward.

Mechanisms that allow effective access are an issue of database design that ideally should remain hidden from users. Effective access cannot be provided by bolting a query system onto an unstructured archive. Instead, the logical organization of the storage of the information must be designed with access in mind, considering what kinds of questions users will want to ask. The structure of the archive must mesh smoothly with the information-retrieval software.

A variety of database queries arise in bioinformatics. Compare the following typical examples:

1. Given a sequence, or fragment of a sequence, find sequences in the database that are similar to it. This is a central problem in bioinformatics. We share such string-matching problems with many fields of computer science. For instance, word processing and editing programs support string-search functions.
2. Given a protein structure, or fragment, find protein structures in the database that are similar to it. This is the generalization of the string-matching problem to three dimensions.
3. Given a sequence of a protein of unknown structure, find other proteins that adopt similar three-dimensional structures. One might be tempted to cheat, to look in the sequence databanks for proteins with sequences similar to the probe sequence. For if two proteins have sufficiently similar sequences, they will have similar structures. However, the converse is not true, and one can hope to create more powerful search techniques that will find proteins of similar structure even though their sequences have diverged beyond the point where they can be recognized as similar by sequence comparison. Comprehensive and reliable collections of protein structures predicted from amino-acid sequences—notably the AlphaFold database—can make this possible.
4. Given a protein structure, find *sequences* in the databank that correspond to similar structures. Again, one can cheat by using the structure to probe a structure databank, but formerly this could give only limited success because there are so many more sequences known than experimentally determined structures. The comprehensive databases of predicted protein structures provide an effective approach to this problem.

Points 1 and 2 are solved problems; such searches are carried out thousands of times a day. Points 3 and 4 are active fields of research.

Database search methods involve a tradeoff between *sensitivity* and *selectivity*. Does the method find all or most of the examples that are actually present, or does it miss a large fraction *(sensitivity)*? Conversely, how many of the 'hits' that it reports are incorrect *(selectivity)*? Suppose a database contains 1000 globin sequences. A search of this database for globins reported 900 results, 700 of which were really globin sequences and 200 of which were not. This result would be said to have 300 false negatives (misses) and 200 false positives. There is a tradeoff between sensitivity and selectivity: Lowering a tolerance threshold will increase the numbers of *both* false negatives and false positives. Often one is willing to work with low thresholds to make sure of not missing anything that might be important; but this requires detailed examination of the results to eliminate the resulting false positives.

Database access

Many databases in molecular biology permit general, free-of-charge public access to the data (see Box 3.3). Users can in general read the data, but almost never make changes. 'Reading' the data usually means seeing a presentation of the data through some program running in a browser. Many 'front ends' may exist for the same database, with individual appearances and different sets of links.

Some but not all databases permit users to extract entries in bulk. For this to be worthwhile, the data must be in a generally accessible format. To this end some databases maintain a version in which each entry appears as plain text (called a *flat file*). This is not necessarily the most useful internal format but facilitates general data exchange. Other collections are maintained using widely available database management systems. These are easily distributable among installations running equivalent software.

Data mining

The examples we have discussed of information retrieval from databases have involved the framing by a user of a specific set of criteria, and the return of relevant entries selected according to these criteria. Consider alternatively a scientific field in the exploratory phase, where a large amount of data

Public access to scientific data

Open and free access to articles in journals, and open and free access to the data the articles contain, are related but distinct issues.

Scientists in the academic world who determine novel data, such as gene sequences or protein structures, are expected to deposit the data in publicly accessible databases. To do this is at least potentially to sacrifice commercial rights, or the intellectual advantages of unshared knowledge in a competitive field of research. The commercial sector of research in molecular biology—prominently including but not limited to the pharmaceutical and biotechnology industries—generally regards as proprietary the results that its scientists generate.

Even within the academic world, this is not a new conflict. Early in the eighteenth century, Isaac Newton demanded access to data collected by the Astronomer Royal, John Flamsteed, in order to prepare a new edition of his *Principia*. Flamsteed refused, claiming ownership of the data despite their having been collected while he occupied an official government post.

Today, journals and granting agencies require deposition of certain kinds of data, including genome sequences and macromolecular crystal structures. Journals will not accept papers without confirmation of deposition from an appropriate database. Although these rules now have general acceptance, their establishment was controversial.

Science made an exception to its mandatory-deposition policy in publishing the Human Genome Draft Sequence by J.C. Venter and co-workers in 2001. There was criticism of this waiver.* A similar waiver applied to the publication of the genome of one of the strains of rice, eliciting similar criticisms.† Conversely, *Science did* require deposition in publicly accessible databases of the genome sequence of the strain of influenza virus active in the 1918–1919 pandemic. R. Kurzweil and W. Joy criticized the *non-withholding* of this sequence, on the grounds that terrorists might use the information to recreate the virus and use it as a weapon.*†

* Powledge, T.M. (2001). Changing the rules? EMBO Reports 2: 171–172.

† Petsko, G.A. (2002). Grain of truth. Genome Biology 3: comment1007.1-comment1007.2.

*† *The New York Times*, 17 September 2005.

has become available, and the challenge is to understand what underlying patterns exist. A first step is to generate hypotheses about those patterns. Perhaps experts might guess what to look for. Testing and refining the experts' hypotheses then requires probing information archives with sets of queries, seeking relationships and correlations in the data. This is the traditional way that science has made progress.

Now, the power of programs permits them to take the initiative in data exploration, to some extent. For example, programs can be adapted to assign data to classes on the basis of 'training' with examples, even if it is not possible explicitly to specify the rules that define the classes. It is even possible for a program to suggest hypotheses about patterns implicit in our data. This amounts to a partial automation of scientific research.

Machine learning is a computational approach to data analysis in which, through analysis of relevant information resources, computer programs achieve the ability to infer properties of data. Two complementary aspects are:

(1) *knowledge discovery*—descriptions, or even *explanations,* of regularities in the data, and

(2) *successful forecasting,* or predictive modelling.

Sophisticated numerical methods to implement these goals include:

- *Statistical techniques,* including clustering and classification algorithms, and principal component analysis (identification of a small number of possibly composite parameters that account for most of the variation in a set of data). Hidden Markov Models are the most powerful non-structure-based methods for detecting homologous amino-acid sequences of proteins.
- *Artificial Neural Networks.* Neural networks are the method of choice for prediction of structures of proteins.
- *Support Vector Machines* are algorithms for classification, that outperform neural networks in a number of applications.

Both Artificial Neural Networks and Support Vector Machines are data structures and algorithms for *supervised learning*. In supervised learning, the general framework of a program is constructed, but the details depend on choices of parameters. By exposing the program to a number of objects of known classification, and telling the program whether its prediction was correct or not (the supervision phase), the program can tune its parameters to give optimal performance.

The computer programs that implement some machine-learning techniques, including Artificial Neural Networks, have complex internal structures. Large numbers of variable parameters give them versatility; optimization of the parameters by training can achieve impressive accuracy in classifying input data. A disappointing aspect is that it is usually impossible to 'pick apart' a trained network, to harvest any insights into the structure of the data, expressible in a simple, understandable form. Some statistical methods—for instance, Singular Value Decomposition—do provide such insight, at least by identifying which are the important variables, or combinations of variables. R. Hamming wrote, 'The goal of computing is insight, not numbers.' Today people want both.

Survey of databases in molecular biology

Primary data collections related to biological macromolecules include:

- nucleic-acid sequences, including whole-genome projects;
- amino-acid sequences of proteins;
- protein and nucleic acid structures;
- small-molecule crystal structures;
- protein functions;
- expression patterns of genes;
- metabolic pathways and networks of interaction and control; and
- the scientific literature.

It is tempting to regard certain databases as *primary*—those that originally gather the data, and are responsible for curation and archiving. *Secondary* databases, or derived databases, would then be those that take data from the primary databases, recombine them, reannotate and reformat them, represent them, and provide different informational environments, different facilities, and different links. However, the distinction between primary and secondary databases is not as clear-cut as it used to be.

Nucleic-acid sequence databases

The worldwide nucleic-acid sequence archive, The International Nucleotide Sequence Database Collaboration, is a partnership of EMBL-Bank at the European Bioinformatics Institute (EBI), the DNA Data Bank of Japan (DDBJ) at the Center for Information Biology (CIB), and GenBank at the National Center for Biotechnology Information (NCBI).

The groups exchange data daily. As a result, the raw data are identical, although the format in which they are stored and the nature of the annotation vary. These databases curate, archive, and distribute DNA and RNA sequences collected from genome projects, scientific publications, and patent applications. Entries have a life cycle in the database. Because of the desire on the part of the user community for rapid access to data, new entries are made available before annotation is complete and checks are performed. Entries mature through the classes: *unannotated* → *preliminary* → *unreviewed* → *standard*. Rarely, an entry 'dies'—a few have been removed when they are determined to be erroneous.

KEY POINT

In addition to the International Nucleotide Sequence Database Collaboration, there are numerous other DNA sequence databanks:

- Genome browsers—databanks organized around one or more genome sequences, with links to other information sources about the organism.
- The International HapMap Consortium—database of SNPs.
- Forensic human DNA databases—for which law enforcement agencies in various countries are responsible.
- Databases of DNA sequences of animals contribute to a number of social and legal matters (see Chapter 2).

Protein sequence databases

In 2002, three protein sequence databases—the Protein Information Resource (PIR, at the National Biomedical Research Foundation of the Georgetown University Medical Center in Washington DC, US), and SWISS-PROT and TrEMBL (from the Swiss Institute of Bioinformatics in Geneva and the European Bioinformatics Institute in Hinxton, UK)—coordinated their efforts to form the UniProt consortium. The partners in this enterprise share the database but continue to offer separate information-retrieval tools for access.

The PIR grew out of the very first sequence database, developed by Margaret O. Dayhoff—the pioneer of the field of bioinformatics. SWISS-PROT was developed at the Swiss Institute of Bioinformatics. TrEMBL contains the translations of genes identified within DNA sequences in the EMBL Data Library. TrEMBL entries are regarded as preliminary. They mature—after curation and extended annotation—into full-fledged UniProt entries.

Today, almost all amino-acid sequence information arises from translation of nucleic-acid sequences. Information about ligands, disulphide bridges, subunit associations, post-translational modifications, glycosylation, splice variants, effects of mRNA editing, etc., are not available from gene sequences. For instance, from genetic information alone, one would not know that human haemoglobin is a tetramer binding haem groups, or that human insulin is a dimer linked by disulphide bridges. Protein databanks collect this additional information from the literature and provide suitable annotations.

Databases of genetic diseases—OMIM and OMIA

Online Mendelian Inheritance in Man™ (OMIM™) is a database of human genes and genetic disorders. Its original compilation, by V.A. McKusick, M. Smith, and colleagues, was published on paper. The NCBI of the US National Library of Medicine has developed it into a database accessible from the web and introduced links to other archives of related information, including sequence databanks and the medical literature. OMIM is now well integrated with the National Center for Biotechnology Information information-retrieval system ENTREZ.

A related database, the OMIM Morbid Map, deals with human genetic diseases and their chromosomal locations.

OMIA (Online Mendelian Inheritance in Animals) is a corresponding database for disease and other inherited traits in animals—excluding human and mouse.

Databases of structures

Structure databases archive, annotate, and distribute sets of atomic coordinates. Approximately 200 000 protein structures have been determined experimentally, by X-ray crystallography, nuclear magnetic resonance (NMR), and cryo electron microscopy (cryo-EM). Recent technical developments are tipping the balance towards cryo-EM. The AlphaFold database contains comprehensive prediction of structures corresponding to entries in amino-acid sequence databanks.

The Worldwide Protein Data Bank (wwPDB) now comprises four collaborating primary archival projects to integrate the archiving and distribution of experimentally determined biological macromolecular structures:

- The Protein Data Bank in Europe (PDBe), at EBI, UK
- The Research Collaboratory for Structural Bioinformatics (RCSB), in the US
- The Protein Data Bank Japan (PDBj) based in Osaka, Japan
- The Biological Magnetic Resonance Data Bank (BMRB), in the US.

The wwPDB sites accept depositions, process new entries, and maintain the archives. The Electron Microscopy Data Bank collects density maps and derived models of macromolecular structures. As a result of recent advances, cryo-electron microscopy (cryoEM) can now produce atomic-resolution models. These appear in the wwPDB. As cryoEM grows in importance, coordination between the Electron Microscopy Data Bank and the wwPDB will respond appropriately. These and many other websites organize and provide access to these data, including pictorial displays. Naturally, there is considerable overlap among them (see Box 3.4). Each offers search facilities to identify structures of interest, based on the presence of keywords (or a logical combination of keywords), or numerical values such as the year of deposition. Different sites differ also in their 'look and feel', and users will discover their own preferences.

In addition to the results of experimental structure determinations, EBI and RCSB contain links to collections of predicted structures. AlphaFold2, from Google subsidiary DeepMind, has been applied comprehensively to create a database of predicted structures now containing 2×10^8 entries. In addition, the RCSB has a link to ModelArchive, based at the Swiss Institute of Bioinformatics, containing predictions from Homology Modelling and from RoseTTAFold, a program developed at the University of Washington, US.

BOX 3.4 Features of the partners in the Worldwide Protein Data Bank

The PDBe, RCSB, and PDBj have the same set of structural data at their core. They share a common basic set of search and retrieval facilities, and links. However, they differ in infrastructure; in particular, in the facilities they support. A few examples:

The PDBe, under the direction of G. Kleywegt, has been especially creative and technically skilful in designing tools for structure retrieval and analysis. These include:

- PDBsum provides an overview of each entry.
- PDBeMotif combines protein sequence, chemical structure and protein-structural data in a single search.
- PDBeFold allows searches for structural similarity and alignment, and for probing the databank with a substructure.
- PDBePISA (Proteins, Interfaces, Structures and Assemblies) allows searching and exploration of macromolecular interfaces.

The RCSB offers Gene View, which maps experimental structures onto the human genome, and many other software tools.

PDBj offers a large number of software tools, including SeSAW, allowing identification of putative functional sites in a query via a structural match to the PDB, based either on an experimentally determined structure or even a model (if only a sequence is known).

wwPDB members also offer short, topical courses over the web.

The wwPDB overlaps in scope with several other databases. The Cambridge Crystallographic Data Centre (CCDC) archives the structures of small molecules. This information is extremely useful in studies of conformations of the component units of biological macromolecules, and for investigations of macromolecule–ligand interactions, including but not limited to applications to drug design.

The Nucleic Acid Knowledgebase (www.nakb.org) complements the wwPDB. It collects nucleic-acid-containing three-dimensional structures. It offers two-dimensional fold and three-dimensional structure visualization, education and standards pages (including a musical atlas!), and a web and software resource list. The NAKB replaces the older Nucleic Acid Database (established in 1991).

Classifications of protein structures

Several websites offer hierarchical classifications of the entire PDB according to the folding patterns of the proteins. These include:

- SCOP and SCOP2: *Structural Classification of Proteins*
- CATH: *Class/Architecture/Topology/Homologous superfamily*
- DALI: based on extraction of similar structures from distance matrices
- ECOD: hierarchical classification of protein based on evolutionary relationships.

These sites are useful general entry points to protein structural data. For instance, SCOP offers facilities for searching keywords to identify structures, navigation up and down the hierarchy, generation of pictures, access to the annotation records in the PDB entries, and links to related databases.

Databases of metabolic pathways

The Kyoto Encyclopedia of Genes and Genomes (KEGG) collects individual genomes, gene products, and their functions, but its special strength lies in its integration of biochemical and genetic information. KEGG focuses on interactions: molecular assemblies, and metabolic and regulatory networks. It has been developed under the direction of M. Kanehisa.

KEGG organizes five data types into a comprehensive system:

1. catalogues of chemical compounds in living cells
2. gene catalogues
3. genome maps
4. pathway maps
5. orthologue tables. (Orthologues are genes in different species that evolved from a common ancestral gene.)

The catalogues of chemical compounds and genes contain information about particular molecules or sequences. Genome maps integrate the genes themselves according to their chromosomal location. In some cases, knowing that a gene appears in an operon can provide clues to its function.

Pathway maps describe potential networks of molecular activities, both metabolic and regulatory. A metabolic pathway in KEGG is an idealization corresponding to a large number of possible metabolic cascades, combining reactions occurring in different organisms. It can generate a real metabolic pathway of a particular organism, by matching the proteins of that organism to enzymes within the reference pathways.

One enzyme in one organism would be referred to in KEGG in its orthologue tables, which link the enzyme to related ones in other organisms. This permits analysis of relationships between the metabolic pathways of different organisms.

Expression and proteomics databases

Recall the Central Dogma: ***DNA makes RNA makes protein***. Genomic databases contain DNA sequences. Expression databases record measurements of RNA levels, describing patterns of gene transcription. Proteomics databases record measurements on proteins, describing patterns of gene translation.

Comparisons of expression patterns give clues to (1) the function and mechanism of action of gene products; (2) how organisms coordinate their control over metabolic processes in different conditions—for instance, yeast under aerobic or anaerobic conditions; (3) the variations in mobilization of genes in different tissues, or at different stages of the cell cycle, or during the development of an organism; (4) mechanisms of antibiotic resistance in bacteria and consequent suggestion of targets for drug development; (5) the response to challenge by a parasite;

and (6) the response to medications of different types and dosages, to guide effective therapy.

Although of course there is a close relationship between patterns of transcription and patterns of translation, direct measurements of protein contents of cells and tissues—proteomics—provides additional valuable information. Because of variable splicing, RNA editing, different mRNA lifetimes, and different rates of translation of different mRNAs, measurements of proteins directly give a more accurate description of patterns of gene expression than measurements of transcription.

Post-translational modifications, and formation of complexes, including but not limited to oligomers, can be detected only by examining the proteins.

The literature as a database

Medline® (Medical Literature Analysis and Retrieval System Online) is the bibliographic database of the US National Library of Medicine. Medline has been integrated into PubMed, the bibliographical component of the US National Center for Biotechnology Information database ENTREZ.

Medline covers the scientific literature of fields related to research, teaching, and delivery of health care. It includes relevant areas of fundamental science. Medline is not primarily patient-oriented; a counterpart, MedlinePlus®, is a less-technical information resource about health care. For instance, a search in MedlinePlus for allopurinol, a drug used in treating gout and related diseases, returns a page linking to descriptions of the diseases for which this drug can be used, recommendations for dosage, side effects, etc. A search for allopurinol in MedLine (via PubMed) returns ~8000 technical articles, most of which do not deal directly with the prescription of allopurinol in current clinical practice.

One very effective feature of PubMed is the option to retrieve related articles. This is a very quick way to 'get into' the literature of a topic. Here is a tip: if you are trying to start to learn about an unfamiliar subject, try selecting Review from the ARTICLE TYPE menu in PubMed, or adding the keyword tutorial or review to your search in a general search engine. In many cases, PubMed also lists papers, published subsequently, that cite the article in the entry. This allows tracking the development of a topic: from the past, through the references within the paper itself; to the future, through the papers that refer to it.

In addition to the use of PubMed by individuals to search for articles to read (or, at any rate, to cite[3]), several projects 'mine' PubMed to create derived databases, usually combining the bibliographical results with other types of relevant data.

Lists of molecular biology databases and servers

It is difficult to explore any topic in molecular biology on the web without quickly bumping into a list of this nature. Lists of web resources in molecular biology are very common. They contain, to a large extent, the same information but vary widely in their 'look and feel'. The real problem is that, unless they are curated, they tend to degenerate into lists of dead links.

> Each year the January issue of the journal *Nucleic Acids Research* contains a set of articles on databases in molecular biology. This is an invaluable reference.

This book does not contain a long annotated list of relevant and recommended sites, for the following reasons. First, you do not want a long list; you need a short one. Second, the web is too volatile for such a list to stay useful for very long. It is much more effective to use a general search engine to find what you want at the moment you want it. My advice is to spend some time browsing; it will not take you long to find a site that appears reasonably stable and has a style compatible with your methods of work. A tip, if desired: a site that is comprehensive and shows signs of a commitment to keeping it up to date is: http://www.expasy.org. It is a suitable site for starting a browsing session.

[3] Swift wrote: 'Some people use books as they use Lords: They memorize their titles and boast of their acquaintance.'

Computers and computer science

Bioinformatics would not be possible without advances in computing hardware and software. Fast and high-capacity storage media are essential even to maintain the archives. Information retrieval and analysis require programs: some fairly straightforward and others extremely sophisticated. Distribution of the information requires the facilities of computer networks and the World Wide Web.

Computer science is a relatively young and flourishing field with the goal of making the most effective use of information technology hardware. Certain areas of computer science impinge on bioinformatics most directly. Consider a specific biological problem, that of retrieving from a database all sequences similar to the human *PAX-6* sequence, a gene essential for development (or any other probe sequence). A good solution to this problem would appeal to computer science for:

- *Analysis of algorithms.* An algorithm is a complete and precise specification of a method for solving a problem. For the retrieval of similar sequences, we need to measure the similarity of the probe sequence to every sequence in the database. It is possible to do much better than the naive approach of checking every pair of positions in every possible juxtaposition, a method that even without allowing gaps would require a time proportional to the product of the number of characters in the probe sequence times the number of characters in the database. A speciality in computer science, known colloquially as 'stringology', focuses on developing efficient methods for this type of problem, and analysing their effective performance.
- *Data structures* and *information retrieval.* How can we organize our data for efficient response to queries? For instance, are there ways to index or otherwise preprocess the data to make our sequence-similarity searches more efficient? How can we provide interfaces that will assist the user in framing and executing queries?
- *Software engineering.* Hardly ever anymore does anyone write programs in the native language of computers. Programmers work in higher-level languages, such as C, C++, PYTHON, PERL, JAVA, or even FORTRAN. The choice of programming language depends on the nature of the algorithm and associated data structure, and the expected use of the program.

Of course, most complicated software used in bioinformatics is now written by specialists. Which brings up the question of how much programming expertise a bioinformatician needs.

Programming

Programming is to computer science what bricklaying is to architecture. Both are creative; one is an art and the other a craft.

Many students of bioinformatics ask whether it is essential to learn to write sophisticated computer programs. My advice (not agreed upon by everyone in the field) is: 'Don't. Unless you want to specialize in it.' To work in bioinformatics, you will need to develop expertise in using tools available on the web. You will need facility in the use of your computer's operating system, including general-purpose application programs such as word processors and presentation tools. Learning how to create and maintain a website is essential.

Some skill in writing simple scripts in a language like PYTHON, PERL, or RUBY provides an essential extension to the basic facilities of the operating system. These are very powerful, and available for all computer systems. They make it very easy to carry out many very useful simple tasks, but can also be effective in projects demanding heavy computation, as much of the 'heavy lifting' can often be carried out by subroutine libraries, which can be written in any language and tuned for high efficiency.

Facility with a graphics program to display molecular structures is an important skill (for sources of software see: https://www.science.co.il/biomedical/software/Molecular-graphics.php).

On the other hand, the size of the data archives, and the growing sophistication of the questions we wish to address, demand respect. Truly creative programming in the field is best left to specialists, with advanced training in computer science. Nor does *using* programs, via highly polished (not to say flashy) web interfaces, provide any indication of the nature of the activity involved in writing and debugging programs. Attributed to Bismarck is the comment: 'Those who love sausages or the law should not watch either being made.' Perhaps computer programs should be added to his list.

Database organization

The internal structure of a database must reflect the interrelationships of the contents, in a way that facilitates responding to queries. Types of database organization in common use include:

- In a *hierarchical structure* items are classified, and clustered, at multiple levels. The original Linnaean taxonomy, and its many descendants including the Tree of Life (http://www.tolweb.org/tree/), are examples. Military chains of command are also hierarchies. The databases SCOP, CATH, and ECOD present hierarchies of protein structures based on evolutionary relationships and structural similarity. We shall see that the markup language XML provides a natural format for databases of information with hierarchical structure.
- In a famous 1970 paper, E.F. Codd of the IBM Corporation described the *relational database*. The basic unit of a relational database is a set of correspondences between different features of the database contents, called *tables*. Codd showed how set-theoretic operations (union, intersection, difference, Cartesian product) on tables facilitate processing of logically complex queries. Mature software is available, both open source and commercial, for managing relational databases, and for processing queries.

An example: a relational database of properties of amino acids

A relational database of amino acids might contain a table that associates with each amino acid its name, its three- and one-letter codes, its volume, its solvent-accessible surface area, and the chemical nature of the distal atoms in its sidechain, at pH7. (See Box 3.5, first table.) This organization makes it easy to answer queries of the form: What are the three-letter codes of all amino acids, the sidechains of which have distal carboxylate groups (a -COO$^-$ at the end of the sidechain)? The operation required is to select, from the first table, all rows in which the *Distal group* has the value carboxylate; and to report the three-letter-code column from those rows. Or, it would be easy to extract a subtable showing the surface areas of all amino acids with distal methyl groups (a -CH_3 at the end of a sidechain). This is called a *view*.

A second table might specify the hydrogen-bonding potential of the distal atom group of each amino acid. The column common to both tables—the distal atom group—allows queries that reflect correlations of these properties, for instance: What are the three-letter codes of all amino acids, the sidechains of which can serve as hydrogen-bond donors? This question requires combining information from both tables. In Codd's terms, it would involve a *join* of the two tables.

The general join operation amounts to forming the Cartesian product of the two tables. (The Cartesian product of two sets is the set of ordered pairs of elements, one from each set. If the sets contain n and m elements, respectively, the Cartesian product will contain $n \times m$ elements.) There are 20 entries in the first table and 13 in the second. The Cartesian product would contain 260 rows, combining information from the first table, appearing at the left, and from the second table, appearing on the right. Here are just a few illustrative lines.

Fragments of the join of the two tables from a relational database of properties of amino acids:

Alanine	Ala	A	88.6	115	methyl	amide	yes	yes
Alanine	Ala	A	88.6	115	methyl	amino	yes	yes
				…				
Alanine	Ala	A	88.6	115	methyl	methyl	no	no
				…				
Aspartic Acid	Asp	D	114.1	160	carboxylate	carboxylate	no	yes

To report the three-letter codes of amino acids that have sidechains that could serve as hydrogen-bond acceptors, we want to do what is called a *natural join*. This retains from the combined tables those rows in which columns 6 and 7 are equal. Of the

BOX 3.5 Two tables from a relational database of properties of amino acids

Amino acid	3-letter code	1-letter code	Volume	Surface area (Å²)	Distal group
Alanine	Ala	A	88.6	115	methyl
Arginine	Arg	R	173.4	225	guanidinium
Asparagine	Asn	N	111.1	150	amide
Aspartic acid	Asp	D	114.1	160	carboxylate
Cysteine	Cys	C	108.5	135	sulphydryl
Glutamic acid	Glu	E	138.4	190	carboxylate
Glutamine	Gln	Q	143.8	180	amide
Glycine	Gly	G	60.1	75	hydrogen
Histidine	His	H	153.2	195	imidazole
Isoleucine	Ile	I	166.7	175	methyl
Leucine	Leu	L	166.7	170	methyl
Lysine	Lys	K	168.6	200	amino
Methionine	Met	M	162.9	185	methyl
Phenylalanine	Phe	F	189.9	210	phenyl
Proline	Pro	P	112.7	145	pyrrolidine
Serine	Ser	S	89.0	115	hydroxyl
Threonine	Thr	T	116.1	140	hydroxyl
Tryptophan	Trp	W	227.8	255	indole
Tyrosine	Tyr	Y	193.6	230	phenol
Valine	Val	V	140.0	155	methyl

Distal group	H-bond donor	H-bond acceptor
amide	yes	yes
amino	yes	no
carboxylate	no	yes
guanidinium	yes	yes
hydrogen	no	no
imidazole	yes	yes
hydroxyl	yes	yes
indole	yes	yes
methyl	no	no
phenol	yes	yes
phenyl	no	no
pyrrolidine	no	yes
sulphydryl	yes	no

four rows shown, one contains methyl in both columns 6 and 7; and another contains carboxylate in both columns 6 and 7. The others would be rejected. The survivors form a new table containing columns specifying the hydrogen-bonding potential associated with each amino acid.

We could then select from this combined table the rows containing hydrogen-bond acceptors, and report their three-letter codes. One such row would appear as follows, after merging the two equal Distal group columns following the natural join:

Aspartic acid	Asp	D	114.1	160	carboxylate	no	yes

From this row we could extract the three-letter code Asp.

The relational database organization lends itself naturally to processing complex queries constructed as logical compositions of simpler queries. A somewhat artificial example: What are the three-letter codes of amino acids with volumes between 100 and 120 AND ((that can serve as hydrogen bond donors AND NOT serve as hydrogen bond acceptors) OR (that have surface areas greater than 120 Å^2 and have distal methyl groups)).

The Structured Query Language (SQL) is a fairly well-standardized syntax for probing relational databases with queries of this type. Complex queries containing logical connectives are translatable into Codd's set of operations on tables. For a flavour of SQL, the query in the preceding paragraph would appear:

```
SELECT <3_letter_code> FROM <amino_acid_table>
WHERE (sidechain_volume between 100 AND 120)
AND
(H-bond_donor = "no" AND H-bond_acceptor = "yes")
OR
(surface_area > 100 AND distal_group = "methyl"))
```

The program MySQL is an open-source relational database management system. It is popular with the bioinformatics community.

Markup languages

'Algorithms + Data Structures = Programs' – N. Wirth

Markup languages implement data structures, which are as essential a component of programs as executable instructions. Data structures are the organization of the information on which a program acts. Choice of the proper data structure is a crucial aspect of programming.

The term markup originally described editors' annotations to manuscripts, that control the appearance of the final published text without explicitly appearing in it. An example would be designation of certain words to appear in italics by underlining them in a manuscript. Computer-typesetting systems include formatting commands: the UNIX facilities of the 'roff' family were an early example, and D. Knuth's TeX system is an 'all possible bells and whistles' development. HTML (hypertext markup language)—the language in which websites are written—is primarily a presentational markup language.

The utility of the close-coupling of annotation with contents extends, beyond presentation markup, to organization of data in files. Such a structure provides an alternative to traditional *positional formatting*. Positional formatting is specifying how to interpret an item in a file through rigid rules specifying *where* the item appears. Typical examples of positional formatting are: 'The number of bases in the sequence appears in columns 5–16' or 'Items, separated by white space, appear in the order: gene name, source organism, number of bases, sequence'.

The markup approach achieves greater flexibility by associating each item with a *local* descriptor. The line:

<number of bases>5386</number of bases>

could appear *anywhere* in a file. A program, or a human reader, would recognize what the number 5386 signified. The syntax <*descriptor*>value</*descriptor*> is common to many markup languages, including HTML. The descriptor is called a *tag*. The material enclosed by the beginning and end of the tag is called the *element*. Standardization of the syntax simplifies the construction of the software to interpret it.

Tag-element combinations provide *self-describing* data. Moreover, the data description is *local;* that is, contiguous with individual data items. In contrast, the Learning Goals that appear at the beginning of the chapters of this book are summaries of contents that are *not* local to the sections to which they refer.

Flexibility of format comes at a price, most obviously in a rather cumbersome and bloated appearance of the files. Nor is adult supervision entirely unnecessary: A set of rules must specify acceptable ranges of values. Programs could not be asked to swallow:

<number of bases>Tuesday</number of bases>

Therefore, any file in a markup language requires a *schema*: a list of allowed element and attribute names, and allowed ranges of values. This permits validating a file for proper formatting and consistency. A Document Type Definition, itself written in a standardized language, specifies the schema. Note that

<number of bases>Tuesday</number of bases>

is valid *syntax* but invalid with respect to any reasonable *schema*.

XML

There are many markup languages, specialized for different types of data. One of the most general is XML (extensible markup language) used in many databases and information-retrieval systems. XML assumes a tree-based, or hierarchical, structure of the material. Lower-level tags and elements can appear within higher-level ones.

An XML database of mammalian species might contain the following:

```
<mammals>
  <genus>Homo
    <species>sapiens
common_name='human'</species>
      <species>neandertalis common_
    name='neandertal man'</species>
  </genus>
</mammals>
```

Note the three nested levels of tags: mammals, genus, species. The species elements include the common name as an *attribute*. In an alternative schema, the common name might be a separate tag within the species.

It would be more difficult to construct an XML database of information that is non-hierarchical. Consider a database of information about movies. It would be possible to define an XML schema in which the movie title was at a higher level in the hierarchy than the list of performers. Then it would be easy to probe the database with a movie title, and retrieve the cast. In contrast, it would be more difficult to retrieve all the movies in which Peter Sellers acted. In an alternative schema the performers could be at a higher level than the movies, making it easy to search for an actor or actress, but then it would be difficult to probe with a title and retrieve the cast. A relational database would be a more natural way to organize the data if one wanted to be able to query with *either* movie title or performer.

However, facilities for such queries are not completely incompatible with XML. Even if a database is structured hierarchically with an XML schema, it is possible to index it in different ways to support versatile approaches to retrieval, including non-hierarchical ones.

XML, unlike HTML, is not directly concerned with appearance or presentation. On the other hand, it is perfectly possible to write formatting programs that control the presentation of the contents of an XML file such as the mammal-genus-species example. Such a program could follow convention to display genus and species names in italics. Different programs could impose independent decisions about how to display common names. One program could display common names in boldface, another in plain Roman type. (Programs are available to convert XML to HTML.)

In contrast, in an HTML file, the decision to display common names in boldface type would be *irrevocably* implemented by tags: <b>neandertal man</b>, which would *force* '**neandertal man**' to appear in boldface. Moreover, it is impossible to make up novel tags for HTML format (without approval by an international commission). In other words, the schema of HTML has been fixed. This has the advantage of complete portability and the disadvantage of inflexibility.

Markup languages in general, and XML and HTML in particular, are becoming standard in database construction and distribution:

- *Archiving and curating data.* XML provides a general and flexible structure compatible with organizing information from many different fields and applications. Data validation—checking that the values of the elements are consistent with the schema—is straightforward. The results provide a format for data interchange, facilitating database interoperability.
- *Providing data to programs.* Insertion of a parser between an XML data file and an application program can simplify the input phase of a calculation.
- *Ease of data extraction and presentation.* Selection of data and formatting into an HTML file can be a natural and fluent mapping that facilitates conversion of data into a form that is both human-friendly and distributable over the web. Other markup languages provide facilities for describing graphics. These are profoundly concerned with *both* data structure and presentation.

Programming languages and tools for database construction and access

A computer program is a set of orders that a computer will execute. At the moment of execution, the orders must be specified in a form that can activate the computer; that is, the orders must be in a form that corresponds to the computer's limited repertoire of basic operations. Human beings would like to specify the orders in a human language. This has led to the development of 'pidgin' languages that allow people to write computer programs in languages as close as possible to natural mathematical discourse, but followed by translation into the computer's operation set. FORTRAN was the first of these.

Programming languages differ from natural human languages in many respects, including a restricted horizon of possibility of expression, and very strict intolerance to error.

A similar intolerance to error affects the preparation and formatting of data to be read by computer programs. To serve as input to a program, data must be (1) presented according to specific rules—for example, terms restricted to a controlled vocabulary, and (2) properly formatted. There is a tension between user-friendliness and program-friendliness in establishing the requirements.

Another distinction, which is not as sharp as it used to be, classifies programs into *systems programs* and *applications programs*. Applications programs are generally specific to one or more users. They solve a particular problem in a particular field. They are active in a computer for limited times, after which they report an answer and disappear. In contrast, systems programs govern the overall workflow of the computer, are common to all users, operate continuously, and are consistent with the use of the computer to solve a wide variety of problems (by means of individual applications programs). For instance, a program to superpose two or more protein structures would be an application program. The programs that create the general operating environment—for instance Linux or Microsoft Windows—are systems programs.

Operating systems offer many specific facilities in addition to their overall 'housekeeping' functions. To create lists of orders invoking the facilities of the operating system is to write a program called a *script*.

The boundaries between systems and applications programs are becoming fluid. All the features of the editor with which I am typing this paragraph are specific to the problem of accepting and editing text. However, many people use it, it was distributed with the operating system, and it remains active (in the 'background') even when I am finished with this passage. Conversely, many programmers who put together large and powerful packages that address a variety of problems—for example, retrieval of genetic sequences from a database—boast of having written 'program systems' (not quite systems programs).

Traditional programming languages

If a programmer writes code in C, C++, or FORTRAN, a separate program called a *compiler* translates a program in these languages into the appropriate set of computer instructions. The maturity of compiler technology, together with the understanding of algorithms provided by computer scientists, and the experience and skill of the community of programmers, combine to make languages such as C++ suitable for large-scale computations which strain the available resources.

Another advantage of not writing in native machine language is code *portability;* that is, the ability of one program, written in C, C++, or FORTRAN, to run on a large variety of platforms. It is true that each target machine language requires its own compiler. But writing a compiler needs to be done only once per machine, and there is mature software that facilitates compiler construction. Then an entire literature of programs becomes executable.

Rarely if ever does any one write a program in a computer's native language any more, despite the fact that the resulting code can execute an order-of-magnitude faster than an optimized compiled version can. This may not be enough of a difference given how inexpensive computer power is. The basic problem is lack of portability of the result. Even LAPACK, a widely used subroutine library for linear algebra, is written in FORTRAN.[4]

Scripting languages

Many extremely useful tasks require only minimal computer resources. For instance, the translation of a gene sequence into an amino-acid sequence requires only a straightforward table-look-up of each codon (see Figure 1.2). For these, a simple program achieves adequate throughput: what is important is to save programmer time. The computer time required is often negligible.

Indeed, there has been a steady trend in the relative costs of hardware and software. The balance is tipping, steeply, in the direction of high costs of creating software relative to purchasing and maintaining hardware. Programming practice has reacted with tools and languages that streamline the effort required to write code that works correctly, even at some cost in efficiency of execution.

We have discussed scripting languages such as PYTHON, PERL, and RUBY. At least in their initial versions they were *interpreted* languages. This means that the systems program that carried out the commands skipped the step of compilation to machine language, but implemented the stated operations on a line-by-line basis. In principle, this makes for less efficient execution. But in most cases, this is a legitimate price to pay for the ease of writing the program and the sharp curtailment of the 'debugging' phase. Often the difference in execution time is unnoticeable.

Some languages can be run either in interpretive or compiled mode, for instance LISP. Demonstration by an interpreted language, such as PYTHON, PERL, or RUBY, of significant advantages and popular appeal will elicit writing of a compiler, or at least a more efficient interpreter. Often compilation will not speed up execution significantly, if the original interpreter can call upon programs written and optimized in a compiled language for computationally intensive operations; for instance, multiple calls to a matrix-diagonalization subroutine.

Useful skills in using a scripting language such as PYTHON are relatively easy to attain, relative to languages such as C or C++. Learning some PYTHON or PERL or RUBY is a good compromise for a research scientist who does not intend to specialize in creation of very complex software.

Program libraries specialized for molecular biology

Programmers usually construct new programs by combining well-established components. For instance an algorithm may contain a step that requires sorting a list, or solving linear equations. Subprograms for these steps are widely available. And all programs depend on standard libraries for input/output. Almost never does one write a program completely 'from scratch'.

In addition to standard libraries for numerical analysis and text-processing, there are libraries specialized for molecular biology. Different libraries are associated with different programming languages. For example, Biopython (http://biopython.org) contains

[4] Some people preferred the complete control over their code that programming directly in machine language affords; it was asserted that 'The difference between programming in machine language and programming in FORTRAN is the difference between playing a piano and playing a phonograph.' This point of view is obsolete.

modules that implement common computational tasks in bioinformatics, written in PYTHON. (Similar libraries exist for PERL and RUBY.) Typical modules translate nucleic-acid sequences to protein sequences, or perform sequence alignments. Modules can be integrated smoothly into a new program.

Java—computing over the web

The Java language has a syntax with many similarities to the languages C and C++. Its operating environment is designed to address the following problem:

Suppose the creator of a website wants to provide a program which users can run interactively from a browser. If the program is run on a computer at the website, and if many users simultaneously avail themselves of the facilities, the hardware on which the website is running will come under pressure. An example of this mode is the server at the European Bioinformatics Institute, which handles between 10 and 13 million job requests per month, running them on two large clusters, in Hinxton and in Hemel Hempsted. Each has ~1500 cores with 128 Gb RAM.

An alternative is to ask each user to provide the computer power. Without leaving the website, the browser will dynamically download programs (called *applets*). The programs will be run on the user's computer.

This, in turn, creates a security problem: The user must give the website access to resources on the user's computer. A website that can download executable code and gain access to the local files can do considerable harm, including crashing the computer, or snooping around the file system to steal or damage confidential information, or carrying out unwanted invasive activity such as displaying unsolicited advertising material.

The basic idea of the way to protect the user is as follows: The downloaded Java program is not run directly by the user's operating system, but involves an intermediate agent. The user's system simulates an internal computer—called a virtual machine—which runs the Java program. The virtual machine carefully restricts the resources to which the Java program running under its auspices has access. The local virtual machine imposes the rules; the distant website programmer must follow them.

Natural language processing

Biomedical research depends crucially on the quality of the data and annotations in databases. Some annotations are generated from the data; others are extracted from articles in the scientific literature. Unlike most input prepared for computers in strictly defined formats, the literature, aimed at human-to-human communication, appears in a natural language, although of course many articles contain equations, tables, and figures. Extraction of data from the literature is a labour-intensive activity that will not be able to keep up with the increasing rate of published articles. Will it be possible for computers to take over this task?

Natural language refers to the oral or written or signed forms of human–human communication. Natural language processing by computer means: *at least* the analysis of a stream of spoken or written words that a human could interpret, and *at best* a suitable reaction—acting on a command, or providing a suitable response in the natural language.

Natural language processing has been a goal of computing for decades. Early hopes, during the 1950s and 1960s, for achieving automatic language translation were unfulfilled (see Box 3.6).

A major difficulty in natural language processing is the ambiguity of words and even phrases (see Box 3.7). If a man married to a lawyer asks his wife to 'press his suit', does he want sartorial or forensic action? Human beings extract the meaning from such phrases by using contextual clues to resolve ambiguities. No reader would interpret the third line of Keats's 'Ode to a Nightingale':

My heart aches, and a drowsy numbness pains
My sense, as though of hemlock I had drunk,
Or emptied some dull opiate to the drains
One minute past, and Lethe-wards had sunk

Automatic translation?

An apocryphal story about automatic translation concerns a program that converted English to Russian and back. From the input 'The spirit is willing but the flesh is weak' came back 'The vodka is fine but the meat is rotten'. (That this occurred in a computer translation system is an urban myth. The first traceable publication of this joke actually is in a newspaper over a century ago: The Decatur, Illinois, *USA Herald*, 20 January 1903, p. 5.) A true computer translation howler was the rendering of '. . . la Cour de Justice considère la création d'un sixième poste d'avocat général' as '. . . the Court of Justice is considering the creation of a sixth general avocado station'.*

* Wheeler, P.J., & Lawson, V. (1982). Computing ahead of the linguists. Ambassador International (March): 21–22.

Ambiguous headline

Headlines, because of their enforced concision, are common sources of ambiguity. A machine parser got this one wrong:

British left waffles on Falklands

It interpreted 'left' as a verb and 'waffles' as a noun.

as signifying that the poet had just poured his opiate down his kitchen sink. Keats was deliberately using archaic senses of words.

Computer programs have access neither to life experience nor to context-related clues—is the lawyer's husband holding a garment or a folder of papers? Therefore, ambiguities are difficult to circumvent. A simplification is to restrict the field of discourse. For instance, an early natural language processing system provided an interface to a database of information about the Boston Red Sox (a US baseball team).

A relatively successful approach to the specific problem of machine translation has been Google Translate. It works by searching a large corpus of paired documents produced by human translators. However, it is much more than just a 'look-up' reporter. Sophisticated artificial intelligence algorithms greatly improve the disambiguation of possible translations.

For standard tourist sentences, such as 'Where is the nearest police station?' or 'Is there a good restaurant nearby?' it is adequate although the translations are often not idiomatic. It gets the translation of 'la Cour de Justice considère la création d'un sixième poste d'avocat général' into English correct. An earlier version mistranslated (English to French):

Stolen painting found by tree.

to

Peinture volé trouvé par l'arbre.

suggesting that the tree found the painting, not that the site of the discovery was near a tree. The current version of Google Translate gets it right: 'Tableau volé retrouvé près d'un arbre.'

Unsurprisingly, it translates Cockney rhyming slang literally: 'apples and pairs → des pommes et des poires', even in a sentence with obvious contextual 'tip-offs', for instance: 'My grandfather held onto the bannister while descending the apples and pears.'

There are other on-line natural-language translators, also data-driven and using artificial intelligence; for instance DeepL: https://www.deepl.com/translator. Needless to say, claims for superiority are abundant for each (and available in many languages).

Natural language processing in mining the biomedical literature

The goal of natural language processing in bioinformatics is the extraction of information from the scientific literature and databases. Applications of textual analysis of databases of biomedical literature include:

Identifying keywords and combinations of keywords. Given a list of names of genes and a list of names of diseases, it should be possible to identify papers that contain references to combinations of genes and diseases, and to produce a list of gene-disease combinations based on co-occurrences in one or more papers. Several aspects of this problem make it more challenging than simple keyword search. Many biological entities have multiple synonyms. Conversely, many terms appear in several technical categories and are used also as colloquial terms. As an extreme example, consider: 'common cold', 'cold sore', 'cold shock protein', 'kept in a cold room', 'cold finger', 'paroxysmal cold haemogloburia', 'cold turkey', 'cold compresses', 'colicigonenic plasmid Cold-CA23', and 'Cold Spring Harbor Laboratory', all of which appear in technical articles.

Knowledge extraction—protein–protein interactions. There are several approaches to compiling a database of protein–protein interactions. One is to write a computer program to extract information automatically from the scientific literature. For instance, an article entitled: 'Calnuc binds to Alzheimer's β–amyloid precursor protein and affects its biogenesis' appeared in the *Journal of Neurochemistry*.[5] (Of course, it makes no difference whether the sentence is in the title or the text of the article.) A human reader could harvest for a protein interaction database the pair: calnuc and Alzheimer's β–amyloid precursor protein.

To extract this information automatically, it would help to have a list of protein names. The challenge is to write a program that can identify, within processed text, patterns of the form:

<protein name>...<binds or some equivalent verb>...
<protein name>

The ... allows for various kinds of intervening material. For instance, another article has the title: 'Ubiquitin binds to and regulates a subset of SH3 domains'.[6] The program should recognize the verb 'binds' and ignore 'to and regulates'. Alternatively if one were trying to deduce regulatory networks, then a different verb would form part of the pattern.

With respect to the proteins that bind ubiquitin, the title of the ubiquitin paper is relatively general. A sentence in the abstract of that paper: 'The yeast endocytic protein Sla1, as well as the mammalian proteins CIN85 and amphiphysin, carry ubiquitin-binding SH3 domains', would, if properly parsed, permit extraction of three specific SH3 domains that bind ubiquitin.

One word within the ... that the pattern should not ignore is 'not': The sentence 'Auxin-binding protein 1 does not bind auxin within the endoplasmic reticulum despite this being the predominant subcellular location for this hormone receptor'[7] satisfies the original pattern but is a false positive. It is not enough to check for the presence of 'not'; consider: 'The human anti-apoptotic proteins cIAP1 and cIAP2 bind but do not inhibit caspases.'[8]

Automatic text-mining by software is a very challenging problem. Some people believe that there are fundamental limitations that will never be overcome. Nevertheless, for extracting information from the literature to create complete and high-quality annotations in the databases on which research crucially depends, what else is there? Annotation by human action is labour-intensive and error-prone.

[5] Lin, P., Li, F., Zhang, Y.W., *et al.* (2007). Calnuc binds to Alzheimer's β–amyloid precursor protein and affects its biogenesis. J. Neurochem. 100: 1505–1514.

[6] Stamenova, S.D., French, M.E., He, Y., *et al.* (2007). Ubiquitin binds to and regulates a subset of SH3 domains. Mol. Cell 25: 273–284.

[7] Tian, H., Klämbt, D., & Jones, A.M. (1995). Auxin-binding protein 1 does not bind auxin within the endoplasmic reticulum despite this being the predominant subcellular location for this hormone receptor. J. Biol. Chem. 270: 26962–26969.

[8] Eckelman, B.P., & Salvesen, G.S. (2006). The human anti-apoptotic proteins cIAP1 and cIAP2 bind but do not inhibit caspases. J. Biol. Chem. 281: 3253–3260.

RECOMMENDED READING

There are many fine books and articles on bioinformatics. A few suggestions:

Tramontano, A. (2005). *The Ten Most Wanted Solutions in Protein Bioinformatics*. London: Chapman & Hall/CRC.

Jensen, L., Saric, J., & Bork, P. (2005). Literature mining for the biologist: From information retrieval to biological discovery. Nat. Rev. Genet. 7: 119–129.

Antao, T. (2022). *Bioinformatics with Python Cookbook*. 3rd edn. Birmingham: Packt Publishing.

Lin, S., Scholtens, D., & Datta, S. (2023). *Bioinformatics Methods: From Omics to Next Generation Sequencing*. London: Chapman & Hall.

Raja, K., ed. (2022). *Biomedical Text Mining (Methods in Molecular Biology, 2496)*. New York: Humana Press.

The transition to electronic publishing

Winograd, S., & Zare, R.N. (1995). 'Wired' science or whither the printed pages. Science 269: 615. [The authors, among the most distinguished of contemporary scientists, raise questions that are still not answered after almost 30 years.]

Lesk, M. (2004). *Understanding Digital Libraries,* 2nd edn. San Francisco: Morgan Kaufmann.

Lesk, M.E. (2012). One in a million: Information vs. attention. Int. J. Commun. 6: 907–919.

Development of the Public Library of Science

Neylon, C. (2015). Open Access Pioneer: The Public Library of Science. In: D. Bollier & S. Helfrich (eds), *Patterns of Commoning*. Amherst, MA: The Commons Strategies Group.

Ahmed, A., Al-Khatib, A., Boum, Y., *et al*. (2023). The future of academic publishing. Nat. Hum. Behav. 7: 1021–1026.

Sources for history and current status of databases in molecular biology

Imker, H.J. (2018). 25 years of molecular biology databases: A study of proliferation, impact, and maintenance. Front. Res. Metr. Anal. 3: 18.

Waman, V.P., Orengo, C., Kleywegt, G.J., *et al*. (2022). Three-dimensional structure databases of biological macromolecules. Methods Mol. Biol. 2449: 43–91.

PDBe-KB consortium. (2022). PDBe-KB: Collaboratively defining the biological context of structural data. Nucl. Acids Res. 50: D534–D542.

XML

Bosak, J., & Bray, T. (1999). XML and the second-generation web. Sci. Amer. 280, 5: 89–93. [An introduction to XML, including descriptions of the problems that motivated its development, and the solutions it provides.]

To learn the 'nuts and bolts' of XML, many websites contain tutorials; for example: https://www.infoworld.com/article/2076396/xml-for-the-absolute-beginner.html

There are also many books; for example: Harold, E.R., & Means, W.S. (2004). *XML in a Nutshell,* 3rd edn. Sebastopol, CA: O'Reilly.

EXERCISES, PROBLEMS, AND WEBLEMS

Exercises

Exercise 3.1 According to the data in Box 3.5, which amino acids satisfy the compound query 'What are the three-letter codes of all amino acids, the sidechains of which contain distal methyl groups?'

Exercise 3.2 Construct a third table to be added to the two tables in Box 3.5 that specifies which amino acid sidechains are non-polar (or hydrophobic), which are polar, which are positively charged, and which are negatively charged. (See https://quizlet.com/231723269/amino-acids-diagram/)

Exercise 3.3 Suppose a university library purchases electronic access to a very broad spectrum of scientific journals. Information about usage patterns of different journals by members of the university are recordable at the *Publishers'* websites. (a) How could a university librarian make use of this information to help make difficult choices in the face of budgetary pressure? (b) Is it to a publisher's financial advantage to make this information available to university librarians?

Exercise 3.4 Consider a database of audio clips (for example, recordings of broadcasts of speeches by Winston Churchill). You want to create software to make this database searchable by computer, using *spoken* English sentences as search objects. (a) Suppose you had software that would perform accurate speech recognition; that is, conversion of speech to text. How could you use this to solve the problem? (b) How, in general terms, *might* you try to solve the problem *without* using speech → text conversion?

Exercise 3.5 Rewrite the XML fragment containing a database of mammals in the section Markup languages, converting common name from an attribute to a tag.

Exercise 3.6 The sentence 'Time flies like an arrow' is ambiguous.

(a) Explain three potential meanings of this sentence, treating Time as (1) a noun, (2) a verb, (3) an adjective (modifying flies). (b) Could you reject any of these meanings because they do not correctly obey rules of grammar? (c) Could you reject any of these meanings because they are not consistent with ordinary experience?

Exercise 3.7 Compose a search pattern to detect regulatory interactions between proteins analogous to <protein name> . . . <binds or some equivalent verb> . . . <protein name>.

Exercise 3.8 A naive way to try to find enzyme names in text is to search for words that end in -ase. Think of ten English words ending in -ase that are NOT names of enzymes. What is the longest word ending in -ase that you can find? Of the words you suggest, would any of them be likely to appear in an article in the biomedical literature? (Two obvious words ending in -ase that are not enzyme names but which appear frequently in the biomedical literature are: disease and case. To turn this exercise into a Weblem, look for an on-line rhyming dictionary.)

Exercise 3.9 (a) Write an XML fragment of a database with the schema illustrated for the human and Neandertal (in the section on XML), containing three subspecies of leopard *(Panthera pardus)*: African leopard *(Panthera pardus ssp. Pardus)*, Indian leopard *(Panthera pardus ssp.fusca)*, and Siberian leopard *(Panthera pardus ssp. orientalis)*. (b) Extend the answer to part (a) to include the snow leopard, *Panthera uncia*. Although the common name suggests leopard, the snow leopard is classified as a different species.

Problems

Problem 3.1 Consider the query: What are the three-letter codes of all amino acids that have volumes greater than 120 Å^2 with distal carboxylate or amide groups? Draw a Venn diagram showing, separately, the distributions of three-letter codes of sidechains, distal functional groups, volumes. Show the overlaps of the distributions and indicate the residues that satisfy the query.

Problem 3.2 Write a program to take the example of the fragment of an XML database of mammalian species (see section, Markup languages) and display it in some conveniently readable format. (One approach would be to use one of the available programs to convert XML to HTML.)

Weblems

Weblem 3.1 Which human chromosomes contain genes or pseudogenes for ubiquitin?

Weblem 3.2 Horse myoglobin contains 154 residues. A 108-residue fragment, comprising residues 32–139, folds to a structure that binds haem, and the haem complex binds oxygen and carbon monoxide. (a) What is the sequence of the 108-residue fragment? (b) Mapping it onto the complete structure, which exons or portions of exons does it contain? (c) Which helices does it contain? (d) Draw a picture of the corresponding fragment of the full structure, based on the structure of the complete molecule, for example wwPDB entry 5D5R. (It is highly unlikely that the fragment retains the precise structure of the corresponding substructure of the intact protein.) (e) Which haem-contact residues does the fragment contain?

Weblem 3.3 Pick a database from the last annual Database Issue of the journal Nucleic Acids Research. Report on its contents and describe the contents of a sample entry. (This is suitable for a class project, such that each student picks a different choice and presents the results to the class.)

Weblem 3.4 (a) Open the entry [https://www.uniprot.org/uniprotkb/P00512/entry] for the amino-acid sequence of *Geobacillus stearothermophilus* phosphofructokinase. How many amino acids are there in the sequence? What are the first (N-terminal) ten amino acids in this sequence?
(b) Open the entry [https://www.ebi.ac.uk/pdbe/entry/pdb/3pfk] for the structure of *Geobacillus stearothermophilus* phosphofructokinase. By what method was this structure determined? What is the primary publication describing this structure determination? What reaction does this enzyme catalyse? How is this reaction regulated? How many helices does the secondary structure of this protein contain? How many strands of sheet does the secondary structure of this protein contain? Have any structures of homologous proteins from other species been determined? (Note: *Geobacillus stearothermophilus* is another name for *Bacillus stearothermophilus*.)

Weblem 3.5 From PubMed, determine how many papers related to breast cancer were published in 2024. Read the abstract of one of them. How long did it take you? Multiply the time required by the number of papers. If you wanted to be an expert on the subject of breast cancer, how much time would it take you to read all the abstracts of all the papers on the subject published in 2024?

Weblem 3.6 What kinds of data does each of the following databases contain? (a) InterPro, (b) Gene3D, (c) Pfam, (d) SUPFAM, (e) PROSITE, (f) GO.

Weblem 3.7 How many open-access journals are there categorized as in (a) the Health Sciences, (b) Biology and Life Sciences, (c) Agriculture and Food Sciences, and (d) Bioinformatics (see the Directory of Open Access Journals: http://www.doaj.org/)

Weblem 3.8 For what types of data are the following markup languages specialized? (a) VRML, (b) CML, (c) BSML, (d) LOGML

Weblem 3.9 When will the works of Dylan Thomas go out of copyright?

Weblem 3.10 The home page of the RCSB [http://www.rcsb.org] includes links to query facilities that permit identification and retrieval of particular macromolecular structures from its contents, by specifying desired features. Find at least two other sites that permit identification of structures from the Protein Data Bank. There will be substantial overlap in the features of these query systems. What if any unique facilities does each have?

Weblem 3.11 (a) Give examples of three types of information about fruit flies that appear in *both* the European Nucleotide Archive (ENA) and FlyBase. ENA is based at the European Bioinformatics Institute (EBI). (b) Give examples of three types of information about fruit flies that appear in FlyBase but *not* in the ENA.

Weblem 3.12 Find, in the Protein Data Bank, two structures of the same protein, determined by X-ray crystallography at different resolution: one at low resolution (≥ 2.9 Å) and the other at high resolution (≤ 1.9 Å). Remember: the lower the number specifying the resolution, the higher the potential quality of the result. Check the PDBREPORT entries for the two entries, in http://swift.cmbi.ru.nl/gv/pdbreport/. Compare the evaluations of the two structures by the software underlying the reports.

Weblem 3.13 (a) Search PubMed for the article: Levitt, M., & Chothia, C. (1976). Structural patterns in globular proteins. Nature 261: 552–558. Can you get the full text of the article via PubMed? Print a copy of the page that provides the most detailed information about this article that is accessible to you from PubMed.

(b) Do a Google search for this article. (It is enough to enter search terms Levitt Chothia Structural Patterns). Try the first two or three links returned. Did any of those give you access to the full text of the article?

(c) Search Google Scholar for the same article. Under this entry, select web search. Can you easily find a site containing a complete text of the article? If so, print the page containing the link to the site that provides access to the full text, and highlight the link to the full-text site.

(d) Compare the ease of getting to the full text through PubMed, Google, and Google Scholar.

(e) What conclusions can you draw?

(f) Now suppose you knew or suspected that if the full text were available on the web it would be as a pdf file (= Adobe public document format). Do a Google search for keywords:

Levitt Chothia Structural patterns pdf

Do the results suggest that you should reconsider your conclusions?

Weblem 3.14 Submit to https://www.bioinformatics.org/textknowledge/genetag.php (a) the title + abstract and (b) the entire text of the following paper: Zhu, Y., Culmsee, C., Klumpp, S., *et al.* (2004). Neuroprotection by transforming growth factor-beta1 involves activation of nuclear factor-kappaB through phosphatidylinositol-3-OH kinase/Akt and mitogen-activated protein kinase-extracellular-signal regulated kinase signalling pathways. Neuroscience 123: 897–906. What gene and protein names does the program find in the title + abstract that were not found in the title? What gene and protein names does it find in the full paper that were not found in the title + abstract?

Weblem 3.15 CbiT is a protein the function of which was originally hypothesized, on the basis of homology to proteins of known function, to be a precorrin-8w decarboxylase in the biosynthesis of vitamin B12. The crystal structure, solved in 2002, showed it to be a methyltransferase. (Keller, J.P., Smith, P.M., Benach, J., *et al.* (2002). The crystal structure of MT0146/CbiT suggests that the putative precorrin-8w decarboxylase is a methyltransferase. Structure 10: 1475–1487.) In many databases, the annotations were corrected. How many sites with incorrect annotation as a precorrin-8w decarboxylase can you still find?

Weblem 3.16 Your research director is scheduled to give a one-hour lecture to beginning graduate students at your institution on 'Information retrieval from the scientific literature'. A month before the lecture, he or she receives an invitation to a prestigious international conference that conflicts with the lecture, and asks you to prepare and give the lecture. What sources on the web do you find useful for preparation? What specific examples would you include? What sites would you recommend to your audience for following up on the lecture? Create a fairly detailed outline of the lecture you propose to give.

Weblem 3.17 Same as previous problem, but change 'A month' to 'Two days'. (Perhaps your supervisor is replacing someone who cancelled.)

Weblem 3.18 Find a website containing ambiguous sentences. Examples: ambiguous garden path [http://www.fun-with-words.com/ambiguous_garden_path.html], or ambiguous headlines [http://www.fun-with-words.com/ambiguous_headlines.html]. For a few selected examples, submit them to a syntactic analyser (for example http://www.link.cs.cmu.edu/link/) and decide whether it parsed the sentence correctly. (If in a classroom context, for this and the following problem, it would be possible for different students to take different examples, and present the results together.)

Weblem 3.19 Find a website containing amusing ambiguous headlines. Examples: ambiguous headlines [http://www.fun-with-words.com/ambiguous_headlines.html], or humour headlines [http://www.ling.upenn.edu/~beatrice/humor/headlines.html]. Pick a foreign language, and for a few selected examples, submit them to Google Translate, and back-translate the result. Did you recover the original?

CHAPTER 4

Classical Genetics, Mapping, and Sequencing

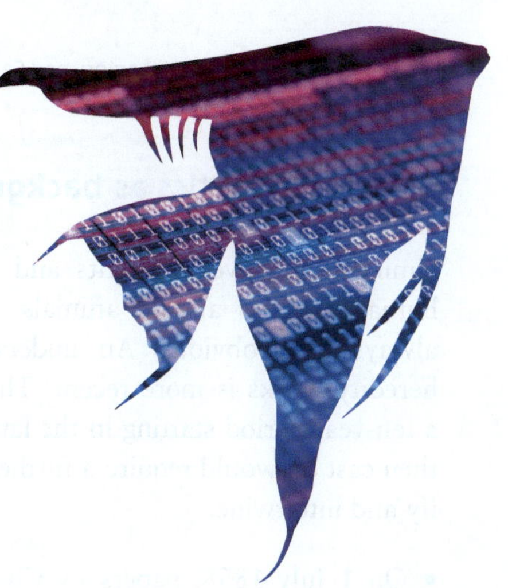

LEARNING GOALS

- *Know some of the important landmarks in the history of genomics*, from the classical work of Darwin and Mendel, through Morgan and Sturtevant, to the more recent research leading to the discovery of the double-helical structure of DNA and the development of the Human Genome Project.
- *Distinguish different types of maps*—genetic linkage maps, chromosome banding patterns, restriction maps, and DNA sequences—and understand the relationships among them.
- *Understand the relationships among linkage, linkage disequilibrium, haplotypes, and the collection of single-nucleotide polymorphisms (SNPs).*
- *Appreciate how the basic techniques of DNA sequencing developed* from the initial pioneering breakthroughs to the current automated high-throughput systems.
- *Understand the primer extension reaction catalysed by DNA polymerase* and its termination by dideoxynucleoside triphosphates.
- *Be familiar with the original sequencing gels, visualized using autoradiography, that were the basis of the original Sanger sequencing method*, and appreciate the advantages of using fluorescent chain-terminating dideoxynucleoside triphosphates.
- *Understand the basis and importance of the polymerase chain reaction (PCR) as a method for amplification of selected DNA sequences within a mixture.*
- *Grasp the significance and relationships of reads, overlaps, contigs, and assemblies* as parts of an overall strategy and organization of a sequencing project.
- *Be able to contrast hierarchical strategies of whole-genome sequencing based on maps and BAC clones*, with the whole-genome shotgun approach.
- *Know about the techniques that underlie contemporary 'next-generation' and 'third-generation' high-throughput sequencing instruments*.
- *Anticipate new developments in high-throughput sequencing*, and the goals set for the next few years of development.

Classical genetics as background

Similarities between parents and offspring, within human families and in animals and plants, have always been obvious. An understanding of how heredity works is more recent. The story begins in a ten-year period starting in the late 1850s. Threads then cast on would require a further century to ramify and intertwine.

- On 1 July 1858, papers by Charles Darwin and Alfred Russel Wallace were presented to the Linnaean Society of London, the first statements of their ideas on the development of species through natural selection of inheritable traits. Darwin's book, *The Origin of Species*, appeared on 22 November 1859.[1]
- In 1860, Louis Pasteur published the observation that the mould *Penicillium glaucum* preferentially metabolized the L-form of tartaric acid, one of two mirror-image molecules that have identical structures except that one is left-handed and the other right-handed. This work, based on Pasteur's earlier separation of racemic tartaric acid (a mixture of left- and right-handed molecules) by manual selection of crystals of different shape, brought the idea of three-dimensional molecular structure into biology: *To understand how a biological process works, we must know the detailed spatial structure of the relevant molecules.* Thus began a long courtship between biology and molecular structure that has flowered into an intimate and fruitful marriage.
- On 8 February and 8 March 1865, the monk Gregor Mendel read his paper, Experiments on Plant Hybridization, to the Natural History Society of Brünn in Moravia. (US President Abraham Lincoln was assassinated on 15 April 1865.) Mendel's paper was published the following year in the Society's Proceedings.

[1] Several earlier writers, including William Charles Wells and Patrick Mayhew, had suggested the idea of natural selection. It appears, for example, in the appendix to Mayhew's 1831 book, *On Naval Timber and Arboriculture*, explicitly alluding to the possibility of creating novel species. (Mayhew's interest was in optimizing the growth of trees for building warships for the Royal Navy.) Mayhew complained when *The Origin of Species* first appeared, and Darwin gave credit to Wells and Mayhew in subsequent editions.

A lack of understanding of the mechanism of heredity was a barrier to the development of Darwin's insights. Mendel's work supplied the crucial missing ideas: the discreteness and persistence of the elements of hereditary transmission, later called genes. Nevertheless, although copies of the Proceedings of the Natural History Society of Brünn were distributed around the scientific community, Mendel's work went largely unnoticed until it was rediscovered early in the twentieth century.

There are several ironic aspects of this situation:

1. Among the recipients in Britain of the Brünn Society Proceedings were the Royal Society of London and the Linnaean Society. Darwin was a member of both and had access to their libraries. Even more, Darwin personally owned a book by W.O. Focke, *Plant Hybridisation,* published in 1880, which included a section describing Mendel's work and its implications. When J.G. Romanes, preparing an article on hybrids for the *Encyclopaedia Britannica*, appealed to Darwin for help in making his review complete, Darwin sent his copy of Focke's book. But, in one of the nearest near-misses in scientific history, neither Darwin nor Romanes read the section on Mendel's work. (How do we know this? The relevant pages were never cut open! Darwin's copy of Focke's book is now in the Cambridge University Library, with the pages still intact.)
2. Darwin came close to an independent statement of Mendel's conclusion that traits that differ between parents persist in the offspring, rather than blend. In a letter to Thomas Huxley in 1857, Darwin wrote:

> 'I have lately been inclined to speculate, very crudely and indistinctly, that propagation by true fertilisation will turn out to be a sort of mixture, and not true fusion, of two distinct individuals, or rather of innumerable individuals, as each parent has its parents and ancestors. I can understand no other view of the way in which crossed forms go back to so large an extent to ancestral forms. But all this, of course, is infinitely crude.'

Indeed, Darwin himself hybridized pea plants and observed segregation of traits! In 1866, he wrote to Wallace:

> 'I crossed the Painted Lady and Purple sweetpeas, which are very differently coloured varieties, and got, even out of the same pod, both varieties perfect but none intermediate.'

In the same letter, Darwin pointed out, as an obvious example of discrete rather than blending inheritance, that male and female parents give rise to male and female offspring.

The legacy of this decade—Darwin, Pasteur, Mendel—was completed by the discovery of DNA (then called nuclein) by Friedrich Miescher in 1869. The structure and function of DNA were equally unknown. However, microscopic observations of the role of chromatin in fertilization led quite early on to suggestions that Miescher's substance was 'responsible ... for the transmission of hereditary characteristics'.[2] The cell biologists got there first, and got it right.

The idea of DNA as the hereditary material then vanished for many years. It encountered considerable resistance when it re-emerged.

What is a gene?

In 1931, Frederick Griffith studied virulent and non-virulent strains of *Streptococcus pneumoniae*, showing that the virulent strain, even if killed, contained a substance that could transform a non-virulent strain into a virulent one, and that the induced virulence is heritable.

In 1944, Oswald Avery, Colin MacLeod, and Maclyn McCarty tested different chemical components of the cells' transforming activity. They identified the DNA from the virulent strain as the molecule that induced the transformation. (We now interpret bacterial transformation as 'horizontal gene transfer'.) As controls, they showed that transformation was inhibited by enzymes that destroy DNA but not by enzymes that destroy proteins. However, their contemporaries were not receptive to their conclusions.[3]

General acceptance of the idea that DNA is the hereditary material awaited the 1952 experiments of Alfred Hershey and Martha Chase, who showed that when bacteriophage T2 replicates itself in *Escherichia coli*, it is the viral DNA and not the viral protein that enters the host cell and carries the inherited characteristics of the virus.

DNA was discovered in 1869. Avery, McLeod, and McCarty showed in 1944 that DNA was the active substance in bacterial transformation. Hershey and Chase showed that during bacteriophage infection, DNA but not protein entered the host cell.

Maps and tour guides

Maps tell us where things are. More specifically, they tell us where things are in relation to other things. In genomics, maps have been essential in revealing the organization of the hereditary material.

Different types of map describe different types of observation:

1. Linkage maps of genes,
2. Banding patterns of chromosomes,
3. Restriction maps—DNA cleavage fragment patterns,
4. DNA sequences.

[2] Hertwig, W.A.O. (1885). Das Problem der Befruchtung und der Isotropie des Eies, eine Theorie der Vererbung. Jenaische Zeitsch. f. Medizin u. Naturwiss. 18: 276–318. For an article about Miescher's scientific career, see Dahm, R. (2005). Friedrich Miescher and the discovery of DNA. Dev. Biol. 278: 274–288.

[3] It is a mystery, why in 1944 the field was unreceptive to the idea that DNA contains hereditary material (the Avery, MacLeod, and McCarty experiments) but in 1952, only eight years later, there was general acceptance of the same conclusion from the Hershey–Chase experiment. Some of the reasons were technical—including the widespread but completely erroneous belief that DNA did not have variable nucleotide sequences. Other reasons were personal; certainly, in 1944 people's attention was largely focused on the Second World War. Also, not until the work of Luria and Delbrück was it widely recognized that prokaryotes have heredity similar to that of other species. A statement attributed to Max Planck, 'Science makes progress one funeral at a time,' is certainly relevant.

Genes, as suggested by Mendel's results, were initially entirely abstract entities. Chromosomes are physical objects, with banding patterns as their visible landmarks. Only with DNA sequences are we dealing directly with stored hereditary information in its physical form. Restriction maps are in effect partial DNA sequences—they give the positions of particular oligonucleotides within DNA molecules (see Figure 2.4).

It was the great achievement of the last century of biology to forge connections among these maps.

A crucial idea that emerged from mapping is that the organization of hereditary information is *linear.* The first steps—and giant strides indeed they were—proved that, within any chromosome, linkage maps are one-dimensional arrays. In fact, all of these types of map are one-dimensional, and in fact they are collinear.

The complete sequences of genomes are the culmination of the entire mapping enterprise. But it is worth emphasizing, again, that genome sequences describe the hereditary information of organisms in *only* a one-dimensional and static form. What they do not tell us is (a) how this information is implemented in space and time; (b) how gene expression is choreographed by orderly developmental and other regulatory programmes; and (c) the influence of surroundings and experience and epigenetics on the structure and activities of the organism, beyond the genome itself. To borrow a phrase, the genome has the roles 'to be consulted, to encourage and to warn'.

BOX 4.1 Vocabulary inherited from classical genetics

Here are traditional meanings of these terms. We must reconsider how they should be defined in the light of recent understanding.

gene	A bearer of hereditary information.*
trait	An observable property or feature of an individual organism.
phenotype	The collection of observable traits of any individual, other than genomic sequence.
genotype	The sequence of any individual's genome.
allele	One of the set of possible gene variants at a particular locus that govern a particular trait. (If there has been a gene duplication in an autosomal gene, the organism would be said to have alleles at each of two loci.)
homozygote	A diploid individual that has two identical alleles at some locus.
heterozygote	A diploid individual that has two different alleles at some locus.
segregation	The separation of corresponding alleles on homologous chromosomes during the reproductive process.
independent assortment	The uncorrelated choices of genes for different characters that each parent transmits to children.
linkage	Absence or reduction of independent assortment of parental genes, which are usually transmitted together because they lie on the same chromosome.

* With apologies to A.S. Eddington.

Genetic maps

Gene maps were classically determined from patterns of inheritance of phenotypic traits (see Box 4.1). Mendel discovered that inherited traits are discrete and persist through a lineage. He observed recessive characteristics that disappear in one generation but re-emerge in a later one. Expressed in modern terminology, he observed traits that depend on a single locus, with a dominant and a recessive allele—denote them *D* and *d*. Mendel further observed that the distribution of phenotypes followed simple statistical rules. The offspring of *Dd* and *Dd* parents showed the dominant phenotype three times as often as the recessive. He inferred that the genotypes of the offspring, *DD, Dd, dD* (all three producing the dominant phenotype) and *dd* (producing the recessive phenotype) occur at the frequencies expected from segregation in the gametes, and independent transmission to the offspring, of the elements of heredity.

Studying the simultaneous inheritance of several traits, Mendel found further statistical regularities later understood to be consistent with the existence of stable, independent, and persistent elements of heredity that distribute themselves randomly. The cards are shuffled and a new hand is redealt to each offspring.

Mendel had no concept of the physical nature of the elements of heredity that he was studying and was content to describe their behaviour in abstract terms. It is interesting to compare the experience of Mendel's contemporary, the physicist James Clerk Maxwell. Despite the success of Maxwell's kinetic theory—based on an abstract model of particles in random motion—the model was for many years taken as a metaphor. The idea that matter is *actually* composed of tiny particles gained widespread acceptance only with Perrin's studies of Brownian motion, over half a century later.

Linkage

In some cases, genes for different traits do not show independent assortment but are *linked*; i.e., their alleles are co-inherited.

Linked traits are governed by genes on the same chromosome. However, in many cases linkage is incomplete. During gamete formation, alleles on different chromosomes of a homologous pair can recombine. This occurs as a result of crossing over, the exchange of material between homologous chromosomes during copying in meiosis (see Figure 4.1).

Thomas Hunt Morgan, at Columbia University in New York City, US, observed varying degrees of linkage in different pairs of genes. His student Alfred Sturtevant proposed that the extent of recombination could be a measure of the distance between the genes on the chromosome.

Morgan's student Alfred Sturtevant, then an undergraduate, made a crucial observation: that the data were consistent with a *linear* distribution of genes. What he found was that genetic distance, as measured by recombination frequency, was *additive*. Consider three genes *A*, *B*, and *C*. Suppose that the distance—that is, the inverse of the recombination rate—from *A* to *B* is 5 and the distance from *B* to *C* is 3. Then, if the distance from *A* to *C* is 8 = 3 + 5, the observations are consistent with a linear and additive structure with gene order *A*–*B*–*C*. (Alternatively, the distances would also be additive if the distance from *A* to *C* were 2, implying gene order *A*–*C*–*B*.) Note that additivity of distances does not hold for points at the vertices of a triangle rather than on a line.

Sturtevant's analysis made it possible to determine the order of genes along each chromosome and to plot them along a line at positions consistent with the distances between them. The unit of length in a gene map is the Morgan, defined by the relation that 1 cM corresponds to a 1% recombination frequency. We now know that 1 cM is ~1 Mb (= 10^6 bp) in humans, but it varies with the location in the genome, the distance between genes, and the gender of the parent: for males, 1 cM is ~1.05 Mb; for females, 1 cM is ~0.88 Mb. Crossing over is reduced in pericentromeric regions; other regions are 'hot spots' for crossing over. It is estimated that 80% of genetic recombination takes place in no more than 25% of our genome.

Linkage guides the search for genes. To identify the gene responsible for a disease, look for a marker of known location that tends to be co-inherited with the disease phenotype. The target gene is then likely to be on the same chromosome, at a position near to the marker.

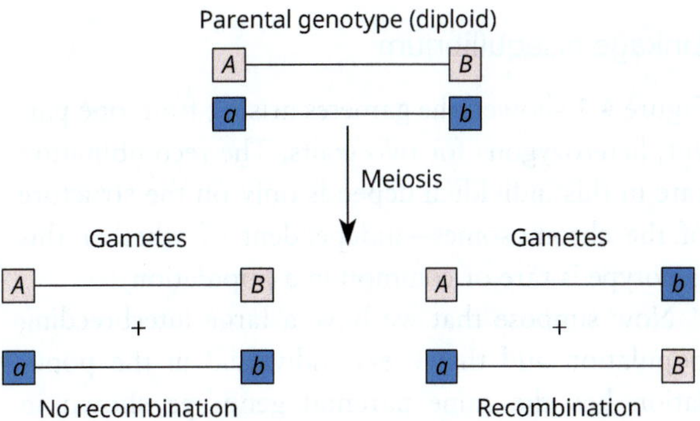

Figure 4.1 Consider two loci on the same chromosome. One locus has alleles *A* and *a*, the other has alleles *B* and *b*. One individual has alleles *A* and *B* on one chromosome (lighter shade) and alleles *a* and *b* (darker shade) on the other (top). Gametes from this individual may form without recombination to give a haploid gamete containing alleles *A* and *B* and another haploid gamete containing alleles *a* and *b* (lower left). Each gamete contains a chromosome identical (at least as far as these loci are concerned) to one of the parental chromosomes. Alternatively, crossover between the two loci may produce recombinant gametes: one haploid gamete containing alleles *A* and *b* and another haploid gamete containing alleles *a* and *B* (lower right). Neither gamete contains a chromosome identical to a parental one. The fraction of gametes showing recombination depends on several factors, notably the distance between the loci. (The same fractions of recombinants and non-recombinants would be produced even if the parent were homozygous, although they would be likely to be indistinguishable.)

In the absence of selection, the fraction of viable recombinant and non-recombinant gametes produced depends only on the genotype of the individual that produced them. It has nothing to do with the allele distribution in the population. If a remote descendant had the parental genotype shown here, its gametes would show the same fractions of recombinants and non-recombinants.

Linkage disequilibrium

Figure 4.1 showed the gametes arising from one parent, heterozygous for two traits. The recombination rate in this individual depends only on the structure of the chromosomes—independent of whether this genotype is rare or common in a population.

Now suppose that we have a large interbreeding population and that every individual in the population has the same parental genotype shown in Figure 4.1. How will the genotype distribution in the population develop? (Assume that no combination of alleles for these two traits has superior fitness or produces any preferential mating pattern.) Recombination at meiosis, followed by zygote formation, can in principle produce individuals of three genotypes: *AB/ab*, *Ab/aB* = *aB/Ab*, and *ab/ab* (where, for instance, *AB/ab* signifies an individual with alleles *AB* on one chromosome and *ab* on the other; this is the parental genotype in Figure 4.1). Starting from the original completely *AB/ab* population, eventually recombination will randomize the allelic correlation, producing a population in which the ratio of genotypes: $AB/ab : Ab/aB = aB/Ab : ab/ab$ is 1:2:1. (This assumes that the overall gene frequency in the population is $A = a$ and $B = b$; see Problem 4.1.) This is what would be observed in a single generation if *A* and *B* were on different chromosomes.

How fast the randomization occurs depends on the recombination rate, which depends on the genetic distance between the loci. In the short term, if recombination is infrequent, the parental genotype *AB/ab* will continue to predominate. The deviation of the genotype distribution in the population from the ultimate 1:2:1 ratio is called **linkage disequilibrium**.

KEY POINT

Two markers are in linkage disequilibrium if the observed distribution of different combinations of alleles differs from that expected on the basis of independent hereditary transmission of the individual alleles.

In the absence of linkage disequilibrium, the frequencies of allelic combinations will be proportional to the products of the frequencies of the individual alleles. Linkage disequilibrium measures the deviation from this equilibrium distribution.

Suppose the overall fractions of the alleles at two loci are p_A, $p_a = 1 - p_A$, p_B, $p_b = 1 - p_B$. (Because p_A, p_a, p_B, and p_b are all fractions, they all have values between 0 and 1.) Then:

- the equilibrium value of $p_{AB} = p_A \times p_B$
  the equilibrium value of $p_{Ab} = p_A \times p_b$
  the equilibrium value of $p_{aB} = p_a \times p_B$
  the equilibrium value of $p_{ab} = p_a \times p_b$
- there is no necessary relationship between p_A and p_B
- at equilibrium:

  $$p_{AB} \times p_{ab} = p_{Ab} \times p_{aB} = p_A \times p_B \times p_a \times p_b$$

- a measure of linkage disequilibrium is:

  $$D = p_{AB} \times p_{ab} - p_{Ab} \times p_{aB}$$

  where $D = 0$ implies that the system is at equilibrium, $D > 0$ implies that chromosomes with *AB* and *ab* are more common than expected, $D < 0$ implies that chromosomes with *Ab* and *aB* are more common than expected.

Suppose there is no selection or gene import into the population, i.e. the overall frequencies of individual alleles in the population remains constant. Then linkage disequilibrium will decay as a result of recombination, at a rate depending on the distance between the loci (see Figure 4.2).

With each successive generation, the value of *D* will become closer to 0. Therefore, linkage disequilibrium can be a useful measure of which of several related populations is older. For instance, the relatively low linkage disequilibrium among African populations is evidence for the 'Out of Africa' hypothesis of human origins.

Linkage and linkage disequilibrium are closely related but distinct concepts

Linkage is about the distribution of loci among chromosomes. Linkage disequilibrium is about the distribution of allelic patterns in populations. Close linkage of two loci on a chromosome is a common source of long-term persistence of linkage disequilibrium. Two genes at opposite ends of the same chromosome, although formally linked, may not show significant

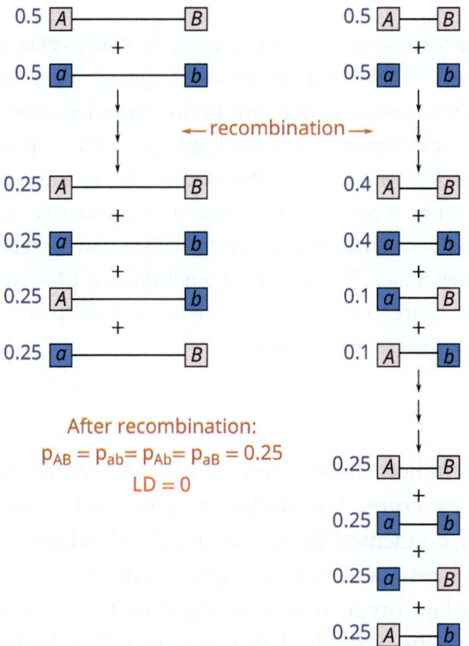

Figure 4.2 Dependence of rate of decay of linkage disequilibrium on separation between linked loci. Shown here are individual chromosomes. Any individual in the population will have a homologous pair. Left: widely separated genes; right, closely separated genes.

Top: Population contains only chromosomes with AB and ab alleles. (Every individual in the population has genotype AB/AB, AB/ab, or ab/ab.) In subsequent generations, recombination brings different pairs of alleles onto a single chromosome. Each chromosome may be AB, ab (parental types) or Ab, aB (recombinants). Because the pair on the left is separated more widely, recombination is more effective. After a certain number of generations of recombination, the allelic combinations are randomized. Linkage disequilibrium measured by $D = p_{AB} \times p_{ab} - p_{Ab} \times p_{aB} = 0$.

On the right, the same number of generations produces fewer recombinants, because the crossing-over rate is lower for two closely spaced loci. After a number of generations sufficient to eliminate linkage disequilibrium for the sites on the left, the sites on the right still retain a majority of parental allele combinations. Further recombination will be necessary to eliminate linkage disequilibrium, finally producing the same random distribution of allelic combinations.

linkage disequilibrium, because crossing over is frequent. Conversely, it is possible—although rare—to observe linkage disequilibrium between two genes on different chromosomes. (This can happen in two ways: (1) a community of immigrants imports a particular set of single-nucleotide polymorphisms (SNPs) into a larger population and these individuals preferentially intermarry for many generations, or (2) theoretically by interactions between gene products that permit only certain combinations of alleles to be viable.)

Classical linkage maps typically involved markers no less than 1 cM apart (~1 Mb in humans) (this is the situation shown in Figure 4.3. Linkage disequilibrium is detectable between markers ~0.01–0.02 cM apart (~10–20 kbp). Therefore, linkage disequilibrium is a much finer tool for localizing a target gene.

Chromosome banding pattern maps

Banding patterns are visible features on chromosomes (see Box 4.2). The most commonly used pattern is G-banding, produced by Giemsa stain. The

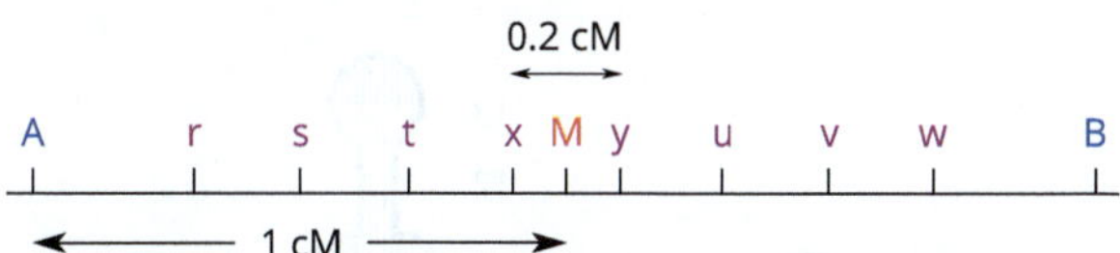

Figure 4.3 Suppose a mutation to a disease gene, *M*, occurred in a human population 50 generations ago. Consider a portion of the genome that includes the site of the disease mutation, *M*, plus genes for two known phenotypic traits, A and B, and two closely spaced markers, *x* and *y*. The genes for traits A and B are 1 cM away from the mutated locus. The markers *x* and *y* are 0.1 cM from *M*. It is highly probable that markers *x* and *y* will be co-inherited with *M* in any pedigrees for which records exist, as the probability of recombination between *x* and *M* or between *y* and *M* in any generation is small: 0.1% = 0.001. The probability of recombination in 50 generations is approximately $0.001 \times 50 = 0.05$. On the other hand, the probability that markers A and *M* or B and *M* have been separated by recombination is very high. The probability of recombination between A and *M* in any generation is 1%. The probability of recombination in 50 generations is approximately 0.4 (see Problem 4.2).

In the history of transmission of this disease gene over 50 generations, markers *x* and *y* are likely to be co-inherited with the disease, but genes for traits A and B will not be reliably coupled with *M*. Therefore, genes A and B, separated by 1 cM (~1 Mb in the human), will not be a reliable guide to localizing the target gene *M* by looking in family pedigrees for genes co-inherited with *M*. However, a distribution of markers such as *x* and *y*, separated by 0.2 cM (~200 kbp) or less, is likely to provide a reliable guide to localizing the target gene *M* through correlation of disease occurrence and genetic markers in family pedigrees.

For humans, we do not have access to 50 generations of records and DNA samples (which would amount to about 1000 years). However, the effects of recombination during the 50 generations since the mutation filter out all but the most closely linked genes from the co-inheritance pedigree. Haplotype groupings simplify the identification of gene-marker correspondences (see Chapter 1).

Nomenclature of chromosome bands

In many organisms, chromosomes are numbered in order of size, 1 being the largest. The two arms of chromosomes, separated by the centromere, are called the p (petite = short) arm and q (queue) arm. Regions within the chromosome are numbered p1, p2 . . . and q1, q2 . . . outward from the centromere. Additional digits indicate band subdivisions. For example, certain bands on the q arm of human chromosome 15 are labelled 15q11.1, 15q11.2, and 15q12. Originally, bands 15q11 and 15q12 were defined; subsequently, 15q11 was divided into 15q11.1 and 15q11.2.

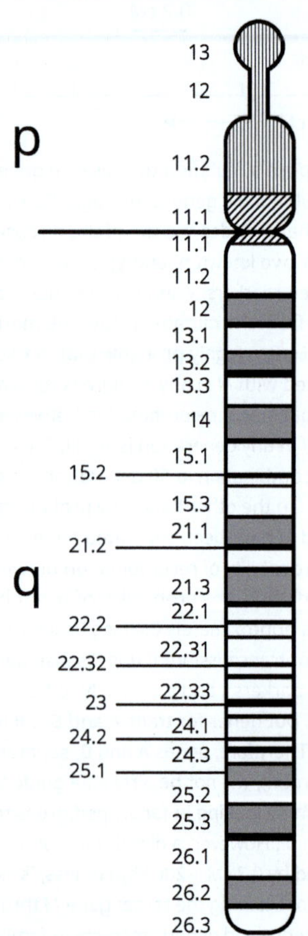

Deletions in the region 15q11–13 are associated with Prader–Willi and Angelman syndromes. These syndromes have the interesting feature that alternative clinical consequences depend on whether the affected chromosome is paternal (leading to Prader–Willi syndrome) or maternal (leading to Angelman syndrome). This observation of **genomic imprinting** shows that the genetic information in a fertilized egg is not simply the bare DNA sequences contributed by the parents. Chromosomes of paternal and maternal origin have different states of methylation, signals for differential expression of their genes. The process of modifying the DNA, which takes place during differentiation in development, is already present in the zygote.

bands reflect base composition and chromosome loop structure. The darker regions tend to contain highly condensed heterochromatin of relatively low GC/AT ratio and sparse in gene content.

The karyotype of an individual comprises the structures of the individual chromosomes. The karyotype is largely constant for all individuals within a species, but varies between species. This is the result of chromosome rearrangement during evolution. The inability of most cells with incongruent karyotypes to pair chromosomes properly is one barrier to fertility that contributes to species divergence.

Although most individuals of a species have the same karyotype, occasionally aberrant chromosomes appear. Some of them are lethal and others are correlated with disease. Examples include Down's syndrome (arising from three copies of chromosome 21); and Prader–Willi or Angelman syndromes (arising from defective chromosome 15 structure) (see Box 4.2).

Studies of chromosome banding patterns support several types of investigations.

Correlation between genetic linkage maps and chromosome structure

Chromosome aberrations include deletions, translocations (of material from one chromosome to another), and inversions. The genetic consequences of a short deletion in only one of a pair of homologous chromosomes—the presence of only one allele instead of two for traits that map to the deleted region—allowed for direct mapping of genes to positions on chromosomes. In this way, the abstract genetic linkage maps could be superimposed onto the chromosome.

This was first done in the 1930s after the discovery of the very large chromosomes in the *Drosophila melanogaster* salivary glands. The correlation of chromosome aberrations with changes in genetic linkage patterns proved the co-linearity of the two maps. Together with the mapping of sex-linked traits to the X chromosome, the genetic consequences of chromosomal deletions confirmed what had previously been a hypothesis—that chromosomes carry hereditary information (see Figure 4.4).

KEY POINT

Complete sequencing of yeast chromosome III in 1992 gave the first opportunity for direct comparison of a genetic linkage map and positions in the DNA sequence.

The modern technique for mapping genes onto chromosomes is **fluorescent *in situ* hybridization (FISH)**. A probe oligonucleotide sequence labelled with fluorescent dye is hybridized to a chromosome. The location where the probe is bound shows up directly in a photograph of the chromosome. Typical resolution is ~10^5 bp, but specialized new techniques can achieve high resolution, down to 1 kbp. Simultaneous FISH with two probes can detect linkage and even estimate genetic distances. FISH can also detect chromosomal abnormalities. Use of chromosome-specific dyes of different colour visualizes the karyotype by 'chromosome painting' (see Figure 4.5).

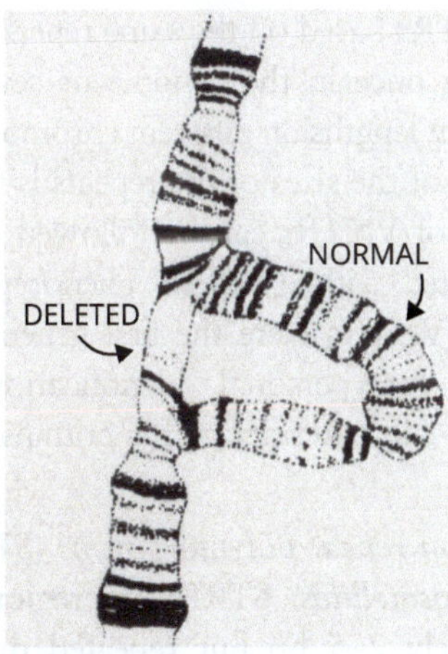

Figure 4.4 Loss of part of one of a pair of homologous chromosomes leads to a structure during chromosome replication called a 'deletion loop'. This drawing, from original work carried out shortly after the discovery in the 1930s of the large chromosomes in the salivary gland of *Drosophila*, shows two chromosomes that are paired except in a region in which one chromosome has suffered a deletion. Genes that map to this region can, in affected individuals, show 'pseudodominance'—the appearance of a recessive phenotype as a result of deletion of a masking dominant allele.

(From: Painter, T.S. (1934). Salivary chromosomes and the attack on the gene. J. Hered. XXV: 465–476.)

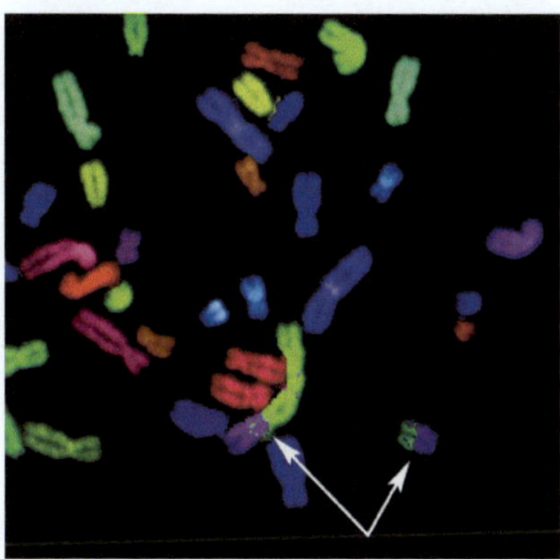

Figure 4.5 Fluorescence *in situ* hybridization: 'chromosome painting' to visualize a metaphase chromosome spread. All chromosomes are monochromatic, except for chromosomes 2 and 8 with regions of purple and green, showing interchange of material.

From: Cornforth, M.N., Greulich-Bode, K.M., Loucas, B.D., *et al.* (2002). Chromosomes are predominantly located randomly with respect to each other in interphase human cells. J. Cell. Biol. 159: 237–244.

Studies of evolutionary changes in karyotype

If we compare our chromosomes with those of a chimpanzee, we see that some large-scale rearrangements have taken place. Human chromosome 2 corresponds to two separate chromosomes in the chimpanzee (see Figure 4.6). In our ancestors, two chromosomes fused to produce modern human chromosome 2. However, most regions in the corresponding chromosomes of the two species show conservation of banding patterns. Such regions are called syntenic blocks. For human and chimpanzee, full genome sequences are available. They confirm that the relationships suggested by the comparisons of banding patterns reflect conservation at the level of DNA sequences.

Applications to diagnosis of disease

Many diseases are caused by—or at least correlated with—chromosomal abnormalities, including many forms of cancer. DNA sequencing is superseding the cytogenetic detection of large-scale mutations.

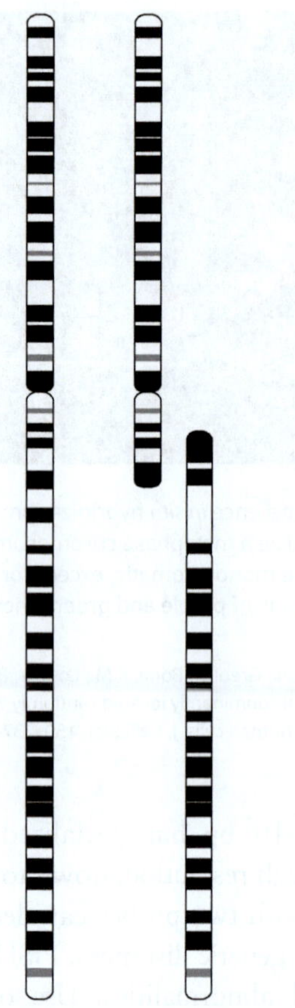

Figure 4.6 Left: human chromosome 2. Right: matching chromosomes from a chimpanzee.

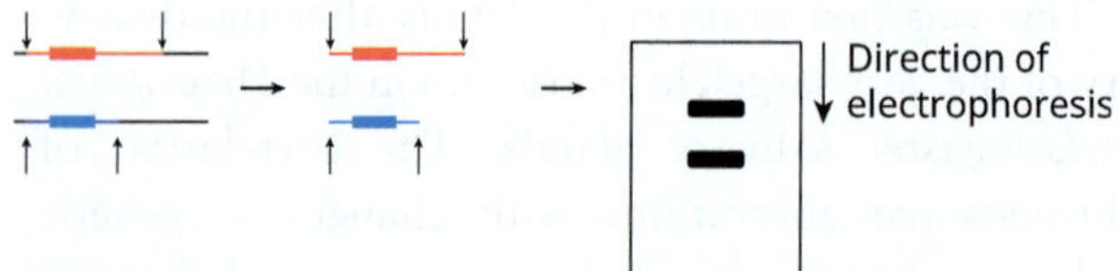

Figure 4.7 Restriction enzymes cleave DNA at the sites of specific sequences to produce fragments separable, according to size, by gel electrophoresis (see Chapter 2). Left: Regions of two homologous chromosomes showing two alleles that differ in the length of a restriction fragment. The fragment contains a marker (red or blue rectangle; red top row, blue bottom row), a region that has *the same sequence* on both chromosomes. Vertical arrows show the cut sites for one restriction enzyme. The common fragment is 'labelled' red on one chromosome and blue on the other. After cleavage (middle), the fragments are loaded on a gel (right) and separated according to size. Cut sites for the restriction enzyme may appear in many places in the genome, on different chromosomes, not just at the positions of the allele illustrated. In order to see only those fragments that originate at the locus of interest, a radioactive or fluorescent probe complementary to the common marker (the rectangle) may be used via Southern blotting to make visible only the selected fragments. (Which band on the gel on the right corresponds to the fragment coloured red on the left, and which band corresponds to the fragment coloured blue?)

High-resolution maps, based directly on DNA sequences

Formerly, we could see genomes only by the reflected light of phenotypes. However, markers are no longer limited to genes with phenotypically observable effects, which are anyway too sparse for an adequately high-resolution map of the human genome. Now that we can interrogate DNA sequences directly, any features of DNA that vary among individuals can serve directly as markers.

The first genetic markers based directly on DNA sequences, rather than on phenotypic traits, were restriction fragment length polymorphisms (RFLPs). The genetic marker is the size of the restriction fragments that contain a particular sequence within them (see Figure 4.7).

Other useful types of marker include:

- *Variable number tandem repeats (VNTRs)*, also called *minisatellites*. VNTRs contain regions 10–100 bp long, repeated a variable number of times—same sequence, different number of repeats. In any individual, VNTRs based on the same repeat motif may appear only once in the genome or several times, with different lengths on different chromosomes. The distribution of the sizes of the repeats is the marker. Inheritance of VNTRs can be followed in a family and correlated with a disease phenotype like any other trait. VNTRs were the first genetic sequence data used for personal identification—'genetic fingerprints'—in paternity and in criminal cases (see Chapter 2).
- *Short tandem repeat polymorphisms (STRPs)*, also called *microsatellites*. STRPs are regions of only approximately 2–5 bp but repeated many times, typically 10–30 consecutive copies. They have several advantages as markers over VNTRs, one of which is a more even distribution over the human genome.

The DNA sequence itself can also be used as a marker. Physically, it is a sequence of nucleotides in the molecule, computationally a string of characters A, T, G, and C. Collection and application of sequence data directly formerly made use of mitochondrial

DNA or haplotypes. Now, whole-genome sequencing has replaced many of the older techniques.

There is no reason why these markers need to lie within expressed genes and usually they do not. The CAG repeats in the gene for Huntington's disease and certain other disease genes are exceptions (see Box 1.7).

Markers have applications in several different fields. Medical applications of markers include their use in tissue typing to identify compatible donors for transplants, detection of disease susceptibility, and prediction of individual drug-response variation (pharmacogenomics). Anthropologists use them to trace migrations and relationships among populations.

Development of methods of sequencing required detailed knowledge of the structure of DNA and the mechanism of its replication.

Discovery of the structure of DNA

Life involves the controlled manipulation of matter, energy, and information. Biochemists were familiar with the structures and mechanisms of enzymes, living molecules catalysing conversions of *matter* and *energy*. What kind of molecule could store and manipulate *information?*

Chemical analysis of DNA during the early part of the twentieth century characterized the constituents—the bases and the sugars—and the nature of their linkage. DNA is a polynucleotide chain, containing a repetitive backbone of sugar-phosphate units, with small organic bases—adenine (A), thymine (T), guanine (G), and cytosine (C)—attached to each sugar (see Figure 4.8). Not only was the distribution of the bases along the chain unknown, but also its significance was entirely unsuspected.[4] Indeed, many people accepted P.A. Levene's idea that DNA contained a regular repetition of a constant four-nucleotide unit. It was, therefore, considered that DNA simply lacked the versatility required to convey hereditary information, compared, for example, with proteins. We now know that eukaryotic DNA does contain some repetitive sequences, but is not limited to them.

An understanding of how DNA worked in biological processes required a detailed three-dimensional structure. In the 1950s, the method of choice for determination of molecular structure was X-ray crystallography. The pioneer of X-ray structure determination, Sir Lawrence Bragg, Cavendish Professor of Physics at Cambridge, was an enthusiastic supporter of the efforts in his department by Max Perutz and his group to extend methods of X-ray crystallography to biological molecules as large as proteins.

However, DNA molecules are flexible and do not form classic crystals. They can be drawn into fibres, which can be thought of as crystalline in two dimensions, and disordered around the fibre axis. This disorder severely impoverishes the available data. X-ray diffraction patterns of three-dimensional crystals permit an objective determination of the individual atomic positions in a structure. In contrast, fibre diffraction data present a trickier puzzle. Guided by all available clues, the molecular biologist must conceive a model that is consistent with the known covalent structure and with the X-ray diffraction pattern.

X-ray diffraction patterns of DNA fibres were measured at King's College, London, by a group that included Rosalind Franklin and Maurice Wilkins. In May 1951, Wilkins spoke at a conference in Naples. In the audience was a young American postdoctoral scientist named James Watson. Watson was converted. He took his newly kindled interest in X-ray structure determination to Cambridge, where he met a graduate student named Francis Crick, and together they sought the structure of DNA.

The clues included: (a) Alexander Gulland's DNA titration curves that showed the bases to be involved in internal hydrogen-bonding interactions; and (b) Erwin Chargaff's observations that, although the

[4] Except for a prescient comment by physicist Erwin Schrödinger—based on ideas of Max Delbrück—in his influential 1944 book, *What is Life?*: 'We believe a gene—or perhaps the whole chromosome fibre—to be an aperiodic solid.' Do not be misled by the reference to a solid. Schrödinger himself included a footnote: 'That it [the chromosome] is highly flexible is no objection, so is a thin copper wire.'

Figure 4.8 The chemical structure of one strand of DNA. The backbone consists of 2′-deoxyribose sugars linked by phosphodiester bonds. In its chemical bonding pattern, the backbone is a regular repetitive structure, independent of the bases. Each base, one attached to each sugar, can be one of four choices. In the sequence of bases, DNA is logically equivalent to a linear message written in a four-letter alphabet.

amounts of the four bases differ among samples of DNA from different organisms, the amounts of adenine and thymine are always equal and the amounts of guanine and cytosine are always equal. Chargaff published his results in 1949 and described them to Watson and Crick during a visit to Cambridge in July 1952.[5]

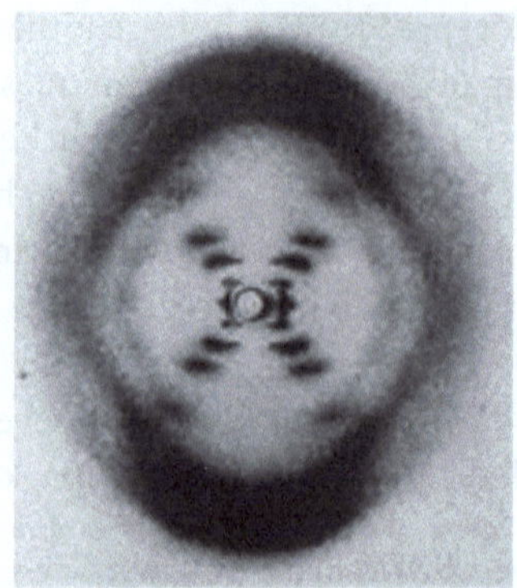

Figure 4.9 X-ray diffraction pattern of DNA fibre, by R. Franklin. The X-shaped pattern at the centre of the picture is diagnostic of a helical structure. From the distribution of intensity, it is possible to deduce the number of residues per turn and the symmetry of the structure.

Franklin, R., & Gosling, R.G. (1953). The structure of sodium thymonucleate fibres. II. The cylindrically symmetrical Patterson function. Acta Crystallographica 6, 8: 678–685. https://doi.org/10.1107/s0365110x53001940

It was recognized as a race. Lined up against Watson and Crick were Franklin and Wilkins in London and Linus Pauling in California, in the US. Pauling was a frightening contender, with many major discoveries already to his credit. He had recently beaten the Cambridge group to the structure of the α–helix in proteins.

Watson and Crick did no experimental work themselves, but confined their efforts to rationalizing all of the data and available clues, including the fibre diffraction photos. Watson attended talks in which the King's group's fibre diffraction photographs were shown. Franklin continued to improve the data, producing in 1952 the photograph in Figure 4.9. This photograph appeared in a Medical

[5] Chargaff was one of the more acerbic scientists of his time. He dismissed molecular biology as 'the practice of biochemistry without a license'. It is likely that he felt that he deserved more credit than he received for providing information crucial to solving the DNA structure; in any event he was certainly no fan of Watson and Crick: Once when asked whether students at his university were taught the central dogma of molecular biology, he replied, 'Of course. The central dogma of molecular biology is: Always use clean glassware and write everything down in ink in a hard-covered notebook.'

Research Council report, which was shown to Watson and Crick without Franklin's knowledge. The fibre diffraction data implied that the chains formed helices, with a repeat distance of 34 Å, containing ten residues per turn (i.e., 3.4 Å between successive bases). Although manuscripts left by Franklin upon her death suggest that she was close to solving the structure herself, Watson and Crick beat her to the goal.

What animated Watson and Crick's model was the idea of the complementarity of specific pairs of bases, together with the recognition that AT and GC pairs had compatible stereochemistry; that is, they could fit interchangeably into a regular structure. There had been several premonitory hints at this crucial idea. A Cambridge mathematician, John Griffith (nephew of the Frederick Griffith who discovered bacterial transformation), told Crick of a preferential attraction between A and T, and between G and C, that he had deduced from theoretical calculations. This, together with Chargaff's rules, immediately suggested the idea of complementarity.[6] (A colleague has commented: In physics experiment suggests, and theory proves; in biology, theory suggests and experiment proves.[7] Even A.S. Eddington, an astrophysicist, once warned against putting 'too much confidence in observational results until they have been confirmed by theory'.)

What in retrospect screams hydrogen-bonded base pairing (see Figure 4.10) only whispered, before the model was imagined. The leap from energetic and compositional complementarity to the correct structural complementarity was *not* (as it may perhaps now appear) an easy step. Even the fact that the model rationalized the hints from Griffith and Chargaff and the titration curves was more confirming evidence than proof. The famous sentence from Watson and Crick's paper, 'It has not escaped our notice that the specific pairing we have postulated immediately suggests a possible copying mechanism for the genetic material' distracts by its coyness from the underlying logic: it was the clear structural basis for the biological activity that made the model immediately and utterly convincing (see Figure 4.10).

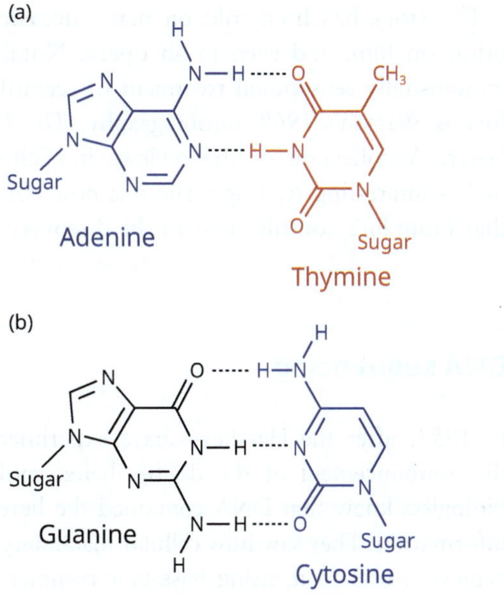

(a)

(b)

Figure 4.10 The complementary base pairings: adenine–thymine and guanine–cytosine.

Once the base pairing was recognized, the pieces of the puzzle fell quickly into place (see Figure 4.11). On 28 February 1953—in a possibly apocryphal story—Crick and Watson strode into the Eagle, a pub in downtown Cambridge near the laboratory, and Crick announced to anyone listening that he and Watson had discovered the secret of life. The 25 April issue of Nature contained three papers on the structure of DNA, one by Watson and Crick, and two from the King's group. A second paper by Watson and Crick, on the genetic implications of the structure, appeared in the 30 May issue of Nature.

It was an eventful spring: Stalin died on 5 March; Edmund Hillary and Tenzing Norgay reached the summit of Mount Everest on 29 May; Queen Elizabeth II was crowned on 2 June.

Rosalind Franklin died in 1958. The 1962 Prize for Physiology and Medicine was awarded to Watson, Crick, and Wilkins.

[6] See: Lagnado, J. (2005). From pabulun to prions (via DNA): Tale of two Griffiths. The Biochemist 27: 33–35.

[7] Perhaps not any longer; think of the Higgs boson.

The story has been told on many occasions, in print, on film, and even in an opera. Notable for its unusually sensational treatment of scientific history is Watson's 1968 autobiography, *The Double Helix*. A collection of the reviews it elicited still makes interesting reading.[8] There is now consensus that Franklin's contributions to the discovery of the structure of DNA were underestimated, not least by Watson, who in *The Double Helix* disparages her mercilessly, both professionally and personally. Of numerous attempts to redress the balance, Sir Aaron Klug's lecture is the most authoritative.[9]

Franklin has been honoured by appearing on postage stamps, coins, and banknotes, in several countries.

DNA sequencing

In 1953, after the Hershey–Chase experiment and the announcement of the double helix, molecular biologists knew that DNA contained the hereditary information. They saw how cellular machinery could achieve access to it, using base-pair complementarity. But if the sequence of the bases was like a text everyone wanted to read, not only was it a text in an unknown language, but there were not even any examples of the language, because the sequences were unknown. The importance of the problem of determining DNA sequences was obvious, but the difficulties frightened most people away from confronting the challenge.

Frederick Sanger and the development of DNA sequencing

Cambridge biochemist Frederick Sanger devoted his career to sequence determination of biological macromolecules. First, he determined the amino-acid sequence of the protein insulin, proving for the first time that proteins had definite sequences. This work won him the Nobel Prize in Chemistry in 1958. He next turned his attention to sequencing of RNA, and then to DNA. His methods for sequencing of DNA won a second Chemistry Nobel Prize, in 1980, shared with Walter Gilbert and Paul Berg. Sanger's professional legacy to the field in general, and his personal influence on his many students and colleagues, is unsurpassed.

To appreciate the state of the art during the early 1970s, it is important to recognize that it was very difficult even to prepare pure samples of single-stranded DNA for attempts at sequencing. Restriction enzymes were a new discovery. They did not provide the mature and versatile technology that we take for granted today. Bacteriophage ϕX-174, providing 5386 bp of purifiable, single-stranded DNA, was a natural source of material for Sanger's early work, yielding the first whole genome sequence.

Even ϕX-174 was too long to sequence 'in one go'. It depended, as did subsequent *de novo* genome sequencing projects—until very recently—on solving the 'Humpty Dumpty problem': sequencing fragments and putting them back together again. (Box 4.3 introduces the sequence assembly problem.) Sanger's achievement was a reliable and convenient method for sequencing relatively long fragments, making unambiguous assembly possible.

Sanger's method used DNA polymerase to synthesize a new strand of DNA, incorporating, as part of the synthesis, reactions that reveal the sequence. DNA polymerase is a replication enzyme that synthesizes the strand complementary to a piece of single-stranded DNA. It requires a primer: a short stretch of complementary strand to be extended by successive addition of nucleotides (see Figure 4.12). The polymerase requires a supply of nucleoside triphosphates. In successive steps, the enzyme adds to the growing primer strand the nucleotide complementary to the next unpaired base in the template. This reaction also forms the basis of the polymerase chain reaction (PCR) (see Figure 4.13).

[8] Stent, G., ed. (1980). *The Double Helix: Text, Commentary, Reviews, Original Papers*. New York & London: W.W. Norton.

[9] Klug, A. (2003). The discovery of the DNA double helix. In: T. Krude (ed.), *Changing Science and Society*, pp. 5–43. Cambridge: Cambridge University Press.

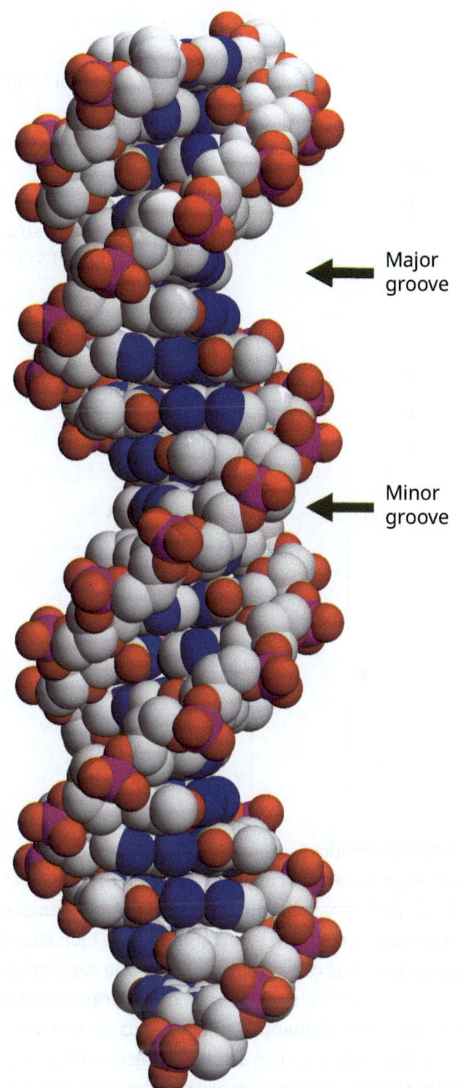

Figure 4.11 The structure of DNA. The two helical strands wind around the outside of the structure. The bases are inside, stacked like the treads of a staircase with their planes perpendicular to the axis of the double helix. The bases are visible through the major and minor grooves and are thereby accessible for interaction with proteins.

KEY POINT

Loop-Mediated Isothermal Amplification (LAMP) is an alternative to PCR. A major advantage is that it does not require thermal cycling steps, simplifying the apparatus. RT-LAMP combines LAMP with reverse transcription of RNA in a test sample to cDNA which is then amplified. RT-LAMP is the basis of portable, robust, and inexpensive kits suitable for use in the field, for rapid detection of viruses such as Ebola and Zika, and, of course, SARS-CoV-2, the virus that causes COVID-19.

The problem of sequence assembly from short fragments

To illustrate what is involved in assembling a long sequence from overlapping short fragments, consider the first two verses of Shakespeare's Richard III:

Now is the winter of our discontent
Made glorious summer by this sun of York

Breaking this up into overlapping ten-letter fragments (treat the line-break as a space), and presenting them in random order, gives:

ntent Made

this sun o

f York

discontent

ious summe

r by this

er of our

glorious

summer by

our disco

winter of

Made glor

Now is the

s the wint

It is easy to reassemble the pieces, and would be easy even if you didn't know the answer. For instance, the first fragment 'ntent Made' overlaps, and must therefore precede, the third fragment from the end of the list, 'Made glor'. Observe that the longer the fragments, the easier the challenge (see Problem 4.3). Recognize also, that repetitive sequences are harder. In Act 2 of Hamlet, Polonius says:

'tis true, 'tis true 'tis pity,
And pity 'tis 'tis true

It would be more difficult to reassemble this from fragments, because the repetitions create ambiguities. Indeed, the very large amounts of repetitive sequence in eukaryotic genomes do create severe problems for assembly algorithms.

If a reference genome for the species, or even a close relative, is known, sequences of fragments can be mapped onto the reference genome, bypassing the assembly problem. The idea would be to search the complete text of Richard III for 'Made glor', and the other fragments.

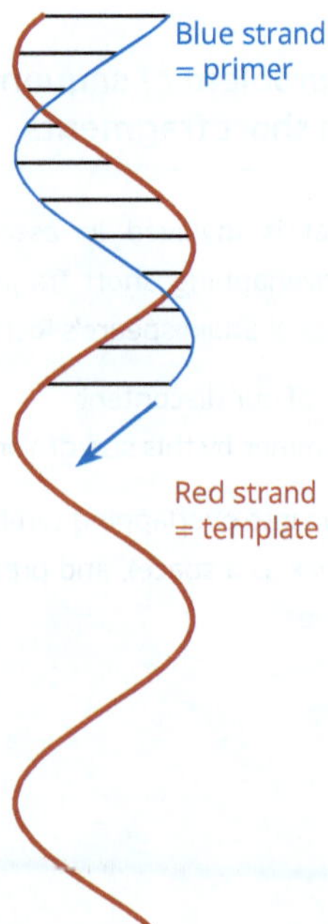

Figure 4.12 Primer extension. Duplication of a piece of single-stranded DNA, the template (red; bolder line), occurs by extension of a short primer (blue) by addition of complementary nucleotides to the primer. The arrow shows the direction of synthesis. The basis of the Sanger method for DNA sequencing is the termination of the extension by 'poisoning' the reaction with a modified base—a dideoxynucleoside triphosphate—that lacks the reactive group necessary for continuing the extension.

DNA sequencing by termination of chain replication

DNA polymerase links the 5'-hydroxyl of the newly added nucleotide to the 3' position of the end of the chain. Hydrolysis of the activating triphosphate provides the energy. Sanger's idea was to include in the mixture of nucleoside triphosphates a fraction of molecules containing no reactive 3' position, a dideoxynucleoside triphosphate (see Figure 4.14). A reaction mixture contains normal deoxyATP, normal deoxyGTP, normal deoxyTTP, and a mixture of normal deoxyCTP plus dideoxyCTP (Figure 4.14). As chain extension proceeds, when the next unpaired base in the template is a guanine, the enzyme will randomly incorporate either a normal deoxycytidine, after which the extension will continue, or a dideoxycytidine, after which no further extension will occur. (Three other reactions are run in parallel containing all

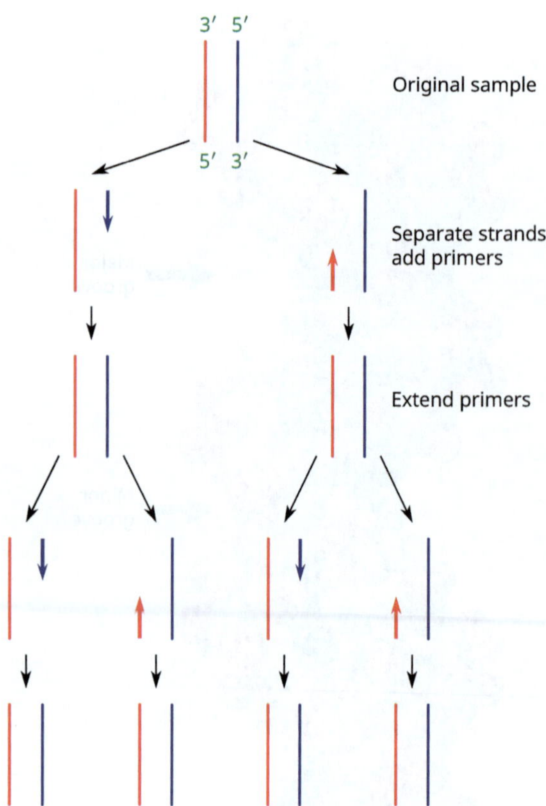

Figure 4.13 Schematic diagram of the polymerase chain reaction (PCR), a molecular copying technique that can selectively amplify very small quantities of DNA in a sample. In this figure, red strands are complementary to blue strands. The target sequence is called the amplicon. One cycle doubles the amount. First, separate the strands of the original sample and add primers complementary to both ends of the target region. (The short red and blue arrows represent the primers.) Primer extension by DNA polymerase reconstitutes two copies of the original sample. Repetition of these operations doubles the amount of amplicon at each step. Only the first and second cycles are shown in this diagram. Exponential growth leads quickly to the production of large quantities of the desired sequence. Specificity is achieved by tailoring the primers to the target sequence. Amplification is effective even if the target sequence appears in a small amount within a large background of other DNA.

A development called real-time PCR allows the quantitative measurement of levels of mRNA in a sample by first converting it to cDNA and then following the rate of the amplification reaction by measurement of incorporation of a fluorescent label. The basic idea is that the more abundant a sequence in a mixture, the faster the build-up of the product of its PCR amplification. Extremely useful in understanding PCR are animations available on the web. Searching will turn up a number of sites, for example: http://users.ugent.be/~avierstr/principles/pcrani.html

deoxycytidine triphosphate

dideoxycytidine triphosphate

Figure 4.14 Left: normal deoxycytidine triphosphate, containing both a reactive 5′-hydroxyl, activated as the triphosphate derivative permitting it to be added to the growing primer strand, and a 3′-hydroxyl (red), to which the *next* nucleotide will be attached. Right: dideoxycytidine triphosphate, in which the 3′ position is unreactive. Dideoxycytidine can be incorporated into the growing strand, but no subsequent extension is possible. **P-P-P** is a simplified representation of the triphosphate group.

four normal triphosphates plus dideoxyATP, dideoxyGTP, or dideoxyTTP.)

This reaction produces a mixture of fragments of different lengths, each of which ends in a (modified) deoxycytidine:

```
3′-atacagagaatctagatacagagttgttcgag        Template
5′-tatgtctcttagat →                 Primer extension
5′-tatgtctcttagatctatgtctcaacaagctc
5′-tatgtctcttagatctatgtctcaacaagc          Fragments
5′-tatgtctcttagatctatgtctcaac               produced
5′-tatgtctcttagatctatgtctc
5′-tatgtctcttagatctatgtc
5′-tatgtctcttagatc
5′-tatgtctc
5′-tatgtc
```

The crucial point is that the fragments are *nested*. To determine the positions of the cytosine residues in the extended primer, it is sufficient to determine the lengths of the fragments ending in dideoxycytidine. For instance, if a fragment ending in a dideoxycytidine has length 26, the base at position 26 in the original sequence must be C.

Polyacrylamide gel electrophoresis can separate oligonucleotides according to their lengths accurately enough for sequence determination. It is possible to separate polynucleotides differing in length by a single base, up to molecules about 1000 bases long.

KEY POINT

Almost all DNA sequencing methods, from the earliest until quite recently, depended on breaking a long DNA molecule into fragments, sequencing the fragments, and then assembling them, via overlaps, into the complete sequence (see Box 4.3). The methods of Sanger and of Maxam and Gilbert worked by creating a set of nested fragments, each terminating in a known nucleotide, that could be separated on a gel. Knowing the length of the fragment and the terminal nucleotide allowed the sequence to be 'read off' the gel.

In Sanger's original procedure, the dideoxycytidine carried a radioactive label and the gel was developed by autoradiography. To determine the positions of the other three nucleotides, it was necessary to run four reactions, in parallel, each containing a radioactive dideoxy analogue of one of the four nucleoside triphosphates. Four separate reactions are necessary because each nucleotide gives the same signal—the darkening of the film from the radioactive spot. Running the products of the four reactions in parallel lanes on the same gel allows the sequence to be read from the autoradiograph (see Figure 4.15).

The Maxam–Gilbert chemical cleavage method

The Maxam–Gilbert method for DNA sequencing, also developed in the mid-1970s, also worked by comparing nested fragments. A sample of single-stranded DNA, labelled at its 5′-end, is cleaved by base-specific reagents. As in Sanger's method, polyacrylamide gel electrophoresis separates the fragments by size. The separate cleavage reactions produce fragments that share the 5′-labelled end, but differ in the bases at which the specific 3′-cleavage occurs. The sequence can be read from an autoradiograph.

The Maxam–Gilbert method had early successes, for which Gilbert shared the Nobel Prize. It was Sanger's method that spawned subsequent developments, however. Because the Maxam–Gilbert method does not use primed DNA synthesis, its applicability is inherently limited to sequences adjacent to restriction sites or other fixed termini. Another

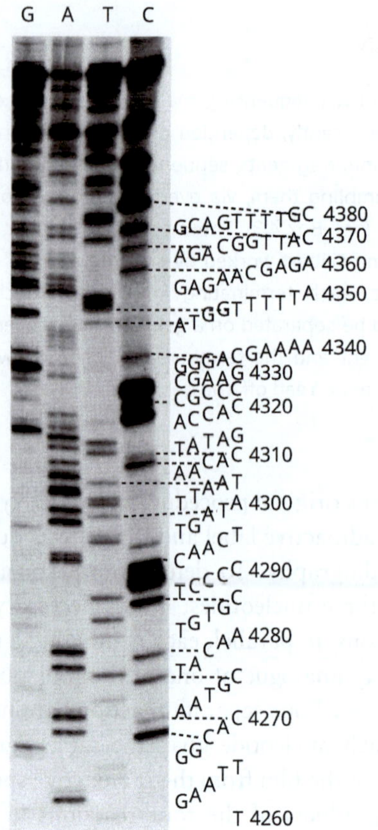

Figure 4.15 A gel from early in the history of DNA sequencing, showing part of the genome of bacteriophage ϕX-174.

Sanger, F., Nicklen, S., & Coulson, A. (1977). DNA sequencing with chain-terminating inhibitors. Proceedings of the National Academy of Sciences of the United States of America 74, 12: 5463–5467. https://doi.org/10.1073/pnas.74.12.5463

disadvantage of the Maxam–Gilbert method was reagent toxicity, notably of hydrazine, a neurotoxin.

Automation of DNA sequencing

Using fluorescent dyes as reporters, rather than radioactivity, was an important technical advance in sequencing. Radioisotopes present health hazards in both use and disposal, and are expensive. (One of Sanger's 'postdocs', who wore his hair long in the fashion of the 1970s, was obliged to have a haircut because of contamination resulting from his frequently pushing his hair out of his eyes.) By attaching different fluorescent dyes to the four dideoxynucleoside triphosphates, each fragment produced gives a different signal, depending on which dideoxynucleotide terminated the extension. All four reactions can be done 'in the same pot', and electrophoresis separates them in a single lane. A laser focused at a fixed point identifies the fragments as they pass. The result can be displayed as a four-colour chromatogram in which successive peaks correspond to successive bases in the sequence (see Figure 4.16).

KEY POINT

Labelling the four dideoxynucleoside triphosphates with different fluorescent dyes, with distinguishable colours, allows 'one-pot-one-lane' sequencing reactions.

Quality is important too: The phred score q measures sequencing accuracy. $q = 20$ implies a probability of $\leq 1\%$ error per base (see Box 4.4). A common unit of sequencing cost is US$ per 1000 bases determined to $q = 20$ accuracy. A sequence of 1000 bases determined at an accuracy of $q = 20$ would be expected to contain no more than ten errors.

The obvious importance of high-quality, high-throughput sequencing motivated the search for improvements in the methods—a quest that continues today.

In 1986, L. Hood, L. Smith, and co-workers described an instrument, based on detection of base-specific fluorescent tags, that allowed automation of DNA sequencing. Instruments produced by Applied Biosystems implemented the work of Hood, Smith, and co-workers, with improvements by J.M. Prober

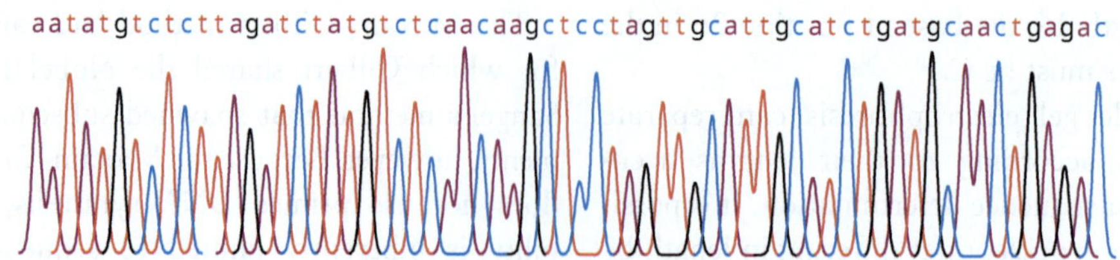

Figure 4.16 A tracing of a sequencing fluorescent chromatogram (simulated). Different-coloured peaks correspond to different bases.

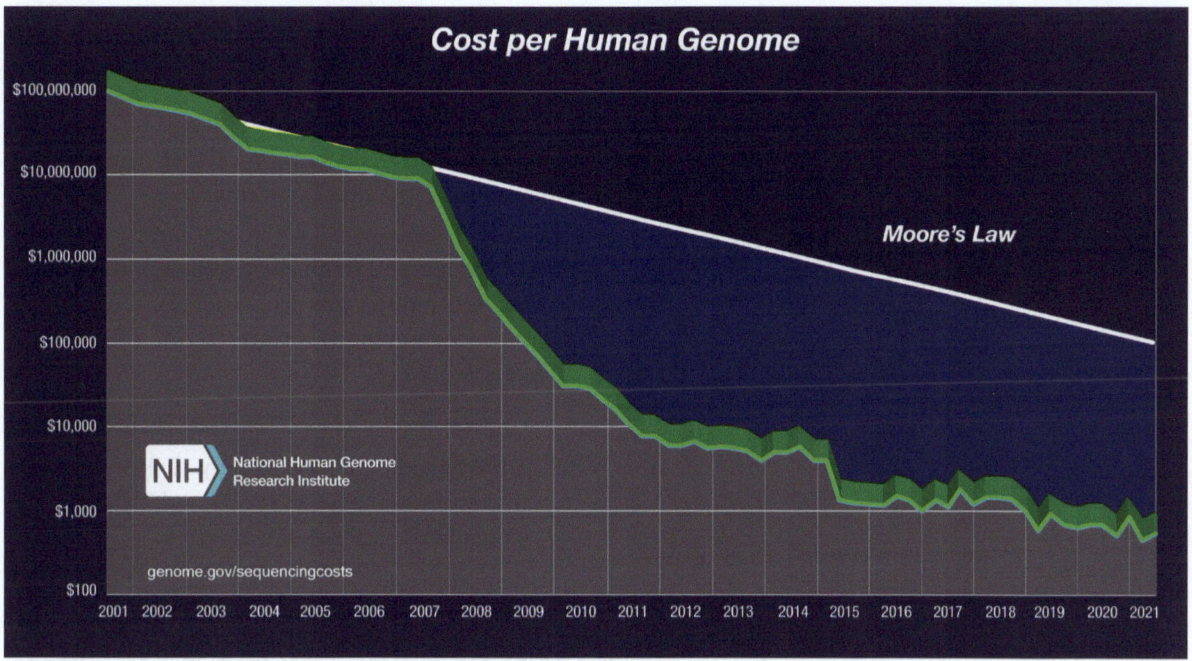

Figure 4.17 Fall in the cost of sequencing over time. Note logarithmic scale on the cost axis. The sea change in 2007 from the introduction of next-generation sequencing platforms is striking.

Moore's law states—very roughly—that the cost of computing power falls by half every two years. In several eras, improvements in sequencing outpaced this.

Wetterstrand KA. DNA Sequencing Costs: Data from the NHGRI Genome Sequencing Program (GSP) Available at: www.genome.gov/sequencingcostsdata. Accessed February 2024.

Phred scores: a measure of quality of sequence determination

The phred score of a sequence determination is a measure of sequence quality. It specifies the probability that the base reported is correct. If p = the probability that a base is in error, then the corresponding phred score $q = -10\log_{10}(p)$.

Quality score q (phred score)	Probability of error	Error rate
10	0.1	1 base in 10 wrong
20	0.01	1 base in 100 wrong
30	0.001	1 base in 1000 wrong
40	0.0001	1 base in 10 000 wrong

and co-workers at DuPont. Capillary systems for fragment separation to replace flat-sheet gels were another essential advance. A population explosion in automatic DNA sequencing machines filled the high-throughput installations that produced the human and other complete genomes. Instruments available in 1998 were able to produce ~1 Mb of sequence per day.

The technology improved dramatically. The US National Institute of Health (NIH) set goals for a US$100 000 human genome by 2009 and a US$1000 genome by 2014 (see Figure 4.17). The US$1000 genome has been achieved, and costs continue to decrease. Preston, VanZeeland, and Peiffer wrote recently 'Today, a human genome can be sequenced for $600, with some predicting that the $100 genome is not far behind.'[10] Readers will scoff: 'Hype springs eternal ...', but the claim is not incredible.[11]

[10] https://www.nature.com/articles/d42473-021-00030-9

[11] See also Pennisi, E. (2022). Upstart DNA sequencers could be a 'game changer'. Science 376: 1257–1258; https://www.nytimes.com/2022/10/12/business/gene-sequencing-ultima-cheaper.html, and https://www.genomeweb.com/sequencing/complete-genomics-unveils-new-platform-ultra-high-throughput-sequencing-market.

Classical methods for organizing large-scale sequencing projects

The first full-genome sequencing projects applied either:

1. The hierarchical method, in which the whole genome is first fragmented and cloned into bacterial artificial chromosomes (BACs) (see Box 4.5), and the order of the fragments is established before sequencing them.
2. The whole-genome shotgun method, which works directly with large numbers of smaller fragments, with a concomitantly more challenging assembly problem.

Bring on the clones: hierarchical—or 'BAC-to-BAC'—genome sequencing

One approach to organizing the sequencing of a large DNA molecule involves dividing the sample into pieces *of known relative position.*

- First, cut the DNA into fragments of about 150 kbp. Clone them into BACs. For example, *Arabidopsis thaliana* has a haploid genome size of about 10^8 bp. A 3948 clone BAC library for *A. thaliana* contained ~100 kbp inserts per clone, giving approximately fourfold coverage.
- Identify a series of clones in the library that contains overlapping fragments. Although referred to as 'fingerprinting', this process depends on *shared* (rather than unique) features of overlapping clones, including:
  1. overlap of restriction fragment size patterns;
  2. amplification of single-copy DNA between interspersed repeat elements and checking for similar size patterns of fragments;
  3. mapping sequence-tagged sites (STSs) and looking for fragments sharing STSs.
- Using the overlaps, order the clones according to their position along the original large target DNA molecule.
- Subfragment each clone, sequence the fragments, and assemble them. The idea is that the clones are small enough that the ~1500 bp sequenced subfragments can be assembled to give the complete sequence of the ~150 kbp BAC clones. Then the clones can be assembled using their known order in the original sequence.

BACs: bacterial artificial chromosomes

A plasmid is a small piece of double-stranded DNA in a bacterial cell, in addition to the main genome. A bacterial artificial chromosome, or BAC, is a plasmid containing foreign DNA—for instance, a fragment of the human genome. A typical BAC, in an *Escherichia coli* cell, can carry about 250 000 bp.

The idea of a BAC is to provide storage for relatively small DNA molecules. The storage should be stable and, by growing up the cells, replicable.

The number of copies per cell varies from plasmid to plasmid. A high copy number would be advantageous if we wanted to use a bacterial culture as a factory to produce large amounts of protein. However, multiple copies of plasmids in cells can undergo recombination and are therefore not suitable for stable storage of DNA fragments. The BACs in common use in genome sequencing projects are based on choosing an *E. coli* plasmid that maintains a low copy number and shows low recombination rates.

Whole-genome shotgun sequencing

The idea of the whole-genome shotgun approach is to sequence random pieces of the DNA and then put them together in the right order (see Figure 4.18). If this can be done, one can skip the laborious stage of creating a map as the basis for assembling partial sequences.

In the whole-genome shotgun sequencing of the *D. melanogaster* genome, the DNA was sheared into random pieces of approximately 2 kbp, 10 kbp, and 150 kbp. For each piece, the sequences of approximately 500 bp from each end were determined; these are called **reads**. A computer program then assembled the results into a maximal set of contiguous sequence, or a **contig** (see Figure 4.18). The fully assembled genome sequence, built by coalescence of contigs, is known as the **'Golden Path'**.

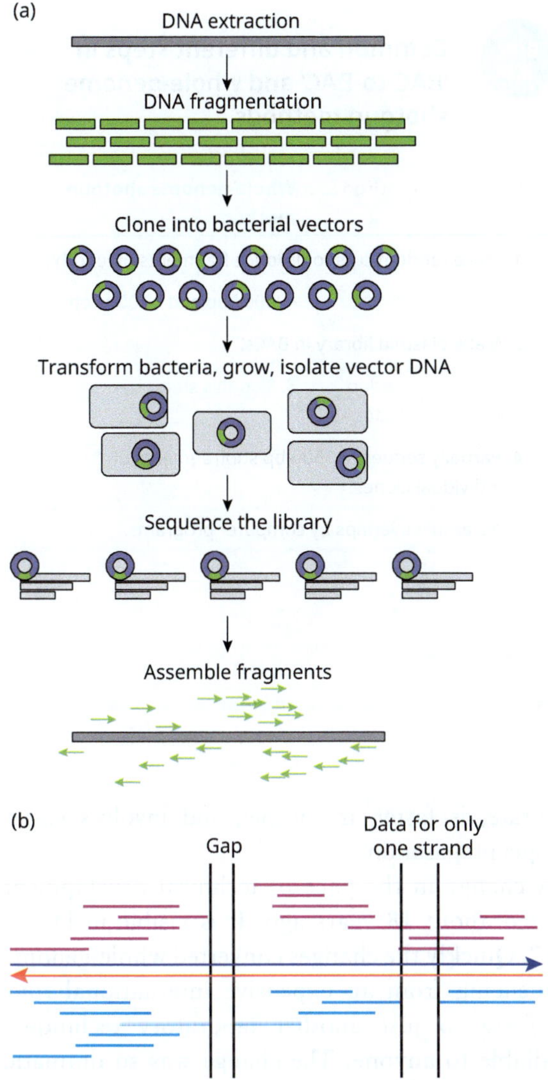

Figure 4.18 (a) Shotgun sequencing of a long DNA molecule. The DNA is broken into shorter fragments, and amplified by cloning into separate bacterial artificial chromosomes (BACs). Sequencing of the individual fragments allows assembly by overlapping subsequences. (b) Details of assembly of a region of DNA from shotgun fragments. Here the two strands of DNA are shown in red and blue. 'Reads' of fragments from the red strand are shown in cyan; those from the blue strand are shown in magenta. In practice, sequences are determined only for about 500 bp at both ends of longer fragments. This assembly, based on overlaps, leaves one gap. Although for one region all data arise from fragments of only one strand, this does not leave a gap.

The coverage is the average number of times each base appears in the fragments. If G = genome length, N = number of reads and L = length of a read, then the coverage = NL / G.

E. Lander and M. Waterman derived formulas for the number of gaps expected as a function of coverage and genome size. For instance, for a 1 Mbp genome, eight-to-tenfold coverage should permit assembly of the reads into five contigs. Completion of the process, called finishing, involves synthesis and sequencing of specific fragments to close the gaps.

Whole-genome shotgun sequencing worked smoothly for prokaryotes, which contain relatively less internal repetitive sequence. Repeats create problems in assembly, and this led initially to scepticism about the feasibility of the shotgun approach for a complex eukaryotic genome. In fact, the *Drosophila* genome has fewer repeats than mammalian genomes and this contributed to the success of sequencing by shotgun methods. In the event, the publication of the *Drosophila* genome in 2000 contained 120 Mb of finished sequence, with about 1600 gaps. In the latest release of the *Drosophila* genome, the gaps have been reduced to less than 0.01% of the total sequence.

Highly skewed base composition, as in *Plasmodium falciparum*—which contains ~80 mol% AT—also complicates application of whole-genome shotgun techniques.

Even if the ultimate goal is a complete, finished genome sequence, it may be possible to identify genes in a partly assembled genome that still show many gaps, provided that the genes are contained within contigs.

Celera took the success of whole-genome shotgun sequencing of the fruit fly as 'proof of principle', justifying its use in their Human Genome Project. Despite the simultaneous announcement of academic and commercial human genomes, on 25 June 2000, the academic group made its results publicly available (largely to preclude patenting) and Celera made use of these data in their assembly.[12]

Comparison of BAC-to-BAC and whole-genome shotgun approaches

Suppose the sample of DNA for sequencing comes from a diploid organism. Fragments arising from homologous regions of two chromosomes of a pair may have sequence differences. Correct assembly

[12] See Sulston, J., & Ferry, G. (2002). *The Common Thread: A Story of Science, Politics, Ethics and the Human Genome*. London: Bantam Press; and Ashburner, M. in Jobling, M.A., Hurles, M.E., & Tyler-Smith, C. (2003). *Human Evolutionary Genetics: Origins, Peoples and Disease*, 27. New York: Garland Science.

must place them at the same location, noting the discrepancies, and must not split these reads into different contigs because of the imperfect matches.

BAC-to-BAC methods are more robust than whole-genome shotgun methods with respect to this problem. Each BAC is a clone and, therefore, has a unique sequence. Successive clones may come from different chromosomes and may, therefore, show mismatches, but this will not affect the order of assembly of the clones.

Note that an unambiguous success of whole-genome shotgun methods, the *Drosophila* genome, was based on a highly inbred laboratory strain. Sequencing the DNA of an individual from a natural, outbred population—or, even worse, DNA pooled from several individuals from such a population—would present a more severe challenge.

Box 4.6 contrasts the alternative approaches.

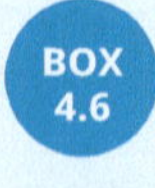

BOX 4.6 Common and different steps in 'BAC-to-BAC' and whole-genome shotgun methods

'BAC-to-BAC' method	Whole-genome shotgun method
1. Make random cuts to produce fragments of length:	
~150 kbp	~2000 kbp and 10 000 kbp
2. Make plasmid library in BACs.	
3. Fingerprint, overlap, and order BAC clones.	3. Skip this step.
4. Partially sequence 1500-bp subfragments of individual clones.	
5. Assemble overlaps by computer programs.	

Second-generation sequencing

Goals for improvement in sequencing methods

Intellectually, all of modern biology rides on the back of DNA sequencing. Genome sequences have no rivals in laying bare an organism's secrets. High quantities of high-quality data raise investigations to higher levels of subtlety. Think of better coverage, or being able to treat a larger sample cohort, as a magnifying glass that brings details into even sharper focus.

Sequencing has proved that it can enhance human welfare. Clinical applications have already entered medical practice; sequencing supports research that will make the connection ever more intimate. Other applications include more effective and safer production of food, biotechnological approaches to generating safe and abundant consumable energy, and environmental remediation.

There is also financial motivation: Sequencing is a multi-billion-pound industry. Substantial rewards await the developer of a breakthrough in DNA sequencing technology. If you want to approach a venture capitalist with a novel sequencing technique, be prepared to argue that your method, compared to the state of the art: gives longer read lengths, is more accurate, is faster, is cheaper, and involves easier sample preparation.

A change in the pace of technical development started about 18 years ago. It is visible in Figure 4.17. Quickly the changes converted whole-genome sequencing from an expensive international *tour de force* to just another laboratory technique, available to anyone. The change was so dramatic as to be called a 'next generation'. Still newer advances have earned the title 'Third-generation sequencing'.

General approaches to improving the throughput/cost ratio include miniaturization; and parallelization, or multiplexing. Common to many but not all of the 'next-generation' methods are preparation steps in which the target DNA is fragmented, common adaptors attached to one or both ends, and the results amplified. This generates a library of short regions that cover the original sequence. The elements of the library are separated spatially, and sequenced in parallel. Assembly of the fragments follows, or mapping the fragments onto a reference sequence. 'Third-generation methods' produce long reads that greatly simplify or even eliminate the assembly process.

One important advance was **sequencing by synthesis:** the real-time detection of primer extension. A basic idea is to expose the growing strand plus polymerase to, *in succession,* the four nucleoside triphosphates (often fluorescently labelled derivatives). The base call depends on detection of which base gives the signal, for instance by a burst of fluorescence; hence the name **pyrosequencing** (see Box 4.7). This eliminates the need for creating overlapping fragments as in the Sanger method.

BOX 4.7

Pyrosequencing

Pyrosequencing detects incorporation of a specific nucleotide into a growing strand by detecting the pyrophosphate (PP_i) released in a step of the reaction catalysed by polymerase:

```
Template: 3′-ATACAGAGAATCTAGAT
                         | T added to primer
Primer:   5′-TATGTCTCT + TTP → TATGTCTCTT + PPi
```

In each cycle, the reaction is separately exposed to each of the four triphosphates.[13] Incorporation of the complementary nucleotide releases pyrophosphate. ATP sulphurylase converts the pyrophosphate to ATP:

adenosine 5′-phosphosulphate + $PP_i \rightarrow$ ATP

which is detected by a light-producing luciferase reaction (see Figure 4.19). Apyrase destroys unincorporated nucleoside triphosphates and ATP produced upon successful incorporation.

Each cycle of successive exposure to the four nucleoside triphosphates produces one sequenced base.

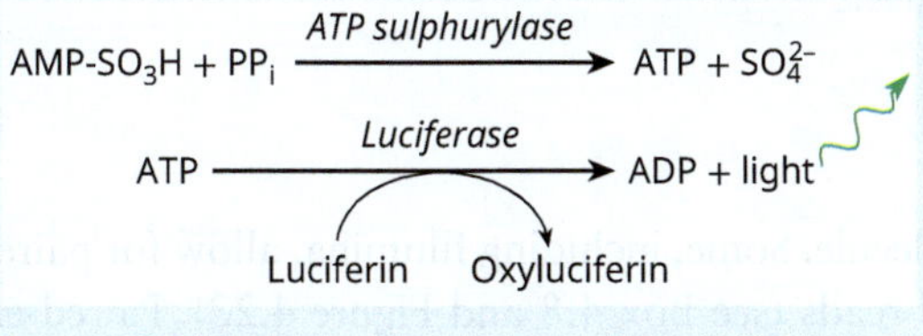

$$\text{AMP-SO}_3\text{H} + \text{PP}_i \xrightarrow{\textit{ATP sulphurylase}} \text{ATP} + \text{SO}_4^{2-}$$

$$\text{ATP} \xrightarrow{\textit{Luciferase}} \text{ADP} + \text{light}$$

Luciferin → Oxyluciferin

Figure 4.19 Detection of matching nucleotides by the luciferase reaction to signal the appearance of PP_i released when a nucleotide is incorporated.

[13] As in many plays, including *The Merchant of Venice, Turandot,* etc., in which an eligible woman is offered several suitors in turn.

Originally developed in Sweden in 1993, the method went through a succession of licencings and acquisitions ultimately to be the basis of instruments produced by 454 Life Sciences. '454 Sequencing' was in common use for genome sequencing because it provided relatively long read lengths (typically ~500 nucleotides; somewhat smaller than a typical Sanger run) and high throughput—a 4-hour run produced 25 million bases, at 99% or better accuracy. It was superseded in 2013.

The next major player in next-generation sequencing was Illumina technology, introduced in 2006. It also involves parallel simultaneous sequencing of a library of fragments, by synthesis. Successive bases are added and detected by fluorescence signals. The workflow (see Figure 4.20) involves:

1. *Generation of a library,* by fragmenting the DNA sample and ligating a common adaptor sequence to both ends (Figure 4.20a). Typical fragment lengths are 300–800 bp.
2. *A flow cell*—a glass slide with lanes—can contain billions of nanowells. The wells present a 'lawn' of attached oligonucleotides complementary to the terminus of the adaptor. Library fragments bind to oligos (Figure 4.20b). Bridge amplification *in situ* (see Figure 4.21) generates a monoclonal cluster of replicated fragments around each original library fragment, no more than one cluster per well.
3. *Sequencing*: In successive cycles, the fragment clusters are exposed to a mixture of nucleoside triphosphates. At each step, the polymerase adds a nucleotide. All four reagents contain a blocking group so that no more than one nucleotide is added at each step. Each of the four nucleotides bears a different fluorescent tag, excited by a laser pulse. The distribution of colours, in an image of the field, identifies which base was added to each cluster. Figure 4.21c, middle, shows a view looking down on the array. The colour of the signal from each cluster gives the base call at that position of that cluster. Next, the fluorescent tags and blocking groups are removed, and the cycle repeated. The result is a kaleidoscopic movie of shifting colours, one frame per position, one colour per cluster, generating the separate sequences of the library fragments.

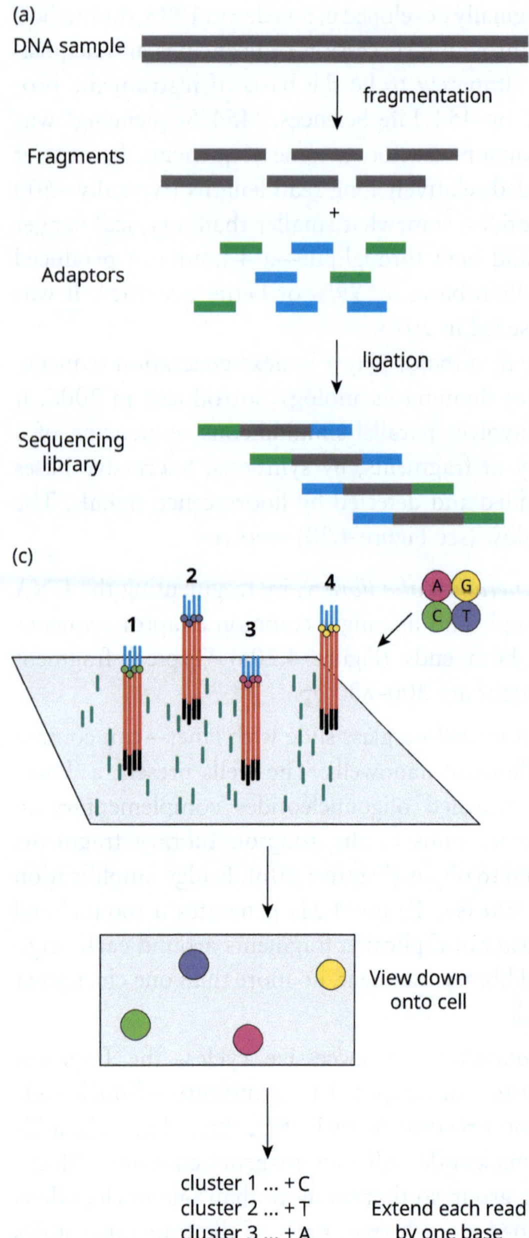

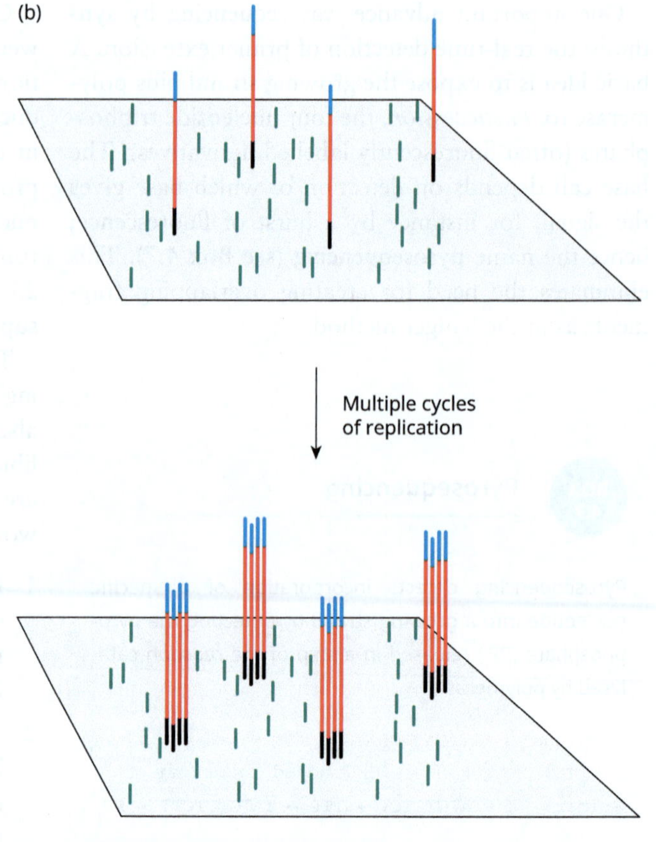

Figure 4.20 Simplified picture of generation of a fragment library, binding to a surface displaying a 'lawn' of fixed oligonucleotides complementary to the adaptor sequences, and sequencing by primer extension, one base at a time. (a) The original DNA sample is cleaved into fragments, and different adaptors are ligated to the 5′ and 3′ ends. (b) Through the adaptors, fragments bind to oligonucleotides fixed on a substrate in the sequencing instrument. Multiple cycles of replication create a cluster of fragments at each position. A clever mechanism for the replication is bridge amplification (see Figure 4.21). (c) Sequencing is carried out by exposing the primed fragments to a mixture of oligonucleotide triphosphates. Incorporation in any cluster gives a unique fluorescent signal. Because each cluster is fixed in position, the position of the signal can be correlated with the fragment that emitted it, and the base assigned as the extension of the determined sequence of that fragment.

Illumina markets a range of instruments, of different power and different costs. In this way they provide different throughput rates to cater to the needs of different installations. Illumina has a ~75% global market share.

Methods differ in the length and accuracy of the fragments sequenced. This determines the parameters of the libraries constructed from a long starting molecule. Some, including Illumina, allow for paired-end reads (see Box 4.8 and Figure 4.23). Paired-end DNA sequencing reads:

- improve assembly of regions containing repetitive sequences,
- produce long contigs for *de novo* sequencing,
- facilitate detection of DNA rearrangements.

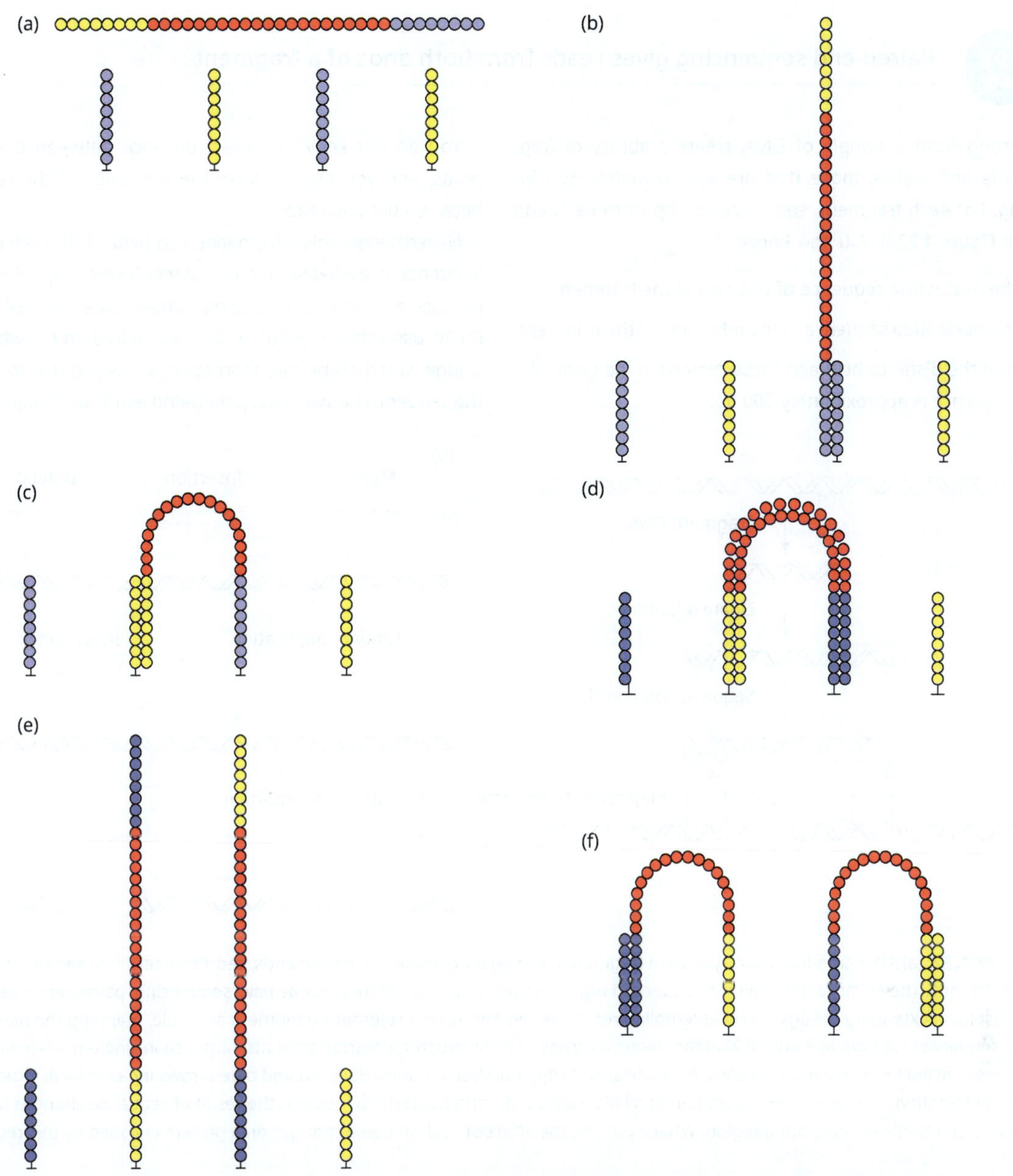

Figure 4.21 In bridge amplification, DNA fragments of about 500 bp, with appropriate adaptors ligated to each end, are repeatedly replicated. In this figure, opposition of two identical yellow or blue or red regions signifies complementary base pairing between them. (a) A series of short adaptors are fixed to a plane. Shown here as a linear array, in reality a two-dimensional 'lawn' (see Figure 4.20). There are two types of adaptors, with different sequences, shown here in blue and yellow. A fragment to be amplified (red) has both adaptors, one affixed to either end. (b) The adaptor at one end of the fragment binds to its complement fixed on the slide. (It would equally be possible for the other end, the yellow end, shown free in this frame, to bind to one of the fixed yellow adaptors.) Next, a polymerase extends the fixed strand to replicate the fragment, to form a blue-red-yellow double-stranded structure. Note that the replicated fragment is fixed to the slide; the original is held only by complementary hydrogen-bonding. By denaturing the double-stranded structure, and washing away the original fragment, one is left with the replicated copy of the fragment, fixed to the slide. (In frame (b) the full-length single-stranded fragment is not fixed to the slide, but in frame (c) the full-length single-stranded copy of the fragment is fixed to the slide (see Exercise 4.21). (c) The replicated fragment arches over, and its distal end binds to another fixed adaptor, with the other sequence (shown here as yellow rather than blue). (d) A polymerase replicates the bridged fragment. The original fragment is fixed to the slide at its blue end, and the replicated fragment is fixed to the slide at its yellow end. (e) Because both ends are fixed to the slide, after denaturation both are still present, each fixed to the slide at one end. Frame (e) is like frame (b) except that the fragment has been replicated. (f) Each fragment can again arch over, to begin another replication cycle. Frame (f) resembles frame (c), with the fragments replicated. Repeating this process creates a cluster of cloned fragments, as in Figure 4.20c.

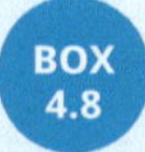

BOX 4.8 Paired-end sequencing gives reads from both ends of a fragment

Starting from a sample of DNA, create a library of fragments and isolate those that are approximately 500 bp long. For each fragment, sequence ~50 bp from each end (see Figure 4.22a). You now know:

- the nucleotide sequence of one end of the fragment,
- the nucleotide sequence of the other end of the fragment,
- that the distance between these regions in the original sequence is approximately 500 bp.

You do not know the exact distance between the two reads, and you do not know the sequence of the region between the two reads.

Nevertheless, this information can help: (1) If a reference sequence is available, one can detect insertions, deletions, repeats, and other structural rearrangements (Figure 4.22b). (2) To assemble a genome *de novo,* paired-end reads can bridge, and thereby link, short contigs. It is possible to 'fill in' the sequence between the paired-end reads, with sequences

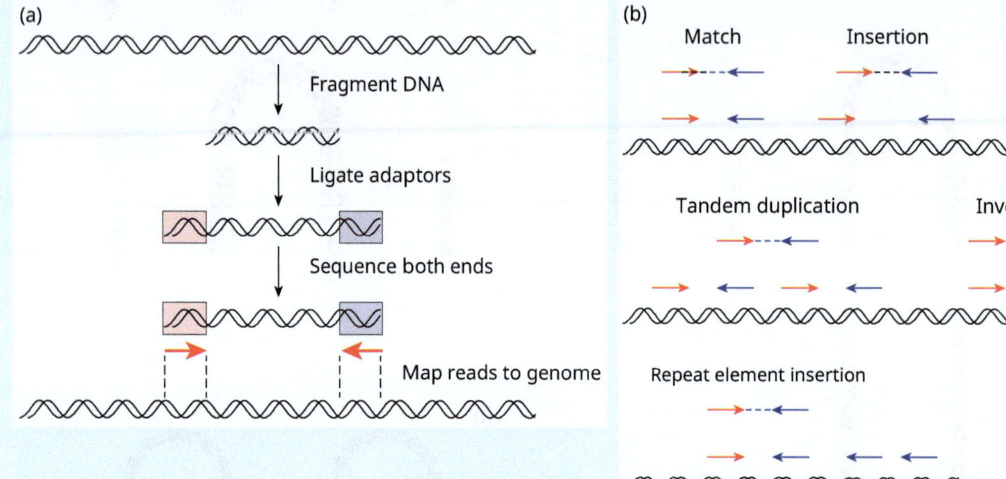

Figure 4.22 (a) After creating a library of DNA fragments, and ligating adaptors to both ends, sequence terminal segments of both strands. The distance between the sequenced regions is known approximately. For *de novo* sequencing, paired-end reads are useful in extending contigs and in assembling repetitive regions. (b) If a reference genome is available, mapping the paired-end sequences (bottom of Figure 4.22a) can detect a variety of local structural rearrangements, larger than single-nucleotide changes, including insertions, deletions, inversions, and copy-number variants. Here red and blue arrows indicate local regions. The combination ⟶ - - - ⟵, common to all diagrams, is an original state; beneath is the result of sequence changes that affect, not a single residue, but a region. What would be the effect of each of these changes on a protein encoded by this region?

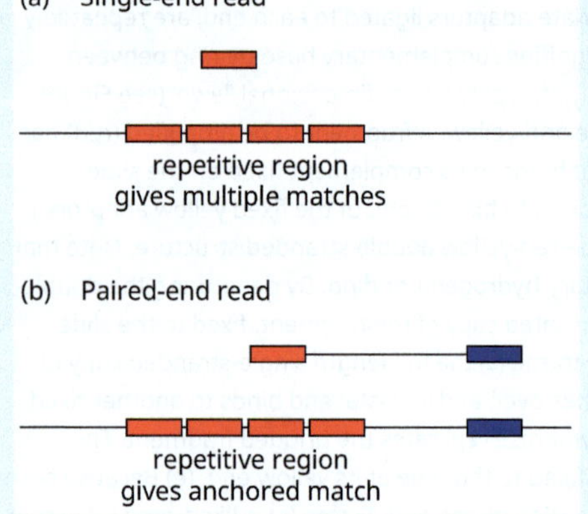

Figure 4.23 Paired-end reads help with the difficult problem of assembling regions containing repetitive sequences. (a) A single-end read that matches part of a repetitive region cannot be mapped unambiguously. (b) If one end of a paired-end read matches a non-repetitive region (blue), it serves as an anchor to help identify where the other end, matching the repetitive region (red), should be mapped. If the ambiguity in the length of the paired-end region is longer than the length of the repeat, the position of the match to the red end cannot be assigned precisely, without other information.

of shorter fragments that overlap the ends. Given that repetitive sequences are particularly difficult to assemble, if one end of the fragment is part of a repeat sequence such as a LINE or SINE, but the other is anchored firmly in a non-repetitive sequence, you know approximately how far the repeat sequence is from the anchored one (see Figure 4.23). You need additional information from other fragment sequences to fix precisely how far apart they are.

In later stages of assembly, paired-end reads can indicate how contigs can be ordered (see Figure 4.24).

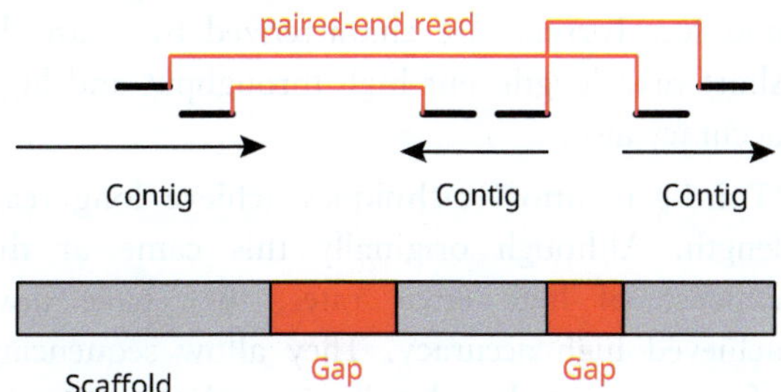

Figure 4.24 Consider a stage in assembly at which substantial contigs have been determined, but gaps make the *order* of the contigs uncertain. Paired-end fragments can join the contigs into a combined scaffold.

KEY POINT

Next-generation sequencing methods had the following features (green = good, red = not so good!):

- *highly parallel:* sequences of many samples carried out simultaneously.
- *miniaturized:* many reactions possible on a single chip.
- *speed:* as a consequence of parallelization, combined results ready quickly.
- *low-cost:* much less expensive than Sanger sequencing.
- *BUT: low read length:* 50–600 bp

The octopus genome

One example of the power of second-generation methods, as of about ten years ago, was the sequencing of the genome of the California two-spot octopus (*Octopus bimaculoides*), by a relatively small cohort of authors.[14] These organisms have a number of interesting features, notably a rich and sophisticated behavioural repertoire, including problem-solving ability and the ability to adopt camouflage.[15] According to Sydney Brenner, one of the investigators, 'They were the first intelligent beings on the planet'. Supporting this intelligence is the largest nervous system known in invertebrates: The octopus has six times as many neurons as the mouse. However, the neurons of the octopus are far more delocalized than those of vertebrates. Three-quarters of the neurons are distributed among the tentacles.

How the octopus achieves its abilities, and comparison of the answer with the corresponding features of vertebrates, were major goals of the investigation. Sequencing the genome was a prominent component of the project.

A whole-genome shotgun approach, using Illumina HiSeq2000 instruments, produced 264 654 660 990 raw base pairs of sequence, giving 92× coverage. The assembly covered 83% of the genome (estimated at 2.7 Gbp). The remaining 17% is primarily repetitive sequence. The sequenced portion includes 33 638 protein-coding genes, estimated at 97% of the total.

One particularly interesting family of proteins is the protocadherins. Homologues in vertebrates are responsible for neuronal development and synaptic specificity. The octopus genome shows an expansion of this family relative to other invertebrates. Vertebrates also show expansion of protocadherins, but by a different mechanism. In the octopus the family shows tandem gene duplication; vertebrates show a complex splicing pattern involving many genes. It is very interesting to see two different ways to achieve a parallel result. It was the availability for comparative analysis of the genomes that made this observation possible.

[14] Albertin, C.B., Simakov, O., Mitros, T., *et al.* (2015). The octopus genome and the evolution of cephalopod neural and morphological novelties. Nature 524: 220–224.

[15] See: https://www.youtube.com/watch?v=qLqB0394V3I

Third-generation sequencing

It is conventional to divide the development of sequencing techniques into 'generations' (see Figure 4.25):

- Sanger sequencing was the founder generation.
- 'Second-generation' techniques, including Illumina and Ion Torrent, are characterized by relatively short read length, but high throughput and high accuracy also.
- 'Third-generation' techniques achieve long read length. Although originally this came at the expense of high error rates, they have now achieved high accuracy. They allow sequencing of unfragmented molecules in real time. Pacific Biosciences and Oxford Nanopore now provide the current 'state-of-the-art' technologies in widespread use. But progress is very rapid, and it is entirely possible that some startup will overtake them in the near future.

A major achievement was to sequence single DNA molecules without amplification. The new methods produce very long reads, typically up to hundreds of kilobase pairs, and sometimes much longer. This overcomes many stubborn assembly problems of earlier, short-read-length methods, especially for repeat-rich low-complexity regions. This was the problem that limited the human genome reported in 2003 to covering only 92% of the sequence; the new methods have brought the project to completion.

Before third-generation methods, a problem inherent in short-read sequencing of diploid or polyploid genomes was **phasing**: the separation of sequences from different homologous chromosomes. The danger is that assembly will produce a chimeric genome, conflating or even collapsing into a single mixed sequence corresponding to regions of the paternal and maternal (in diploid organisms) chromosomes. One possible but imperfect approach to the problem was to sequence, separately, mother–father–child trios. But third-generation sequencing, with its long read lengths, makes it possible to assign confidently the different contributions to each chromosome of a pair. Correct phasing facilitates accurate detection of structural variants, including length variations in highly repeated motifs, insertion and deletion events, duplications, inversions, and translocations.

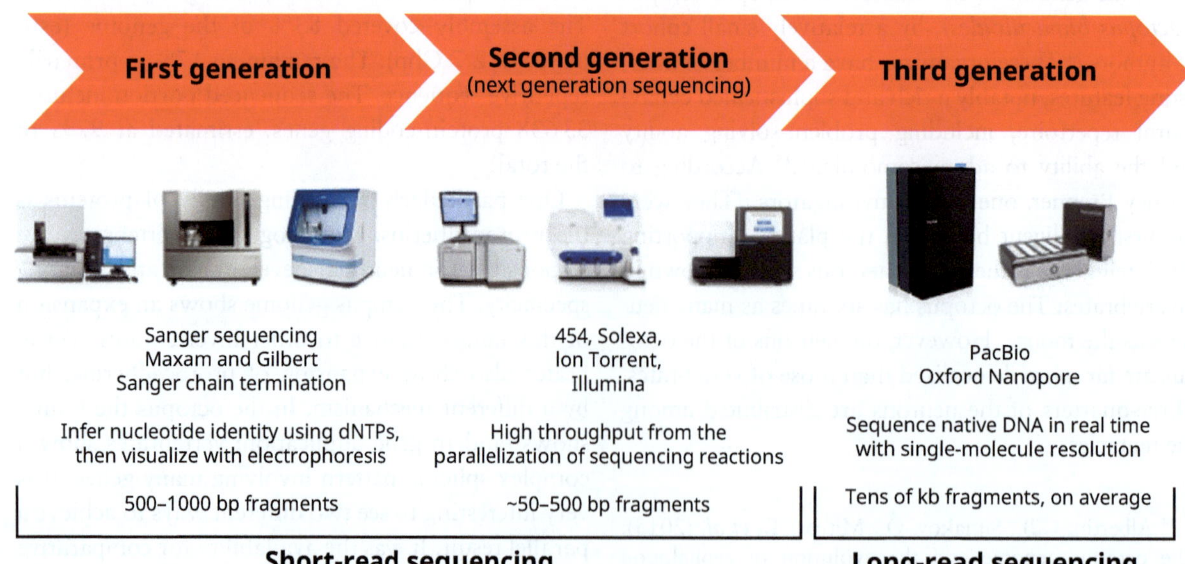

Figure 4.25 Evolution of sequencing technology.

PacBio. (2023). Sequencing 101: The evolution of DNA sequencing tools. PacBio. (20 September 2023), https://www.pacb.com/blog/the-evolution-of-dna-sequencing-tools/

Single-Molecule Real-Time sequencing (SMRT)

Pacific Biosciences (PacBio) offers several sequencing platforms, based on a fluorescent pulse arising from base addition to a replicating chain, called Single-Molecule Real-Time sequencing (SMRT). It is an accurate and high-throughput method: The PacBio Revio system can analyse 100 million molecules in parallel, to achieve 360 Gb of reads per day.

In SMRT sequencing, libraries prepared by ligating adaptors to the ends of DNA molecules are loaded into a large array of nanowells (~70 nm in diameter and ~100 nm deep). A polymerase immobilized at the bottom of each well catalyses the successive addition of fluorescently labelled bases. Each incorporation produces a *transient* base-specific emission signal recorded as a video for real-time accumulation of sequence information, called the Continuous Long Read (see Figure 4.26). After a base is incorporated,

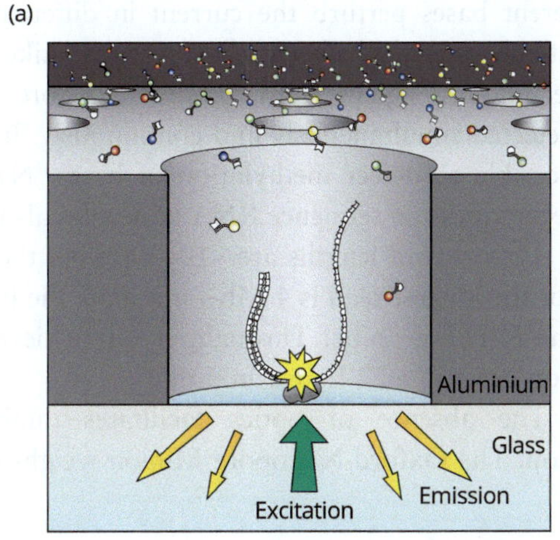

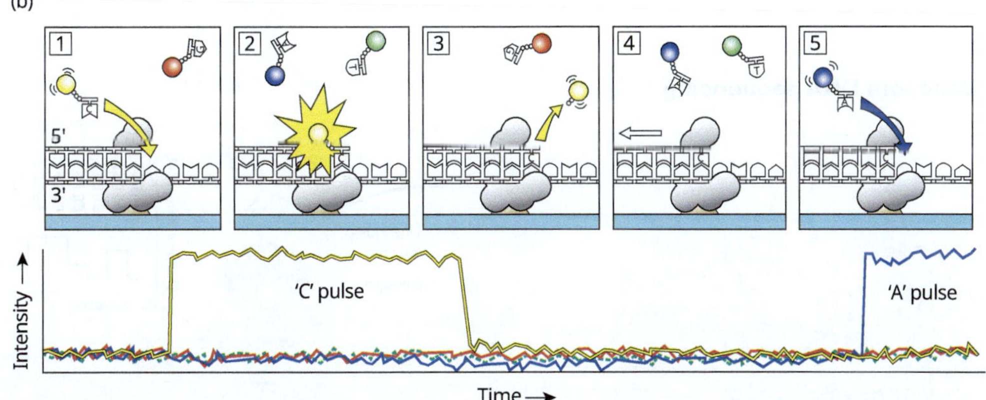

Figure 4.26 (a) An element of the prepared library diffuses into a nanowell. The attached adaptor binds to a polymerase immobilized at the bottom.

(b). Each of the four nucleotides is labelled with a different fluorescent dye emitting a different colour (red = G, yellow = C, green = T, blue = A).

(b)1. A molecule of CTP, labelled with a yellow-fluorescent dye, approaches the polymerase.

(b)2. Concomitant with incorporation of the base is a transient yellow fluorescent emission signal.

(b)3. The dye is cleaved and diffuses out of the well.

(b)4. The polymerase translocates, ready to incorporate the next base.

(b)5. The next nucleotide, an ATP labelled with a blue fluorophore, is incorporated next.

Bottom: The signals for incorporation of first the C and then the A appear at successive times, and are recorded as a video. (The interpulse duration is variable but typically tenths of a second.)

[Rhoads, A., & Au, K.F. (2015). PacBio sequencing and its applications. Genomics Proteomics Bioinformatics 13, 5: 278–289. doi: 10.1016/j.gpb.2015.08.002. Epub 2015 Nov 2. PMID: 26542840; PMCID: PMC4678779. Made available under Creative Commons Attribution 4.0 International]

and a burst of fluorescent emission observed, the dye is cleaved off the base and diffuses out of the well. Recent advances have lowered the error rate; achieving >99.999% consensus accuracy. Read lengths are typically 10–20 kbp, but can be over 30 kbp.

Localization of the signal to the individual nanowell makes possible assembly of very large arrays of wells (up to 8 million in current instruments), and *simultaneous* and *continuous* detection of signals from different wells. The current PacBio instrument can produce up to 4 000 000 reads of a median length 15 kbp, in a runtime of up to 30 hours. (See https://www.pacb.com/videos/video-overview-of-smrt-technology/)

Nanopore sequencing

Oxford Nanopore Technologies Ltd offers desktop and even portable sequencing devices.

A nanopore is a small-diameter molecular tunnel. Many proteins containing small pores are known from their roles in membrane transport. Nanopores used in DNA sequencing include modified *Staphylococcus aureus* α-haemolysin, and *Mycobacterium smegmatis* porinA (MspA). Many copies of the proteins are inserted into an artificial lipid bilayer, distributed in an array of individual microwells ~10 nm in diameter. One flow cell may contain several thousand microwells.

For sequencing, a DNA molecule traverses a channel, at a rate of ~400–450 bases per second. A voltage across the membrane induces a current of ions through the pores. The passage of each base alters the current through the pore (see Figure 4.27). Different bases perturb the current in different, identifiable ways, which can be recorded. Like SMRT techniques, signals from the different pores can be detected simultaneously and continuously. It is even possible to detect methylation of bases. Nanopore techniques can sequence RNA molecules also.

Typical read lengths are ~10–300 kbp; the record for the longest read is 4 Mb—not quite the length of the *E. coli* genome! Throughput can achieve 4 million bases in a 72-hour run time.

The absence of optics facilitates miniaturization. The Oxford Nanopore Minion weighs <450 g,

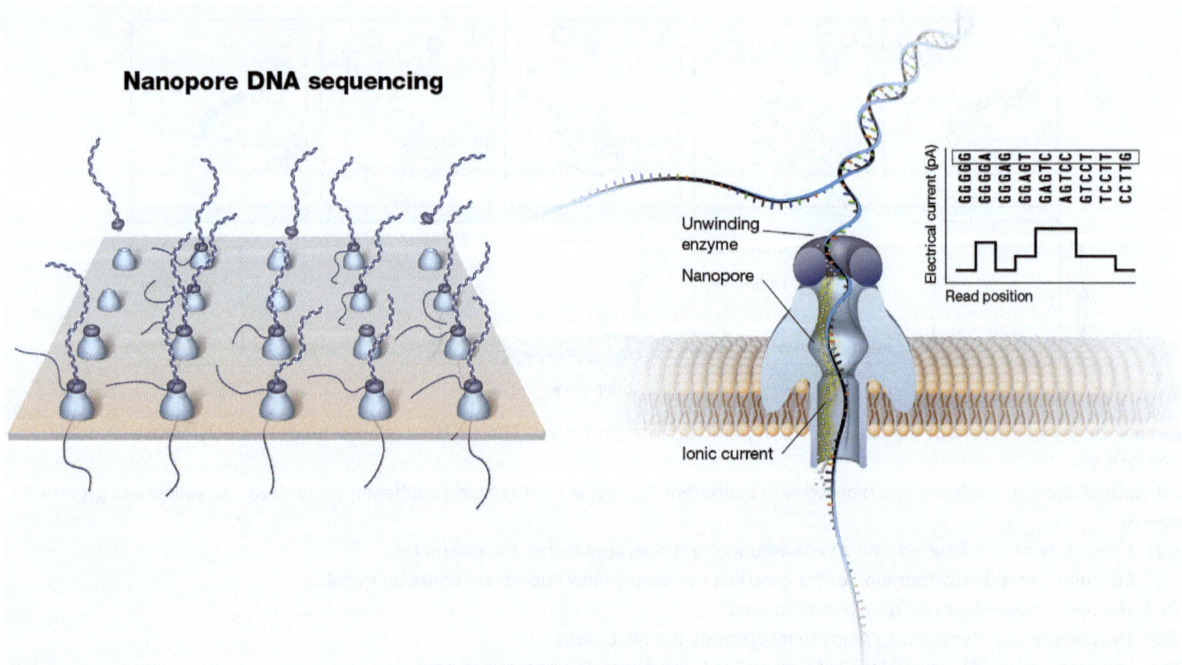

Figure 4.27 Individual DNA molecules are unwound and traverse nanopores embedded in a membrane. Passage of each base produces an individual electrical signal, which can be recorded to collect sequence information.

Nanopore DNA Sequencing: NHGRI Available at: https://www.genome.gov/genetics-glossary/Nanopore-DNA-Sequencing. Accessed February 2024.

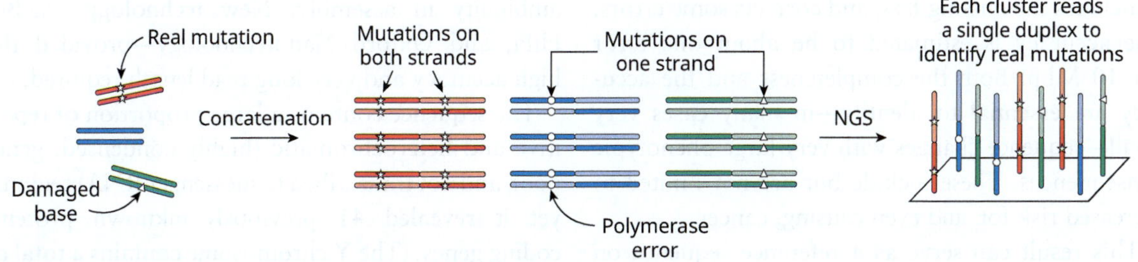

Figure 4.28 Consider two loci on the same chromosome. One locus has a genuine mutation (red strand), the other a damaged base in one of the strands (green), or is a position subject to a PCR error (blue). Using CODEC, it would be possible to distinguish these cases, as a genuine mutation would appear in the *two* corresponding positions in the concatenated strands.

Bae, J.H., Liu, R., Roberts, E., *et al.* (2023). Single duplex DNA sequencing with CODEC detects mutations with high sensitivity. Nat. Genet. 55: 871–879. https://doi.org/10.1038/s41588-023-01376-0

easily portable for field work: An example was the sequencing of Ebola virus isolates in West Africa during an outbreak, and Zika in Brazil (https://www.chronicle.com/paid-content/university-of-birmingham/devising-international-strategies).

Concatenating Original Duplex for Error Correction (CODEC)

To pick up a single site mutation requires perfect accuracy in sequencing. Suppose sequencing a region reports an isolated single-base change relative to a reference sequence. Is this a mutation in the sample, an error in PCR amplification, or a sequencing error? To help resolve this problem, CODEC links the one strand in a DNA sample with the reverse complement of the other strand, and sequences both in tandem (see Figure 4.28). Genuine mutations would show up in the *two* corresponding positions in the concatenated strands (red in Figure 4.28). Incompatibilities between the two concatenated strands would, instead, indicate a sequence change confined to one strand, or an error introduced during PCR amplification or sequencing (blue and green in Figure 4.28).

Completion of the human genome

The original Human Genome Project took two decades to complete, involved hundreds of scientists around the world, and cost over US$3 billion (more than US$1 per base pair).

After the famous transatlantic press conference on 25 June 2000 by UK Prime Minister Tony Blair and US President William Clinton, the original Human Genome Project continued for a couple of years. In September 2003, it had produced a sequence accounting for ~92% of the total. This pushed the then-available technology as far as possible. The remaining ~8% contained highly repetitive regions, at the time impossible to assemble.

A decade later, new methods have allowed completion of a human genome sequence. Notably, long read length shotgun sequencing allowed confident assembly of repetitive regions. PacBio's single-molecule multikilobase reads resolved variations and gaps in the previous sequence; Oxford Nanopore's >100 kbp 'ultralong' read-length permitted full-chromosome assembly.

The sequencing was based on DNA from a **complete hydatidiform mole.** This is a human cell that arises from an abnormal fertilization containing a paternal-only genome: An anuclear empty ovum is in most cases fertilized by a haploid 23,X sperm (monospermy); this then replicates its chromosomes to produce a 46,XX karyotype that is EVERYWHERE homozygous. It does not, however, contain a Y chromosome.

The newly reported sequence comprises complete gapless assemblies for all 23 somatic chromosomes + X.

It includes the missing 8%, and corrects some errors. The accuracy is estimated to be about one error per 10 Mbp. Both the completeness and the accuracy are essential to identify—in many cases very small—sequence changes with very large phenotypic consequences. These include but are not limited to increased risk for, and even causing, cancer.

This result can serve as a reference sequence on which studies of human variation, in health and especially in disease, can be based.

Of course, there are many human genome sequences. Small differences in DNA sequences confer many of our individual features. (Of course identical siblings initially share the same sequence.) Indeed, there is even more than one sequence per individual, as we accumulate mutations over the course of our lives, and cancers lead to proliferation of cells with significantly altered sequences. Somatic mutation in the maturation of the immune response generates very large numbers of B cells, with different sequences in the regions encoding antibodies.

The Y chromosome

Separate projects have sequenced the Y chromosome completely, and explored its variation in different populations. The chromosome sequenced contained 62 460 029 base pairs.

The Y chromosome presents difficulties in sequencing, because it contains 66% repeated sequences, including extensive palindromic regions. If the overall strategy involves sequencing fragments and piecing them together via overlaps, repeated sequences create ambiguity in assembly. New technology—PacBio HiFi, and Oxford Nanotechnology—provided the high accuracy and very long read length required.

The sequence contains a large proportion of repetitive and heterochromatic (highly condensed, gene-poor, and not transcribed to messenger RNA) regions, yet it revealed 41 previously-unknown protein-coding genes. (The Y chromosome contains a total of 106 protein-coding genes.)

A companion study reported Y chromosome variation among 43 men. Populations sampled included South America, West Africa, and East Asia, which have been historically underrepresented in genomic studies. The results revealed wide differences in copy number—for instance the TSPY gene, involved in sperm formation, appeared in 23 copies in one case and 39 copies in another. The lengthy repetitive region of the Y chromosome varies between 17.6 million and 37.2 million base pairs in different cases.

Y-chromosome sequences provide an important resource for studies of variation and relationships in human populations. Because most of the Y chromosome is unpaired, it undergoes very little recombination, relative to autosomes. It is in effect a forced haplotype. Also, the Y chromosome is inherited from father to son, in an exclusively male line. In this respect, it complements mitochondrial DNA, inherited exclusively in the female line.

The general power of sequencing technology has led to very ambitious projects, to sequence very large numbers of human genomes,[16] and comprehensive sequencing of different species.[17]

Life in the fast lane

High-coverage genome sequencing has become standard practice in research in biology and medicine. Available technology includes methods for automated sample preparation, generation and sequencing of fragment libraries, and powerful software for assembly. A reference genome, from the same or a closely related species, can simplify or even replace the assembly step. The methods can support many types of sequencing goals (see Box 4.9).

Many newly determined genomes emerge from large, specialized, national and even international institutions. However, individual research groups,

[16] Khamsi, R. (2022). A more-inclusive genome project aims to capture all of human diversity. Nature 603: 378–381.

[17] See, for example: Hotaling, S., Kelley, J.L., & Fradsen, P.B. (2021). Toward a genome sequence for every animal: Where are we now? Proc. Nat. Acad. Sci. (USA) 118: e2109019118.

BOX 4.9 Applications of high-throughput sequencing

Whole-genome sequencing—both *de novo* and resequencing (= assembling short reads by mapping them onto a reference genome). Active projects include comprehensive sequencing of genomes of many humans, and sequencing of tumours to determine sequence changes between healthy and cancer cells. Another goal is extended or even comprehensive genome sequencing of different species.

Metagenome sequencing—sequencing the genomes of the mixture of organisms occurring in natural samples, such as soil, or sea water.

Transcriptome sequencing, or *RNASeq*—sequencing the RNA in a sample, including messenger RNA, structural RNA, and other types of non-protein-coding RNAs. Classically, RNA sequencing is carried out by reverse transcribing the RNAs, and sequencing the cDNA, but direct RNA sequencing is possible. Measurement of the amounts of different RNAs can reveal transcription patterns of genes.

Small RNA sequencing—RNA sequencing of size-selected RNA transcripts; for instance, sequencing of microRNAs (miRNAs), small RNA molecules that modulate protein expression.

CHiPSeq (chromatin immunoprecipitation sequencing)—sequencing of isolated regions of DNA that interact with transcription factors.

Epigenetic or *methylation sequencing*—determination of which bases in a genome are methylated. Methylation of the 5 position of cytosine, in a CpG dinucleotide, reduces gene expression. N6-methyladenine is another widespread epigenetic signal. (Methylation of bases by chemical reagents can lead to mutation if the base-pairing is affected and the error unrepaired.)

and even beginning research students, need not be daunted by the prospect of undertaking full-genome sequencing. This is not to be dismissive of the skills and industry of the investigators, but rather to emphasize the power and widespread accessibility of the infrastructure of which they can avail themselves. The octopus genome was an example.

Suppose you discover a new animal unknown to science, and want to determine its genome sequence. Assume that you are based at a typically equipped university. What would be an estimate of the calendar time between walking into the lab leading the animal on a leash, and printing out the genome sequence?

The steps would be:

1. *DNA isolation:* From a blood sample, isolate DNA using a kit, such as one offered by ThermoFisher. Elapsed time: probably no more than 30 minutes.
2. *Library preparation:* Convert the genomic DNA to fragments of 300–500 bp, and attach adaptors. The adaptors bind the fragments to a flowcell, and contain priming sites for forward, reverse, and index reads. Elapsed time: probably no more than 3 hours.
3. *Sequence the fragment library:* an Illumina NextSeq 2000 P2 150 x 150 paired-end sequencing run would produce ~40× coverage of a 3 GB genome. Elapsed time: probably no more than 30 hours. (You would not need to be present in the lab while the instrument was running.)
4. *Assemble the sequence:* This would involve taking the reads produced in step 3 and running them through a computer program.

Conclusion: We are talking about a couple of days: It could be a weekend project, and even allow for attending a sports event Saturday afternoon.

What is the World Record? A group at Stanford Medical School recently sequenced a full human genome in 5 hours, 2 minutes![18] They used several Oxford Nanopore sequencers in parallel. The elapsed time from sample acquisition, through sequencing, to actionable interpretation for clinical application was 7 hours, 18 minutes. A patient could enter a critical-care unit in hospital in the morning, and information from his or her sequence applied to treatment by the evening.

How much sequencing power is there in the world?

One measure of overall sequencing power is the size of the databanks. The holdings of assembled and annotated sequences in The International Nucleotide Sequence Database Collaboration (INSDC;

[18] https://med.stanford.edu/news/all-news/2022/01/dna-sequencing-technique.html

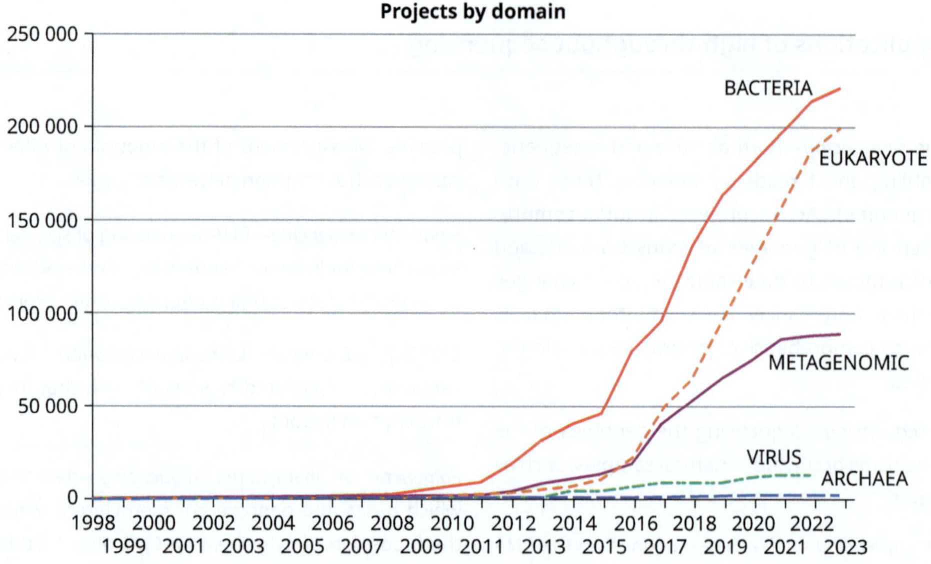

Figure 4.29 Growth in total number of projects, according to group of organism.

Project Totals in GOLD (by year and Domain Group). Supratim Mukherjee, Dimitri Stamatis, Cindy Tianqing Li, Galina Ovchinnikova, Jon Bertsch, Jagadish Chandrabose Sundaramurthi, Mahathi Kandimalla, Paul A. Nicolopoulos, Alessandro Favognano, I-Min A. Chen, Nikos C. Kyrpides and T.B.K. Reddy. Twenty-five years of Genomes OnLine Database (GOLD): data updates and new features in v.9. Nucl. Acids Res. (2022) doi: doi.org/10.1093/nar/gkac974

http://www.insdc.org) grew from 450 481 663 919 bases in 2012, to 19 004 610 381 126 in August 2022. The growth is exponential, with a current doubling time of ~18 months. The Sequence Read Archive, containing 'raw'—i.e., pre-assembly—sequence data, contains over 3.6 petabases (= 3.6×10^{15} bases).

Figure 4.29 shows the growth in sequencing projects.

Another approach is a census of the infrastructure. The Genohub Blog (https://bloggenohub.files.wordpress.com/2013/07/omicsmaps1.png) shows a distribution of major sequencing centres throughout the world. There are some very large concentrations; for instance, in China at the BGI (formerly the Beijing Genomics Institute); the Wellcome Trust Sanger Institute, in Hinxton, UK; and Washington University and Baylor College of Medicine in the US. But most universities and many hospitals have their own sequencing equipment, admittedly of lower throughput capacity. There are even portable devices for field work (as well as for the lab); for genomic surveillance of epidemics, including but not limited to COVID-19. (The challenge for field work is not only to miniaturize, but to produce a device that does not require regular calibration and adjustment by engineers.)

It is possible to integrate the data associated with the omicsmaps.com website. Stephens et al.[19] estimate total worldwide sequencing capacity in 2015 at greater than 35 petabytes per year (1 Pb = 10^{15} bytes). This is the equivalent of approximately 11 million human genomes. Predictions of the future are not trustworthy, but an extrapolation by Illumina from current trends suggests that in 2026 sequencing capacity will be 10^9 human genomes per year (a relatively conservative estimate). Whether or not this is achieved, it is unquestionable that storage and distribution of large data sets will very soon become a severe technological challenge to molecular biologists.

LOOKING BACK AND FORWARD

In the previous chapter, we discussed the history and applications of the Human Genome Project. In this chapter, we have gone into more detail, about classical genetics, the structure of DNA, and the methods for determining sequences. Technical progress has been extremely rapid, enabling many novel lines of

[19] Stevens, Z.D., *et al.* (2015). Big Data: Astronomical or genomical? PLoS Biology 13: e1002195.

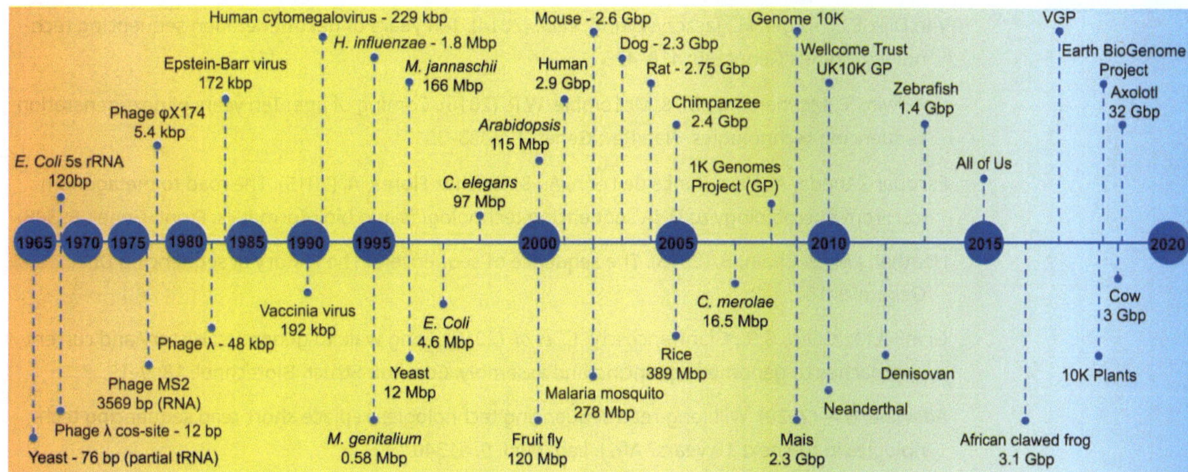

Figure 4.30 History of nucleic acid sequencing, from beginnings with small RNAs and fragments (1965), through the beginnings of DNA sequencing (φX-174, 1977) to the current large-scale full-genome projects. Colour coding of background indicates sequencing method: Light red: early sequencing methods, yellow: Sanger-based shotgun sequencing, green: next-generation sequencing, light blue: third-generation sequencing.

From: Gianni, A.M., Gallo, G.R, Gianfranceschi, L., *et al.* (2020). Long walk to genomics: History and current approaches to genome sequencing and assembly. Comput. Struct. Biotechnol. J. 18: 9–19. [open access]

research that depend on high-throughput sequencing. The 'next-generation' and 'third-generation' infrastructure is providing a deluge of data.

The data demand powerful and sophisticated computational support and synergy with the experiments.

Figure 4.30 gives a synoptic view of the entire history.

In the next chapter, we begin to treat the applications of the data. The focus is evolution, the major organizing principle of biology. Genomics reveals the relationships among species, and allows delineating the pathways that evolution has traversed. With this as background we shall then turn attention to genomes of the organisms with which we share the Earth.

RECOMMENDED READING

Some historical perspectives

Portin, P., & Wilkins, A. (2017). The evolving definition of the term 'Gene'. Genetics 205: 1353–1364.

Smith, T.F. (1990). The history of the genetic sequence databases. Genomics 6: 701–707.

Linkage disequilibrium

Neale, B.M. (2010). Introduction to linkage disequilibrium, the HapMap, and imputation. Cold Spring Harb Protoc. doi:10.1101/pdb.top74.

Slatkin, M. (2008). Linkage disequilibrium—understanding the evolutionary past and mapping the medical future. Nat. Revs. Genet. 9: 477–485.

Mardis, E. (2008). The impact of next-generation sequencing technology on genetics. Trends Genet. 24: 133–141.

The following articles describe the recent advances that have produced the high-throughput sequencing platforms on which contemporary sequencing depends.

Davies, K. (2010). *The $1000 Genome: The Revolution in DNA Sequencing and the New Era of Personalized Medicine*. New York: Free Press.

Van Dijk, E.L., Auger, H., Jaszczyszyn, Y., *et al.* (2014). Ten years of next-generation sequencing technology. Trends Genet. 30: 418–426.

Goodwin, S., McPherson, J.D., & McCombie, W.R. (2016). Coming of age: Ten years of next-generation sequencing technologies. Nat. Rev. Genet. 17: 333–351.

Escobar-Zapeda, A., Vera-Ponce de León, A., & Sanchez-Flores, A. (2015). The road to metagenomics: From microbiology to DNA sequencing technologies and bioinformatics. Front. Genet. 6: 348.

Heather, J.M., & Chain, B. (2016). The sequence of sequencers: The history of sequencing DNA. Genomics 107: 1–8.

Giani, A.M., Gallo, G.R., Gianfranceschi, L., *et al.* (2019). Long walk to genomics: History and current approaches to genome sequencing and assembly. Comput. Struct. Biotechnol. 18: 9–19.

Adewale, B.A. (2020). Will long-read sequencing technologies replace short-read sequencing technologies in the next 10 years? Afr. J. Lab. Med. 9: a1340.

Logsdon, G.A., Vollger, M.R., & Eichler, E.E. (2020). Long-read human genome sequencing and its applications. Nat. Rev. Genet. 21: 597–614.

Hu, T., Chitnis, N., Monos, D., *et al.* (2021). Next-generation sequencing technologies: An overview. Hum. Immunol. 82: 801–811.

Athanasopoulou, K., Boti, M.A., Adamopoulos, P.G., *et al.* (2021). Third-generation sequencing: The spearhead towards the radical transformation of modern genomics. Life (Basel) 12: 30.

Zhang, T., Zhou, J., Gao, W., *et al.* (2022). Complex genome assembly based on long-read sequencing. Brief Bioinform. 23: bbac305.

Au, K.F. (2022). The blooming of long-read sequencing reforms biomedical research. Genome Biology 23: 21.

Pervez, M.T., Hasnain, M.J.U., Abbas, S.H., *et al.* (2022). A comprehensive review of performance of next-generation sequencing platforms. Biomed. Res. Int. 2022: 3457806.

Lorenzo, D., Esquereda, M., Bofarull, M., *et al.* (2023). The reuse of genetic information in research and informed consent. Eur. J. Hum. Genet. 31: 1393–1397.

D'Gama, A.M., & Agrawal, P.B. (2023). Genomic medicine in neonatal care: Progress and challenges. Eur. J. Hum. Genet. 31: 1357–1363.

Warburton, P.E., & Sebra, R.P. (2023). Long-Read DNA Sequencing: Recent advances and remaining challenges. Ann. Rev. Genomics and Hum. Gen. 24: 109–132.

Satam, H., Joshi, K., Mangrolia, U., *et al.* (2023). Next-generation sequencing technology: Current trends and advancements. Biology (Basel) 12: 997.

EXERCISES, PROBLEMS, AND WEBLEMS

Exercises

Exercise 4.1 Two loci with alternative alleles *A/a* and *B/b,* respectively, are 1 cM apart. A cross between parents of genotype *AB/AB* and *ab/ab* produces a large number of offspring. Assuming no selective difference between genotypes, estimate the fraction of the next generation that has genotype *Ab/aB.*

Exercise 4.2 Gene *A* has two alleles, A_1 and A_2. Gene *B* has two alleles, B_1 and B_2. In a population, the following haplotype frequencies are observed: $A_1B_1 = 0.2, A_2B_2 = 0.45, A_1B_2 = 0.15, A_2B_1 = 0.2$. Calculate *D*, the extent of linkage disequilibrium.

Exercise 4.3 Consider two traits controlled by separate genetic loci. One locus has alleles *A* and *a*; the other has alleles *B* and *b*. *A* is dominant over *a* and *B* is dominant over *b*.

Consider two parents each heterozygous for both traits. What would be the statistical distribution of the phenotypes of offspring of these parents if (1) the loci are on different chromosomes; that is, unlinked. (2) the loci are on the same chromosome, 1 cM apart, and in each parent the alleles *A* and *B* are on the same strand, and the alleles *a* and *b* are on the other. (3) the loci are on the same chromosome, 20 cM apart, and in each parent the alleles *A* and *b* are on the same strand, and the alleles *a* and *B* are on the other.

Exercise 4.4 A fruit fly with a chromosome deletion shows pseudodominance for a trait. On a photocopy of Figure 4.4, indicate with an 'X' where the locus for this trait might be.

Exercise 4.5 The Philadelphia translocation, a swapping of material between chromosomes 9 and 22, occurs in a bone marrow cell, and causes chronic myeloid leukaemia. Would the patient transmit this leukaemia to his or her offspring?

Exercise 4.6 On a photocopy of Figure 4.6, indicate two positions in the human chromosome where one would look for genes that are linked in humans but unlinked in chimpanzees.

Exercise 4.7 (a) On a photocopy of Figure 2.4a, indicate with an 'A' the band on the gel that corresponds to the *Bam*HI fragment in Figure 2.4b which begins at 1 kbp and ends at 5 kbp. (b) On a photocopy of Figure 2.4b, indicate with a 'B' the fragment that gives rise to the band in the *Eco*RI lane of the gel which corresponds to the lowest molecular mass fragment.

Exercise 4.8 The lengths of blocks that define human haplotypes vary, partly because of the variation of recombination rates along the genome. Would you expect the haplotype blocks to vary more if their sizes are measured in terms of number of base pairs or in terms of genetic distance in centimorgans?

Exercise 4.9 Suppose that there are ten SNPs in a 10 kbp region. (a) If the region is on the Y chromosome, how many possible haplotypes are there? (b) If the region is on a diploid chromosome, how many possible haplotypes are there?

Exercise 4.10 On a photocopy of Figure 4.8, indicate, by crossing atoms out and writing atoms in, how the structure would have to be changed to illustrate an RNA molecule with the equivalent base sequence.

Exercise 4.11 From a photocopy of Figure 4.8, cut out or redraw the individual bases and show that guanine and thymine could form a non-canonical base pair containing two hydrogen bonds. By comparing with a copy of the standard base pairs, show the extent to which a guanine–thymine pair would not match the correct relative position and orientation of the sugars to be stereochemically compatible with standard base pairs in a double helix of standard structure. Uracil has the same hydrogen-bonding specificity as thymine. (Guanine–uracil 'wobble' base pairs appear in most non-messenger RNAs, and are implicated in codon–anticodon interactions between tRNA and mRNA.)

Exercise 4.12 The tetranucleotide illustrated in Figure 4.8 is self-complementary. (a) What does this mean? (b) Make two photocopies of Figure 4.8. From one of them trim off the names of the bases. From the other, cut out the individual bases and mount them adjacent to the first, in position to form Watson–Crick base pairs. Draw in the hydrogen bonds between bases.

Exercise 4.13 If all of the DNA in all of the cells of your body were laid end to end, would you be surprised if it were longer than the diameter of the solar system? Calculate the result and compare. The semi-major axis of Pluto's orbit is 5 906 376 272 km. The number of cells in an adult human body has been estimated as 10^{13}.

Exercise 4.14 In Figure 4.12, the region of the template strand not complexed with the primer is shown as continuing the helical structure. Although justifiable pedagogically for clarity, why is this not a structurally correct representation?

Exercise 4.15 On a photocopy of Figure 4.18b, indicate the longest contig available from the data given.

Exercise 4.16 In Figure 4.18b, what is the minimal coverage of any position (this is obvious)? (b) What is the maximal coverage of any position (i.e., the largest number of fragments in which the same position appears)? (c) Estimate the average coverage of the entire region. (Hint: Measure the total lengths of the fragments and divide by the length of the region.)

Exercise 4.17 The International Human Genome Mapping Consortium fingerprinted 300 000 BAC clones. Assuming an average insert size of 150 kbp and a 3.2 Gbp genome size, what coverage would be expected?

Exercise 4.18 In determining paired-end reads, using sequence methods that involve primer extension by DNA polymerase, why must the raw sequence data for the two ends come from different strands? (Of course, it is trivial to compute the reverse complement and end up with data from both ends of the same strand.)

Exercise 4.19 Here is a fragment of length 200 from a mitochondrial genome. If this fragment were processed by a sequencer giving read length 35, what would be (a) the result of a single-end read? (b) the result of a paired-end read? (See Figures 1.6 and 4.22.)

```
1 5'-gttaatgtag cttaaactaa agcaaggcac tgaaaatgcc tagatgagtc tgcctactcc
61 ataaacataa aggtttggtc ctagcctttc tattagttga cagtaaattt atacatgcaa
121 gtatctgcct cccagtgaaa tatgccctct aaatccttac cggattaaaa ggagccggta
181 tcaagctcac ctagagtagc tcatgacgcc ttgctaaacc acgcccccac gggatacagc-3'
```

Exercise 4.20 One difficulty in extracting reads that correspond to mitochondrial DNA from sequencing mixed fragments of nuclear and mitochondrial DNA is that the nuclear genome contains segments homologous to regions of the mitochondrial genome, called numts. Mammalian genomes contain 50–450 kbp of numts. (The human genome contains 1005 such segments, of average length 446 bp.) Estimate the fraction of reads from fragments of mammoth DNA that are likely to be numts. The mammoth genome is approximately 4.7×10^9 bases in length.

Exercise 4.21 (a) On a photocopy of Figure 4.21c, draw the appearance after replication of the fragment but before denaturation, and after denaturation. Remember that the links to the slide, shown in the figure as pedestals (upside-down Ts), cannot be broken by denaturation of double-helical structures. (b) Draw the next steps in bridge amplification after the frames of Figure 4.21. That is, Figure 4.21b is analogous to Figure 4.21f. Draw the analogues of Figure 4.21c and 4.21d that would follow Figure 4.21f.

Exercise 4.22 Suppose in the mapping procedure shown in Figure 4.23 two nicks appear on opposite strands quite near each other—the two seven-base sequences could even overlap. What problem for the procedure would this create?

Exercise 4.23 The equivalent length of how many human genomes was generated as raw sequence reads for the octopus genome-sequencing project?

Problems

Problem 4.1 Consider two linked traits in a population in which half of the individuals are double heterozygotes with genotype *AB/ab* and the other half are double homozygotes *(AB/AB)*. Assuming no selective advantage of any combination of alleles for these traits and no preferential mating, after recombination brings the population to equilibrium, what will be the ratio of *AB/AB, Ab/aB* = *aB/Ab*, and *ab/ab* individuals?

Problem 4.2 Consider two markers 1 cM apart. (a) What is the probability that there will be recombination between them in one generation? (b) What is the probability that there will not be recombination between them in one generation? (c) What is the formula for the probability that there will not be recombination between them in n generations? (d) Evaluating this formula, what is the probability that there will not be recombination between them in n = 10, 20, 30, 40, and 50 generations?

Problem 4.3 As a simplified but illustrative example of sequence assembly, we saw the first two verses of Richard III chopped into overlapping ten-character fragments. (a) Chop these lines into consecutive overlapping five-character fragments, and scramble these fragments into random order. Is it still possible to reconstruct the lines without ambiguity? Why is it more difficult to do so than to reconstruct the lines from ten-character fragments? (b) Try generating, and then trying to reassemble, ten and five-character fragments, presented in random order, of the lines of Polonius (ignore punctuation marks):

> . . . 'tis true, 'tis true 'tis pity,
> And pity 'tis 'tis true

or the last two of the following lines

> . . . to expostulate
> What majesty should be, what duty is,
> Why day is day, night night, and time is time,
> Were nothing but to waste night, day, and time.

In each case, is it still possible to reconstruct the lines without ambiguity?

(c) Try the same with Richard II's speech:

> Your cares set up do not pluck my cares down.
> My care is loss of care, by old care done;
> Your care is gain of care, by new care won:
> The cares I give I have, though given away;
> They tend the crown, yet still with me they stay.

Problem 4.4 Extend Problem 4.3 to simulate the effect of paired-end reads. Take any text of approximately 120 words in length (a sonnet is about the right length) and—either by hand or with a program—create fragments with a distribution of lengths distributed roughly normally around 30 ± 5 characters. Print reads of eight characters from each end. Tabulate the data from different fragments in random order. Try to reassemble the text. Study how the difficulty of the assembly depends on the read length, fragment length, and coverage.

Problem 4.5 Lander and Waterman[18] derived formulas for the expected completeness of an assembly as a function of coverage (G = genome length, N = number of reads, L = read length, $c = NL / G$ = coverage):

> probability that a base is not sequenced $= e^{-c}$
> total expected gap length $= G \times e^{-c}$
> total number of gaps $= Ne^{-c}$.

(a) What fraction of a genome could you expect to assemble from eightfold coverage? (b) What total gap length would you expect in an assembly of a 2 Mb target genome size from eightfold coverage? (c) How many gaps would you expect in an assembly of a 2 Mb target genome size from an eightfold coverage of fragments with a read length of 500? (d) You want to sequence a 4 Mb genome by the shotgun method, by assembling random fragments with read length 500. What coverage would you require, to expect no more than four gaps, assuming no complications arising from repetitive sequences or far-from-equimolar base composition?

[18] Lander, E.S., & Waterman, M.S. (1988). Genomic mapping by fingerprinting random clones: A mathematical analysis. Genomics 2: 231–239.

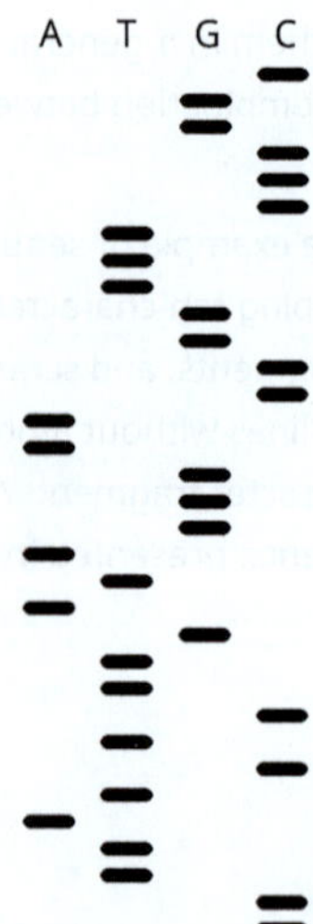

Figure 4.31 Autoradiograph of a sequencing gel (simulated). The shortest fragment travels the farthest. In this diagram, the direction of travel is down the page.

Problem 4.6 Figure 4.31 shows a sequencing gel. (a) What is the sequence of this fragment? (b) Can you see any self-complementary regions in this fragment that might form hairpin loops? (c) On the basis of your answer to part (b), would you guess that this region encodes RNA or protein?

Problem 4.7 How many water molecules are there in the PacBio observation volume of 10^{-21} litres?

Problem 4.8 Many people have asked the author whether he is related to Filippo Brunelleschi, who was born in 1377, died in 1446, and is buried beneath the cathedral in Florence (the dome of which he famously created). Assume that you were granted permission to exhume his body and collect a tissue sample. (a) Estimate the number of generations between Brunelleschi and the author. (b) Assuming that the author is a direct descendant of Brunelleschi, would you expect to be able to prove it by DNA sequencing? Explain your answer.

Problem 4.9 You are asked to devise a Master's degree programme to train annotators for databases in molecular biology. (a) What background would you require for entry to the programme? (b) What courses would you require students to take during the programme?

Weblems

Weblem 4.1 Which of the following families have genes appearing in tandem arrays in the human genome and which have genes dispersed among several chromosomes? (a) Actin, (b) tRNA, (c) all globins, (d) HOX genes, (e) the major histocompatibility complex.

Weblem 4.2 In which countries has Rosalind Franklin been honoured by appearing on postage stamps and banknotes?

Weblem 4.3 What institution currently has the highest sequencing throughput power? How many human genome equivalents per week can this institute produce?

Weblem 4.4 For the major companies providing sequencing equipment: What is the current throughput rate and typical read length of their state-of-the-art instrument?

Weblem 4.5 On a photocopy of Figure 4.17, bring the figure up-to-date.

CHAPTER 5

Evolution and Genomic Change

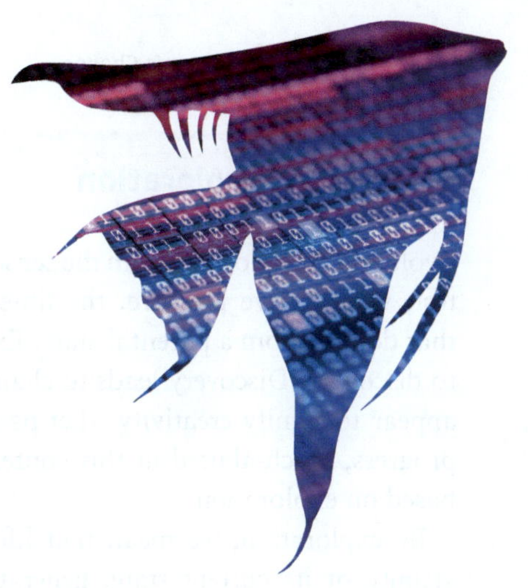

LEARNING GOALS

- *Understand the coordination of changes in genotype and phenotype during evolution.*
- *Know the principles of biological classification and the grammar of biological nomenclature.*
- *Appreciate the distinction between similarity and homology*; and accept the idea that homology, usually unobservable, is an inference from similarity.
- *Recognize that measurements of similarity between gene or protein sequences* offer our best insight into relationships between individuals and between species.
- *Understand the general idea of pattern recognition*, and be able to use tools such as the dotplot to recognize similarities among sequences.
- *Appreciate the generalization of ideas of sequence alignment to alignment of protein sequences* by structural superposition.
- *Know the relationship between divergence of sequence and divergence of structure* in protein evolution.
- *Understand the basis for classification and clustering*, and some of the techniques that can be applied.
- *Understand the basis of constructing phylogenetic trees*, methods for calculating them, and the information different types of trees contain.
- *Appreciate the relationships among amino-acid sequences, protein structures, and protein functions.*

In Chapter 1, we described some of the differences observed in comparing closely related species, and distantly related ones, at various levels from molecules to karyotypes. In this chapter, we seek to understand how these changes came about. The general answer is of course evolution. The availability of the data from genomics and related fields—such as protein structure determinations—challenges us to probe the detailed mechanism by which evolutionary changes occur. To illuminate these changes we rely on measurements of the nature and extent of the differences between related sequences, structures, and functions. Tools for analysing similarities include sequence-alignment algorithms, structural superpositions, and methods for generating phylogenetic trees. These tools are part of an essential skill set for anyone working in the field of genomics.

Evolution is exploration

Evolution is exploration—in the sense that mutations test, via selective pressure, the fitness of sequences that deviate from a parental state. Exploration leads to discovery. Discovery leads to change. Change can appear to signify creativity. (Let us avoid the word progress, much-abused in this context.) But it is all based on exploration.

By exploration, we mean that life can probe the vicinity of its current state, generating and testing variations. Evolution involves exploration and change at many levels. These include genome sequences, and phenotypic traits at the molecular and macroscopic levels. What makes biology so complex is that all these changes at different levels are intimately linked.

- The most fundamental level of exploration is *mutation of genome sequences*. It is through mutations—and, in sexually reproducing organisms, allelic reassortment—that life explores the neighbourhood of a current genotypic and phenotypic state.

A mutation can affect a transcribed molecule, changing either the amino-acid sequence of a protein, or a base in a non-protein-coding RNA. Alternatively, changes in splice sites or regulatory sequences can change protein expression levels. Other variations involving insertions and deletions, copy-number changes, and transpositions can also affect RNAs and proteins themselves, and patterns of expression levels.

A mutation that causes loss of an essential function will be lethal—a blind alley of evolutionary exploration. Conversely, certain other mutations might be expected to have little or no effect. Putatively 'silent' mutations include changes among synonymous triplets in protein-coding genes, or mutations in pseudogenes, or mutations in the large portions of some genomes currently described as junk. However, even mutations producing synonymous codons in protein-coding genes could interact with transfer RNA (tRNA) levels to affect translation rates, and thereby affect protein structure (see Chapter 12).

To allow evolution to proceed, the system is robust to some small changes, in order that large change is achievable by a succession of small steps. If the system were 'locked into' the sequence of a particular region such that any mutation would be lethal, evolution would be in handcuffs.

- Altered *proteins* explore possibilities of altered structure, including altered post-translational modification; and altered function, including changes in enzymatic activity or changes in regulatory signals. Changes in splice sites can affect exon combination patterns.

A single conservative amino acid substitution at a site on the surface of a protein distant from the active site would be expected to have only minor localized, self-contained effects on protein structure and function.

However, some mutations send ripples through the system and have far-reaching effects. Small changes in single *HOX* genes have immense leverage in creating the overall body plans of animals. For example, a major change in body plan in metazoans occurred about 400 million years ago, with the emergence of insects with six legs from arthropod ancestors with larger numbers of legs. Experiments of W. McGinnis and co-workers showed that changes in one protein, Ubx, a *HOX* homologue, are sufficient to achieve this large-scale anatomical transition.

- *At the chromosomal level*, evolution can explore distributions of genes. This can involve local or global gene duplication and transposition of either small segments of chromosomes or large-scale blocks. Degradation of synteny can lead to infertility. (Synteny = conservation of order of regions in two sets of chromosomes.) This is one of the mechanisms of speciation.
- *At the cellular level*, evolution has explored different kinds of organization, notably the prokaryote–eukaryote division. Some but not all cells have cell walls. Some, but not all, have chloroplasts. Complex organisms develop many types of specialized cells, tissues, and organs.
- *Individuals* explore different possible life histories. The context and interactions of our lives shape our development, in ways consistent with our genomes.

For humans, more than other species, cultural heritage and experience have a great effect on physical as well as on mental development. We as individuals also have a certain amount of personal control over how we explore the potential inherent in our genomes. This freedom is, of course, incomplete, for our environment as well as our genome constrains our development and activities. This environment includes the societies in which we live.

- *Within populations of individuals of the same species, evolution explores varying distributions of allele frequencies*. Natural populations show genotypic and phenotypic variation. Even in the absence of mutations, populations can 'react' to changing conditions by varying gene frequencies: Industrial melanism is a classic example (see Box 5.1).
- *At the level of body plan*, a visit to a zoo or botanical garden, or suitable web browsing, reveals life's stunning variety. Comparative anatomy reveals the underlying similarities among different animals and plants, but also the different design

BOX 5.1 Industrial melanism and its reversal

One variety of the British pepper moth, *Biston betularia* (variety *typica*), has a mottled black-and-white colouring. The moths are nocturnal; during the day, they roost on tree trunks. Before the rise of industrial pollution, light-coloured lichens encrusted the trees and the moths were protected from birds by camouflage.

Another variety, *carbonaria*, has a uniformly dark colour. It was first observed, as a mutant, in the mid-nineteenth century, near Manchester in the north of England. Historical collections show the rarity of the dark variety at that time. The difference between the two varieties is mediated by a single gene that controls the amount of the black pigment, melanin, produced.

As the Industrial Revolution advanced, soot killed the lichens and blackened the bark of the trees. Against the darkened trees, the dark moths were better camouflaged. Within a century, the population had become 90% variety *carbonaria*. Figure 5.1 shows, quite convincingly, the differences in appearance of both trees and moths. The difference is a shift within the population of the allelic frequency distribution of a single gene.

Let it not be said that England has not taken steps to curb air pollution. Coal fires were banned in London by Edward I in 1273, albeit only temporarily. Parliament followed up with the Clean Air Act of 1956. At present, *B. betularia* populations in areas recovering from soot deposits are shifting *back* to higher proportions of the mottled *typica* variety.

(a)

(b)

Figure 5.1 Industrial melanism. Left: light- and dark-coloured pepper moths (*Biston betularia*) on normal trees, with lichens growing on the bark. Right: light- and dark-coloured pepper moths on lichen-free trees encrusted with soot. Each picture contains two moths, one camouflaged and the other easily visible. Can you spot the camouflaged moths?

Reproduced with permission; left from: Kettlewell, H.B.D. (1956). Further selection experiments on industrial melanism in the *Lepidoptera*. Heredity 10: 287–301; right from: Science Photo Library.

solutions for structural, locomotory, and sensory systems between vertebrates and invertebrates.

- *At the level of ecosystems*, different populations explore their modes of interaction. Many pairs, or groups, of species co-evolve. Some examples of co-evolution of species are known from *cooperating species*; for instance, the correlation between the anatomy of flowers and their insect pollinators—a subject studied by Darwin himself—or the correlation between colour changes in fruit ripening and the development of colour vision in animals. Other examples of co-evolution involve *species in competition or conflict*; for example, predator–prey relationships. These include the war between humans and pathogenic bacteria and viruses.

The mechanism of evolutionary change is now understood, in general terms. Genetic reassortment and mutation generate inheritable phenotypic variation. Phenotype-dependent differential rates of reproduction—that is, natural selection—govern which alleles, at which frequencies, are passed on to succeeding generations. Alternatively, even in the absence of selection, genetic drift can lead to alterations in genome contents and distributions in populations.

The two elements of exploration—variation from the current state of the system and change to a new one—occur at many levels, from individual genomes to proteins to cells to ecosystems. A long-standing challenge of biology has been to understand the relationships among these different levels of evolution.

Biological systematics

Classically, the unit of large-scale evolution is the species. Species represent nature's experiments in structure, metabolism, and lifestyle. Both Linnaeus and Darwin recognized the importance of species, and made them the focus of their work. The concept of species remains essential, despite its attendant difficulties (see Chapter 1). In view of the theoretical problems, biologists present the analysis of known species, and higher order taxa, in terms based on tradition and convention. We shall explore how modern analytic methods based on genomic and structural data mesh with the classical approaches.

Study of the vast variety of living organisms requires that we organize what we observe and measure. We have to agree on what we call things. Biological taxonomy encompasses identifying different life forms, deciding where they fit in, and assigning them a name—based on some 'real or fancied characteristic of the form described' (A.S. Romer), and equipped with a proper description and deposition of specimen material.

Biological nomenclature

Two problems in organizing biological nomenclature are what must be named and how to assign the names. The taxonomic hierarchy introduced by Linnaeus—kingdom, phylum, class, order, family, genus, and species—is still in use, although with modifications. (Another legacy is the continued use of classical languages.) Members of more restrictive categories, for instance, several species in the same genus, have more shared features and higher degrees of similarity, than the members of more inclusive categories, such as phylum or class. As Nature is not as 'neat' as the eighteenth-century scientists would have liked, boundaries between taxa are fuzzy. Other levels have been interpolated into the classical hierarchy, such as subspecies or superfamilies.

To identify a species, it usually suffices to specify the lowest two levels of the hierarchy, as a *binomial*: genus and species. For instance, *Homo sapiens, Bos taurus, Drosophila melanogaster, Solanum lycopersicum*. Each binomial uniquely identifies a species that may also be known by one or more common names;

for instance, human, cow, fruit fly, tomato. Conversely, many common names refer to whole groups of species. For example, there are many species of whales, not all in the same genus or even the same family. Of course, most species do not have common names at all.

Classifications of human, cow, fruit fly, and tomato

	Human	Cow	Fruit fly	Tomato
Kingdom	Animalia	Animalia	Animalia	Plantae
Phylum	Chordata	Chordata	Arthropoda	Spermatophyta
Class	Mammalia	Mammalia	Insecta	Magnoliopsida
Order	Primata	Artiodactyla	Diptera	Solanales
Family	Hominidae	Bovidae	Drosophilidae	Solanaceae
Genus	*Homo*	*Bos*	*Drosophila*	*Solanum*
Species	*sapiens*	*taurus*	*melanogaster*	*lycopersicum*

Common names can result from, and cause, confusion. Settlers on new continents often applied common names to animals similar in appearance to familiar ones, but not in fact closely related to them. For instance, early Europeans in Australia called a native animal the koala bear, although marsupial koalas are not closely related to European bears.

Discoveries of new species are more common than discoveries of new genera, families, etc. During the years 1970–1998, five times as many new species and subspecies were described than new genera. New families, orders, classes, and phyla appear much more infrequently (Table 5.1).

The Earth no longer contains vast unexplored territories that supported the cornucopia of novelties that emerged during the classic era of exploration in the eighteenth and nineteenth centuries. However, it remains true that the species described by scientists are a small selection from all those that exist. Approximately 18 000 new species are discovered and named each year. Nor are these all merely minor variations on well-known themes. The echinoderm *Yorkicystis haefneri*, discovered in 2022, is a new genus. A new primate species was reported in 2020, the Popa langur *(Trachypithecus popa)* from Myanmar.

Table 5.1 New taxa described from 1970 to 1998

Taxa	Numbers described
New phyla	11
New classes	44
New orders	100
New families	731
New genera	8 579
New species	38 590
New subspecies	3231

From: Winston, J.E., & Metzger, K. (1998). Trends in taxonomy revealed by published literature. BioScience 48: 125–128.

Biological names usually describe features of a species; e.g., giant kangaroo, *Macropus giganteus*, which means 'gigantic large-foot' (not to be confused with the 'bigfoot' primate alleged to inhabit the northwest US and adjacent regions of Canada). Names may indicate location (e.g., Virginia opossum, *Didelphis virginiana*), or recognize the discoverer (e.g., Darwin's rhea, *Rhea darwinii*, a large bird encountered by Darwin on his visit to South America. J. Gould named the species in Darwin's honour in 1837. P.H.G. Mohring had named the genus in 1752, to reflect the large size of the birds: the Greek goddess Rhea, mother of Zeus, was a female titan, member of a mythological race of giants.) The scientific name of Père David's deer, *Elaphurus davidianus*, is another example of an animal named after its discoverer.

Other sources of names include expedition sponsors, thesis supervisors, public figures such as kings and queens, politicians, artists, musicians, sporting figures, and fictional characters, e.g., *Phytotelmatrichis osopaddington* (osopaddington is Latin for Paddington Bear, a nod to Peruvian origin; but the species is actually a beetle, not a bear). And even rude slang.

A marine mollusc, *Rotaovula hirohitoi*, was named for the former Emperor of Japan, who was himself a serious marine biologist. Some scientists have named creatures with loathsome features after rivals, as insults. Finally, J.E. Winston has written: 'These days, it would be considered pretty tacky . . . to name a species after yourself'.[1] For unusual and in some cases amusing examples, see (among other places): https://www.fnps.org/assets/pdf/palmetto/v13i4p8austin.pdfhttps://www.australiangeographic.com.au/topics/science-environment/2014/03/weirdest-species-names/, and www.faculty.ucr.edu/~heraty/menke.html.

Biological nomenclature is governed by international agreements, adopted by consent among professional scientists. The International Codes of Zoological and Botanical Nomenclature separately offer rules for the naming of animals and plants. Nomenclature of bacteria grew out of, and eventually split off from, the botanical code. Virologists have developed their own classification.

Currently the International Union of Biological Sciences, a member of the International Council of Scientific Unions, concerns itself with biological nomenclature. It is a sponsor of the Species 2000 project, an effort to curate a complete and integrated database of the world's species, including plants, animals, fungi, and microbes. The Species 2000 project coordinates its activities with other similar international efforts, including the Interagency Taxonomic Information System (ITIS) and the Global Biodiversity Information Facility (GBIF). The related projects of the Tree of Life (http://tolweb.org/tree/) and ARKive (http://www.arkive.org/) include pictorial databases. The Catalogue of Life (COL) has the goal of maintaining a comprehensive list of recognized species: https://www.catalogueoflife.org/

The World Conservation Union, usually known by its former name, the International Union for the Conservation of Nature and Natural Resources (IUCN), maintains a 'Red list' of endangered species.

Measurement of biological similarities and differences

Ultimately, comparisons in biology involve observations of differences between individual organisms. With access to living populations, it is possible to get a sense of the variability among individuals and to observe a variety of features, including physiology and lifestyle, and trajectories of development, in addition to 'static' anatomy.

In contrast, for extinct species, palaeontologists are often limited to fragmentary samples of hard parts—bones and teeth—sometimes from a single individual. Indeed, in the classical era of natural history exploration, a museum in Europe would often receive only a preserved body, even from species that still thrived elsewhere in the world. (Père David's deer is a typical example.) Despite these handicaps in data collection, biologists in the pre-molecular era built their taxonomic edifice on studies of comparative anatomy, embryology, and stratigraphy for dates.

Understanding biological diversity requires observation and measurement of similarities and differences. What features should one compare? W.E. Le Gros Clark wrote,

> While it may be broadly accepted that, as a general proposition, degrees of genetic relationship can be assessed by noting degrees of resemblance in anatomical details, it needs to be emphasized that morphological characters vary considerably in their significance for this assessment. Consequently it is of the utmost importance that particular attention should be given to those characters whose taxonomic relevance has been duly established by comparative anatomical and palaeontological studies.
>
> —Le Gros Clark, W.E. (1971). *The Antecedents of Man*, 3rd edn. Chicago: Quadrangle Books, pp. 11–12.

This approach works fine in the hands of a professional with expertise and as distinguished as Le Gros Clark, but it has also elicited attempts to make classification methods more quantitative and objective. These attempts include the development

[1] Winston, J.E. (1999). *Describing Species*. New York: Columbia University Press.

of computational methods for interpreting similarities of a wide spectrum of features, some but not all based on sequence data.

Molecular techniques

Many molecular properties have been used for phylogenetic studies, some surprisingly long ago. Serological cross-reactivity was applied to detect relationships from the beginning of the last century until superseded by the direct use of sequences. E.T. Reichert and A.P. Brown published, over a century ago (in 1909), a phylogenetic analysis of fishes based on haemoglobin crystals. Their work was based on Stenö's law (1669), which states that although different crystals of the same substance have different dimensions—some are big, some small—they have the same interfacial angles. We now understand that this law reflects the similarity in microscopic arrangement and packing of the atomic or molecular units within the crystals. Reichert and Brown showed that the interfacial angles of crystals of haemoglobins isolated from different species showed patterns of similarity and divergence parallel to the species' taxonomic relationships.

Reichert and Brown's results are replete with significant implications, which can be appreciated only in retrospect. They demonstrate that proteins can adopt definite, fixed shapes, an idea by no means recognized at the time. They imply that, as species progressively diverge, the structures of their haemoglobins progressively diverge also. In 1909, no one had a clue about nucleic acid or protein sequences. In principle, therefore, the recognition of evolution of protein structures preceded, by half a century, the idea of evolution of nucleotide and amino-acid sequences. Reichert and Brown even saw a structural difference between oxy- and deoxyhaemoglobin.

The Reichert and Brown work has my nomination for the most premature scientific result ever.

Today, DNA sequences provide the best measures of similarities among species for phylogenetic analysis. Many genes are available for comparison. Genes vary widely in their rates of change. This is fortunate, because, given a set of species to be studied, it is necessary to find genes that vary at an appropriate rate. Some genes remain almost constant among the species of interest. These provide no discrimination. Other genes that vary too rapidly cannot be aligned. (The situation is somewhat similar to dating by radioactive decay. It is necessary to choose an isotope with a half-life of the same order of magnitude as the age of the material.)

The mammalian mitochondrial genome, a circular, double-stranded DNA molecule typically ~16 568 bp long, provides a useful fast-changing set of sequences for the study of evolution among closely related species. In contrast, slowly changing ribosomal RNA (rRNA) sequences were used by C. Woese to identify the three major divisions of life: archaea, bacteria, and eukarya (see Figure 1.14).

The idea that we can choose any gene that varies at the desired rate assumes that different genes would give similar phylogenetic relationships. This is often true, but not always.

KEY POINT

In order to develop a clear picture of the relationships between species, it is necessary to pick a molecule that is changing at a reasonable rate. There must be enough change such that the signal does not sink below the noise level, but not too much change as to obscure common features.

Homologues and families

Products of evolution retain similarities. The similarities appear at many levels—related people, recently diverged species, different tissues within an organism containing related cell types but varying protein expression patterns, amino-acid sequences and structures of proteins, DNA sequences. A major theme of biology has traditionally been to recognize and classify such similarities, with a view to understanding how they arose and, when appropriate, to what purpose (i.e., with what selective advantage).

However, it is necessary to keep clearly in mind that similarity, which is observable, is a surrogate for relationship, which usually is not. Related biological objects are homologues, or families. In many cases, such as mammalian haemoglobins, the similarities are sufficient to give us confidence that we are analysing a family of related—i.e., homologous—molecules. Ideally, we have a spectrum of similarities, including some close relatives and some distant ones, with the distant relatives linked by chains of close ones. For the comparisons of haemoglobins from different species, the congruence of the degrees of similarity of the molecules, with measurements of similarities of other sets of molecules, and with the classical taxonomic relationships among species, is reassuring. For globins in the human genome, the common conserved features argue that we are dealing with a single diverged family.

But divergence does not stop within the scope of our ability to detect homologues. There are many cases where (a) a tantalizing tenuous degree of similarity suggests homology, but we remain unsure whether or not the inference of relationship is valid, or (b) there is sufficient dissimilarity between two molecules or structures that homology is unsuspected, but a series of missing links clearly connects the two.

Our most precise tools measure similarity between sequences or between molecular structures. These tools are mature methods. They have been calibrated to allow us to decide, in all but the hardest cases, whether or not we are dealing with homologues.

Pattern matching—the basic tool of bioinformatics

Given suitable data, computer programs can measure similarities and extract common patterns. Programs to extract patterns in sequences are powerful and readily available. Indeed, sequence comparisons are a problem common to many fields, including the text editors available in all computer systems. For protein structures also, it is possible to detect and measure similarities and common patterns. This makes it possible to study sequence–structure relationships quantitatively.

Other types of biological information—such as protein function, expression patterns, metabolic pathways, information about characteristics that distinguish species—do not present themselves quite so naturally in forms adapted for computational analysis. They require foundational work to create models of the information, including identification of the important categories of data, and controlled and carefully defined vocabularies for their description. The Gene Ontology Consortium's classifications of protein functions allowed development of tools for quantitative measurement of similarity and divergence of function (see Chapter 13).

Rules and regulations stipulating how to express data are also essential for database integration. They provide the basis on which independent databases in related or overlapping fields can communicate and cooperate with one another, and reconcile differences. They allow writing information-retrieval software to handle queries requiring coordinated access to several databases.

Sequence alignment

Given two or more sequences, we wish to:

- measure their similarity;
- understand how the residues match up;
- observe patterns of conservation and variability; and
- infer evolutionary relationships.

If we can do this, we will be in a good position to go fishing in databanks for related sequences, and measuring relative degrees of similarity among genes or proteins.

Sequence alignment is the identification of residue–residue correspondences. Any assignment of correspondences that preserves the *order* of the residues within the sequences is an alignment. Alignments may contain gaps. For example,

```
Given two text strings: first string  a b c d e
                        second string a c d e f
a reasonable alignment would be:      a b c d e -
                                      a - c d e f
```

Some alignments are better than others. For the sequences gctgaacg and ctataatc:

```
An uninformative
  alignment:   - - - - - - - g c t g a a c g
               c t a t a a t c - - - - - - -
An alignment
  without gaps:  g c t g a a c g
                 c t a t a a t c
An alignment
  with gaps:   g c t g a - a - - c g
               - - c t - a t a a t c
And another:   g c t g - a a - c g
               - c t a t a a t c -
```

Most readers would consider the last of these alignments of gctgaacg and ctataatc the best of the four. To decide whether it is the best of *all* possibilities, we need a way of examining all possible alignments systematically. We need to compute a score reflecting the quality of each possible alignment, and to identify an alignment with the optimal score. The optimal alignment may not be unique: several different alignments may give the same best score. Moreover, even minor variations in the scoring scheme may change the ranking of alignments, causing a different one to emerge as the best.

KEY POINT

To measure similarity between two sequences, find their optimal alignment—the best matching up of the individual characters—and produce a cumulative score of the similarities between the characters at each position.

The dot plot

The **dot plot** is a simple picture that gives an overview of pairwise sequence similarity. Less obvious is its close relationship to alignments.

The dot plot is a table or matrix. The rows correspond to the residues of one sequence and the columns to the residues of the other sequence. In its simplest form, the positions in the dot plot are left blank if the residues are different and filled if they match. Stretches of similar residues show up as diagonals in the upper left–lower right (northwest–southeast) direction (see Figure 5.2).

Dot plots give quick pictorial statements of the relationship between two sequences. Obvious features of similarity stand out. Figure 5.3 shows a

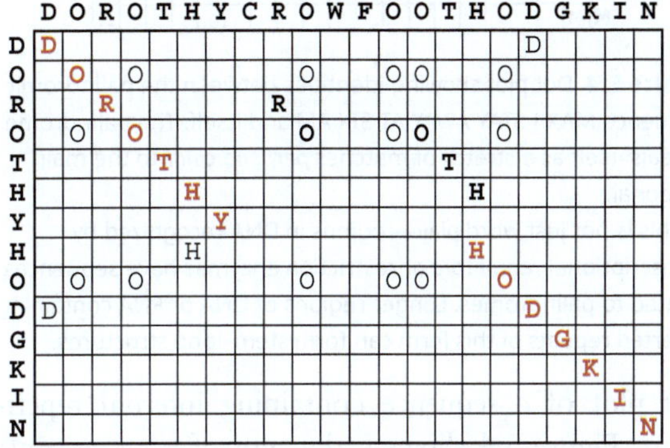

Figure 5.2 Dot plot showing identities between the short name (DOROTHYHODGKIN) and full name (DOROTHYCROWFOOT HODGKIN) of a famous protein crystallographer.

Letters corresponding to *isolated* matches are shown in non-bold type. The longest matching regions, shown in red, are the first and last names DOROTHY and HODGKIN. Shorter matching regions, such as the OTH of dorOTHy and crowfoOTHodgkin, or the RO of doROthy and cROwfoot, are noise. Note the effect of the 'insertion' of Crowfoot in interrupting and displacing the matching.

	A	B	R	A	C	A	D	A	B	R	A	C	A	D	A	B	R	A
A	A			A		A		A			A		A		A			A
B		B							B							B		
R			R							R							R	
A	A			A		A		A			A		A		A			A
C					C							C						
A	A			A		A		A			A		A		A			A
D							D							D				
A	A			A		A		A			A		A		A			A
B		B							B							B		
R			R							R							R	
A	A			A		A		A			A		A		A			A
C					C							C						
A	A			A		A		A			A		A		A			A
D							D							D				
A	A			A		A		A			A		A		A			A
B		B							B							B		
R			R							R							R	
A	A			A		A		A			A		A		A			A

Figure 5.3 Dot plot showing identities between a repetitive sequence (ABRACADABRACADABRA) and itself. The repeats appear on several subsidiary diagonals parallel to the main diagonal.

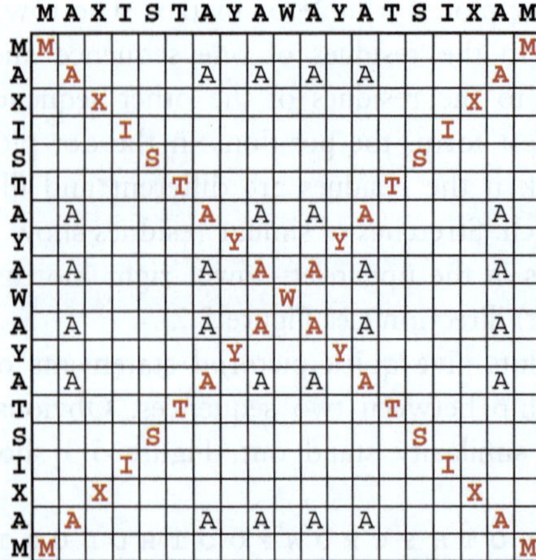

Figure 5.4 Dot plot showing identities between the palindromic sequence MAX I STAY AWAY AT SIX AM and itself. The palindrome reveals itself as a stretch of matches *perpendicular* to the main diagonal.

This is not just word play—regions in DNA recognized by transcriptional regulators or restriction enzymes have sequences related to palindromes. Longer regions of DNA or RNA containing inverted repeats of this form can form stem-loop structures.

dot plot of a sequence containing internal repetitions. Figure 5.4 shows a dot plot of a **palindrome** (a sequence that is identical to its reversal).

A dot plot relating real amino-acid sequences—human and *Xenopus laevis* ephrin B3—shows that the similarity is stronger in the N-terminal part of the protein. Figure 5.5 shows the dot plot and the corresponding sequence alignment. It is useful to look at these together and see how the regions of high and low similarity appear in the two figures.

KEY POINT

Ephrins are proteins that guide axons in the developing nervous system, and play a number of other important roles in development.

A disadvantage of the dot plot is that its 'reach' into the realm of distantly related sequences is poor. In analysing sequences, one should always look at a dot plot to be sure of not missing anything obvious, but be prepared to apply more subtle tools.

Dot plots and alignments

How can we derive an optimal alignment of two sequences? Conceptually, any alignment—that is, any assignment of residue–residue correspondences—is equivalent to a path through a dot plot. A diagonal move corresponds to an equivalence between two residues. Horizontal and vertical moves correspond to insertions and deletions. Because any allowable alignment assigns residues uniquely, and in order along the sequences, the *only* allowable moves are southeast (diagonal), east (horizontal), and south (vertical).

Figure 5.6 shows the optimal path through the Dorothy Hodgkin dot plot. The path passes through the largest number of matching residues. The horizontal segment of the path corresponds to the insertion of Crowfoot.

In this example, the optimal alignment and optimal path are obvious. In general, a computer program must examine all possibilities. How to do that effectively is a matter of some delicacy. Algorithms for relating *locally* optimal moves to integrated optimal pathways—that is, for constructing full alignments—depend on a mathematical technique called **dynamic programming**. (For details, see Lesk, A.M. (2019). *Introduction to Bioinformatics*, 5th edn. Oxford: Oxford University Press.)

KEY POINT

A dotplot shows perspicuously the quality and distribution of the pattern of similarity between two sequences. Each possible alignment of the two sequences corresponds to a path through the dotplot, from upper left to lower right.

Defining the optimum alignment

To go beyond 'alignment by eyeball' via dot plots, we must define quantitative measures of sequence similarity and difference.

Given two character strings, two measures of the distance between them are as follows:

- the **Hamming** distance, defined between two strings of equal length, is the number of positions with mismatching characters.
- the **Levenshtein**, or edit, distance between two strings of not necessarily equal length is the minimal number of 'edit operations' required to change one string into the other, where an edit operation is a deletion, insertion, or alteration of a single character in either sequence.

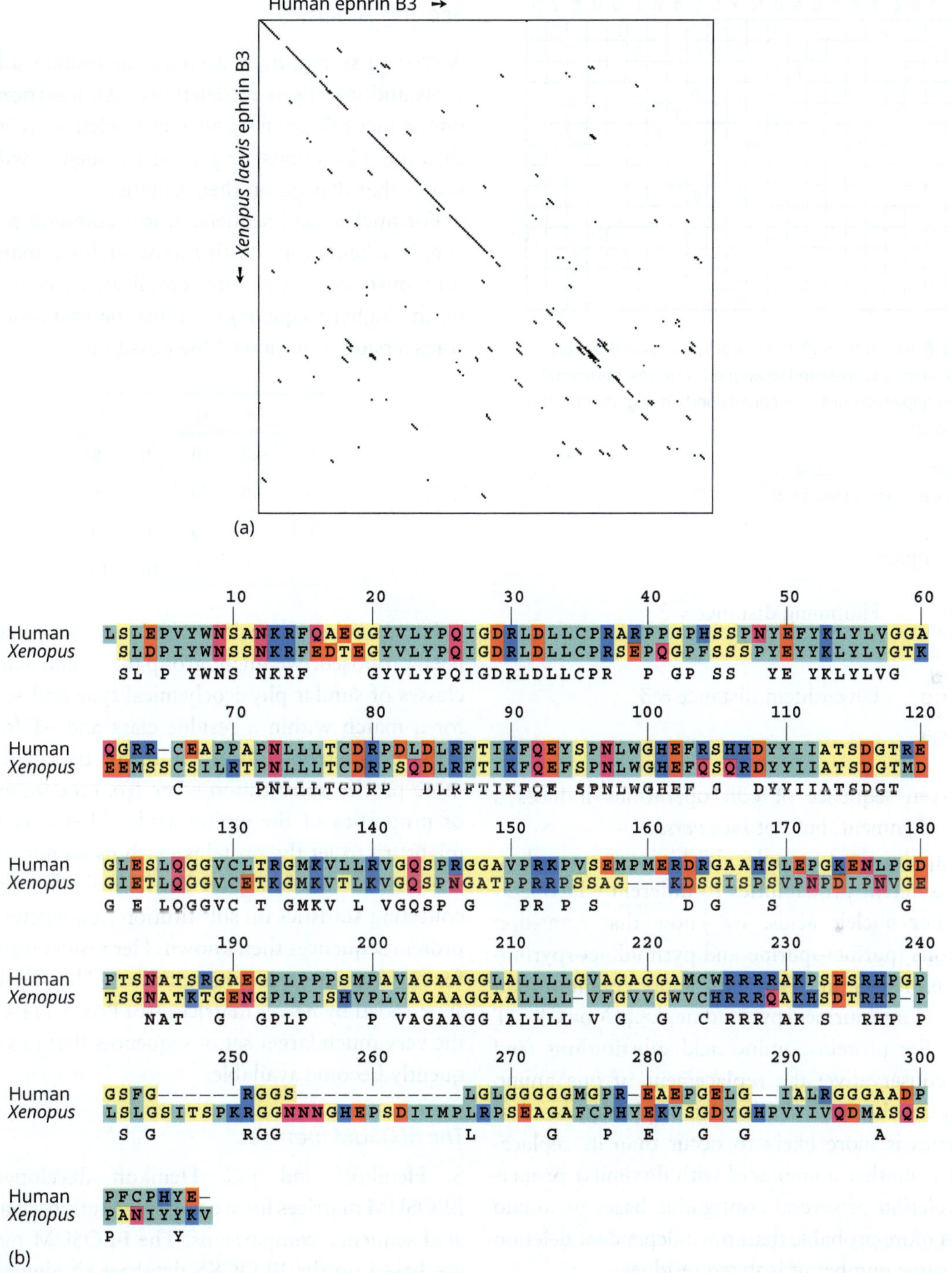

Figure 5.5 Relationships between the sequences of ephrin B3 proteins from human and *Xenopus laevis*. (a) Dot plot. The major signal is along the main diagonal, interrupted by occasional divergent regions, and showing the substantially weaker similarity near the C terminus. (b) Sequence alignment. Amino acids are colour coded by physicochemical type. Letters under the sequences indicate positions occupied by the same residue in both sequences. The sequence similarity is stronger at the N-terminus than at the C-terminus.

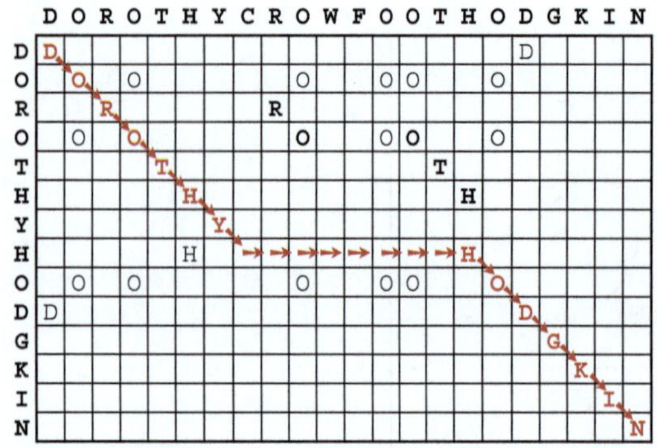

Figure 5.6 A path through the Dorothy Hodgkin dot plot. Diagonal arrows correspond to aligned residues, horizontal arrows to gap insertions. The corresponding alignment is the obvious one:

```
DOROTHY--------HODGKIN
DOROTHYCROWFOOTHODGKIN
```

For example:

`agtc` `cgta`	Hamming distance = 2
`ag-tcc` `cgctca`	Levenshtein distance = 3

A given sequence of edit operations induces a unique alignment, but not *vice versa.*

For applications to molecular biology, we wish to assign different probabilities to different edit operations. For nucleic acids, we know that transition mutations (purine↔purine and pyrimidine↔pyrimidine; i.e., A↔G and T↔C) are more common than transversions (purine↔pyrimidine; i.e., (A or G)↔(T or C)). For proteins, amino acid substitutions tend to be conservative: the replacement of one amino acid by another with similar size or physicochemical properties is more likely to occur than its replacement by another amino acid with dissimilar properties. Deletion of several contiguous bases or amino acids is more probable than the independent deletion of the same number of isolated residues.

These relative probabilities allow assigning scores to edit operations. A computer program can score each path through the dot plot by adding up the scores of the individual steps. For each substitution, it adds the score of the mutation, depending on the pair of residues involved. For horizontal and vertical moves, it adds a suitable gap penalty.

Scoring schemes

A scoring system must account for residue substitutions and insertions or deletions. (An insertion from one sequence's point of view is a deletion as seen by the other.) Deletions, or gaps in a sequence, will have scores that depend on their lengths.

For nucleic-acid sequences, it is common to use a simple scheme for substitutions, +1 for a match, –1 for a mismatch, or a more complicated scheme based on the higher frequency of transition mutations than transversion mutations. One possibility is:

	A	G	T	C
A	20	10	5	5
G	10	20	5	5
T	5	5	20	10
C	5	5	10	20

For proteins, a variety of scoring schemes have been proposed. We might group the amino acids into classes of similar physicochemical type and score +1 for a match within a residue class and –1 for residues in different classes. We might try to devise a more precise substitution score from a combination of properties of the amino acids. Alternatively, we might try to let the proteins teach us an appropriate scoring scheme. Margaret Dayhoff did this first, by collecting statistics on substitution frequencies in the protein sequences then known. Her results were used for many years to score alignments. They have been superseded by newer matrices (see Box 5.2) based on the very much larger set of sequences that has subsequently become available.

The BLOSUM matrices

S. Henikoff and J.G. Henikoff developed the BLOSUM matrices for scoring substitutions in amino-acid sequence comparisons. The BLOSUM matrices are based on the BLOCKS database of aligned protein sequences; hence the name: BLOcks SUbstitution Matrix. From regions of closely related proteins alignable without gaps, Henikoff and Henikoff calculated the ratio of the number of observed pairs of amino acids at any position to the number of pairs expected from the overall amino acid frequencies. In order to avoid overweighting closely related sequences, the Henikoffs replaced groups of proteins that had sequence identities higher than a threshold by either a single representative or a weighted average. The threshold of 62% similarity produces the commonly used BLOSUM62 substitution matrix. This is offered by all programs as an option and is the default in most.

The BLOSUM62 matrix is shown in Box 5.2. It expresses scores as *log-odds* values:

$$\text{Score of mutation } i \leftrightarrow j = \log_{10} \frac{\text{observed } i \leftrightarrow j \text{ mutation rate}}{\text{mutation rate expected from amino acid frequencies}}$$

The numbers are multiplied by 10, to avoid decimal points.

The matrix entries reflect the probabilities of mutational events. A value of +2 (e.g., leucine ↔ isoleucine) implies that in related sequences the mutation would be expected to occur 1.6 times more frequently than random. The calculation is as follows: the matrix entry 2 corresponds to the actual value 0.2 because of the scaling. The value 0.2 is $\log_{10}$ of the relative expectation value of the mutation. As $\log_{10}(1.6) = 0.2$, the expectation value is 1.6.

BOX 5.2

The BLOSUM62 matrix used for scoring amino-acid sequence similarity

Rows and columns are in alphabetical order of the three-letter amino acid names. Only the lower triangle of the matrix is shown, as the substitution probabilities are taken as symmetric (not because we are sure that the rate of any substitution is the same as the rate of its reverse, but because it is difficult to determine the differences between the two rates).

	A	R	N	D	C	Q	E	G	H	I	L	K	M	F	P	S	T	W	Y	V
Ala (A)	4																			
Arg (R)	−1	5																		
Asn (N)	−2	0	6																	
Asp (D)	−2	−2	1	6																
Cys (C)	0	−3	−3	−3	9															
Gln (Q)	−1	1	0	0	−3	5														
Glu (E)	−1	0	0	2	−4	2	5													
Gly (G)	0	−2	0	−1	−3	−2	−2	6												
His (H)	−2	0	1	−1	−3	0	0	−2	8											
Ile (I)	−1	−3	−3	−3	−1	−3	−3	−4	−3	4										
Leu (L)	−1	−2	−3	−4	−1	−2	−3	−4	−3	2	4									
Lys (K)	−1	2	0	−1	−3	1	1	−2	−1	−3	−2	5								
Met (M)	−1	−1	−2	−3	−1	0	−2	−3	−2	1	2	−1	5							
Phe (F)	−2	−3	−3	−3	−2	−3	−3	−3	−1	0	0	−3	0	6						
Pro (P)	−1	−2	−2	−1	−3	−1	−1	−2	−2	−3	−3	−1	−2	−4	7					
Ser (S)	1	−1	1	0	−1	0	0	0	−1	−2	−2	0	−1	−2	−1	4				
Thr (T)	0	−1	0	−1	−1	−1	−1	−2	−2	−1	−1	−1	−1	−2	−1	1	5			
Trp (W)	−3	−3	−4	−4	−2	−2	−3	−2	−2	−3	−2	−3	−1	1	−4	−3	−2	11		
Tyr (Y)	−2	−2	−2	−3	−2	−1	−2	−3	2	−1	−1	−2	−1	3	−3	−2	−2	2	7	
Val (V)	0	−3	−3	−3	−1	−2	−2	−3	−3	3	1	−2	1	−1	−2	−2	0	−3	−1	4

The probability of two independent mutational events is the product of their probabilities. By using logarithms, we have scores that we can add up rather than multiply, a computational convenience.

Scoring insertions and deletions, or 'gap weighting'

To form a complete scoring scheme for alignments, we need, in addition to the substitution matrix, a way of scoring gaps. How important are insertions and deletions, relative to substitutions? We need to distinguish gap initiation:

```
aaagaaa
aaa-aaa
```

from gap extension:

```
aaaggggaaa
aaa----aaa
```

For aligning DNA sequences, the popular alignment software package **CLUSTAL Omega** recommends use of the identity matrix for substitution (+1 for a match, 0 for a mismatch) and gap penalties of 10 for gap initiation and 0.1 for gap extension by one residue. For aligning protein sequences, the recommendations are to use the BLOSUM62 matrix for substitutions, with gap penalties of 11 for gap initiation and 1 for gap extension by one residue.

KEY POINT

To define optimal alignment, we must assign scores for each possible substitution, and for gap initiation and extension.

Varieties and extensions

Global alignment takes into account *all* residues in both sequences. If one sequence is shorter than the other, the difference in length must be made up by insertions/deletions.

Local alignment is a pattern-matching technique for identifying a match for a short probe sequence within a much longer text. Gaps outside the local match are not penalized. This is a common task in text searching and editing. Finding all instances of the word 'dream' in the text of Hamlet is an example of local pattern matching. (Allowing a mismatch and a gap would pick up the word 'drum' as well as 'dream'.) If you consider the DNA sequence of an entire human chromosome as a long string of characters, searching for a particular gene sequence within the chromosome is a local matching problem.

A very important extension of pairwise sequence alignment is **multiple sequence alignment**, the mutual alignment of three or more sequences. Usually, we can find large families of similar sequences by identifying homologues in many different species. Multiple sequence alignment reveals the underlying patterns contained in a set of related sequences much more clearly than pairwise sequence alignments.

Programs for all of these different alignment problems are available on the web. There are many different programs, each with many different mirrors. The European Bioinformatics Institute sites are one trustworthy source.

Global alignment (pairwise and multiple)

CLUSTAL Omega	http://www.ebi.ac.uk/Tools/msa/clustalo/
T-Coffee	http://www.tcoffee.org/Projects/tcoffee/, http://www.ebi.ac.uk/Tools/msa/tcoffee/
EMBOSS	http://emboss.sourceforge.net, http://www.ebi.ac.uk/Tools/emboss/http://www.ebi.ac.uk/

Local alignment

SSEARCH	http://pir.georgetown.edu/pirwww/search/pairwise.shtml
EMBOSS Water	http://www.ebi.ac.uk/Tools/psa/emboss_water/

(there are separate versions for protein and nucleic-acid sequences).

Approximate methods for quick screening of databases

It is routine to screen genes from a new genome against databases, to find similarities to other sequences. Databases have grown so large that programs based on exact local alignments are too slow. Approximate methods can detect close relationships well and quickly but are inferior to the exact ones in picking up very distant relationships. In practice, they give satisfactory performance in the many cases in which the probe sequence is fairly similar to one or more sequences in a databank. Therefore, they are certainly worth trying first.

A typical approximation approach such as BLAST (basic local alignment search tool) takes a small integer k and determines all instances of each 'word' of length k (i.e., each set of k consecutive characters, with no gaps) of the probe sequence that occur in any sequence in the database. A candidate sequence is a sequence in the databank containing a large number of matching k-tuples, with equivalent spacing in probe and candidate sequences. For a selected set of candidate sequences, *approximate* optimal alignment calculations are then carried out, with the time- and space-saving restriction that the paths through the matrix considered are restricted to bands around the diagonals containing many matching k-tuples. It is clearest to show the procedure in terms of a dot plot (see Figure 5.7).

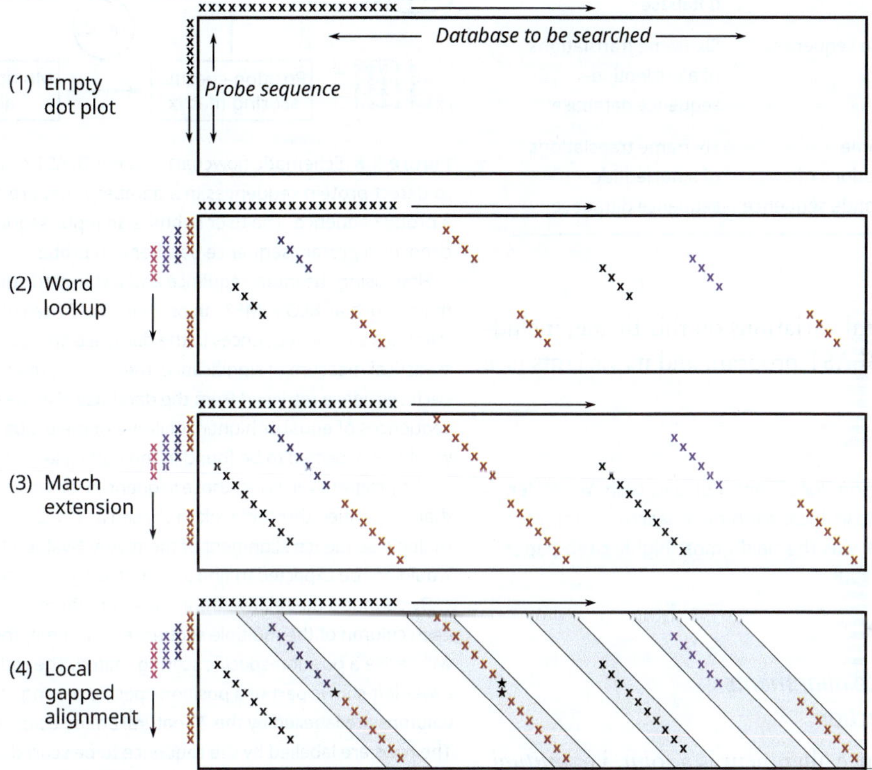

Figure 5.7 Schematic diagram showing the mechanism of a BLAST search. BLAST solves the problem of finding close matches of a probe sequence in a full genome or a full database that are much longer than the probe sequence.

1. The 'playing field' of the algorithm is the outline of a dot plot, just as if the problem were going to be solved by application of an exact-alignment method.
2. BLAST first divides the probe sequence into fixed-length words of length k; here $k = 4$. It then identifies all exact occurrences of these words in the full database, with no mismatches or gaps. Note that the same four-letter word may occur several times in the probe sequence (shown here in red), and of course each four-letter word may match many times within the database. It is possible to do this step quickly after pre-processing the database to record the sites of appearance of all four-letter words.
3. Starting with each match, BLAST tries to extend the match in both directions, still with no mismatches or gaps allowed.
4. Given the extended matches, BLAST tries to put them together by doing alignments *allowing* mismatches and gaps, but only within limited regions containing the preliminary matches (grey areas). The result of this step is to add to the matches the positions shown as *. This produces longer matching regions.

It is the restriction of the more complex matching procedure to relatively small regions, rather than applying it to the entire matrix, that gives the method its speed. The price paid is that the method will miss a combined match lying outside the grey area. In the example illustrated, the matching regions coloured red and purple, at the right of the matrix, will not be combined but reported as separate hits.

BOX 5.3 Different 'flavours' of BLAST search different databases

Program	Searches for:	In:
BLASTN	Nucleotide sequence	Nucleotide-sequence database
BLASTX	Six-frame translations of a nucleotide sequence	Protein sequence database
BLASTP	Protein sequence	Protein-sequence database
TBLASTN	Protein sequence	Six-frame translations of a nucleotide-sequence database
TBLASTX	Six-frame translations of a nucleotide sequence	Six-frame translations of a nucleotide-sequence database

There are several variations on this theme, including the original BLAST program and its variants (see Box 5.3).

The original paper on BLAST, Altschul, S.F., Gish, W., Miller, W., *et al.* (1990). Basic local alignment search tool. J. Mol. Biol. 215: 403–410, was the field's most highly cited paper published in the 1990s.

Multiple sequence alignments and pattern detection

A multiple sequence alignment is a *joint*, i.e. *mutual*, alignment of more than two sequences.

Multiple sequence alignments are rich in information about patterns of conservation. They help us to understand the common features of structure and function of a family of sequences, by showing us which residues are conserved (and therefore crucial). They also help us to identify distant homologues with greater confidence than a pairwise sequence alignment could.

The most powerful methods of protein structure prediction are based on multiple sequence alignments (see Chapter 12).

The patterns inherent in a multiple sequence alignment are not merely inferences from the alignment table—this is leaving it too late—but can actively contribute to *creating* a high-quality alignment. The idea is for an algorithm to learn the underlying patterns *while* it is assembling the multiple sequence alignment.

One very powerful program based on this approach is **PSI-BLAST**, an extension of BLAST for multiple sequence alignment (see Figure 5.8). PSI-BLAST constructs a **profile**, i.e. a conservation

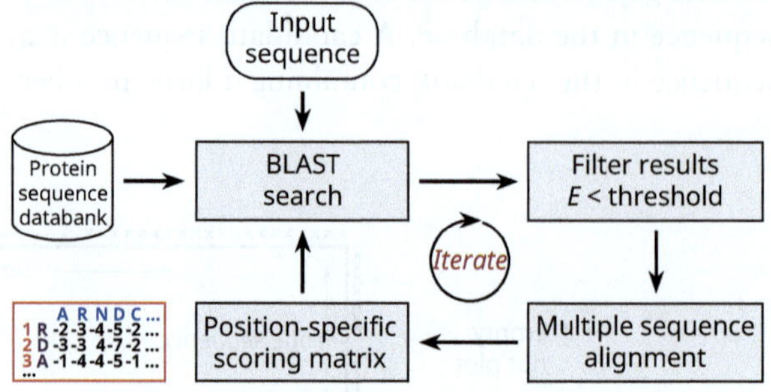

Figure 5.8 Schematic flowchart of a PSI-BLAST calculation to detect protein sequences in a database that are similar to a probe sequence. The user submits an input sequence and chooses a protein sequence databank to probe.

First, using the input sequence and a standard substitution matrix such as BLOSUM62, an ordinary BLAST calculation identifies similar sequences in the database and assigns a statistical measure of significance, the ***E-value***, to each 'hit'. For each sequence retrieved from the database, *E* is the number of sequences of equal or higher similarity to the probe sequence that would be expected to be found in the database, just by chance.

The program will select those sequences for which *E* is no greater than a specified threshold, often chosen as 0.005, and perform a multiple sequence alignment of them. Any *E*-value <1 means that it would not be expected to find *any* matches, just by chance.

By counting the relative frequencies of different amino acids in each column of the multiple sequence alignment, the program will derive a position-specific scoring matrix. The red box at the lower left shows part of a position-specific scoring matrix. The columns are labelled by the 20 natural amino acids, shown in blue. The rows are labelled by the sequence to be scored by the matrix, residue numbers in red and amino acids in purple. In this case, the N-terminal sequence of the sequence to be scored is RDA . . . The entries in the column are the log-odd scores of finding any amino acid at any position in the multiple alignment. For instance, the entry under A in row 3 is −1; therefore, the probability of finding an A at the third position is proportional to 10^{-1}.

To find the score of the sequence, add up the values in the R column of the first row, the D column of the second row, the A column of the third row, etc., to give: $10^{-3} + 10^{-7} + 10^{-1}$. In this example, the probabilities are expressed unscaled and as logarithms to the base 10. Note that the sequence being scored may contain gaps.

This matrix can be used as an alternative to the input sequence and substitution matrix in a BLAST search. Each subsequent BLAST search, based on the matrix derived in the previous step, will return a different set of 'hits'. With a sensible choice of input parameters, the procedure will usually converge to produce a more reliable set of similar sequences than was returned by the simple BLAST search of the input sequence performed in the first step.

pattern, in an initial multiple alignment of the 'hits' derived from a preliminary BLAST search. Armed with the profile, the method returns to the database and does a more sensitive search, giving higher weight to well-conserved positions; it then realigns what it finds and refines the profile. Several such cycles of refinement of the profile give PSI-BLAST the power both to detect distant relationships and to create high-quality multiple sequence alignments.

Perhaps the most powerful pattern analysis algorithms are based on hidden Markov models. A hidden Markov model is a mathematical construct that generates sequences according to internal probabilistic rules. For example, successive rolls of a pair of dice generate a sequence of numbers between 2 and 12, with different probabilities. However, for rolling dice there is not even a probabilistic link between the successive numbers that come up. Similarly, typing with one finger at random generates a sequence of characters, again with no correlation between successive letters. However, if you insist on typing ten keys per second, so that you can only move your hand a limited extent between successive keystrokes, then Q is more likely to be followed by W than by M. The result would be a probabilistically generated sequence with the observed distribution of each character dependent on what preceded it.

To represent a set of sequences with a hidden Markov model, imagine a computer program that generates sequences of nucleotides, or amino acids, according to rules that govern the probability distributions of *successors* to each letter. For example, an adenine would be assigned a set of probabilities for being followed by another adenine, or a thymine, or a cytosine, or a guanine, or a gap. A different set of probabilities would govern the successors of a thymine, a guanine, a cytosine, or a gap. The enhanced power of hidden Markov models over position-specific scoring matrices stems from their incorporation of statistics about the correlation between successive positions.

For example, it is observed that the dinucleotide frequency CpG is lower in higher organisms than would be expected from the overall mole fractions of C and G in the genome. A hidden Markov model would reflect this in a lowered probability of G in a position following a C, relative to the probabilities of a G following A, T, or another G.

Given a set of sequences, the process of *training* a hidden Markov model involves adjusting all of the probability distributions so that the sequences generated by the model have a high probability of reproducing the set of sequences analysed.

KEY POINT

A multiple sequence alignment is much richer in information than a pairwise sequence alignment. A hidden Markov model is a method for capturing the information.

Pattern matching in three-dimensional structures

Given two or more structures—perhaps of several homologous proteins—we can frame questions generalized from sequence alignment. How can we measure structural similarity quantitatively? Can we derive a sequence alignment from structural comparisons?

In the native state of a protein, the mainchain follows a curve in space. The general spatial layout of this curve defines a folding pattern. The backbones of related proteins show recognizably similar but not identical folding patterns. A letter of the alphabet in different type fonts—for instance, b and *b*—illustrates the topological similarities, and differences, in detail, that are seen among proteins related by evolutionary divergence that share a common folding pattern. A better analogy for widely divergent proteins might be the letters B and R, which share the letter P as a common core substructure but in addition have either a loop (B) or a stroke (R) that differ. Homologous protein structures typically contain fairly large, well-fitting substructures. Figure 5.9 shows a superposition of local regions of two proteins and an overall superposition of two entire structural domains (see also Figure 1.13).

Extraction of the maximum common substructure induces an alignment of the sequences. This is called a structural alignment. Because structure changes

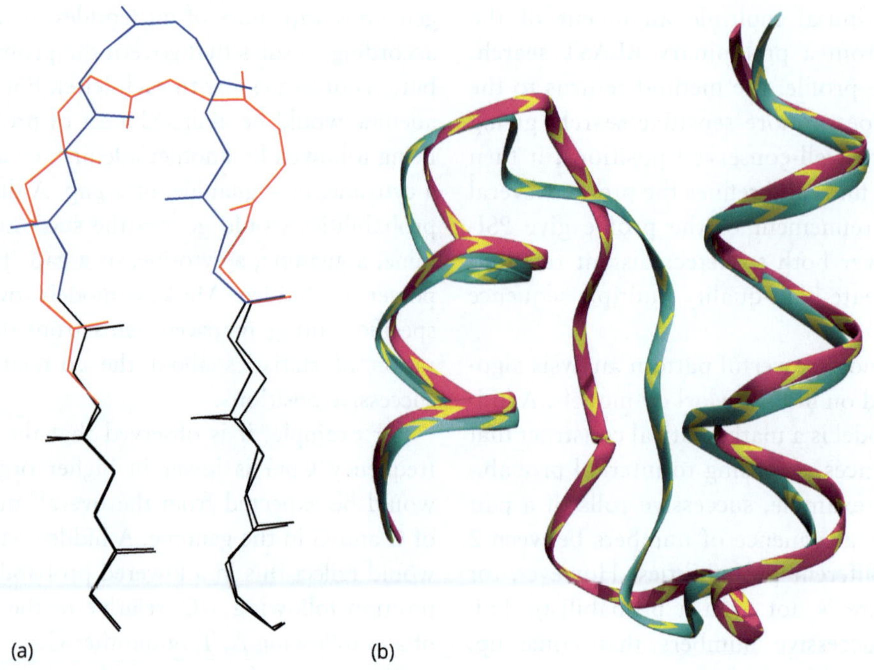

Figure 5.9 Local and global superpositions of protein structures.

(a) Two β-**hairpins** from the antigen-binding site of an antibody [1VFA], [2FBJ]. Only mainchain atoms are shown. The 'stems' of the loops, parts of strands of β-sheet, superpose well (regions in black, at bottom of picture). The connections have different lengths and conformations, and do not superpose well (red and blue regions, at top of picture). This is an example of a *local* well-fitting substructure. It involves only small contiguous regions of the chains.

Observing which residues are superposed in the black regions induces an alignment of the residues in these regions.

(b) Superposition of regions with a folding pattern called the HeH (helix–extended loop–helix). Magenta: domain from RNA-binding domain of transcriptional terminator protein *r* (from *E. coli*) [1a62]. Cyan: domain from KU heterodimer (human) [1JEQ].

The helices at either end of the chains superpose well. The extended regions between the helices do not. The structural alignment—that is, the sequence alignment induced by the structural superposition—is this:

```
RNA-binding domain    TPVSELITLGENMGLEN–LARMRKQDIIFAILKQH
KU heterodimer        FTVPMLKEACRAYGL–KSG-L-KKQELLEALTKHF
```

more conservatively than sequence during evolution, for distantly related proteins it may be possible to align the sequences on the basis of the structures even if methods based purely on sequences cannot recognize the relationship.

KEY POINT

A structure alignment is nevertheless an alignment = an assignment of residue–residue correspondences. Instead of assigning the correspondence by matching the characters in two or more sequences, a structural alignment assigns the correspondence to residues that occupy similar positions in space, relative to the molecular framework.

Evolution of protein sequences, structures, functions

Extending Crick's classic 'central dogma' gives us the basic paradigm:

DNA → RNA → amino-acid sequence of a protein → protein structure → protein function

Transcription of DNA to RNA, and translation of mRNA by ribosomes, takes us as far as the amino-acid sequence. Under appropriate physicochemical conditions, the amino-acid sequence dictates the protein structure by a spontaneous folding process (see Figure 1.2). Folding produces a native state. For most proteins, the native state structure contains an active site with the proper geometry, charge distribution, and hydrogen-bonding potential to interact specifically with other proteins, or with small-molecule ligands. In many cases, the active site contains catalytic residues that produce enzymatic activity.

During evolution, selection acts on protein function to alter gene frequencies in populations, closing the loop back to DNA.

The effects of single-site mutations

The native states of proteins are the cumulative effect of many inter-residue interactions. What then should we expect to be the result of a perturbation in the amino-acid sequence?

Consider an SNP leading to a single amino acid substitution. Will the structure stay the same? Changing one amino acid without otherwise altering the structure would leave most interactions intact, except for those involving the mutated residue itself (and conservative mutations may preserve even these). Nevertheless, sometimes changing a single residue is enough to blow the original structure apart. An example is the mutation A174D in human aldolase (see Box 5.4). In other cases, changes to residues providing specific interactions with ligands may alter activity. Some mutations do not alter the structure but destabilize it; frequently this is enough to cause disease. Mutations in human aldolase illustrate several effects of single amino acid substitutions.

Many small changes in amino-acid sequence leave the basic structure intact, producing only small conformational changes. In this sense, protein structures are *robust* to mutation—not to all mutations, but to enough mutations to allow variability. This is essential—and sufficient—for evolution.

BOX 5.4 Hereditary fructose intolerance and mutants of aldolase B

The enzyme aldolase catalyses the cleavage of two substrates:

fructose-1,6-bisphosphate → glyceraldehyde-3-phosphate + dihydroxyacetone phosphate

fructose-1-phosphate → glyceraldehyde + dihydroxyacetone phosphate

Fructose-1,6-bisphosphate is a mainstream metabolite in glycolysis and gluconeogenesis, classic pathways of glucose metabolism. Fructose-1-phosphate arises in metabolism of dietary fructose. Different isozymes of aldolase have different relative activities towards fructose-1, 6-bisphosphate and fructose-1-phosphate.

Approximately 1 in 20000 people suffer from *hereditary fructose intolerance*, a defect in the liver isozyme, aldolase B. The gene encoding this protein maps to locus 9q22.3 in the human, giving the trait an autosomal recessive inheritance pattern. For affected individuals, ingestion of fructose, a monosaccharide common in fruits and honey, leads to vomiting, discomfort, and hypoglycaemia. Problems often first appear in infancy as fructose and sucrose are added to the diet upon weaning. The condition can be fatal if unrecognized and untreated; however, for most patients it is sufficient to adopt a diet free of fructose and sucrose.

Numerous mutations have been associated with aldolase B dysfunction, including amino acid substitutions, nonsense mutations producing truncated protein, insertions and deletions including frameshifts, and changes in splice sites. The most common mutations are A149P (over 50% of cases worldwide) and A174D.

T. Cox and co-workers* characterized the proteins corresponding to several known mutants. Normal aldolase B is a tetramer of four 363-amino-acid subunits. Because all mutants were discovered in patients presenting with hereditary fructose intolerance, all had reduced or absent enzymatic activity. Two classes of mutants were:

- *Catalytic mutants:* these can be expressed as intact tetramers, retaining some activity at 37°C. These include W147R and R303W.
- *Structural mutants:* these are destabilized, and show catalytic activity (if at all) only after expression at 22–23°C. These include N334K, A149P, L256P, and A174D.

The substitutions in the catalytic mutants occur in or near the active site. The substitutions in the structural mutants occur either in a residue buried in the monomeric structure (A174D, which does not fold at all, as a result of burying a charged sidechain), or in the subunit interface, which causes the protein to dissociate into monomers (N334K, L256P, and A149P).

* Rellos, P., Sysgusch, J., & Cox, T.M. (2000). Expression, purification and characterization of natural mutants of human aldolase B. J. Biol. Chem. 275: 1145–1151.

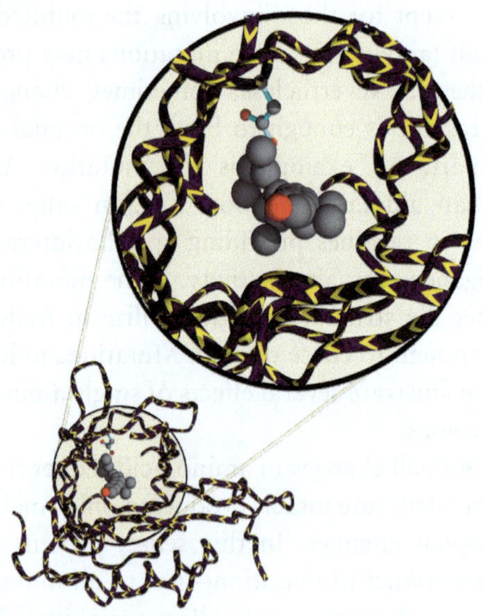

Figure 5.10 Retinol-binding protein transports vitamin A around the bloodstream, bound in a deep hydrophobic cavity within the protein. This figure shows a model of mutant 75Gly → Asp of retinol-binding protein. Note that the model was built *solely* by inserting the sidechain, to observe the structural consequences. No attempt was made to try to predict the structural deformation produced. What this model shows is that there are steric and electrostatic incompatibilities between the sidechain of 75Asp and the ligand. This explains the observation of decreased affinity for retinol, producing vitamin A deficiency and night blindness.

Z. Wang and J. Moult[2] described some of the kinds of structural effects of single amino acid substitutions related to human diseases (see Figures 5.10, 5.11, and 5.12).

The protein structure prediction program AlphaFold is applicable to modelling effects of mutations.[3]

A sequence that so lacked robustness that any mutation would destroy it, could not exist. It could have no neighbouring precursor and processes of evolution could never find its sequence.

[2] Wang, Z., & Moult, J. (2001). SNPs, protein structure, and disease. Hum. Mut. 17: 263–270. See also: Peterson, T.A., Doughty, E., & Kann, M.G. (2013). Towards precision medicine: advances in computational approaches for the analysis of human variants. J. Mol. Biol. 425: 4047–4063.

[3] Pak, M.A., Markhieva, K.A., Novikova, M.S., Petrov, D.S., Vorobyev, I.S., *et al.* (2023). Using AlphaFold to predict the impact of single mutations on protein stability and function. PLoS One 18: e0282689.

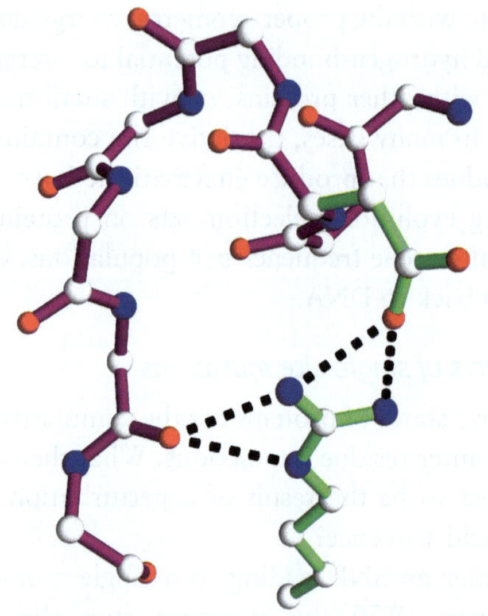

Figure 5.11 Factor XIIIa is the enzyme at the final step of the blood coagulation cascade. It cross-links fibrin molecules, stabilizing clots. The normal protein contains an arginine sidechain that forms a salt bridge and multiple hydrogen bonds (broken lines), to a neighbouring asparate sidechain and to mainchain carbonyl atoms. The arginine and aspartate sidechains are shown in green. Mutation of the arginine to an isoleucine (not shown) removes these interactions, destabilizing the protein and resulting in poor clot formation [1F13].

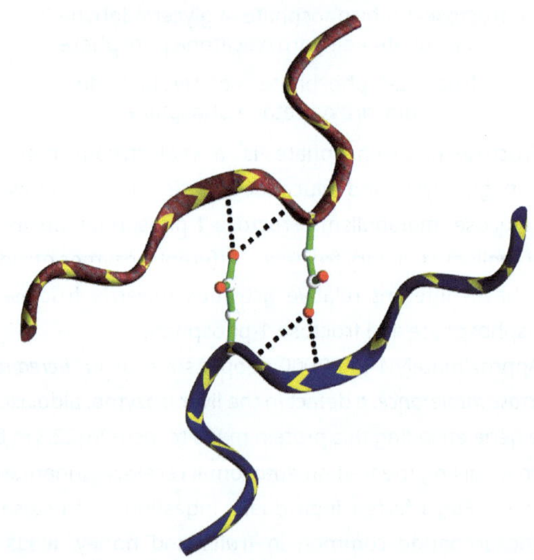

Figure 5.12 Aldolase A is an enzyme in the glycolytic pathway. It is an isozyme of aldolase B, the protein involved in hereditary fructose intolerance. Aldolase is normally a tetramer, stabilized by hydrogen bonds involving aspartate residues at the inter-subunit interface. Mutation of aspartates to glycines destabilizes the tetramer. Although this has no detectable effect on most cell types, red blood cells show weakened cell membranes, causing a congenital form of haemolytic anaemia [2ALD].

Evolution of protein structure and function

Sequences and structures of related proteins show coordinated evolutionary divergence. As sequences progressively diverge, structures progressively deform. Typically, a core of the structure, including the major elements of secondary structure and, usually, the active site, retains its folding pattern. Other, peripheral regions of the structure can refold entirely (see Box 5.5).

Structure changes more conservatively than sequence. That is, protein families can show relatively

BOX 5.5 The relationship between divergence of amino-acid sequence and the divergence of protein structure

Systematic studies of the structural differences between pairs of related proteins have defined a quantitative relationship between the divergence of the amino-acid sequences of the core of a family of structures and the divergence of structure. As the sequence diverges, there are progressively increasing distortions in the mainchain conformation, and the fraction of the residues in the core usually decreases. Until the fraction of identical residues in the sequence drops below about 40–50% these effects are relatively modest. Almost all the structure remains in the core, and the deformation of the mainchain atoms is on average no more than 1.0 Å. With increasing sequence divergence, in most cases some regions refold entirely, reducing the size of the core, and the distortions of the residues remaining within the core increase in magnitude.

A correlation between the divergence of sequence and structure applies to all families of proteins. Figure 5.13 shows the changes in structure of the core, expressed as the r.m.s. deviation of the mainchain atoms after optimal superposition; plotted against the sequence divergence, expressed as the percentage conserved amino acids of the core after optimal alignment. The points correspond to pairs of homologous proteins from many related families. (Those at 100% residue identity are proteins for which the structure was determined in two or more crystal environments, and the deviations show that crystal packing forces—and, to a lesser extent, solvent and temperature—can modify slightly the conformation of the proteins.) Figure 5.13b shows the changes in the fraction of residues in the core as a function of sequence divergence. The fraction of residues in the cores of distantly related proteins can vary widely: In some cases, the fraction of residues in the core remains high, in others it can drop to below 50% of the structure.

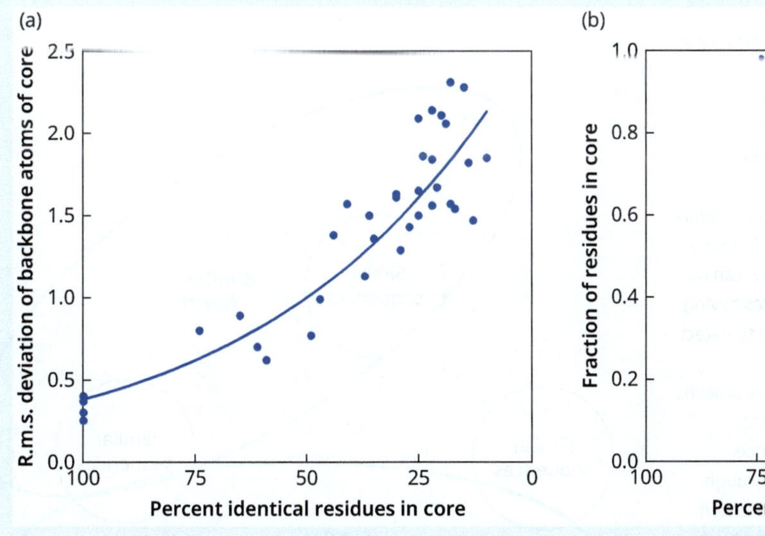

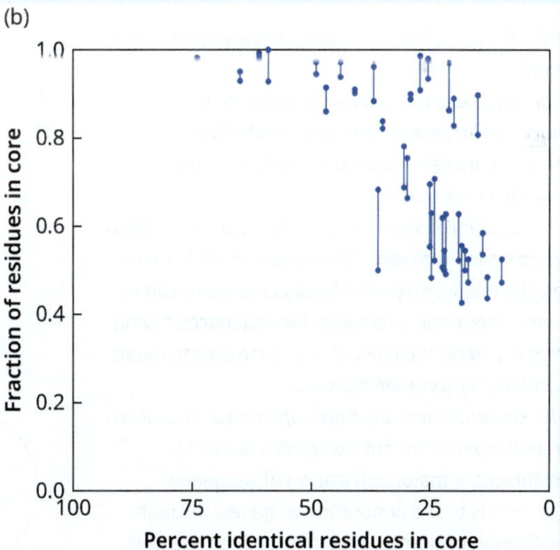

Figure 5.13 Relationship between divergence of amino-acid sequence and three-dimensional structure of the core, in evolving proteins. (a) Variation of r.m.s. deviation of the core with the percent identical residues in the core. (b) Variation of size of the core with the percent identical residues in the core. This figure shows results calculated for 32 pairs of homologous proteins of a variety of structural types.

large amino-acid sequence changes while retaining high structural (and functional) similarity. Indeed, in many families of proteins, we can recognize structural similarity in relatives so distant that there is no easily visible signal of the similarity in the sequence.

A common reason for retention of protein conformation in general, and the structure of the active site in particular, is selection for maintenance of function. Evolutionary constraints may preclude large changes in protein stability or structure—a resulting loss of function may be disfavoured by selection. It is easier to see the effects of these constraints than to understand their mechanism. In many cases, certain specific residues are directly involved in function—for example the iron-linked histidine of the globins—and these are immutable. In contrast, constraints that maintain the overall folding pattern are dispersed around the sequence, and it is only by studying patterns of residue conservation in large-scale alignments of homologous proteins that we can begin to understand the constraints imposed by structure on sequence.

When a protein evolves to change its function, many of these constraints are released—or, more precisely, replaced by alternative constraints required by the new function. The relationship between sequence and function is much more complex than the relationship between sequence and structure. Small changes in sequence, during evolution, usually make only small changes in structure. Often they make only small changes in function also. But changes in function do not necessarily require large changes in sequence or structure—function can jump.

Indeed, a protein can change function without any sequence changes at all. For instance, in the duck, an active lactate dehydrogenase and an enolase serve as crystallins in the eye lens, although they do not encounter the substrates *in situ*. In other birds, crystallins are closely related to enzymes, but some divergence has already occurred, with loss of catalytic activity. (This proves that the enzymatic activity is not necessary in the eye lens.) Many other such examples are known of 'recruitment', or 'moonlighting', by proteins—adapting to a novel function with relatively little sequence change.

Conversely, proteins with very different sequences and structures can have the *same* function. For instance, many families of proteinases differ in sequence and structure, sharing only a common general catalytic activity. Figure 5.14 summarizes, in a schematic way, the landscape of protein space with respect to the relations among sequence, structure, and function.

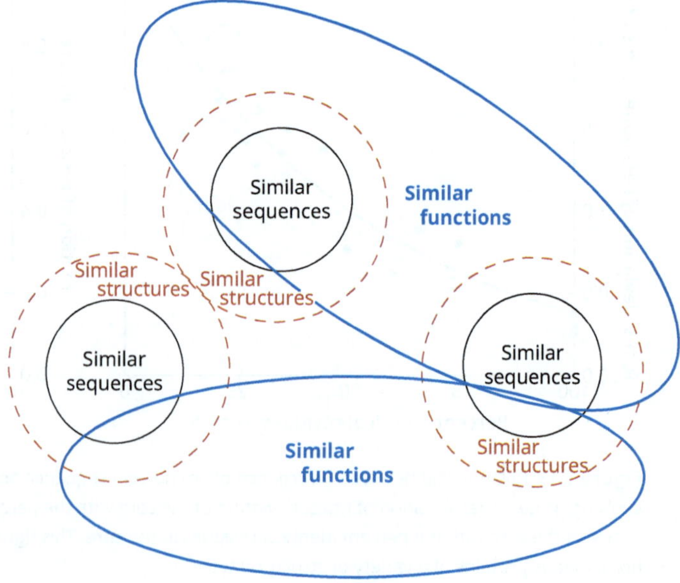

Figure 5.14 Relationships among sequence, structure, and function:

- *similar sequences can usually be relied on to produce similar protein structures*, with divergence in structure increasing progressively with the divergence in sequence;
- conversely, *similar structures are often found in distantly related proteins, with very different sequences*: in many cases, the relationships in a family of proteins can be detected *only* in the structures, the sequences having diverged beyond the point of our being able to detect the underlying common features;
- *similar sequences and structures often produce proteins with similar functions*, but exceptions abound; nevertheless, a major application of sequence **alignment** is to the annotation of genes, through identification of homologues. The goal is to assign structure and function to as many genes as possible;
- conversely, *similar functions are often carried out by non-homologous proteins with dissimilar structures*, e.g. the different families of proteinases, sugar kinases, and lysyl-tRNA synthetases.

All three features of proteins—sequence, structure, and function—are potentially useful in interpreting new genome sequences. We expect that many regions of the new genome encode proteins similar to relatives known in other species. We can find them by looking for similar patterns in the sequences. We can expect that the structures will be similar, and indeed can calibrate the expected difference in structure from the extent of divergence in sequence. As Figure 5.14 shows, however, we cannot be as confident in assuming that function will be conserved.

Even if sequence divergence gives rise fairly smoothly to progressive structure divergence, tiny structural changes in sensitive regions can have profound effects on function (see Figure 5.15). Such regions obviously include active sites, but are not necessarily limited to them. It is possible for function to make discrete jumps with minimal sequence change.

During evolution it is possible for different branches of a protein family to invent a new function independently, or for a function to be lost and rediscovered. In such cases, similarity of function does not imply close similarity of sequence. The evolutionary tree of the sequences of members of the family, and the graph of the relationships between different functions, can show very different topologies.

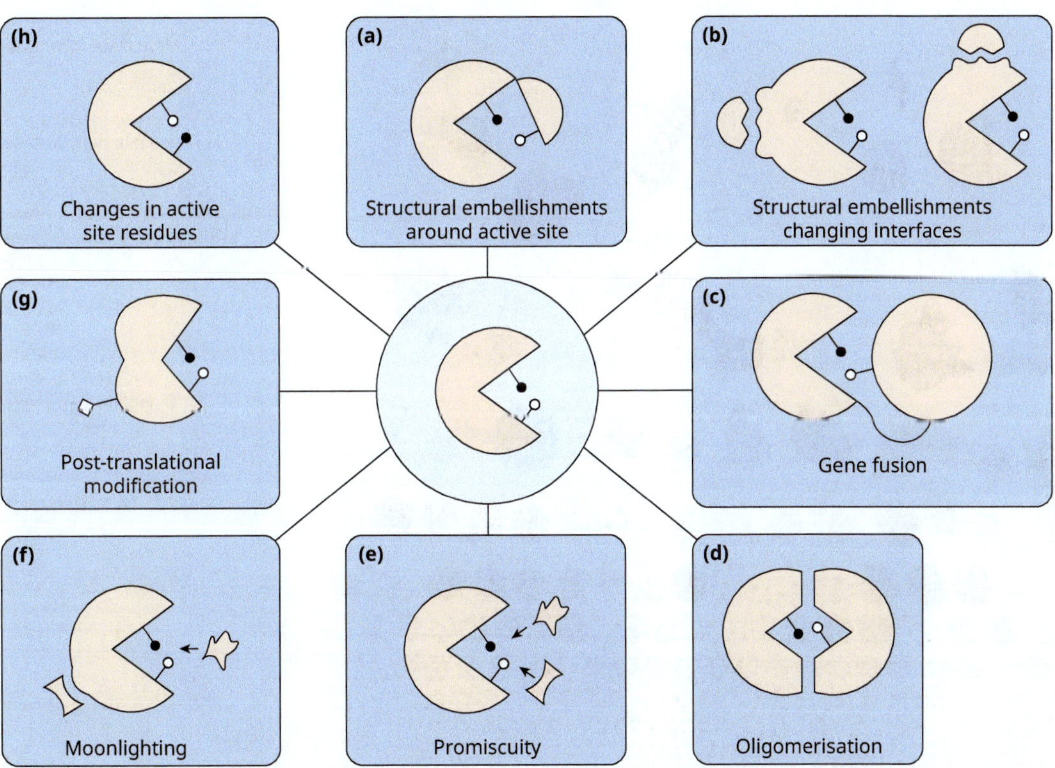

Figure 5.15 From an original structure shown at the centre of this figure, mechanisms of functional change in protein evolution may include one or more of the following: (a) Mutations in residues around the active site, (b) Mutations in residues at interfaces, (c) Gene fusion, (d) Oligomerization (changes in quaternary structure), (e) Promiscuity (= relaxation of specificity), (f) Moonlighting (adopting an additional function), (g) Post-translational modification, and (h) Changes in active site residues. For mechanisms (a), (c), and (d), there are contributions to the new active site from a partner domain or subunit.

Das, S., Dawson, N.L., & Orengo, C.A. (2015). Diversity in protein domain superfamilies, Current Opinion in Genetics & Development 35: 40–49, ISSN 0959-437X, https://doi.org/10.1016/j.gde.2015.09.005. (https://www.sciencedirect.com/science/article/pii/S0959437X15000982)

Divergence of function in the enolase superfamily

The enolase superfamily shows wide functional divergence. Enolase itself catalyses the isomerization of 2-phosphoglycerate and phosphoenolpyruvate in glycolysis; however, there are large numbers of homologues, showing at least 20 different catalytic activities.

Perhaps surprisingly, some enolase superfamily enzymes that catalyse *different* reactions are more similar to each other than to other pairs of enzymes that catalyse the *same* reaction. (These might appear in the region in Figure 5.14 where the two blue ellipses overlap.) Conversely, if the activities evolved more than once, independently, some proteins catalysing a common reaction may in fact be distant rather than close relatives. (Indeed, they may not be related at all: there are many evolutionarily unrelated families of proteases, for example.) These discrepancies frustrate attempts to classify the functions of enzymes on the basis of sequence or even of structure, or to assign functions to all the sequences that have never set foot in a wet lab.

Gerlt and Babbitt, and their co-workers, have described the relationships of the known proteins of the enolase superfamily, in terms of a network (see Figure 5.16). Neighbourhoods in the network depend on sequence similarity. Closely related proteins form clusters.

Absent from the figure are the enolases themselves, which would loom overwhelmingly large, because so

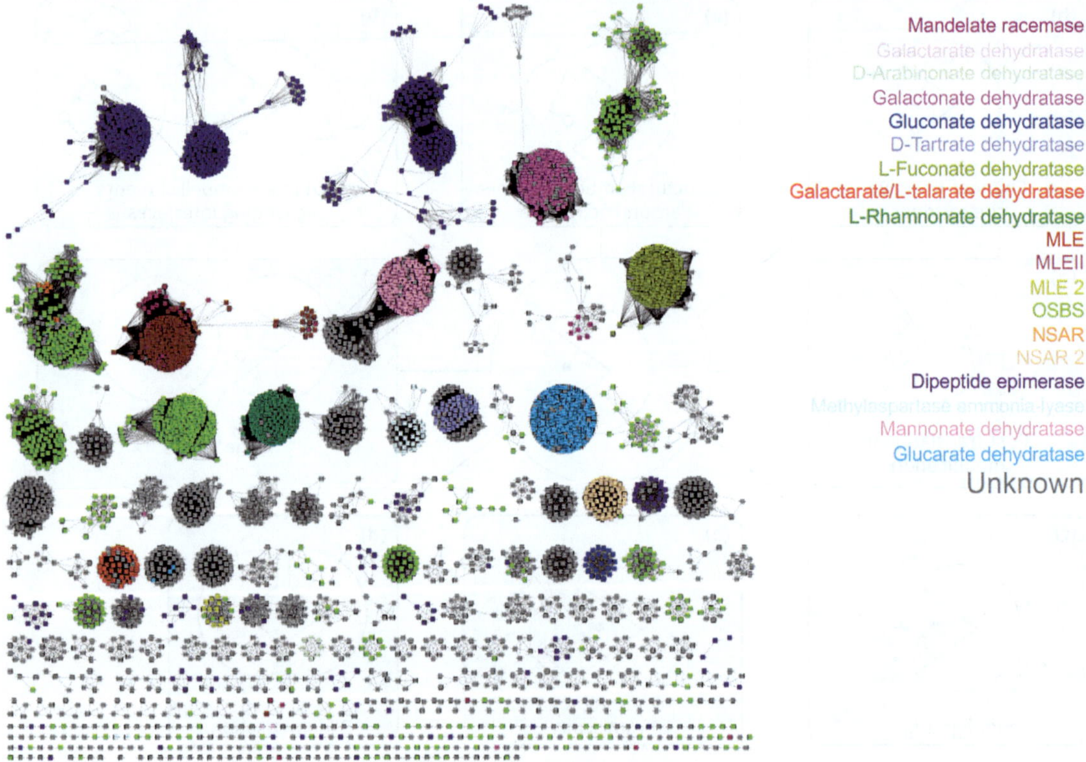

Figure 5.16 Network based on sequence similarity for enolase superfamily members, excluding the enolase subgroup itself. Each node, shown as a circle, corresponds to a protein in the superfamily. Lines connect pairs of nodes which have high sequence similarity; shorter edges indicate more similar sequences. Thus the graph contains clusters of closely related proteins. The threshold for drawing an edge is such that the similarity of the sequences is within homology-modelling range. (Homology modelling is the prediction of a structure from a close homologue, as suggested by Figure 5.13.) Therefore, knowledge of the structure of any member of a cluster allows prediction of the structures of other cluster members.

If the function of a protein is unknown, the corresponding node is grey. Nodes corresponding to proteins of known function are coloured. A list of known functions appears at the right of the figure.

Gerlt, J.A., Babbitt, P.C., Jacobson, M.P., *et al.* (2012). Divergent evolution in enolase superfamily: Strategies for assigning functions. The Journal of biological chemistry 287, 1: 29–34. https://doi.org/10.1074/jbc.R111.240945

many sequences are known. Those superfamily members that appear in Figure 5.16 segregate into about a dozen major clusters; some of which have themselves almost broken into separate groups, with only tenuous links between them. In almost all cases, the elements of a cluster are all either grey (unknown function) or a single colour (corresponding to a known function). It would be justifiable to assume that all nodes of these clusters have the same function. There are exceptions however! For the clusters that are entirely grey, no function of any member of the cluster is known.

Were the enolases themselves added, this figure would display a comprehensive view of the relationships between sequence and function in the superfamily. It would also provide a guide to focus future investigations on interesting and significant questions. A cluster for which no member has a known function (all nodes grey) would be a good candidate for investigation. If a cluster has the unusual property of containing members with different functions, it would be interesting to see how these minimal sequence changes alter function. If two (or more) separated clusters both show a common function, it may be that this function was invented twice (or more times), independently, in different lineages; it would be interesting to sort this out.

To allow comprehensive and facile access to the data and the analytic tools, Babbitt, Holliday, and co-workers have developed the **Structure-Function Linkage Database (SFLD)**, at http://sfld.rbvi.ucsf.edu/django/. The data are organized around the relationships between sequences and structural features, and reaction mechanisms. The website supports browsing, information retrieval, following links to other databases, and interactive analysis, such as computation of networks such as the one appearing in Figure 5.16. Doing this interactively allows selecting different colour schemes, or clicking on nodes to retrieve annotation information for particular molecules. For instance, an operation that would be impossible on the fixed picture in Figure 5.16 but could be done using the interactive tools, would be to identify within the network the particular molecule that corresponds to a node, and determine its most similar homologues. The site is also rich in documentation and tutorial materials, including introductory videos.

Phylogeny

Once we have measured the similarity of one or more properties of a set of individuals, or species, we can try to arrange them according to their apparent pattern of divergence. If in fact the similarities do arise during descent from a common ancestor, it should be possible to depict the relationships in a family tree (for related individuals), or phylogenetic tree (for species and higher taxa). The goal is to present the pattern of similarities and divergences in a consistent diagram such that close relationships within the diagram correspond to high degrees of similarity.

The basic principle is that *the origin of similarity is common ancestry*. Although there are many exceptions, arising from convergent evolution or horizontal gene transfer, this basic principle is crucial both for rationalizing contemporary observations and for opening a window onto the history of life.

The goal of phylogeny is a *logical arrangement* of the relationships among members of a set of species, populations, individuals, or genes. The arrangement is derived from observed similarities. Commonly, the assumption is that the organization will have the form of a tree. (We saw an example of a tree in Figure 1.12.) The computations that derive the optimal phylogenetic tree from a matrix of similarities are not trivial, and the problem has been a challenge in research for some time.

We assume that phylogeny reflects relationships—among species, populations, individuals, or genes. Relationship is taken in the literal sense of kinship or genealogy, i.e. assignment of a scheme of ancestors and descendants (see Box 5.6).

The basis for determining and presenting relationships between individuals, populations, and species is clustering and classification. We need to discuss these tools in some considerable detail.

KEY POINT

A phylogenetic tree is a diagram showing ancestor-descendant relationships, that captures a pattern of similarities, in that individuals or species closely linked in the tree have high similarity.

BOX 5.6 Concepts related to biological classification and phylogeny

- **Homology** means, specifically, descent from a common ancestor.
- **Similarity** is the measurement of resemblance or difference, independent of the source of the resemblance. Similarity is observable *now* and involves no historical hypotheses. In contrast, assertions of homology require inferences about historical events, which are almost always unobservable.
- *Similarity and dissimilarity*. Data suitable for phylogenetic analysis may be specified equivalently in terms of similarities between objects, or by dissimilarities. In comparing two DNA sequences, we may count the percentage of identical residues in an optimal alignment. This is a measure of similarity—the higher the value, the more similar the sequences. Alternatively, we could count the number of mutations separating the sequences. The Hamming and Levenstein distances are measures of dissimilarity.
- **Clustering** is bringing together similar items, distinguishing classes made up of objects that are more similar to one another than they are to other objects outside the classes. Most people would agree about degrees of similarity, but clustering is more subjective. When classifying objects, some people prefer larger classes, tolerating wider variation; others prefer smaller, tighter classes. They are called, respectively, groupers and splitters.
- **Hierarchical clustering** is the formation of clusters of clusters of . . .
- The distinction between **clustering** and **classification**: Clustering is the determination of the set of classes into which a group of samples should be divided. Classification is the assignment of a sample to its proper place in a known set of classes (see Box 5.7).
- **Phylogeny** is the description of biological ancestor–descendant relationships, usually expressed as a tree. A statement of phylogeny among objects *assumes* homology and *depends* on classification.

BOX 5.7 Clustering and classification

A clustering is a partition of a set of objects, such that the similarity of pairs within each cluster is higher than the similarity of pairs from different clusters. Classification may be based on clustering: Given a set of clusters, assignment of objects to one of the clusters is a **classification**:

Clustering	Classification
Classes to be determined from data	Prior knowledge of classes
Goal: determine classes	Goal: assign sample to one class
Need set of samples, no class assignment	Need set of samples with class assignment
In a machine learning setting:	
Unsupervised learning	Supervised learning
Clustering can be a basis for classification of subsequently presented samples.	

Clustering is simplest if the classes do not overlap and there is a separating line (or, in higher dimensions, plane or hyperplane). In Figure 5.17, a line separates two half-planes, each occupied exclusively by red or black points. Given any new point, it is easy to calculate onto which side

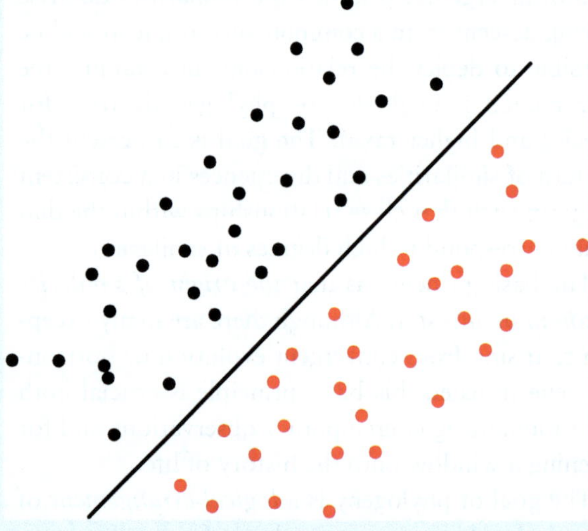

Figure 5.17 A distribution of points separable by a line.

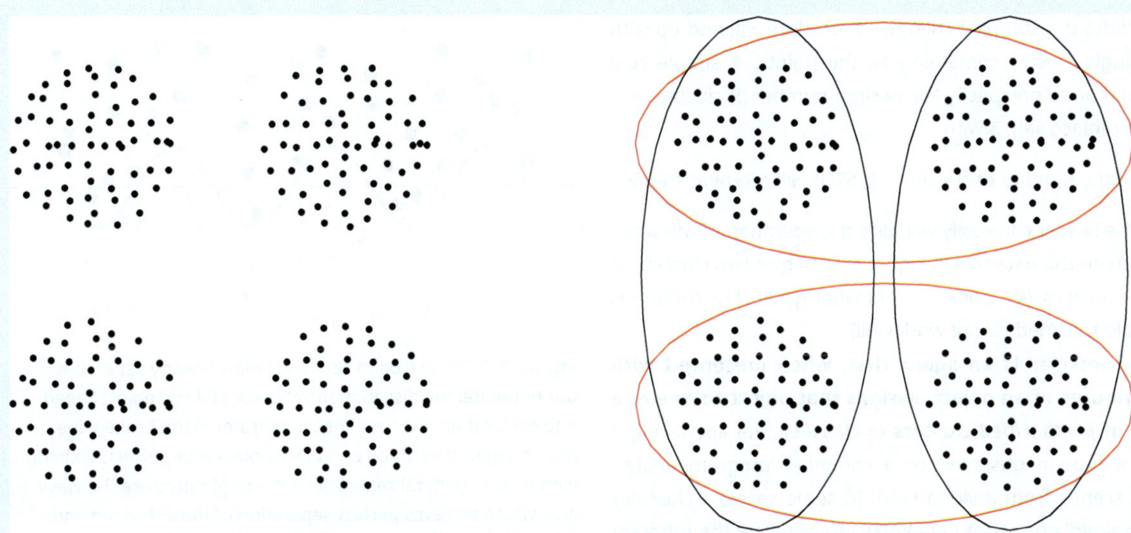

Figure 5.18 How many clusters? Four? Two? (Which two?)

of the line it falls, and, thereby, classify it. However, not all classifications depend on geometric clusterings: it is possible to classify the squares on a chessboard into black and white sets but it is not possible to cluster them spatially.

Often, clustering does not have a unique solution. In Figure 5.18, the set of points on the left can be clustered in two valid ways as shown on the right. One could also divide this distribution into four clusters.

In some cases, multiple clusterings of the same data are useful. Figure 5.19 shows overlapping clusterings of the amino acids, according to different properties.

From a measure of distance between points, we can proceed to define clusters. One approach is to build up clusters from nearest neighbours, called *single-linkage clustering:*

1. Begin by regarding each point as a cluster.
2. Find the pair of clusters with the minimum distance. Initially, when each cluster contains only one point, the distance between clusters is the distance between the points. The distance between two sets containing more than one point may be defined as the minimum distance between points in the two sets, or the average distance between points in the two sets. (In discussing phylogenetic trees, we shall describe the UPGMA method, Unweighted Pair Group Method with Arithmetic mean, which uses the average.) Join the minimum-distance clusters into a single cluster.
3. Repeat step 2.

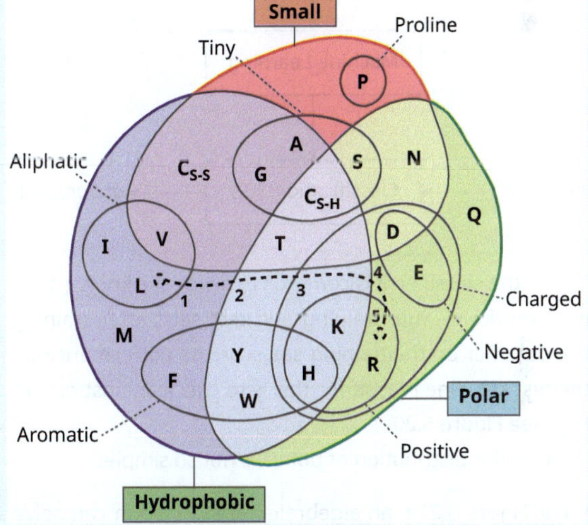

Figure 5.19 Multiple clusterings of the canonical 20 amino acids, according to different physicochemical properties, such as charge and size. A possible measure of distance between two amino acids is the number of boundaries that must be crossed to get from one amino acid to another. For instance, the path from L to R involves crossing five boundaries: (1) Aliphatic, (2) Polar, (3) Charged, (4) Hydrophobic, (5) Positive.

Without a stopping rule, this procedure will end up with a single cluster containing all the points. A simple rule would be to prespecify the desired number of clusters = k, and replace step 3 with

3. If the number of clusters = k, STOP; else Repeat step 2.

The reader can easily visualize the operation of this algorithm on the data sets in Figure 5.18 to give two clusters or four clusters. ($k = 2$ or $k = 4$.) For what types of distributions would this method *not* work well?

A *classifier* is an agent that, when presented with attributes of an object, assigns that object to one of a set of *prespecified* clusters (= classes). The agent might be a human observer, or a computer program analysing scenes from a video feed. In some cases, a classifier is 'trained' on sets of data known to populate the different classes.

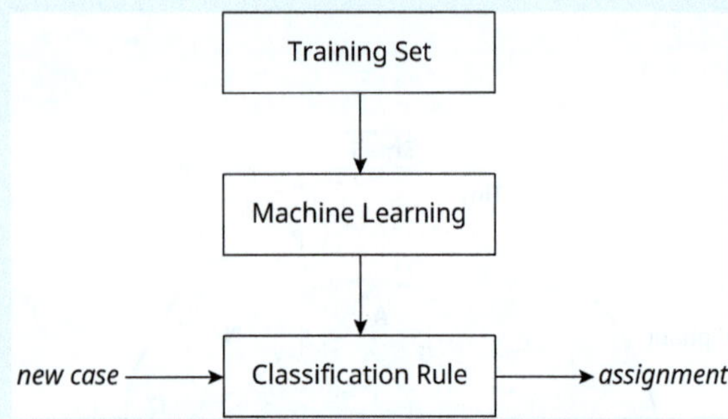

To test the classifier, it is common to divide the known samples, randomly, into training and test sets. After training the classifier on the training set, score its performance on the test set. One reason for this is to check against *overfitting* (see Figure 5.20).

Often the distribution of points is not so simple.

- For Figure 5.21a, an algebraic transformation can solve the problem. Converting to polar coordinates and plotting radius and θ (Figure 5.21b) again gives us non-overlapping sets of points that could be separated, for example by the line radius = 7.
- The points distributed along a line at the top of Figure 5.22 are not linearly separable. Mapping the points into two dimensions: $x \rightarrow (x, x^2)$ would allow a separating line.
- Figure 5.23 shows another simple example. On the left, the four points are not linearly separable into black and red classes. Map the two-dimensional points (x, y) into three dimensions by the function: $\Phi(x, y) = (x^2, \sqrt{2}xy, y^2)$, shown on the right. (In effect, (x, y) is mapped to (x', y', z'), where $x' = x^2, y' = \sqrt{2}xy, z' = y^2$). Because all the original points are on one of the original axes, either $x = 0$ or $y = 0$ and therefore $xy = 0$. Hence, in three dimensions, on the right, all points in fact map into a two-dimensional subspace, $\sqrt{2}xy = 0$. Moreover, because of the squared terms in Φ, both red points on the left map into the same position on the right, and both black points on the left map into the same position on the right. But nevertheless, there is now a plane that separates them.

In this case, the transformation took us to three dimensions. A program can easily deal with three, rather than

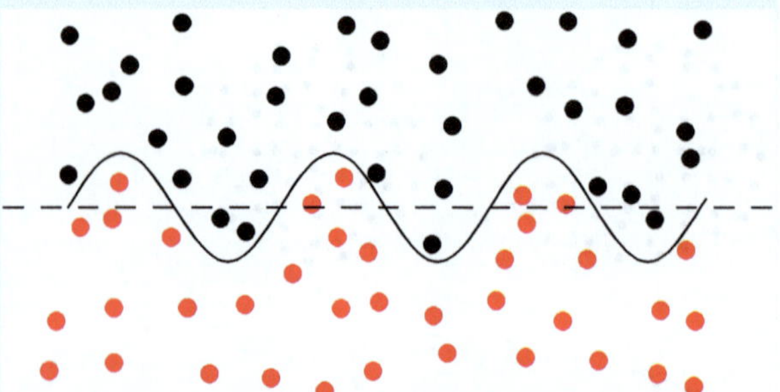

Figure 5.20 These point sets are *almost* linearly separable, but not quite: the distributions of black and red points 'bleed' into each other. The best linear separator is the broken line, but its separation of red and black points is imperfect. Training with a more general separator curve might produce the wavy line, which achieves perfect separation of *this* set of red and black points.

Suppose however that this training set were a sample from a large population. Selection of a different subset would in general produce an alternative distribution. The new distribution might have the same degree of overlap; that is, of invasions across the broken line, but these might have a different distribution. The wavy line works for the particular training set from which it was derived, but may not work well on other sets of points selected from the population. Such a procedure is said to be *overtrained*. Overtraining is a common problem in machine-learning problems.

A defence against overtraining is to test for it, by dividing the known points, randomly, using half as a 'training' set and holding back the other half as a 'test' set. Suppose the figure shows the results for the training set. The wavy line classifies the training set perfectly. If it also classifies perfectly the points in the test set, fine. But if it does not, the procedure was overtrained—that is, tuned to specific details of the training set rather than features of the entire population. Perhaps the broken line might give the best performance over the entire population.

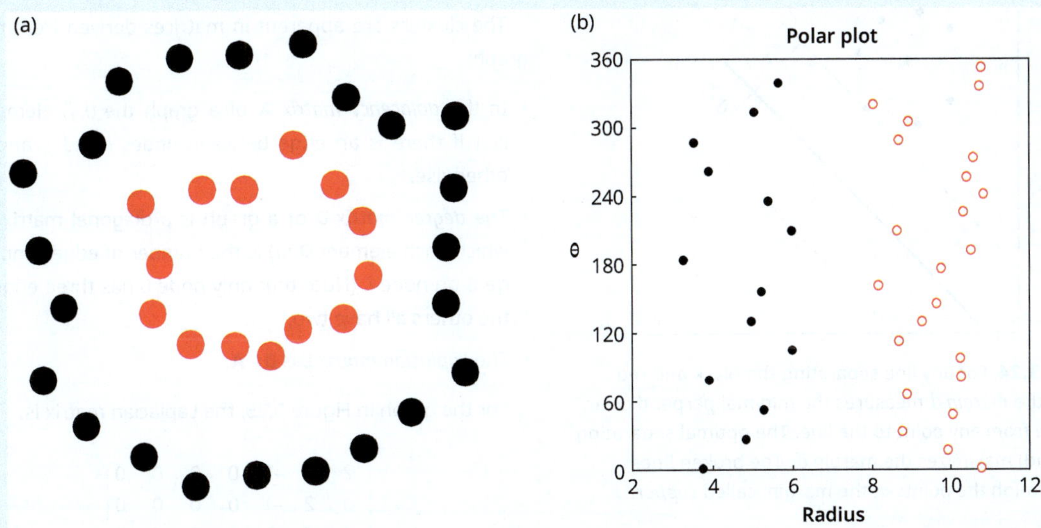

Figure 5.21 (a) A set of points not separable by a line. (b) A polar plot of the same points: convert $(x, y) \to (r, \theta)$ and then plot r as the abscissa and θ as the ordinate. This permits separation of the points by a line.

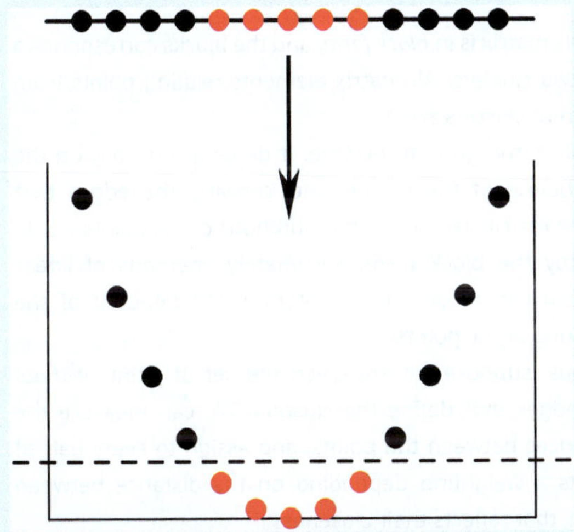

Figure 5.22 Another transformation to achieve separability. In some cases, mapping into a higher dimension allows separation by a line. Converting the one-dimensional distribution x_i into the two dimensions: $x_i \to (x_i, x_i^2)$ allows separation.

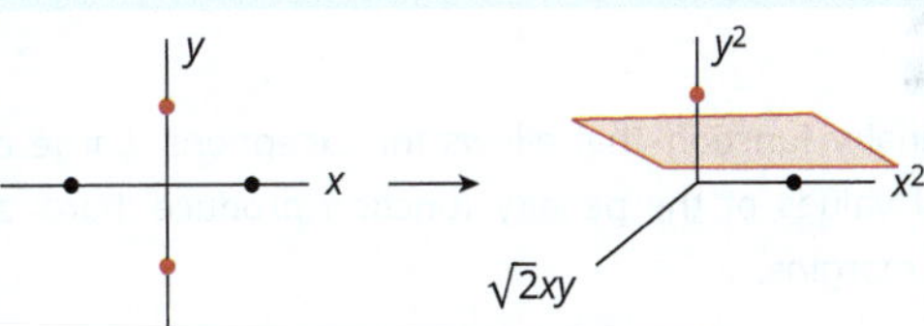

Figure 5.23 Mapping of a non-linearly-separable set of points into higher dimensions, where they are separable by a plane. The transformation from two to three dimensions is: $\Phi(x, y) = (x^2, \sqrt{2}xy, y^2)$.

two, numbers per point. But in other cases, a useful transformation might map the points into a space of very high dimensions. It would no longer be convenient to compute and store the results of the mapping. And of course it would be impossible to draw a simple graph, like the one on the right of Figure 5.23, to display them.

Support Vector Machines (SVMs)

Even in a simple case of well-separated classes (see Figure 5.17) there are many ways to draw the separating line. For each such line there is a 'margin': the minimum perpendicular distance from any point to the line.

Support Vector Machines provide an approach to choosing an optimal dividing line, by maximizing the margin (Figure 5.24). This is the most robust choice: Should additional points submitted for classification invade the margin, the chance is greatest that they will be classified correctly.

The closest points to the line are called the support vectors. Moving a support vector will shift the decision boundary. Moving other vectors will not, provided that their new positions do not invade the margin.

What if the distributions overlap? An optimization procedure to choose the dividing line can be altered to include

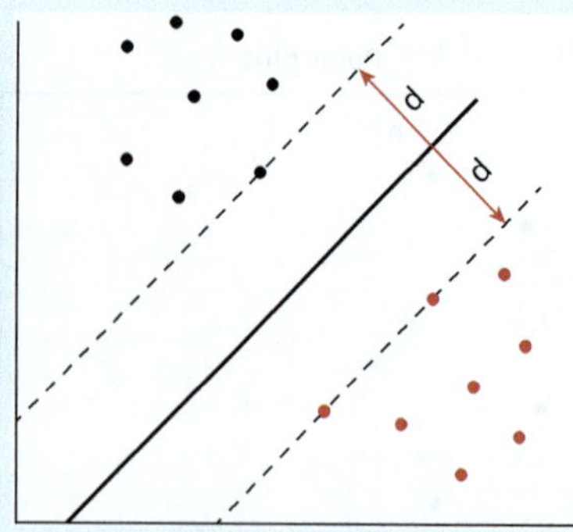

Figure 5.24 For any line separating the black and red points, the *margin d* measures the minimal perpendicular distance from any point to the line. The optimal separating line (solid) maximizes the margin *d*. The broken lines pass through the points *at* the margin, called *support vectors*.

Add a black point at the position of the letter d that appears nearer the red points. Could you still draw a separating line? Estimate the ratio of the new margin to *d*. It might be preferable to keep the old dividing line and accept the margin violation.

a 'penalty function' that allows for exceptions. Large and small values of the penalty function produce 'hard' and 'soft' margins.

Clustering by graph spectral theory

A graph is a collection of nodes, and edges between pairs of nodes.

There is a close relationship between clusters and graphs. A *path* in a graph between two nodes is a succession of edges beginning at one node and ending at the other. A *connected subset* of a graph has a path between any pair of points in the subset. A dissection of a graph into connected subsets creates a partition, or clustering, of the set. Figure 5.25 shows a graph with two connected subsets.

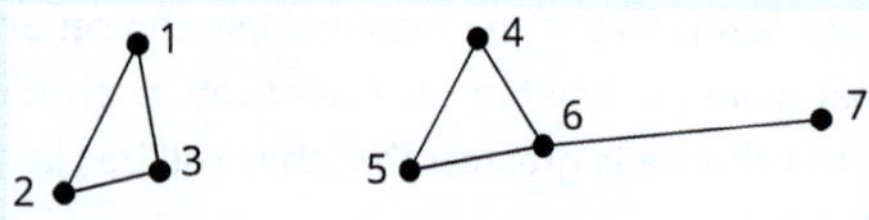

Figure 5.25 Two clusters of points.

(From: Kannan, N., & Vishveshwara, S. (1999). Identification of side-chain clusters in protein structures by a graph spectral method. J. Mol. Biol. 292: 441–464.)

The clusters are apparent in matrices derived from the graph:

- In the *adjacency matrix* **A** of a graph the (i, j) element is 1 if there is an edge between nodes i and j, and 0 otherwise.
- The *degree matrix* **D** of a graph is a diagonal matrix in which each element $\mathbf{D}(i,i)$ is the number of edges impingent on node i. (Note that only node 6 has three edges; the others all have two.)
- The *Laplacian matrix* **L** is $\mathbf{D} - \mathbf{A}$.

For the graph in Figure 5.25, the Laplacian matrix is:

$$\mathbf{L} = \begin{bmatrix} 2 & -1 & -1 & 0 & 0 & 0 & 0 \\ -1 & 2 & -1 & 0 & 0 & 0 & 0 \\ -1 & -1 & 2 & 0 & 0 & 0 & 0 \\ 0 & 0 & 0 & 2 & -1 & -1 & 0 \\ 0 & 0 & 0 & -1 & 2 & -1 & 0 \\ 0 & 0 & 0 & -1 & -1 & 3 & -1 \\ 0 & 0 & 0 & 0 & 0 & -1 & 1 \end{bmatrix}$$

This matrix is in *block form,* and the blocks correspond to the two clusters. All matrix elements relating points from different clusters are 0.

This is too good to be true. It depends crucially on the numbering of the points, and knowing the edges that define the clusters. Most permutations of the points would destroy the block form. Fortunately, methods of linear algebra can recover the clustering, independent of the indexing of the points.

Thus, suppose we are given the set of points *without* the edges that define the clusters. We can measure the distances between the points, and assign to every pair of points a weighting depending on the distance between them, that reflects their closeness:

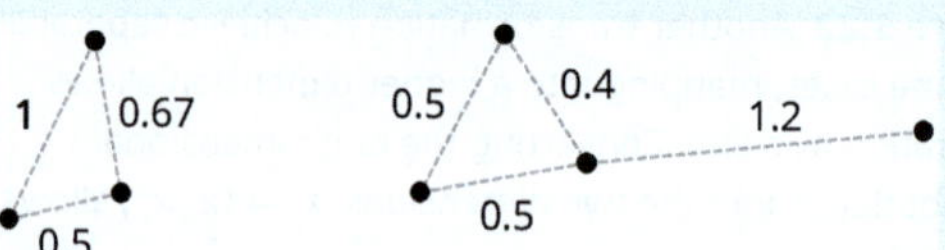

The Weight matrix **W** is a generalization of the adjacency matrix **A**. If $d(i, j)$ is the distance between points i and j,

$$\mathbf{W}(i,j) = \begin{cases} 1/d(i,j) & \text{if } d(i,j) \le 1.3 \\ 0.01 & \text{if } d(i,j) > 1.3 \end{cases}$$

Let a_i be the sum of the elements in row i of **W**: $a_i = \sum_j \mathbf{W}(i,j)$. Then a generalization of **D** is:

$$\mathbf{D}'(i,j) = \begin{cases} a_i & \text{if } i = j \\ 0 & \text{if } i \neq j \end{cases}$$

D′ is still a diagonal matrix.

The new Laplacian, **D′** – **W**, is:

$$\mathbf{L} = \begin{bmatrix} 1.71 & -1 & -0.67 & -0.01 & 0 & -0.01 & 0 \\ -1 & 1.54 & 0.5 & -0.01 & 0 & -0.01 & 0 \\ 0.67 & -0.5 & 1.21 & -0.01 & 0 & -0.01 & -0.01 \\ -0.01 & -0.01 & 0 & 0.94 & -0.5 & -0.4 & -0.01 \\ -0.01 & -0.01 & 0 & -0.5 & 1.21 & -0.67 & -0.01 \\ -0.01 & -0.01 & 0 & -0.4 & -0.67 & 2.3 & -1.2 \\ -0.01 & -0.01 & -0.01 & -0.01 & -0.01 & -1.2 & 125 \end{bmatrix}$$

Note that the clustering is clear from the distribution of off-diagonal values. But suppose it were not obvious, if for example the points were renumbered.

The Laplacian matrix can be diagonalized. This means solving the equation:

$$(\mathbf{L} - \lambda\mathbf{I})X = 0$$

where **I** is the identity matrix ($\mathbf{I}(i,j) = 1$ if $i = j$ and 0 if $i \neq j$), and X is a vector. In general, if the matrix to be diagonalized has n rows and n columns, there will be n values of λ, called the eigenvalues, and n corresponding n-vectors X, its eigenvectors.

The first two eigenvalues (in descending numerical order) and the corresponding eigenvectors are:

Eigenvalues:	0.0000	0.7000	…	(7 eigenvalues in all)
Eigenvectors:	0.3780	0.4364	…	(7 eigenvectors in all)
	0.3780	0.4364		
	0.3780	0.4364		
	0.3780	−0.3273		
	0.3780	−0.3273	…	
	0.3780	−0.3273		
	0.3780	−0.3273		

The first eigenvector does not distinguish the points. But the second eigenvector has different values for points in different clusters. It has extracted the clusters for us. It was not necessary to specify a target number of clusters.

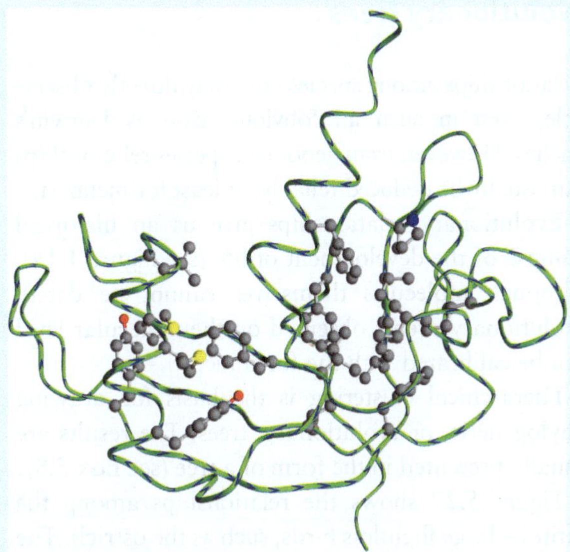

Figure 5.26 Bovine angiogenin [1AGI], hydrophobic clusters determined by Kannan & Vishveshwara, using graph spectral methods.

What if the numbering of the points were permuted? The eigenvectors and eigenvalues are invariant, although the order of the values would change to reflect the renumbering. This solves the problem that the clusters were evident in the original adjacency matrix only because points in the same clusters were numbered consecutively in Figure 5.25.

> Graph-spectral clustering projects data into a space of fewer dimensions. In contrast, the kernel method for Support Vector Machines projects data into a higher dimensional space.

Kannan and Vishveshwara have applied graph spectral theory to detecting clusters of sidechains in proteins. Figure 5.26 shows their result for bovine angiogenin [1AGI].[4]

Vishveshwara and colleagues have applied graph spectral techniques to a number of other problems in bioinformatics; for instance, the detection of clusters in metabolic reaction networks.

[4] Acharya, K.R., Shapiro, R., Riordan, J.F., *et al.* (1995). Crystal structure of bovine angiogenin at 1.5-Å resolution. Proc. Natl. Acad. Sci. USA 92: 2949–2953.

Evolutionary trees

Relationships among species are rarely directly observable, even in such an 'obvious' case as Darwin's finches. However, from genomics, species relationships can usually be deduced reliably, at least for metazoa.

Evolutionary relationships give us an historical glimpse of the development of life (see Figure 1.14). Although molecules themselves cannot be dated, evolutionary events observed on the molecular level can be calibrated with the fossil record.

Hierarchical clustering is the basis for drawing phylogenetic, or evolutionary, trees. The results are usually presented in the form of a tree (see Box 5.8).

Figure 5.27 shows the relationships among the ratites—large flightless birds, such as the ostrich. The ancestor of the ratites is believed to be a bird that could fly, probably related to the extant tinamous. Such a tree, showing descendants of a single original ancestral species, is said to be **rooted**. (The root of the tree usually appears at the top or the side; botanists will have to get used to this.)

Alternatively, we may be able to specify relationships but not order them according to a history. The relationships among the finches of the Galapagos Islands, studied by Darwin, plus a related species from the nearby Cocos Island, are shown in an unrooted tree (Figure 5.28). Addition of data from a species on the South American mainland ancestral to the island finches would allow us to root the tree.

KEY POINT

The idea of phylogeny is to observe different degrees of similarity among species or higher taxa, assume that the species are related by descent from a common ancestor and that higher degrees of similarity correspond to closer relationships, and try to capture the relationships in a tree diagram showing ancestor–descendant relationships such that species more closely related according to the tree do have higher degrees of similarity.

Structure and contents of an evolutionary tree

In computer science, a tree is a particular kind of **graph**. A graph is a structure containing nodes (abstract points) connected by edges (represented as lines between the points). A path between two nodes in a graph is a series of consecutive edges that begins at one node and ends at the other. In a general graph, there may be many paths between any two nodes. We shall discuss graphs in more detail in Chapters 13 and 14.

A tree is a special kind of graph. First of all, a tree must be connected, meaning that there is a path through the graph between any two points. Second, in a tree there is *only* one path between every two points. We have already seen several trees; for example, Figure 1.12.

A particular node may be selected as a **root** of a tree. However, this is not necessary—abstract trees may be **rooted** (for instance, Figure 5.27) or **unrooted** (for instance, Figure 5.28).

In phylogenetic trees, the root is the earliest common ancestor of all of the other nodes. Rooted phylogenetic trees explicitly show ancestor–descendant relationships, as from any node there is a connected path up through successive ancestors terminating at the root. Unrooted trees show the topology of relationship but not the pattern of descent.

It may be possible to assign numbers to the edges of a graph to indicate some kind of 'length' of the edges, corresponding to a 'distance' between the nodes that the edges connect. These lengths are not necessarily geometric distances, but may be abstract values. Given edge lengths, the graph may be drawn to scale, with the sizes of the edges proportional to the assigned lengths.

In phylogenetic trees, edge lengths signify either some measure of the dissimilarity between two taxa, or the length of time since their separation. The assumption that differences between properties of living species reflect their divergence times will be true only if the rates of divergence are the same in all branches of the tree. Many exceptions are known. For instance, among mammals, many proteins from rodents show relatively fast evolutionary rates.

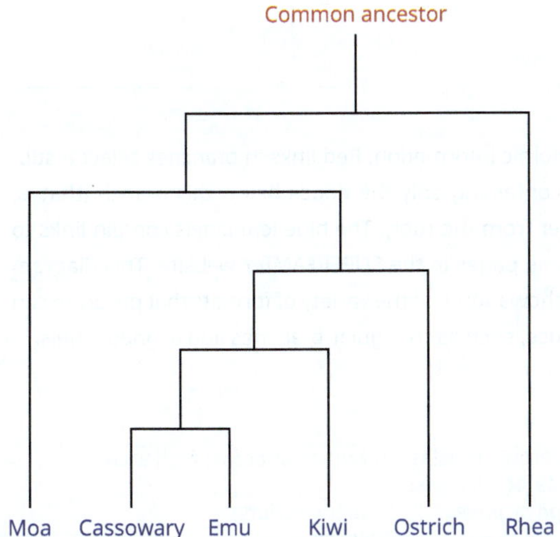

Figure 5.27 Phylogenetic tree of ratites (large flightless birds), based on mitochondrial DNA sequences. The common ancestor is at the **root** of this tree, appearing at the top of the graph. A surprising implication of these DNA sequences is that the moa and kiwi are not the closest relatives and, therefore, New Zealand must have been colonized twice by ratites or their ancestors. In terms of geography, this is less surprising if one looks at a map of the ancient continent Gondwanaland prior to its break-up, rather than at a contemporary map on which the distance between Africa and New Zealand is large.

The statement of a tree of relationships may reveal only the connectivity or topology of the tree, in which case the lengths of the branches are arbitrary. A more ambitious goal is to show the distances between taxa quantitatively, for instance to label the branches with the time since divergence from a common ancestor (see Box 5.9).

A phylogenetic tree tells us the organization of a set of taxa. It does not tell us how they should be grouped or partitioned. For example, the tree of Darwin's finches *looks* as if it could reasonably be partitioned into two or three clusters. The guiding principle in selecting a partition is that the intracluster similarities be relatively high and the intercluster similarities be relatively low. If the data are intrinsically well grouped, then the clustering is obvious. If there is no clear separation, proper clustering is ambiguous and difficult.

If a hierarchical clustering is achievable, the results can be represented as a tree. Trees have many favourable features for describing relationships. However, other possible representations, with more general, and complicated, geometric structure, may in some cases be worth consideration. (Most evolutionary biologists are 'tree-huggers' and are reluctant to consider any alternative.) H.-J. Bandelt and A.W.M. Dress developed a more general graphical clustering method, called *split decomposition,* that can better account for a set of distances than a tree structure.[5]

[5] Bandelt, H.-J., & Dress, A.W.M. (1992). Split decomposition: A new and useful approach to phylogenetic analysis of distance data. Molecular Phylogen. and Evol. 1: 242–252.

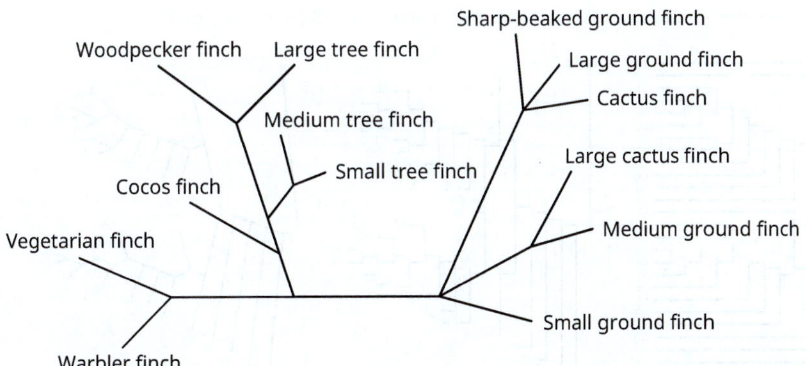

Figure 5.28 Unrooted tree of relationships among finches from the Galapagos and Cocos Islands. Darwin studied the Galapagos finches in 1835, noting the differences in the shapes of their beaks and the correlation of beak shape with diet. Finches that ate fruits had beaks like those of parrots, whereas finches that ate insects had narrow, prying beaks. These observations were seminal to the development of Darwin's ideas. As early as 1839 he wrote, in *The Voyage of the Beagle*, 'Seeing this gradation and diversity of structure in one small, intimately related group of birds, one might really fancy that from an original paucity of birds in this archipelago, one species had been taken and modified for different ends'.

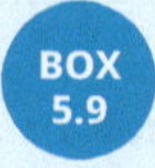

BOX 5.9 Drawing phylogenetic trees

There are many programs available for drawing phylogenetic trees; some static, some interactive (https://en.wikipedia.org/wiki/List_of_phylogenetic_tree_visualization_software). The diagram shows output from TreeVector. On the web it creates interactive diagrams: moving the mouse over the red circles displays extended taxonomic information. Red links in branches select a subtree containing only the nodes below this branch (that is, farther from the root). The blue leaf labels contain links to genome pages in the SUPERFAMILY website. This diagram also shows some of the variety of formats that program can produce, such as triangular branches and aligned labels.

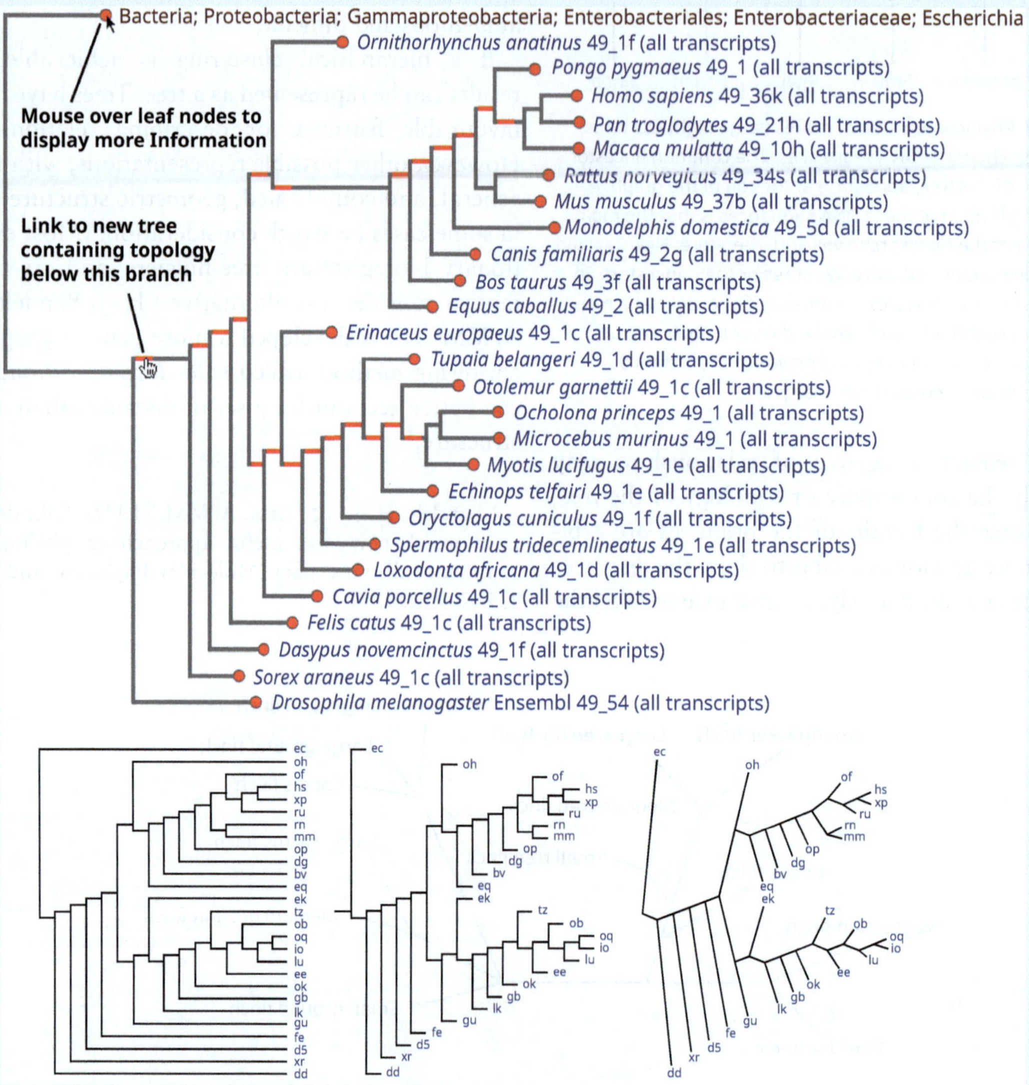

Pethica, R., Barker, G., Kovacs, T., *et al.* (2010). TreeVector: Scalable, interactive, phylogenetic trees for the web. PLoS ONE 5, 1: e8934. https://doi.org/10.1371/journal.pone.0008934 © 2010 Pethica et al.

An example where split decomposition is useful appears in mammalian evolution. The early radiation of placental mammals was rapid, occurring during a period of both climate warming and fragmentation of the continents. The nodes in a phylogenetic tree lie very close together, making it difficult to determine the branching order. Incomplete separation of lineages, and genomic mosaicity, leads to confounding phylogenetic signals.

B.M. Hallström and A. Janke have studied the early evolution of placental mammals, and present the results in the form of a split decomposition (Figure 5.29a). Major clades of placental mammals are:

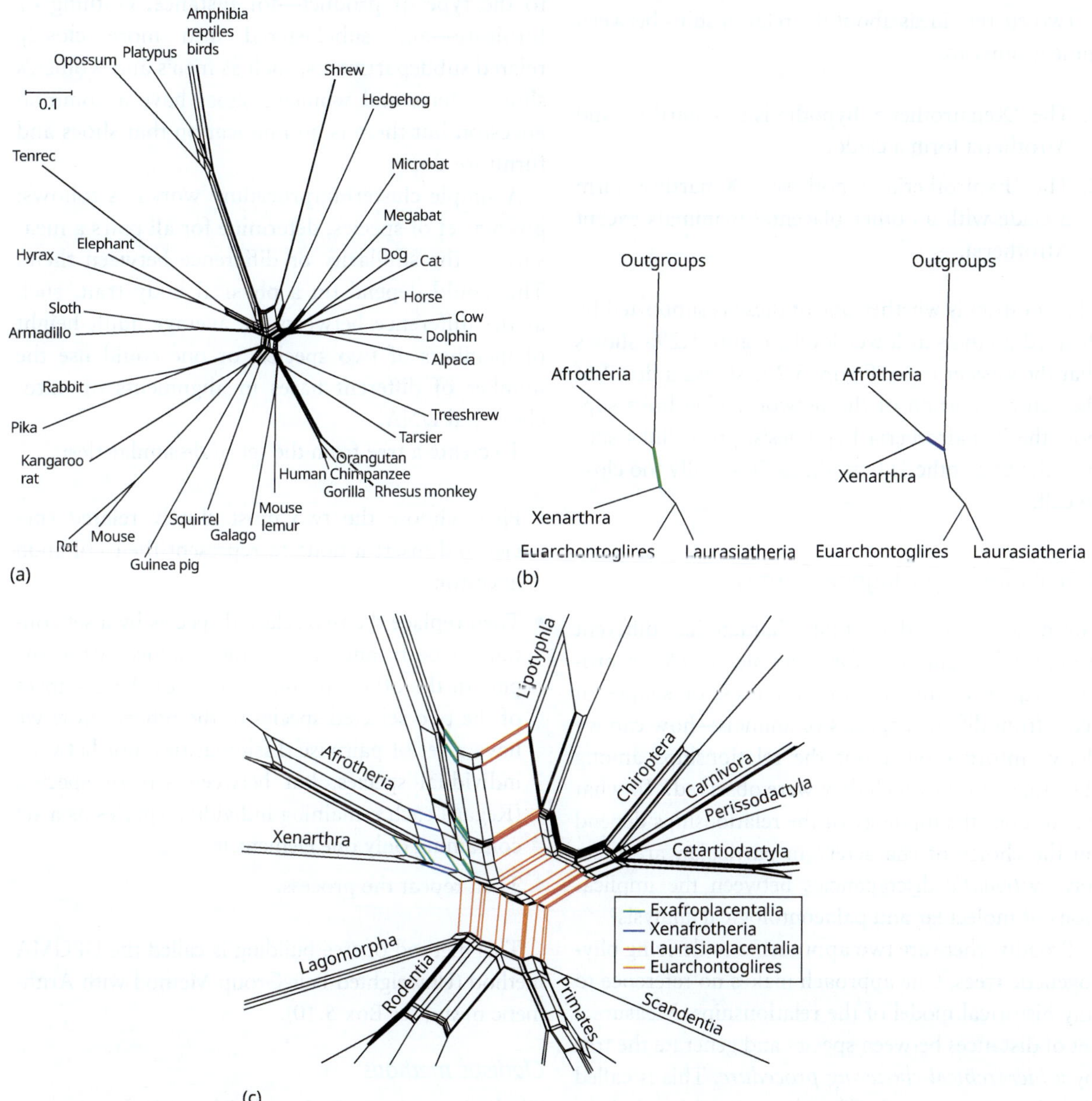

Figure 5.29 (a) Split-decomposition network of mammalian species, plus outliers, based on alignment protein sequences containing a total of 954 000 amino acids, from 37 species. (b) Trees illustrating the Xenafrotheria hypothesis (left) and Exafrotheria hypothesis (right). Blue and green links point out the difference, and correspond to the same colours in Figure 5.29c. (c) Central part of the network from (a), highlighting major splits. (a) and (c) from Hallström, B.M., & Janke, A. (2010). Mammalian evolution may not be strictly bifurcating. Molecular Biology and Evolution 27, 12: 2804–2816. https://doi.org/10.1093/molbev/msq166; (b) redrawn from personal communication from B.M. Hallström.

Clade	Extant examples
Afrotheria	elephants, dugongs, aardvarks
Euarchontoglires	rodents, rabbits, treeshrews, primates
Xenarthra	anteaters, tree sloths, armadillos
Laurasiatheria	shrews, pangolins, bats, whales, ungulates

Two current ideas about the relationships between these groups are:

1. The 'Xenafrotheria hypothesis': Xenarthra and Afrotheria form a clade.
2. The 'Exafrotheria hypothesis': Xenarthra form a clade with all other placental mammals except Afrotheria.

The question is whether one of these is supported by detailed analysis at lower levels. Figure 5.29a shows that the answer is no. Figure 5.29c shows a detail of the central portion of the network. Blue links support the Xenafrotheria hypothesis; green links support the Exafrotheria hypothesis. It is really too close to call.

Calculation of phylogenetic trees

Given a set of data that characterize different groups of organisms—for example, DNA or protein sequences, or protein structures, or shapes of teeth from different species of animals—how can we derive information about the relationships among the organisms in which they were observed? To what extent does the topology of the relationships depend on the choice of character? In particular, are there any *systematic* discrepancies between the implications of molecular and palaeontological analysis?

Broadly, there are two approaches to deriving phylogenetic trees. One approach makes no reference to any historical model of the relationships. Measure a set of distances between species and generate the tree by a *hierarchical clustering procedure*. This is called the *phenetic* approach. The alternative, the *cladistic* approach, is to consider possible pathways of evolution, infer the features of the ancestor at each node, and choose an optimal tree according to some model of evolutionary change. Phenetics is based on similarity; cladistics is based on genealogy.

Clustering methods

Phenetic, or clustering, approaches to determination of phylogenetic relationships are explicitly non-historical. Indeed, hierarchical clustering is perfectly capable of producing a tree even in the absence of evolutionary relationships. A departmental store has goods clustered into sections according to the type of product—for instance, clothing or furniture—and subclustered into more closely related subdepartments, such as men's and women's shoes. Men's and women's shoes have a common ancestor, but there is no implication that shoes and furniture do.

A simple clustering procedure works as follows: given a set of species, determine for all pairs a measure of the similarity or difference between them. This could depend on a physical body trait, such as the difference between the average adult height of members of two species, or one could use the number of different bases in alignments of mitochondrial DNA.

To create a tree from the set of dissimilarities:

- First, choose the two most closely related species and insert a node to represent their common ancestor.
- Then replace the two selected species by a set containing both, and replace the distances from the pair to the others by the average of the distances of the two selected species to the others. Now we have a set of pairwise dissimilarities, not between individual species, but between sets of species. (Regard each remaining individual species as a set containing only one element.)
- Then repeat the process.

This method of tree building is called the UPGMA method (Unweighted Pair Group Method with Arithmetic mean; see Box 5.10).

Cladistic methods

Cladistic methods deal explicitly with the patterns of ancestry implied by the possible trees relating a set of taxa. Their aim is to select the correct tree by utilizing an explicit model of the evolutionary process. The most popular cladistic methods in molecular phylogeny are the *maximum parsimony*

BOX 5.10 Calculation of phylogenetic trees by clustering

Consider four species characterized by homologous sequences ATCC, ATGC, TTCG, and TCGG. Taking the number of differences as the measure of dissimilarity between each pair of species, we will use a simple clustering procedure to derive a phylogenetic tree.

The distance matrix is:

	ATCC	ATGC	TTCG	TCGG
ATCC	0	**1**	2	4
ATGC		0	3	3
TTCG			0	2
TCGG				0

(As the matrix is symmetric, we need fill in only the upper half.)

The smallest distance is **1** (in boldface), between ATCC and ATGC. Therefore, our first cluster is {ATCC, ATGC}. The tree will contain the fragment:

The reduced distance matrix is:

	{ATCC, ATGC}	TTCG	TCGG
{ATCC, ATGC}	0	$\frac{1}{2}(2+3)=2.5$	$\frac{1}{2}(4+3)=3.5$
TTCG		0	**2**
TCGG			0

For example, the number 3.5 in the upper right was calculated by averaging the distances between ATCC and TCGG = 4 and between ATGC and TCGG = 3.

The next cluster is {TTCG, TCGG}, with distance **2**. Finally, linking the clusters {ATCC, ATGC} and {TTCG, TCGG} gives the tree:

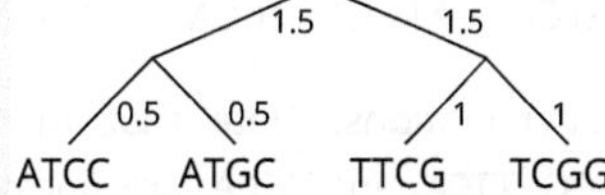

Branch lengths have been assigned according to the rule:

$$\text{branch length of edge between nodes X and Y} = \frac{1}{2}\text{distance between X and Y}$$

Whether the branch lengths are truly proportional to the divergence times of the taxa represented by the nodes must be determined from external evidence.

and *maximum likelihood* approaches. They are specialized to sequence data, starting from a multiple sequence alignment. Neither maximum parsimony nor maximum likelihood could be applied to anatomic characters such as average adult height.

The maximum parsimony method of W. Fitch defines an optimal tree as the one that postulates the fewest mutations (see Box 5.11).

The maximum likelihood method assigns quantitative probabilities to mutational events, rather than merely counting them. Like maximum parsimony, maximum likelihood reconstructs ancestors at all nodes of each tree considered; however, it also assigns branch lengths based on the probabilities of the mutational events postulated. For each possible tree topology, the assumed substitution rates are varied to find the parameters that give the highest likelihood of producing the observed sequences. The optimal tree is the one with the highest likelihood of generating the observed data.

Both maximum parsimony and maximum likelihood methods are superior to clustering techniques. This has been demonstrated with cases where independent evidence—for instance, from palaeontology—provides a correct answer, and also with simulated data—computer modelling of evolving sequences.

Calculation of phylogenetic trees by maximum parsimony

Given species characterized by homologous sequences ATCG, ATGG, TCCA, and TTCA, the tree:

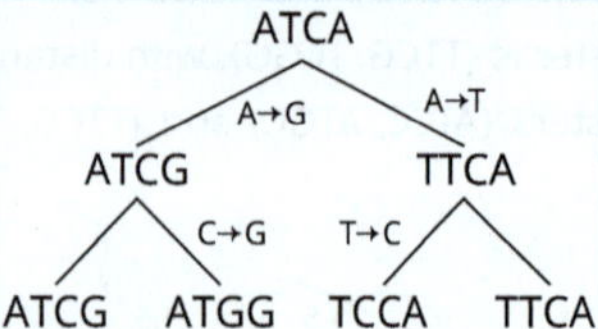

postulates four mutations. Note that the ancestral sequences, ATCG, TTCA, and ATCA, are not part of the observable data.

An alternative tree:

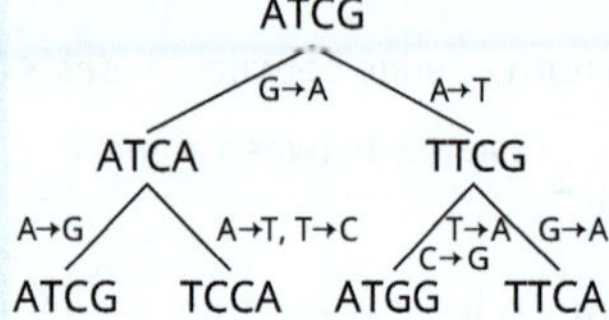

postulates seven mutations. Note that the second tree implies that the $G \rightarrow A$ mutation in the fourth position occurred twice independently. The former tree is optimal according to the maximum parsimony method, because no other tree involves fewer mutations. In many cases, several trees may postulate the same number of mutations, fewer than any other tree. For such cases, the maximum parsimony approach does not give a unique answer.

The problem of varying rates of evolution

Suppose that four species, A, B, C, and D, have the phylogenetic tree:

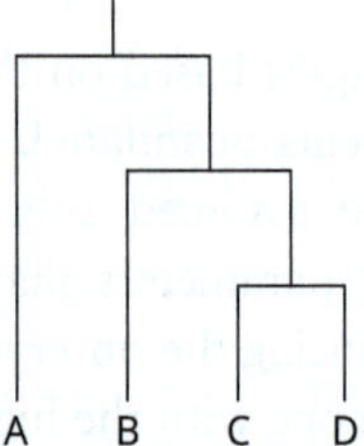

This tree is consistent with the dissimilarity matrix:

	A	B	C	D
A	0	3	3	3
B		0	2	2
C			0	1
D				0

Suppose, however, that taxon D is changing very fast, although the phylogeny is unaltered. The dissimilarity matrix might then be observed to be:

	A	B	C	D
A	0	3	3	20
B		0	2	20
C			0	20
D				0

from which we would derive the *incorrect* phylogenetic tree:

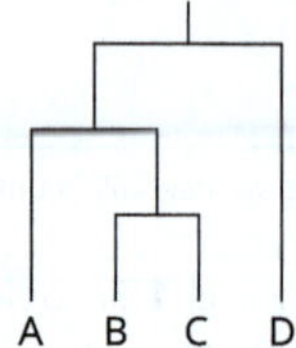

All of the methods discussed here are subject to errors of this kind if the rates of evolutionary change vary along different branches of the tree. To test for varying rates, compare the species under consideration with an **outgroup**—a species more distantly related to all of the species in question than any pair of them is to each other. For instance, if we are studying species of primates, a non-primate mammal such as the cow would be a suitable outgroup. If the rates of evolution among the primate species were constant, we would expect to observe approximately equal dissimilarity measures between all primate species and the cow. If this is not observed, the suggestion is that evolutionary rates have varied among the primates, and the character being used may well not provide the correct phylogenetic tree.

Bayesian methods

The problem we are trying to solve is: Of all possible phylogenetic trees organizing the relationships among different species, based on a given multiple sequence alignment, which one has the highest probability of generating the observed multiple sequence alignment, under some model of evolutionary change? The model might be specified in terms of probabilities of mutation rates, etc. For the moment it would seem that a weakness of the approach is the difficulty of knowing how to specify the model explicitly and accurately.

Nevertheless, from any such model of evolutionary change, we can compute the probability that any tree would produce the observed multiple sequence alignment. Suppose we begin the problem in a state of complete ignorance, meaning that we consider that initially—for all we know—all potential phylogenetic trees must be regarded as equally probable. Then Bayes' rule states that we want to choose the tree with the highest probability of producing the observed multiple sequence alignment.

What makes this approach so powerful is that we can optimize the probability of producing the observed data not only over possible trees, but *over different models of evolutionary change*. This releases us from making overly constricting assumptions such as constancy of molecular clock rates over different branches of the tree, identical mutation probabilities at all sites, etc. The calculations are nevertheless feasible. There is consensus that programs based on the Bayesian approach are the most powerful tools for deriving phylogenetic trees from multiple sequence alignments.

There are many available software packages. The user should be aware that these calculations are quite sophisticated, and the programs are very powerful but require an investment in understanding what they do to make them work for you. MEGA is perhaps the simplest for casual use. Phylip had the field to itself for many years.

Program	Approach	Website
Phylip	Various choices	http://evolution.genetics.washington.edu/phylip.html
MEGA	Various choices	http://www.megasoftware.net/
PhyML	Maximum likelihood	http://www.atgc-montpellier.fr/phyml/
RAxML	Maximum likelihood	http://sco.h-its.org/exelixis/software.html
MrBayes	Bayesian	http://mrbayes.sourceforge.net/
BEAST	Bayesian	http://beast.bio.ed.ac.uk/

Short-circuiting evolution: genetic engineering

Evolutionary divergence arises in nature through generation of variation by random mutation, followed by either selection or genetic drift to alter allele frequencies in populations or to create novel species. Contemporary techniques allow deliberate transfer of genes, to create organisms with altered characters directly, or to use CRISPR/Cas methods to edit genomes.

Biochemists used to be like astronomers—we could observe our subjects but not modify them. This is no longer true.

Another line of research has the goal of modifying individual enzymes—for instance, to enhance thermostability—or even to design and create novel ones. We shall treat this topic in Chapter 12.

In addition to gene therapy for disease, and genetically modified crop plants, many other applications are available or under development:

Use of microorganisms as protein factories. Microorganisms are routinely used in the laboratory to express proteins—human or otherwise—for research. An example with clinical application is the microbial synthesis of human growth hormone. Formerly, the only source of the hormone was by post-mortem extraction from pituitary glands. This carried the risk of transmitting prion disease. Other microbiologically produced human proteins with clinical applications include insulin, and many monoclonal antibodies. Still other applications include manufacture of fuels, or plastics, and dissolving oil spills.

Genetically modified animals. Higher animals are also used as protein factories, in cases where the active protein requires post-translational modifications of which microorganisms are incapable. Genetically engineered goats secrete an anticoagulant, human antithrombin III, in their milk. This product has been approved for clinical use in the US. Production of drugs by this route is called 'pharming'.

Other goals of genetically modified animals include:

(a) enhancing the nutritional value of food; e.g., pork enriched in ω–3 fatty acids.

(b) pigs lacking the cell-surface antigens that produce rejection by the human immune system, as a source of organs for transplant.

(c) animals that grow faster and/or require less expensive feed; for example, fast-growing salmon.

(d) protecting livestock against disease; e.g., cows lacking prion proteins and thereby not susceptible to bovine spongiform encephalopathy (BSE).

(e) allergen-free pets.

(f) fish that glow in colours by virtue of genes for fluorescent proteins.

Genetically modified plants. Many crop plants are targets for genetic modification. Goals include:

(a) pesticide-resistant plants, that allow treatments to kill weeds or insect pests without damaging the plants.

(b) a related approach is a plant that makes its own insecticide. Bt-corn (maize) contains a natural insect-killing gene transferred from *Bacillus thuringiensis*.

(c) crops with enhanced nutritional value. An example is 'golden rice' enriched in vitamin A (see Chapter 9).

(d) fruit with longer shelf life, such as the 'flavr-savr' tomato.

(e) crops that produce only sterile seeds.

(f) a blight-resistant genetically modified chestnut tree (see Box 5.12).

There are a number of controversial aspects to these activities. In the case of genetically modified plants,

BOX 5.12 Blight-resistant chestnut trees

In nature, there are four major species of chestnut trees: the American chestnut (*Castanea dentata*), the European chestnut (*C. sativa*), the Chinese chestnut (*C. mollissima*), and the Japanese chestnut (*C. crenata*), plus nine minor species.

Even today, much of the mountain regions of Italy are chestnut forest, cultivated for nut production and used for timber. Orchards with 500-year-old trees are still harvested every year. Among famous individual trees is the 3000-year-old *Chestnut of a Hundred Horses,* in Sicily on the Eastern slope of Mt. Etna, a few km from the crater. Legend has it that Queen Joanna of Aragon (reigned 1516–1555), together with one hundred knights and horses, took shelter during a storm under the branches of this huge tree (trunk 58 m in circumference!).

American chestnut trees (*Castanea dentata*) used to populate forests of the Eastern US very extensively. Because it was so fast-growing and could attain huge sizes, it often dominated both rural and urban ('Under the spreading chestnut tree the village smithy stands'—Longfellow) landscapes.

Chestnut wood is rot-resistant, straight-grained, and suitable for building and for furniture. (It is likely that US President Abraham Lincoln's celebrated 'log cabin' was built from chestnut trees.)

The edible nuts made a major contribution to agriculture.[6] Chestnut fodder fattened hogs and cattle. Roasted chestnuts are traditional at a Christmas dinner. They are sold from carts on streets in many cities (see: a contemporary photograph https://anamericaninrome.com/2016/12/roasted-chestnuts-in-rome-a-winter-time-snack/ and a sketch by Toulouse-Lautrec https://www.moma.org/collection/works/65562). Historically, during catastrophes—after the fall of the Roman Empire, the Middle Ages, and the First and the Second World Wars—chestnuts staved off

[6] 'When chestnuts were ripe I laid up half a bushel for winter. It was very exciting at that season to roam the then boundless chestnut woods of Lincoln . . . amid the rustling of leaves and the loud reproofs of the red squirrels and the jays, whose half-consumed nuts I sometimes stole, for the burs which they had selected were sure to contain sound ones. Occasionally I climbed and shook the trees. They grew also behind my house, and one large tree, which almost overshadowed it, was, when in flower, a bouquet which scented the whole neighborhood, but the squirrels and the jays got most of its fruit; the last coming in flocks early in the morning and picking the nuts out of the burs before they fell, I relinquished these trees to them and visited the more distant woods composed wholly of chestnut. These nuts, as far as they went, were a good substitute for bread.'—Henry David Thoreau, *Walden*.

starvation. Chestnuts are still used in cooking, for instance for soup, and for a cake called *Pattona,* made of chestnut flour, pine nuts, and raisins. (Incidentally, chestnut flour is gluten-free.)

Now the bad news: Chestnut trees are host to a bark fungus, *Cryphonectria parasitica* (see Figure 5.30).

C. parasitica originally infected Asian chestnut trees, but those evolved resistance to its most severe effects. The fungus was accidentally introduced to Europe and the US in the early twentieth century. Its effects have been devastating: At least until recently, chestnut blight was the greatest ecological disaster in US history. The blight killed off almost the entire population of American chestnut trees on the East coast, from Maine to Georgia (see Figure 5.31). Between 1904 and 1940, an estimated four billion trees died. Large American chestnuts all but disappeared from forests and cities.

Figure 5.30 Canker on American chestnut tree caused by *Cryphonectria parasitica* infection. The pathogen causes the collapse of stems. If the canker girdles the stem completely, it disrupts nutrient and water transport, and the parts of the tree above the canker die.

(Photo courtesy of Michigan State University Extension.)

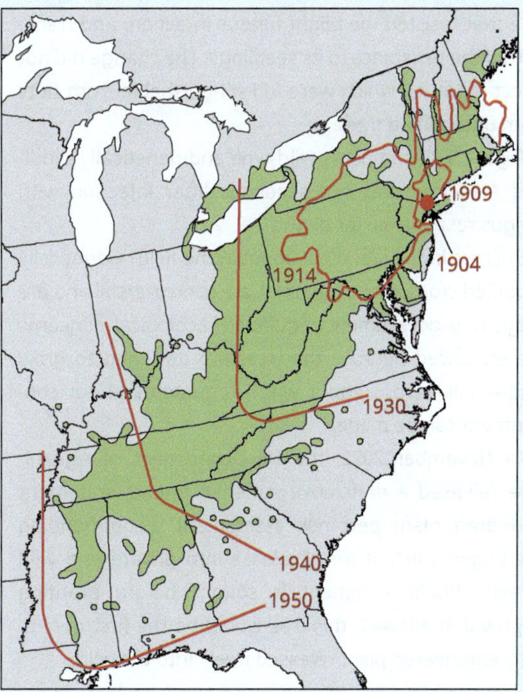

Figure 5.31 Map of spread of chestnut blight fungus in Eastern US after introduction in 1904.

Gurney, K. (2014). The American Chestnut: Restoring a Native Tree. [online] The American Chestnut Foundation. Available at: https://legislature.vermont.gov/Documents/2014/WorkGroups/House%20Agriculture/American%20Chestnut%20Tree%20Restoration/W~Kendra%20Gurney~American%20Chestnut%20Tree%20Restoration%20Hearing~4-10-2014.pdf.

European chestnut trees also suffer from the blight, first observed in 1938. Attempts, in both Europe and the US, to introduce resistant Asian chestnut trees, or hybrids produced by traditional breeding methods, were not successful.

In 1990, two tree geneticists, William Powell and Charles Maynard, at the State University of New York College of Environmental Science and Forestry, sought to create blight-resistant chestnut trees by genetic engineering.

The fungus kills the trees by secreting oxalic acid. Powell and Maynard introduced into the chestnut genome a gene from wheat that encodes an enzyme, oxalate oxidase, that degrades the toxin to CO_2 and H_2O_2 (hydrogen peroxide). It was also necessary to modify the promoter sequence to enhance expression levels. Powell and Maynard created a tree that resisted the blight fungus infection, and transmitted the resistance to its seedlings. The change did not affect the nuts, which were indistinguishable from nuts from unmodified trees.

Figure 5.32 compares wild-type and genetically modified American chestnut seedlings after infection with fungus responsible for blight.

Success in the lab! What about in the field? Genetically modified crops, including trees, are controversial and are subject to government regulation. Ecological concerns are enhanced because the trees are designed to grow wild—unlike crops about which arguments about containment can be made.

In November 2022, the US Department of Agriculture released a draft environmental impact statement and draft plant pest risk assessment recommending the blight-resistant trees' release into private and wild forests. Public comment is sought before granting approval. If allowed, this tree would be the first genetically engineered plant released freely into the wild.

Figure 5.32 A month after infection with chestnut blight fungus *Cryphonectria parasitica*, wild-type American chestnut seedlings (left) are wilted but transgenic trees are healthy.

Photograph by A. Newhouse (slightly cropped). From: https://theconversation.com/new-genetically-engineered-american-chestnut-will-help-restore-the-decimated-iconic-tree-52191

there is concern over the spreading of genes from the crops to undesired hosts. For instance, the purpose of introducing a gene for herbicide resistance into a crop plant is to make it easier selectively to kill weeds without affecting the crop plant. However, it has been observed that the gene can spread to the weeds.

A concern about sterile seeds is economic. Use of sterile seeds requires farmers to purchase new seeds each year. It precludes the traditional agricultural practice of holding back a portion of a crop for replanting. In addition to the specific economic implications, there is a widespread feeling that biotechnology might alter the relationship between people and nature that has been a common cultural heritage for thousands of years. It would be wrong to dismiss these feelings as irrational or as characterizing only a fringe.

There are many examples of natural genetic engineering. We discussed the CRISPR/Cas system in Chapter 1. Box 5.13 describes another example.

BOX 5.13 Tom and Jerry, and toxoplasmosis

Parasites modify their hosts in many ways. Interference with higher mental processes of mammalian hosts, an effect of infection by the protozoan *Toxoplasma gondii*, is unusual and interesting.

Tom and Jerry cartoons feature a mouse (Jerry) unnaturally unafraid of a cat (Tom), that a mouse would normally shun as a threatening predator. Jerry is showing symptoms of toxoplasmosis. The protozoan parasite *T. gondii* alters the perception and behaviour of a rodent host. Mice and rats infected with *T. gondii* lose their normal fear and avoidance of cats, and instead show a 'fatal attraction' to cat odours. The altered response is not a generalized fear suppression—a tranq'—but a very narrow and precisely directed one.

There is a good reason for this from the point of view of the protozoan. Only in the primary host, a cat, can it complete its life cycle, and reproduce sexually. Secondary hosts such as rodents promote the proliferation of the protozoan only upon transfer to cats, through killing and ingestion of prey.

(*T. gondii* is capable of infecting all mammals. Of course, only those that are smaller and living in the same community are possible prey to cats. Tom and Jerry cartoons are usually set in a city with no felidae other than moggies like Tom. However, the entire family can serve as primary hosts; large cats such as lions and tigers can acquire *T. gondii* from prey larger than small rodents.)

How does the parasite achieve its effect? It combines a number of mechanisms:

(1) It is likely that dopamine is involved. There have been reports of increased dopamine production in infected rodent brains. The *T. gondii* genome contains two genes encoding tyrosine hydroxylase, that catalyses the conversion of L-tyrosine to L-3,4-dihydroxyphenylalanine. This reaction is the rate-limiting step in dopamine synthesis. Support for this idea is that, in *T. gondii*-infected rodents, drugs that interfere with the dopamine-receptor binding tend to restore the normal aversion to cats.

(2) *T. gondii* upregulates synthesis of testosterone, in male hosts.

(3) Rewiring neural circuits. Normally in rodents, exposure to cat odour activates neural circuits responsible for perception of fear. One way that *T. gondii* changes this response is by an epigenetic change in the host.

The targets of *T. gondii* genetic engineering are the amygdalae, two symmetrically disposed clusters of cells in the temporal lobes of the brain. They are involved in various aspects of the fear response, including recognition of fear-inducing stimuli and evincing of fear-related behavioural responses. (People with the rare genetic condition Urbach–Wiethe disease show bilateral lesions of the amygdalae, and fail to exhibit fear-related behaviour.) The amygdalae are also involved in response to sexual stimulation, which in rodents is primarily olfactory.

Two relevant regions of the amygdala are the posteroventral part of the medial amygdala (MePV), which receives information from the olfactory system, and the adjacent posterodorsal medial amygdala (MePD) that responds to reproductive pheromones. In uninfected rats, cat odour activates the MePV more strongly than the MePD. Infection enhances the activation of MePD neurons. This causes the reinterpretation of the odour as a sexual stimulant, overriding the defensive response.

How does *T. gondii* control the response to cat odour by the different segments of the amygdalae? The rodent MePD contains large numbers of neurons potentiated by the hormone arginine vasopressin. Vasopressin is a nine-amino-acid peptide that modulates fear, aggression, and anxiety, and male social and sexual behaviour, through stimulation of neural circuits involving the amygdala. (It has other, better-known functions, *viz.*, control of water retention and constriction of blood vessels.)

T. gondii effects an epigenetic change in the host to upregulate the synthesis of arginine vasopressin. Infected rats show hypomethylation of the promoter of arginine vasopressin, leading to increased expression of the peptide. The potentiated vasopressinergic system enhances the activation of MePD neurons in response to cat odour.

Two types of evidence support this scenario. Systematic hypermethylation (by subcutaneous injection of L-methionine) reverses the loss of fear of cats in *T. gondii*-infected animals. Conversely, the change from fear of cats to attraction is inducible in *non-infected* rats by hypomethylation (by infusion of an inhibitor of DNA methyl transferase into the medial amygdala).

Toxoplasmosis is not limited to rodents and cats. It may infect up to 350 000 people in the UK. It is estimated that approximately 30% of the world's population is infected. *T. gondii* infection of a pregnant mother increases the risks of miscarriage or stillbirth or developmental defects, as well as of passing on the infection to the baby. Toxoplasmosis is also an important cause of abortion and stillbirth in sheep, goats, and pigs, although very rarely in cattle.

T. gondii infection has been implicated for enhanced risk for neuropsychiatric illness in humans. The evidence for an association with schizophrenia is the clearest. There is some specific evidence for changed attitudes towards cats, but because most humans do not normally have an aversion to cats, the effect is harder to measure than in rodents. Also, one might naively have thought that increased dopamine synthesis might be protective against Parkinson's disease, but this does not seem to be true.

The idea of a shortcut to alter perception and thought is a staple of movies; without offering a credible mechanism. *T. gondii* has developed one.

LOOKING BACK AND FORWARD

This chapter and its predecessor have treated the types and amounts of data available to us, and introduced the approaches to their analysis in some detail. These data, linked to the central dogma, include genome sequences, transcriptomes, proteomes. Later we shall add metabolomes and control cascades. We shall next embark on a survey of types of life forms: viruses, prokaryotes, and eukaryotes. What do the data tell us about their evolution in general? What can we learn by looking in detail into particular cases?

RECOMMENDED READING

General presentations of phylogeny (Felsenstein is particularly strong on the history of the subject, expected from a scientist who has had such a long-term commitment to the field)

Whelan, S., Liò, P., & Goldman, N. (2001). Molecular phylogenetics: State-of-the-art methods for looking into the past. Trends Genet. 17: 262–272.

Baldauf, S.L. (2003). Phylogeny for the faint of heart: A tutorial. Trends Genet. 19: 345–351.

Huson, D.H., Rupp, R., & Scornavacca, C. (2010). *Phylogenetic Networks: Concepts, Algorithms and Applications*. Cambridge: Cambridge University Press.

Yang, Z. (2014). *Molecular Evolution: A Statistical Approach*. Oxford: Oxford University Press.

Felsenstein, J. (2003). *Inferring Phylogenies*. Sunderland, MA: Sinauer Associates.

The basis of sequence analysis

Gusfeld, D. (1997). *Algorithms on Strings, Trees and Sequences*. Cambridge: Cambridge University Press.

Doolittle, R.F. (1986). *Of URFS and ORFS: A Primer on How to Analyze Derived Amino Acid Sequences*. Mill Valley, CA: University Science Books.

Evolution of protein structure

Chothia, C., Gough, J., Vogel, C., *et al.* (2003). Evolution of the protein repertoire. Science 300: 1701–1703.

Chothia, C., & Gough, J. (2009). Genomic and structural aspects of protein evolution. Biochem. J. 419: 15–28.

Weatheritt, R.J., & Babu, M.M. (2013). Evolution. The hidden codes that shape protein evolution. Science 342: 1325–1326.

Phylogenetic relationships in eukaryotes

Baldauf, S.L. (2003). The deep roots of eukaryotes. Science 300: 1703–1706.

Jackson, J.B.C., & Johnson, K.G. (2001). Paleoecology: Measuring past biodiversity. Science 293: 2401–2405.

It is possible to represent phylogenetic relationships in forms more general than tree structures

Bandelt, H.-J., & Dress, A.W.M. (1992). Split decomposition: A new and useful approach to phylogenetic analysis of distance data. Molec. Phylog. Evol. 1: 242–252.

Huson, D.H., & Scornavacca, C. (2011). A survey of combinatorial methods for phylogenetic networks. Genome Biol. Evol. 3: 23–35.

Natural genetic engineering by *Toxoplasma gondii*

Flegr, J., & Markossuperv, A. (2014). Masterpiece of epigenetic engineering—how *Toxoplasma gondii* reprogrammes host brains to change fear to sexual attraction. Mol. Ecol. 23: 5934–5936.

Fabiani, S., Pinto, B., Bonuccelli, U., *et al.* (2015). Neurobiological studies on the relationship between toxoplasmosis and neuropsychiatric diseases. J. Neurol. Sci. 351: 3–8.

EXERCISES, PROBLEMS, AND WEBLEMS

Exercises

Exercise 5.1 On two photocopies of Figure 5.28, indicate a reasonable division of the species into (a) three clusters; (b) five clusters.

Exercise 5.2 What is the Hamming distance between the words DECLENSION and RECREATION?

Exercise 5.3 What is the Levenshtein distance between the words BIOINFORMATICS and CONFORMATION?

Exercise 5.4 The Levenshtein distance between the strings agtcc and cgctca is 3, consistent with the following alignment:

```
ag-tcc
cgctca
```

Provide a sequence of three edit operations that convert agtcc to cgctca.

Exercise 5.5 To what alignment does the path through the following dot plot correspond?

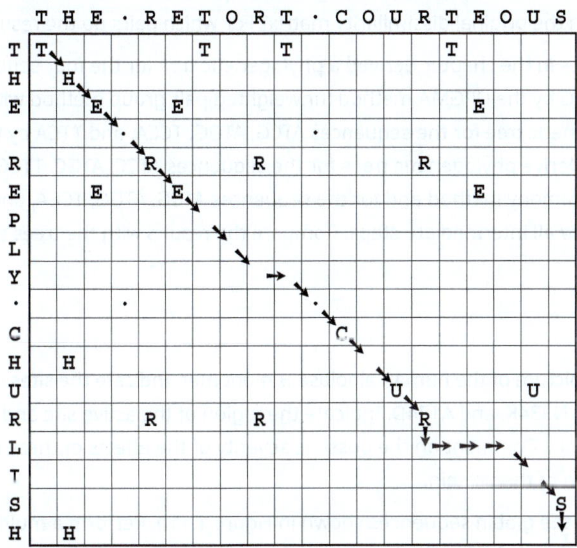

Exercise 5.6 In the dot plot appearing in Figure 5.5, there is an interruption of the matching at a height approximately at the level of the downward-pointing arrow at the left that precedes the words *Xenopus laevis*. On a photocopy of Figure 5.5b, indicate where in the sequence this region appears.

Exercise 5.7 How would you use a dot plot to pick up palindromic DNA sequences of the type that appear partly on each strand, as in the specificity sites of restriction endonucleases?

Exercise 5.8 According to the BLOSUM62 matrix: (a) is a histidine (H) more likely to change to an asparagine (N) or to an aspartic acid (D)? (b) What is the ratio of the probability that a histidine will be *observed* to change to an asparagine to the probability *expected* on the basis of the amino acid composition of the protein that it will change to an asparagine?

Exercise 5.9 Consider the box (red outline) showing a part of a position-specific scoring matrix in Figure 5.8. Suppose you were scoring a protein with 225 residues according to this matrix. (a) How many columns would you expect there to be in the position-specific scoring matrix? (b) How many rows would you expect there to be?

Exercise 5.10 On photocopies of Figure 5.14, indicate points (a) where a pair of highly diverged homologous proteins with similar structure but without obvious sequence similarities might lie; (b) where a pair of non-homologous proteins with similar structure might lie; (c) where a pair of enzymes that share a function but not a structure (for instance, serine and cysteine proteinases) might lie.

Problems

Problem 5.1 Draw a dot plot of the following sequence from the wheat dwarf virus genome: ttttcgtgagtgcgcggaggctttt against itself. In what respects is it not a perfect palindrome?

Problem 5.2 How would you adapt the dot plot formalism to search for regions of DNA or RNA that form local double-helical hairpin regions? Assume that the two hydrogen-bonded regions are separated by only a short unpaired loop, as for example in tRNA (see Figure 1.8).

Problem 5.3 (a) How might the course of the BLAST calculation shown in Figure 5.7 differ if the word length were chosen as 3 instead of 4? (b) How might the course of the BLAST calculation shown in Figure 5.7 differ if the word length were chosen as 7 instead of 4?

Problem 5.4 The phylogenetic tree in Box 5.10 is derived from a complete dissimilarity matrix, i.e. a specification of a measure of the dissimilarity between *every* pair of tetranucleotides. The numbers associated with each edge reproduce the measures of dissimilarity between connected nodes: i.e., the sum of the edges in the path between ATCC and ATGC is $0.5 + 0.5 = 1$, which is the value in the matrix corresponding to row ATCC and column ATGC. For every pair of tetranucleotides, calculate the sum of the numbers associated with the edges in the path between them. For which pairs do the results agree with the original dissimilarity matrix? For which pairs do the results disagree?

Problem 5.5 Examples in the chapter derived a phylogenetic tree for the four sequences ATCC, ATGC, TTCG, and TCGG by the UPGMA method (unweighted pair group method with arithmetic mean) and a phylogenetic tree for the sequences ATCG, ATGG, TCCA, and TTCA by the maximum-parsimony method. Derive phylogenetic trees for the sequences ATCC, ATGC, TTCG, and TCGG by the maximum-parsimony method and for the sequences ATCG, ATGG, TCCA, and TTCA by the UPGMA method. Show all intermediate steps. Compare the results with the trees derived in the chapter.

Weblems

Weblem 5.1 Draw a picture of the human aldolase B monomer. Indicate the sites of the mutations N334K, L256P, A149P, N334K, and A174D. Indicate the region of the active site and the regions of intersubunit contacts. Comment on the possible severity of the effects of these mutations on structure and function of the protein.

Weblem 5.2 Retrieve the globin sequences shown in Figure 1.11b. Perform a multiple sequence alignment, and draw the phylogenetic tree. Comment on ways in which the tree seems biologically reasonable; and—if any—ways in which it does not.

Weblem 5.3 (a) Identify five new mammalian species discovered since 2000. (b) Identify five new bird species discovered since 2000. (c) Identify five new flowering plants discovered since 2000. (d) Identify ten new species discovered in 2024.

CHAPTER 6

Comparative Genomics

LEARNING GOALS

- *Know the three major divisions of living things*—archaea, bacteria, and eukaryotes—based on analysis of the sequences of 16S rRNA genes.
- *Recognize the prevalence of horizontal gene transfer*, especially among prokaryotes, and understand that horizontal gene transfer is inconsistent with the hierarchical 'tree of life' picture that the Linnaean classification scheme suggests.
- *Be familiar with major events in the history of life.*
- *Appreciate the general distribution of genome sizes* and numbers of genes among species.
- *Distinguish the characteristics of different types of genome organization* in viruses, prokaryotes, and eukaryotes.
- *Know the effects of gene duplication on genome evolution.*
- Be able to distinguish the meanings of homologue, orthologue, and paralogue.
- *Understand the mechanism of genome change* at the levels of individual bases, genes, chromosome segments, and whole genomes.
- *Recognize the limits of what genomes determine and what they do not determine*, and the limits of what we can currently explain on the basis of genomics and what we cannot.
- *Appreciate, as far as possible, what makes us human.*
- *Understand the idea of a model organism in the study of human disease.*

Unity and diversity of life

The diversity of life fascinates everyone. The macroscopic life forms most familiar to us come in discrete types called species (see Chapter 1). Linnaeus, an eighteenth-century Swedish naturalist, first organized the characteristics of different species into a logical framework. He introduced the system of nomenclature still used today. (We dealt with biological systematics in more detail in Chapter 4.)

For macroscopic organisms, the Linnaean classification is reinterpretable as a phylogenetic tree—a set of ancestor–descendant relationships between species. A thread of continuous family history unites all organisms. We have already alluded to the dissonance between the concept of a continuous evolutionary pathway from a common ancestor to a daughter species and the idea that species are fundamentally discrete. Nevertheless, these ideas remain a cornerstone of biological thought. The concept of a species remains a useful one, even though it has proved very difficult to define precisely, even for macroscopic organisms.

Genome sequences provide the most general, detailed, and consistent approach to definition of species. Sequences rule microbial taxonomy, but jostle for power with traditional methods in the classification of plants and animals.

Microbiologists use Linnaean nomenclature for bacteria but find themselves uncomfortable doing so. Structural characteristics of bacteria lend themselves less well to distinguishing species than the physical features of higher organisms. Selected regions of mitochondrial DNA provide signatures of different species, called the Barcode of Life (see Box 6.1).

Traditional methods for bacterial classification were based on features of morphology (cell size and shape), biochemistry (uptake of stains, carbon and nitrogen sources, fermentation products), and physiology (growth temperature range and optimum, osmotic tolerance). Gram-positive and Gram-negative bacteria differ in their ability to take up crystal violet or methylene blue stain: Gram-positive bacteria contain a thick (20–80 nm) peptidoglycan layer in their cell wall that binds the stain. Immunological cross-reactivity has also been a basis for classification, especially among infectious species that elicit a clinical motivation for their classification. Before sequencing, hybridization of DNA from two different bacteria was a criterion for similarity. Most bacterial DNAs will form hybrid double-helical structures provided that the similarity in base sequence is >80%.

Later, following the seminal work by C. Woese, prokaryotic species were defined in terms of variations in 16S ribosomal RNA (rRNA) and other sequences. Bacteria for which the 16S rRNA sequences are more than about 2.5–3% different are considered different species. Typically, this corresponds to no more than 70% similarity in overall genome sequence. If humans and chimpanzees were bacteria, we would *easily* be considered as the same species!

DNA barcoding

Field biologists now often characterize populations by DNA sequences. For higher animals, the sequence of the cytochrome *c* oxidase subunit I mitochondrial region (COI) provides a compact index for species identification. In most groups, this region is 648 bp long. The sequence variation within a species is small compared with differences between species.

Biologists describe the identification of species by the sequence of this region as 'barcoding'. The Barcode of Life database (BOLD) collects the information, currently containing approximately 15 million entries. Its query system converts COI sequences to taxonomic assignments (http://www.ibol.org).

Barcoding is a focus of the debate over traditional morphology-based taxonomy versus reliance on sequences. Of course, for classification of long-extinct organisms for which no DNA sequences are available, there is no choice. Degradation of samples collected from fossils could degrade the quality of the data. And, despite their utility, barcodes tell us very little about the organisms they identify. No one would deny the phenotypic richness observable in living specimens—not least in their behaviour—much of which we cannot yet infer, even from complete genome sequences.

It was by sequencing the barcode regions that the New York secondary school students investigated the species attributions in sushi restaurants (see Chapter 1).

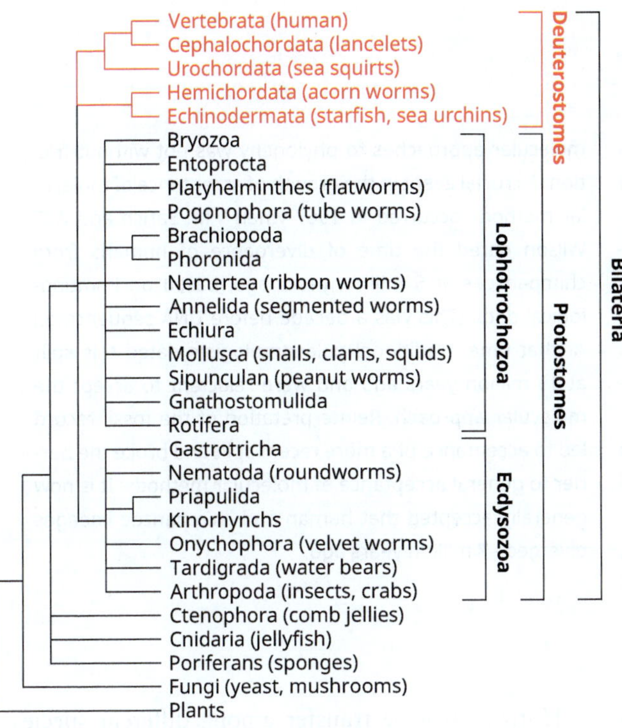

Figure 6.1 Phylogenetic tree of metazoa (multicellular animals). *Bilateria* include all animals that share a left-right symmetry of body plan. *Protostomes* and *deuterostomes* (red) are two major lineages that separated at an early stage of evolution, estimated at 670 million years ago. They show very different patterns of embryological development, including different early cleavage patterns, opposite orientations of the mature gut with respect to the earliest invagination of the blastula, and the origin of the skeleton from mesoderm (deuterostomes) or ectoderm (protostomes). Protostomes comprise two subgroups distinguished on the basis of the sequences of an RNA from the small ribosomal subunit and *HOX* genes. (*HOX* genes govern the development of body plans.) Morphologically, *ecdysozoa* have a moulting cuticle—a hard outer layer of organic material. *Lophotrochozoa* have soft bodies. Figure 6.2 shows in more detail the group that includes us—the deuterostomes (red).

Taxonomy based on sequences

Protein, RNA, and DNA sequences have illuminated relationships between species, both for macroscopic organisms and microbes. The sequences have clarified some relationships but have exposed others as simplistic. Major results include the following:

- *All life on Earth has enough general similarity to show that all life forms had a common origin.* Evidence includes the universality of the basic chemical structures of DNA, RNA, and proteins, the universality of their general biological roles, and the near-universality of the genetic code.
- *On the basis of 16S rRNAs, C. Woese divided living things into: bacteria, archaea, and eukarya.* Figure 1.14 showed the major divisions of the tree of life. At the ends of the eukaryote branch are the metazoa, including yeast and all multicellular organisms—fungi, plants, and animals (see Figure 6.1). We and our closest relatives are in the vertebrate branch of the deuterostomes (see Figure 6.2).

Although archaea and bacteria are both unicellular organisms that lack a nucleus, at the molecular level archaea are in some ways more closely related to eukarya than to bacteria. The recent discovery of the Lokiarchaeum, the most complex known

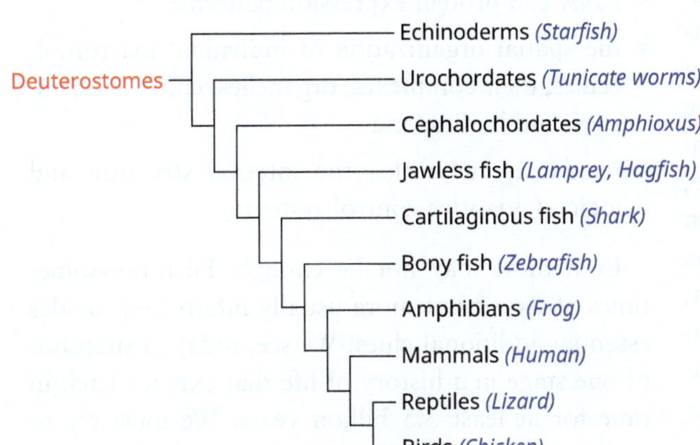

Figure 6.2 Phylogenetic tree of vertebrates and our closest relatives. Chordates, including vertebrates, and echinoderms are all deuterostomes. Examples are shown in blue.

BOX 6.2

Molecular phylogeny and chronology

Molecular approaches to phylogeny developed against a background of traditional taxonomy, based on a variety of morphological characters, embryology, geographical distribution, and, for fossils, information about the geological context (stratigraphy). The classical methods have some advantages. Traditional taxonomists have much greater access to extinct organisms via the fossil record. They can date the appearance and extinction of species by geological methods (see Figure 1.17).

Molecular biologists, in contrast, have limited access to extinct species, although 'palaeosequencing' has achieved some very impressive results.

It must be admitted that, before DNA sequencing, the relationship between proponents of classical and molecular approaches to phylogeny was not without friction. A crucial event in the growth of acceptance of molecular methods occurred in 1967 when V.M. Sarich and A.C. Wilson dated the time of divergence of humans from chimpanzees at 5 million years ago, based on immunological data. (This was a decade before DNA sequencing.) At that time, traditional palaeontologists dated this split at 15 million years ago and were reluctant to accept the molecular approach. Reinterpretation of the fossil record led to acceptance of a more recent split and broke the barrier to general acceptance of molecular methods. It is now generally accepted that human and chimpanzee lineages diverged ~7 million years ago.

prokaryote, bridges the gap between archaea and eukaryotes. It is also likely that the archaea are the closest living organisms to the root of the tree of life.

- *Dating of historical events from sequence differences.* As species diverge, their sequences diverge. L. Pauling and E. Zuckerkandl suggested that if sequence divergence occurred at a constant rate, it would provide a 'molecular clock' that would allow dating of the splits in lineage between species.

Although the clock is not universal, judicious calibration of rates of sequence change with palaeontological data permits dating of events in the history of life (see Box 6.2).

- *The importance of horizontal gene transfer.* This is the acquisition of genetic material by one organism from another by natural rather than laboratory procedures, through some means other than descent from a parent during replication or mating (see Box 6.3). Several mechanisms of horizontal gene transfer are known, including direct uptake, as in Griffith's pneumococcal transformation experiments, or via a viral carrier. Arrangements of species into phylogenetic trees, in contrast, depends on strict ancestor–descendant relationships between different organisms during evolution (see Figure 1.15).

Horizontal gene transfer among different species has affected most genes in prokaryotes. It requires a change in our thinking from ordinary 'clonal' or parental models of heredity. Indeed, microorganisms do not easily fit into the structure of the 'tree' of life but require a more complex organizational chart.

Sizes and organization of genomes

We appeal to genomes to help us to understand ourselves as individuals, and our relationships with all of the other organisms that march in the pageant of life. To make progress, we must integrate several data streams, including:

- genome sequences;
- RNA and protein expression patterns;
- the spatial organization of individual macromolecules, their complexes, organelles, entire cells, tissues, and bodies; and
- regulatory networks, the internal structure and logic of adaptive control systems.

Even these may not be enough. History—sometimes observed but more usually inferred—provides essential additional clues. We see, today, a snapshot of one stage in a history of life that extends back in time for at least 3.5 billion years. We must try to

BOX 6.3 Please pass the genes: horizontal gene transfer

On learning that *Streptomyces griseus* trypsin is more closely related to bovine trypsin than to other microbial proteinases, Brian Hartley commented in 1970 that '... the bacterium must have been infected by a cow'. This was a clear example of lateral or horizontal gene transfer—a bacterium picking up a gene from the soil in which it was growing, that an organism of another species had deposited there. The classic experiments on pneumococcal transformation by Griffiths, and those by O. Avery, C. MacLeod, and M. McCarthy that identified DNA as the genetic material, are other examples.

Horizontal gene transfer in bacteria may occur through direct uptake of DNA (as in the Griffiths, and Avery, MacLeod, and McCarthy experiments), conjugation (plasmid-mediated gene transfer), and transduction (phage-mediated transfer).

Evidence for horizontal transfer includes (1) discrepancies among evolutionary trees constructed from different genes; and (2) direct sequence comparisons between genes from different species.

- In *Escherichia coli*, about 25% of the genes appear to have been acquired by transfer from other species.
- In microbial evolution, horizontal gene transfer is more prevalent among operational genes—those responsible for 'housekeeping' activities such as biosynthesis—than among informational genes—those responsible for organizational and regulatory activities in transcription and translation. For example:
  - *Bradyrhizobium japonicum*, a nitrogen-fixing bacterium, symbiotic with higher plants, has two glutamine synthetase genes: One is similar to those of its bacterial relatives; the other is 50% identical to those of higher plants;
  - rubisco (ribulose-1,5-bisphosphate carboxylase/oxygenase), the enzyme that first fixes carbon dioxide at entry to the Calvin cycle of photosynthesis, has been passed around between bacteria, mitochondria, and algal plastids, as well as undergoing gene duplication.
  - many phage genes appearing in the *E. coli* genome provide further examples and point to a mechanism of transfer.

Nor is the phenomenon of horizontal gene transfer limited to prokaryotes. Eukaryotes originated as chimaeras, probably about 2 billion years ago. Eukaryotes derive their informational genes primarily from an organism related to *Methanococcus*, and their operational genes primarily from proteobacteria, with some contributions from cyanobacteria and methanogens. (The exact role of Lokiarchaeum is not yet established.) Almost all informational genes from *Methanococcus* itself are similar to those in yeast. At least eight human genes appeared in the *Mycobacterium tuberculosis* genome. *S. griseus* trypsin is an example of eukaryote → prokaryote transfer.

In eukaryotes, gene transfer may involve nuclear DNA, mitochondrial DNA, and chloroplast DNA. Intracellularly, there has been extensive gene transfer from organelles to and from the nucleus, and even between mitochondria and chloroplasts. Transfer of organelle DNA among different cells of a single organism is common, and can contribute to tissue repair *in vivo*. Gene transfer between species can involve nuclear or organelle DNA.

Leigh's syndrome is a neurodegenerative disorder arising from mutations in mitochondrial DNA, typically resulting in early death. Clinically, substitution of healthy mitochondria from a donor in a Leigh's syndrome carrier mother's egg cell allows development of a child unaffected by the mother's mutations. (The popular press has referred to this as a 'three-parent baby'.)

The observations hint at the model of a 'global organism'—or a genomic World Wide DNA Web from which organisms download genes at will! How can this be reconciled with the fact that the discreteness of species has been maintained? I offered the conventional explanation, that the living world contains ecological 'niches' to which individual species are adapted: the discreteness of niches, together with mechanisms of reproductive isolation, explains the discreteness of species.

But this explanation depends on the stability of normal heredity to maintain the fitness of the species. Why would the global organism not break down the lines of demarcation between species, just as global access to pop culture threatens to break down lines of demarcation among national and ethnic cultural heritages? Perhaps the answer is that it is the informational genes, which appear to be less subject to horizontal transfer, that determine the identity of the species.

I do not pretend to have a satisfactory answer.

read the past in contemporary genomes, which contain records of their own development.

US Supreme Court Justice Felix Frankfurter wrote that '. . . the American constitution is not just a document, it is a historical stream.' Like a genome!

This programme requires development of novel methods. New fields of study require new approaches. Salvador Luria once suggested that to determine common features of all life one should not try to survey everything, but, rather, identify the organism most different from us and see what we have in common with it. Let us combine this with a complementary idea: to take the most *closely related* organisms and identify the *differences*. That is, consider the following *two* questions:

- How do the human genome and the *E. coli* genome express our *common* heritage?
- How do genomes that are 99% identical create the *differences* between humans and chimpanzees?

If we could answer these questions, we would have achieved a lot.

KEY POINT

We appeal to genomes to help us to understand evolutionary relationships. With a few exceptions, our access to genome sequences is limited to organisms alive today, that is, to a single snapshot in time. However, genomes contain internal records of their history which gives us a window onto the past.

Genome sizes

One reason for resistance to Darwin's theory of evolution is its denial to human beings of a special status relative to animals. Genomics threatens to do this all over again. Humans *do* have unique features. Many people, not excluding molecular biologists, expect these features to be reflected in the genome. And so they must be, although, quite frankly, not in any obvious way.

KEY POINT

The term C-value has been used to refer to the amount of DNA in a haploid cell, i.e. a gamete; the letter C refers to the *constancy* of the amount of DNA per cell in a species.

The overall size of the human genome is not special. Different organisms have different total amounts of DNA per cell (see Figure 6.3 and Table 6.1).

On a broad scale, there is a general correlation between complexity of organism and amount of DNA per cell. Prokaryotes have less DNA per cell than eukaryotes, and yeast has less than mammals. However, although humans have more DNA per cell than certain other organisms popular in molecular biology, including *Caenorhabditis elegans* and *Drosophila melanogaster*, many organisms have even greater amounts than we do. The genome of *Amoeba dubia* is 200 times larger than the human genome. The genome of the marbled lungfish (*Protopterus aethiopicus*), a closer relative, is 43 times as large as ours.

Why the different amounts of DNA? As far as we know, most of the human genome does not encode protein or RNA. Regions of genomes without known function are often referred to as 'junk DNA'. Of course, the fact that we may not know the function of much of our genome does not mean that it has none. (Maybe it is junk, but it is certainly not all transcriptionally inert. Recent discoveries have revealed many new types of RNA molecules, mostly involved in control processes. It would be naïve to doubt that many more types will come to light.) Moreover, the amount of space between genes affects the rate of crossing over and recombination and, thereby, rates of evolution. Indeed, the large amount of repetitive sequence between our genes enhances recombination rates by promoting homologous recombination.

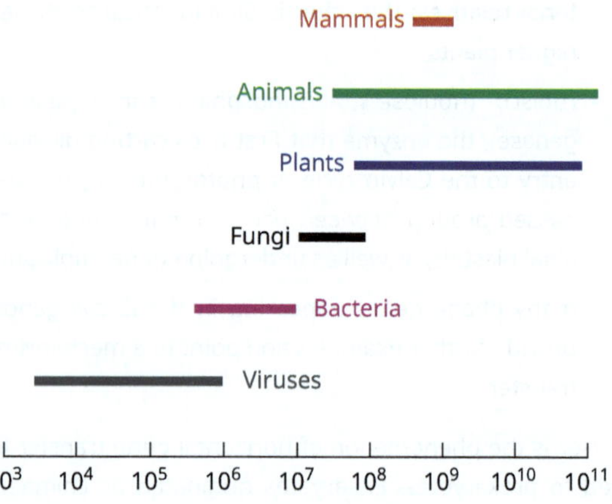

Figure 6.3 Distribution of genome sizes in different groups of living things. The horizontal scale gives the number of bases or base pairs.

Table 6.1 Genome sizes

Organism	Number of base pairs	Number of genes	Comment
φX-174	5386	10	Virus infecting *E. coli*
Influenza A	13590	10	Strain A/goose/guandong/1/96(H5N1)
Human mitochondrion	16569	37	Subcellular organelle
Epstein–Barr virus (EBV)	172282	80	Cause of mononucleosis
Nanoarchaeum equitans	490885	552	Archaeon, smallest known genome of a cellular organism
Mycoplasma pneumoniae	816394	680	Cause of cyclic pneumonia epidemics
Rickettsia prowazekii	1111523	834	Bacterium, cause of epidemic typhus
Mimivirus	1181404	1262	Virus with the largest known genome
Borrelia burgdorferi	1471725	1738	Bacterium, cause of Lyme disease
Aquifex aeolicus	1551335	1749	Bacterium from hot spring
Thermoplasma acidophilum	1564905	1509	Archaeal prokaryote, lacks cell wall
Helicobacter pylori	1667867	1589	Chief cause of stomach ulcers
Methanococcus jannaschii	1664970	1783	Archaeal prokaryote, thermophile
Haemophilus influenza	1830138	1738	Bacterium, cause of middle-ear infections
Thermotoga maritime	1860725	1879	Marine bacterium
Archaeoglobus fulgidus	2178400	2437	Another archaeon
Deinococcus radiodurans	3284156	3187	Radiation-resistant bacterium
Synechocystis	3573470	4003	Cyanobacterium, 'blue-green alga'
Vibrio cholera	4033460	3890	Cause of cholera
Mycobacterium tuberculosis	4411532	3959	Cause of tuberculosis
Bacillus subtilis	4214814	4779	Popular in molecular biology
Escherichia coli	4639221	4485	Molecular biologists' all-time favourite
Saccharomyces cerevisiae	12495682	5770	Yeast, first eukaryotic genome sequenced
Caenorhabditis elegans	100258171	19099	'The worm'
Arabidopsis thaliana	135000000	25498	Flowering plant (angiosperm), 'the weed'
Drosophila melanogaster	122653977	13472	The fruit fly
Takifugu rubripes	3.65×10^{8}	23000	Pufferfish (fugu fish)
Human	3.2×10^{9}	23000	
Wheat	16×10^{9}	30000	
Salamander	10^{11}	?	
Psilotum nudum	2.5×10^{11}	?	Whisk fern, a simple plant
Amoeba dubia	6.7×10^{11}	?	Protozoan
Protopterus aethiopicus	1.33×10^{11}	?	Marbled lungfish

Rate of evolutionary change is a characteristic of a species that is certainly subject to selective pressure. Features of the genome that affect rate of evolution cannot be dismissed entirely as junk.

If genome size *per se* does not single out humans, what about numbers of genes? Again there is a general correlation between complexity of organism and estimated numbers of genes.

- Viral genomes encode only a few proteins.
- Prokaryote genomes contain hundreds or thousands of genes.

  Genomes of prokaryotes tend to be compact. The 4.6 Mb chromosome of *E. coli* encodes approximately 4500 genes, distributed on both strands. The absence of introns and the shorter intergenic regions account for the higher coding densities. A very large fraction of the DNA, 87.8%, codes for proteins, 0.8% codes for structural RNAs, and only 0.7% has no known function.
- The simple eukaryote yeast has almost 6000 genes, fewer than twice as many as *E. coli*.
- Metazoa have tens of thousands of genes.

However, within groups of related organisms, including vertebrates, there is no simple correlation between apparent complexity of organism, or even genome size, and numbers of genes (see Table 6.2). Two vertebrates, the pufferfish and humans, appear to have roughly the same number of genes but differ by almost an order of magnitude in genome size. It was also unexpected to find that the worm *C. elegans* appears to have more genes than the fruit fly. In fact, to a first approximation many groups of metazoa have roughly the same numbers of genes.

The phenomena of variable splicing and RNA editing show the situation to be more complicated than simple gene estimates make it appear. This is one reason why it has been difficult to get an accurate count of the number of genes in humans and other higher organisms. In eukaryotes, estimates of gene number refer to maximal sets of exons in units that are coordinately transcribed and translated. In fact, variation in splicing may create many proteins from each gene. As an extreme example, in the mammalian immune system, billions of distinct antibodies arise from regions in the genome containing fewer than ~100 exons. (However, the immune system is special: splicing occurs at the DNA, not the RNA, level.)

RNA editing is the alteration of bases in mRNA, after transcription. The changes are usually either C→U or A→I (I = inosine, has the coding properties of G). If only some mRNA from the same gene is edited, an extra degree of variability in the proteins arises. Investigation of RNA editing is a relatively new field, and many more implications of the process in health and disease remain to be revealed.

For instance, in the human intestine, two isoforms of apolipoprotein B appear. To produce one of these forms, mRNA editing alters a single nucleotide, changing a 'c' to 'u'. This turns a Gln codon (caa) into a Stop codon (uaa), truncating the altered isoform to half the size of the longer form. The longer form is the translation of the full unedited mRNA.

Defective RNA editing contributes to the pathology of sporadic amyotrophic lateral sclerosis (a neurodegenerative disease, of which the most famous sufferers have been US baseball player Lou Gehrig, Cambridge physicist Stephen Hawking, and

Table 6.2 Distribution of genome sizes and gene densities

Species	Genome size (Mb)	Coding (%)	Approximate number of genes	Estimated gene density (kb/gene)
E. coli	4.64	88	4485	1.03
Yeast	12.5	70	6000	2.1
Pufferfish	365	15	23000	10
A. thaliana	115	29	23000	6
Human	3289	1.3	23000	143

Mao Zedong, founder and president of the People's Republic of China.)

Still another complication is that many mature proteins are complexes of multiple polypeptide chains, the genes for which are typically not closely linked. Haemoglobin is a good example: the gene for the α chain is on chromosome 16, the gene for the β chain is on chromosome 11.

The conclusion is that it is very different to estimate the size—to say nothing of the complexity—of a eukaryote's proteome from its genome.

KEY POINT

Even taking protein assembly from separate polypeptide chains, alternative splicing, and RNA editing into account, these figures give only a static idea of proteome complexity. Cells control gene expression patterns by complex and dynamic regulatory networks. Conclusion: it is difficult to correlate numbers of expressed genes with organismal complexity if one has no good way of measuring either.

Genome organization in eukaryotes

Genomic information in eukaryotic cells is divided between the main nuclear genome and cytoplasmic organelles: mitochondria and chloroplasts.

In the nucleus, DNA is complexed with proteins to form chromosomes (see Figure 1.3). We have already noted that chromatin remodelling is an important component of regulation of gene expression. DNA in mitochondria or chloroplasts also forms nucleoprotein complexes. These resemble bacterial nucleoids, reflecting the endosymbiont origin of organelles. Typically, organelle genomes are circular, double-stranded DNA molecules. In some species, organelles contain more than one DNA molecule.

With a few exceptions, the amount of nuclear DNA per cell is constant in all cells of an organism except for gametes. In contrast, organelles vary in number of copies of the DNA that they contain. Moreover, cells of different tissues contain different numbers of organelles. Mitochondria are more numerous in cells that consume large amounts of energy, such as brain, heart, and eye (about 10 000 mitochondria per cell), than in skin cells (only a few hundred). Some cells of parasites have no mitochondria at all. In plants, a leaf cell may contain up to 100 chloroplasts, the number varying among species. Leaves may contain 10^6 chloroplasts per mm^2 of surface area. Unsurprisingly, root cells have none.

Mitochondrial genomes vary in size among species. Human mitochondrial DNA is 16 568 bp long. Yeast mitochondrial DNA is 75 kb and that of plants is considerably larger: the DNA of muskmelon (cantaloupe) mitochondria is 2.4 Mb! Chloroplast genomes range from about 110 to 160 kb, larger than animal mitochondrial DNAs. In some species, such as the protozoan *Cryptosporidium*, the mitochondria contain no DNA at all! All mitochondrial proteins are encoded in the nuclear DNA, synthesized in the cytoplasm, and imported into the mitochondria.

Mitochondria and chloroplasts carry out their own protein synthesis. (With a few exceptions such as *Cryptosporidium*.) Chloroplasts and plant mitochondria translate their genes according to the standard genetic code, but animal mitochondria use variants.

There is active traffic between organelle and nuclear genomes (see Box 6.4). Approximately 90% of chloroplast proteins are encoded by nuclear genes, and gene transfer is still going on. However, the differences in genetic code inhibit mitochondrial → nuclear transfer in animals (see Table 6.3). In the *Arabidopsis thaliana* genome, a 620 kb insertion of mitochondrial DNA in nuclear chromosome 2 contains much of the mitochondrial genome including

Table 6.3 Numbers of RNAs and proteins encoded in organelle genomes

Organelle	RNAs encoded	Proteins encoded
Animal mitochondria	Two ribosomal RNAs, 22 tRNAs	12 or 13
Plant mitochondria	Three ribosomal RNAs, ~22 tRNAs	30–39
Chloroplast	Four ribosomal RNAs (two copies), 37 tRNAs	50–57

BOX 6.4 Traffic between the mitochondrial and nuclear genomes

Rps14, a protein from the small subunit of the rice mitochondrial ribosome, is encoded by a nuclear gene, *rps14*, on chromosome 8. In fact, by alternative splicing, the five exons of this region (centre strip) encode both Rps14 (ribosomal protein 14 of the small subunit) and SdhB (the B subunit of succinate dehydrogenase).

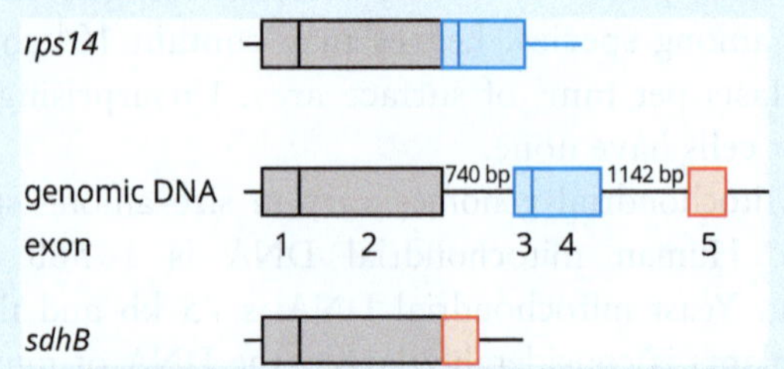

Both Rps14 and SdhB are synthesized in the cytoplasm and transported into the mitochondria. It is likely that both genes were originally in the mitochondrial genome. A region similar to the nuclear gene *rps14* remains in the rice mitochondrial genome. This mitochondrial gene is translated, but the product has become non-functional as a result of four single-nucleotide deletions that destroy the reading frame. Certain other higher plants have functional mitochondrial *rps14* genes (broadbean, rapeseed), whereas others resemble rice in containing non-functional mitochondrial *rps14* genes, but functional nuclear genes (potato, *Arabidopsis*).

Genes similar to *sdhB* have not been observed in plant mitochondrial genomes. It is likely, therefore, that the move of the *sdhB* gene from the mitochondrial to the nuclear genome is an old event and the move of the *rsp14* gene is a relatively recent one.

Moving a mitochondrial gene to the nucleus moves the site of its expression to the cytoplasm. It needs a leader sequence containing a proper targeting signal to direct translocation of the protein into mitochondria. The protein encoded by the rice nuclear gene for mitochondrial *rps14* appears to have borrowed a mitochondrial targeting signal from *sdhB*, part of an earlier generation of immigrants, by alternative splicing. Thus, compared with products of mitochondrial genes for Rps14, the nuclear-encoded version has an N-terminal extension derived from the *sdhB* exons. This extension is cleaved off in the mitochondria.

some duplicated material. The total insertion is even longer than the mitochondrial genome itself, which is 366 924 bp long.

Photosynthetic sea slugs: chloroplasts in animals

The endosymbiotic origin of mitochondria and plant chloroplasts is a well-accepted theory about events that happened 1–2 billion years ago. Acquisition of endosymbiotic chloroplasts by sea slugs is observable today (see Figure 6.4). The slugs, which are molluscs—i.e., animals—eat algae. They open the algal cells and discard the contents—including the nucleus—except for the chloroplasts. The chloroplasts are taken up into host cells, where they carry out photosynthesis. The slug can live for months on the molecules synthesized using solar energy.

Figure 6.4 A lettuce sea slug (*Elysia crispata*) on a patch of the alga *Bryopsis*. The slug eats algae, extracts and endocytoses the chloroplasts, and then basks in the sun, as in the picture, while the chloroplasts photosynthesize organic compounds.

Photograph by William Capman, Augsburg College, Minneapolis, MN, US.

The mollusc does not get an entirely 'free lunch'. In algae typical of the slug's food, the chloroplast genome encodes only 13% of the organelle proteins. During the active life of the chloroplast within the animal's cells, its proteins turn over and must be synthesized. Genes from the algal chloroplast have entered the mollusc nuclear genome and are expressed by the host.

How genomes differ

There is a growing consensus that the dynamics of expression patterns embodies the most interesting features of genomes. It is nevertheless prudent to begin less ambitiously, with static aspects—the sequences themselves. Similarities and differences among genome sequences appear (1) at the levels of individual bases; (2) at the level of genes (see Figure 6.6); (3) in larger-scale blocks; and (4) at the level of whole genomes that have undergone complete duplications.

Variation at the level of individual nucleotides

Closely related genomes tend to contain regions encoding closely related proteins. Alignments of the sequences of homologous genes reveal differences, many in the form of single-site mutations or insertions and deletions. Typically, there is reasonable correlation between overall species divergence and divergence of sequences of individual genes and the corresponding proteins. Comparisons of amino-acid and gene sequences of thioredoxins provide a typical example (see Figure 6.5).

Compare these protein sequences with the corresponding gene sequences from human, chicken, and *Staphylococcus aureus* (see Figure 6.6). Note the large gap in the bacterial gene corresponding to the intron in the human and chicken genes. The asterisks under the sequences indicate positions containing the same base in all three genomes. The colons indicate positions containing two identical bases among the three; in most cases, two common bases appear in the human and chicken sequences, even in the non-coding regions. Note the frequent occurrence of patterns '**:' and '**-blank'. What is the likely reason for this?

Other types of changes affecting local regions of the genome

Proteins have other routes of evolution than the wriggling around sequence space by successive point mutations. Larger-scale *local* changes in genes include insertions and deletions of multiple residues, inversions and translocations (see Figure 4.22b), and duplications.

Post-transcriptional events

- One mechanism that breaks the fidelity of the gene→mRNA→protein translation is the enzymatic modification of certain bases between transcription and translation, a process known as **mRNA editing.**
- The normal function of ubiquitin is to link to proteins as a signal for their degradation. Ubiquitin+1 (UBB+1) is an aberrant protein arising from deletion of a dinucleotide from mRNA, with 19-amino-acid extension resulting from the frameshift. UBB+1 cannot target proteins for

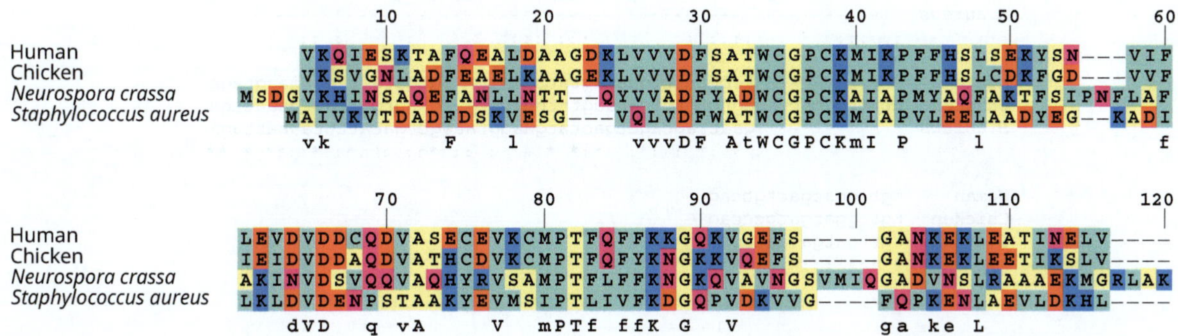

Figure 6.5 Thioredoxins are proteins that catalyse disulphide-exchange reactions, contributing to the speed and accuracy of the protein-folding process. The human thioredoxin gene extends over 13 kb and consists of five exons. This figure shows the alignment of amino-acid sequences of thioredoxins from two vertebrates (human and chicken), a fungus (*Neurospora crassa*), and a bacterium (*Staphylococcus aureus*).

```
Human       actgcttttcaggaagccttggacgctgcaggtgataaacttgtagtagttgacttctca
Chicken     gctgattttgaggcagaactgaaagctgctggtgagaagcttgtagtagttgatttctct
S.aureus    gcagattttgattcaaaagtagaatctggtgtacaa---------ttagtagattttgg
             *:* **** *:: *:   *: * :***: *:::* :: :::::::****:** **:*:

Human       gccacgtggtgtgggccttgcaaaatgatcaagcctttctttcat---------------
Chicken     gccacatggtgtggaccatgtaaaatgatcaagccattttccatgtaagtagcctttgt
S.aureus    gcaacatggtgtggtccatgtaaaatgatcgctccggtattagaa---------------
            **:** ******** ** ** *********:::** :* ** :*:

Human       --------------gtgagtattaaacaatgtctgctttgtaagagatttgtgtttttg
Chicken     ttttcacagtaacagtaagtat-acacaaatacttctgtgcaacttgtcagtaatatg-g
S.aureus    ------------------------------------------------------------
                          :: ::::: : ::::   :: :: :: ::    :  ::  : :  :

Human       agttggtggtcacagtggtaggaaagaaagacagtt----aaaggatttggtttcggtg
Chicken     aggaaacctcctttgtctgtggggtggatggtatttcttgaaggagaatttgtagaagta
S.aureus    ------------------------------------------------------------
            ::        :   ::    ::   : : :  : ::    :: :    :: ::     ::

Human       gg----gggatttctttggctccatctttggtctaaaagtagtagtataacaaataatt
Chicken     tgtgattggtaactattgaataaagtacttggatacacagcagggaacagacatgctgtt
S.aureus    ------------------------------------------------------------
             :     :: :   : ::        :  ::::    : :: ::       :::     ::

Human       taggtttgatacatgtagcccattgaaa-acaaatttagaagttaatttgtcttaaat
Chicken     gcatttgtctgtcctggtgctctgatgcacagtctgtaggtacctagcttccctcaaga
S.aureus    ------------------------------------------------------------
                ::::       : :  :  :::   :::   : :::      :  ::  :: ::

Human       agttcttttttccccacattgaaaca-----tgggcctta--tttgaaatcccagccta
Chicken     a---ctggtaacagtggagttgaaacagtgtgtgtacactggctctgattattaaaactg
S.aureus    ------------------------------------------------------------
            :   ::  :          ::::::::     ::  :  :   : :::      :  ::

Human       gaatttgatatgccaaactgtttt---atactaa--gaaaaatttgatttagagaaaatt
Chicken     cattcagataagctgtagcatctttctgtagtgatggggaggctgtaaggaaggaaagg
S.aureus    ------------------------------------------------------------
             : :  :::: ::   :   : ::    :: : :  :  :   :  :     :: :::

Human       tatgtctcttagatctatgt-ctccaaaaga----tctaaattttggatctttaattag
Chicken     cctttcccttgtgcttaggtgcttcagcagactcttccaggggatggagctgaaaattaa
S.aureus    ------------------------------------------------------------
              : :: :::     :: :: :: ::  :::    :: :     : :        :::::

Human       tctctagtttattaagtttccatttaagaagcttaagcttgggtatgttgcattgccat
Chicken     tttctggtca-gtaaagtagctgctttaaggtaacgagtcaag---tctcacagcaccag
S.aureus    ------------------------------------------------------------
            : ::: ::    : ::::  :   :: :        ::    :   : :  ::   :::

Human       tacctagttctaaatctttt-------tggattttcatttaaattttccag-------
Chicken     tttctatgaatgcatcttttaaagaagtggctttcctggagcagtaactacataatttg
S.aureus    ------------------------------------------------------------
            :  :::    :  :::::::       ::: :::        :     : ::

Human       -------------tccctctctgaaaagtattccaac--gtgatattccttgaagtaga
Chicken     tttttcattctagagtctgtgtgacaagtttggtgat---gtggtgttcattgaaattga
S.aureus    -------------gaattagcagctgactatgaaggtaaagctgacattttaaaattaga
                            :* : :*: :*:* *    : :::*:: : :*: *::** * **

Human       tgt ggatgactgtcag
Chicken     tgt ggatgatgcccag
S.aureus    tgttgatgaaaatcca
            ***:*::::*    **:
```

1

Figure 6.6 Alignment of partial thioredoxin gene sequences from the genomes of human, chicken, and *Staphylococcus aureus*. The region shown contains exons 2 and 3 from the vertebrate genes. * = same base in all three sequences; : = same base in two of the three sequences.

degradation. Even worse, it can accumulate in cells and may contribute to neuronal cell death in Alzheimer's disease.

- Another effect applies at the level of translation. In *E. coli* one gene codes for both the τ and γ subunits of DNA polymerase III. Translation of the entire gene forms the τ subunit. The γ subunit corresponds approximately to the N-terminal two-thirds of the τ subunit. A frameshift on the ribosome leads to premature chain termination 50% of the time, yielding a 1:1 expression ratio of τ and γ subunits. (**Ribosomal frameshifting** is an exception to the processing of mRNA by the ribosome exactly three bases at a time. There is a slippage of the ribosome along the mRNA that changes the reading frame. All organisms, but especially many viruses, use this as a mechanism to synthesize multiple proteins from a single gene. It is not an error but is programmed in the gene.)

Duplications

Duplications of individual genes, of regions containing many genes, and of complete genomes have been an important mechanism of evolution. They are a prolific source of variation, the raw material of both selection and genetic drift.

Duplications are seen in archaea, bacteria, and eukaryotes. Estimates of amounts of duplication vary, but there is agreement that it is substantial (see Table 6.4). The *A. thaliana* genome, for instance, contains over 60% duplications.

Duplication of genes

Organisms in all three domains of life show duplication of individual genes.

After duplication, both copies of a gene may survive and diverge. Alternatively, one copy may turn into a pseudogene or be deleted, leaving only one functional copy (see Box 6.5). Or, both could be lost.

S. Ohno proposed in 1970 that duplication followed by divergence could be an important source of proteins with novel functions. Duplication of a gene allows one copy to retain an essential function whilst the other copy can diverge to acquire modified or even novel functions. It is generally easier to 'recruit' and adapt an already active molecule to a new function than to evolve a new protein 'from scratch'.[1]

The course of evolution of proteins descended from a common ancestor will differ, depending on whether they are retaining or changing their function (see Box 6.6).

Table 6.4 Percentages of duplicated genes

Species	Duplicate genes (%)
Bacteria	
Mycoplasma pneumoniae	44
Helicobacter pylori	17
Haemophilus influenzae	17
Archaea	
Archaeoglobus fulgidus	30
Eukarya	
Saccharomyces cerevisiae	30
Caenorhabditis elegans	49
Drosophila melanogaster	41
Arabidopsis thaliana	65
Homo sapiens	38

From: Zhang, J. (2003). Evolution by gene duplication: An update. Trends Ecol. Evol. 18: 292–298.

What can happen to a gene?

During evolution:

1. A gene may pass to descendants, accumulating favourable (or unfavourable) mutations or drifting neutrally.
2. A gene may be lost.
3. A gene may be duplicated, followed by divergence or by loss of one of the pair.
4. A gene may undergo horizontal transfer to an organism of another species.
5. A gene may undergo complex patterns of fusion, fission, or rearrangement, perhaps involving regions encoding individual protein domains.

[1] 'Walls supply stones more easily than quarries, and palaces and temples will be demolished to make stables of granite, and cottages of porphyry.'—Johnson, *Rasselas*.

BOX 6.6 Homologues, orthologues, and paralogues

Homologues are regions of genomes, or portions of proteins, that are derived from a common ancestor. Because only in rare cases can we actually observe the ancestor-descendant relationship, most assignments of homology are inferences from similarity in sequence, structure, and/or genomic context.

Paralogues are related genes that have duplicated and diverged to provide *separate* functions in the *same* species.

Orthologues, in contrast, are homologues that perform the *same* function in *different* species, and have diverged during speciation. (For instance, the α and β chains of human haemoglobin are paralogues, and human and horse myoglobin are orthologues.)

Other related sequences may be pseudogenes, which may have arisen by duplication or by retrotransposition from mRNA, followed by the accumulation of mutations to the point of loss of function or expression.

In analysing the divergence of related genes, how can we distinguish the effect of selection from genetic drift? Given two aligned gene sequences, we can calculate K_s, the number of synonymous substitutions, and K_a, the number of non-synonymous substitutions. Most but not all synonymous substitutions are changes in the third position of codons. (The calculation of K_a and K_s involves more than simple counting because of the need to estimate and correct for possible multiple changes.) The ratio of K_a/K_s distinguishes the role of selective pressure and drift in the divergence of genes after duplication:

$K_a/K_s \sim 1$	**Neutral evolution**: silent and substitution mutations have occurred to approximately equal extents.
$K_a/K_s > 1$	**Positive selection**: substitution mutations are more prevalent than silent mutations, implying that selective pressures are active and the substitutions are advantageous.
$K_a/K_s < 1$	**Purifying selection**: substitution mutations are underrepresented, implying that the sequence is optimized fairly rigidly, with relatively little tolerance for mutation.

KEY POINT

A common mechanism of evolution is the duplication of a gene corresponding to a protein subunit, or of an entire gene, or even of a whole genome. Creating two copies of any of these entities means that one copy can continue to provide an essential function, while the other can diverge to explore other possibilities.

The many globins in the human genome provide a good example of gene duplication and divergence (see Figure 6.7). Genes for several versions of the

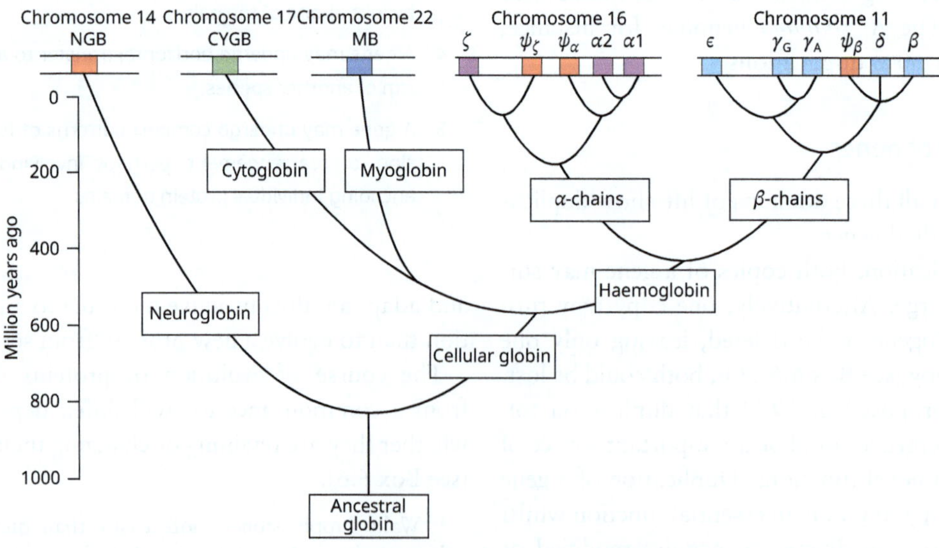

Figure 6.7 Duplication and dispersal through the genome of globin genes during animal evolution. Top line shows current human complement of globin genes.

Burmester, T., Ebner, B., Weich, B., et al. (2002). Cytoglobin: A novel globin type ubiquitously expressed in vertebrate tissues. Mol. Biol. Evol. 19, 4: 416–421, https://doi.org/10.1093/oxfordjournals.molbev.a004096

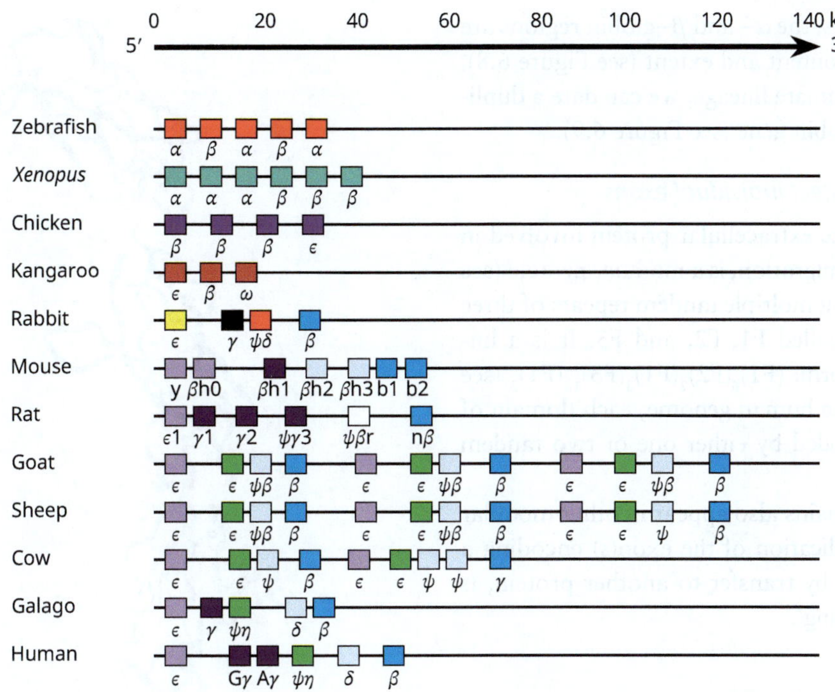

Figure 6.8 Layout of the β–globin locus in selected vertebrates. ψ signifies pseudogene. Colour coding: Light brown, fish; green, amphibian; purple, avian; magenta, marsupial; dark brown, ε–like; dark blue, γ–like; orange, δ–like; cyan, β–like; white, rat-specific pseudogene; red, η–like.

The zebrafish and *Xenopus* regions illustrate the organization of the region prior to the separation of α– and β–globins. They alone of the species illustrated here contain only α– and β–globin genes.

Goat and sheep are closely related species that show similar patterns. Rat and mouse are closely related species that show different patterns.

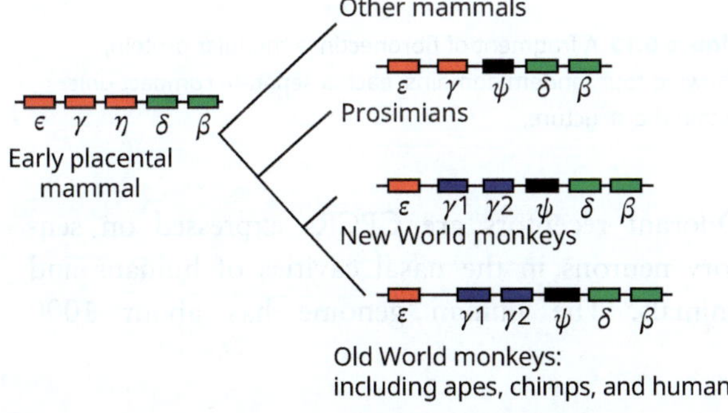

Figure 6.9 Evolution of primate β–globin region (not drawn to scale). Red boxes, embryonically expressed genes; green boxes, post-embryonically expressed genes; blue boxes, foetally expressed genes; black boxes, pseudogenes. Mammals ancestral to the groups in this chart had an embryonic ε–globin and a post-embryonic β–globin gene. Marsupials retain this pair (see Figure 6.8).
By the time this diagram begins, the ε gene had duplicated to form three embryonic genes, ε, γ, and η, and the β gene had duplicated to form δ and β. The η gene fell into desuetude, mutating into a pseudogene. Subsequent duplication of the γ gene in anthropoids produced γ_1 and γ_2, and a change in the control of expression to convert the γ genes from embryonic to foetal expression.

haemoglobin α and β chains form clusters on chromosomes 16 and 11. Other, isolated, loci contain genes for neuroglobin, cytoglobin, and myoglobin. Closely linked genes, as in the α– and β–globin regions, suggest relatively recent divergence. Yet even within these clusters, the proteins encoded have diverged in function, showing small but significant variations in oxygen affinity and responses to allosteric effectors. Also, these proteins appear at different stages of our development, implying divergence in the control of their expression.

We can date the globin duplications by looking back into evolutionary history (see Figure 6.7). Neuroglobin split off from other globins before the last common ancestor of the vertebrates, perhaps 10^9 years ago. The divergence of myoglobin and cytoglobin from haemoglobin occurred before the emergence of the jawless fishes, during the Cambrian about 500 million years ago. The divergence of α– and β–globins occurred early in the vertebrate lineage, approximately 450 million years ago.

Within mammals, the α– and β–globin regions are quite variable in content and extent (see Figure 6.8). Even within the primate lineage, we can date a duplication of the γ–globin gene (see Figure 6.9).

Duplication can affect individual exons

Fibronectin, a large extracellular protein involved in cell adhesion and migration, is a **modular protein** (see Box 6.7) containing multiple tandem repeats of three types of domain called F1, F2, and F3. It is a linear array of the form: $(F1)_6(F2)_2(F1)_3(F3)_{15}(F1)_3$ (see Figure 6.10). In the human genome, each domain of fibronectin is encoded by either one or two tandem exons.

Fibronectin domains also appear in other modular proteins. The duplication of the exon(s) encoding a domain, followed by transfer to another protein, is called 'exon shuffling'.

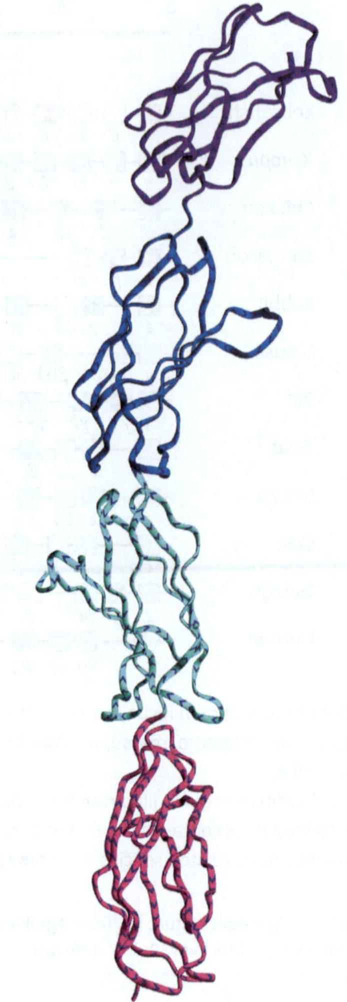

Figure 6.10 A fragment of fibronectin, a modular protein, showing four tandem domains, each a separate compact unit within the structure.

Family expansion: G-protein-coupled receptors

Repeated duplications can generate large numbers of homologues. **G-protein-coupled receptors (GPCRs)** are a large superfamily of eukaryotic cell-surface receptors active in signal recognition and processing, including the senses of sight, taste, and smell.

The human genome contains about 700 active GPCRs. They are integral membrane proteins with a common structure comprising seven transmembrane helices (see Figure 6.11). GPCRs interact with **G proteins** within the cell. Some GPCRs mediate responses to extracellular chemical signals. Reception of the signal triggers *intracellular* signalling cascades, which may reach as far as the nucleus to affect gene expression.

The interaction patterns of large families of functionally related proteins can create great complexity. Odorant receptors are GPCRs expressed on sensory neurons in the nasal cavities of humans and animals. The human genome has about 1000

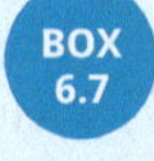

BOX 6.7 Modular proteins

A **modular protein** contains a linear string of compact units, called **domains**. Domains appear to have independent stability and can be 'mixed and matched' with one another in different proteins. Domains sometimes, but by no means always, correspond to single exons.

Modular proteins are common in eukaryotes. Individual domains of eukaryotic modular proteins are often separately homologous to single-domain prokaryotic proteins (and, less commonly, *vice versa*).

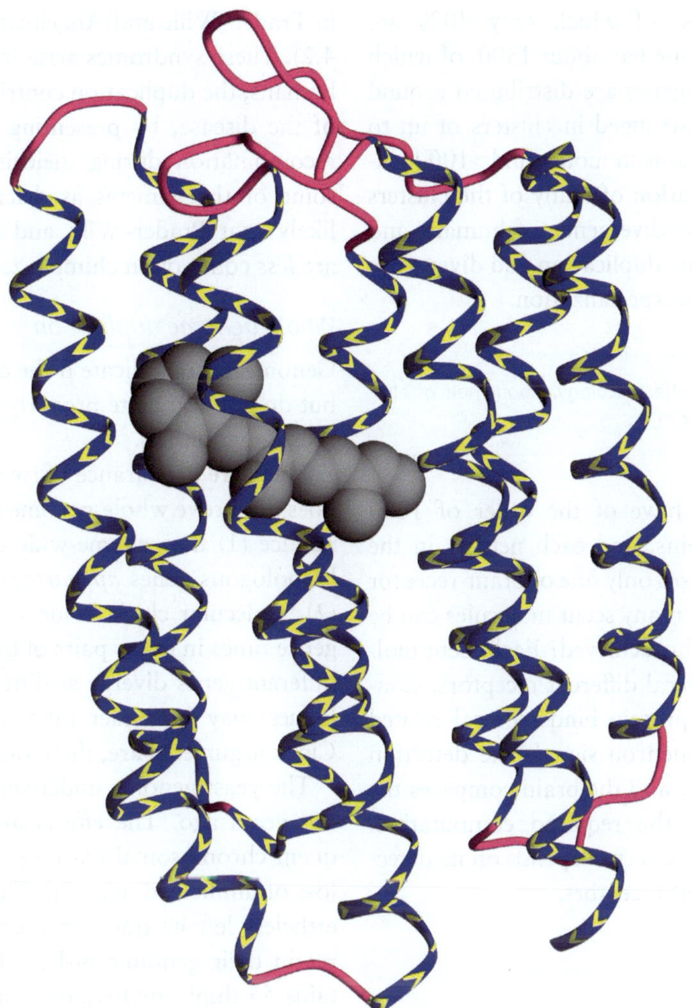

Figure 6.11 G-protein-coupled receptors (GPCRs) are a large family of transmembrane proteins involved in signal transduction into cells. They share a substructure containing seven transmembrane helices, arranged in a common topology. This figure shows the first experimentally determined mammalian GPCR structure, bovine opsin [1H68]. This molecule senses light and generates a nerve impulse.

The seven-helical structure is common to the family of GPCRs. The helices traverse the membrane, with loops protruding outside and inside the cell. This figure shows a view parallel to the membrane, with the extracellular side at the top. The transmembrane region is generally flanked by N- and C-terminal domains. The N-terminal domain is always outside the cell and the C-terminal domain always inside.

GPCRs constitute the largest known family of receptors. The family is as old as the eukaryotes and is large and diverse. Mammalian genomes contain ~1500–2000 GPCRs, accounting for about 3–5% of the genome. A similar fraction of the *C. elegans* genome codes for GPCRs.

Some GPCRs are involved in sensory reception, including vision, smell, and taste. Some, like opsin and bacteriorhodopsin, bind chromophores. (Bacteriorhodopsin is not a signalling molecule but a light-driven proton pump.) Others respond to extracellular ligands including hormones and neurotransmitters.

The common mechanism of function of GPCRs is a conformational change, induced by receptor binding or light absorption. The activated state of the GPCR interacts with an intracellular G protein, triggering a signal cascade. As there are substantially more GPCRs than G proteins, many GPCRs must interact with a single G protein. For instance, all odorant receptors interact with the same G protein α-subunit.

GPCRs are the targets for many drugs used in the treatment of high blood pressure, asthma, allergies, and other conditions. The large number of related GPCRs is a challenge to the design of drugs that bind to a unique target. Many drugs have undesired side effects because of imperfect specificity.

odorant-receptor genes, of which only 40% are active. The mouse genome has about 1300, of which 80% are active. These genes are distributed around mammalian genomes, arranged in clusters of up to 100 genes, in >40 locations in mouse and >100 locations in humans. Formation of many of the clusters appears to antedate the divergence of humans and mice, but individual gene duplication and divergence within clusters has led to specialization.

GPCRs are important in pharmacology. About half of all prescription drugs target GPCRs.

Although mammals have of the order of 1000 odorant-receptor proteins, and each neuron in the nasal epithelium expresses only one odorant-receptor allele, over ten times as many scent molecules can be distinguished. How is this achieved? Each scent molecule interacts with several different receptors. Conversely, each receptor protein binds several related scent molecules. Each neuron signals the detection of one *group* of odours, and the brain compares the outputs and performs the required computation. Identification of a specific scent depends on its detection by a *combination* of receptors.

KEY POINT

We depend less on our sense of smell than mice and dogs do, but we have about half the number of expressed odorant receptor proteins. Would you have expected a larger discrepancy?

Large-scale duplications

The genomes of many species contain duplications of multigene regions, the length varying from species to species.

Large-scale segmental duplications are an important component of the difference between human and chimpanzee genomes, affecting about 2.7% of the genome. Some duplications are found in chimpanzees but not humans, some in humans but not chimpanzees, and some in both.

Some duplications that appear in the human but not in the chimpanzee genome involve segments associated with developmental disorders, including a region on human chromosome 15 involved in Prader–Willi and Angelman syndromes (see Box 4.2). These syndromes arise from microdeletions. In humans, the duplication contributes to the frequency of the disease, by presenting sites for homologous recombination during meiosis, which show up in some of the gametes as deletions. It is, therefore, likely that Prader–Willi and Angelman syndromes are *less* common in chimpanzees than in humans.

Whole-genome duplication

Genomes can duplicate if the chromosomes replicate but do not segregate properly into separate progeny cells upon mitosis.

The mere appearance of two copies of many genes does not prove whole-genome duplication. One must adduce (1) the genome-wide occurrence of pairs of homologous genes *appearing in the same order*; or (2) 'molecular clock' evidence showing equal divergence times in many pairs of homologues. (However, different genes diverge at different rates, and homologues may be under different selective pressures. Clock arguments are, therefore, relatively weak.)

The yeast genome underwent a duplication about 10^8 years ago. The effects are obscured by subsequent chromosomal rearrangements and by massive loss of duplicated material. The duplication has nevertheless left its traces in multiple homologues that retain their genomic order. The yeast genome contains 55 duplicated regions, on average 55 kb long, together covering ~50% of the genome and including 376 pairs of homologous genes.

In a seminal 1970 book, *Evolution by Gene Duplication*, S. Ohno proposed that the vertebrate genome is the product of one or more complete genome duplications. Genome sequences confirm his prescient insight. Nor are whole-genome duplications only limited to vertebrates. They have occurred frequently in plant lineages.

In the lineage leading to vertebrates, the genomes of the cephalochordate Amphioxus (*Branchiostoma floridae*), and the urochordate *Ciona intestinalis* (see Figure 6.2), showed evidence for two rounds of whole genome duplication. Individual genes in these relatives corresponded to multiple genes in vertebrates, with enough synteny preserved to show that the process happened in parallel on a large scale. (Synteny = conservation of order of regions in two sets of chromosomes.) It appears that two whole-genome duplications in the vertebrate line occurred after the split between the primitive chordate relatives—urochordates and cephalochordates—from vertebrates, about 400–600 million years ago. More recently, a third duplication occurred in the lineage leading to ray-finned fishes, such as zebrafish and medaka.

What happens to all those extra genes? Most metazoa have roughly the same number of protein-coding genes, in the range 20000–25000. Whole-genome duplication is followed by massive gene loss. Some genes do form paralogous groups, and can even duplicate further, individually, creating gene and protein families of various sizes. The globins and GPCRs are examples.

It is interesting to see which kinds of genes do take advantage of the duplication. The comparison of the genome of the primitive chordate *B. floridae* with genomes of vertebrates shows that the set of duplicates that is retained after whole-genome duplication is enriched in genes for signal transduction, transcriptional regulation, neuronal activity, and development. Precisely the features with which early chordates were experimenting.

> Evolution subscribes to the advice, attributed to Yogi Berra: 'When you come to a fork in the road—take it.' We may venture to add: Even if you don't come to a fork in the road—take it anyway.

Plant genomes are very susceptible to duplication. The sequence of the *Arabidopsis* genome reveals at least three and probably five successive duplication events.

Most plants are **polyploids**, i.e. they contain multiple sets of entire chromosomes. **Autopolyploids** contain multiple copies of genomes from the same parent. **Allopolyploids** contain multiple copies of genomes from different parents. Many crop species are polyploids, relative to the wild species from which they were domesticated, including wheat, alfalfa, oats, coffee, potatoes, sugar cane, cotton, peanuts, and bananas. Often polyploidy increases the size of the fruit or grain, a useful property for agriculture (see Box 6.8).

Polyploidy may have other advantages. In studies of Arctic flora, it is observed that the fraction of diploid and polyploid plant species increases towards higher latitudes. Many Arctic plants tend to exist in small, separated populations and frequently go through 'bottlenecks' of marginal survival, for instance during glaciations. After recession of the ice, deglaciated areas may be repopulated by a few or even one dispersed seed. Carrying many copies of the genome in the cells of each individual may help to preserve genetic diversity, even in tiny populations.

Although polyploidization is much more common in plants than in animals, related species of frogs (genus *Xenopus*) are diploid, tetraploid, octaploid,

Polyploidy in wheat

The wheat first used in agriculture, in the Middle East at least 10000–15000 years ago, is a diploid called **einkorn** (*Triticum monococcum*), containing 14 pairs of chromosomes. Emmer wheat (*T. dicoccum*), also cultivated since palaeolithic times, and durum wheat (*T. turgidum*), are merged hybrids of relatives of einkorn with other wild grasses to form tetraploid species. Additional hybridizations, to different wild wheats, gave hexaploid forms, including spelt (*T. spelta*) and modern common wheat (*T. aestivum*). Triticale, a robust crop developed in modern agriculture and currently used primarily for animal feed, is an artificial genus arising from crossing durum wheat (*T. turgidum*) and rye (*Secale cereale*). Most triticale varieties are hexaploids.

Variety of wheat	Classification	Chromosome complement
Einkorn	*Triticum monococcum*	AA
Emmer wheat	*Triticum dicoccum*	AABB
Durum wheat	*Triticum turgidum*	AABB
Spelt	*Triticum spelta*	AABBDD
Common wheat	*Triticum aestivum*	AABBDD
Triticale	*Triticosecale*	AABBRR

A, genome of original diploid wheat or a relative; B, genome of a wild grass, *Aegilops speltoides* or a relative; D, genome of another wild grass, *T. tauschii* or a relative; R, genome of rye *S. cereale*.

All of these species are still cultivated—some to only minor extents—and have their individual uses in cooking. Spelt, or *farro* in Italian, is the basis of a well-known soup; pasta is made from durum wheat; and bread is made from *T. aestivum*.

and dodecaploid. One tetraploid mammal is known, the rat *Tympanoctomys barrerae* from the Monte Desert in west-central Argentina. These species provide a model for control of expression of duplicated genes. For example, in polyploid frogs, silencing is non-syntenic. Each copy of the genome contains some expressed and some silenced genes. This is a different model of dosage compensation from the silencing of an entire X chromosome in cells of mammalian females (see Chapter 1).

There are a number of examples of tissue-specific polyploidization. The endosperm of maize kernels undergoes repeated cycles of endoreplication (replication of nuclear DNA in the absence of mitosis) to produce cells that can have as many as 96 copies of the haploid genome. In mammals, it is observed that the number of polyploid cells in the liver increases with age, or in response to disease or surgery even in children. This may be a defence against oxidative stress. In the bone marrow of mammals, very large polyploid cells called megakaryocytes 'bud off' portions of their cytoplasm to form platelets. (Platelets are enucleate cells in the blood involved in clotting.) In a related condition, called polyteny, replicated chromosomes remain in alignment rather than separate as in polyploids. This is the origin of the giant salivary gland chromosomes of *Drosophila*, which played such an important role in the history of cytogenetics (see Figure 4.4).

Comparisons at the chromosome level: synteny

Comparison of chromosome banding patterns provides snapshots of similarities and differences in large-scale organization among eukaryotic genomes. Synteny literally means 'on the same band', that is, on the same chromosome. (The chromosome exchange that causes chronic myeloid leukaemia (see Chapter 10) is a *breaking* of synteny.) Closely related species generally show a correspondence between large syntenic blocks. The similarity of the banding patterns reveals the underlying similarity of the patterns in the DNA sequences themselves.

Figure 6.12 shows the relationships among the karyotypes of human, chimpanzee, gorilla, and orangutan.

What makes us human?

It is too difficult to look only at the human genome and try to deduce . . . ourselves. Three approaches help in understanding the genome.

- *Variation within our species*. Humans vary in appearance, traits, and personality. Some of the differences are the result of differences in genome sequence, others the result of different life history. Typically, a random pair of humans will differ in ~0.1% of the genome sequence; many of these changes have no phenotypic consequence.
- *Comparative genomics*. We can compare the human and chimpanzee or bonobo genomes and ask how *differences* between these genomes might give rise to *differences* between the species. The human and chimpanzee genomes differ by ~1%; the human and bonobo genomes differ by ~1.2%, about 12 times the intraspecific variation among humans. We call attention again to the paradox that the differences between human and chimpanzee or bonobo DNA sequences are, overall, quite small, but the small differences have tremendous phenotypic consequences.
- *Study of human disease*. Many mutations cause disease and give clues to the functions of the affected regions. These regions may encode enzymes or regulatory proteins or RNAs, or they may be DNA sequences that are targets of regulatory mechanisms.

Understanding the effects of mutations both illuminates human biology and, often, has immediate clinical applications.

Genomes of chimpanzees and humans

The genome sequence of Clint, a male chimpanzee from the Yerkes National Primate Research Center at Emory University, in Atlanta, Georgia, US, was reported in 2005. Clint represented the West African subspecies *Pan troglodytes verus*.

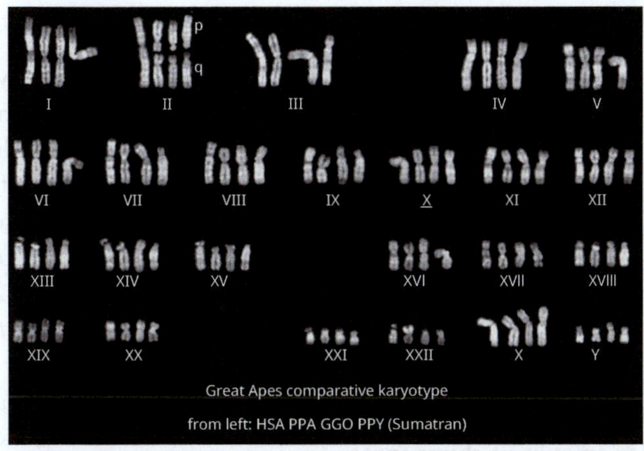

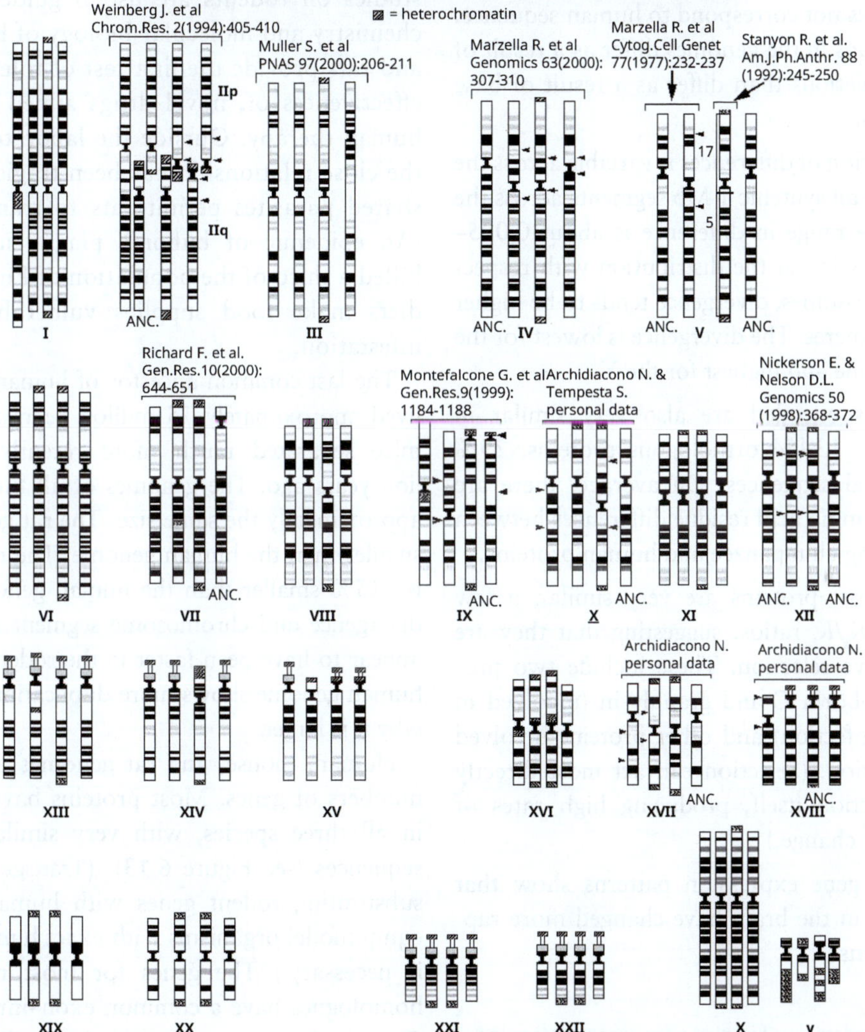

Figure 6.12 Top: photograph of banding patterns. Bottom: ideograms. HSA, *Homo sapiens*; PTR, *Pan troglodytes* (chimpanzee); GGO, *Gorilla gorilla*; PPY, *Pongo pygmaeus* (orangutan).

Photographs courtesy of Prof. M. Rocchi, Università di Bari, Italy.

As expected from so closely related a species:

- There is close alignment of the genome: 99% is alignable with the human genome.
- The sequences of the alignable regions differ at 1.23% of the positions. Recognizing that there is intraspecies divergence among humans and among chimpanzees, it is likely that the true interspecies difference amounts to about 1% of the alignable sequence.
- The non-alignable regions represent insertions and deletions. It is estimated that about 45 Mb of human sequence do not correspond to chimpanzee sequence and a similar amount of chimpanzee sequence does not correspond to human sequence. More positions in the genomes differ as a result of insertions/deletions than differ as a result of base substitutions.
- The distribution of differences is variable across the genome. For all syntenic 1 Mb segments across the genomes, the range in difference is about 0.005–0.025%. Looking at the distribution with respect to the chromosomes, divergence tends to be higher near the telomeres. The divergence is lowest for the X chromosome and highest for the Y.
- The proteins encoded are also very similar in sequence. Of 13454 orthologous proteins, 29% have identical sequences. On average, there are one to two amino acid residue differences between corresponding chimpanzee and human proteins.
- Although most proteins are very similar, a few show large K_a/K_s ratios, suggesting that they are under positive selection. These include two proteins, glycophorin C and granulysin (involved in combating infection) and other proteins involved in reproduction. (Selection can act most directly on reproduction itself, producing high rates of evolutionary change.)
- Changes in gene expression patterns show that genes active in the brain have changed more rapidly in humans.

Sadly and unexpectedly, Clint died a few weeks before the paper on his genome was submitted to *Nature*. He was 24 years old. Even in the wild, chimpanzees can live for 40–45 years. Cheeta, a chimp that appeared in Tarzan movies in the 1930s, died in 2011 at age 80.

Genomes of mice and rats

The mouse and rat are by far the most common mammalian laboratory animals. Knowledge of these species and correlation with human biology is encyclopaedic. Determination of mouse and rat complete genome sequences was clearly a high-priority goal. The genome of the laboratory mouse (*Mus musculus*) appeared in December 2002. The genome of the brown or Norway rat (*Rattus norvegicus*) appeared in April 2004.

Mice, rats, and humans are closely related mammals and illuminate one another. Laboratory studies on rodents are useful guides to the biochemistry and molecular biology of humans. Mice and rats provide the first test of tolerance to, and effectiveness of, novel drugs aimed ultimately at human therapy. Outside the laboratory, however, the close relationship has been tragic for humans: shared parasites permit rats to transmit disease. (An epidemic of bubonic plague in 1347–1352 killed a third of the population of Europe.) Shared diets make food supplies vulnerable to rodent infestation.

The last common ancestor of humans and rodents lived approximately 75 million years ago. Rats and mice separated much more recently: 12–24 million years ago. The genomes of all three species are approximately the same size. The rat genome is ~5% smaller than the human genome. The mouse genome is ~15% smaller than the human genome. Sequence divergence and chromosome segment rearrangement appear to have been faster in the rodent lineage. The human genome shows more duplication—one reason why it is larger.

Human, mouse, and rat genomes encode similar numbers of genes. Most proteins have homologues in all three species, with very similar amino-acid sequences (see Figure 6.13). (Transgenic animals—substituting rodent genes with human genes—can equip model organisms with exact human sequences if necessary.) The genes for most rodent–human homologues have a common exon–intron structure. Gene duplications create protein families, which may be of different sizes in different species. For instance, consistent with their greater dependence on the sense of smell, rodents have more odorant receptors than we do.

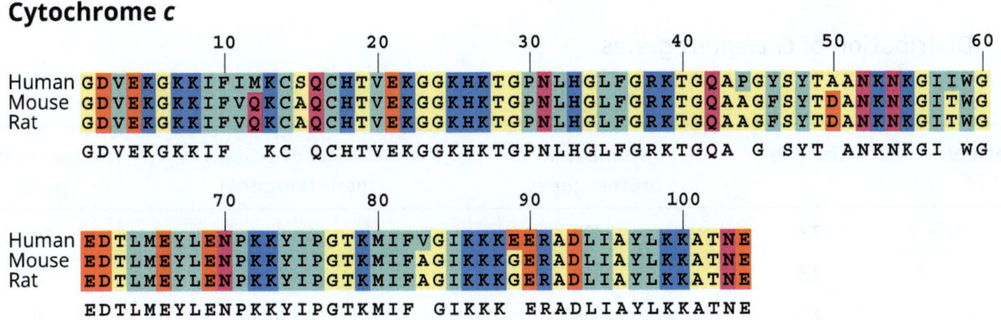

Figure 6.13 Alignment of the amino-acid sequences of cytochromes *c* from human, mouse, and rat. All have 104 residues. The mouse and rat sequences are identical. Letters below the sequences show residues conserved in all three species. The human sequence shows nine substitutions, most of which are conservative. For colour coding, see Figure 1.11.

Some genomic variation is observable at the chromosomal level. The mouse has 19 chromosomes, plus X/Y. The rat has 20 chromosomes, plus X/Y. Synteny between the mouse and rat genomes is high. Synteny between the mouse and human genomes is variable. Most human chromosomes contain many small blocks that correspond separately to regions distributed among several mouse chromosomes. However, almost all of human chromosome 20 corresponds almost continuously with a region of mouse chromosome 2. Almost all of human chromosome X appears in mouse chromosome X, differing by rearrangement of nine contiguous blocks, including some reversals. Almost all of human chromosome 17 appears within a region of mouse chromosome 11; however, the sequence is broken into 16 segments that are rearranged, including some reversals.

Alignment of genetic maps is less stringent than alignment of sequences. The rearrangements within and among chromosomes make it impossible to align the three genomes as one long linear sequence. However, at the nucleotide level, 40% of human and mouse genomes are block-alignable, about 1 Gb in all.

The blocks are identified by matching sequences of genetic markers. Even though most of the correspondences are distributed among different chromosomes, most of the genomes can be partitioned into syntenic blocks, making up a total of 2.35 Gb (over 90% of the mouse genome). These regions include almost all known exons and regulatory regions.

The differences among the human, mouse, and rat genomes arise from selection or neutral drift. Regions changing under selection rather than drift are likely to be functional. This is a powerful way to search for regulatory regions, which are harder to identify in genomes than protein-encoding genes are.

Genomics confirms the utility of the rat and mouse for clinical research. Of a set of ~1000 genes for which known mutations are associated with human disease, almost all have homologues in rodents. In certain interesting cases, the sequence of the human disease-associated mutant is identical to the mouse and rat wild type. There has probably been co-evolution, with a compensatory change in some other gene or genes. This can be a source of clues to the function and interactions involved in the disease.

Model organisms for study of human diseases

Above the molecular level, the differences between humans and flies and between humans and worms are more obvious than the similarities. At the biochemical and genomic level, the situation is reversed. The underlying common features of the structure, organization, and development of different species is of both academic interest and practical importance, as flies and worms provide models for human diseases.

The genome of *Caenorhabditis elegans*

The nematode worm *C. elegans* entered molecular biology in 1963, at the invitation of Sydney Brenner. Brenner recognized its potential as a sufficiently complex organism to be interesting but simple enough to permit complete analysis of its development and neural circuitry, *at least* at the cellular level.

BOX 6.9

Distribution of *C. elegans* genes

Chromosome	Size (Mb)	Number of protein genes	Density of protein gene (kb/gene)	Number of tRNA genes
I	7.9	2803	5.06	13
II	8.5	3259	3.65	6
III	7.6	2508	5.40	9
IV	9.2	3094	5.17	7
V	9.8	4082	4.15	5
X	10.1	2631	6.54	3

From: The *C. elegans* Sequencing Consortium (1998). Genome sequence of the nematode *C. elegans*: A platform for investigating biology. Science 282: 2012–2018.

The *C. elegans* genome, completed in 1998, was the first full DNA sequence of a multicellular organism. *C. elegans* contains ~97 Mb of DNA distributed on six paired chromosomes (see Box 6.9). There is an X but no Y chromosome: different genders in *C. elegans* are a self-fertilizing hermaphrodite, genotype XX, and a male, genotype XO (i.e., a single unpaired X chromosome).

The *C. elegans* genome is about eight times larger than that of yeast, and its 19099 predicted genes are approximately three times the number in yeast. Exons cover ~27% of the genome. Protein-coding genes contain an average of five introns. The gene density is relatively low, for a eukaryote, with ~1 gene/5 kb of DNA. Approximately 25% are in clusters of related genes.

The genome of *Drosophila melanogaster*

The total chromosomal DNA of *D. melanogaster* contains about 180 Mb. Approximately two-thirds is euchromatin, a relatively uncoiled and non-compact form, containing most of the active genes. The euchromatic portion, about 120 Mb, was the first segment of the sequence released. The other one-third of the *Drosophila* genome appears as heterochromatin, highly compact regions flanking the centromeres. Heterochromatin contains many tandem repeats of the sequence AATAACATAG and relatively few genes.

The genome is distributed over five chromosomes: three large autosomes, a tiny chromosome containing only ~1 Mb of euchromatin, and an X/Y chromosome pair, of which the Y chromosome is heterochromatic and relatively gene poor (see Box 6.10). The fly's ~14 000 genes are approximately double the number in yeast, but fewer than in *C. elegans*, perhaps a surprise. The average density of genes in the euchromatin sequence is 1 gene/9 kb; about half that of *C. elegans* (see Box 6.9). The genes of the metacentric chromosomes 2 and 3 are reported separately for the two arms, arbitrarily designated left (L) and right (R). The other chromosomes are telocentric (see Figure 6.14).

Determination of the *D. melanogaster* genome sequence was a collaboration between industry (Celera Genomics) and the academic *Drosophila* Genome Projects based in Berkeley, California, US, and in Europe. The project was a methodological testbed.

- First, it showed that a relatively large eukaryotic genome could be completed by the method of whole-genome shotgun sequencing (see Chapter 4).
- Second, the annotation of the sequence took place in a burst of activity: the intensive sessions of an 11-day 'jamboree' meeting held at Celera in November 1999. The ~45 participants included experts representing the inherited knowledge of a century of fruit-fly biology and local computer experts involved in the sequencing. The flavour of the meeting is well described from the personal point of view by one of the participants, Michael Ashburner.[2]

[2] Ashburner, M. (2006). *Won for All: How the* Drosophila *Genome Was Sequenced*. Cold Spring Harbor, New York: Cold Spring Harbor Laboratory Press.

BOX 6.10 Distribution of *D. melanogaster* genes

Chromosome arm	Size (Mb)	Number of protein genes	Density of protein genes (kb/gene)	Number of tRNA genes
X	22.2	2279	9.7	25
2L	22.4	2537	8.8	40
2R	20.8	2947	7.1	100
3L	23.8	2718	8.8	49
3R	27.9	3501	8.0	80
4	1.28	83	15.4	0

Data from Release 5.22: http://www.ncbi.nlm.nih.gov/mapview/mapsearch.cgi?taxid=7227.

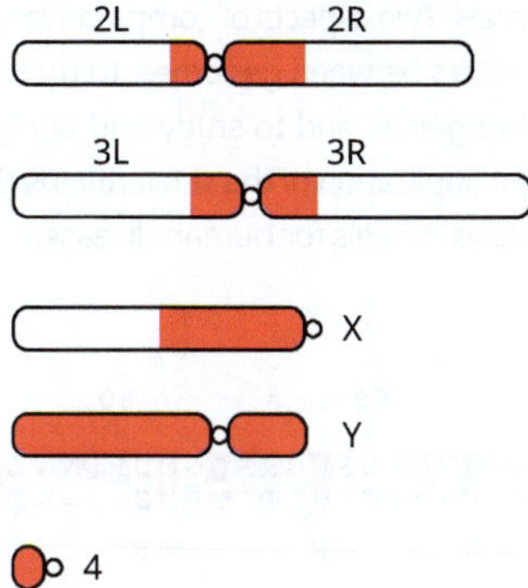

Figure 6.14 The chromosomes of *Drosophila melanogaster*. Heterochromatin is shaded/red.

Homologous genes in humans, worms, and flies

Once the genomes were sequenced and annotated, comparisons showed that homologues of many genes appear in all three species. Forty-four per cent of protein-coding genes from *D. melanogaster* have human homologues, 25% of protein-coding genes from *C. elegans* have human homologues, and 23% of protein-coding genes from *D. melanogaster* have homologues in *C. elegans*.

Proteins with common functions are expected to be quite similar in the different species (see Figure 6.15). In other cases, different species have modified and adapted homologous proteins for slightly different functions.

C. elegans and *D. melanogaster* are favourite subjects of developmental biologists. It is therefore of interest that a number of transcription regulators involved in developmental control have homologues in human, *D. melanogaster*, and *C. elegans*. These include PAX (paired box domain) and HOX (homeobox domain) proteins.

Models of human disease

A model organism is a species in which an interesting feature of human biology—especially a disease—can be studied (Table 6.5). Ideally, a model organism is small and robust, has a relatively simple genome, is easy to maintain and manipulate (both physically and genetically) in the laboratory, has a short generation time, is safe to humans, and comes with an extensive knowledge of its biology. As each model organism has its own strengths and limitations, different organisms are useful for different investigations.

In principle, biologists will use the simplest organism that illustrates the human feature of interest. It is easier to do experiments with yeast than with fruit flies. However, sometimes there is no choice. The only animal other than humans that is susceptible to leprosy is the armadillo, which satisfies few, if any, of the criteria of an ideal laboratory organism.

There are two ways in which model organisms can contribute to understanding and treatment of human disease. The first is to observe homologues in the model organisms of genes implicated in human diseases. One can then study the effect on the model organism of mutation or knockout of the homologues. The second is to introduce a human gene into a model organism and discover its phenotypic effect. A model animal containing an active human gene makes it possible to screen libraries of compounds for potential drugs.

Table 6.5 shows some of the human disease-associated genes with homologues in *D. melanogaster*, *C. elegans*, and *Saccharomyces cerevisiae*.

Despite the fact that insects are not very closely related to mammals, even fruit flies are useful in the study of human disease. The *D. melanogaster* genome contains homologues of human genes implicated in cancer and in cardiovascular, neurological, endocrinological, renal, metabolic, and haematological diseases. Some of these homologues have different functions in humans and flies. Other human disease-associated genes can be introduced into, and studied in, the fly. For instance, the gene for human spinocerebellar ataxia type 3, when expressed in the fly, produces similar neuronal cell degeneration. There are now fly models for Parkinson's disease and malaria.

C. elegans also provides human disease models. Mutations in the human gene for presenilin-1 (*PS1*) are associated with familial early-onset Alzheimer's disease. Mutations in the homologous gene in *C. elegans, sel-12* (Figure 6.15) do show neurological defects, but in only a few neurons. Mutants do show more profound defects in egg laying, but this may be a secondary effect.

Although there are greater differences between the nervous systems of humans and *C. elegans* than between their machineries for respiratory energy transduction, the sequences of these homologues are actually quite similar. The relationship between sequence and function in proteins is full of surprises.

KEY POINT

We have discussed selected genomes—chimpanzee, mouse, rat, worm, and fly—the first because it is the closest extant relative we have, and the others because of their importance as laboratory animals. Two aspects of comparative genomics are to study *differences* between genomes, to try to account for phenotypic divergence; and to study and apply *similarities*. One important application of the similarities is the use of laboratory animals as models for human diseases.

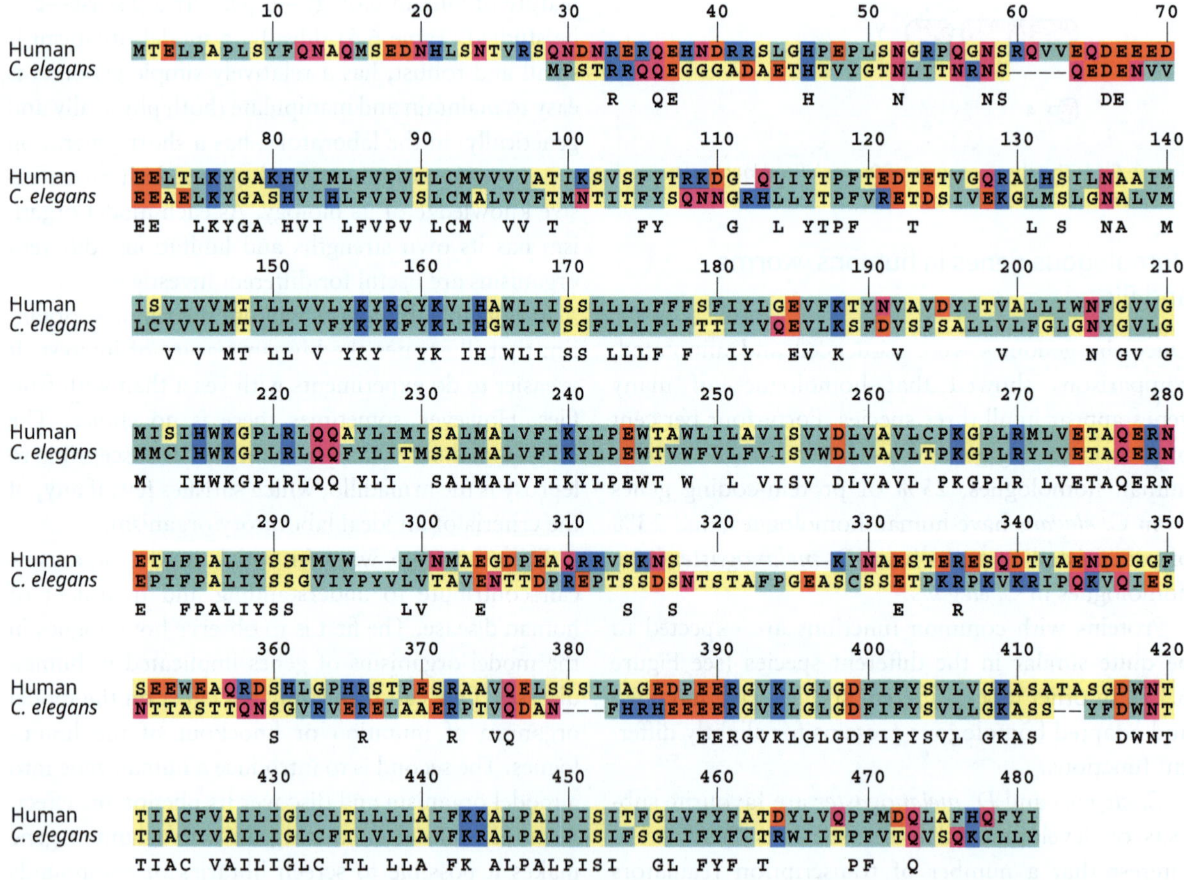

Figure 6.15 Alignment of the amino-acid sequences of the human protein presenilin-1 and the *C. elegans* homologue SEL-12.

Table 6.5 Human disease-associated genes shared with worms, flies, and yeast

Affected area	Disease	Description	Gene	Similarity in		
				Worm	Fly	Yeast
Bones	Multiple exostoses	Ossification at tips of femur, pelvis, or ribs	*EXT1*	***	**	–
Blood	Leukaemia	Chronic myelogenous leukaemia, a blood cell cancer	*ABL1*	***	***	*
	Bruton agammaglobulinaemia	Lack of mature B cells	*BTK*	***	**	*
	Glucose-6-phosphate dehydrogenase deficiency	Drug and stress-induced rupture of red blood cells	*G6PD*	****	****	****
Brain	Early-onset Alzheimer's disease	Common cause of mental retardation	*PS1*	**	**	–
	Fragile X syndrome		*FMR1*	**	–	–
	Juvenile Parkinson's disease		*PARK2*	***	**	*
Colon	Hereditary non-polyposis cancer	Polyps that become malignant	*MSH2*	***	***	***
	Adenomatous polyposis		*APC*	***	*	–
Ears	Hereditary deafness		*MYO15*	***	***	***
Eyes	Retinoblastoma	Cancer of the eye	*RB1*	*	*	–
Heart	Familial cardiac myopathy	Inherited cardiac disease	*MYH7*	***	***	***
	Long QT syndrome	Sometimes fatal cardiac arrhythmias	*3-SCN5A*	***	**	*
Kidney	Polycystic kidney disease 2		*PKD2*	**	**	–
Liver	Wilson's disease	Build-up of copper in cells, causing liver disease and other symptoms	*ATP7B*	***	***	***
Lung	Cystic fibrosis	Progressive disease of lungs and pancreas	*CFTR*	***	***	–
	Lung cancer	Caused by defects in p53 gene, which can also cause cancer of the oesophagus, colon, brain, lung, breast, and skin	*p53*	*	–	–
Muscles	Duchenne's muscular dystrophy	Progressive atrophy of muscles	*DMD*	***	***	–
Pancreas	Pancreatic cancer		*MADH4*	***	*	–
	Pancreatic cancer		*RAS*	**	**	**
Prostate	Advanced cancer of the prostate	Caused by mutations in the PTEN gene, which can also cause cancer of the brain, endometrium, and breast	*PTEN*	**	**	*
Skin	Xeroderma pigmentosum D	Early-onset skin cancer	*XPD*	***	**	***
	Neurofibromatosis 1	Soft tumours at many sites, plus skeletal and neurological defects	*NF1*	***	*	**
Thyroid	Cancer of the thyroid	Multiple endocrine neoplasia type 2	*MEN2*	***	**	*

Based on data from Rubin, G.M., Yandell, M.D., Wortman, J.R., *et al.* (2000). Comparative genomics of the eukaryotes. Science 287: 2204–2215. Presentation adapted from http://www.hhmi.org/genesweshare/e400.html.

LOOKING BACK AND FORWARD

Living things are a finite selection of choices from an infinite range of possibilities. The choices are partly the result of environmental pressures—including the biosphere itself as a prominent part of the environment with which every organism and every species has to deal, and to the shaping of which it contributes in important ways—and partly historical accident.

It is likely that life originated on Earth about 3.5 billion years ago. The first cellular life forms were undoubtedly prokaryotes. Eukaryotes appeared about 2 billion years later. But there are enough residual similarities among living things to suggest a common ancestor of us all. The great diversity of living forms is, therefore, the result of divergence.

In the next two chapters, we shall survey some features of a number of representative prokaryotes, viruses, and eukaryotes.

RECOMMENDED READING

Origin of eukaryotes

Nobs, S.J., MacLeod, F.I., Wong, H.L., *et al.* (2022). Eukarya the chimera: Eukaryotes, a secondary innovation of the two domains of life? Trends Microbiol. 30: 421–431.

Papers treating some of the current debate in biological taxonomy and the strengths and weaknesses of barcoding, and a description of the database

Ratnasingham, S., & Hebert, P.D.N. (2007). BOLD: The Barcode of Life Data System. Molecular Ecology Notes 7: 355–364.

Stoeckle, M.Y., & Hebert, P.D. (2008). Barcode of life. Sci. Am. 299, 4: 82–86, 88.

Antil, S., Abraham, J.S., Sripoorna, S., *et al.* (2023). DNA barcoding, an effective tool for species identification: A review. Mol. Biol. Rep. 50: 761–775.

Detailed review of work on genome comparisons and what they tell us about genome contents and evolution

Miller, W., Makova, K.D., Nekrutenko, A., *et al.* (2004). Comparative genomics. Annu. Rev. Genomics Hum. Genet. 5: 15–56.

Liberles, D.A., Chang, B., Geiler-Samerotte, K., *et al.* (2020). Emerging frontiers in the study of molecular evolution. J. Mol. Evol. 88: 211–226.

Stearns, S.C. (2020). Frontiers in molecular evolutionary medicine. J. Mol. Evol. 88: 3–11.

Barik, S. (2022). Special issue: Structure, function and evolution of protein domains. Int. J. Mol. Sci. 31: 6201. (Introduction to a collection of articles.)

Buda, K., Miton, C.M., Fan, X.C., *et al.* (2023). Molecular determinants of protein evolvability. Trends Biochem. Sci. 48: 751–760.

How modern developments in biological data collection affect the use of model organisms in the study of human biology and disease

Barr, M.M. (2003). Super models. Physiol. Genomics 13: 15–24.

No authors listed. (2020). Model organisms: Nature's gift to disease research. Genetics 214: 233–234.

Stephan, T., Burgess, S.M., Cheng, H., *et al.* (2022). Darwinian genomics and diversity in the tree of life. Proc. Natl. Acad. Sci. (USA) 119: e2115644119.

The significance of duplications

Levasseur, A., & Pontarotti, P. (2011). The role of duplications in the evolution of genomes highlights the need for evolutionary-based approaches in comparative genomics. Biol. Direct. 18: 11.

Copley, S.D. (2020). Evolution of new enzymes by gene duplication and divergence. FEBS J. 287: 1262–1283.

Mallik, S., Tawfik, D.S., & Levy, E.D. (2022). How gene duplication diversifies the landscape of protein oligomeric state and function. Curr. Opin. Genet. Dev. 76: 101966.

Molecular clocks

Alvarez-Ponce, D. (2021). Richard Dickerson, molecular clocks, and rates of protein evolution. J. Mol. Evol. 89: 122–126.

Variable splicing

Park, J.W., & Graveley, B.R. (2007). Complex alternative splicing. Adv. Exp. Med. Biol. 623: 50–63.

Verta, J.P., & Jacobs, A. (2022). The role of alternative splicing in adaptation and evolution. Trends Ecol. Evol. 37: 299–308.

Wright, C.J., Smith, C.W.J., & Jiggins, C.D. (2022). Alternative splicing as a source of phenotypic diversity. Nat. Rev. Genet. 23: 697–710.

RNA editing

Maas, S., Kawahara, Y., Tamburro, K.M., *et al.* (2006). A-to-I RNA editing and human disease. RNA Biol. 3: 1–9.

Farajollahi, S., & Maas, S. (2010). Molecular diversity through RNA editing: A balancing act. Trends Genet. 26: 221–230.

Eisenberg, E. (2021). Proteome diversification by RNA editing. Methods Mol. Biol. 2181: 229–251.

EXERCISES, PROBLEMS, AND WEBLEMS

Exercises

Exercise 6.1 On a photocopy of Figure 6.3, (a) indicate on the abscissa the position of the human genome; (b) indicate the position of the pufferfish genome; (c) indicate the position of the rice (*Oryza sativa*) genome; (d) indicate the position of the yeast *S. cerevisiae* genome; (e) indicate the position of the *E. coli* genome; (f) indicate the range of sizes of mitochondrial genomes; (g) indicate the range of sizes of chloroplast genomes; and (h) indicate the positions of human and canteloupe mitochondria.

Exercise 6.2 In Figure 6.15, which substitutions are conservative?

Exercise 6.3 Which of the following pairs are orthologues? Which are paralogues? Which are neither? (a) Human trypsin and horse trypsin; (b) human trypsin and horse chymotrypsin; (c) human trypsin and human elastase; (d) *Bacillus subtilis* subtilisin and horse chymotrypsin.

Exercise 6.4 Human mitochondrial DNA is 16 568 bp long. A brain cell may contain 10 000 mitochondria. What fraction of the DNA in a brain cell is mitochondrial? Assume 1 mtDNA/mitochondrion.

Exercise 6.5 Some antibiotics, for example streptomycin, block protein synthesis in bacteria but not cytoplasmic protein synthesis in eukaryotes. Mitochondria and chloroplasts contain their own protein-synthesizing machinery. Would you expect streptomycin to block mitochondrial protein synthesis? Explain your answer.

Exercise 6.6 On a photocopy of Figure 1.14, indicate by an arrow linking them the approximate positions of the source and destination organisms for the endosymbiotic origins of mitochondria and chloroplasts.

Exercise 6.7 Why could a mollusc that extracts chloroplasts from algae that it eats not simply let the chloroplasts reside in some body cavity (like the symbiotic bacteria in the guts of ruminant animals) rather than endocytosing them?

Exercise 6.8 The main *E. coli* chromosome contains 4 639 221 bp. The cell is roughly a cylinder about 0.1 μm in diameter and 0.2 μm long. If the length of an extended segment of DNA in the B conformation is 3.4 Å per base pair (0.34 nm), what would be the diameter of the chromosome if it were geometrically a circle?

Exercise 6.9 On a photocopy of Figure 6.5, mark the positions that (a) contain the same amino acid in all three eukaryotes but differ in *S. aureus*; (b) contain the same amino acid in humans and chickens but are not the same in *Neurospora* and/or *S. aureus*; (c) contain the same amino acid in humans and *S. aureus*, but this amino acid does not appear at this position in either of the other two species.

Exercise 6.10 On a photocopy of Figures 6.5 and 6.6, indicate which regions of the amino-acid sequences are encoded by which blocks of the nucleotide sequences in Figure 6.6.

Exercise 6.11 On a photocopy of Figure 6.7, draw horizontal lines at the beginning and end of the Devonian (see Figure 1.17). What is the earliest type of species to emerge after the split between neuroglobin and cytoglobin?

Exercise 6.12 Which animals in Figure 6.8 have the largest number of functional β-globin genes?

Exercise 6.13 Which animals in Figure 6.8 show a triplication of part of the β-region? Which genes make up the repeating unit?

Exercise 6.14 Which animals in Figure 6.8 show a pseudogene most closely related to a δ-globin?

Exercise 6.15 The human and chimpanzee genomes are 96% identical. (a) Approximately how many individual bases of the human genome differ from the corresponding positions of the chimp genome? (b) Assume that the human genome is 3% protein coding and contains about 20 000 genes, and that all of the sequence differences are independent, single-base changes distributed randomly throughout the genome (these assumptions are definitely *not true*). Estimate the fraction of protein-coding genes mutated between humans and chimpanzees.

Exercise 6.16 On a photocopy of Figure 6.7, indicate where three whole-genome duplications are believed to have occurred.

Problems

Problem 6.1 From the partial sequences in Figure 6.6, (a) how many positions (necessarily in the coding regions) contain the same base in all three genomes? What percentage of positions contain the same base in all three genomes? (b) How many positions in the coding regions are common to human and chicken but different in *S. aureus*? To what percentage of coding positions does this correspond? (c) How many positions in the non-coding regions are common to human and chicken? To what percentage of non-coding positions does this correspond?

Problem 6.2 For the coding regions of the genes for human and chicken thioredoxins (Figure 6.5), calculate the K_a/K_s ratio. What can you infer about the relative importance of selection and drift in accounting for the difference in the corresponding amino-acid sequences?

Problem 6.3 To what time frame can you date the duplication of the γ gene in the β–globin locus of primates? (See Figures 1.17, 6.8, and 6.9.)

Problem 6.4 Figure 6.16 shows an evolutionary tree of six primates. With reference to Figure 6.12, can you find similarities in the chromosome structures that confirm these relationships between human, common chimpanzee, gorilla, and orangutan?

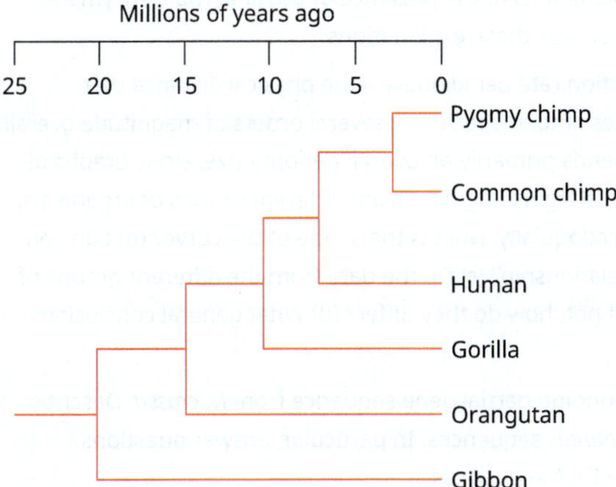

Figure 6.16 Phylogenetic tree of six primates.

Problem 6.5 Which substitutions between rodent and human cytochrome *c* would *not* be considered conservative mutations?

Problem 6.6 For the presenilin-1 homologues of humans and *C. elegans* (Figure 6.15): (a) What are the percentages of identical amino acids in an optimal alignment? (b) What are the percentages of identical residues *or* conservative substitutions?

Weblems

Weblem 6.1 Identify a virus and a prokaryote such that the genome of the virus is larger than the genome of the prokaryote (see Figure 6.3).

Weblem 6.2 Find examples of viruses with (a) a double-stranded circular DNA genome; (b) a double-stranded linear DNA genome; (c) a single-stranded (+)sense RNA genome; and (d) a single-stranded (-)sense RNA genome.

Weblem 6.3 Give an example of a species that represents the simplest known form of (a) metazoan; (b) deuterostome; (c) placoderm; and (d) eutherian.

Weblem 6.4 Mitochondrial DNA is often edited before translation. When a mitochondrial gene is transferred to the nucleus, there are two possibilities: (1) the copying is DNA→DNA. In this case, the nuclear version will initially have the mitochondrial DNA sequence and will require mutation to effect the changes introduced in the mitochondrion by the editing; or (2) the nuclear gene is reverse transcribed from the edited mitochondrial mRNA. Compare the sequences of the nuclear gene for cytochrome oxidase II from mung bean (*Vigna radiata*) with mitochondrial sequences in related legumes, before and after editing. Do the data support either hypothesis?

Weblem 6.5 Align the following sequences: (1) the nuclear-encoded *rps14* gene from rice (*Oryza sativa*); (2) the mitochondrial-encoded *rps14* gene from broadbean (*Vicia faba*); (3) the nuclear-encoded *sdhB* gene from rice. Identify the leader sequence in the rice *rps14* gene, targeting mitochondrial import, that appears to have been borrowed from the rice *sdhB* gene.

Weblem 6.6 Many people believe that Rickettsiae are the closest extant relatives of the organism that, after endosymbiosis, gave rise to mitochondria. Rickettsiae are obligate aerobes. (a) Can you identify an enzyme that (1) catalyses an anaerobic function in mitochondria and (2) lacks a known homologue in Rickettsiae? (b) If you can find one, and if the rickettsial origin of mitochondria is a valid hypothesis, what are reasonable explanations of the presence of the anaerobic enzymatic function in mitochondria? (c) How would you test these explanations?

Weblem 6.7 In eukaryotes, the recombination rate per kilobase—the physical distance that corresponds to the genetic distance—varies among species by several orders of magnitude overall. (It also varies within each genome.) It depends primarily on overall genome size. Draw graphs of recombination rate against genome size, distinguishing data from different groups of organisms. (a) What relationship do you observe, i.e. colloquially, what is the shape of the curve? (b) Can you plot the data in a way that gives a linear relationship? (c) Do the data from the different groups of organisms follow the same relationship? If not, how do they differ? (d) What general conclusions can you draw?

Weblem 6.8 Add to Figure 6.6 the corresponding partial gene sequence from *N. crassa*. Describe its relationship to human, chicken, and *S. aureus* sequences. In particular, answer questions analogous to those in Exercise 6.10 for the DNA sequences.

Weblem 6.9 What full-genome project is in progress, but not yet complete, for an organism in each of the following categories: (a) fungus; (b) amphibian; (c) land plant; (d) insect (*not* a species of *Drosophila*); (e) primate.

Weblem 6.10 Collect and align sequences of the protein HSP70 from about six (of each) Gram-positive bacteria, proteobacteria, other Gram-negative bacteria, and archaea. Identify an insertion common to proteobacteria and other Gram-negative bacteria but absent from Gram-positive bacteria and archaea. On this basis, sketch the topology of a phylogenetic tree relating Gram-positive bacteria, proteobacteria, other Gram-negative bacteria, and archaea. Where do these results suggest placing the root of the prokaryotic tree?

Weblem 6.11 The genetic code used in translation of genes in animal mitochondria differs from the standard one. Was this feature simply inherited from the original symbiont that gave rise to the organelle, or did it arise subsequently by divergence? Determine (a) what genetic code is used by the living organism that appears to be the closest extant relative of the original symbiont, *Rickettsia prowazekii*. (b) Do all animal mitochondria use the *same* variant of the standard genetic code? Based on these observations, suggest an answer to the question.

Weblem 6.12 What is significant about the following species that justified sequencing their entire genome sequences? (a) *Plasmodium falciparum*; (b) *Aspergillus fumigatus*; (c) *Tropheryma whipplei*; (d) *Sulfolobus tokodaii*; (e) sea urchin; (f) *Ciona intestinalis*.

Weblem 6.13 In 1976, swine flu killed Private David Lewis, a soldier at Fort Dix, in central New Jersey south of Princeton, US. What was the response of the US Government, under President Gerald Ford? What was the course of the potential epidemic? In retrospect, does the response appear to have been the right one? Could you have advised President Ford to make a different response? If so, based on what facts known at the time of Private Lewis's illness?

Weblem 6.14 Compute K_a/K_s ratios for the genes for the two isoforms of *D. melanogaster* cytochrome *c*. Does the divergence appear to have arisen from selection or drift?

Weblem 6.15 Compute K_a/K_s ratios for the genes for human presenilin-1 and the *C. elegans* homologue *sel-12*. Does the divergence appear to have arisen from selection or drift?

Weblem 6.16 The classification of tarsiers among primates is ambiguous. Palaeontology places tarsiers with lemurs, lorises, and galagos, in the suborder prosimians. Some genetic relationships place tarsiers with monkeys, apes, and humans. It is also possible that tarsiers form a separate prosimian infraorder. Find the structure of the β-globin region of the tarsier genome and explain what these data suggest about where, in Figure 6.9, the tarsiers belong.

Weblem 6.17 The human homologues of *C. elegans* NPR-1 are the neuropeptide Y receptors NPY1R, NPY2R, . . . What mutations are known in these human receptors and what, according to Online Mendelian Inheritance in Man (OMIM™), are their effects?

Weblem 6.18 In humans, cholesteryl ester transfer protein is important in controlling blood levels of high-density lipoproteins. Do homologues of this protein exist in (a) mouse; (b) rat; and (c) hamster?

Weblem 6.19 Mouse and rat cytochrome *c* have identical amino-acid sequences. Do they have identical gene sequences also? If not, show the differences between the cytochrome *c* gene in mouse and rat. Distinguish between exons and introns.

Weblem 6.20 How many cytochrome *c* pseudogenes are there in the human genome?

Weblem 6.21 Which of the following genes have the same number of exons in human and mouse orthologues? (a) Haemoglobin A (human gene *HBA*); (b) cytochrome *c* (human gene *HCS*); (c) spermidine synthase (human gene *SRM*).

Weblem 6.22 Are any regions containing globin genes, other than haemoglobin α and β, included in the choices of ENCODE regions studied by the HapMap Project? If so, which ones?

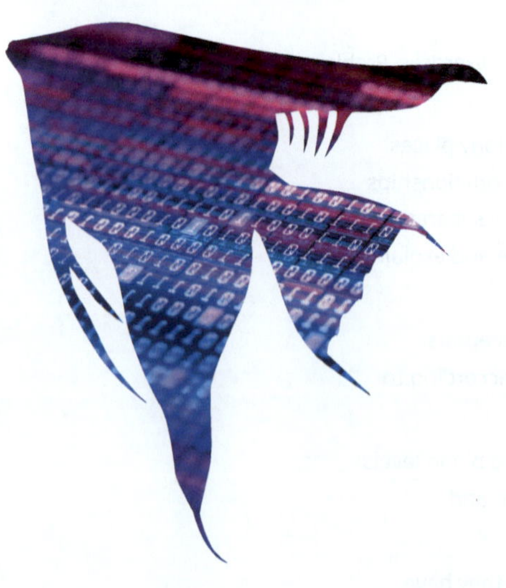

CHAPTER 7

Genomes of Prokaryotes and Viruses

LEARNING GOALS

- *Recognize the relationships among the three major divisions of cellular life*: archaea, bacteria, and eukarya.
- *Know the features that distinguish archaea, bacteria, and eukarya* and appreciate how differences of lifestyle reflect differences in genomes and structures.
- *Understand the molecular basis of adaptations*; for example, to life at high temperatures, or different ocean depths.
- *Appreciate, at the molecular level, the genomic and phenotypic differences among selected related species of prokaryotes.*
- *Understand the basic structure and mechanism of replication of viruses.*
- *Recognize the problem of bacterial and viral pathogenicity*, and the importance of the development of antibiotic resistance.
- *Appreciate developments in metagenomics, metatranscriptomics, and metaproteomics*: the application of methods from the laboratory to natural ecosystems.
- *Appreciate the vast variety of different microorganisms that inhabit, and mutually interact in, environmental samples.* These habitats include oceans and soils, and the human (or animal) body.

Evolution and phylogenetic relationships in prokaryotes

Prokaryotes have several claims on our interest.

- They cause infectious diseases. Some, for instance tuberculosis, are major public health problems. It is a challenge to control these diseases in the face of the development of antibiotic resistance.
- Molecular biologists study prokaryotes as examples of relatively simple cells, to understand fundamental principles of metabolism, genetics, and regulation.
- Historically, prokaryotes represent the earliest forms of life, from which all others are derived. They had the biosphere to themselves for over 2 billion years.
- Prokaryotes are important mediators of ecological processes and geological cycles. Indeed, geological and biological phenomena are linked in an intimate marriage, which has seen its turbulent episodes (Table 7.1). Purely geological events such as asteroid impacts have caused or contributed to mass extinctions. Purely biological events, such as the development of photosynthetic processes that released large quantities of O_2 into the atmosphere, and respiration that released CO_2, have altered flows of matter and energy, affecting the development of the Earth's geochemistry and climate. Microbes respond to human-caused environmental damage. They can aggravate ecological problems, but also hold out hope of remediating them.
- Prokaryotes show a large variety of metabolic processes, many of which are useful in biotechnology.

Table 7.1 Landmarks in history of life

Event	Date	24-hour equivalent
Formation of Earth	$\sim 4.5 \times 10^9$ years ago	24 hours ago
Origin of life	$>3.8 \times 10^9$ years ago	20 hours ago
Cyanobacterial photosynthesis	$>2.7 \times 10^9$ years ago	14 hours ago
Rise of atmospheric O_2	$\sim 2.5 \times 10^9$ years ago	13 hours ago
Cambrian explosion	$\sim 541 - 490 \times 10^6$ years ago	3 hours ago

The major habitats of prokaryotes are the open ocean, surface soils, and subsurface sediments beneath both ocean and soil. The bodies of humans and other animals harbour many microbes, but—important as the consequences for health and disease may be—as an overall reservoir we are a minor player. The total carbon content of prokaryotes is between 60% and 100% of the total carbon found in plants, but the total nitrogen and phosphorus in prokaryotes is probably ten times that of plants.

> The oceans also contain viruses in very great abundance and variety. Most of these are uncharacterized. They have been called the 'dark matter of the biosphere'. It is likely that viruses are an important mediator of gene transfer between marine prokaryotes.

The exploration of potential habitats by prokaryotes approaches saturation. Prokaryote cells divide actively. Production is estimated at 1.7×10^{30} cells per year, the open ocean being the highest contributor. This fecundity gives prokaryotes the opportunity to evolve quickly. The resulting variety of prokaryotes includes the colonists of inhospitable habitats such as hot springs and very salty lakes (Table 7.2). It also includes almost continuous local variations, adaptations to microniches.

Table 7.2 Distribution of prokaryotic cells

Habitat	Number of prokaryotic cells ($\times 10^{28}$)	Total carbon in prokaryotes ($\times 10^{15}$ g)
Ocean subsurface	355	303
Terrestrial subsurface	25–250	22–215
Soil	26	26
Oceans, lakes, and rivers	12	2.2
Within all human bodies	0.00004	

From: Whitman, W.B., Coleman, D.C., & Wiebe, W.J. (1998). Prokaryotes: The unseen majority. Proc. Natl. Acad. Sci. USA 95: 6578–6583.

Major types of prokaryotes

C. Woese divided prokaryotes into archaea and bacteria, on the basis of 16S ribosomal RNA (rRNA) gene sequences (Figure 1.14). Figure 7.1 shows the secondary structures of a region within the 16S rRNA that differs in bacteria, archaea, and eukaryotes. In context, Figure 7.2 shows the tertiary structure of this region within the full *Escherichia coli* 16S rRNA structure in the ribosome.

Numerous other differences between archaea and bacteria have emerged, involving genomic, structural, and metabolic features:

- some genes in archaea but none in bacteria contain introns;
- there are systematic differences in tRNA sequences between archaea and bacteria;
- enzymes involved in DNA replication, such as DNA polymerases, and some of the tRNA synthetases involved in protein synthesis, differ systematically between archaea and bacteria;
- archaea, like eukaryotes, have three RNA polymerases; bacteria have only one.
- archaea but not bacteria contain DNA-associated proteins resembling histones;

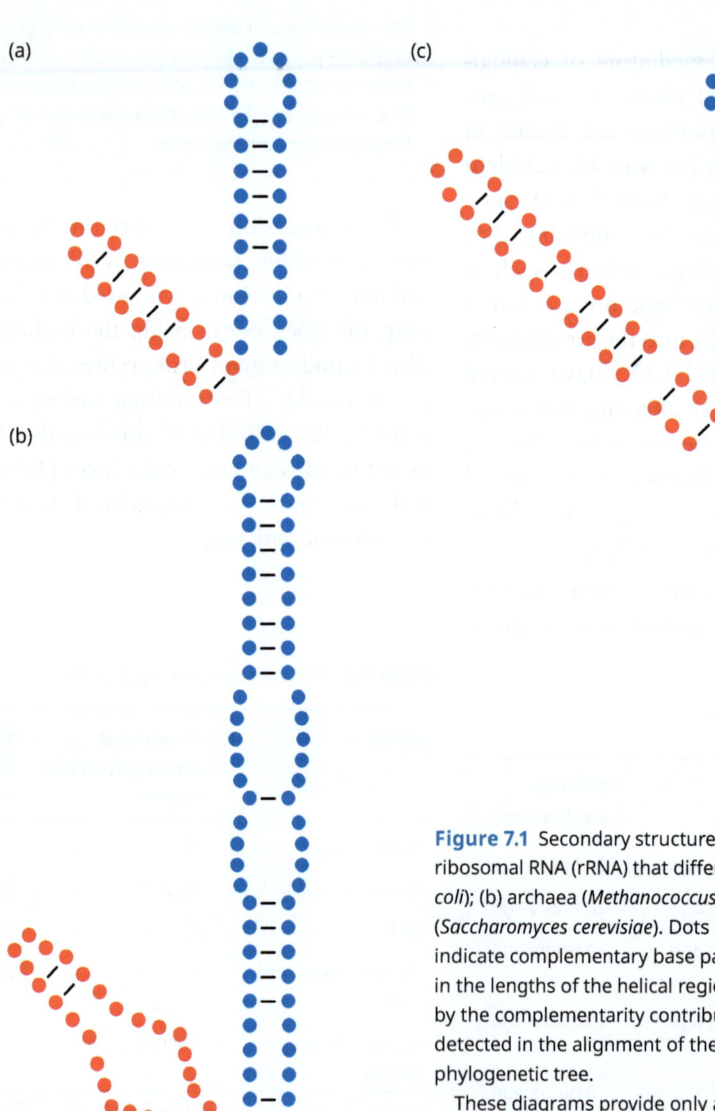

Figure 7.1 Secondary structure patterns of regions of 16S ribosomal RNA (rRNA) that differ among (a) bacteria (*Escherichia coli*); (b) archaea (*Methanococcus vannielii*); and (c) eukaryotes (*Saccharomyces cerevisiae*). Dots represent individual residues. Lines indicate complementary base pairing. The systematic differences in the lengths of the helical regions and the constraints imposed by the complementarity contributed to the patterns that Woese detected in the alignment of the sequences and in the derived phylogenetic tree.

These diagrams provide only a two-dimensional view. Figure 7.2 shows the actual three-dimensional structure of this region within the entire 16S RNA structure of *E. coli*.

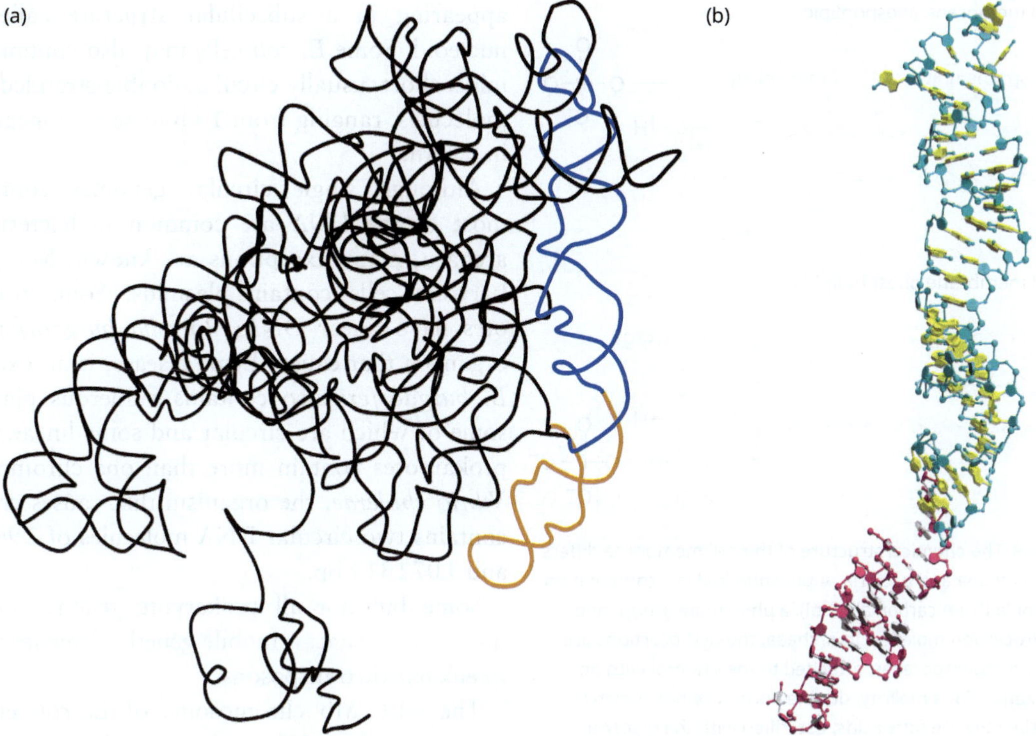

Figure 7.2 (a) Three-dimensional structure of 16S rRNA from the *Escherichia coli* ribosome [2AVY], showing the region of Figure 7.1a highlighted in orange and blue. (b) Detailed structure of the region shown in orange and blue in Figure 7.1a. Note that the acute angle between two helical regions drawn in the conventional representation of the secondary structure in the preceding figure is merely a drafting convention and does not correspond to the true three-dimensional structure.

- membranes of all cells contain phospholipids: compounds combining a glycerol molecule with long-chain organic molecules (see Figure 7.3); however:
  - bacteria and eukaryotes build cell membranes from phospholipids containing D-glycerol; archaea use L-glycerol
  - the organic chains in bacteria and eukaryotes are fatty acids, typically 16–18 carbon atoms long. Archaea instead use polyisoprenes. The branching of the isoprene chains permits the formation of links between different phospholipids in archaeal membranes. This allows the membrane to develop a higher order structure.
  - bacteria and eukaryotes link the organic side-chains to glycerol with an ester linkage; archaea prefer an ether linkage;
- cell wall structures: bacterial but not archaeal cell walls contain peptidoglycan, a combination of sugar derivatives and peptides; and
- archaea and bacteria differ in their complements of metabolic pathways.

Do we know the root of the tree of life?

There is consensus that life on Earth began over 3.5 billion years ago. The earliest remaining evidence for cellular life has the form of microfossils, from South Africa, Australia, and Canada. South African and Australian stromatolites arose from cyanobacteria related to modern prokaryotes. Many people believe that forms of life based on RNA—as both information archive and catalysts—existed before proteins took over the 'executive branch'.

The name archaea suggests that they represent the oldest forms of life. However, there has been extensive gene transfer between archaea and bacteria. It is not possible to assign LUCA—the last universal common ancestor of all known life forms—to either the archaeal or bacterial branch of the evolutionary tree. It is thought that the two lineages split very soon after the origin of cellular life. A branch of the archaea, the Asgardarchaeota, may be the closest extant relatives of LUCA.

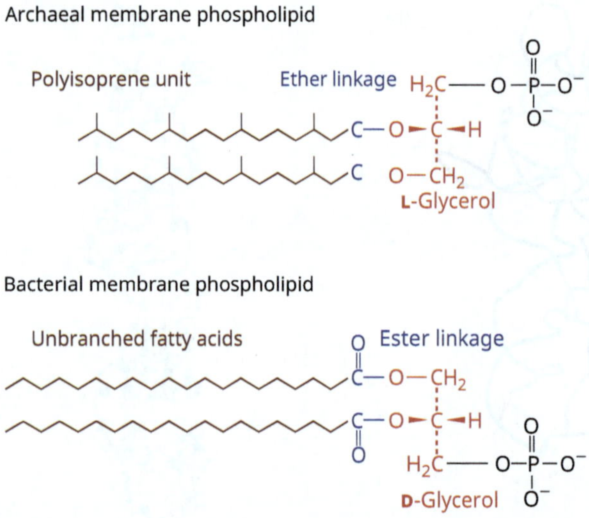

Figure 7.3 The chemical structure of the cell membrane differs between archaea and bacteria. A phospholipid is a combination of glycerol (a three-carbon alcohol), a phosphate group, and long hydrocarbon moieties. In archaea, the hydrocarbons are terpenes, or polyisoprenes, attached to the glycerol with an ether linkage. (For simplicity, double bonds are not shown.) In bacteria, they are fatty acids, esterified with the glycerol. Glycerol has two mirror-image forms. Archaeal and bacterial membranes contain different enantiomers. In the glycerol moieties in the figure (shown in red), triangles indicate bonds to groups in front of the central carbon, whereas broken lines indicate bonds to groups farther away than the central carbon. These differences must have correlates in the genomes, which contain genes that encode alternative sets of synthetic enzymes to produce these structures.

KEY POINT

C. Woese divided all living things into three groups: archaea, bacteria, and eukaryotes. Although we do not have reliable knowledge of the earliest events in life history, it is likely that archaea are closest to LUCA, the last universal common ancestor of us all. However, the split between archaea and bacteria happened quite early on.

Genome organization in prokaryotes

A typical prokaryotic genome has the form of a single circular molecule of double-stranded DNA, between 0.6 and 10 million bp long. For instance, a cell of *E. coli* strain K12 contains a single molecule of double-stranded DNA 4 639 675 bp long, closed into a circle. The DNA is supercoiled and associated with histone-like proteins into a 'chromosome', appearing in a subcellular structure called the nucleoid. Some *E. coli* cells may also contain plasmids: short, usually circular, double-stranded DNA molecules, ranging from 1 kb to several megabases in length.

Although single circular genomes containing most of the DNA are common in bacteria and archaea, many exceptions are known. Many prokaryotic cells contain plasmids. Some prokaryotes have linear DNA; *Borrelia burgdorferi*, the organism that causes Lyme disease, is an example. *B. burgdorferi* also contains numerous plasmids, some of which are circular and some linear. Other prokaryotes contain more than one chromosome. *Vibrio cholerae*, the organism that causes cholera, contains two circular DNA molecules of 2 961 146 and 1 072 314 bp.

Some but not all prokaryote genomes contain **insertion sequences**, mobile genetic elements similar to eukaryotic transposons.

The ~4.6 Mb chromosome of *E. coli* encodes approximately 4500 genes, distributed on both strands. The absence of introns and the shorter intergenic regions account for the high coding densities. A very large fraction of the DNA, 87.8%, codes for proteins, 0.8% codes for structural RNAs, and only 0.7% has no known function (see Table 7.3 and Figure 7.4).

Many prokaryotic genomes have been sequenced. They illuminate, in a somewhat simpler context than the human genome, how these organisms satisfy requirements common to all cellular life forms, such as energy metabolism and regulation. In addition, there are good practical motives for studying features of prokaryotes. Differences between prokaryotic and eukaryotic metabolism—enzymes unique to prokaryotes—are appropriate targets for drugs against infection. Of great importance to clinical medicine is understanding how prokaryotes evolve to develop pathogenicity and antibiotic resistance.

Table 7.3 Coding percentage and average gene density

Species	Coding	Average gene density
E. coli	>90%	1 gene/kb
Pufferfish	15%	1 gene/10 kb
Human	5%	1 gene/30 kb

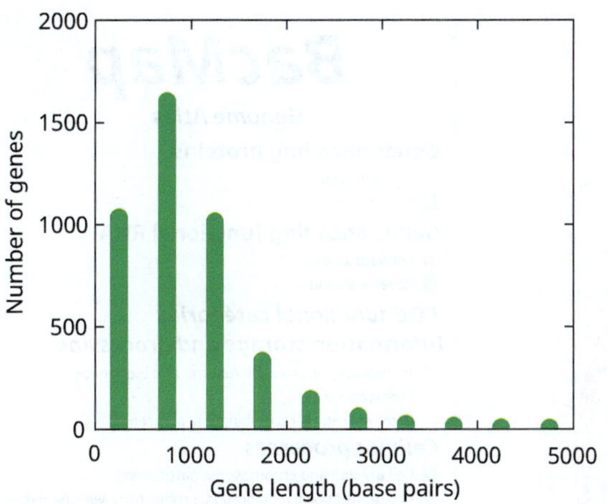

Figure 7.4 Distribution of gene lengths in *E. coli*. Two very long genes for hypothetical proteins, *yeeJ* and *ydbA*, of length 7152 and 8619 bp, respectively, are omitted. The average gene length is 960 bp. Most genes are less than 1500 bp long.

We visualize the contents of bacterial chromosomes as concentric circular diagrams, looking vaguely like 'tie-dyed' patterns (see Figure 7.5).

Replication and transcription

In *E. coli*, replication begins at a specific site called *oriC* and proceeds in both directions. This site is the calibration point from which the genome is indexed. Replication ends at the *terC* site, found almost, but not exactly, half way around the circle. In contrast, archaea often have multiple sites of origin of replication.

In prokaryotes, many mRNA transcripts contain several tandem genes, which require separate initiation of translation. (In this, they are unlike viral polyproteins, which are translated in one piece and then cleaved.) In bacteria, but less frequently in archaea, co-transcribed genes typically have related functions, forming an 'operon'.

Timing illuminates the interrelationships among these processes. Under ordinary conditions, it takes *E. coli* 40 minutes to replicate its genome. The full generation time between cell divisions is about an hour. This explains why genes that require high rates of expression tend to be near the origin of replication: the availability for transcription of partially replicated DNA in effect increases the copy number of such genes. Conversely, the half-life of mRNA is only a few minutes. Therefore, translation must overlap transcription.

Gene transfer

There are three methods of transfer of DNA between prokaryotic cells.

- *Transformation*. The uptake of 'naked' DNA, as in the experiments of Griffith and of Avery, MacLeod, and McCarthy.
- *Conjugation*. Insertion of some or all of the DNA from one cell into another—the prokaryotic equivalent of 'mating', although there is no meiosis or zygote formation. Bacterial conjugation does permit formation of recombinants. The start point for DNA transfer varies with the position in the genome of a mobile site. This is *not* the same as the origin of replication, *oriC*.
- *Transduction*. Transfer of DNA from one cell to another via a bacteriophage. During replication in one cell, a phage can pick up fragments of bacterial DNA and transmit it to another cell infected subsequently by progeny virions.

Bacterial conjugation has proved very useful in genome mapping. Transfer of a complete *E. coli* genome takes 100 minutes. Interrupting the process at different times (by physical agitation) results in partial genome transfer. Identifying which genes have entered the recipient cell after different intervals reveals the order of the genes. Positions in the genetic map of *E. coli*, for example, were classically expressed in minutes. Now, of course, they are specified in terms of the DNA sequence itself.

KEY POINT

Prokaryotes have several mechanisms for sharing genetic material: transformation by naked DNA, conjugation, and transfer via viruses.

Archaea

The first archaea discovered lived at high temperatures near sea-floor hydrothermal vents or in lakes containing very high concentrations of salt, such as the Dead Sea. However, not all archaea are adapted to extreme environments. Indeed, there is some evidence that mesophilic archaea came first and that thermophiles were a later adaptation. Conversely, not all thermophiles are archaea; *Thermus aquaticus*

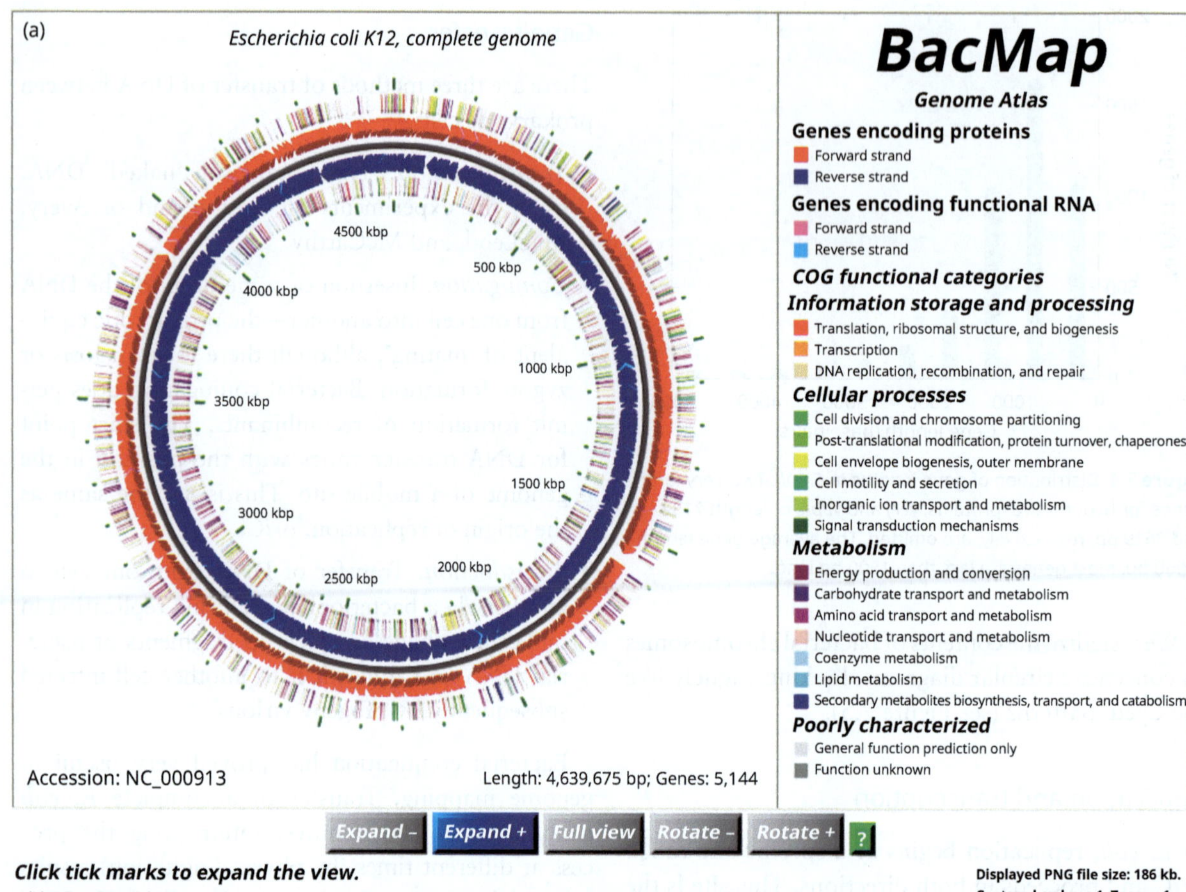

Figure 7.5 Map of the genome of *E. coli* K12. (a) Full view. There are two concentric circles of oppositely oriented arrows. The outer circle, in red, shows protein-coding regions of the forward strand. The inner circle, in blue, shows protein-coding regions of the reverse strand. A few arrows appearing in pink in the outer circle show structural-RNA-encoding regions of the forward strand. A few arrows appearing in cyan in the inner circle show structural-RNA-encoding regions of the reverse strand. Radial ticks identify individual gene products, colour coded according to function. COG categories refer to the Clusters of Orthologous Groups database (http://www.ncbi.nlm.nih.gov/COG/). (b) Expanded view of the region containing the *his* operon. The BacMap site provides access to genomes of bacteria and archaea (http://bacmap.wishartlab.com).

Stothard, P., Van Domselaar, G., Shrivastava, S., *et al.* (2005). BacMap: An interactive picture atlas of annotated bacterial genomes. Nucleic Acids Research 33, suppl_1: D317–D320. https://doi.org/10.1093/nar/gki075. Made available under Creative Commons Attribution 4.0 International.

(the source of *Taq* polymerase, an enzyme in common use at high temperature for polymerase chain reaction (PCR) amplification of DNA) is a bacterium.

Archaea are an abundant component of life in the open ocean, making up ~20% of all marine microbes. They also associate with a variety of metazoan hosts.

The major groupings of archaea (Figure 7.6) are as follows:

- *Crenarchaeota*. Many but not all of these are thermophiles. They include *Sulfolobus* and *Thermoproteus*.
- *Euryarchaeota*. These include methanogens, sulphate reducers, and many extreme halophiles, thermophiles, and acidophiles, including:
  - *Halobacter salinarum*, which can grow in salt concentrations above 4 M! Many people find its photosynthetic abilities even more interesting: *H. salinarum* contains a bacteriorhodopsin with which it captures sunlight energy as ATP without involving chlorophyll.
  - *Picrophilus torridus*, an extreme acidophile first isolated from the sulphurous volcanic springs of northern Japan. It can grow at pH 0.7!
  - Methanogens. These are strict anaerobes, dependent on the reaction:

$$CO_2 + 4H_2 \rightarrow CH_4 + 2H_2O$$

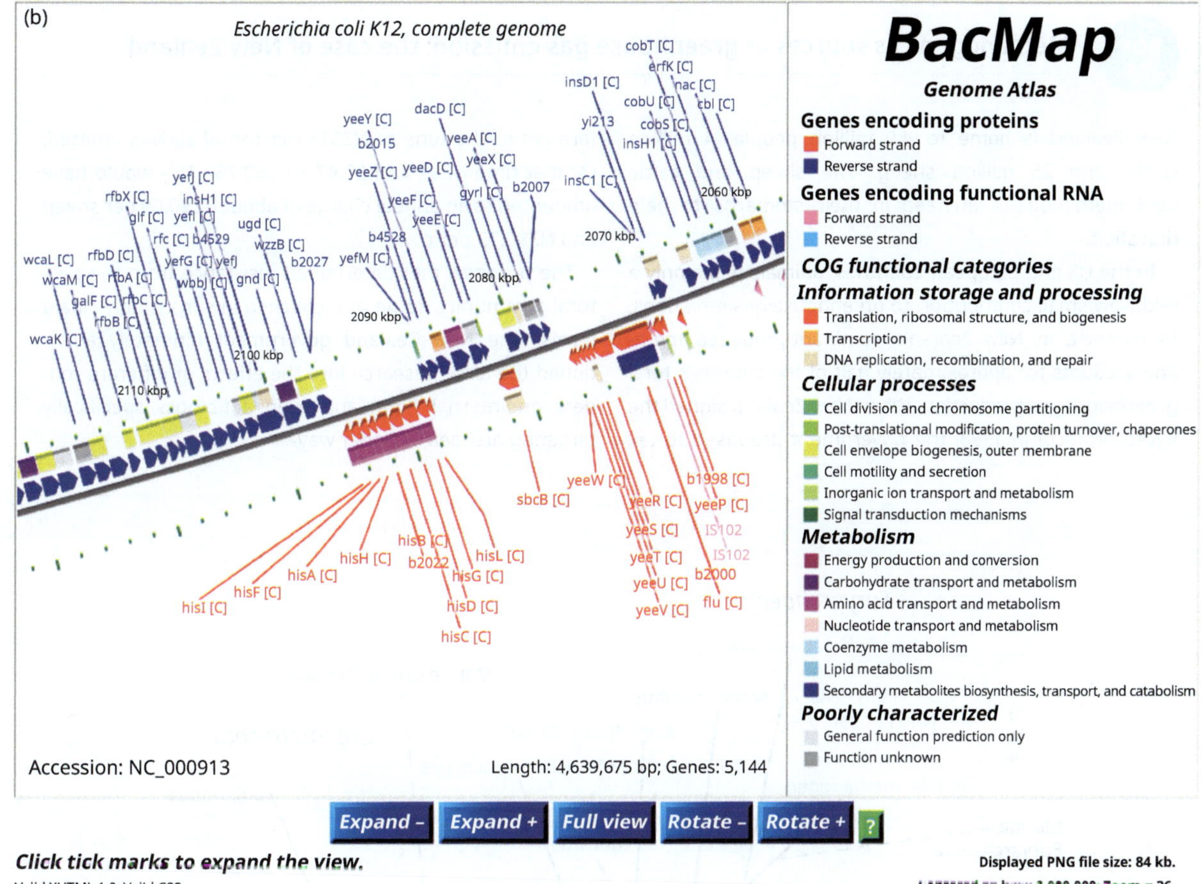

Figure 7.5 (*continued*)

Methanogenic archaea live in the guts of ruminant animals and help to digest cellulose. Archaeal cellulases hydrolyse plant fodder to simple sugars, from which CO_2 and H_2 are produced by fermentation. A cow can produce hundreds of litres of methane per day! (See Box 7.1.)

- *Korarchaeota*. These were discovered by environmental sampling of a hot spring in Yellowstone National Park, Wyoming, in the western US. They are perhaps closest to the root of all archaea (see Figure 7.6).
- *Nanoarchaeota*. These have been identified as a single small (~400 nm diameter) hyperthermophile from a submarine hot vent.

The last two phyla are minor, at least in terms of our current knowledge of them.

The genome of *Methanococcus jannaschii*

The microorganism *Methanococcus jannaschii* was collected from a hydrothermal vent 2600 m deep off the coast of Baja California, Mexico, in 1983. It is a thermophilic organism, surviving at temperatures from 48°C to 94°C, with an optimum at 85°C. *M. jannaschii* is capable of self-reproduction from inorganic components. Its overall metabolic equation is to synthesize methane from H_2 and CO_2. It is a strict anaerobe.

> Hydrothermal vents are underwater volcanoes emitting hot lava and gases through cracks in the ocean floor. They create niches for living communities disconnected from the surface; these use minerals from the vent as nutrients. These communities of microorganisms, and some animals, are the only forms of life not dependent on sunlight, directly or indirectly, for their energy.

BOX 7.1 Methanogens as sources of greenhouse gas emission: the case of New Zealand

New Zealand is home to 4.9 million people, 4 million cattle, and 25 million sheep. The sheep and cattle host methanogenic archaea in their stomachs to help digestion.

In the US and European countries, animals make only a relatively small contribution to greenhouse gas emissions. In contrast, in New Zealand, ruminant-produced methane accounts for approximately half of the country's total greenhouse gas production. When New Zealand signed the Kyoto protocol in 1998, the government proposed to tax farmers to the tune of NZ$11 per ton of carbon emitted. (At that time NZ$1 ≈ UK£0.47 ≈ US$0.76.) This would have amounted to an annual charge of about NZ$0.09 per sheep and NZ$0.72 per cow.

The proposal met determined resistance from the pastoral community, some of it couched in surprisingly ribald terms. The New Zealand government ultimately abandoned the idea. Research into the effects of different fodders on internal flora, and even antibiotics specifically targeting archaea, is under way.

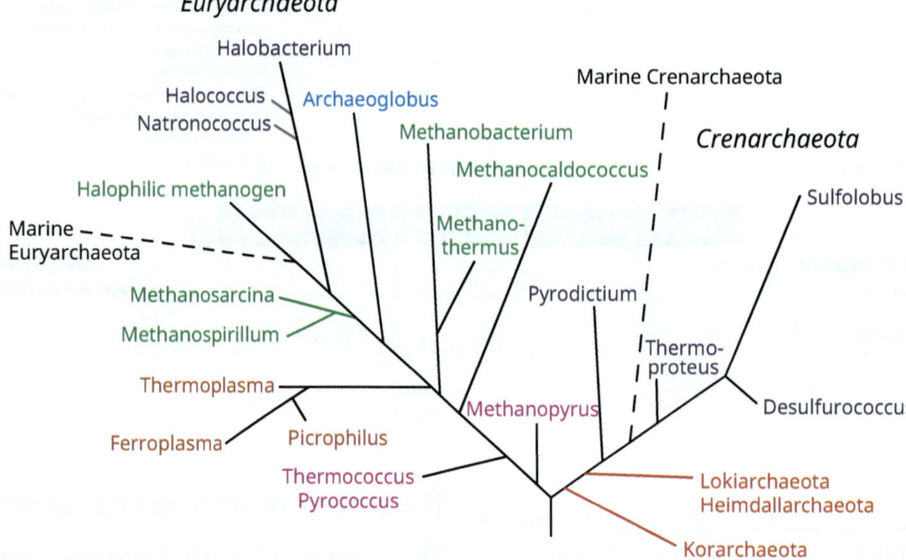

Figure 7.6 Phylogenetic tree of archaea, based on analysis of 16S rRNA sequences. Major archaeal groupings are coloured as follows:

Euryarchaeota:	**Crenarchaeota:**
Archaeoglobali	*Desulfurococcales*
Halobacteria	
Methanobacteria	*Korarchaeota*
Methanococci	*Heimdallarchaeota*
Methanomicrobia	*Nanoarchaeota* (not shown)
Methanopyri	
Thermococci	
Thermoplasmata	

It is widely believed that the Korarchaeota are the closest living organisms to the last universal common ancestor (LUCA), and that Heimdallarchaeota are the closest relatives of eukaryotes.

The genome of *M. jannaschii* was sequenced in 1996 by The Institute for Genomic Research (TIGR). It was the first archaeal genome sequenced. It contains a large chromosome with a circular double-stranded DNA molecule 1 664 976 bp long and two extrachromosomal elements of 58 407 and 16 550 bp. There are 1743 predicted coding regions, of which 1682 are on the chromosome and 44 and 12 are on the large and small extrachromosomal elements, respectively. Some RNA genes contain introns. As in other prokaryotic genomes, there is little non-coding DNA.

Recall Salvador Luria's suggestion that to determine common features of all life one should not try to survey everything, but, rather, identify the organism most different from us and see what we have in common with it.

M. jannaschii would appear to satisfy Luria's goal of finding our most distant extant relative. Comparison of its genome sequence with others shows that it is distantly related to other forms of life. Only 38% of the open reading frames could be assigned a function on the basis of homology to proteins known from other organisms. However, to everyone's great surprise, archaea are in some ways more closely related to eukaryotes than to bacteria! They are a complex mixture. Archaeal proteins involved in metabolism are more similar to those of bacteria than to those of eukaryotes. But archaeal proteins involved in transcription, translation, and regulation are more similar to those of eukaryotes. The recently discovered archaeon, Heimdallarchaea, is a close link between archaea and eukarya. It is genetically closer to eukaryotes than any other known prokaryote.

Life at extreme temperatures

Organisms from the three major divisions of life—archaea, bacteria, and eukaryotes—show wide, overlapping ranges of optimal growth temperatures (see Figure 7.7). Thermophiles and hyperthermophiles survive at elevated temperatures because they synthesize molecules that are stable to heat denaturation. Accordingly, adaptations to high-temperature survival might be observed in DNA, in RNA, and in proteins. Enzymes from hyperthermophiles have applications in laboratory molecular biology and in industry.

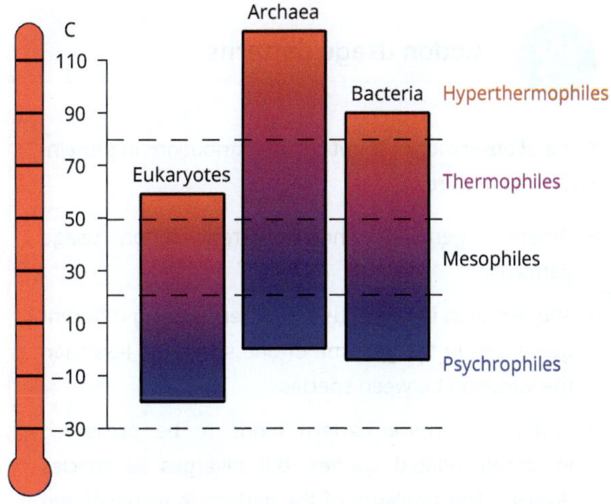

Figure 7.7 'Some like it hot, some like it cold . . .' Distribution of known growth temperatures of eukaryotes, archaea, and bacteria. Ranges defining hyperthermophiles, thermophiles, mesophiles, and psychrophiles are arbitrary.

What choices do organisms have to adjust the thermal stability of their constituents?

Thermophiles can evolve proteins with enhanced stability. Moreover, any set of favourable amino-acid sequence is compatible with many gene sequences. In principle, organisms can take advantage of the redundancy of the genetic code to adjust the thermal stability of their nucleic acids. For double-stranded DNA and RNA, the thermal stability increases linearly with the G+C content. (However, this is only one aspect of the application of the redundancy in the code. See Box 7.2.)

Singer and Hickey compared genome sequences from 40 prokaryotes.[1] The organisms included eight archaeal mesophiles and thermophiles and 32 bacterial mesophiles and thermophiles, with optimal growth temperatures ranging from 18°C to 97°C. This permitted a study focusing on adaptations to high temperature, not biased by archaeal–bacterial

[1] Singer, G.A., & Hickey, D.A. (2003). Thermophilic prokaryotes have characteristic patterns of codon usage, amino acid composition and nucleotide content. Gene 317: 39–47; Hickey, D.A., & Singer, G.A. (2004). Genomic and proteomic adaptations to growth at high temperature. Genome Biol. 5: 117.

Codon usage patterns

General observations about codon distributions in protein-coding genes are:

- Different genomes show different codon usage patterns.
- The variation in codon usage pattern among different genes within the genome of one species is less than the variation between species.
- Codon preference pattern tends to be preserved in closely related species, but diverges as species diverge. The similarity of the pattern in archaeal and bacterial thermophiles is evidence that codon usage patterns can be determined by selection.
- Within genomes, highly expressed proteins show stronger bias in codon usage. Genes for highly expressed proteins are enriched in sets of preferred codons, and these preferred codons are often correlated with greater tRNA abundances. Matching the pattern of codon usage and tRNA abundance can make protein synthetic throughput higher.

differences. Their results show association of thermostability with:

- *DNA*. Perhaps surprisingly, overall genomic G+C content is *not* correlated with growth temperature. However, a correlation with growth temperature *is* observed in the distribution of *dinucleotides*, with thermophiles, mesophiles, and psychrophiles showing different characteristic patterns.

Thermophiles and hyperthermophiles stabilize their DNA structures, not by increasing the G+C content, but by tight binding of special ligands protective against denaturation.

- *RNA*. The G+C content of *non-protein coding RNA* is correlated with growth temperature, especially in double-stranded regions. High G+C content in double-stranded regions enhances their thermostability. *Single-stranded RNAs*, including messenger RNAs (mRNAs), are relatively rich in purines, notably adenine, for reasons that are not clear.

Although the overall G+C content of DNA is not correlated with growth temperature, codon usage patterns are. Coding sequences in thermophiles are enriched relative to mesophiles in synonymous codons ending in C or G. It is the third position of the codons that offers the greatest opportunity for choice among synonymous codons.

- *Proteins*. Comparisons of homologues show that proteins from thermophiles and hyperthermophiles:
  - tend to be shorter than their homologues from mesophiles, most of the residues lost coming from surface loops;
  - have more charged residues at their surfaces, both positive and negative (Asp, Lys, His, Asp, Glu); the formation of stabilizing salt bridges is a common feature of their structures (see Figure 7.8); (A salt bridge is the juxtaposition of a positively charged sidechain with a negatively charged sidechain, producing an electrostatic attraction between them.)
  - contain relatively fewer uncharged polar residues (such as Ser, Thr, Gln, Asn, Cys); some of these have thermolabile sidechains (His, Gln, Thr); and
  - contain higher proportions of hydrophobic β–branched residues (see Box 7.3).

Also, hyperthermophiles have special 'chaperones'—proteins that assist in protein folding. It is likely that this is an adaptation to the challenge of high-temperature growth.

Comparative genomics of hyperthermophilic archaea: *Thermococcus kodakarensis* and *Pyrococci*

A hyperthermophilic archaeon, *Thermococcus kodakarensis* strain KOD1, was isolated from a hot sulphur spring (102°C, pH 5.8) on the shore of Kodakara Island, in the Ryukyu archipelago between Kagoshima in southwest Japan, and Okinawa (29° 12° N, 129° 19° E). *T. kodakarensis* KOD1 is a strict anaerobe, reducing elemental sulphur to H_2S.

General features of the genome

Fukui and co-workers reported the complete genome sequence of *T. kodakarensis* KOD1 (see Figure 7.10). The single, circular chromosome contains 2088737 bp, with a G+C content of 52 mole %. A total of

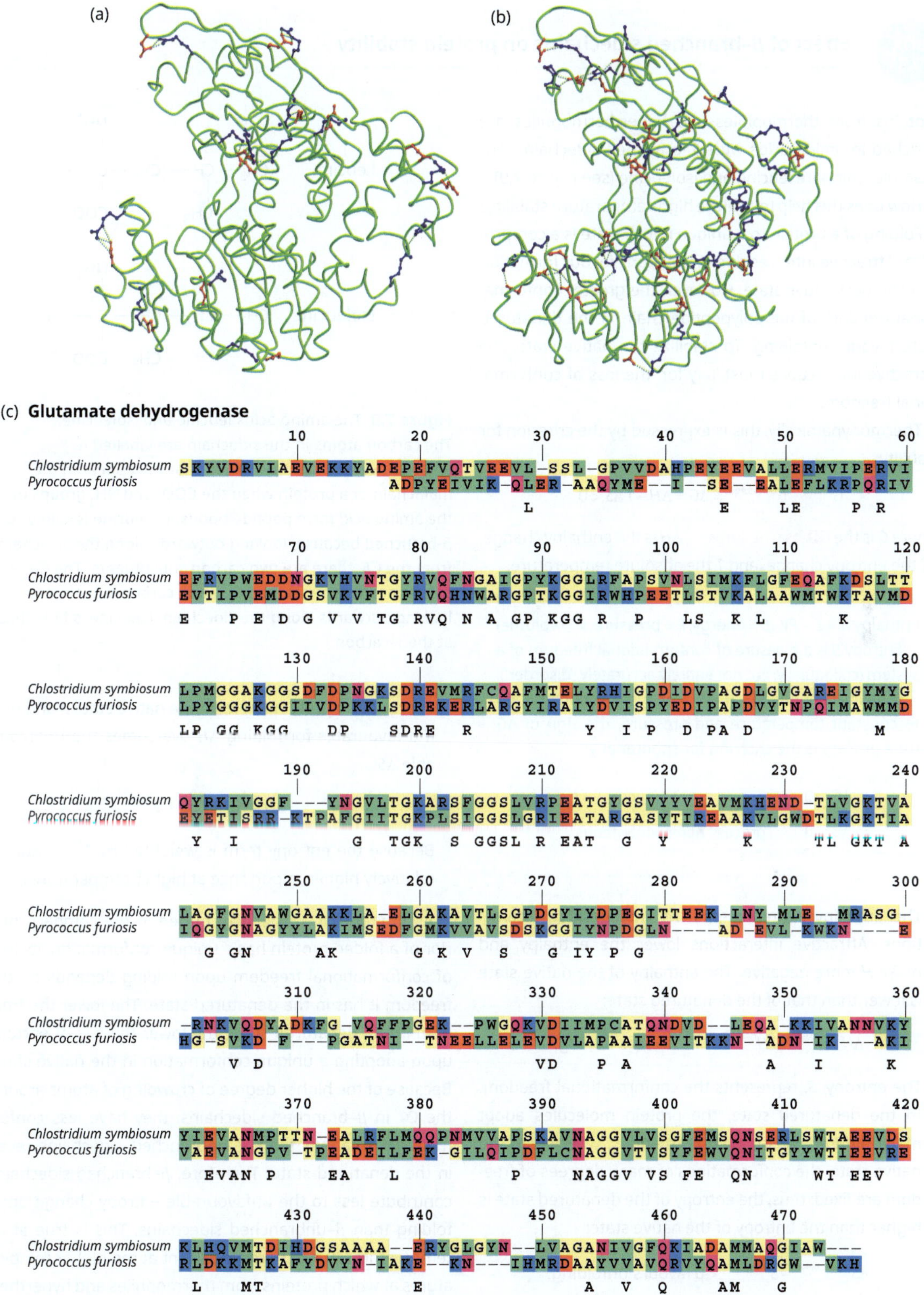

Figure 7.8 Proteins from thermophiles and hyperthermophiles are enriched in salt bridges relative to their mesophilic homologues. Positively charged sidechains are shown in blue. Negatively charged sidechains are shown in red. (a) Subunit of glutamate dehydrogenase from mesophilic archaeon *Clostridium symbiosum* [1HRD]. (b) Subunit of glutamate dehydrogenase from hyperthermophilic archaeon *Pyrococcus furiosus* [1GTM]. (c) Sequence alignment of archaeal hyperthermophilic and mesophilic glutamate dehydrogenase subunits.

BOX 7.3 Effect of β-branched sidechains on protein stability

Proteins from thermophiles and hyperthermophiles are enriched in amino acids with β-branched sidechains. For example, compare leucine and isoleucine (see Figure 7.9).

How does this help to achieve high-temperature stability?

Folding of a protein to a unique native state is a compromise. Attractive inter-residue interactions favour formation of a compact native state. However, the greater conformational freedom of the polypeptide chain in the denatured state favours unfolding. To stabilize the native state, the attractive interactions must 'pay for' the loss of conformational freedom.

Thermodynamically, this is expressed by the criterion for stability:

$$G^{\text{Native}} - G^{\text{Denatured}} = \Delta G = \Delta H - T\Delta S < 0$$

where G is the **Gibbs free energy**, ΔH is the enthalpy change, ΔS the entropy change, and T the absolute temperature.

> Enthalpy $H = E + PV$ (E = energy, P = pressure, V = volume)
>
> Entropy S is a measure of conformational freedom of a system (colloquially, but not entirely accurately, 'disorder').
>
> Gibbs free energy $G = H - TS$. For systems constrained to constant temperature and pressure, the sign of ΔG for a process is the criterion for spontaneity:
>
> $\Delta G < 0$ process spontaneous
>
> $\Delta G > 0$ process non-spontaneous

- The enthalpy, H, represents the inter-residue interactions. Attractive interactions lower the enthalpy and make H more negative. The enthalpy of the native state is lower than that of the denatured state:

$$\Delta H = H^{\text{Native}} - H^{\text{Denatured}} < 0 \text{ favours folding.}$$

- The entropy, S, represents the conformational freedom. In the denatured state, the protein molecules adopt many different possible conformations, whereas in the native state, the conformations of many degrees of freedom are fixed; thus, the entropy of the denatured state is higher than the entropy of the native state:

$$\Delta S = S^{\text{Native}} - S^{\text{Denatured}} < 0 \text{ favours unfolding.}$$

- Because systems (at constant temperature and pressure) come to an equilibrium state of minimum Gibbs free energy (G), a protein will form the native state if and only if a favourable-for-folding ΔH overcomes the unfavourable ΔS:

$$\Delta G = \Delta H - T\Delta S < 0$$

Because the entropy term is weighted by T, it assumes relatively higher importance at higher temperatures.

Figure 7.9 The amino acids leucine and isoleucine. The carbon atoms in the sidechain are labelled α, β, γ, and δ outwards from the carbon that will appear in the mainchain of a protein when the COO^- and NH_3^+ groups of the amino acid form peptide bonds. Isoleucine is said to be β-branched because, looking outwards along the sidechain from the Cβ, there are two carbon substituents. The Cα of leucine, in contrast, has only one carbon substituent looking outwards along the sidechain. (Leucine is branched at the γ-carbon.)

Assuming that a sidechain buried in the compact interior of a folded protein has a unique conformation, its loss of conformational freedom upon folding depends on the freedom it has in the denatured state. The lower the freedom in the denatured state, the lower the *loss* of entropy upon adopting a unique conformation in the native state. Because of the higher degree of crowding of atoms around the Cα in β-branched sidechains, they have less conformational freedom than β-unbranched sidechains, even in the denatured state. Therefore, β-branched sidechains contribute less to the unfavourable entropy change upon folding than β-unbranched sidechains. This is true at all temperatures but is more significant at the higher temperatures at which proteins from thermophiles and hyperthermophiles have to form their native states, because of the factor T in the entropy term.

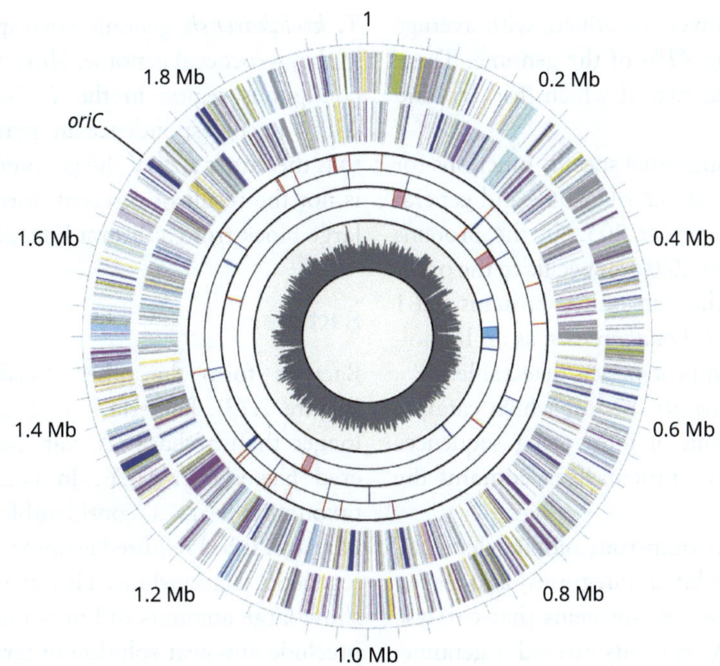

Figure 7.10 Diagram of the genome of *T. kodakarensis* strain KOD1. The contents of the consecutive circles, from the outermost, are:

(1) Scale in 0.2 Mb increments, plus the predicted origin of replication (*ori*C).
(2) Predicted protein-coding regions in clockwise direction.
(3) Predicted protein-coding regions in anticlockwise direction.
(4) Predicted tRNA coding regions in clockwise direction (red).
(5) Predicted tRNA coding regions in anticlockwise direction (blue).
(6) Predicted mobile elements in clockwise direction (red). Lines indicate transposase genes; boxes indicate virus-related regions.
(7) Predicted mobile elements in counter-clockwise direction (blue). Lines indicate transposase genes; boxes indicate virus-related regions.
(8) G+C content (mol. %) in 10 kb window.

Circles (2) and (3) are colour coded according to function:

Functional category	Colour
Translation, ribosomal structure, biogenesis	Magenta
Transcription	Pink
DNA replication, recombination, repair	Pale pink
Cell division, chromosome partitioning	Forest green
Post-translational modification, protein turnover, chaperones	Yellow
Cell envelope biogenesis, outer membrane	Light yellow
Cell motility, secretion	Light green
Inorganic ion transport/metabolism	Pale green
Signal transduction	Medium turquoise
Energy production/conversion	Purple
Carbohydrate transport/metabolism	Light blue
Amino acid transport/metabolism	Cyan
Nucleotide transport/metabolism	Violet
Co-enzyme metabolism	Pale turquoise
Lipid metabolism	Medium purple
Secondary metabolites biosynthesis/transport/catabolism	Light sky blue

Fukui, T., Atomi, H., Kanai, T., *et al.* (2005). Complete genome sequence of the hyperthermophilic archaeon Thermococcus kodakaraensis KOD1 and comparison with Pyrococcus genomes. Genome Research 15, 3: 352–363. https://doi.org/10.1101/gr.3003105

2306 coding sequences were identified, with average length 833 bp, covering 92% of the genome. There are 46 genes for tRNA, two of which (for Trp and Met) contain introns.

Database searching suggested specific functions for half of the proteins (1165 out of 2306), and general functional classes for another 205. Of the proteins with known homologues, 240 are specific to the order Thermococcales. Of the remaining proteins, 261 appear to be unique to *T. kodakarensis*, as no homology to any other known protein was detectable.

Fifteen of the proteins are inteins, which catalyse the excision and splicing of intervening sequences *after* translation; i.e., the protein itself contains the self-splicing activity.

The genome contains numerous mobile elements, including four virus-related integrases and seven homologues of transposases—proteins that catalyse the movement of DNA segments around a genome. However, transposase activity was not observed, suggesting that these proteins may have lost their function.

Molecular physiology of T. kodakarensis

By mapping the proteins of *T. kodakarensis* for which functions can be assigned, it is possible to reconstruct the metabolic and transport pathways of *T. kodakarensis*. (See Figure 7.11 for an overview.)

Comparative genomics of T. kodakarensis

The genome sequences of three close relatives of *T. kodakarensis—Pyrococcus abyssi, Pyrococcus horikoshii*, and *Pyrococcus furiosus*—allowed comparisons and descriptions of the evolutionary relationships between these archaea at the genome and protein levels.

Some of the differences are characteristic of the different genera—*Thermococcus* and *Pyrococcus*. These include the G+C content and the number of coding sequences (Table 7.4).

The loss of synteny between *Pyrococcus* and *Thermococcus* genera illuminates the origin of the difference in protein content (see Figure 7.12). The rearrangement among the *Pyrococcus* species themselves is substantial and involves a major inversion between *P. abyssi* and *P. horikoshii*. In contrast, between the genera, the genome has been completely shuffled and redealt. That is, almost everything in the *T. kodakarensis* genome corresponds to *something* in the pyrococcal genome. However, there is no large contiguous region in the *T. kodakarensis* genome with *no* correspondence in pyrococci. This shows that the larger size of the genome of *T. kodakarensis* is not the result of a recent horizontal transfer of a large block from a distant lineage.

Bacteria

Bacteria form the other division of prokaryotes (Figure 7.13). Bacteria have been known for much longer than archaea. A. van Leeuwenhoek discovered bacteria in 1676. In consequence, bacterial taxonomy bears a considerable load of historical baggage. It has required genomes to sort out the phylogenetic relationships. However, the genomes also show large amounts of horizontal gene transfer that preclude any neat solution in terms of a simple phylogenetic tree.

Figure 7.13 suggests one approach to classifying the main groups of bacteria. Box 7.4 gives some examples of better known organisms in the different groups, with some brief comments. For a classification of bacteria focusing on pathogens, see http://www.mgc.ac.cn/VFs/.

Genomes of pathogenic bacteria

A bacterial pathogen is a bacterium that can cause disease. It can do so by virtue of having virulence factors, which may include toxins, surface proteins that mediate attachment to cells, defensive shields (proteins and carbohydrates), and secreted enzymes. In many cases, closely related strains or species differ in pathogenicity. This suggests that comparison of their genomes could help to identify virulence factors. Knowledge of virulence factors would permit: (a) testing of foods for presence of pathogenic strains, (b) choosing suitable drug targets, and (c) designing vaccines.

Examples include:

- *E. coli*. Because a non-pathogenic strain (K12MG 1655) of *E. coli* occupies such a central role in molecular biology, it is easy to forget that it exists in nature, and that related strains are pathogenic.

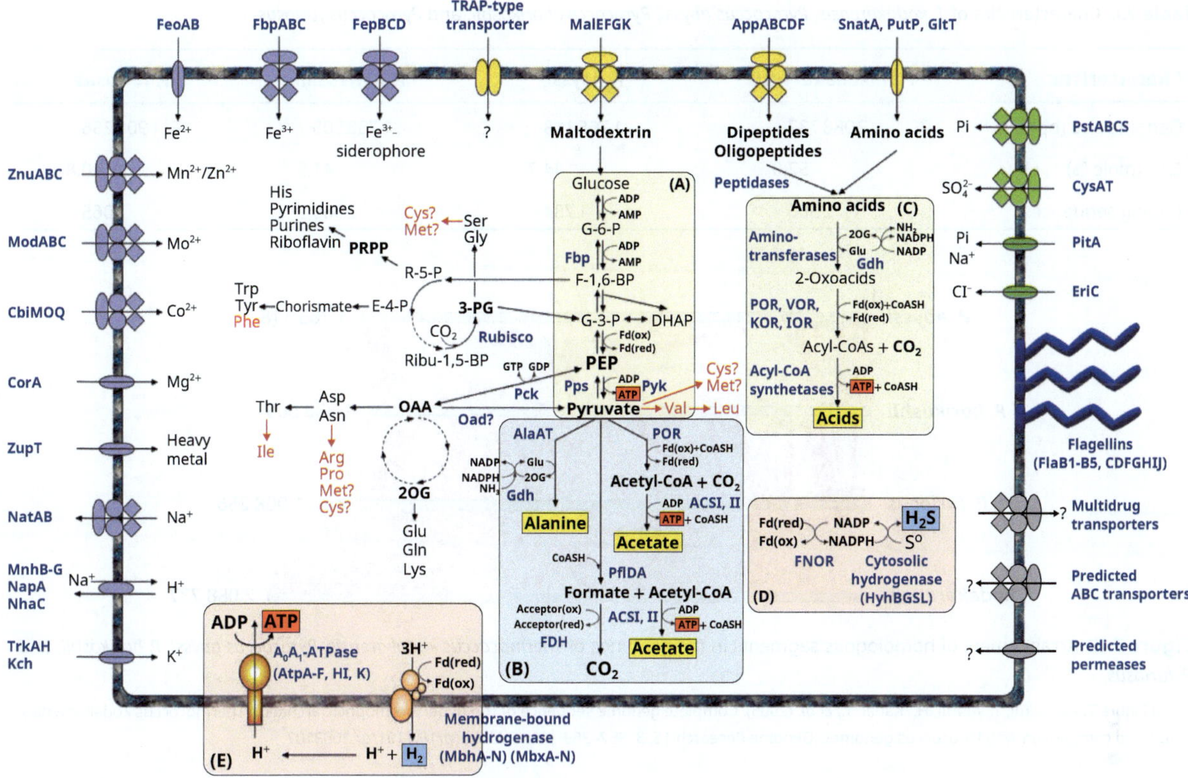

Figure 7.11 Reconstructed scheme of metabolism and solute transport in *Thermococcus kodakarensis.* Components or pathways for which no predictable enzymes could be assigned appear in red.

Each gene product with a predicted function in ion or solute transport is illustrated on the membrane. The transporters and permeases are grouped by substrate specificity, as cations (violet), anions (green), carbohydrates/carboxylates/amino acids (yellow), and unknown (grey).

Metabolic pathways appear in the interior of the cell: (A) glycolysis (modified Embden–Meyerhof pathway); (B) pyruvate degradation; (C) amino acid degradation; (D) sulphur reduction; and (E) hydrogen evolution and formation of proton-motive force, coupled with ATP generation.

Abbreviations: DHAP, dihydroxyacetone phosphate; E-4-P, erythrose 4-phosphate; F-1,6-BP, fructose 1,6-bisphosphate; G-3-P, glyceraldehyde 3-phosphate; G-6-P, glucose 6-phosphate; OAA, oxaloacetate; 2OG, 2-oxoglutarate; PEP, phosphoenolpyruvate; 3-PG, 3-phosphoglycerate; PRPP, 5-phosphoribosyl 1-pyrophosphate; R-5-P, ribose 5-phosphate; Ribu-1,5-BP, ribulose 1,5-bisphosphate; ACS, acetyl-CoA synthetase (ADP-forming); AlaAT, alanine aminotransferase; Fbp, fructose 1,6-bisphosphatase; FDH, formate dehydrogenase; FNOR, ferredoxin:NADP oxidoreductase; Gdh, glutamate dehydrogenase; IOR, indolepyruvate:ferredoxin oxidoreductase; KOR, 2-oxoacid: ferredoxin oxidoreductase; PflDA, pyruvate formate lyase and its activating enzyme; Oad, oxaloacetate decarboxylase; Pck, phosphoenolpyruvate carboxykinase; POR, pyruvate:ferredoxin oxidoreductase; Pps, phosphoenolpyruvate synthase; Pyk, pyruvate kinase; VOR, 2-oxoisovalerate: ferredoxin oxidoreductase; AppABCDF, ABC-type dipeptide/oligopeptide transporter; CbiMOQ, ABC-type Co^{2+} transporter; CorA, Mg^{2+}/Co^{2+} transporter; CysAT, ABC-type sulphate transporter; EriC, voltage-gated Cl^- channel protein; FbpABC, ABC-type Fe^{3+} transporter; FeoAB, Fe^{2+} transporter; FepBCD, ABC-type Fe^{3+}-siderophore transporter; GltT, H^+/glutamate symporter; Kch, Ca^{2+}-gated K^+ channel protein; MalEFGK, ABC-type maltodextrin transporter; MnhB-G, multisubunit Na^+/H^+ antiporter; ModABC, ABC-type Mo^{2+} transporter; NapA and NhaC, Na^+/H^+ antiporter; NatAB, ABC-type Na^+ efflux pump; PitA, Na^+/phosphate symporter; PstABCS, ABC-type phosphate transporter; PutP, Na^+/proline symporter; SnatA, small neutral amino acid transporter; TrkAH, Trk-type K^+ transporter; ZnuABC, ABC-type Mn^{2+}/Zn^{2+} transporter; ZupT, heavy metal cation transporter.

From Figure 7.9 in Fukui, T., Atomi, H., Kanai, T., *et al.* (2005). Complete genome sequence of the hyperthermophilic archaeon Thermococcus kodakaraensis KOD1 and comparison with Pyrococcus genomes. Genome Research 15, 3: 352–363. https://doi.org/10.1101/gr.3003106

Table 7.4 Characteristics of *T. kodakarensis, Pyrococcus abyssi, Pyrococcus horikoshii*, and *Pyrococcus furiosus*

Characteristic	*T. kodakarensis*	*P. abyssi*	*P. horikoshii*	*P. furiosus*
Genome size (bp)	2 088 737	1 765 118	1 738 505	1 908 256
G+C (mole %)	52.0	44.7	41.9	40.8
Coding sequences	2306	1784	2065	2065

Figure 7.12 Arrangement of homologous segments in the genomes of *Thermococcus kodakarensis, Pyrococcus abyssi, P. horikoshii*, and *P. furiosus*.

From Figure 7.9 in Fukui, T., Atomi, H., Kanai, T., *et al.* (2005). Complete genome sequence of the hyperthermophilic archaeon Thermococcus kodakaraensis KOD1 and comparison with Pyrococcus genomes. Genome Research 15, 3: 352–363. https://doi.org/10.1101/gr.3003107

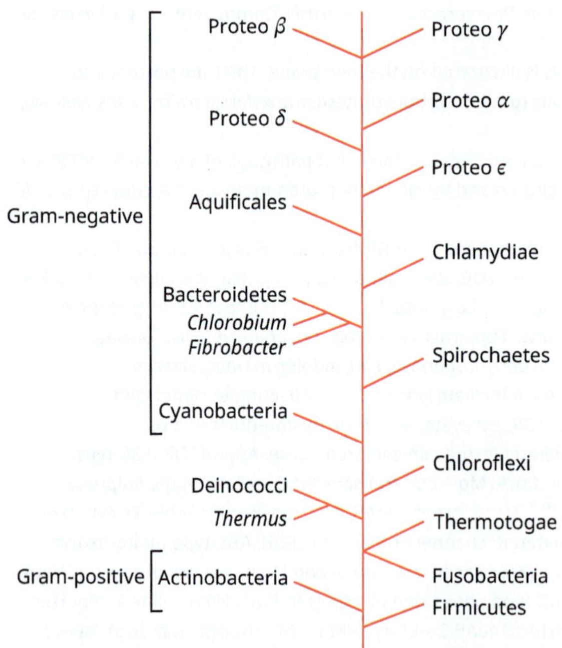

Figure 7.13 Phylogenetic tree of some major bacterial types. This diagram reflects the topology of the tree but not the extent of divergence between and within groups. Some of the groups are phyla, others are genera.

Reference: Bacterial (Prokaryotic) Phylogeny Webpage (2006) (http://www.bacterialphylogeny.com/index.htm).

Strain *E. coli* 0157:H7 causes haemorrhagic colitis, which can be fatal. A comparison of the genomes of this strain with that of K12 strain MG1655 shows that the genome of K12 MG1655, 4639221 bp, is shorter than that of 0157:H7, 5 528 445 bp. Regions amounting to 4.1 Mb are common to both strains. The unshared genes tend to cluster in strain-specific regions. Strain-specific regions in K12MG1655 contain 1.34 Mb, 1387 genes, and in 0157:H7 contain 0.53 Mb, 528 genes.

It is likely that the strains diverged about 4.5 million years ago. A clue to the origin of the differences between the strains is the atypical base composition of the strain-specific regions. This suggests that they entered the respective genomes by horizontal gene transfer.

KEY POINT

Horizontal gene transfer is a common theme in development of virulence and antibiotic resistance.

BOX 7.4

Characteristics of major groups of bacteria

Group	Examples	Comments
Firmicutes	*Bacilli, Staphylococci, Lactobacilli, Clostridia*	*Listeria* and staphylococci can be infectious; *Clostridia* can cause food poisoning; *Lactobacilli* are useful in yoghurt production
Actinobacteria	*Micrococcus, Streptomyces*	Decompose dead plant material; source of antibiotics
Fusobacteria	*Fusobacterium nucleatum*	Live in human gut, involved in periodontal infection
Thermotogae	*Thermotoga subterranea*	Thermophilic or hyperthermophilic; some are anaerobic
Thermus	*Thermus aquaticus*	Thermophilic; source of *Taq* polymerase
Deinococci	*Deinococcus radiodurans*	*D. radiodurans* is unusually radiation resistant
Chloroflexi	*Chloroflexus aurantiacus*	Photosynthetic, but do not produce O_2; may provide clue to early development of photosynthesis
Cyanobacteria	*Prochlorococcus marinus*	Chlorophyll-based photosynthesis; most split H_2O and produce O_2; gave rise to chloroplasts via symbiosis
Spirochaetes	*Leptospira, Borrelia burgdorferi, Treponema pallidum*	Some are pathogenic (leptospirosis, syphilis, Lyme disease)
Fibrobacters	*Fibrobacter intestinalis*	Live in gut; help cattle to digest cellulose
Chlorobium	*Chlorobium tepidum*	Green sulphur bacteria; photosynthetic: reduce sulphide to sulphur
Bacteroidetes	*Bacteroides fragilis*	Some are marine plankton; others are anaerobic, live in the gut and can cause infection. *Porphyromonas gingivalis* causes gum disease, and is also associated with rheumatoid arthritis and with schizophrenia
Chlamydiae	*Chlamydia trachomatis*	Grow intracellularly; major cause of blindness; also cause sexually transmitted infections of the urogenital system
Aquificales	*Aquifex aeolicus*	Extremophiles, autotrophs
α–Proteobacteria	*Rhodospirillum rubrum, Rhizobium, Rickettsia*	*Rhizobium* are symbiotic with legumes and fix nitrogen; *Rickettsia* cause typhus; gave rise to mitochondria via symbiosis
β–Proteobacteria	*Burkholdia, Bordetella, Thiobacillus, Neisseria*	Some live on inorganic nutrients; others are infectious, different species causing pertussis, gonorrhoea, and meningitis
γ–Proteobacteria	*Escherichia coli, Haemophilus influenzae, Pseudomonas aeruginosa, Yersinia pestis, Salmonella typhimurium*	Important in medicine and molecular biology: cause enteritis, typhoid, bubonic plague, and others
δ–Proteobacteria	*Desulfovibrio desulfuricans*	Mostly aerobic; some anaerobic examples reduce sulphur or sulphate
ε–Proteobacteria	*Helicobacter pylori*	*H. pylori* lives in gut, cause of ulcers

- *Helicobacter pylori*. Half the world's population is infected with *H. pylori*. One out of ten people develop clinical disease: gastritis, duodenal and gastric ulcers, and some cancers. Proof that *H. pylori* infection is the cause of ulcers was obtained—over the disbelief of the scientific and medical establishments at the time—by Barry Marshall, who deliberately swallowed a culture of *H. pylori*, and quickly developed symptoms of gastritis.

H. pylori strains are very diverse (they have been applied to tracking of patterns of human migration). Almost 2500 strains have been sequenced completely. Strain 26695 contains about 1.7 Mbp, and about 1550 genes. Virulence appears to be associated with a common 40 kb cag pathogenicity island containing >40 genes that encode a secretion system. The appearance of genes within this island is correlated with virulence. This pathogenicity island is common to many bacteria. It is likely that it has been circulated by horizontal gene transfer.

- *Staphylococcus aureus*. *S. aureus* infections are a growing clinical problem because of the aggressive development of antibiotic resistance. (The development of resistance to vancomycin—the 'antibiotic of last resort'—is discussed in Chapter 11.)

The genomics of *S. aureus* has been pursued vigorously, in order to identify the mechanisms of development and spread of resistance. The *S. aureus* genome is about 2.8–2.9 Mb long. Assignment of approximately 2600 open reading frames accounts for almost 85% of the genome. There is a single plasmid containing about 25 000 bp.

Genes for enhanced antibiotic resistance are encoded by a transposon inserted into the plasmid. Comparison of the sequences—in particular, observation of lack of synteny—has made it clear that the development of methicillin resistance was not a single event, producing a clone that was subsequently selected. Instead, the resistance elements were acquired many times by many strains, via horizontal gene transfer.

A comparison of the sequences of many *S. aureus* strains, encompassing different clinical isolates, showed that 78% of genes were common to all strains, including those from cow and sheep. The remaining 22%, that are at least partially strain-specific, tend to be localized within 18 large regions of difference (RDs), ranging from 3–50 kb long. Figure 7.14 shows the presence of these regions in the different strains, and the correlation of the pattern with methicillin resistance.

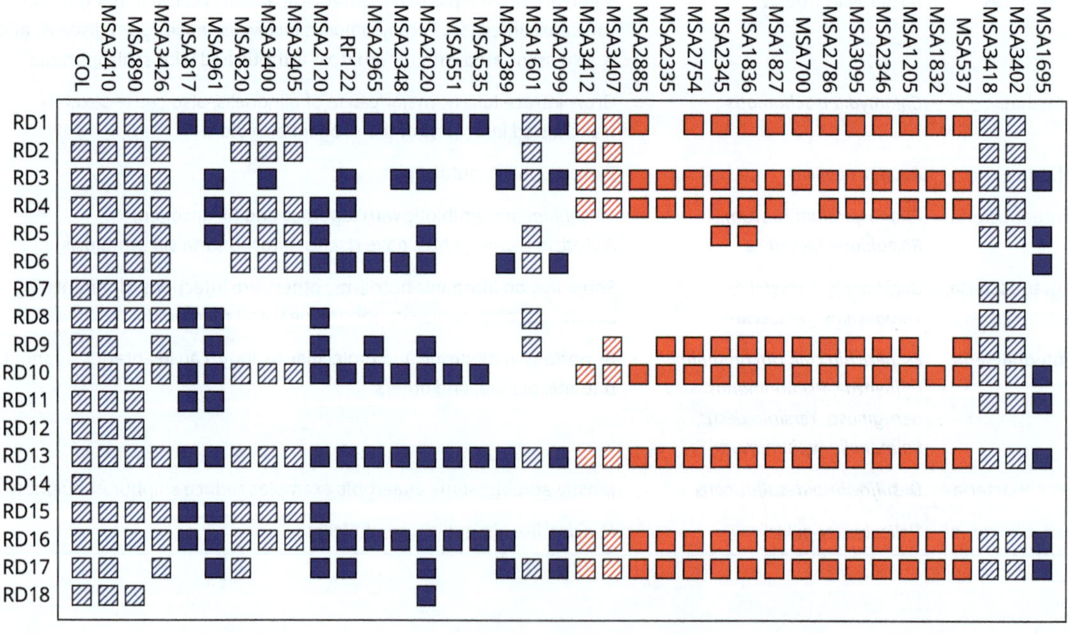

Figure 7.14 Results of comparison of sequences of 36 isolates of *S. aureus*. RD = regions of difference, localized segments of the genome of high variability among strains. Filled squares indicate RD present; empty squares, RD absent. Hatched squares correspond to methicillin-resistant strains. Red indicates isolates of electrophoretic type 234, the predominant type causing toxic shock syndrome.
Fitzgerald, J.R., Sturdevant, D.E., Mackie, S.M., *et al.* (2001). Evolutionary genomics of *Staphylococcus aureus*: Insights into the origin of methicillin-resistant strains and the toxic shock syndrome epidemic. Proceedings of the National Academy of Sciences of the United States of America 98, 15: 8821–8826. https://doi.org/10.1073/pnas.161098098. Reproduced with permission. Copyright (2001) National Academy of Sciences, USA.

Genomics and the development of vaccines

Genomics and recombinant technology have made possible a new generation of approaches to antibiotic development (see Box 7.5).

A vaccine against hepatitis B virus is expressed in yeast cells. The vaccine is a surface antigen, a viral envelope protein, the gene for which was cloned into yeast.

A vaccine against *Bordetella pertussis* (the causative agent of whooping cough) is based on the toxin, a multi-subunit protein. By genetic engineering, the molecule was completely detoxified by introduction of mutants, which removed the enzymatic activity but left the immunological properties intact. Antibodies raised against the detoxified form protect against the native protein.

A more general approach involves comparing the genome sequences of pathogenic and non-pathogenic strains to identify virulence factors that might serve as the basis of vaccines.

Neisseria meningitidis serogroup B is the major cause of meningitis and septicaemia in children and young adults. From the 2272351 bp genome sequence, computational methods predicted 2158 genes. Algorithms predicted that 600 of them would be on the cell surface or secreted. These were candidates for vaccines. Of these, 350 were expressed in *E. coli*, and tested in mice for an immune response that produced bactericidal antibodies. In order to achieve a vaccine that covered as many strains as possible, surveys of different strains for sequence variability of these candidate vaccines revealed which ones were relatively constant. Vaccines are now available: In the UK, they are offered under the NHS childhood vaccination programme to all babies at 8 and 16 weeks, with a booster after their 1st birthday.

KEY POINT

Our descendants may well look back at the second half of the twentieth century as a narrow window during which bacterial infections could be controlled, and before and after which they could not.

BOX 7.5 The future of antibiotic development

The development of resistant strains of pathogens presents a severe challenge to medicine. There is consensus that novel antibiotics will be needed. Some are already in the 'pipeline', currently in the clinical testing phase. However, the research that produced today's 'new' antibiotics was initiated in the early 1990s and pharmaceutical companies are reducing their emphasis on antibiotic research. It is likely that fewer new antibiotics will emerge in the current industrial climate. The paradox of a growing need for new discoveries coupled with the reduction in resources aimed at generating them creates a problem that may become a crisis.

The novel developments associated with genomics, the subject of this book, can in principle contribute to the development of antibiotics. It is possible to identify *targets*—specific proteins, essential for a pathogen, that differ from mammalian proteins sufficiently to suggest that drugs against these proteins would be effective against the pathogen but non-toxic to mammals. Unlike classical antibiotic research practice, experimental methods are now available that can define the mechanism of action of a drug while it is still under development.

A related prospect is to turn to biology as well as chemistry to discover new therapeutic agents, including revival of an old suggestion of using bacteriophages clinically. A number of small biotech companies have started up to explore a number of non-traditional avenues. Given the decade 'lead time' required for a new drug to make its way from the laboratory to approval for clinical use, the process must be set in motion immediately. A problem with this approach is a line of recent US court decisions that impose stricter criteria on the patentability of procedures that might be considered 'natural processes'.

The problems are multidimensional, involving science, economics, long-term forecasting, and regulatory and patent law. Each field presents an individual set of difficulties. These difficulties are compounded by the necessity to solve them all simultaneously in the face of both genuine conflicts between different goals, and boundaries between professions that impede communication and cooperation. There is, however, consensus that the problems *must* be solved. Artificial intelligence may well play an important role.

Viruses

Viruses figure prominently in the study of life, and in the study of death. Einstein's comment, that nature is 'subtle but not malicious' does not apply to viruses—they are emphatically both. It is said that there is no discovery in cell biology that viruses don't already know about.

Most viruses are nucleoprotein particles, with a nucleic acid genome encapsulated within a protein or glycoprotein coat. Viruses infect cells using specialized proteins on cell surfaces that effect attachment and invasion. Viral nucleic acid enters the host cell. In some cases, viral proteins required for replication also enter the host cell. Once inside the host cell, the invading viral molecules must (1) make multiple copies of the viral genome; (2) synthesize viral proteins, including enzymes active only within the host cell, and coat proteins (and others) to be assembled into the progeny virions; and then (3) 'pack up and leave'.

For viruses in eukaryotic cells the burst size can range from thousands to tens of thousands. For viruses in prokaryotic cells burst sizes are typically smaller.

The new viruses go on to infect additional cells. In many cases they kill the infected organism. Despite improved treatments, AIDS killed over half a million people in 2022. Novel epidemics, such as Ebola and Zika virus, and COVID-19, pose grave threats. Deaths of animals also number in the millions, in some cases as a result of culls to prevent spread of disease. In response to the foot-and-mouth disease outbreak in 2001, 10 million sheep and cattle were killed in the UK.

Some viruses can integrate their genomes into those of the host cells. Sometimes this is mandatory for replication; in other cases, it makes the host cell and its descendants a reservoir. In some cases, viruses 'fossilize' to leave inactive regions in the host genome.

Many viruses have short genomes. The first natural DNA sequenced was that of bacteriophage ϕX-174, 5386 bases of single-stranded DNA. The genome that causes chicken pox, varicella zoster virus, has a double-stranded DNA genome 124 884 bp long. The genomes of viruses HIV-1 and influenza A contain 9181 and 13 588 bases, respectively. Both are single-stranded RNA. These sizes were considered typical.

Nucleocytoplasmic large DNA viruses (or giant viruses)

The discovery of giant viruses has broken this generalization. Early in the history of the subject, scientists spoke of 'filterable viruses' because the infectious particles were small enough to pass through a filter of diatomaceous earth or porcelain, which retain bacteria. Giant viruses are not filterable, by the historical criterion.

A number of types of giant virus are known. They have large double-stranded DNA genomes, exceeding in length some cellular genomes, and large particle sizes (see Table 7.5). Their proteomes include more proteins than the repertoires of typical viruses—and, in some cases, of cells. One pandoravirus is predicted to contain over 2500 protein-coding genes.

Table 7.5 Types of giant viruses

Type of virus	Genome length	Size	Where discovered
Mimivirus	1.1 Mb	750 nm	Cooling tower near Paris
Megavirus	1.2 Mb	680 nm	Ocean, off coast of Chile
Pithovirus	0.61 Mb	1 500 nm	30 000-year-old ice core from Siberian permafrost
Pandoravirus	1.9–2.5 Mb	1 000 nm	One example in ocean, off coast of Chile (2.5 Mb genome); another in freshwater pond in Melbourne, Australia (1.9 Mb genome)

These giant viruses challenge the distinction between viruses and cells. It has been suggested that the eukaryotic cell nucleus originated from a virus.

Some large viruses are themselves parasitized by small (15–30 kb) DNA **virophages**. According to current ideas, the giant virus provides the transcription machinery required for replication of the virophage within the cytoplasm of an overall eukaryotic cellular host (*not* the nucleus). The virus contributes enzymes involved in protein biosynthesis, including amino-acyl tRNA synthetases. Some mimiviruses even have a CRISPR/Cas-like defence system.

Viral genomes

Viral genomes contain only relatively short stretches of nucleic acids. These may be DNA or RNA (see Box 7.6). Some, for instance the genome of the virus that causes hepatitis C, encode a single polyprotein. Cleavage produces the few proteins the virus needs to take over the cell. Other viruses, such as human immunodeficiency virus type 1 (HIV-1), contain several genes. The HIV-1 genome is about 9.8 kb long, containing a total of nine genes (see Figure 7.15). One gene encodes the Gag–Pol fusion protein, which is cleaved to release Gag (the HIV-1 protease), reverse transcriptase, and integrase. Other mRNAs expressed by HIV-1 contain introns and are spliced to express Rev and Tat (see Figure 7.15).

Recombinant viruses

Mixed infections of a cell by different viruses permit genetic recombination. It is even possible to package *unaltered* nucleic acid from one strain into an envelope composed of protein from another strain. In that event, the absorption–penetration–surface antigenicity characters are those of the coat proteins, but the hereditary characteristics are those of the nucleic acid. (Infection with such a virus is a kind of natural Hershey–Chase experiment; see Chapter 4.) These effects can alter host specificity.

Types of viral genome

As assembled within the virion, a viral genome may consist of:

Nucleic acid	Examples
Single-stranded DNA	Bacteriophages ϕX-174 and M13
Double-stranded DNA	Adenoviruses, smallpox virus, Epstein–Barr virus, bacteriophage λ
Single-stranded RNA	Bacteriophages MS2, Qβ, tobacco mosaic virus, HIV-1
Double-stranded RNA	Bluetongue virus

Single-stranded DNA viral genomes are generally converted to double-stranded DNA by the host.

Replication of RNA viruses is prone to mutation, because the RNA-dependent RNA polymerases that replicate their genomes usually do not have the proof-reading and error-correcting features of eukaryotic DNA polymerases. This helps viruses to evade host immune systems and facilitates their jumping between host species. (Emerging viral diseases, including but not limited to acquired immunodeficiency syndrome (AIDS) and avian flu, are usually based on viruses with RNA genomes.)

A viral genome consisting of single-stranded RNA can be:
(+)sense = same sequence as protein-translatable mRNA
(–)sense = complementary sequence to mRNA
ambisense = mixture of both.

Inside the cell, (+)sense viral RNAs present themselves as messenger RNA (mRNA) and are translated. (–)sense viral RNAs and double-stranded viral RNAs require specialized polymerases for conversion to mRNA. Retroviral genomes contain (+)sense RNA, which is reverse transcribed into host DNA. These viral polymerases and reverse transcriptases are proteins that are contained in the infecting virion and enter the host cell along with the viral nucleic acid.

Some viral genomes are infectious on their own. For some RNA viruses, a DNA reverse transcript of the viral RNA is infectious (although at a lower rate than the natural virion). This permits preparation of large quantities of viral genomes for vaccines, avoiding the lability and high mutation rate of viral RNA replication.

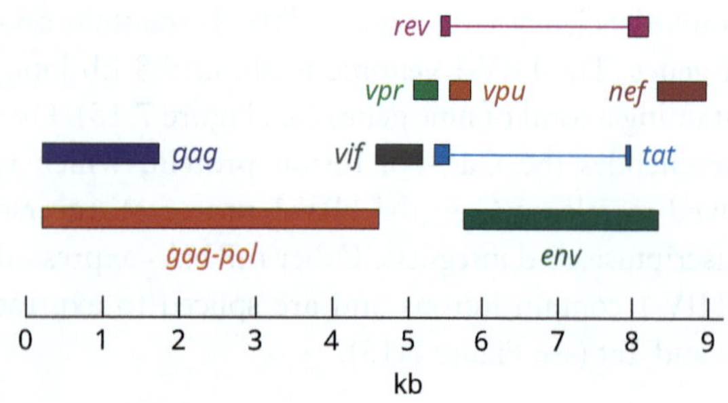

Figure 7.15 Diagram showing the sizes of the individual gene transcripts of HIV-1. The introns of the *rev* and *tat* genes are indicated by a thin line. The proteins encoded by these genes are:

Gene	Proteins	Function
Gag	p24, p6, p7, p17	Structural proteins of capsid and matrix
Pol	Reverse transcriptase, integrase, protease	Integration into host genome and cleavage of viral-encoded polyproteins
Env	Precursors of gp120, gp41	Envelope proteins, active in attachment and fusion to host cells
Tat	Tat	Facilitates transcription of viral RNA
Rev	Rev	Enhances cytoplasmic export of transcripts
Nef	Nef	Interferes with host immune function
Vif	Vif	Interferes with host defence
Vpr	Vpr	Needed for nuclear import of viral nucleic acid
Vpu	Vpu	Promotes assembly and release of progeny virus; also stimulates degradation of host CD4 proteins, disabling the host immune system

KEY POINT

HIV-1, influenza viruses, and SARS-CoV2 (the virus responsible for COVID-19) have become major threats to human health after jumping from animal hosts. Their high mutation rates—with severe clinical consequences—reflect the fact that their genomes are RNA.

In the laboratory, a virus can be constructed as a vector to produce foreign proteins inside a cell. Two applications of this technique are as follows.

1. To produce a vaccine, insert into the vaccinia virus genome a DNA sequence coding for an immunogen (perhaps the HIV-1 surface glycoprotein gp120). (The HIV-1 surface protein itself is of course not infectious: the Hershey–Chase experiment again!).[2] Infection by recombinant virus leads to expression of immunogen and elicitation of an immune response, giving the host protective immunity. The immunity created by such an *intracellular* exposure to the immunogen is much more powerful than that achievable simply by injecting the immunogen into the bloodstream.
2. A recombinant virus carrying a normal human gene can be useful for gene therapy. Can a retroviral vector reintroduce the normal variant into the patient's genome? A successful application has been the treatment of adrenoleukodystrophy (an inherited degenerative disorder leading to progressive brain damage, adrenal gland failure, and early death) by gene delivery using a virus derived from HIV. Treatments for many other diseases are at various stages of research. For instance, cystic fibrosis arises from a mutation in the cystic fibrosis transmembrane regulator (*CFTR*) gene. Gene therapy for cystic fibrosis is now in Phase II clinical trials.

The CRISPR/Cas system has already cured patients with sickle-cell disease and β-thalassemia. Many other applications are waiting in the wings.

Viruses and evolution

Viruses evolve much faster than cellular life forms. Their rates of mutation are a million times faster than those of cellular forms. In addition, there are more of them—at least an order of magnitude more viruses than cells—and viruses have short generation times and large numbers of progeny.

Spurring this evolutionary gallop is the lack of fidelity in genome replication. The enzyme RNA-dependent RNA polymerase replicates genomes of RNA viruses. Unlike cellular DNA-replication systems, equipped with error-correction mechanisms, the error rate of RNA replication is very high. And because viral genomes do not contain much 'junk', a mutation is more likely to have a phenotypic effect. (Some viruses do have error-correcting processes, but

[2] But HIV-1 *does* insert protein as well as RNA into the host cell. Fortunately for Hershey, Chase, and the field, the virus they worked with, bacteriophage T2, does not.

these are rudimentary and sparse.) As a result, viral replication generates very great sequence diversity. In an untreated AIDS patient, every day viral replication will generate mutations, not only at every site in the HIV-1 genome, but at every possible *pair* of sites. Changes in epitopes allow viruses to evade host immune systems, and to develop drug resistance.

However, the accumulation of errors threatens to degrade viral fitness. Essential functions are at risk. In fact, too high a mutation rate can tip the population into an instability, called 'mutational meltdown'. Living with very high mutation rates, viruses skate very close to this line. Chemical mutagens can push the population over the edge. For example, mutagenic nucleoside analogues can drive a poliovirus population to extinction: A fourfold increase in mutation rate produces a 95% reduction in viral titre. Similar results have been obtained with other viruses.

As part of the generated diversity, viral proteomes contain protein domains specific to viruses, not found in cells. Perhaps unsurprisingly, about half of them are involved in capsid structure. Most of the others are involved in nucleic acid metabolism. But the implication that viral evolution can create novel domains has implications for cellular evolution.

For viruses play a role in the evolution of cellular life forms. They are agents of gene transfer. By insertion into crucial sites in the genome they can affect gene expression patterns. Consider also the process of reciprocal gene transfers: host → virus → host. Given that viral evolution is so much more rapid, perhaps some of the heavy lifting in generating diversity in protein structure and function is taken on at the viral stage.

Influenza: a past and current threat

Influenza is a contagious disease caused by a virus that infects the respiratory tract. The virus is passed around a population in droplets created when an infected individual coughs or sneezes. Unlike HIV-1, the virus that causes AIDS, influenza virus can survive outside the host, greatly facilitating its transmission.

Three types of influenza virus are known, of which type A is the most dangerous. The virion contains a spherical lipoprotein coat enclosing eight nucleoproteins containing the RNA genome, encoding a total of ten proteins. Protruding from the envelope are several hundred 'spikes' containing the proteins haemagglutinin (80% of the spikes) and neuraminidase (about 20%) (see Figure 7.16). These proteins are essential to the reproduction of the virus. Haemagglutinin binds to host cell surface glycoprotein receptors to promote viral entry into cells. Neuraminidase helps progeny virions to get out. Both haemagglutinin and neuraminidase are targets of drugs.

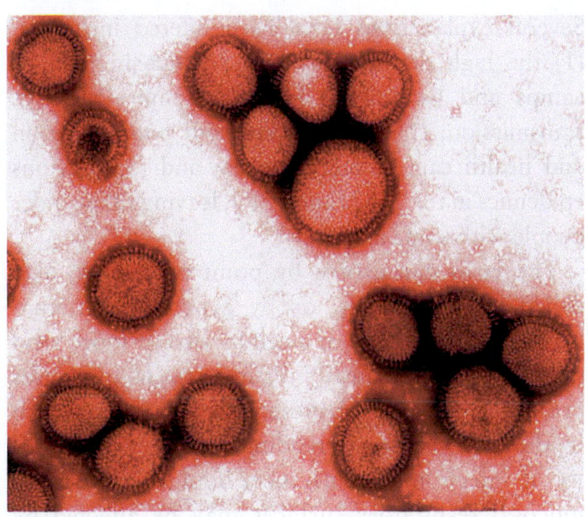

Figure 7.16 Influenza virus.

Picture courtesy Professor Y. Kawaoka, University of Wisconsin, US, and University of Tokyo, Japan.

Every year, influenza seasonally affects many people worldwide. In a typical year, 38 000 people die in the US from influenza or related complications, and 200 000 are hospitalized. The mortality rate is 0.8%, with most fatalities occurring in the very young or in older adults. Worldwide, the usual annual fatality rate is 1–1.5 million.

However, in some years, influenza and associated complications attack more viciously. A famous pandemic occurred at the end of the First World War: within an 18-month period in 1918–1919, influenza killed an estimated 50 million people, far more than had died in the war. The mortality rate was 1% of those affected. Such an influenza pandemic today could have an even higher mortality in regions of the world containing many people immunocompromised by AIDS.

In most influenza seasons, fatalities occur as a result of bacterial infection of lungs weakened by the virus, to which older adults are more vulnerable. The 1918–1919 epidemic was different, in the higher percentage of fatalities among *young* people.

Several explanations have been offered including: (1) the high density of young soldiers in military camps and battlefields, leading to more effective transmission; (2) overcrowding and poor nutrition and health care among refugees; and (3) previous epidemics in the 1850s and 1889, leaving many older people with some immunity.

The virus can evolve by point mutations, also called antigenic drift, or by genetic recombination after coinfection of a cell by two or more viral strains. Immunologically distinct strains of viruses are called serotypes. Different serotypes vary both in the ease of their spreading and in the mortality of infection. For instance, binding of virus to mucosal respiratory surfaces may be affected by amino-acid sequence polymorphisms of both viral proteins and host receptors. Contagion and mortality are also dependent on characteristics of the host population, including density and general health levels. A strain that is both highly contagious and has a high mortality rate would be very dangerous indeed. As part of an effort to understand why the 1918–1919 strain was so dangerous, scientists have recently reconstructed that virus, based on material recovered from contemporary postmortem specimens.

Different strains of influenza virus contain 1 of 13 recognized types of gene for haemagglutinin (H) and one of nine recognized types of gene for neuraminidase (N). These types identify different major strains of viruses. For instance, the strain that caused the 1918–1919 pandemic was H1N1.

Many strains of the virus infect only a restricted range of species, and have different mortality rates in different species. These properties can change as the virus evolves. A strain that infects animals can potentially become infectious to humans. Species range depends on the different forms of sialic acid presented on viral glycoproteins. An important determinant is haemagglutinin residue 226, which is Gln in viruses infectious to birds and Leu in viruses infectious to humans.

Avian flu

In 2006, an H5N1 strain of avian flu characterized by very high mortality infected domestic poultry in several countries. It was considered a particularly dangerous threat to humans because it has a high mutation rate and recombines readily. One way for the virus to jump from birds to humans is for two strains to co-infect pigs and use them as a 'mixing vessel' for recombination.

Avian flu can normally infect only birds and, in some cases, pigs. Domestic poultry stocks raised under conditions of very high population density are particularly vulnerable. Often migratory birds are carriers but do not get sick. (The 2006 avian flu strain was spread from southeast Asia to Russia by migratory birds.)

The H5N1 strain prevalent in 2006 was first identified in Hong Kong in 1997 and traced to ducks from Guandong province. It jumped the species barrier to mammals, becoming infectious to pigs, in April 2004. It then became supervirulent, killing rodents, birds, and humans. This H5N1 strain is

ETHICAL ISSUES

Ethical dilemma: Publication of RNA sequence of the virulent 1918–1919 strain of influenza virus

Recently, scientists were able to recover and sequence the strain of influenza active in the 1918–1919 pandemic. The journal *Science* published the work and, consistent with editorial policy, required that the sequence be deposited in the nucleic acid databanks.

In *The New York Times* on 17 September 2005, R. Kurzweil and W. Joy wrote an article critical of the decision to make the sequence generally available in databanks on the grounds that terrorists might use the information to recreate the virus and use it as a weapon.

Reactions to the publication of the reconstructed pandemic viral sequence illustrate the conflict between the recognition that free and open access to information is a benefit to the progress of science, and the dangers of its misuse. A precedent occurred before the Second World War when physicist Leo Szilard tried, unsuccessfully, to persuade colleagues not to publish results that might prove useful in the development of atomic weapons. He suggested that journals record dates of receipt and acceptance of manuscripts but then sequester the articles for the duration. This occurred well before the strict secrecy imposed after the Manhattan Project was organized.

Science did make an exception to its mandatory deposition policy in publishing the Human Genome Draft Sequence by J.C. Venter and co-workers in 2001.

100% fatal in domesticated chickens and in 54% of reported human cases.

Compared with previous epidemics, the world today is particularly vulnerable because of:

- increased human population densities;
- widespread long-distance travel;
- intensive livestock production (including antibiotic feeding, which may select drug-resistant strains of infectious bacteria); and
- policies of various governments that do not adequately reimburse farmers who must sacrifice animals, creating a disincentive to report disease; the result is a delay or even default of an effective response.

Increased population densities of both humans and animals threaten a greater rate of spread of a dangerous strain of virus, even in comparison with the recent 1968–1969 epidemic (see Table 7.6).

Aggressive approaches to controlling avian flu have involved large-scale culling of stocks. In 1997, the H5N1 strain infected poultry in Hong Kong and caused six human fatalities. The entire poultry population of the island had to be destroyed: 1.5 million birds in three days. An H7N7 epidemic in the Netherlands in 2003 led to the killing of >30 million birds (approximately twice the human population of the country). In Asia in 2004, over 100 million birds were culled. In 2015, 5 300 000 hens were slaughtered in Iowa, in the midwestern US. Now, many authorities take prompt action: In 2015, detection of avian flu led to destruction of 170 000 chickens on one farm in Lancashire. There are plans to cull almost 40 000 birds on a farm in Fife. The smaller numbers involved in the UK reflect both the promptness of the action taken and the fact that poultry farms tend to be smaller than in the US.

Table 7.6 Population in China (millions)

Year	Humans	Pigs	Chickens
1971	800	250	648
2005	1 307	433	4 338
2014	1 376	466	4 630
2023	1 455	453	5 000

https://www.worlddata.info/asia/china/populationgrowth.php
https://www.pig333.com/pig-production-data/china_11/
https://www.youtube.com/watch?v=fOsvT4Uf9yc

Drugs against influenza

Tamiflu (oseltamivir) and Relenza (zanamivir) are the two major drugs against influenza. Both are inhibitors of the viral neuraminidase.

Relenza (Figure 7.17) was designed at the Commonwealth Scientific and Industrial Research Organization (CSIRO) laboratory in Melbourne, Australia. Crystal structures of influenza neuraminidase showed that conserved sequences formed a cavity, suggesting a target site for drugs. By targeting the active site of the enzyme, it is harder for the virus to evolve resistance.

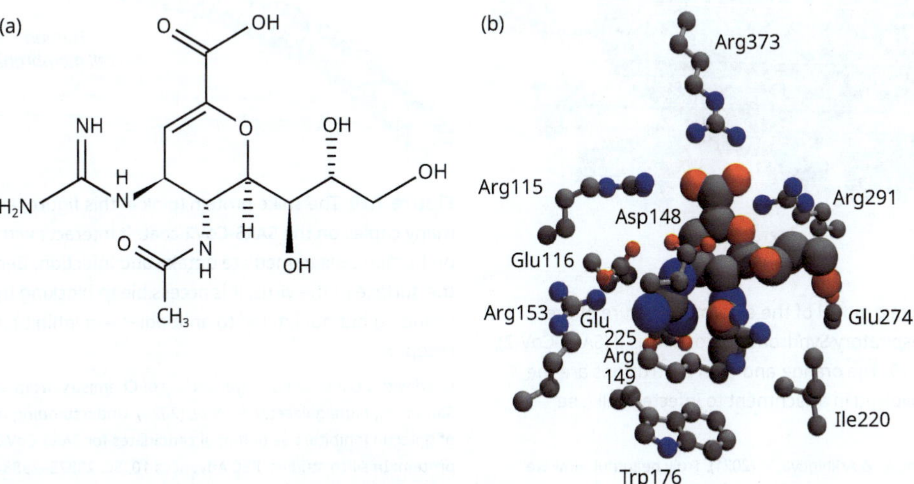

Figure 7.17 (a) The structure of the anti-influenza drug zanamivir (Relenza). (b) Zanamivir is a transition-state analogue that binds to the active site of influenza neuraminidase. Here the atoms of the drug are shown as large spheres and the residues from the neuraminidase are shown in ball-and-stick representation.

COVID-19

A prominent feature of life during the last few years has been the COVID-19 epidemic. Caused by the SARS-CoV2 virus (see Figure 7.18), the disease is both dangerous and highly transmissible. More than 7 million people have died of COVID-19. Many others have been infected: Some have recovered fairly promptly; others have suffered long-term consequences. Defence measures have involved major disruptions to social and economic activities.

Widespread testing for COVID-19 has involved (1) PCR—the most reliable test—involves taking a sample by a health care provider and submission to a laboratory, a procedure taking several days, and (2) an antigen test, a rapid do-it-yourself test that reports results in a few minutes. Positive tests are trustworthy, but negative results may be false, especially if symptoms are not present.

Figure 7.18 Atomic model of the external structure of the Severe Acute Respiratory Syndrome CoronaVirus 2 (SARS-CoV-2), cause of COVID-19. The orange and cyan protrusions are the spike protein, involved in attachment to infected cell (see Figure 7.19).

From: Solodovnikov, A., & Arkhipova, V. (2021). Truly beautiful: How we made the SARS-CoV-2 3D model (in Russian). N+1. https://nplus1.ru/blog/2021/07/29/sars-cov-2-model.

Fortunately, molecular biologists rose to the occasion, and produced effective vaccines in a very short time. The goal of the vaccines is to produce antibodies to the SARS-CoV2 spike protein, a component of the viral coat, involved in attachment to human cells (see Figure 7.19).

One approach is to use purified spike protein itself in a vaccine, eliciting antibodies that will react with intact infecting virus. Another approach that has proved extremely effective is to use the mRNA that encodes the spike protein. Delivery of this mRNA

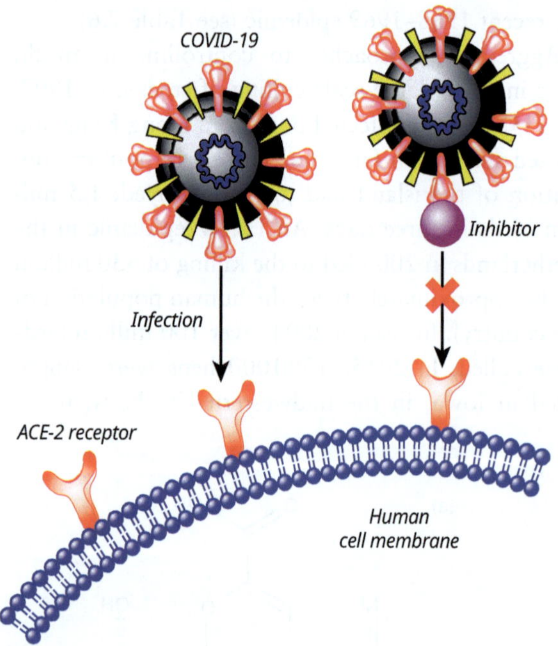

Figure 7.19 The spike protein (pink in this figure) appears in many copies on the SARS-CoV2 coat. It interacts with receptors on human cells to mediate uptake and infection. Because it is on the surface of the virus, it is accessible to blocking by ligands—including but not limited to antibodies—to inhibit binding to the receptor.

Used with permission of Royal Society of Chemistry, from Sethi, A., Sanam, S., Munagalasetty, S., *et al.* (2020). Understanding the role of galectin inhibitors as potential candidates for SARS-CoV-2 spike protein: In silico studies. RSC Advances 10, 50: 29873–29884. https://doi.org/10.1039/d0ra04795c. Permission conveyed through Copyright Clearance Centre, Inc.

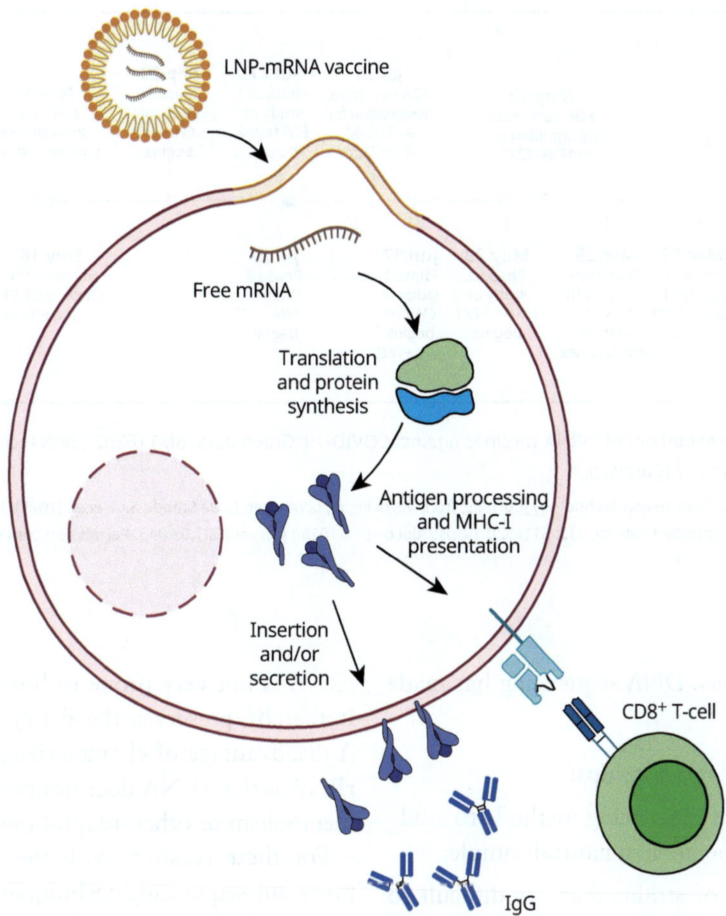

Figure 7.20 Simplified view of mechanism of vaccine action. A messenger RNA (mRNA) molecule corresponding to the spike protein encased in a lipid nanoparticle (LNP) is absorbed by a cell, and releases the mRNA. Ribosomes translate the mRNA to produce, intracellularly, copies of the spike protein. These can be inserted in the cell membrane or secreted, eliciting a B-cell immune response; or cut into peptides presented on the cell surface by MHC proteins, to elicit a T-cell-mediated immune response. The antibodies produced will be active against spike proteins on the surface of an intact infecting virus. (Broken lines indicate cell nucleus.)

From: Piccaluga, P.P., Di Guardo, A., Lagni, A., *et al.* (2022). COVID-19 vaccine: Between myth and truth. Vaccines 10: 349.

will cause the host to express the spike protein intracellularly, where it will elicit an immune response (see Figure 7.20).

Figure 7.21 shows the timeline of the development of mRNA vaccines against COVID-19. It was less than a year between recognition of the outbreak to the approval of a vaccine!

mRNA therapy is being tested for safety and effectiveness in treating melanoma, lung cancer, and other solid tumours.

'Ome, ome, on the range: metagenomics, the genomes in a coherent environmental sample

Classically, microbiologists studied prokaryotes by growing them in culture, isolating pure strains for detailed study. Powerful as the methods were, and useful as they were for clinical applications and research, they were also blinders that prevented full appreciation of the variety and interactions of species

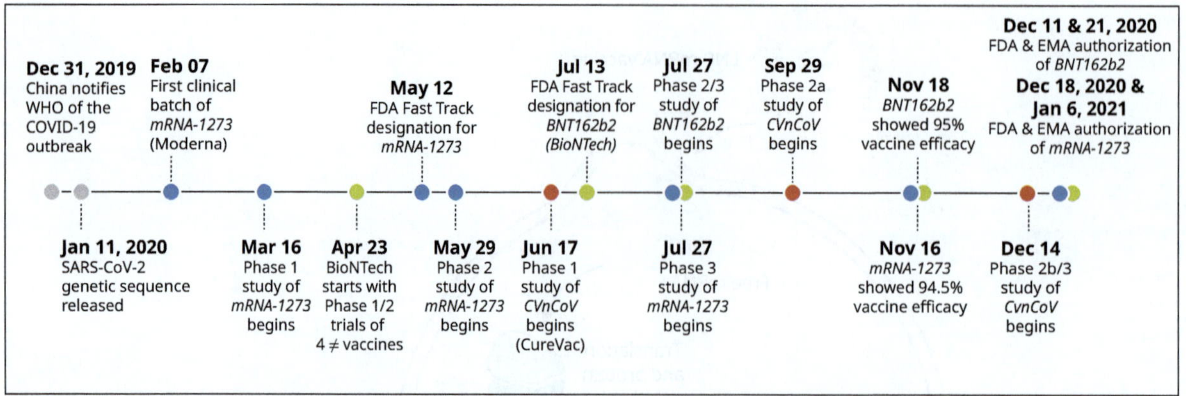

Figure 7.21 Timeline of development of mRNA vaccines against COVID-19. Green dots: BMT162b2 (BioNTech), blue dots: mRNA-1273 (Moderna), orange dots: CVnCoV (CureVac).

Used with permission of Elsevier Science and Technology Journals, from Verbeke, R., Lentacker, I., De Smedt, S.C., *et al.* (2021). The dawn of mRNA vaccines: The COVID-19 case. Journal of Controlled Release 333: 511–520. https://doi.org/10.1016/j.jconrel.2021.03.043. Permission conveyed through Copyright Clearance Centre, Inc.

in natural environments. DNA sequencing has made it possible to:

- clarify evolutionary relationships;
- use high-throughput sequencing methods to study a cross-section of the life in a natural sample;
- study the majority of strains that are difficult to grow in culture; and
- appreciate the relationships and interactions among different species that share an ecosystem.

Metagenomics can explore the microflora of every niche in the world. These include the exterior and interior of the human body. We shall discuss the human microbiome in Chapter 10.

KEY POINT

A millilitre of ocean water may contain 100–200 species. A gram of soil may contain 4000.

From natural samples containing complex mixtures, it is possible to amplify and determine particular sequences directly, without culturing individual strains. The first molecule of choice was 16S rRNA. This is partly because of its traditional role as a molecule that varies at the appropriate rate to distinguish ancient phylogenetic branching patterns. In addition, rRNA is not very prone to horizontal gene transfer. It thereby preserves the distinctions between taxa. A disadvantage of characterizing an organism by its rRNA is that rRNA does not reveal any details of the metabolism or other adaptations of the species.

For these reasons, with the availability of more powerful sequencing techniques, metagenomics has turned to shotgun sequencing—the sequencing of short fragments arising by randomly shearing the DNA in a sample. Clearly, the assembly problem is much more difficult in dealing with samples from a mixture of organisms.

The sequencing of a mixed sample collected from environmental sources (called eDNA) has applications in conservation, monitoring, and ecosystem assessment. It allows detection of rare species, and warnings of appearances of invasive species.

A student once complained to me that biology had become the study, not of life as a whole, but of just a few laboratory organisms. She changed fields, claiming that metagenomics is bringing biology back out into the field, and later sent me a picture of herself in a T-shirt printed with the slogan: 'From *E. coli*-gy to Ecology!'.

An example of metagenomics is the sequencing of 16S rRNA genes from ocean water from the Sargasso Sea. A group led by J.C. Venter sequenced 10^9 non-redundant regions. Many novel sequences were found, although it is difficult to assemble complete genomes and avoid chimaeras.

There are very large numbers of metagenomic projects completed or in progress, including microbes and viruses from hot springs in Yellowstone National Park in Wyoming, US; flora of the mouth and distal gut of humans and digestive tracts of parasites; microbial communities in drinking-water sources; acid mine drainage systems; ocean waters both coastal and pelagic; and soils, including samples from the rhizospheres—the surroundings of roots—of many different types of plants (see http://www.genomesonline.org/). Metagenomics reports sequences from viruses as well as from cells. Viruses are the most abundant and variable life forms on Earth. However, although human pathogens of importance in medicine and the bacteriophages of importance in the molecular biology laboratory have received intensive study, little is known about the rest. Viruses in soils and oceans remain the 'dark matter' of the biosphere.

Metagenomics and metaproteomics are extensions of laboratory techniques to natural environments. This allows study of interacting organisms and species, at the cost of greater complexity of the measurements.

Marine cyanobacteria—an in-depth study

The basic concepts of genome evolution are secure: organisms create, and select for, and against, genome variations. Adaptations drive some changes—via divergence, allelic redistribution within a population, gene loss, or gene acquisition by horizontal gene transfer. Neutral genetic drift accounts for other changes, especially in small populations.

What is more difficult to understand is *how* selective pressures act. If organisms encounter environments varying in space and/or time, into how many populations, or even new species, will they split? Which proteins will diverge—in sequence, in function, or in expression pattern? What novel genes are needed and where will they come from?

In most field situations, the 'topology' of evolutionary space depends on a complicated interaction of physical and ecological variables, such as geographic barriers imposed by landscape, climate, and interspecies cooperation or competition. Complex environments give rise to complex biological communities.

The distribution of cyanobacteria in the open oceans in temperate regions offers a relatively simple ecological context. The distributions—both of environmental features and of species or strains—depend on a *single* variable, depth. There are many correlates of depth: light intensity and quality, temperature, pressure, ultraviolet light penetration, nutrient availability (notably, sources of nitrogen and iron), and the occurrence of predators and viruses. Yet, for all its complexity, the system is one-dimensional.

The major populations of cyanobacteria in temperate and tropical oceans belong to two related genera, *Prochlorococcus* and *Synechococcus*. They are responsible for a significant fraction of worldwide photosynthesis. *Prochlorococcus* is believed to have diverged from *Synechococcus* fairly recently.

Ocean environments are stratified. Studies of the distribution of *Prochlorococcus* ecotypes in a vertical column in the Sargasso Sea reveal a division into *two* types of strain. Closer to the surface than about 130 m depth, the majority of *Prochlorococcus* strains are adapted to high light levels. Strains prevalent below 130 m depth are adapted to low light levels (see Figure 7.22).

KEY POINT

Prochlorococcus strains are adapted to different ambient light levels, which decrease with increasing depth below the ocean surface. High light levels: ≥200 μmol photons m^{-2} s^{-1}. Low light levels: ≤30–50 μmol photons m^{-2} s^{-1}.

Rocap and co-workers compared the genomes of high-light and low-light-adapted *Prochlorococcus* strains with *Synechococcus*[3] (see Box 7.7). *Prochlorococcus* strain MIT9313 is adapted to growth in low light intensity. Both its distribution with depth (Figure 7.22) and the general features of its genome are similar to *Synechococcus*. In contrast, the high-light-adapted strain MED4 is 'lean and mean': its genome is unusually small and encodes fewer

[3] Rocap, G., Larimer, F.W., Lamerdin, J., *et al.* (2003). Genome divergence in two *Prochlorococcus* ecotypes reflects oceanic niche differentiation. Nature 424: 1042–1047.

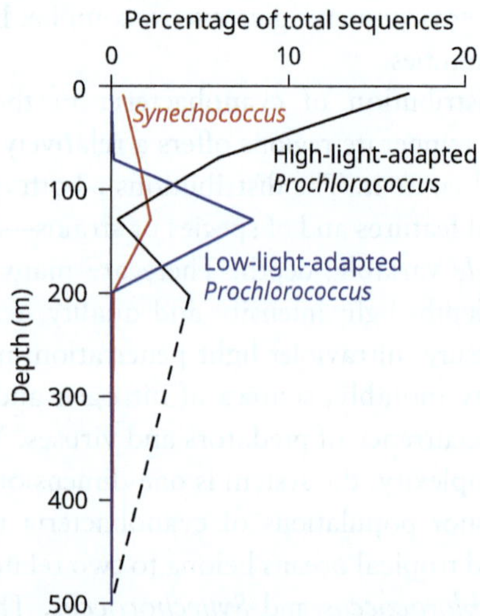

Figure 7.22 Distribution with depth of three types of cyanobacteria: *Synechococcus, Prochlorococcus* strains adapted to high-light-intensity habitats, and *Prochlorococcus* strains adapted to low-light-intensity habitats. The appearance of large amounts of low-light-adapted *Prochlorococcus* at 140 m, below the level where it seemed to have virtually disappeared, is puzzling. The broken line from 200–500 m depth is interpolated; measurements have been made at 200 and 500 m but not in between.

Used with permission of American Association for the Advancement of Science, from DeLong, E.F., Preston, C.M., Mincer, T., *et al*. (2006). Community genomics among stratified microbial assemblages in the ocean's interior. Science 311, 5760: 496–503. https://doi.org/10.1126/science.1120250; permission conveyed through Copyright Clearance Center, Inc.

proteins. *Prochlorococcus* MED4 is the smallest known organism that generates oxygen.

Approximately half of the proteins of *Synechococcus* are common to all three species: 1314 out of 2526. Only 38 genus-specific proteins appear in both *Prochlorococcus* strains but not in *Synechococcus*. Many of these 38 proteins are involved in the synthesis of the light-harvesting complex of *Prochlorococcus*, which has a structure unusual among bacteria (see Box 7.8).

Within the *Prochlorococcus* genus, many genes are strain specific: MIT9313 has 923 proteins that do not appear in MED4 (about half of these do appear in *Synechococcus*). This, together with the observation that the MED4 genome and proteome are substantially smaller, implies that many genes have been lost in the differentiation of the strains.

How are different *Prochlorococcus* strains adapted to differences in ambient light intensities and spectral distributions?

Effective interactions with light require both efficient energy transduction and protection from photochemical damage caused by excitation energy spillover.

- Antenna complexes of photosystem II (PSII) of most cyanobacteria, including *Synechococcus*, contain protein complexes called phycobilisomes. *Prochlorococcus* is unusual among cyanobacteria in using, as PSII antennae, proteins binding unusual modified (divinyl) chlorophylls, called Pcb proteins.

Where did the Pcb proteins come from? They appear to have been recruited from a family called *iron-stress-inducible* (ISI) proteins (see Weblem 7.6).

BOX 7.7

Prochlorococcus and *Synechococcus* genomes

Feature	*Prochlorococcus*		*Synechococcus*
	Strain MED4	Strain MIT9313	Strain WH8102
Preferred light level	High	Low	
Length (bp)	1657990	2410873	2434428
G+C (mol%)	30.8	50.7	59.4
Protein coding (%)	88	82	85.6
Protein coding genes	1716	2273	2526
RNA genes	40	51	44

BOX 7.8

Cyanobacterial photosynthesis

The photosynthetic apparatus of cyanobacteria contains large chromophore-containing macromolecular complexes:

- Two coupled photosystems that carry out energy transduction—the capture of light energy. These are called photosystem I and photosystem II (PSI and PSII).
- Antenna pigments that make light harvesting more efficient by absorbing light and transferring the excitation energy to the reactive chlorophylls.
- Prochlorococcus contains atypical light-harvesting molecules.

ISI proteins are expressed in cyanobacteria under conditions of low iron concentrations. They provide an alternative to the iron-rich protein ferredoxin in the electron transport chain.

- High-light-adapted and low-light-adapted strains differ in the relative amounts of chlorophyll a_2 and b_2. (Chlorophyll a_2 has a methyl group at a position where chlorophyll b_2 has an aldehyde group.) High-light-adapted strains have mostly chlorophyll a_2; low-light-adapted strains have more chlorophyll b_2, which absorbs optimally in the blue region of the spectrum. This is an appro priate match to the colour of the ambient light below the ocean surface. The ratio of concentrations of chlorophylls a_2 and b_2 can vary with habitat conditions. In the low-light-adapted strain MIT9313, the ratio can change by at least a factor of two, producing more chlorophyll a_2 at higher light intensities. The chlorophyll a_2/b_2 ratio in the high-light-adapted strain MED4 is less sensitive to ambient light intensity. The nature of the control mechanism and the set of proteins that change expression patterns are not yet fully understood.

Another adaptation for 'scavenging' light in dark environments is to increase the number of genes for light-harvesting proteins. Low-light-adapted strains have more copies of the genes that code for the chlorophyll-binding antenna protein Pcb. In addition, MED4 contains a phycoerythrin, an antenna protein that binds a chromophore absorbing green light.

Protection against photochemical damage

Ultraviolet light can damage DNA. Major products include thymine dimers, the linkage of adjacent thymine residues in DNA. In response to the threat of mutation, cells contain repair enzymes, including *photolyase*, which recovers thymines from dimers.

Because ultraviolet light does not penetrate far into sea water, *Prochlorococcus* strain MED4, living nearer the surface, is in greater danger from photochemical damage than MIT9313. Indeed, MED4, but not MIT9313, contains a gene for photolyase. Another difference between the strains is also probably related to photo-oxidative stress: MED4 contains perhaps twice as many *high-light-inducible proteins* as MIT9313. From their distribution in the genome, some of these appear to have arisen by recent duplication events.

'Mad dogs and Englishmen go out in the midday sun,' sang Noël Coward. Plants and ocean-surface-dwelling cyanobacteria do too—they have no choice.

Utilization of nitrogen sources

In the ocean, the prevalent form of nitrogen near the surface is ammonium, produced in part by fixation of atmospheric nitrogen. In deep waters, nitrate is more common. Nitrate is produced by degradation of dead organic matter: dead organisms sink.

Different organisms assimilate nitrogen from molecular nitrogen, nitrate, nitrite, or ammonium ions.

The two *Prochlorococcus* strains differ in their ability to assimilate nitrogen from different sources (see Box 7.9). There has been a successive loss in nitrogen-assimilating ability with the divergence from *Synechococcus* to low-light-adapted *Prochlorococcus* (deeper-living) to high-light-adapted *Prochlorococcus* (living nearer the surface). *Synechococcus* has active nitrate and nitrite reductase genes and can use either ion, as well as ammonium, as a nitrogen source. *Prochlorococcus* MED4 has lost nitrate reductase but retains nitrite reductase; it can use nitrite but not nitrate as a nitrogen source. *Prochlorococcus* MIT9313 has neither reductase and must get its nitrogen from ammonium or from other reduced nitrogen compounds such as amino acids.

BOX 7.9

Assimilation of nitrogen

Reaction	Enzyme	*Synechococcus*	*Prochlorococcus*	
		WH8102	MED4	MIT9313
$N_2 \rightarrow NH_4^+$	Nitrogenase	Absent	Absent	Absent
$NO_3^- \rightarrow NO_2^-$	Nitrate reductase	Present	Absent	Absent
$NO_2^- \rightarrow NH_4^+$	Nitrite reductase	Present	Present	Absent
$NH_3 \rightarrow$ glutamine	Glutamine synthetase	Present	Present	Present

Protection against predators and viruses

Other cellular life forms graze on *Prochlorococcus* and *Synechococcus* strains. Viruses infect them.

Defensive adaptations involve genes encoding proteins involved in the synthesis of lipopolysaccharides and polysaccharides, which form the basis of cell surface recognition. Both strains of *Prochlorococcus* have acquired, by horizontal gene transfer, clusters of genes for surface polysaccharides not shared by the other strain or by *Synechococcus*. As evidence of foreign origin, this 40.8 kb cluster in MIT9313 has a G+C content of 42 mole %, substantially lower than that of the MIT9313 genome as a whole, 50.7 mole %.

LOOKING BACK AND FORWARD

The transition from prokaryotes to eukaryotes is one of the most important landmarks in the history of life. In this chapter, we have surveyed some classes of archaea, bacteria, and viruses, and have looked in detail at a few of them. We next turn to eukaryotes. The greater complexity of higher eukaryotes in terms of macroscopic bodies, composed of organized collections of differentiated cells, is obvious. Arguably the nervous system is the crowning achievement. But what is not so obvious is the extent to which the novelties of eukaryotes are developments from genetic and biochemical components of eukaryotes.

RECOMMENDED READING

General discussions of prokaryotic classification

Oren, A., & Papke, R.T. (2010). *Molecular Phylogeny of Microorganisms*. Norfolk, UK: Caister Academic Press.

Last universal common ancestor (LUCA) and related topics

Mat, W.K., Xue, H., & Wong, J.T. (2008). The genomics of LUCA. Front. Biosci. 13: 5605–5613.

Puigbò, P., Wolf, Y.I., & Koonin, E.V. (2009). Search for a 'Tree of Life' in the thicket of the phylogenetic forest. J. Biol. 8: 59.

Watson, T. (2019). The trickster microbes shaking up the tree of life. Nature 569: 322–324.

McGrath, C. (2022). Highlight: Unraveling the origins of LUCA and LECA on the Tree of Life. Genome Biol. Evol. 14: evac072.

Spang, A., Mahendrarajah, T.A., Offre, P., *et al.* (2022). Evolving perspective on the origin and diversification of cellular life and the virosphere. Genome Biol. Evol. 14: evac034.

Forterre, P. (2022) Archaea: A goldmine for molecular biologists and evolutionists. Methods Mol. Biol. 2522: 1–21.

The history of the growth of oxygen in the atmosphere: an intersection between biochemistry and geology

Raymond, J., & Segrè, D. (2006). The effect of oxygen on biochemical networks and the evolution of complex life. Science 311: 1764–1767.

Holland, H.D. (2006). The oxygenation of the atmosphere and oceans. Phil. Trans. R. Soc. Lond. B: Biol. Sci. 361: 903–915.

Kasting, J.F. (2013). What caused the rise of atmospheric O_2? Chemical Geology 362: 1320–1325.

Lyons, T.W., Reinhard, C.T., & Planavsky, N.J. (2014). The rise of oxygen in Earth's early ocean and atmosphere. Nature 506: 307–315.

Fischer, W.W., & Valentine, J.S. (2019). How did life come to tolerate and thrive in an oxygenated world? Free Radic. Biol. Med. 140: 1–3.

Mills, D.B., Boyle, R.A., Daines, S.J., *et al.* (2022). Eukaryogenesis and oxygen in Earth history. Nat. Ecol. Evol. 6: 520–532.

West, J.B. (2022). The strange history of atmospheric oxygen. Physiol. Rep. 10: e15214.

Craig, J.M., Kumar, S., & Hedges, S.B. (2023). The origin of eukaryotes and rise in complexity were synchronous with the rise in oxygen. Front. Bioinform. 3: 1233281.

New approaches to vaccine design

Scarselli, M., Giuliani, M.M., Adu-Bobie, J., *et al.* (2005). The impact of genomics on vaccine design. Trends Biotech. 23: 84–91.

Rossolini, G.M., & Thaller, M.C. (2010). Coping with antibiotic resistance: Contributions from genomics. Genome Medicine 2: 15.

Luciani, F. (2016). High-throughput sequencing and vaccine design. Rev. Sci. Tech. 35: 53–65.

Lu, G., Shan, S., Zainab, B., *et al.* (2021). Novel vaccine design based on genomics data analysis: A review. Scand. J. Immunol. 93: e12986.

Jain, S., Venkataraman, A., Wechsler, M.E., *et al.* (2021). Messenger RNA-based vaccines: Past, present, and future directions in the context of the COVID-19 pandemic. Adv. Drug. Deliv. Rev. 179: 114000.

Hogan, M.J., & Pardi, N. (2022). mRNA vaccines in the COVID-19 pandemic and beyond. Annu. Rev. Med. 73: 17–39.

Fiyouzi, T., & Reche, P.A. (2023). Vaccine design: An introduction. Methods Mol. Biol. 2673: 1–14.

Genomics of viruses

Abroi, A., & Gough, J. (2011). Are viruses a source of new protein folds for organisms?—Virosphere structure space and evolution. Bioessays 33: 626–635.

Rohwer, F., & Barott, K. (2013). Viral information. Biol. Philos. 28: 283–297.

Abergel, C., Legendre, M., & Claverie, J.-M. (2015). The rapidly expanding universe of giant viruses: Mimivirus, Pandoravirus, Pithovirus and Mollivirus. FEMS Microbiol. Revs. 39: 779–796.

Forterre, P. (2016). To be or not to be alive: How recent discoveries challenge the traditional definitions of viruses and life. Stud. Hist. Philos. Biol. Biomed. Sci. 59: 100–108.

Cobián-Güemes, A.G., Youle, M., Cantú, V.A., *et al.* (2016). Viruses as winners in the game of life. Ann. Rev. Virol. 3: 197–214.

Prangishvili, D., Bamford, D.H., Forterre, P., *et al.* (2017). The enigmatic archaeal virosphere. Nat. Rev. Microbiol. 15: 724–739.

Debarbieux, L., Forterre, P., Krupovic, M., *et al.* (2018). Special issue on Centennial Celebration of Bacteriophage Research [collection of articles]. Research in Microbiology 169: 479–541.

Silva, J.M., Pratas, D., Caetano, T., *et al.* (2022). The complexity landscape of viral genomes. Gigascience 11: giac079.

Regulski, K., Champion-Arnaud, P., & Gabard, J. (2021). Bacteriophage manufacturing: From early twentieth-century processes to current GMP. In: D.R. Harper, S.T. Abedon, B.H. Burrowes, & M.L. McConville (eds), *Bacteriophages*. Cham: Springer.

Wise, J. (2024). Antimicrobial resistance: MPs call on UK government to maximise potential of bacteriophages. Brit. Med. J. 384: q12.

Metagenomics

DeLong, E.F., & Karl, D.M. (2005). Genomic perspectives in microbial oceanography. Nature 437: 336–342.

Scanlan, D.J., Ostrowski, M., Mazard, S., *et al.* (2009). Ecological genomics of marine picocyanobacteria. Microbiol. Mol. Rev. 73: 249–299.

Wooley, J.C., Godzik, A., & Friedberg, I. (2010). A primer on metagenomics. PLoS Comput. Biol. 6: e1000667.

Bohlin, J. (2011). Genomic signatures in microbes—properties and applications. Sci. World J. 11: 715–725.

Fernández, L.A. (2005). Exploring prokaryotic diversity: There are other molecular worlds. Mol. Microbiol. 55: 5–15.

Guerrero, R., & Berlanga, M. (2006). Life's unity and flexibility: The ecological link. Int. Microbiol. 9: 225–235.

Yilmaz, P., Yarza, P., & Glöckner, F.O. (2016). Expanding the world of marine bacterial and archaeal clades. Front. Microbiol. 6: 1524.

Joseph, T.A., & Pe'er, I. (2021). An introduction to whole-metagenome shotgun sequencing studies. Methods Mol. Biol. 2243: 107–122.

EXERCISES, PROBLEMS, AND WEBLEMS

Exercises

Exercise 7.1 Which of the differences between archaea and bacteria described in this chapter could you derive from genome sequence alone?

Exercise 7.2 Using the standard genetic code (Box 1.2), are there any amino acids for which a hyperthermophilic organism could *not* preferentially choose a codon with G or C in the third position?

Exercise 7.3 From the table in Box 7.4, what are two groups of bacteria that are (a) photosynthetic, (b) live in the human gut, (c) pathogenic?

Exercise 7.4 What RNA molecule is most closely linked to the *his* operon in *E. coli* (see Figure 7.5)?

Exercise 7.5 Some large viruses contain genes for amino-acyl-tRNA synthetases. What experiments could you perform to determine (a) whether these genes encode a functional protein? (b) whether expression of the viral gene in infected cells is necessary for replication?

Exercise 7.6 What qualitative feature of the genomes accounts for the fact that one infection with the virus causing chicken pox creates lifetime immunity, but we need a new 'flu shot' each year?

Exercise 7.7 On a photocopy of Figure 7.8a, circle salt bridges that appear at (approximately) common positions in both structures.

Exercise 7.8 On a photocopy of Figure 7.8b, identify positions at which there is a charged residue in the *P. furiosus* sequence but an uncharged residue in the *C. symbiosum* sequence. (Charged residues = K, R, H (shown in blue), D, E (shown in red), uncharged residues = all the others.)

Exercise 7.9 On a photocopy of Figure 7.11, (a) circle the glycolysis/gluconeogenesis pathway; (b) circle the Calvin cycle.

Exercise 7.10 On a photocopy of Figure 7.12, circle a region in which synteny is maintained in *P. abyssi*, *P. horikoshii*, and *P. furiosis*.

Exercise 7.11 Why is the recombinant vaccine against hepatitis B expressed in yeast and not *E. coli*?

Exercise 7.12 If 1 ml of seawater contains 200 species, the metagenome would be how big?

Exercise 7.13 In H.G. Wells' 1898 novel *The War of the Worlds*, invaders from Mars are overcome by disease. Assuming that life on Mars developed independently of life on Earth, why is it unlikely that the Martians died of viral infections?

Problems

Problem 7.1 Draw a Venn diagram showing the numbers of genes specific to one, common to pairs, and common to all three of the following: *Prochlorococcus* strain MED4, *Prochlorococcus* strain MIT9313, and *Synechococcus* strain WH8102.

Problem 7.2 In Figure 7.14, hatched squares indicate MRSA strains. (a) Find two RDs (regions of diversity) that are present in all methicillin-resistant strains studied. (b) Find two RDs that are present in some methicillin-resistant strains, but not every strain containing either of them is methicillin resistant. (c) Is there any RD that is present in every methicillin-resistant strain, and absent from every methicillin-sensitive strain?

Problem 7.3 On a copy of Figure 7.17b, indicate the following interactions: (a) the positively charged guanidino group in zanamivir (left in Figure 7.17b) forms salt bridges with Glu116 and Glu225 in the neuraminidase active site; (b) hydroxyl groups of the glycerol moiety of zanamivir (at the right in Figure 7.17b) are hydrogen bonded to Glu274; (c) the carbonyl oxygen of the *N*-acetyl sidechain is hydrogen bonded to Arg149; (d) the methyl group of the *N*-acetyl sidechain makes hydrophobic interactions with Ile220 and Trp176.

Problem 7.4 Read the letters to *The New York Times* (17 September 2005) discussing the decision to publish the genome sequence of the 1918 pandemic strain of influenza virus. Summarize the arguments for and against publication.

Weblems

Weblem 7.1 Print out the complete secondary structures of 16S rRNAs from *E. coli, Methanococcus vannielii, and Saccharomyces cerevisiae*. On each structure, indicate where the region illustrated in Figure 7.1 appears.

Weblem 7.2 Are any archaea implicated in human disease?

Weblem 7.3 What was the most common serotype of influenza in the US during the 2022–2023 season?

Weblem 7.4 (a) Identify a bacterium with a growth temperature below 10°C. (b) Identify an archaeon with a growth temperature below 10°C (see Figure 7.5).

Weblem 7.5 Make two histograms of *E. coli* genes, similar to that of Figure 7.4, showing in one of them the size distribution of genes appearing clockwise in the genome and in the other the size distribution of genes appearing counter-clockwise in the genome. Describe any systematic differences that appear.

Weblem 7.6 Where did *Prochlorococcus* get the chlorophyll-binding proteins of its photosystem II antenna? (a) Using either of the *Prochlorococcus* Pcb proteins (UniProt ID PCBA_PROMM or PCBA_PROMP), search using PSI-BLAST for homologous cyanobacterial proteins with different functions. You should find—among others—ISIA_SYNP6, ISIA_SYNP7, and ISIA_SYNP2. Present the results of this search, editing the output down to the most relevant information. (b) What is the function of the ISI family of proteins? (c) Align the full sequences of PCBA_PROMM, PCBA_PROMP, ISIA_SYNP6, ISIA_SYNP7, and ISIA_SYNP2, using CLUSTAL W or T-Coffee. Comment on the extent of the divergence. (d) Determine the fraction of identical residues between every pair of sequences from the multiple sequence alignment and draw a phylogenetic tree.

Weblem 7.7 The high-light-intensity-adapted *Prochlorococcus* strain MED4 contains a phycoerythrin, but the low-light-intensity-adapted strain MIT9313 does not. Do most *Synechococcus* strains contain phycoerythrins? If so, align the sequences of a prochlorococcal and synechococcal phycoerythrin and comment on the extent of the divergence compared with the divergence of prochlorococcal Pcb proteins from synechococcal ISI proteins (see Weblem 7.6).

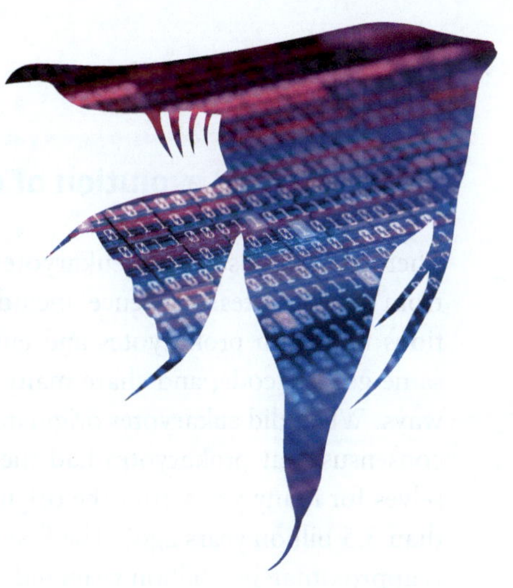

CHAPTER 8

Genomes of Eukaryotes

LEARNING GOALS

- *Have a clear sense of how eukaryotic cells differ from prokaryotic cells.*
- *Understand the relationships among the major types of eukaryotes.*
- *Recognize that fungi in general and yeasts in particular are among the simplest eukaryotes.* They have served as model organisms in the molecular biology laboratory (in addition to applications in agriculture and cooking).
- *Know the unique features of higher plants*, and the focus on *Arabidopsis thaliana*—'the fruit fly of botany'.
- *For animals, appreciate the evolutionary path from early organisms, more distantly related to humans, to mammals*.
- *Respect the power of our ability to recover and sequence DNA from extinct organisms.*

We, like all other species that exist or have existed, are the product of a long evolutionary history. Genomes give us snapshots of landmarks along the route. They allow us to understand much of the topology of the path—where and when it branched. They reveal at which points certain features arose. In some cases, they show us the experiments—some productive, some abortive—that preceded the results that life subsequently adopted.

The origin and evolution of eukaryotes

There is consensus that eukaryotes are descended from prokaryotes. Evidence includes the observations that both prokaryotes and eukaryotes use the same genetic code, and share many metabolic pathways. When did eukaryotes originate? Again, there is consensus that prokaryotes had the world to themselves for many years after the origin of life (no later than 3.5 billion years ago). The first eukaryotic fossil is approximately 2 billion years old. The discovery in datable oil deposits of biochemicals produced only by eukaryotes suggests an earlier origin, almost 3 billion years ago.

The first eukaryotes were unicellular organisms. Multicellularity began with formation of colonies, followed by cell specialization, perhaps first by simple symbiosis. Developmental programmes allowed specialization within clonal clusters.

There followed development of a great variety of body plans, based on differentiation of a variety of tissue types. A major landmark that we know about, in the development of higher organisms, was the division between animals and plants. Some animals adopted body structures with bilateral symmetry. Some of these became vertebrates. Eventually, some became human.

Until relatively recently, our view of the history of life was limited to observations of organisms that are extant, or that have left either descendants or fossils. The Burgess Shale deposits show us that we have missed a lot of interesting possibilities. It is true that it has become possible to recover and sequence ancient DNA. This has opened another window on extinct species. But only in a limited way.

Our only real possibility of reconstructing our history is through sequencing the genomes of extant organisms that have been around for a long time, and to read the story that their sequences tell.

KEY POINT

In order to reconstruct evolutionary history from genomes, we must analyse genomes from species the ancestors of which first arose in the distant past. What do they share with their close relatives, that might offer genomic characterization of a group of species; for instance, what are the defining genomic features of vertebrates? What do ancient species share with their predecessors and successors? What innovations did they achieve? Which of those were dead ends and which did other species, descended from them, adopt and develop?

Evolution and phylogenetic relationships in eukaryotes

Genome sequences provide detailed information about evolutionary relationships among species. So many genomes are now known that here we must pick and choose only a few of the interesting examples. The Genomes On Line Database (GOLD) lists over 50 000 eukaryotic genome sequencing projects.

Figure 8.1 shows the major groups of eukaryotes. Note the 'star' topology—at this level of low resolution, eukaryotes are a bush rather than a tree. In this chapter, however, we shall follow a more directed path, roughly in the direction towards higher complexity. This correlates fairly well with date of origin.

From the comparative genomics of eukaryotes, we can ask questions about features of humans that we consider important. For instance, we can look into the origin of some features of a body plan, or the immune system, or the endocrine system, or features of the nervous system. Phrasing the questions loosely: Who invented them? When? What if anything did they come from? What alternatives were experimented with and how well did they work?

This is an ambitious programme. Let us back off, and begin with one of the simplest eukaryotes, yeast.

The yeast genome

Yeast, like *Escherichia coli*, is an organism better known to many of us from molecular biology laboratories than from nature. It has served as a model eukaryote, because of its relative simplicity, ease of growth, having both haploid and diploid states, and safety. Of course, we are also grateful to yeast for bread, wine, and beer. On the other hand some related fungi are infectious.

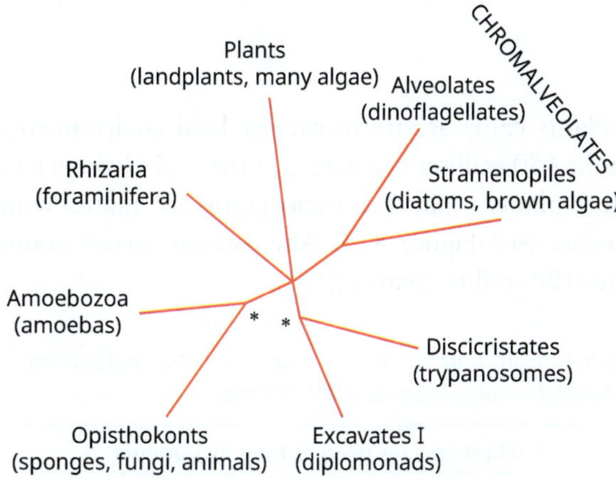

Figure 8.1 Major classes of eukaryote, with examples in parentheses. The asterisks mark possible positions of the root of the tree.

Based on: Baldauf, S.L. (2003). The deep roots of eukaryotes. Science 300: 1703–1706, and personal communication.

Yeast is one of the simplest eukaryotic organisms. Its cells, like our own, contain a nucleus and other specialized intracellular compartments. The sequencing of its genome, by an international consortium comprising ~100 laboratories, was completed in 1992.

The genome of baker's yeast, *Saccharomyces cerevisiae*, contains 12052 kbp, distributed among 16 chromosomes.[1] The chromosomes range in size over an order of magnitude, from the 1352 kbp chromosome IV to the 230 kbp chromosome I.

The *S. cerevisiae* genome is 2.6 times the length of that of *E. coli*, and less than a hundredth of the human genome. Many strains also contain one or more plasmids. The genome is relatively compact, with genes accounting for about 72% of the sequence. There are fewer repeat sequences compared with genomes of more complex eukaryotes.

A duplication of the entire yeast genome appears to have occurred ~150 million years ago. This was followed by translocations of pieces of the duplicated DNA and loss of one of the copies of most (~92%) of the genes.

Approximately 6700 protein-coding genes are predicted. Relatively few contain introns. However, the genes of a related yeast, *Schizosaccharomyces pombe*, are much richer in introns.

Over 5000 of the protein-coding genes correspond to molecules to which a function can be assigned. About 1000 more contain some similarity to known proteins in other species. Another ~800 are similar to ORFs in other genomes that correspond to unknown proteins. Many of these homologues appear in prokaryotes. Only about 40% of yeast proteins have identifiable homologues in the human genome.

S. cerevisiae contains many genes for non-protein-coding RNAs. A single tandem array on chromosome XII encodes 120 copies of ribosomal RNA. There are 40 genes for small nuclear RNAs, and 275 genes for transfer RNA, about one-third of which contain introns.

The classification of yeast protein functions shown in Table 8.1 is taken from the *Saccharomyces* genome database.

[1] 1 kbp = 1 kilo base pair = 1000 base pairs.

Table 8.1 Assignment of *Saccharomyces cerevisiae* gene products to different functional categories

Functional category	Number of proteins
Metabolism	1514
Energy	367
Cell cycle and DNA processing	1012
Transcription	1077
Protein synthesis	480
Protein fate (folding, modification, destination)	1154
Protein with binding function or cofactor requirement (structural or catalytic)	1049
Regulation of metabolism and protein function	253
Cellular transport, transport facilities, and transport routes	1038
Cellular communication/signal transduction mechanism	234
Cell rescue, defence, and virulence	554
Interaction with the environment	463
Transposable elements, viral and plasmid proteins	120
Cell fate	273
Development (systemic)	69
Biogenesis of cellular components	862
Cell type differentiation	452
Unclassified proteins	1393
Functionally classified proteins	4777
Functionally unclassified proteins	1394

The evolution of plants

Plants and animals parted company a long time ago. Although all life forms share much of their molecular biology, plants derive energy from sunlight via photosynthesis. The consequences for their structure, lifestyle, and developmental programmes have been profound. They require many proteins dedicated to their unique biophysical and metabolic activities. Plant genome sequences illuminate the similarities to and differences from other eukaryotes:

- Plants share some functions with animals. At the genomic and proteomic level, are they achieved in similar ways?
- Some functions are unique to plants. At the genomic and proteomic level, where did they come from? Were they evolved *de novo*, adapted, or borrowed?

From a common ancestor living about 800 million years ago, metazoa split into three major groups: fungi, animals, and plants. Higher plants evolved from single-celled organisms formerly classified as green algae. Green algae have now been split into streptophytes and chlorophytes; streptophytes are related to higher plants.

Plants came ashore to occupy land environments about 450 million years ago, in the mid-Ordovician. Most plants today are angiosperms, or plants with flowers (see Figure 8.2). Angiosperms arose about 140–190 million years ago.

Since the introduction of agriculture, humans have become dependent on a few crop plants for food:

Plant species used as food by humans

Human use/classification	Plant species
Total described species	250 000
Edible	30 000
Cultivated	7 000
Important on national scale	120
Making up 90% of world's calories	30

Source: UN Food and Agriculture Organization

The first complete nuclear genome of a higher plant to be sequenced was that of *Arabidopsis thaliana*, common name thale cress. It is related to turnip, cabbage, and broccoli. Its ease of handling and rapid

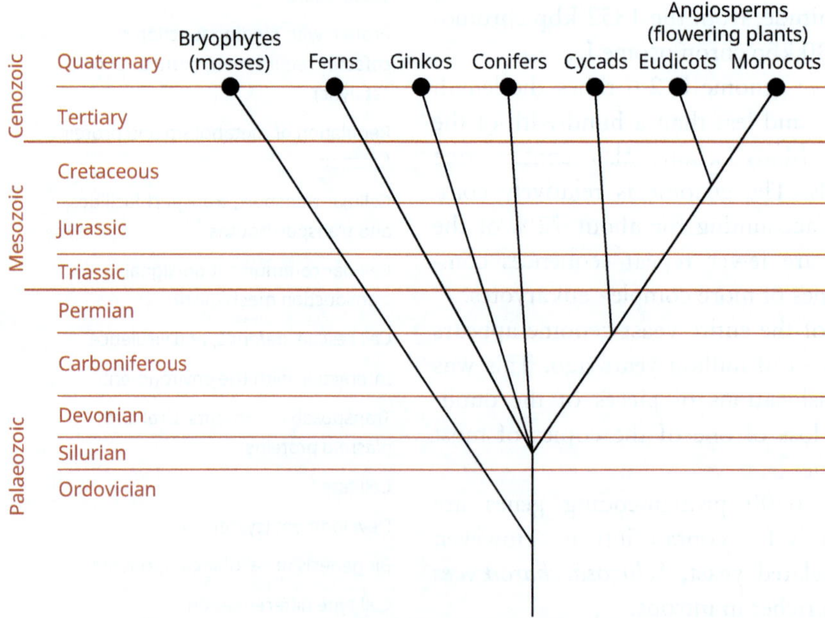

Figure 8.2 Phylogeny of land plants. The picture is limited to groups with extant examples with which readers may be familiar. Monocots and eudicots are named for the difference between single and double cotyledons in the embryo, but many other features separate them.

BOX 8.1 The *Arabidopsis thaliana* genome

Genome size	146 Mb (estimated)
Sequenced nuclear DNA	115 936 794 bp
Predicted protein coding genes	26 732
Pseudogenes	3818
Variably spliced genes	2330
Transposons	>10% of genome

Gene distribution	Length (bp)	Number of genes
Chromosome 1	30 432 563	6905
Chromosome 2	19 705 359	4178
Chromosome 3	23 403 063	5313
Chromosome 4	18 585 042	4088
Chromosome 5	23 810 767	6248
Full nucleus	115 936 794	26 732
Mitochondrion	366 924	135
Chloroplast	154 478	122

generation time have made it a favoured subject for research in plant molecular biology. *A. thaliana* has been called 'the fruit fly of botany'.

The *Arabidopsis thaliana* genome

A. thaliana has a relatively small genome—135 Mb in the laboratory strain—distributed over five chromosomes (see Box 8.1). (The maize genome is almost 20 times as large.)

The compact genome was one reason why the research community adopted *Arabidopsis*.

The *Arabidopsis* nuclear genome is relatively compact. Protein-coding genes contain an average of 5.4 exons, of average length 276 bp, separated by relatively short introns about 165 bp long. The intergenic spacing is also short, about 4.6 kb. A feature of plant genes is that the G+C content of exons (44 mole %) is higher than that of introns (32 mole %).

The structure of the *A. thaliana* genome reveals both local and genome-wide duplications. There were probably *five* polyploidizations. It has been possible to date the latest two whole-genome duplications fairly precisely: they occurred 90 million, and 40 million, years ago. In addition, local duplications have affected ~17% of genes. Close relatives, such as cabbage and cauliflower, have undergone additional polyploidizations during the 12 million years since they diverged from *Arabidopsis*.

A. thaliana has a mitochondrial and a chloroplast DNA sequence as well. Genome analysis must address questions of divisions of labour. Relative to animal cells, organelles in plant cells bear a greater metabolic burden, if only because of the activities of chloroplasts. Chloroplast genomes are relatively gene dense, with well-preserved gene order. In plant mitochondria, genes are more widely spaced and recombination is more common, relative to chloroplasts. Mitochondrial and chloroplast genes contain fewer introns (Table 8.2).

The *Arabidopsis* proteome contains many molecules specific to plants, including those involved in photosynthesis and in the metabolism of components of cell walls. *Arabidopsis* is rich in genes that encode water-transporting channels, peptide-hormone transporters, metabolic and biosynthetic enzymes, and proteins involved in defence, detoxification, and environmental sensing.

Plants have many special metabolic pathways, for photosynthesis and for the metabolism of cell wall components, alkaloids, and growth regulators such as auxins and gibberellins. Complex metabolism requires the genome to encode a large and varied set of enzymes. In keeping with the essential role of light in plant life, *Arabidopsis* has many light sensors that regulate development and circadian responses.

Table 8.2 Genes containing introns

	Genome		
	Nuclear	Chloroplast	Mitochondrial
Genes containing introns (%)	80	18.4	12

Plants are also threatened by pathogens and have evolved defence mechanisms dissimilar from our immune system. One weapon that plants deploy against pathogens involves the production of reactive oxygen species. Plants synthesize other defence molecules against animals. Conversely, they also produce molecules that attract pollinators or seed-dispersal agents (think colour changes upon fruit ripening). These attractants have provided useful sources of flavours, fragrances, and drugs, encompassing traditional 'herbal medicine' and modern pharmacology.

Comparing the proteins encoded in the nuclear genome of *Arabidopsis* with human proteins, the fraction of homologues observed varies with functional category. For protein synthesis, 60% of nuclear-encoded *Arabidopsis* genes have human homologues. For transcription regulation, the figure is only 30%. It is not that transcription is poorly represented in plant genomes; it is just that plants do it differently. In fact, plants have several times as many transcription factors as the fruit fly. Although many components of the signal-transduction pathways familiar from animals are absent in plants, plants have developed specific transcription factor families not present in animals.

Many *Arabidopsis* genes are homologous to human genes implicated in disease. For instance, plants and animals have similar DNA repair systems, and *Arabidopsis* has a homologue of *BRCA2*. For some human disease-associated genes, the plant homologue is more similar to the human protein than those from fruit fly or *Caenorhabditis elegans*. Study of the function of the plant homologues will be illuminating, even though it is unlikely that *Arabidopsis* will be suitable for trials of drugs intended for clinical use in humans!

Genomes of animals

Plants are of course not ancestral to humans. We now turn to species that represent some important branching points in our own ancestry. That is, a succession of species with which we share more and more recent common ancestors: urochordates, fishes, birds, monotremes, and other eutherian mammals. (We discussed the genomes of *C. elegans* and *Drosophila melanogaster* in Chapter 6.)

The genome of the sea squirt (*Ciona intestinalis*)

Vertebrates arose within the chordate phylum, branching off from other lines that led to sea squirts and lancelets (see Figure 8.3). The sea squirt (*C. intestinalis*) represents one of the most primitive of our chordate relatives (Figure 8.4). Its genome provides insight into chordate and vertebrate origins.

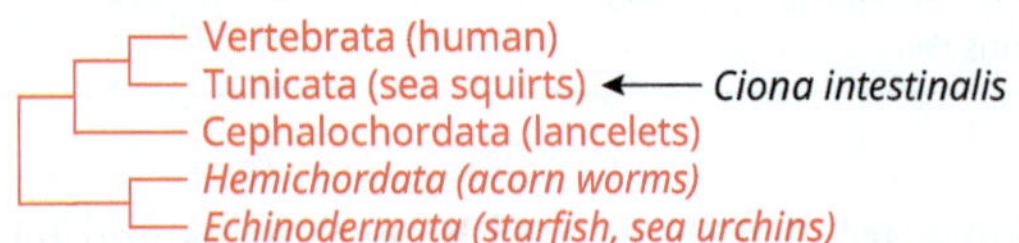

Figure 8.3 The evolutionary position of the sea squirt, *Ciona intestinalis*, compared to other chordates and our nearest non-chordate deuterostome relatives.

KEY POINT

Chordates have a notochord, a rudimentary cartilaginous skeleton running dorsally from head to tail. A nerve cord lies parallel, adjacent, and dorsal to the notochord. Vertebrates retain the notochord during early embryonic development (vertebrates are also chordates) but replace it with the spinal cord.

The *C. intestinalis* genome has approximately 160 Mbp, about 1/20 the size of the human genome. From the initial sequence determined, approximately 16 000 proteins were adduced. This is comparable to invertebrates, and lower than typical vertebrate repertoires. The genes are tightly packed (7.5 kb/gene, compared to 100 kb/gene for humans).

Because of the position of *Ciona* in the evolutionary tree, it is of interest to analyse its genes to determine in which species homologues appear.

- Almost 60% of the genes have homologues in *C. elegans* and/or *D. melanogaster*. These represent genes shared by species ancestral to both invertebrates and chordates.
- A few genes look more similar to genes from worm and/or fly than to vertebrate genes. It is likely that these are vestiges of the common ancestor lost in the lineage leading to vertebrates. An example is

Figure 8.4 Adult *Ciona intestinalis.*
Photo by Andrew Martinez in Cañestro, C., Bassham, S., & Postlethwait, J.H. (2003). Seeing chordate evolution through the Ciona genome sequence. Genome Biol. 4: 208. https://doi.org/10.1186/gb-2003-4-3-208. Springer Nature.

the gene for haemocyanin, the oxygen carrier in many invertebrates. *Ciona* also contains genes for four globins, the vertebrate oxygen carrier.

Some haemocyanins also have enzymatic activity, such as phenoloxidases, enzymes which convert monophenols to diphenols, and/or diphenols to *o*-quinones. These are defence mechanisms: activation of phenoloxidase activity to produce reactive quinones, which contribute to the inactivation of foreign organisms.

Indeed, one system carefully looked for in *Ciona*, but conspicuous by its absence, is an adaptive immune system. *Ciona* does not appear to have genes for immunoglobulins, T-cell receptors, or MHC proteins. This appears then to be a later invention, by vertebrates. The characteristic molecules are absent from even the primitive jawless vertebrates—lamprey and hagfish.

- A fifth of the genes have no apparent homologue in vertebrates or invertebrates. It is likely that homologues will be discovered—perhaps when the protein structures are determined—or they may be very highly diverged within the urochordate lineage.

Some *Ciona* proteins carry out functions specific to urochordates. The urochordate body is surrounded by a 'tunic', made of fibrous cellulose-like polysaccharide. (Tunicates is another name for this group.) *Ciona* contains enzymes for synthesis of cellulose, and endogluconases, which degrade cellulose. Cellulose as a structural material is unusual in organisms other than plants and bacteria. There is evidence that the last common ancestor of urochordates acquired the cellulose synthase gene by lateral transfer from bacteria. Cellulose *degradation* is more widespread. The endogluconases of *Ciona* are most similar to homologues in animals that express proteins that digest cellulose, such as termites and some cockroaches.

The genome of the pufferfish (*Tetraodon nigroviridis*)

Tetraodon nigroviridis is a freshwater pufferfish (see Figure 8.5). Ancestors of humans and fish parted company 450 million years ago. Comparing the genome of *T. nigroviridis* with other vertebrate genomes should therefore reveal defining properties of vertebrates.

The *T. nigroviridis* genome is about 340 Mb in length, just over double that of *Ciona*, and an order

Figure 8.5 *Tetraodon nigroviridis* (common name: spotted green pufferfish).
Photo by Starseed on Wikimedia Commons. Made available under CC BY-SA 3.0 DEED Attribution-ShareAlike 3.0 Unported.

of magnitude smaller than the human. Contributing to the compactness are a relative paucity of repetitive transposable elements, and shorter introns and intergenic regions. Forty per cent of the genome is protein coding! Approximately 28 000 protein-coding genes have been identified.

The *T. nigroviridis* genome strikingly illuminates the large-scale structure of vertebrate genomes. First, there is evidence for whole-genome duplication. Chromosome rearrangements have complicated what might originally have been a simple pattern. There remains, however, a considerable degree of synteny between groups of paralogous genes on different chromosomes. These common syntenic blocks *within* the *T. nigroviridis* genome arose by whole genome duplication followed by chromosome rearrangement.

Second, it is possible to map syntenic groups between *T. nigroviridis* and human. Figure 8.6 shows reciprocal maps. Consider Figure 8.6a. Imagine each human chromosome coloured a constant separate colour; for example, colour human chromosome 2 pink. Then for each gene on human chromosome 2, find homologues on *T. nigroviridis* chromosomes, and colour them pink also. The large blocks of pink in *T. nigroviridis* chromosome 2 indicate long blocks that are syntenic with human chromosome 2. Of course, some of human chromosome 2 appears elsewhere in the *T. nigroviridis* karyotype; for instance, at the top of *T. nigroviridis* chromosome 3. Figure 8.6b is the reciprocal map: colour the *T. nigroviridis* chromosomes a solid colour, and map them onto the human set.

It cannot be seen in Figure 8.6 directly, but typically one human region aligns with *two* regions in *T. nigroviridis*. The explanation is whole-genome duplication in *T. nigroviridis* but not in human.

Figure 8.7 shows this in more detail. Here Hsa = *Homo sapiens*; human chromosomes are numbered Hsa1–Hsa22 plus HsaX. Tni = *T. nigroviridis*, and Anc stands for ancestral vertebrate chromosomes. This pattern is very interesting. First, as mentioned, to most regions of the human chromosomes there correspond *two* regions from *T. nigroviridis*.

Moreover, there are strange interleaving patterns within the mapping. This is expanded in two examples, from human chromosomes 16 and X. Each small box represents a gene. The expanded region of human chromosome 16 is a combination of *T. nigroviridis* chromosomes 13 and 15. It is the pattern you

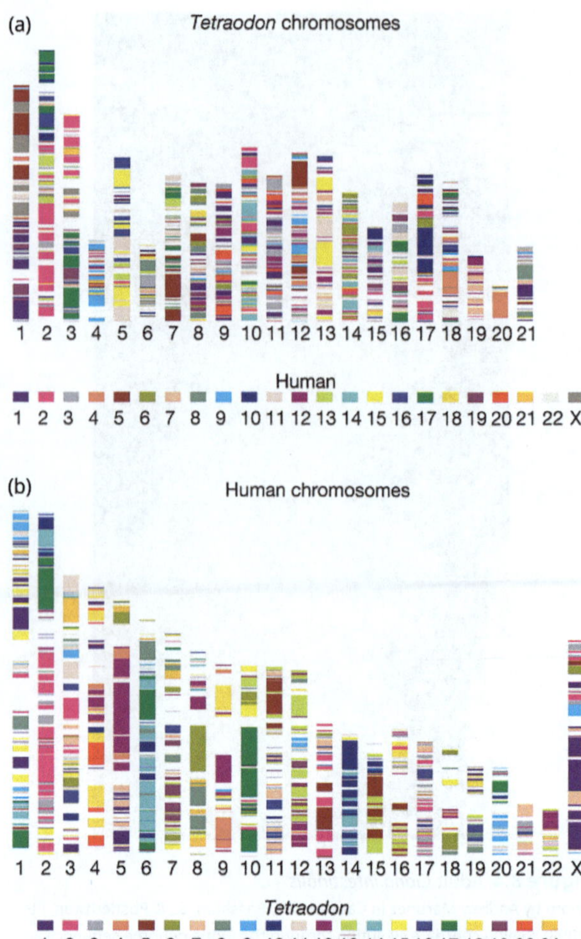

Figure 8.6 Mapping of syntenic blocks between human and *Tetraodon nigroviridis* chromosomes. The definition of synteny is not a very strict one. A synteny is recorded if a grouping of two or more genes in one species has an orthologue on the same chromosome in the other species, independent of order and orientation.

(a) Each coloured band in each of the *T. nigroviridis* chromosomes corresponds to a conserved syntenic block in the human chromosome of the same colour. For instance, *T. nigroviridis* chromosome 2 has many pink areas, indicating extensive relationship with human chromosome 2. *T. nigroviridis* chromosome 17 has many purple areas, indicating relationship with human chromosome 10.

(b) Reciprocal map, showing mapping of *T. nigroviridis* blocks onto human chromosomes. Here the close relationship between human chromosome 10 and *T. nigroviridis* chromosome 17 appears in green.

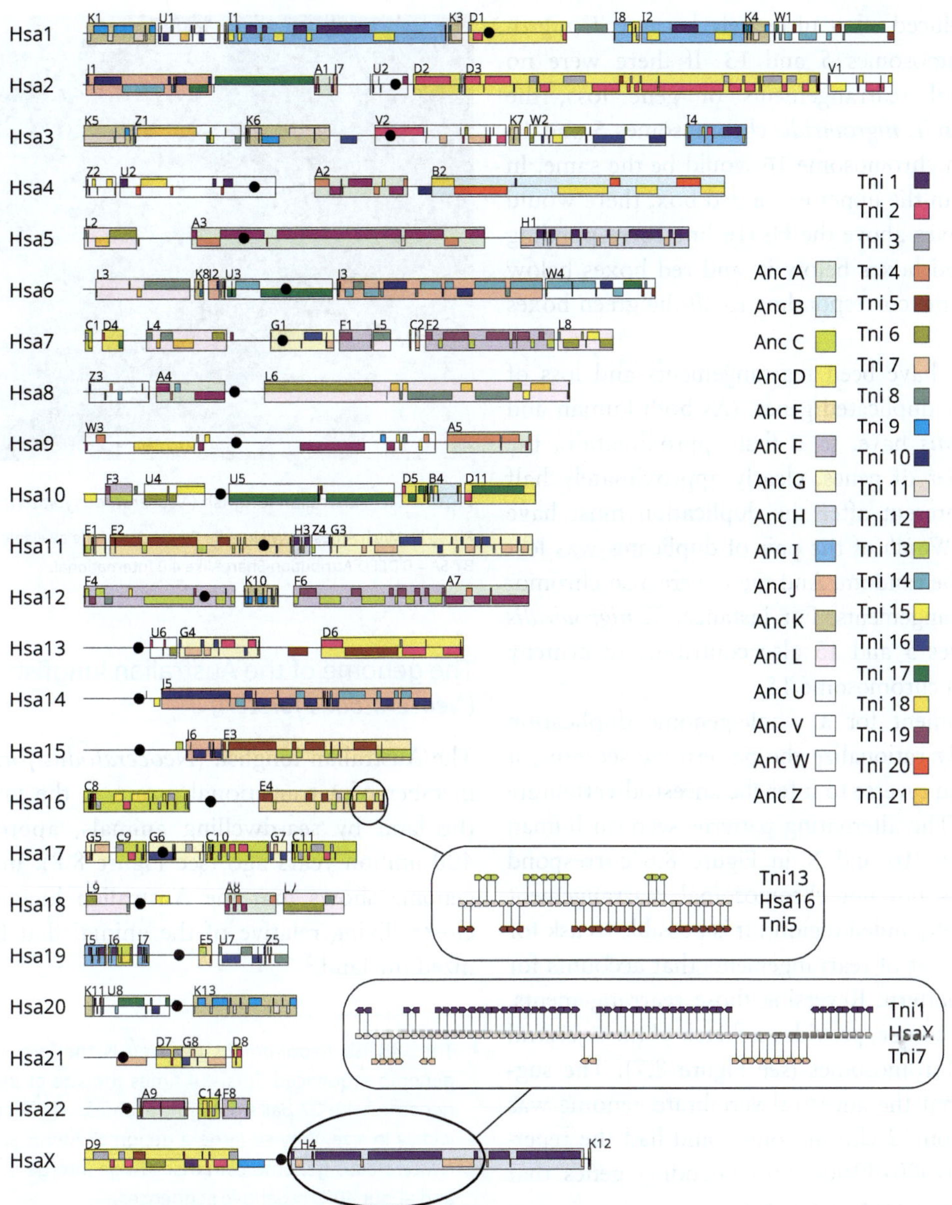

Figure 8.7 More-detailed mapping of syntenic blocks between *T. nigroviridis* and human chromosomes. The two 'blown-up' regions show matches between individual genes. Notice that in general *two T. nigroviridis* regions map to one human one, evidence for whole genome duplication. In the detailed regions, there is an alternation of matches, arising from random loss of one copy of each pair of genes produced by the whole genome duplication. Hsa = *Homo sapiens;* human chromosomes are numbered Hsa1–Hsa22 plus HsaX. Tni = *T. nigroviridis*. Anc = ancestral vertebrate. (Blocks AncU, AncV, AncW, and AncZ contain small amounts of sequence that could not be assigned to the 12 ancestral chromosomes.)

Used with permission of Springer Nature BV, from Jaillon, O., Aury, J.M., Brunet, F., *et al.* (2004). Genome duplication in the teleost fish *Tetraodon nigroviridis* reveals the early vertebrate proto-karyotype. Nature 431: 946–957. https://doi.org/10.1038/nature03025. Permission conveyed through Copyright Clearance Centre, Inc.

would expect from a pack of cards if you started with the red cards in one hand, the black cards in the other hand, and shuffled the deck.

The history that explains this pattern—and for which these observations provide compelling evidence—is that there was a whole-genome duplication in the *T. nigroviridis* lineage, but not in the common ancestor, nor in the human lineage after divergence. The chromosomes duplicated in an ancestor of *T. nigroviridis*, *after* divergence from the lineage leading to humans.

This produced what ultimately became *T. nigroviridis* chromosomes 5 and 13. If there were no chromosomal rearrangements or gene loss, the matchings of *T. nigroviridis* chromosomes 5 and 13 with human chromosome 16 would be the same. In Figure 8.7, in the upper expanded box, there would be green boxes above the Hsa16 line corresponding to *all* the red boxes below it, and red boxes below the Hsa16 line corresponding to *all* the green boxes above it.

But there have been rearrangements and loss of many of the duplicated genes. (As both human and *T. nigroviridis* have, to a first approximation, the same number of genes, clearly approximately half the genes present after the duplication must have been lost.) *Which* of the pair of duplicates was lost appears to be random. And there were also chromosomal rearrangements. For instance, *T. nigroviridis* chromosomes 5 and 13 also contribute to synteny with human chromosome 15.

The argument for a whole-genome duplication can not only rationalize the pattern we see now, it can be run in reverse to infer the ancestral vertebrate karyotype. The alternating patterns seen on human chromosomes 16 and X in Figure 8.6 correspond to gene loss but not chromosomal rearrangement (within the expanded region). It is possible to ask for the minimal set of rearrangements that accounts for the entire pattern. Reversing those rearrangements, on paper of course, provides a sketch of the ancestral vertebrate chromosomes (see Figure 8.7). The suggestion is that the ancestral vertebrate genome was distributed on 12 chromosomes, and had the repertoire of ~20 000–30 000 protein-coding genes that are typical of extant vertebrates.

KEY POINT

The pufferfish is a vertebrate. At least three-quarters of its genes have human homologues. By comparison of its genome with humans, it is possible to reconstruct the ancestral vertebrate karyotype.

[2] ... a process which many Australian humans enthusiastically reverse.

Figure 8.8 Australian lungfish (*Neoceratodus forsteri*).

Photo by Mitch Ames on Wikimedia Commons. Made available under CC BY-SA 4.0 DEED Attribution-ShareAlike 4.0 International.

The genome of the Australian lungfish (*Neoceratodus forsteri*)

The Australian lungfish (*Neoceratodus forsteri*) is a member of a transitional group in the invasion of the land by sea-dwelling animals, approximately 400 million years ago (see Figure 8.8). In fact, the genome shows that the Australian lungfish is the closest living relative of the animal that first colonized the land.[2]

The >43 Gb *Neoceratodus* genome is the largest animal genome sequenced. It is ~14 times the size of the human genome, with 27 pairs of chromosomes, which differ very widely in size. Why so large a genome? About 50% of the *Neoceratodus* genome comprises huge intergenic regions, and about 40%, repetitive sequences.

The genome of the marbled lungfish (*Protopterus aethiopicus*) is even larger: 130 Gbp. Even this is less than the genome of the flowering plant *Paris japonica*, ~150 Gbp. It is estimated that if the DNA from one cell of *P. japonica* were laid out end-to-end, it would be 100 m long. It would take >9.58 seconds for a human to run that distance (Usain Bolt's current world record).

Obvious developments required for emerging from the sea include—but are not limited to—lungs, for air breathing, and conversion of fins to limbs with digits, initially for locomotion on the ground, later for grasping.

How were the required adaptations achieved?

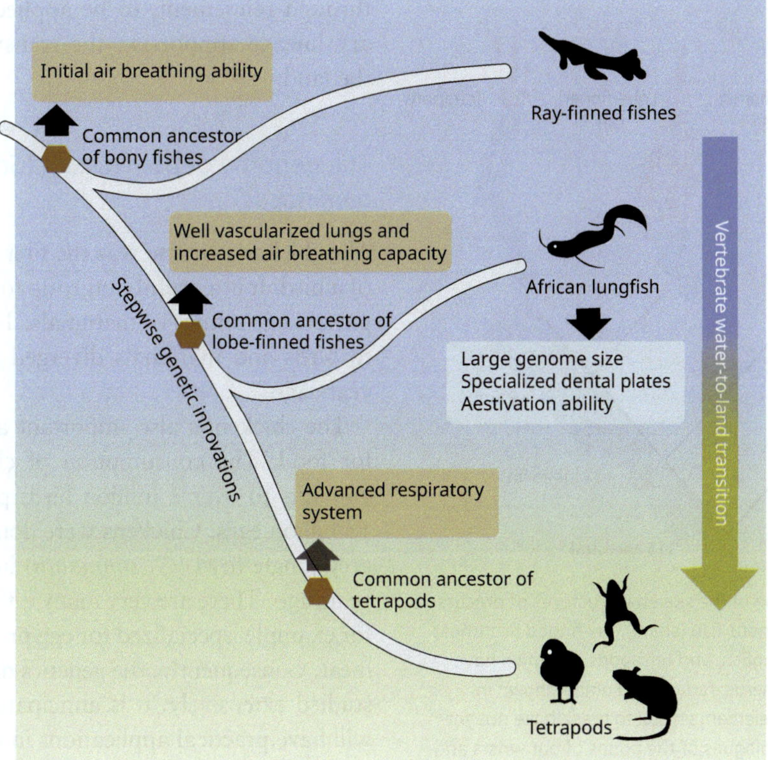

Figure 8.9 Development of lungs from ray-finned fishes, lobe-finned fish—including lungfish—and tetrapods. A few true fishes are air-breathing. The pirarucu (*Arapaima gigas*) is an *obligate* air-breather. It does *not* have a lung.

Aestivation refers to dormancy when confronted with hot, dry environments. Rehydration restores normal activity. Unlike certain other lungfishes which can survive complete dehydration for years, the Australian lungfish can live out of water only for several days, and provided its body is kept moist.

Used with permission of Elsevier Science & Technology Journals, from Wang, K., Wang, J., Zhu, C., *et al.* (2021). African lungfish genome sheds light on the vertebrate water-to-land transition. Cell 184, 5: 1362–1376.e18. https://doi.org/10.1016/j.cell.2021.01.047. Permission conveyed through Copyright Clearance Centre, Inc.

1. Breathing air

Lungfish have a primitive single (*not* paired) lung, derived from the swim bladder, the organ used for buoyancy in bony fishes (see Figure 8.9).

In humans and other animals, the secreted protein sonic hedgehog (SHH) functions as a critical morphogen during embryonic lung development. SHH is expressed strongly in *Neoceratodus* embryos, producing lung development similar to amphibians. Expansion of two other families of genes contribute to adaptation to life on land: lung surfactant proteins (molecules that lower surface tension to prevent alveolar collapse during exhalation), and airborne odorant-receptor genes (with a reduction in some receptor classes for waterborne odour).

Reduced pulmonary surfactant levels are typical in COVID-19 patients, contributing to respiratory distress. The virus kills the alveolar type II cells that produce the surfactants.

2. Anatomically, the Australian lungfish has bones homologous to the humerus, radius, and ulna of our arms. These are *not* present in bony fishes (see Figure 8.10). The skeletal similarities do not extend distally: the lungfish has a fan of bones but no wrist or digits (see Figure 8.11).

'Fish fingers' are products of kitchens, not evolution.

Nevertheless, a gene responsible for development of the hand in tetrapods is also activated in the

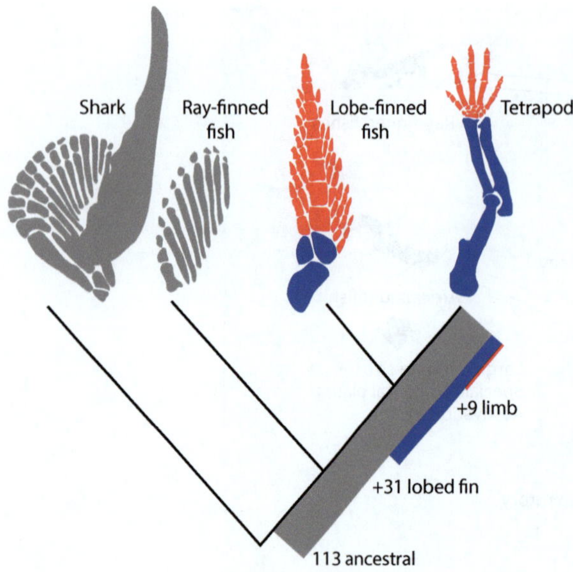

Figure 8.10 Comparison of the skeletal structures of precursors of our limbs, in cartilaginous fish (shark), ray-finned fish (most ocean-dwelling fish), lungfish, and tetrapods. Lungfish have homologues of our humerus, radius, and ulna—shown in blue in this figure—but elements distal to these have not yet differentiated into homologues of the bones of our wrists and digits, which do appear in tetrapods.

Meyer, A., Schloissnig, S., Franchini, P., *et al.* (2021). Giant lungfish genome elucidates the conquest of land by vertebrates. Nature 590: 284–289. https://doi.org/10.1038/s41586-021-03198-8]

homologous skeletal domain of the lungfish. Differential regulation of genes involved in digit development could explain changes between the anatomy of fish fins and that of tetrapod limbs with fingers and toes.

Indeed, Hawkins, Henke, and Harris[3] have shown that zebrafish with mutations in regulatory pathways develop novel bones in the fin, indicating *latent* developmental programmes normally unimplemented in wild-type zebrafish and other teleost fishes. Hawkins *et al.* are careful *not* to claim homology between these novel bones in mutant zebrafish and the normal bones of the tetrapod limb. Moreover, the observed mutations in regulatory genes are not observed in tetrapods. Nevertheless, the fact that the general mechanism of skeletal patterning appears already in teleosts suggests that it was available, through refinement, to be applied in the evolutionary lineage supporting the transition to occupying the land.

[3] Hawkins, M.B., Henke, K., & Harris, M.P. (2021). Latent developmental potential to form limb-like skeletal structures in zebrafish. Cell 184: 899–911.

The genome of the chicken (*Gallus gallus domesticus*)

The chicken genome was the first complete sequence of a bird. It is a useful outgroup for study of the comparative genomics of mammals. The lineages leading to birds and mammals diverged about 310 million years ago.

The chicken is also important as an animal raised for food. The consumption of chickens in the UK amounts to over 1 million birds per year, and about 13 billion eggs. Chickens were domesticated from the grey jungle fowl (*G. sonneratii*) in Asia, about 4000 years ago. There are very many different breeds; some, for example, specialized for egg production, others for meat. Consequently, the genetics of chickens has been studied extensively. It is anticipated that the genome will have practical applications in food production.

The chicken has also been a popular laboratory animal. It has contributed to research in developmental biology, virology, immunology, and cancer.

At ~1.2 Gbp, the chicken genome is substantially smaller than most mammalian genomes. However, it contains approximately the same number of genes. The chicken has 38 autosomes and one pair of sex chromosomes, called Z and W. Like other birds, but different from mammals, females are heterogametic (ZW) and males are homogametic (ZZ). About 90% of the 1.05 Gb of assembled sequence was anchored to its proper chromosome location. It is interesting that the synteny between human and chicken is more conserved than between human and mouse.

Why is the chicken genome smaller than the human and other mammalian genomes? The chicken genome is relatively poor in interspersed repeats and pseudogenes. Expansion of many gene families is greater in mammalian genomes.

To try to extract a 'core' set of vertebrate proteins, comparison of human, pufferfish (*Takifugu rubripes*), and chicken genomes exposes a common gene set. Approximately 7000 protein-coding genes from the chicken have orthologues in both pufferfish and human. These are likely to implement common necessary functions, and one expects to find them in

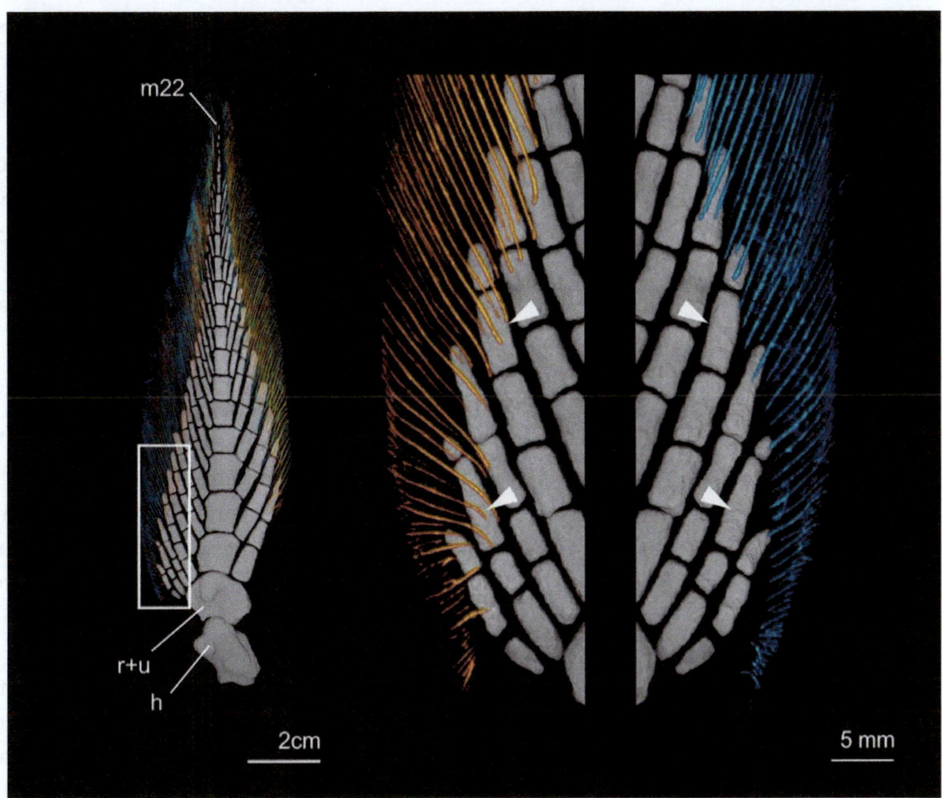

Figure 8.11 Australian lungfish fin resembles tetrapod limbs in homologues of the humerus, radius, and ulna, but, distally, with a fan of bones rather than a wrist and digits. (h = humerus, r + u = fused radius and ulna, m22 = the 22nd and distal-most mesomere, or fin ray bone.) In the adult the radius and ulna are fused, but are separate at earlier developmental stages.

Stewart, T.A., Lemberg, J.B., Taft, N.K., *et al.* (2020). Fin ray patterns at the fin-to-limb transition. Proceedings of the National Academy of Sciences of the United States of America 117, 3: 1612–1620. https://doi.org/10.1073/pnas.1915983117

most higher vertebrates. These common genes are expressed in many different tissues. This is another typical signature of a gene that is not rapidly evolving for a lineage-specific function.

The next step in this line of enquiry would be to determine which of these genes are also expressed in primitive vertebrates and invertebrates. This will define a higher vertebrate common core gene repertoire.

Comparing chicken and human only, about 60% of the ~23 000 chicken protein-coding genes have unique human homologues. Whereas such pairs between human and mouse or rat show an average of 88% sequence conservation, between human and chicken this drops, albeit only modestly, to 75.3%. Proteins with different classes of function differ in degree of conservation with human homologues: transport proteins are more highly conserved than average, and proteins of the immune response show only 60% sequence conservation.

The chicken has some avian-specific proteins. These include one family of keratins, which in chickens form feathers; mammals have expanded a different family, to form hair. Chickens have genes for avidin, a protein appearing in the egg whites of reptiles, amphibians, and birds. The very strong binding of biotin to avidin $\left(K_D \sim 10^{-15}\right)$ has been applied in the laboratory for protein purification. What is its natural function? It is thought to protect eggs from bacteria, which require free biotin as a cofactor in numerous reactions.

Conversely, some human genes that chickens lack are:

- milk proteins such as casein
- enamel proteins, associated with loss of teeth in the bird lineage subsequent to Archaeopteryx, a primitive bird that did have teeth
- vomeronasal receptors. The vomeronasal system is a secondary chemosensory or odour detection

system, that appears in many vertebrates, including humans, fish, reptiles, and others. Its absence from chickens and other birds signifies a loss in the avian lineage, rather than an invention in the mammalian one.

The genome of the platypus (*Ornithorhynchus anatinus*)

Extant mammals form a class divided into three orders:

monotremes: only the platypus and two species of echidna

marsupials: kangaroos, opossums, koalas, and many others, including all mammals native to Australia and New Guinea

placentals: all other mammals, including humans.

The platypus (*Ornithorhynchus anatinus*) is as distant a relative as we have among mammals (see Figure 8.12). Startling to its discoverers, and still to us now, is its mixture of mammalian and reptilian characteristics. Like mammals, the platypus has hair, and nurses its young (it has mammary glands, but not teats—the milk is released through localized pores in the skin, modified sweat glands). Like reptiles it lays eggs; and has venom, delivered by males through ankle spurs. Careful study of the anatomy revealed many other unusual characteristics. For instance, the name monotreme (= single aperture) refers to the single orifice serving both the urogenital and digestive systems.

Figure 8.12 Platypus (*Ornithorhynchus anatinus*), one of the three extant species of monotremes.
Photo by Stefan Kraft on Wikimedia Commons. Made available under CC BY-SA 3.0 DEED Attribution-ShareAlike 3.0 Unported.

An unusual sensory capacity is electroreception, the ability of the platypus to perceive electrical impulses. The platypus can locate and catch prey through use of a combination of mechano- and electroreceptors in its bill. A platypus will attack a battery immersed in water in the dark.

Study of the molecular biology of the platypus revealed other surprises, including ten sex chromosomes (males are always XYXYXYXYXY). However, the sex determination system is closer to that of birds than of most mammals. Whereas in marsupials and placental mammals, the primary locus for sex determination is SRY, a gene on the Y chromosome that encodes a transcription factor, the platypus lacks this gene.

The closest to an extant reptile–mammal transitional form that we have, the platypus has offered us the chance to see how the basic distinctive features of mammals originated.

The platypus's was the first monotreme genome sequenced. It has 2.2 billion base pairs. 18 527 protein-coding genes were identified. Not unexpectedly, a majority of these have orthologues in opossum (a marsupial), human, dog, and mouse (placentals), and even chicken. Of particular interest are genes *not* found in other mammals. Like its anatomy, the platypus genome shows a mixture of mammalian and non-mammalian features. These include:

Odour receptors: the platypus odorant receptor genes are for the most part recognizable homologues of those in other mammals. The repertoire is more akin to that of other mammals than to reptiles. There are roughly half the number of odorant-receptor genes as in other mammals, but this may reflect the animal's aquatic lifestyle.

Milk: although true milk is unique to mammals, non-mammal animals that incubate their eggs secrete fluids that protect eggs from desiccation and/or infection. However, unlike those primitive precursors, platypus milk resembles that of other mammals. It is a complex mixture with both nutritive and antimicrobial functions.

Eggs: unlike the eggs of marsupials and placental mammals, which are nourished internally by the mother, the platypus lays eggs that contain yolk. Common to the yolks of eggs of fish, amphibians, reptiles, birds, and most invertebrates is the protein vitellogenin. Vitellogenin is the precursor of the lipoproteins and phosphoproteins that are major protein components of egg yolk. Moreover, vitellogenins are not restricted to eggs. For instance, bees use it as a food store also. Vitellogenin genes were lost in the lineages leading to marsupials and placental mammals, but retained in the monotremes.

In other mammals, the placenta became the locus of embryonic development, and the mother supplied nutrients. Monotremes have a primitive form of placenta, called a yolk-sac placenta.

Platypus eggs do not have a typical calcium carbonate hard shell, but a leathery protein envelope.

Venom: venom is one of those evolved traits that has proved useful to a variety of species, including monotremes and reptiles. Platypus venom is a complex cocktail containing proteins evolved by duplications of genes with other functions. However, although there are some biochemical features common to platypus and snake venoms, there is evidence that they developed independently. The enlistment of defensin-like peptides as components of venom is an example of convergent, or at least parallel, evolution (see Figure 8.13). (In convergent evolution, ancestors are less similar than descendants. In parallel evolution, ancestors and descendants differ, but in similar ways.)

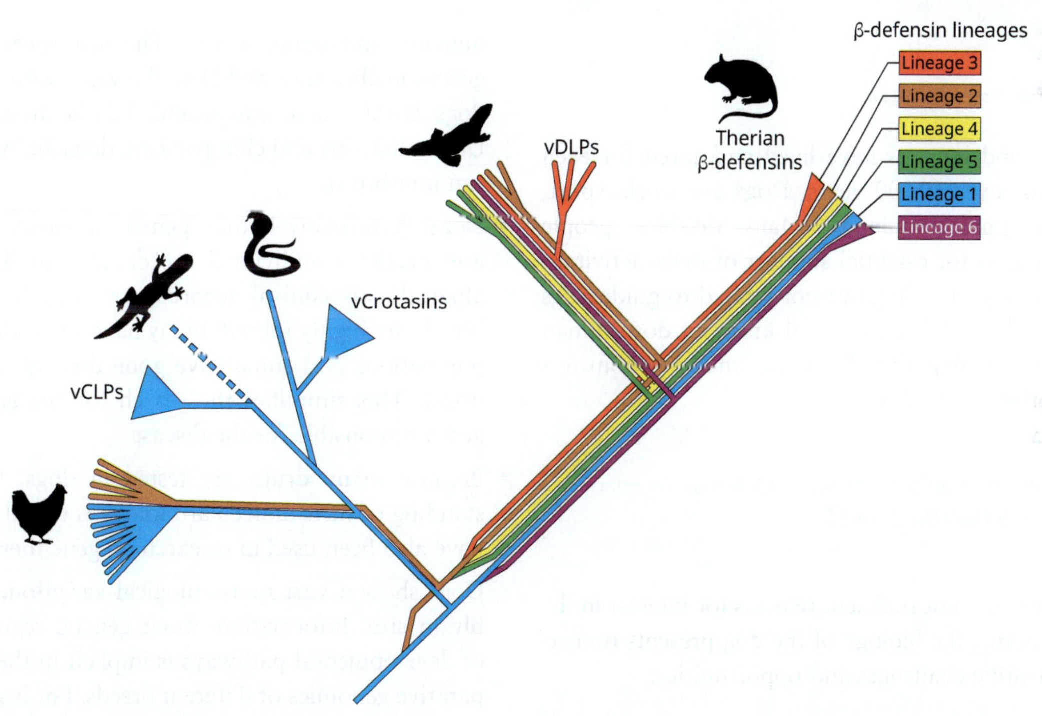

Figure 8.13 Evolutionary tree and points of gene duplication of defensins in birds, reptiles, platypus, and therians (= marsupial + placental mammal). Defensins are a group of families of small proteins found in a variety of vertebrate and invertebrate species. The therian molecules are not components of venom. They have antibacterial activity, generally functioning by forming pores within the microbial cell membrane, allowing cell contents to leak out.

A class of venom defensin-like proteins (vDLPs) has arisen independently in several lineages, including reptiles and platypus. In *Crotalus* snake venoms, the vDLPs are neurotoxins affecting voltage-gated sodium channels. The mechanism of action of the vDLP in platypus venom is still unknown.

KEY POINT

Platypus anatomy combines reptilian and mammalian features. Analysis of its genome also reveals these characteristics of a transitional form. However, there are some details that appear only from detailed sequence information; for instance, that the defensin homologue in platypus venom is not retained from reptiles, but convergently evolved.

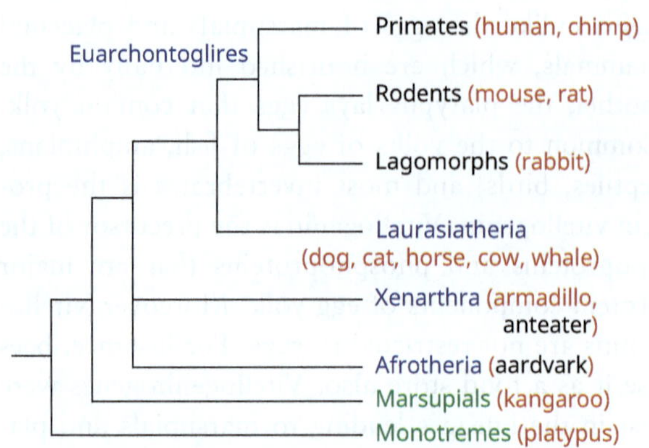

Figure 8.14 Phylogeny of mammals, showing monotremes and marsupials (green) and the four major groups of eutherian mammals: Euarchontoglires, Laurasiatheria, Xenarthra, and Afrotheria (blue). Human, chimpanzee, mouse, and rat are all Euarchontoglires. Dogs belong to the Laurasiatheria. Complete genome sequences are known for all example species shown in red.

The genome of the dog (*Canis lupus familiaris*)

Used with permission of V.E. Lesk.

Dogs and humans have lived and cared for each other for over 10 000 years. Dogs are work, sport, and companion animals. Many disabled people rely on dogs for essential support of daily activities, prominently including but not limited to guide dogs for the blind. All readers will know of dog–human partnerships that rival human–human relationships in emotional intimacy.

According to tradition, Romulus and Remus, founders of Rome, were suckled by a wolf.

Were those not sufficient reasons for interest in the dog genome, the biology of the dog presents numerous scientific challenges and opportunities.

- The dog is an outgroup of the Euarchontoglires, a set of mammals that includes humans (see Figure 8.14).
- Dogs are an ideal species in which to study domestication. To a far greater extent than other genera, dogs and their relatives offer both a variety of inbred populations—the different breeds—and corresponding wild populations. The genomes of dogs and wolves are much closer than those of humans and chimpanzees. The sequence divergences in chromosomal DNA between wolves and dogs is 0.04% in exons and 0.21% in introns. Unlike humans and chimpanzees, dogs and wolves can interbreed.
- Dogs share many human genetic diseases. Many are specific to individual breeds, and good genealogical and clinical records are available. The breeds are highly inbred: many have small founder populations and some have gone through bottlenecks. This simplifies the search for the gene or genes responsible for the disease.
- Because many drugs are tested in dogs, understanding of their molecular biology is useful. Dogs have also been used in research on gene therapy.
- Dogs show a vast morphological variation, notably in size. Information about genetic regulation of developmental pathways is implicit in the comparative genomics of different breeds. For instance, a single mutation controls breadth of skull and shortness of face. In humans, mutation in the homologous protein is responsible for Treacher Collins syndrome, a developmental disorder affecting the skull and face.

Different breeds also vary generally in personality traits, providing an opportunity to identify genes for aggressiveness and passivity.

History of the dog

The order Carnivora, to which domestic dogs and cats belong, originated during the Palaeocene, ~60 million years ago (see Figure 8.15). Dog-like carnivores are known from fossils from 40 million years ago. The current closely related species—wolf, coyote, jackal, and red fox—split off about 3–4 million years ago. The wolf lineage gave rise to the domesticated dog, *Canis familiaris*.

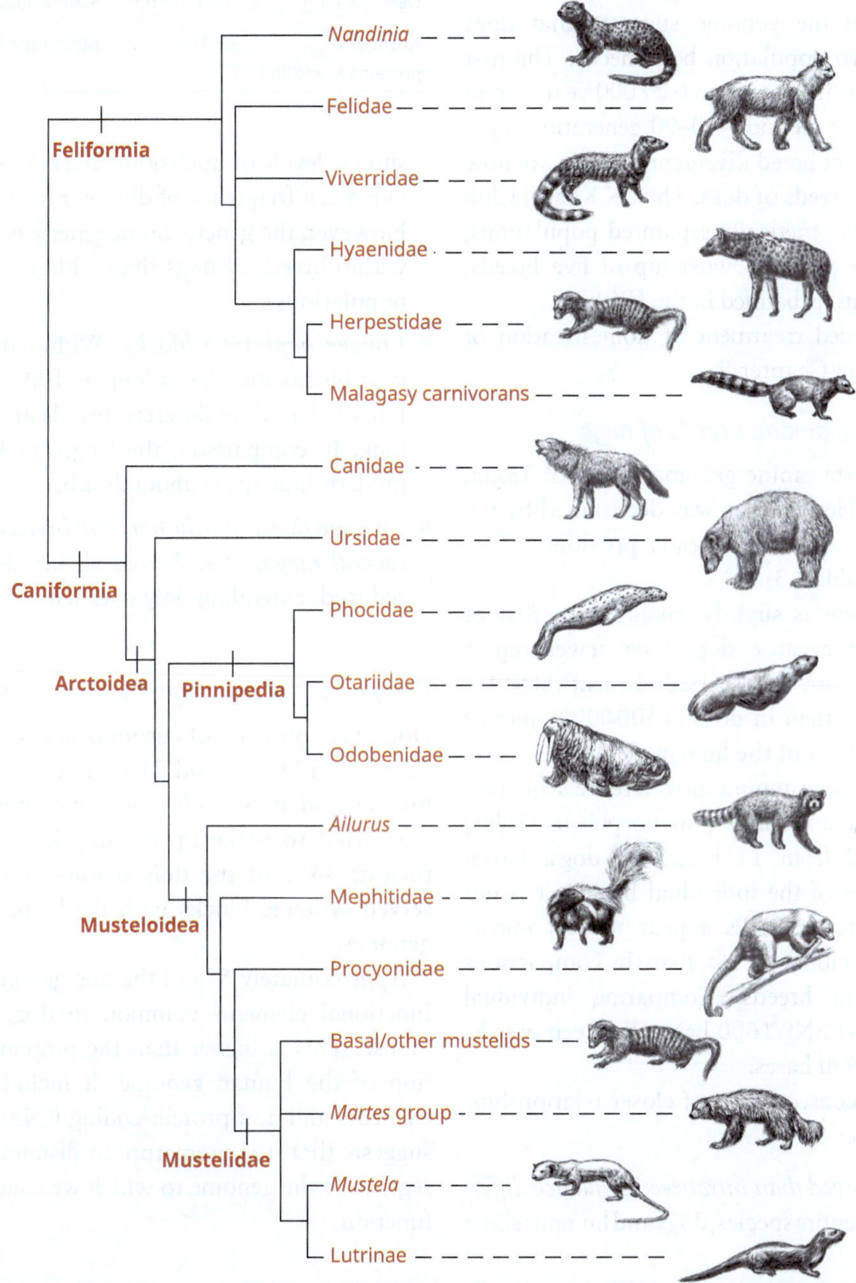

Figure 8.15 Domestic dogs and cats fall into the two main suborders of the order Carnivora. The two lineages split about 48 million years ago. Note that the pictures of the animals are not drawn to scale.

Used with permission of Oxford University Press, from Flynn, J.J., Finarelli, J.A., Zehr, S., *et al*. (2005). Molecular phylogeny of the carnivora (Mammalia): Assessing the impact of increased sampling on resolving enigmatic relationships. Systematic Biology 54, 2: 317–337. https://doi.org/10.1080/10635150590923326 Permission conveyed through Copyright Clearance Centre, Inc.

Domestication of dogs is recorded in archaeological artefacts 14000–15000 years old, but probably took place much earlier. The first colonists of North America, who came across the Bering Strait about 20000–15000 years ago, brought domesticated dogs with them.

Evidence from the genome suggests that dogs went through two population bottlenecks. The first occurred ~9000 generations ago (~27000 years) upon domestication. The second, ~30–90 generations ago, signals the origin of breed divergence. There are now about 300–1000 breeds of dogs. The UK Kennel Club recognizes 222 as genetically separated populations, with closed gene pools. (Ownership of five breeds, deemed dangerous, is banned in the UK.)

A more extended treatment of domestication of the dog appears in Chapter 9.

Genome variation among breeds of dogs

The most complete canine genome is that of Tasha, a female boxer. Her genome was determined by the shotgun method, with 31.5 reads providing ~7.5-fold coverage (Table 8.3).

Table 8.3 The dog genome

Feature	Value	Comment
Number of chromosomes	39 pairs	More than humans
Genome length	2.4×10^9 bp	Slightly less than humans
Number of proteins identified	37774	More than humans

The dog genome is slightly smaller than that of humans, in part because dogs have fewer repeat sequences. The short, interspersed element (SINE) is longer in humans than in dogs (1500000 copies of SINEs make up 13% of the human genome).

In addition to determining the reference sequence, 2.5 million single-nucleotide polymorphisms (SNPs) were determined from 11 breeds of dogs. Given the inbred nature of the individual breeds, it is not surprising that fewer SNPs appear when comparing individuals within breeds than in comparisons amongst different breeds. Comparing individual boxers, there is ~1 SNP/1600 bases. Between breeds, there is ~1 SNP/900 bases.

Concomitant consequences of closer relationships within breeds are:

- *Greater interbreed than intrabreed sequence difference.* Over the entire species, dogs and humans show similar levels of nucleotide diversity between individuals: a frequency of different bases of $\sim 8 \times 10^{-4}$. However, the genetic homogeneity is much greater within breeds of dogs than within distinct human populations.
- *Longer haplotype blocks.* Within breeds, haplotype blocks may be as long as 100 kb. Haplotype blocks shared by different breeds are about 10 kb long. In comparison, the length of haplotypes in modern humans is about 20 kb.
- *Linkage disequilibrium within breeds extends over several megabases.* Across all breeds it is greatly reduced, extending only over tens of kilobases.

Comparison of dog, human, and mouse genomes

Dogs have 39 pairs of chromosomes compared with 23 pairs in human and 21 in mouse. Therefore, the human and mouse chromosomes must have been reassorted to make up the dog karyotype. Nevertheless, 94% of the dog genome appears in conserved syntenic blocks with the human and mouse genomes.

Approximately 5% of the dog genome constitutes functional elements common to dog, human, and mouse. This is higher than the protein-coding fraction of the human genome. It includes regulatory elements and non-protein-coding RNAs, and further suggests that it is premature to dismiss as 'junk' the regions of the genome to which we cannot yet assign function.

Palaeosequencing—ancient DNA

Recovery of DNA from ancient samples

The recovery and sequencing of DNA from extinct species offers us a window onto evolutionary history. Source material includes fossils collected from their deposits, mummified samples, eggshells in the subfossil state (it was even possible, by testing DNA from the outer surface of moa shells, to show that *male* birds were responsible for incubating the eggs), preserved seeds or other plant material, specimens from museums, and clinical collections of pathogens. For instance, there are repositories of samples of influenza virus dating back almost a century.

Although often only minuscule amounts of material are available, PCR amplification can produce reasonable quantities for sequencing.

Like other biological material, DNA degrades after death, unless it is preserved. The best protection is to exclude liquid water. This can occur if the samples are frozen, or desiccated by heat, or if sequestered within compartments such as teeth, bone, or hair. Because DNA from ancient samples is usually present in only microscopic quantities, contamination is a serious danger. Contaminants can be microbial, or human from the scientists handling the specimens. Even without contamination, samples suffer from fragmentation, and from chemical change resulting in sequence changes. The most common is deamination of cytosine to uracil.

However, advances in isolation methods can reduce further damage during the extraction phase, and careful technique can exclude contamination by scientist DNA. This research puts wide-ranging demands on the personalities of its practitioners: Paleosequencers must 'rough it' during field work; but apply unusually painstaking care in handling samples, both in the field and back in the lab. A speciality practised by few, but of interest to many.

DNA from extinct birds

The moas of New Zealand

In the absence of terrestrial mammals, New Zealand's largest animals were flightless birds. The moas ranged in size up to 3 m tall, 300 kg giants (Figure 8.16). They became extinct after the arrival of human settlers from Polynesia.

Moas are ratites, an order of birds that also includes the New Zealand kiwi, the ostriches of Africa, the emu and cassowary of Australia and New Guinea, the rhea in South America, and the extinct elephant bird of Madagascar (*Aepyornis maximus*): This monster was 3.3 m tall, and weighed 450 kg!

The taxonomy of moas has been difficult to resolve by classical methods, in the absence of living examples. Mitochondrial DNA sequences from 29 specimens produced a phylogenetic tree comprising nine species,

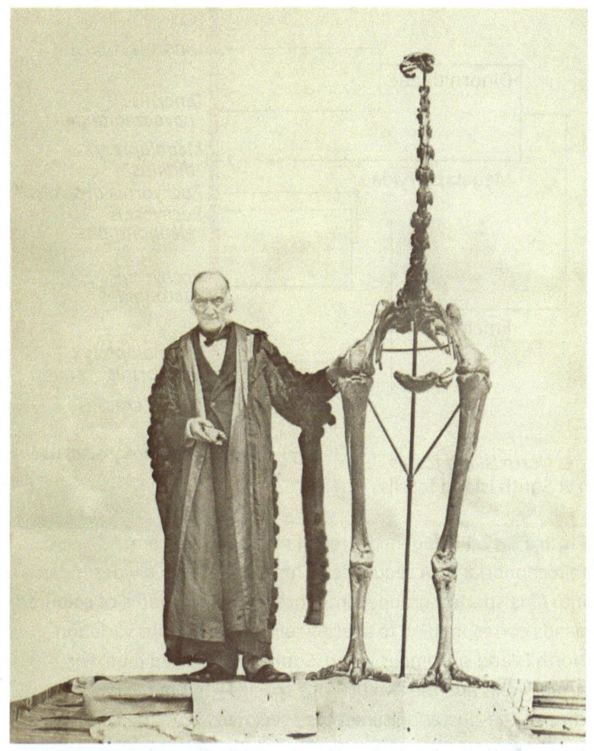

Figure 8.16 Photograph of an assembled moa skeleton, of *Dinornis novaezealandiae*, towering over celebrated nineteenth-century anatomist Sir Richard Owen. Based in London, Owen was the recipient of many interesting specimens discovered in the far reaches of the Empire, including the bones of moas, and a preserved specimen of the platypus. He was the driving force behind the establishment of the British Museum of Natural History in South Kensington.

The first specimen from a moa to reach Owen was a 15-cm fragment of bone. Owen correctly identified it as coming from the femur of a giant bird. In this photograph, taken in 1877, Owen is holding the original fragment in his right hand. With his left hand, he is indicating its position within the full skeleton, discovered and assembled later.

Plate XCVII of Owen and a Moa Skeleton, Owen, R. (1879). *Memoirs on the Extinct Wingless Birds of New Zealand*, Vol. 2. London: John van Voorst.

grouped into three families: the Dinornithidae, Megalapterygidae, and Emeidae (Figure 8.17). The data suggest a separation date of the Emeidae from the other two of 5.27 million years ago, and more copious recent radiation within the last 2 million years.

It is interesting to correlate the estimated divergence times of the species with the geological history of New Zealand. Today, the North and South Islands are separated by Cook Strait, 23 km across at its narrowest. But, approximately 30–21 million years ago, higher sea levels reduced New Zealand to

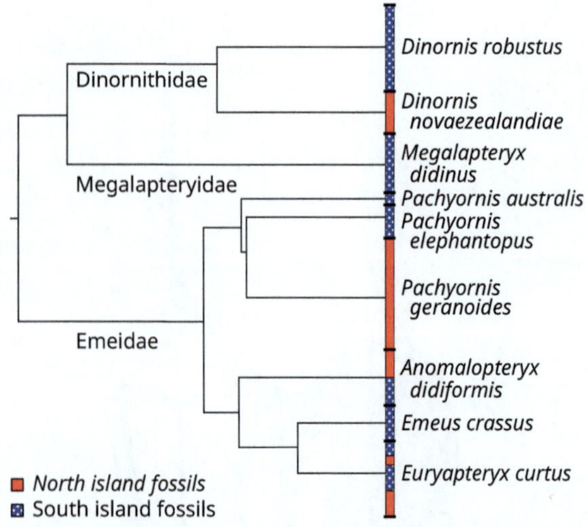

Figure 8.17 Phylogenetic tree of moa species, from mitochondrial DNA sequences. This classification divides moas into nine species, grouped into three families. Widths of coloured bands corresponding to species reflect intraspecies variation. North Island specimens in red, South in patterned blue. For *A. didiformis* and the two *Dinornis* species—but not *Euryapteryx curtus*—the clustering separates specimens from the two islands.

Bunce, M., Worthy, T.H., Phillips, M.J., *et al.* (2009). The evolutionary history of the extinct ratite moa and New Zealand Neogene paleogeography. Proceedings of the National Academy of Sciences of the United States of America 106, 49: 20646–20651. https://doi.org/10.1073/pnas.0906660106

a few scattered islands. The somewhat larger South Island was isolated from the North, until about 2–1.5 million years ago.

Approximately 8.5–5 million years ago the New Zealand 'alps' formed. This mountain chain runs roughly parallel to the main axis of the South Island. It divides the habitats into wet rainforest on the west, and dry, warmer regions on the east.

The data on divergence of sequences is consistent with the historical geology. The suggested scenario is that the divergence of the major groups took place on the South Island, after the alps formed. When the land links arose, during glaciations, birds began to inhabit the North Island, taking advantage of the new surroundings to generate a new round of divergence (Figure 8.18).

The dodo and the solitaire

The dodo (*Raphus cucullatus*) was a large, flightless bird that inhabited the island of Mauritius, in the Indian Ocean, east of Madagascar (Figure 8.19). It was a large, robust bird, about a metre in height and weighing about 20 kg (about four times the size of a typical family Christmas turkey in the UK). It is common for island species to be either significantly larger or significantly smaller than their mainland counterparts.

Sailors stopping at the island found the dodo easy prey—it could not fly, and lacked appropriate fear of human predators. As a result, the dodo became extinct. The last survivor was shot in 1681.

The solitaire (*Pezophaps solitaria*) was a related bird from a neighbouring island east of Mauritius, Rodrigues. It also became extinct, outliving the dodo by perhaps a century.

Even museum specimens of the dodo are rare. The Oxford University Museum of Natural History had one. It was seen by don Charles Dodgson, who turned it into a character in Alice in Wonderland. Only partially saved from a fire that started during a tidying-up exercise, the Oxford specimen is the only known source of soft tissues from a dodo. What remains now comprises a head, and a leg and foot, each with some skin attached. There are many bones on the tropical island, but conditions are unpropitious for preservation of DNA.

In 2022, Beth Shapiro of the University of California, Santa Cruz, US, achieved complete sequencing of the dodo genome, from DNA extracted from a skull in the collection of Zoological Museum of Copenhagen. The genome of the Rodrigues solitaire has also been sequenced.

A phylogenetic tree constructed from sequence data fixed the taxonomic position of the dodo. It is a pigeon, of the family Columbidae. The closest extant relative of the dodo and solitaire is the Nicobar pigeon (*Caloenas nicobarica*), which lives in Southeast Asia. Shapiro and colleagues are now involved in a project to recover the dodo from extinction, starting from the Nicobar pigeon. The same group is working on the mammoth and the thylacine.

High-throughput sequencing of mammoth DNA

About 20 000 years ago, before the peak of the last ice age, the most abundant large animals were the mammoths. Adapted to the cold, they died out as a result of global warming, hunting by humans, and changes in their preferred habitats.

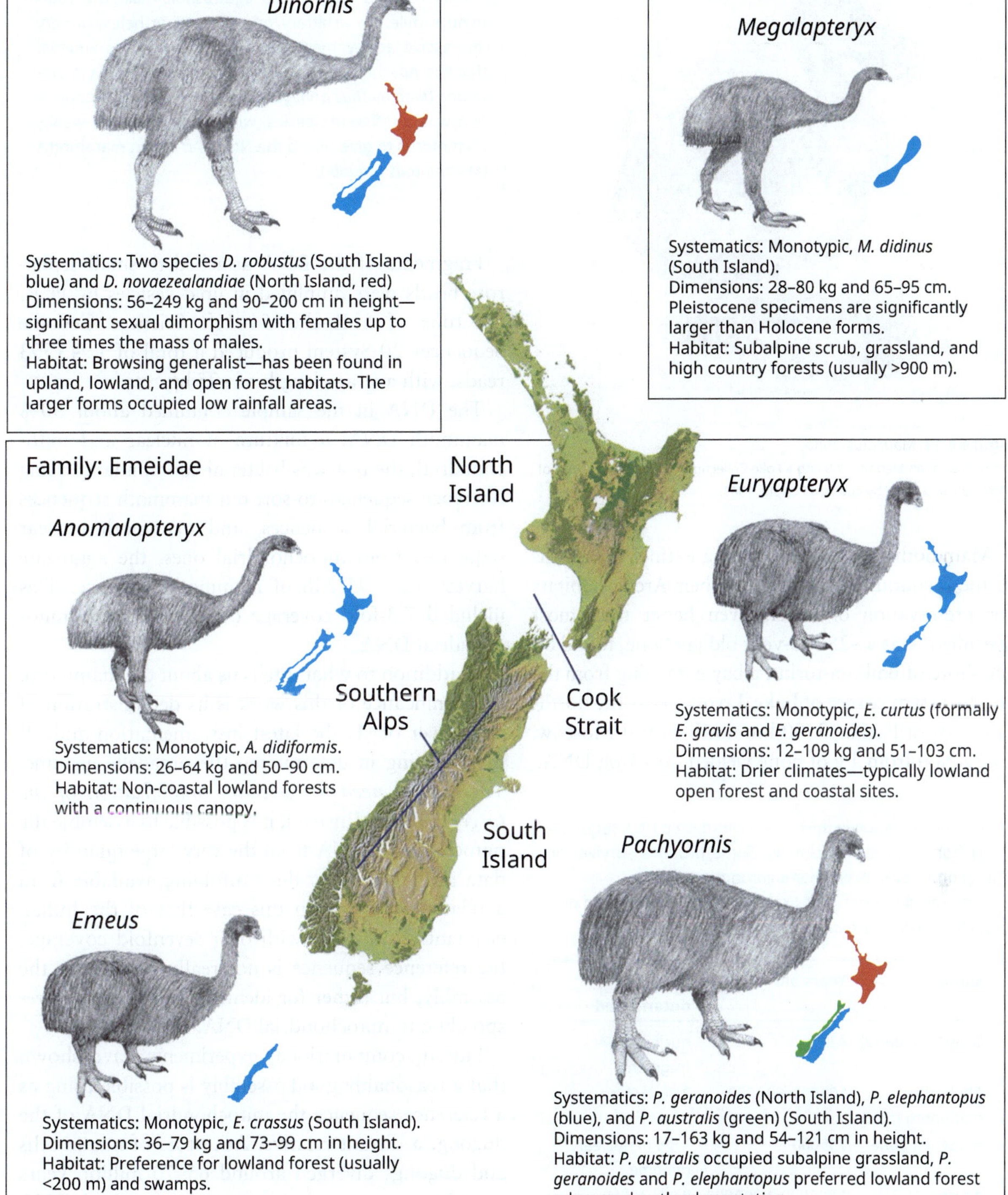

Figure 8.18 Reconstructions, classification, and estimated geographical distributions of species of moa, extinct flightless birds from New Zealand.

Bunce, M., Worthy, T.H., Phillips, M.J., *et al.* (2009). The evolutionary history of the extinct ratite moa and New Zealand Neogene paleogeography. Proceedings of the National Academy of Sciences of the United States of America 106, 49: 20646–20651. https://doi.org/10.1073/pnas.0906660107

Figure 8.19 Mauritius dodo.
From: 'A German Menagerie Being a Folio Collection of 1100 Illustrations of Mammals and Birds' by Edouard Poppig, 1841.

Mammoths are unusual among extinct organisms in the favourable conditions of their Arctic habitats for preservation of DNA. Even better than most specimens was a ~28 000-year-old jawbone, found on the shore of Baikura-turku, a bay extending from the south-eastern corner of Lake Taymyr, near the northern coast of Russia, 3500 km north-east of Moscow. Extraction from 1 g of bone yielded ~0.73 μg DNA.

Good preservation in the cold has permitted DNA sequencing from very old specimens. Some proteins survive for long times even in warmer environments.

In most instances, soft tissue does not survive, and one is left with *'Nature read in tooth and claw'*.

Sample	Years old	Sequence determined
Early Neandertal	430 000	nuclear DNA fragments
Chukochya mammoth tooth	500 000–600 000	3.66×10^6 bp
Horse leg bone	735 000	~ 3% of nuclear genome
Adycha mammoth tooth	1 000 000–1 300 000	8.84×10^8 bp
Krestovka mammoth tooth	1 100 000–1 650 000	4.9×10^7 bp
Ostrich egg shell	3 800 000	protein sequences
Rhinoceros tooth	1 770 000	protein sequences

Sequencing of DNA of mammoth teeth from more than 1 million years ago breaks a threshold—like the four-minute mile, an arbitrary one—but nevertheless a very impressive achievement indeed. The sequences suggest that the Adycha population was ancestral to woolly mammoths (*Mammuthus primigenius*) but the Krestovka population was a different species, which interbred with woolly mammoths to give rise to the North American mammoths (*Mammuthus columbi*).

Fragments of the DNA attached to small sepharose beads were amplified in lipid vesicles by PCR. Six runs of a Roche/454 Life Sciences Genome Sequencer 20 System produced a total of 1 943 593 reads, with average length of ~95 bp.

The DNA in the sample contained about 50% mammoth DNA, a mixture of nuclear and mitochondrial; the rest was bacterial contaminant. Using reference sequences to sort out mammoth sequences from bacterial sequences, and mammoth nuclear sequences from mitochondrial ones, the aggregate harvest was ~95 Mb of mammoth sequence. This included 7.3-fold coverage of the 16 770 bp mitochondrial DNA.

In addition to what it tells us about the mammoth, the significance of this work is its demonstration of the power of (1) the latest instrumentation and (2) resequencing in determining the organelle genome. *There is no need for selective amplification of the target sequence.* Instead, it is possible to assemble the mitochondrial DNA from the very large quantity of data available, given the scaffolding available from a related sequence, in this case that of the Indian elephant. Note that, with over sevenfold coverage, the reference sequence is not really needed for the assembly, but rather for identifying the reads corresponding to mitochondrial DNA.

Indeed, computational experiments have shown that a reasonably good assembly is possible using as a reference sequence the mitochondrial DNA of the dugong, a distant relative. Ancestors of mammoths and dugongs diverged around 65–70 million years ago. Mammoth and dugong mitochondrial DNA sequences are only 75.3% identical. In practice, a reference sequence from a closer relative than the dugong is to mammoth would in most cases be available. Therefore, the dugong–mammoth assembly offers a 'worst-case' analysis.

For the sake of argument, suppose one's interest were limited to the mitochondrial DNA sequence. One might choose to amplify the mitochondrial DNA and sequence that only. Sequencing fragments from all of the DNA is, comparatively, very inefficient in its use of the data produced, as far as determining the mitochondrial sequence is concerned. However, comparisons with other ancient-DNA sequencing projects suggest that it *is* efficient in terms of the amount of precious sample used. For the study of extinct species, this is an overriding consideration.

The mammoth nuclear genome

The mammoth nuclear genome presented a harder problem. It is estimated to be approximately 4.17 Gbp in length, longer than the human genome. Its sequencing depended on finding better samples, and on advances in sequencing technique.

DNA was extracted from hair samples from a Siberian animal, denoted M4, that died about 20 000 years ago. It is interesting that because the DNA is fragmented, the relatively short read lengths of the sequencer were not a great problem. The average read length produced was about 150 bp. This yielded 3.6 Gb of sequence. Combining this with additional sequences determined from other individuals produced a total of 4.17 Gb of sequence. Calibration of error rates suggest an average of 6 errors out of 10 000 bases arising from DNA damage, and 8 out of 10 000 bases from sequencing.

The genome of the African elephant (*Loxodonta africana*) provided a reference sequence. The *L. africana* genome had been sequenced at 7× coverage, and assembled. Alignment of the reads from the mammoth samples, to *L. africana* and to other genomes representing potential contaminants, showed that over 90% of the reads were mammoth DNA, for a total of 3.3 Gb of mammoth sequence.

It was possible to compare amino-acid sequences of mammoth proteins with the orthologues in elephants and other species. The results suggest that mammoth and African elephant differ, on average, in one residue per protein. It is difficult to assign selective or even functional significance to these changes, in general. Even in the cases of residues unique to mammoth compared to a wide spectrum of other placental mammals, the sequence is only the starting point for investigation of the proteins thereby identified as interesting candidates for follow-up studies.

The phylogeny of elephants

Genome sequences have allowed resolution of two problems in elephant phylogeny:

(a) How many species of African elephants are there?

    There are two populations of elephants in Africa, living in the savannah and in the forest. Some authorities have described them as separate species: *L. africana* (savannah) and *L. cyclotis* (forest). Others have considered them a single species, or regard *L. cyclotis* as a subspecies of *L. africana*.

(b) Are mammoths more closely related to African elephants or Indian elephants (*Elephas maximus*)?

Both questions have engendered considerable debate in the relevant specialist literature.

In order to assess phylogenetic relationships among African and Indian elephants and mammoths, the extinct American mastodon provided an outgroup. DNA from a tooth, estimated to be between 50 000 and 130 000 years old, provided 1.76 Mb of mastodon sequence. The corresponding regions from elephants and mammoths were also sequenced.

The data set for analysis contained, from each species, approximately 40 000 bp of sequence from 375 loci. This allowed comparison of the divergences between the groups.

The results showed that the variation between *L. africana* and *L. cyclotis* is approximately the same as between mammoths and Asian elephant (*E. maximus*). Reliance on similarities of DNA sequences can be confusing: Mammoths and Asian elephants are not even in the same genus, but many people have believed that *L. africana* and *L. cyclotis* are the same species! Nevertheless, the conclusion is that mammoths are more closely related to Asian elephants than to African elephants, and that *L. africana* and *L. cyclotis* should be considered as separate species.

The splitting of a species into two has implications for conservation, because in principle splitting would require a separate decision about whether each were endangered. In fact, all world elephant species are endangered.

Returning to the question of what defines species boundaries, it is clear that this distinction was drawn at least primarily on the basis of similarity of DNA sequence. What about the classical biological definition of whether the species hybridize in nature? (In captivity, even African and Indian elephants are fertile.) *L. Africana* and *L. cyclotis* do give rise to hybrids in the Uganda–Congo border region where their ranges overlap. However, as implied by the divergence of the DNA sequences, hybrids are in fact rare, and the two populations do maintain separate gene pools.

LOOKING BACK AND FORWARD

In the last two chapters, we have surveyed selected individual species of viruses, prokaryotes, and eukaryotes. We asked what their genomes tell us about their capabilities and lifestyles, their interrelationships, and their places in the history of life.

Conspicuous by their absence from the chapter about eukaryotes is the human genome itself, and also details of the species with which we most intimately interact, including domesticated animals and plants. In the next chapter, we shall discuss several applications of genomics to human biology and human history, including our extinct relatives such as Neandertal man. Chapter 10 will treat applications of genome sequencing to human health. (This was, after all, the impetus that launched the project and continues to drive it forward.)

RECOMMENDED READING

General discussions of topics in eukaryotic evolution, some in the form of collections of papers

Hirt, R.P., & Horner, D.S., eds. (2004). *Organelles, Genomes and Eukaryote Phylogeny: An Evolutionary Synthesis in the Age of Genomics*. Boca Raton, FL: CRC Press.

On 29 June 2006, The Royal Society held a discussion meeting, major steps in cell evolution: palaeontological, molecular, and cellular evidence of their timing and global effects. The meeting was organized by T. Cavalier-Smith, M. Brasier, & T.M. Embley, and published in Philosophical Transactions of the Royal Society B, volume 361, issue 1470.

Baldauf, S.L. (2008). An overview of the phylogeny and diversity of eukaryotes. *Journal of Systematics and Evolution* 46: 263–273.

Telford, M.J., & Littlewood, D.T.J., eds. (2009). *Animal Evolution: Genomes, Fossils and Trees*. Oxford: Oxford University Press.

Dunn, C.W., Giribet, G., Edgecombe, G.D., *et al.* (2014). Animal phylogeny and its evolutionary implications. Annual Review of Ecology, Evolution, and Systematics 45: 371–395.

Burki, F., Roger, A.J., Brown, M.W., *et al.* (2019). The new tree of eukaryotes. Trends Ecol. Evol. 35: 43–55.

Rhie, A., McCarthy, S.A., Fedrigo, O., *et al.* (2021). Towards complete and error-free genome assemblies of all vertebrate species. Nature 592: 737–746.

Christmas, M.J., Kaplow, I.M., Genereux, D.P., *et al.* (2023). Evolutionary constraint and innovation across hundreds of placental mammals. Science 380: eabn3943.

Whole-genome duplications

Cañestro, C., Albalat, R., Irimia, M., *et al.* (2013). Impact of gene gains, losses and duplication modes on the origin and diversification of vertebrates. Semin. Cell. Dev. Biol. 24: 83–94.

Soltis, P.S., Marchant, D.B., Van de Peer, Y., *et al.* (2015). Polyploidy and genome evolution in plants. Curr. Opin. Genet. Dev. 35: 119–125.

Del Pozo, J.C., & Ramirez-Parra, E. (2015). Whole genome duplications in plants: An overview from Arabidopsis. J. Exp. Bot. 66: 6991–7000.

Soltis, D.E., Misra, B.B., Shan, S., *et al.* (2016). Polyploidy and the proteome. Biochim. Biophys. Acta 1864: 896–907.

Tekaia, F. (2016). Inferring orthologs: Open questions and perspectives. Genomics Insights 9: 17–28.

Moriyama, Y., & Koshiba-Takeuchi, K. (2018). Significance of whole-genome duplications on the emergence of evolutionary novelties. Brief. Funct. Genomics 17: 329–338.

Onimaru, K., & Kuraku, S. (2018). Inference of the ancestral vertebrate phenotype through vestiges of the whole-genome duplications. Brief. Funct. Genomics 17: 352–361.

Clark, J.W., & Donoghue, P.C.J. (2023). Constraining whole-genome duplication events in geological time. Methods Mol. Biol. 2545: 139–154.

An atlas of life forms, showing phylogenetic relationships and dates of divergence

Hedges, S.B., & Kumar, S. (2009). *The Timetree of Life*. Oxford: Oxford University Press.

Sequencing of ancient DNA

Shapiro, B. (2015). *How to Clone a Mammoth: The Science of De-Extinction*. Princeton, NJ: Princeton University Press.

Johanson, Z., Joss, J., Boisvert, C.A., *et al.* (2007). Fish fingers: Digit homologues in sarcopterygian fish fins. J. Exp. Zool. B. Mol. Dev. Evol. 308: 757–768.

Card, D.C., Shapiro, B., Giribet, G., *et al.* (2021). Museum genomics. Annu. Rev. Genet. 55: 633–659.

Orlando, L., Allaby, R., Skoglund, P., *et al.* (2021). Ancient DNA analysis. Nat. Rev. Methods. Primers 1: 14.

EXERCISES, PROBLEMS, AND WEBLEMS

Exercises

Exercise 8.1 What fraction of the intergenic space in the nuclear genome of *A. thaliana* is occupied by transposons?

Exercise 8.2 On a photocopy of Figure 8.2, mark the approximate dates of the duplications in the *A. thaliana* genome on the branch leading to eudicots.

Exercise 8.3 Calculate the average gene density in the nuclear, mitochondrial, and chloroplast genomes of *A. thaliana*.

Exercise 8.4 Describe how you would test the assertion that the last common ancestor of urochordates acquired the cellulose synthase gene by lateral transfer from bacteria. What sequence information would you gather, and how would you analyse it?

Exercise 8.5 In Figure 8.6b, which *T. nigroviridis* chromosomes have substantial regions of synteny with human chromosome 10?

Exercise 8.6 On a photocopy of Figure 8.6a, indicate the regions in which *T. nigroviridis* chromosomes 5 and 13 contribute to human chromosome 15.

Exercise 8.7 Figure 8.18 shows that specimens of *Euryapteryx curtus* appear on both North and South Islands. On which island is it likely that the species arose? What reasoning leads you to this conclusion?

Exercise 8.8 In the dog, the variation in mitochondrial DNA sequences is lower than the variation in nuclear DNA sequences. What does this suggest about the breeding behaviour of domesticated dogs?

Exercise 8.9 For which of the following domesticated species could a population survive if released into the wild? Dog, cat, chicken, parakeet, maize, rice, and wheat.

Exercise 8.10 It is much easier to study mitochondrial DNA than nuclear—it is smaller, and more abundant in cells. What is the danger of assigning phylogeny by comparing populations through sequencing mitochondrial DNA of various individuals, in species that are matrilocal (that is, females remain within a herd, males leave)? An example was a study of elephants, based on mitochondrial DNA sequences. What criticism might be raised?

Problem

Problem 8.1 Figure 8.20 shows an alignment of globins from *Ciona intestinalis* and human haemoglobin α– and β–chains, myoglobin, cytoglobin, and neuroglobin. Some N- and C-terminal extensions have been trimmed. (a) Which pair of globins has the largest number of identical residues in this alignment? (b) Are the *Ciona* globins more similar to one another than the human globins are to one another? (c) Which human globin do the *Ciona* globins most resemble?

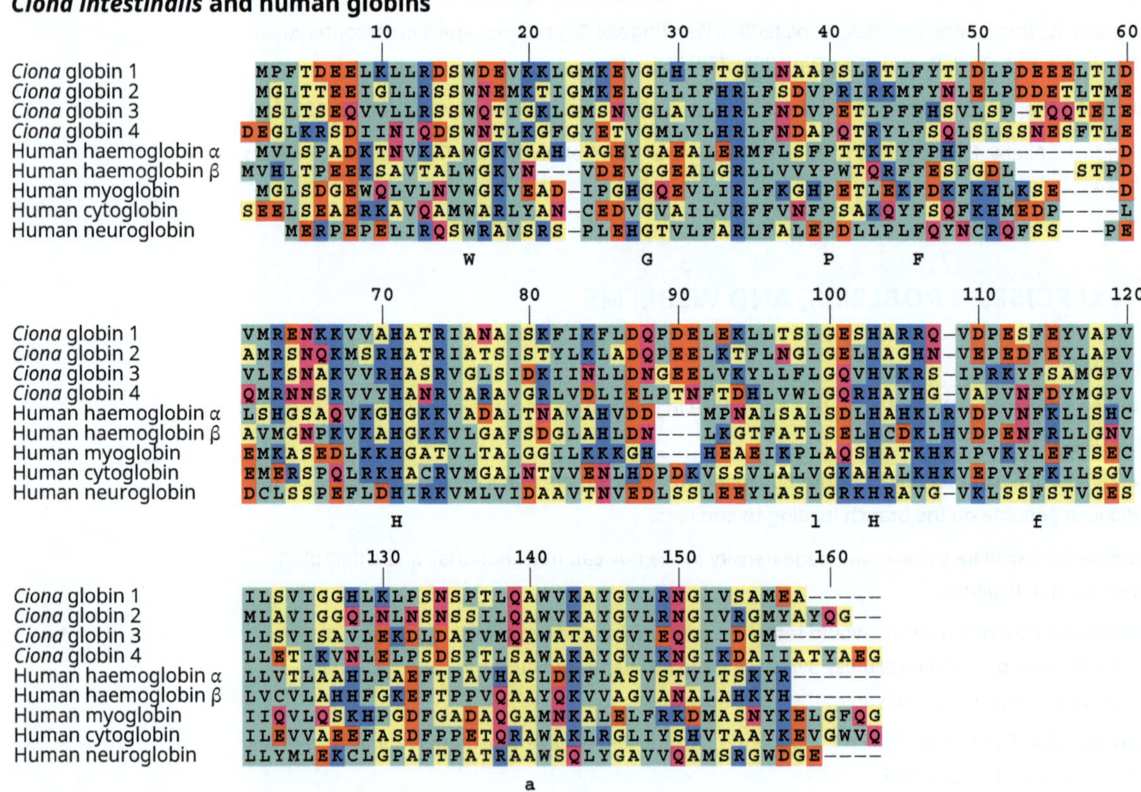

Figure 8.20 Alignment of globins from *Ciona intestinalis* and human haemoglobin α– and β–chains, myoglobin, cytoglobin, and neuroglobin. Some N- and C-terminal extensions have been trimmed.

Weblems

Weblem 8.1 For which eukaryotic animal and plant phyla have at least one species been the subject of a full-genome sequencing project? For each, give an example of a species.

Weblem 8.2 Identify a yeast protein in each of the first ten functional categories in Table 8.1.

Weblem 8.3 What is the latest common ancestor of the human and the aardvark? (Hint: compare the full taxonomy listings in any entry of a human and an aardvark sequence.)

Weblem 8.4 The UniProtKB entry for chicken ovocleidin 116 is Q9PUT1_CHICK. This protein is eggshell-specific in chickens. Does it have any mammalian homologues?

Weblem 8.5 Mammals excrete nitrogenous waste in the form of urea. Birds excrete uric acid. Urea in mammals is formed by the urea cycle, a metabolic pathway that forms urea from ammonia and aspartate. The first enzyme in this pathway is carbamoyl phosphate synthetase 1, which catalyses the reaction:

$$2ATP + HCO_3^- + NH_4^+ \rightarrow 2ADP + \text{Carbamoyl phosphate} + P_i$$

(a) Does the chicken contain a homologue of human carbamoyl phosphate synthetase I?

(b) Is there reason to believe it is not functional?

(c) In what tissues in chicken is it expressed?

(d) What might be the function of this enzyme in the chicken?

CHAPTER 9

Genomics and Anthropology: Human Evolution, Migration, and Domestication of Plants and Animals

LEARNING GOALS

- *Appreciate how genomics, archaeology, and languages combine* to elucidate important transitions in human history.
- *Know the evidence for placing the origin of modern humans in Southern and Eastern Africa.*
- *Understand how sequencing of mitochondrial and nuclear genomes elucidates the relationship of modern humans to Neandertals and Denisovans*, our early European relatives.
- *Appreciate how DNA sequencing can map patterns of human migration.*
- *Recognize that the domestication of animals and plants are experiments in directed genome change.*
- *Know how, in many important cases, we can analyse the genetic differences between the wild progenitor and domesticated varieties*. For example, we can study the genome changes associated with domestication of the dog and the horse.
- *Understand the phenotypic qualities required for domestication of animals.*
- *Understand the phenotypic qualities required for domestication of plants.*
- *Know the genome changes associated with domestication of rice and maize.*
- *Understand the effects of domestications on human civilization, and on human genomes.*

Today, humans occupy most of the world's land. We have diverse civilizations, characterized by different languages; systems of belief; tastes in food, sports, art, and music; and types of social organization, including systems of government and laws. We depend on domesticated animals, plants, and microbes (our microbiomes are also domesticates); natural resources including minerals and access to fresh water; industry and commerce, and, to an ever-increasing extent, on technology. With pathogenic bacteria and viruses, we are ever at war. Only very rarely now are we at risk from large vertebrate predators, such as sharks, alligators, and tigers.

Progress has become so fast that it is easy to lose sight of the fact that most of the present depends on the past. Genomics has a role in telling the story of how this came about. That is the subject of this chapter.

Ancestry of *Homo sapiens*

Human evolution excites the curiosity and imagination of scientists and the public alike.

Humans are primates, emerging from the same gene pool that gave rise to monkeys and apes; including our closest extant relatives, the chimpanzees and bonobos. Fossils give us only a bare glimpse at the variety of early hominins. (Hominins comprises modern humans, previous human species, and all ancestral populations since the divergence of our lineage from that of the great apes). Many anatomical, biochemical, developmental, and behavioural features distinguish modern humans from other species. The most important are our conceptual and linguistic abilities—arguably a redundancy—that create a cumulative culture, growing over successive generations. Humans are the only species that wonders about its evolution, or sequences its genomes.

Two tools for studying human evolution are the shovel and the computer. For only a few fossils do we have the corresponding genomes, but the ones we have are spectacularly informative. These include genomes of Neandertals and Denisovans, two related groups that inhabited what is now Europe and Western Eurasia.

What made us human? We can compare genome sequences from humans and other species to identify unique aspects of the human genome. The closer the relationship, the fewer the genomic differences. Our closest extant relatives are the chimpanzees and bonobos. Two randomly chosen humans differ in ~0.1% of their sequences. A randomly chosen human and a randomly chosen chimpanzee differ at ~1% of sites. This amounts to about 64 million bases. Humans and chimpanzees show differences at larger scales also, in chromosomal rearrangements (see Figure 6.12). Sequences from closer relatives, Neandertals and Denisovans, sharpen the focus on the differences. A random modern human and a random Neandertal or Denisovan would differ at ~0.15% of sites. Neandertals and Denisovans are more similar to each other than either group is to modern humans.

Where did we come from? Modern humans evolved in Southern and Eastern Africa, and spread out throughout the world (see Box 9.1). Hominins left Africa for Europe more than once (indeed there was probably two-way traffic). Premodern humans in Europe were the progenitors of Neandertals and Denisovans. A snapshot taken 140 000 years ago would show a variety of hominin lineages in Africa and Eurasia, already with some cultural/technological capacities, such as tool making.

A second wave, of modern humans, entered Europe about 50 000 years ago. They challenged the Neandertals and Denisovans for the same ecological niche. Successfully. Interactions with modern humans were at least a major contributing cause of Neandertal and Denisovan extinction. A snapshot taken 30 000 years ago would show a single modern human lineage throughout the old world. Figure 9.1 shows a timeline of the current model of human evolution.

Neandertals, Denisovans, and modern humans all inhabited Europe and Central Asia at the same time. Articles in scientific journals enquire about similarities and differences, and dates. The popular press focuses on the possibility of sexual congress between Neandertals and humans.

BOX 9.1 What is the evidence that modern humans originated in Africa?

Genetic variation is highest in African populations, both in number of single-nucleotide polymorphisms (SNPs) and range of repeat lengths of microsatellites. The farther from Africa, the less the variation. This is consistent with the idea that most genetic change is random drift. African populations show the least persistence of linkage disequilibrium (see Chapter 4).

Microbiomes confirm the evidence from the genomes. *Helicobacter pylori* lives in the human gut (where it can cause ulcers). Different populations host divergent strains of *H. pylori,* the result of genetic drift after splitting of the human populations. As a result comparisons of *H. pylori* sequences allow tracing of relationships and migration patterns of human populations. Indeed, because of their greater variability, analysis of *H. pylori* sequences can in principle provide more detailed information about human migrations than analysis of the corresponding human genome sequences. Also, *H. pylori* provides information about migrations of males, as opposed to the mitochondrial sequences described in Figure 9.4. *H. pylori* sequences collected from Africans are more variable than those from inhabitants of other regions. In general, the larger the distance from Africa, the lower the genetic diversity.

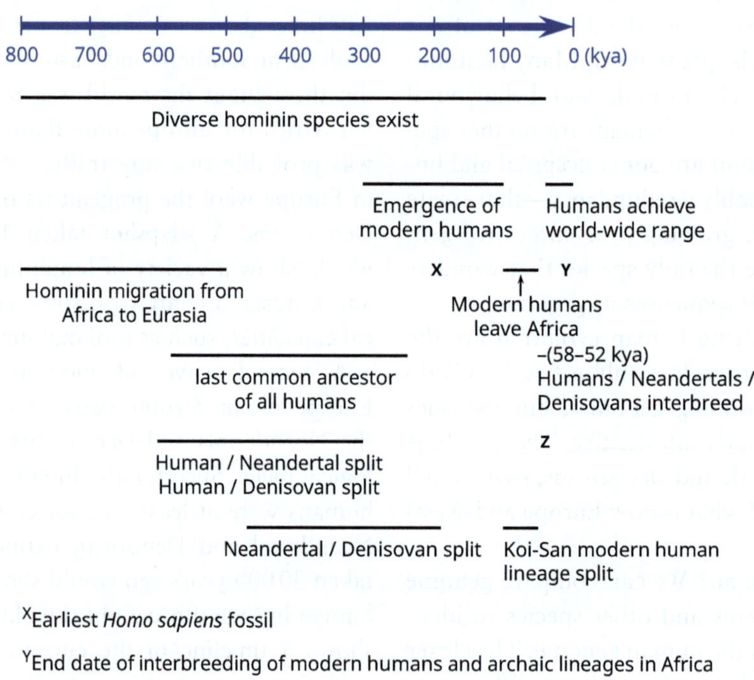

Figure 9.1 Trajectory of the evolution of modern humans, Neandertals, and Denisovans. kya = thousands of years ago. Many of the time frames are quite broad. In some cases, this reflects a lengthy process. In other cases, it is an admission of our being at the mercy of an extremely sparse set of available fossils. The date of extinction of the Neandertals is estimated at 40 kya, based on dating of the last known remains. But how long, could how many, Neandertals have survived in obscure, isolated refugia, of which we know nothing? In contrast, we know that the passenger pigeon (*Ectopistes migratorius*) became extinct at 12 noon on 1 September 1914, when the last one died in a zoo in Cincinnati (central US).

Data from Marean, Curtis W., An Evolutionary Anthropological Perspective on Modern Human Origins (October 2015). Annual Review of Anthropology, Vol. 44, pp. 533-556, 2015, Available at SSRN: https://ssrn.com/abstract=2679288 or http://dx.doi.org/10.1146/annurev-anthro-102313-025954

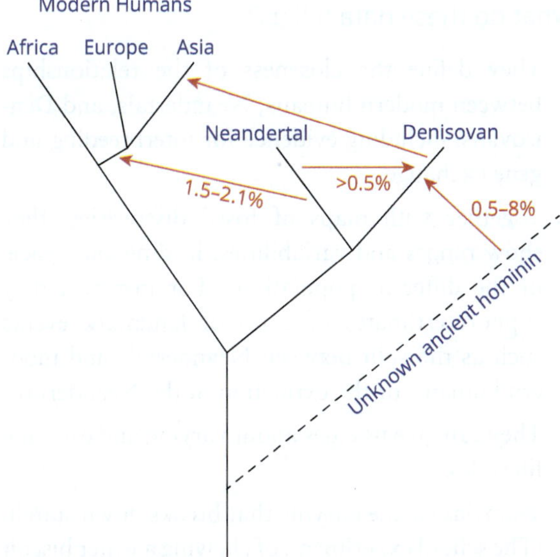

Figure 9.2 Evolutionary tree of modern humans, Neandertals, and Denisovans, and extents of transfer of DNA sequence. Neandertals and Denisovans are more similar to each other than either is to modern humans. In some cases, the sources of gene transfer can be more precise. The population of Neandertals that contributed genes to humans is more closely related to Neandertals from the Caucausus (Mezmaiskaya) than from Croatia (Vindija) or Siberia (the Altai mountains).

Adapted from Prüfer, K., Racimo, F., Patterson, N., *et al.* (2014). The complete genome sequence of a Neanderthal from the Altai Mountains. Nature 505: 43–49. https://doi.org/10.1038/nature12886, Springer Nature.

Interbreed they did, and exchanged genetic material. All of these populations were interfertile. Figure 9.2 shows an evolutionary tree that includes modern humans, Neandertals, and Denisovans; and estimated amounts of gene transfer among these three groups and from more ancient hominin species. Neandertals contributed 1–4% of the genomes of humans from outside sub-Saharan Africa. Denisovans contributed 1–6% of the genomes of people in Southeast Asian islands and Oceania, including the indigenous peoples of Papua New Guinea and Australia. Neandertals and Denisovans share a distal branch of the tree, their sequences share >99.5% similarity. More exotically, a few percent of the Denisovan genome appears to retain traces of a more primitive species, probably *Homo erectus*.

The arrows in this figure are based on only a few cases of proven interbreeding, and it is likely that gene flow was bidirectional.

The Neandertal genome

In 1856, about a dozen bones and fragments were found in a limestone quarry in the Neander Valley, near Düsseldorf (Valley = Tal in German). They were recognized as being anatomically different from modern humans. Although, in retrospect, Neandertal bones had appeared earlier, the 1856 discovery was the first to be recognized—in 1864—as remains from a new species: *Homo neandertalensis*. (Why did it take eight years? *The Origin of Species* was published in 1859.)

Neandertals were closely related to modern humans. They were approximately our height, but substantially more robust. Their cranial capacity was as large as modern humans or perhaps slightly larger, although this in itself does not guarantee comparable cognitive abilities. Neandertals did fashion and use tools, and produce at least decorative arts. It has been suggested that they wore makeup. They engaged in complex funerary practices.

The group of Svante Pääbo at the Max Planck Institute for Evolutionary Anthropology in Leipzig has pioneered the genomic sequencing of ancient hominin remains. The Neandertal mitochondrial genome was sequenced in 2008. Because cells can contain thousands of copies of mitochondrial DNA, and it is so short (usually 16 565 bases in Neandertals), mitochondrial DNA is much easier to recover and sequence than nuclear DNA. ENTREZ now contains entries for 25 complete Neandertal mitochondrial genomes, and there have been very many sequence determinations of specific, evolutionarily informative regions. The earliest Neandertal mitochondrial DNA sequence data are from a specimen found in the Scladina cave (Meuse Basin, Belgium), deposited 100 000 years ago. The earliest Neandertal fossils known so far are 130 000 years old.

Pääbo's group published a draft sequence of the Neandertal nuclear genome in 2010, at an average 1.3× coverage. This combined results from three individuals, two of them probably close relatives. The specimens were from a cave in Vindija, Croatia. This project also produced partial sequences of the nuclear genomes of three additional individuals, yielding between 2.2 Mb and 56.4 Mb from each specimen. In 2014, the group published a high-quality (50× coverage) Neandertal genome sequence from a single, different individual. Since then, sequences of multiple individuals from a community in Siberia, including members of a family, have been determined.

It has even been possible to determine the methylation patterns of Neandertal and Denisovan DNA, telling us about their epigenetics, and proteomics.

The speed at which the techniques have increased in power is very impressive.

The Denisovan genome

In March 2010, a 41 000-year-old fragment of a finger bone from a young girl was found in the Denisova cave, in the Altai Mountains in Siberia, 442 km SSE of Novosibirsk, near the border between Russia and Kazakhstan. Sequencing showed that she represented a third population of ancient humans. They are now called Denisovans, from the site of discovery of the first specimen.[1] However, like the Neandertals, their range was wide.

Other Denisovan remains, all extremely fragmentary, have turned up. In comparison with classical palaeontology, genomic sequencing is not handicapped so severely by limitation to fragmentary specimens, provided the DNA has been preserved. Nevertheless one must recognize and respect the very great abilities of classical palaeontologists to draw inferences from limited skeletal material—Cuvier claimed to be able to reconstruct the whole animal from a single bone.

Pääbo's group sequenced mitochondrial and nuclear DNA from three Denisovan individuals. The mitochondrial sequence was complete, and high-coverage. The three specimens also yielded nuclear DNA sequence mappable onto the human genome, of, respectively, 1.86 Gb (average 31× coverage), 265 Mb, and 54.6 Mb.

There are no DNA sequences of archaic hominids in Africa. Conditions are unpropitious for preservation of DNA.

[1] The same cave has been occupied by Denisovans, Neandertals, and modern humans. The unlikely possibility that they were all there at once—and perhaps even snowed in for a time—presages Bret Harte's *The Outcasts of Poker Flat*.

What do these data tell us?

1. They define the closeness of the relationships between modern humans, Neandertals, and Denisovans, including evidence for interbreeding and gene exchange.
2. Together with maps of fossil discoveries, they show ranges and variabilities, in time and space, of the different populations. For instance, they support estimates of dates of landmark events such as the split between Neandertals and modern humans, or the extinction of the Neandertals.
3. They can give us clues about varying and evolving lifestyles.

   α-amylase is the enzyme that breaks down starch. (The school experiment, of chewing a water biscuit [= soda cracker] until it starts to taste sweet, is an observation of the action of salivary α-amylase.) Most primates, including the Neandertals and Denisovans, have a single copy of the α-amylase gene. In modern humans, the copy number is larger ($\geq$2 and often many more: average 6, maximum 20) and quite variable.

   The first duplication was ancient, although coming after the split of human ancestors from Neandertals and Denisovans. A second duplication occurred tens of thousands of years ago, and is likely correlated with the development of agriculture. Many contemporary humans descended from populations with low-starch diets have only two copies of the α-amylase gene. Indeed, because α-amylase is not very effective at hydrolysing raw starches, the multiplication of the gene copy number may be related to the development of cooking. The expression level of α-amylase expression is controlled by copy number, but also by the starch content of the individual's current diet. Physical or psychological stress also increases α-amylase levels. (Dogs have higher levels of α-amylase than wolves, perhaps because dogs scavenge human food.)
4. They allow us to isolate genes unique to humans (including Neandertals and Denisovans), with a view to understanding the origin of our special features. One interesting example is the *FOXP2* gene, involved in the development of language function in modern humans (see Box 2.4). The

protein it encodes is a transcription factor, that governs expression of many proteins in the developing brain. The human FOXP2 protein has two amino-acid differences in the coding region from the homologous genes of extant primates. The Neandertal and Denisovan genomes contain the modern human version.

However, to infer from the FOXP2 sequence that Neandertals and Denisovans had sophisticated languages is far too simplistic (see Box 9.2). The region flanking this gene contains a binding site for the transcription regulator POU3F2. This site differs in humans relative to Neandertals and Denisovans; and indeed between humans and all other tetrapods. The mutation should alter the level of expression of the *FOXP2* gene itself. The observation of mutations both in a gene coding for a protein and in a regulatory sequence shows that the effects of even single-base changes in the genome may involve complex patterns of interaction.

In the case of *FOXP2*, no one would be naïve enough to think that one protein could account for language abilities. Indeed, the protein encoded by FOXP2 switches on, or off, the expression of hundreds of other proteins in the developing brain. On the other hand, mutations in other genes than *FOXP2* are associated with language impairment in humans. Thus a correct *FOXP2* sequence is a *necessary* but not a *sufficient* condition for language competence.

The message, from α-amylase and *FOXP2*, should be clear: *The observation that traits show simple Mendelian inheritance does not imply that one protein, or region of DNA, is acting on its own. The genes may be fulcra of very complex processes.*

What have Neandertals and Denisovans done for us lately?

Among the genes contributed by the Denisovans to the modern human genome is an MHC class I allele. The MHC complex controls the versatility and vigour of the immune response to different antigens. A more varied MHC haplotype allows a stronger immune response to a wider variety of pathogens. A small population of modern humans, migrating from Africa to Europe, might find themselves (a) with a restricted set of MHC alleles, as a result of a population bottleneck associated with the migration, with strong 'founder' effects on allele frequencies; and handicapped, as are all small populations, by the weakness of selective pressure, relative

BOX 9.2

Did the Neandertals have language?

Sources of relevant evidence includes genomics, anatomy, and archaeology.

- *Genomics*: Neandertals and Denisovans share with modern humans—and *only*, among all vertebrates, with modern humans—a version of the *FOXP2* gene. However, the *FOXP2* sequence cannot by itself imply language competence. *FOXP2* is part of a complex pattern of interacting genes and gene products that together create the cognitive substrate of language.
- *Anatomy*: Studies of vocal and auditory features of Neandertals show a modern-human-like laryngeal anatomy and pulmonary airflow control, consistent with an enhanced vocal repertoire and prowess relative to other primates. Perhaps concomitantly, the ear ossicles of Neandertals gave their hearing a frequency spectral sensitivity curve similar to that of modern humans. This is a genuine difference from other primates, notwithstanding the fact that animals can hear and interpret human speech (at least to the extent of recognizing simple commands).
- *Archaeology*: Neandertal technology, ornaments (if you wish to be dismissive) or art (if you wish to be generous), and burial practices suggest a relatively high degree of social interaction, or culture. Many people feel that this is inconsistent with absence of language.

The conclusion, however, is that the evidence for Neandertal language, although highly suggestive, is inconclusive: sufficient to indict, but not to convict beyond reasonable doubt. No obvious method to resolve the question definitively presents itself.

to genetic drift, to fix useful variant alleles and (b) an inappropriate set of MHC alleles, leaving them poorly equipped to deal with unfamiliar pathogens in the new environment. In modern times, the succumbing of indigenous populations in Central and South America and in the South Seas to diseases such as measles, upon first contact with Europeans, may be attributable to immunological 'naiveté' of their MHC haplotypes.

Some modern human genes that present the highest Neandertal ancestry in Europeans and Asians code for Toll-like receptors (*TLR1*, *TLR6*, *TLR10*). These genes encode cell-surface proteins that detect pathogens and trigger inflammatory and immune responses. These genes may enhance the immune response to *Yersinia pestis,* the bacterium responsible for the Black Plague that in the 14th century killed an estimated 30–60% of the population of Europe.

Conversely, it has been suggested that modern humans brought with them into Europe some tropical diseases with which the Neandertals could not easily cope.

Ancient populations and migrations

When people move around, they take their DNA with them. This makes it possible to trace patterns of migration. Many studies focus on mitochondrial DNA sequences, which reveal lines of maternal inheritance (see Box 9.3). Y chromosomes provide complementary paternal information. Although studies of Y sequences are more sparse, in many cases they corroborate the implications of the mitochondrial data.

There is now consensus that our species, *Homo sapiens*, arose in Africa approximately 300 000 years ago. Recent migrations beginning approximately 70 000–50 000 years ago took our ancestors around the world, and continue to do so. Unlike modern population flows, documented in historical records, we depend on archaeological relics, modern genomics, and linguistics to infer the timing, the routes, the numbers of individuals, and even perhaps the

BOX 9.3 Human mitochondrial DNA haplogroups

Human mitochondrial DNA is a double-stranded, closed, circular molecule, usually 16 568 bp long. It is inherited almost exclusively through maternal lines. A fertilized egg contains the mother's mitochondria. Although sperm contain mitochondria—essential to provide energy for their motility—the few paternal mitochondria that enter the egg are selectively eliminated. As a haploid entity, mitochondrial DNA is, therefore, not subject to recombination, and changes only by mutation.

Mitochondrial DNA is estimated to adopt one mutation every 25 000 years. This gives a reasonable rate of divergence to trace human migration patterns. (Nuclear DNA mutates approximately ten times more slowly than mitochondrial DNA because (1) histones protect it; (2) active repair mechanisms edit out some mutations; and (3) the activity of mitochondria in oxidative phosphorylation exposes the mitochondrial DNA to mutagenic oxygen radicals.)

Human mitochondrial DNA contains genes for 22 tRNAs, two ribosomal RNAs, and 13 proteins. The major non-coding region is the control region, or D-loop, involved in regulation and initiation of replication. This region is about 1 kb long. It shows a higher rate of substitution than the rest of the mitochondrial genome, by a factor of about four.

Different mitochondrial DNA sequences are associated with different populations. Mutations are referred to the first human mitochondrial DNA sequence determined, and subsequently corrected, called the revised Cambridge Reference Sequence. Groups of related sequences are called haplogroups. (The distribution of the number of sequence differences between different individuals has a peak at ~70 for Africans and ~30 for non-Africans.) The original classification of sequence variants depended on changes in restriction sites. This was followed by explicit sequencing of the control region, focusing on its two highly polymorphic segments. For finest resolution, contemporary studies

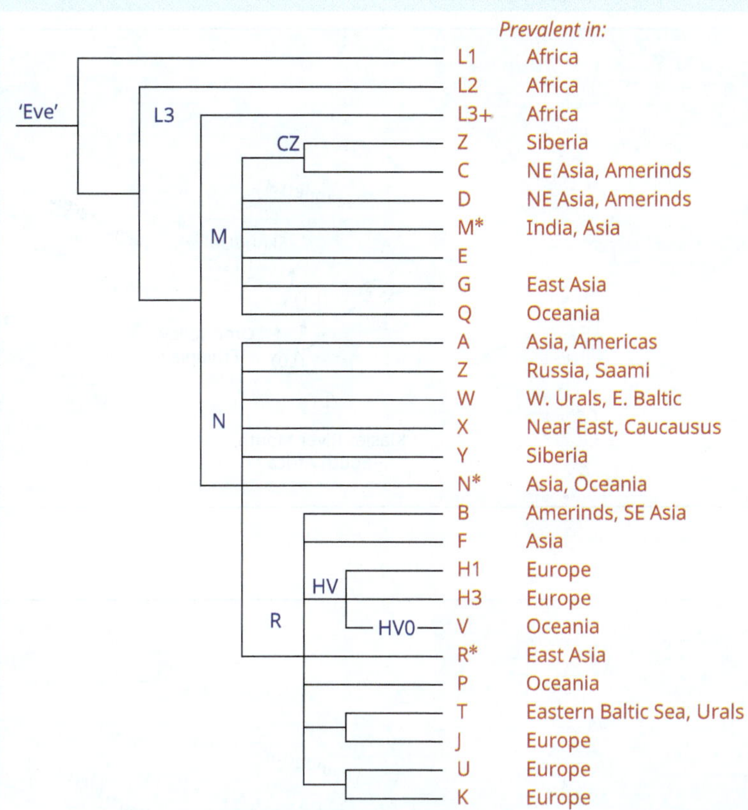

Figure 9.3 Phylogenetic tree of major mitochondrial haplogroups. The nomenclature began with a study of Native Americans, or Amerinds, and the letters A, B, C, and D were assigned to them. Other letters were introduced and were subdivided as needed as more detailed sequencing data appeared. HV0 was formerly called pre-V.

now more frequently determine full mitochondrial DNA sequences, except in cases of ancient DNA where the best recoverable material may be fragmentary (see Figure 9.3).

Several databases focus on human mitochondrial genomes, including MITOMAP (http://www.mitomap.org), hmtDB (https://www.hmtdb.uniba.it/), EMPOP (European Mitochondrial POPulation, https://empop.online/), and AmtDB (Ancient mitochondrial DNA, https://amtdb.org).

P. Forster has created a 'movie' showing successive stages of human dispersal (see Figure 9.4).

The evidence for human origins in Africa is that contemporary genetic diversity is highest there. The mitochondrial DNA haplogroup L1, believed to be the oldest haplotype that survives, is found in the KhoiSan of the Kalahari Desert in southern Africa and in the Biaka pygmies of the central African rainforest (see Figure 9.4a). An expansion of L2 and L3 haplogroups took place within Africa about 80 000–60 000 years ago (see Figure 9.4b). Over two-thirds of contemporary Africans belong to these groups.

The first emigration from Africa occurred about 85 000–55 000 years ago. The participants in this first dispersal carried the L3 mitochondrial DNA haplogroup, which arose ~84 000 years ago. Mutations in L3 gave rise to haplogroups M and N, ~60 000 years ago (see Figure 9.4c). Haplogroup M mitochondrial DNA appears in ancient populations from the Andaman Islands, southern continental India, and the Malaysian Peninsula. This suggests a dispersal via what is now southern Iran along the coast to India, the Malaysian Peninsula, and Australia. Archaeological evidence shows that humans reached Australia by 46 000 years ago. Dating of the earliest human remains in Australia is consistent with the extinction of many large mammals and birds shortly thereafter (see Figure 9.4d).

From 60 000 to 30 000 years ago, human populations expanded in southern Europe and Asia, accumulating mutations in mitochondrial DNA to form new haplogroups (see Figure 9.4e).

Expansion into Europe was delayed and interrupted. Between 20 000 and 30 000 years ago—an interglacial era—the first human Europeans encountered and replaced the Neandertal population (see Figure 9.4f). The same climate conditions permitted a northwards expansion in Asia.

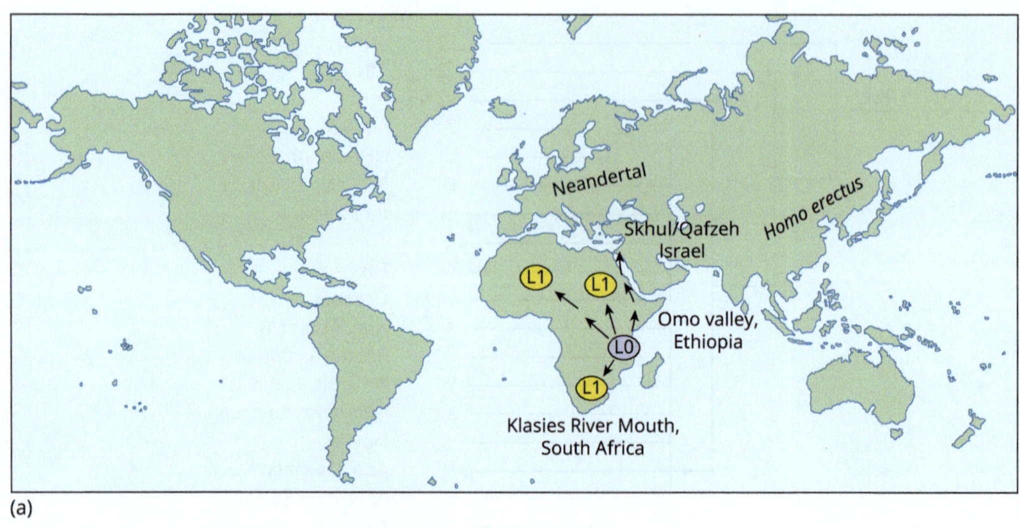

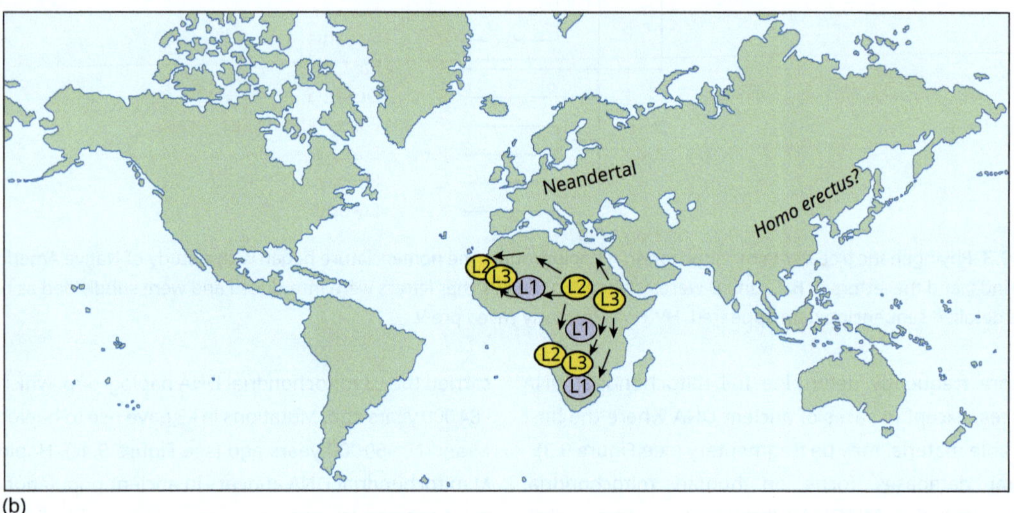

Figure 9.4 'Movie' of human migration patterns, based on mitochondrial DNA sequences. Letters indicate haplotypes. (a) The beginnings, ~150 000 years ago. (b) Expansion and divergence within Africa, 80 000–60 000 years ago. (c) Out of Africa, 60 000–50 000 years ago. (d) Spread through the south coast of the Indian Ocean, reaching Australia, 50 000–30 000 years ago. (e) Expansion in the eastern Mediterranean, India, southeast Asia and Australia, and central Asia, 30 000 years ago. (f) After 30 000 years ago, milder climate conditions permitted expansion northwards in Europe and Asia. (g) Crossing the Bering Strait, first human inhabitants of American continents, 20 000–15 000 years ago. (h) Ice Age, 20 000 years ago, with humans forced to retreat south. (i) Spread to South America, 18 000 years ago. (j) Warmer climate, with resettlement of northern latitudes, 15 000–13 000 years ago. (k) Sudden warming and subsequent stable climate, 11 400 years ago, allowing the spread of agriculture. (l) Expansion to islands, 2000 years ago. (m) The current picture. (Showing only the most prevalent and not the only haplotypes in each place.)

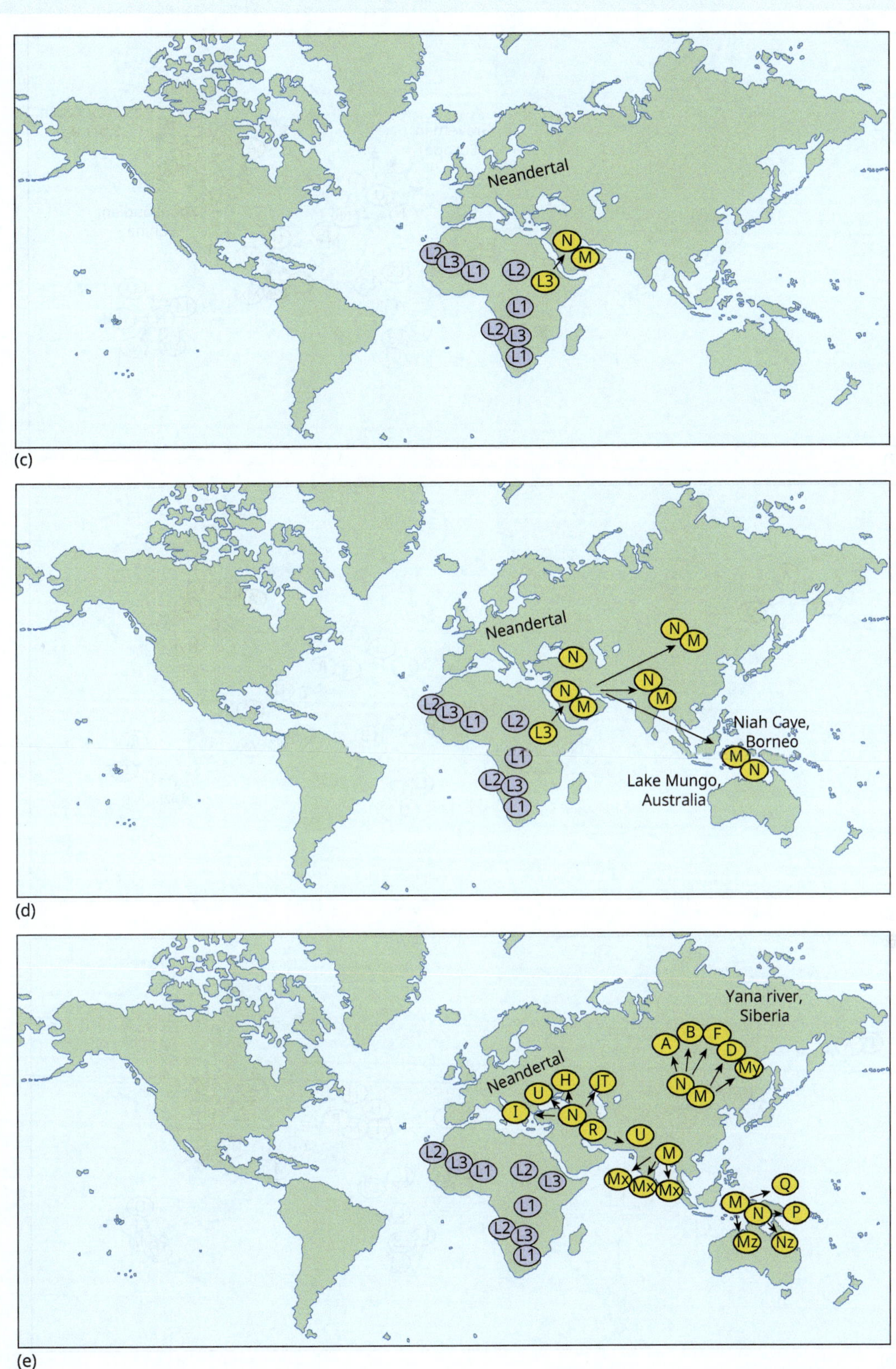

Figure 9.4 *(continued)*

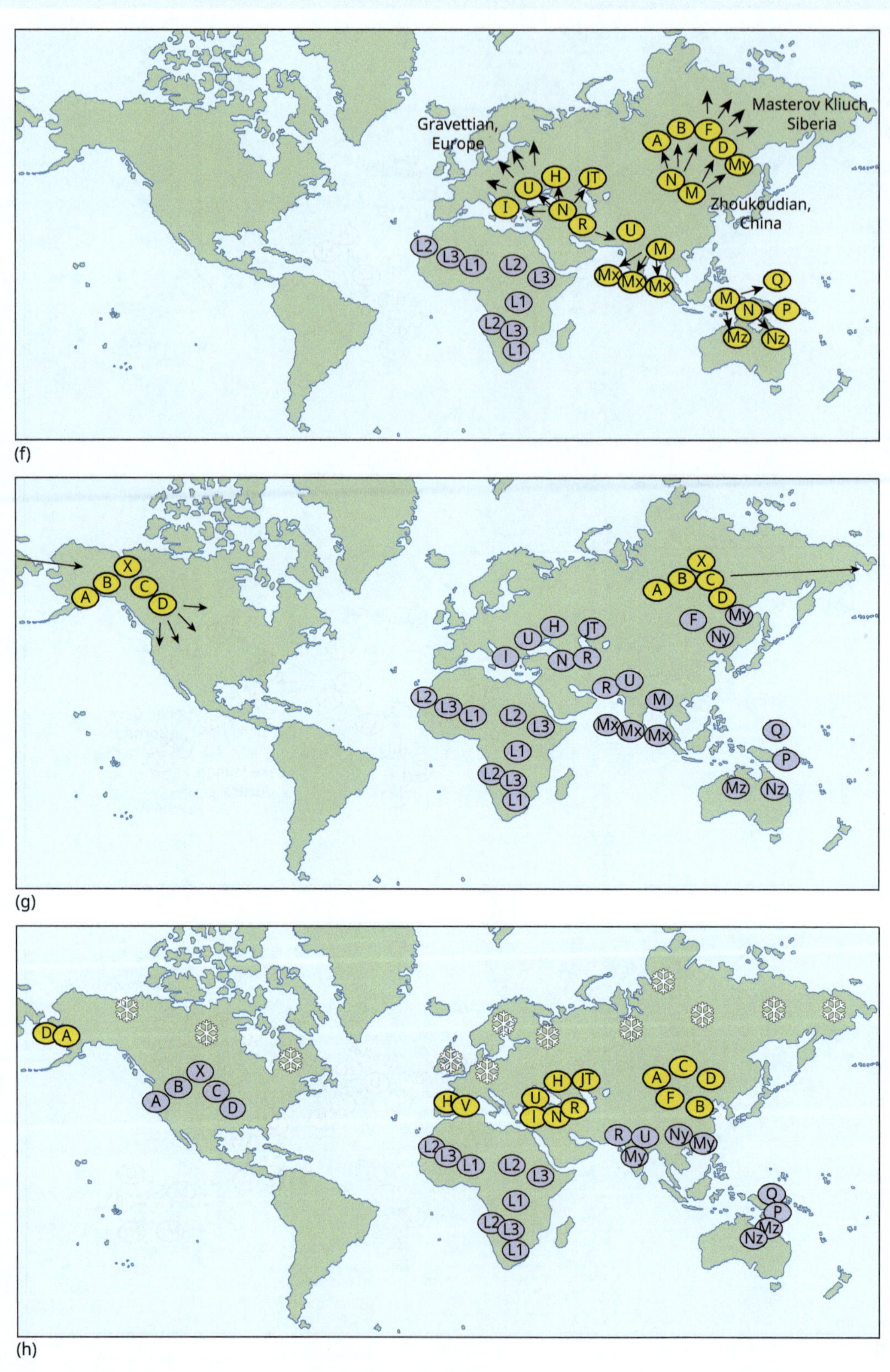

Figure 9.4 *(continued)*

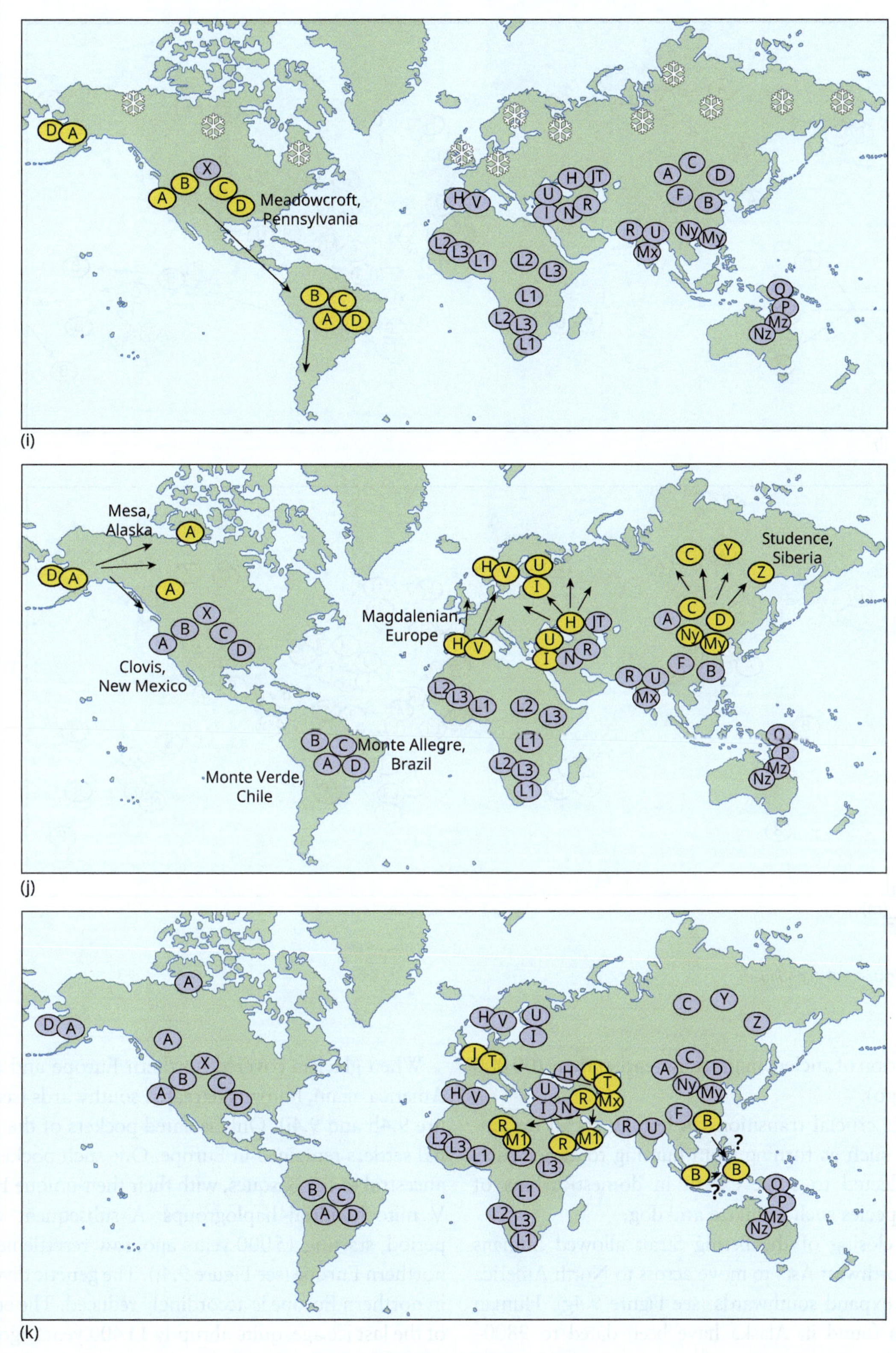

Figure 9.4 *(continued)*

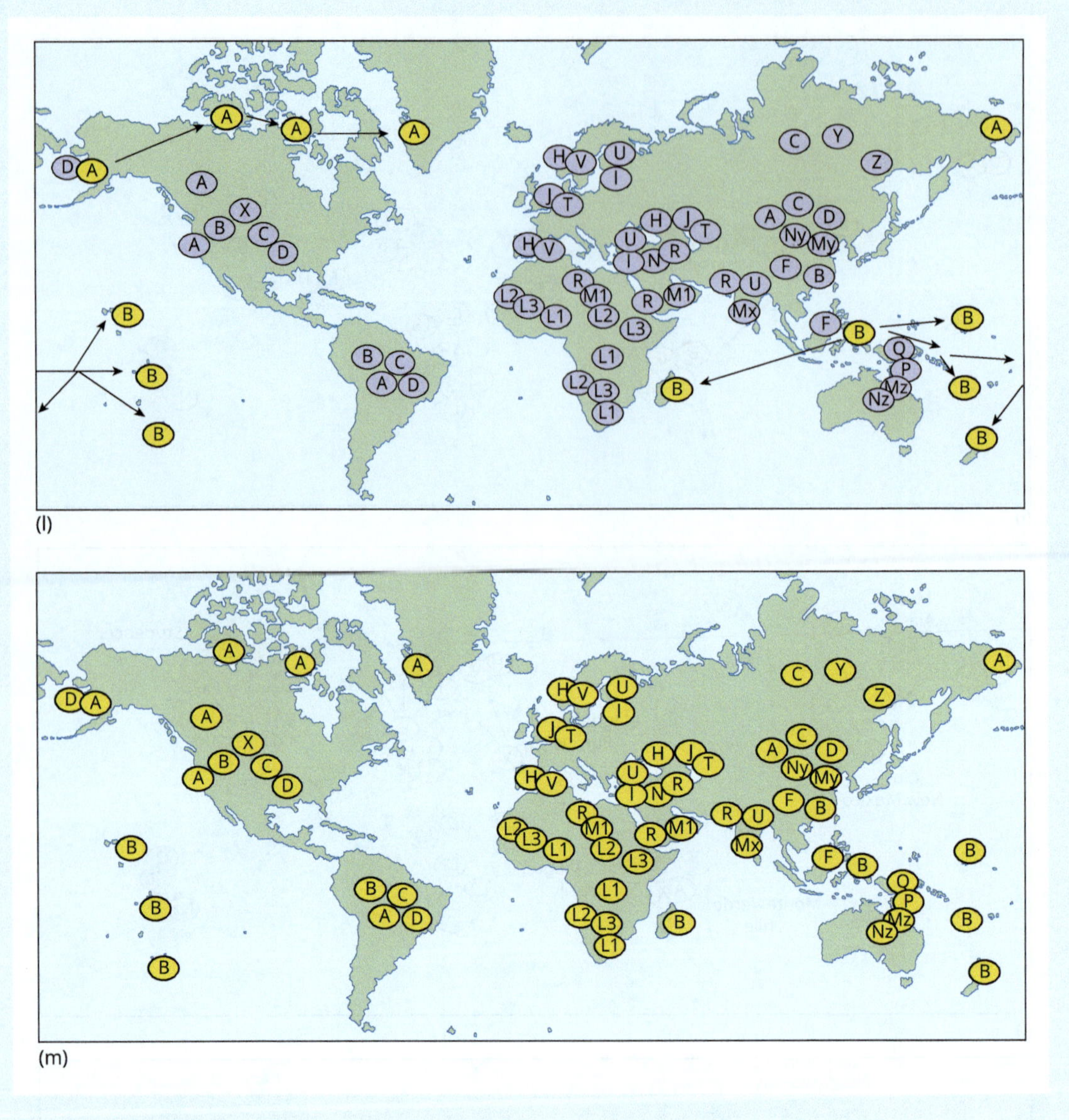

Figure 9.4 *(continued)*

motivation of ancient migrations (earlier than 100 000 years ago).

Other crucial transitions in human social organization, such as turning from hunting to agriculture, are reflected to some extent in domestications of other species such as maize and dog.

The closing of the Bering Strait allowed humans from northwest Asia to move across to North America and to expand southwards (see Figure 9.4g). Human remains found in Alaska have been dated to 9800–9200 years ago. Current evidence suggests that humans first arrived in America 20 000–15 000 years ago.

When glaciers covered northern Europe and arctic America again, humans retreated southwards (see Figure 9.4h and 9.4i). Only isolated pockets of the original settlers remained in Europe. One such pocket was ancestral to the Basques, with their then-unique H and V mitochondrial haplogroups. A subsequent warm period, starting 15 000 years ago, saw resettlement of northern Europe (see Figure 9.4j). The genetic diversity in northern Europe is accordingly reduced. The ending of the last ice age, quite abruptly 11 400 years ago, permitted the expansion of agriculture (see Figure 9.4k and Box 9.4).

Late migrations, during the last 2000 years, populated islands such as Greenland by Inuit from Alaska, Madagascar by people from southeast Asia, and Pacific islands by Austronesians (see Figure 9.4).

The contemporary distribution of mitochondrial haplogroups contains the records of this history (see Figure 9.4m).

Western civilization? 'I think it would be a good idea'[2]

Originally populated by hunter-gatherers, the European human landscape changed with the arrival of domesticated plants and animals, and the inventions they stimulated. There was an interleaving, and synergy, of crop domestication, animal domestication, and technological innovation. An animal pulling a plough reflects all three. Horse-drawn wagons, and chariots, required not only domestication of the horse, but invention of the wheel.

Domestication of plants and animals provided sustenance, labour, and companionship (see Box 9.4). The effects on human social organization were profound. The interval between planting and harvesting crops led to fixed abodes, replacing migration to follow herds of prey. Food production higher than the subsistence level led to population growth, larger

[2] Attributed to Mohandas K. Gandhi.

BOX 9.4 Worldwide domestications of plants and animals

Humans in different regions domesticated available plants and animals. Several examples of multiple, independent domestications are known.

Area	Domesticated plants	Domesticated animals
Middle East	wheat, barley	sheep, goats, cattle, pigs
Mesoamerica	maize, beans, squash	dog
South America	potato	llama, alpaca, guinea pig
North America	amaranth	dog
China	millet, rice	cattle, pigs, dogs
Africa	sorghum	cattle

Candidates for domestication must present favourable features.

Wild forms of some plants are harvestable directly; for instance, chocolate. But for many plants, domestication was a complex, multistage process; typically involving:

- an early, primarily passive, stage, of gathering / collecting, perhaps some protective tending;
- storage of seeds, sowing, weeding, drainage, irrigation;
- land clearance, systematic tillage;
- selection for desired traits.

One characteristic of many crop plants is that the seeds, and the harvests, can be stored. This allows retention of seeds for planting next year's crop. It also makes for a more robust continuity of food supply in the face of variable climatic conditions. Pharaoh's dream, in Genesis 14, as interpreted by Joseph, led to stocking of granaries during seven years of plenty, to bridge the seven lean years that followed.

For animals, features favouring domestication include:

- Docility
- Non-territorial; that is, tolerant to herding

(The distinction: docility = not afraid of humans; tolerant to herding = not afraid of one another)

- Socially, showing hierarchy of dominance, allowing humans to co-opt leadership role
- Unconstrained breeding
- Rapid growth and maturation; short generation time

communities, and trade. Human activities became specialized: Artisans developed the expertise required for technological innovation. Economic inequalities, leading to class distinctions, arose, leading to the need for some form of government.

Three strands of evidence bear witness to these changes: archaeology, genomics, and linguistics. Archaeology can date lifestyles, domestications, and inventions, and reveal diets. Genomics can follow migrations, trace histories of domestications and inventions, and identify selective pressures for novel traits such as adult-persistent lactose tolerance associated with diets that include milk. Linguistics can also trace human migrations, and indicate the history of domestication and invention. The structures and vocabularies of languages control how people perceive the world, and their social organization.

Crops leading to agriculture in Europe appeared first in the Middle East, in the area known as the fertile crescent (modern Western Iran, Southern Iraq, Syria, Lebanon, Jordan, Israel, and Northern Egypt). Approximately 11 500 years ago, people harvested wild barley, wheat, and lentils. This region also saw the domestication of cattle, sheep, goats, and pigs, approximately 10 000–13 000 years ago. By 7000 years ago, farming and domesticated animals were widespread in the Middle East, and beginning to appear in Southern Europe and Northern Africa. Agriculture advanced across Europe, reaching northeast Europe by about 6000 years ago. Trade spread agricultural products in advance of their local cultivation.

Agriculture reached Britain about 6000 years ago.

Surveys of DNA sequences from human remains distributed around Europe have delineated this history. Lazaridis *et al.* (2015) studied DNA sequences from 727 West Eurasians, who lived over the past 11 000 years (see Figure 9.5). By correlating sequences with archaeological evidence, they showed that:

- The first human inhabitants of Europe were the Yamnaya hunter-gatherers based north of the Caucasus, in present-day Ukraine and Russia. It is likely that they spoke an Indo-European language (see Box 9.5), and that they brought the wheel into Europe.

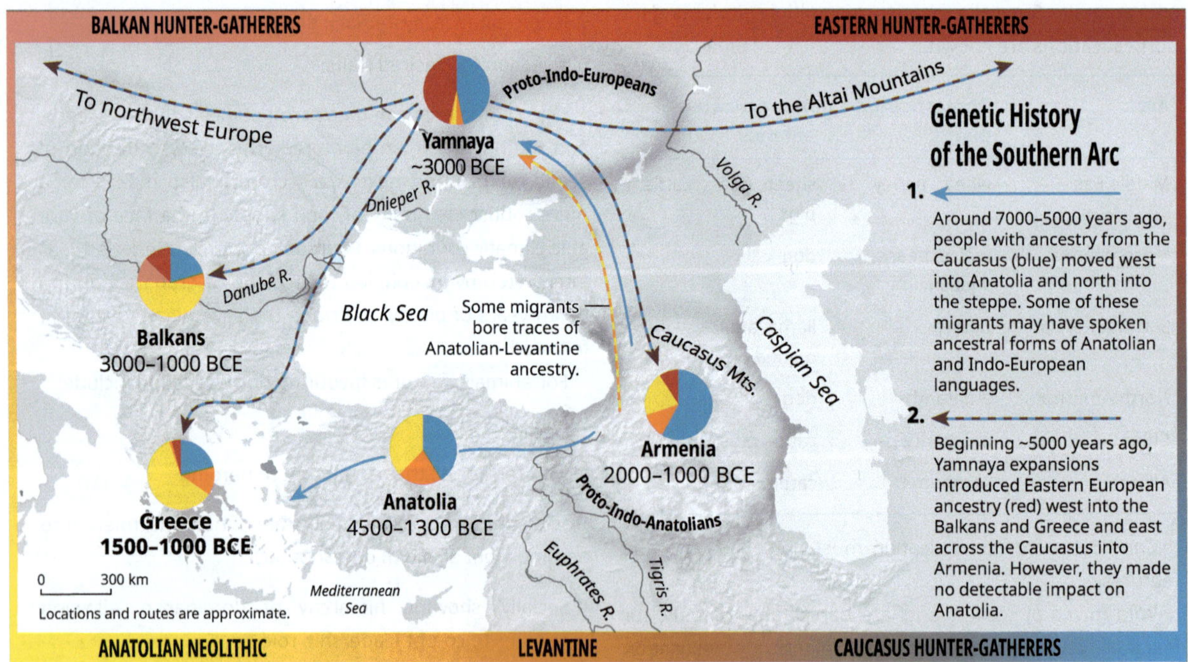

Figure 9.5 One giant steppe for mankind: Migrations of the Yamnaya, based in the West Asian highlands, populated Anatolia, the Balkans, central Europe, South Siberia, and Central/South Asia. They brought with them both Indo-European language and advanced technology including the wheel.

Lazaridis, I., Alpaslan-Roodenberg, S., Acar, A., *et al.* (2022). The genetic history of the Southern Arc: A bridge between West Asia and Europe. Science 377, 6609. https://doi.org/10.1126/science.abm4247 Reprinted with permission from AAAS.

The Indo-European language family

Europeans speak many different languages. To translate texts among the EU's 24 official languages (still including English, plus Russian, Arabic, and Chinese), the European Commission's Directorate-General for Translation has a staff of 2500; a parallel organization handles oral translation.

Many European languages, even if not mutually intelligible, have obvious affinities. The major Romance languages are Italian, French, Spanish, Portuguese, and Romanian. Other languages have diverged much farther, to the point where the relationships cease to be obvious. However, almost all languages spoken in Europe, plus parts of Western, Central, and South Asia, belong to a larger language family called Indo-European (see Figure 9.6). The geographic range of Indo-European languages runs from India (Sanskrit, Hindi) and Pakistan (Urdu) on the East to Icelandic in the West. The exceptions, the non-Indo-European languages spoken in Europe, include the family comprising Finnish, Estonian, Hungarian—which are related to Asian languages—and Basque, for which no affinity to any other language is known. In modern times, colonization and emigration transplanted European languages to the Americas, Africa, Australia, and New Zealand.

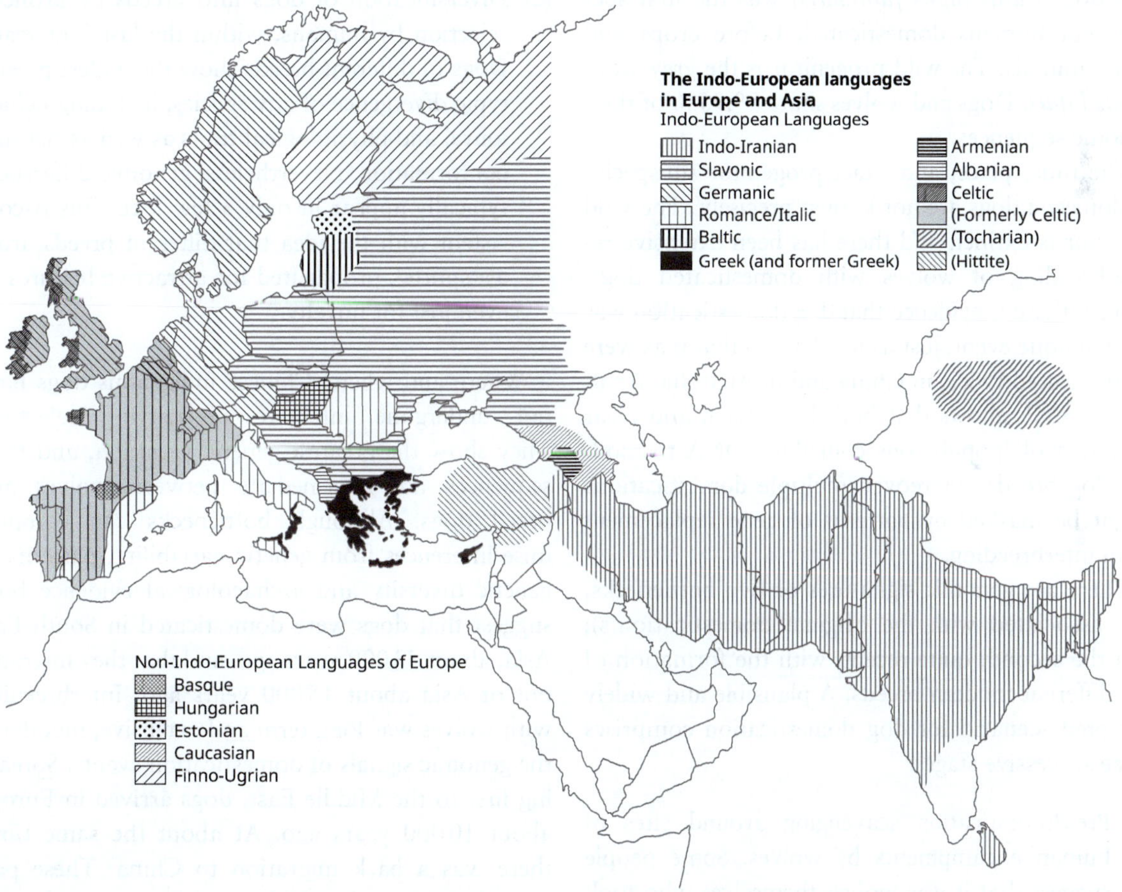

Figure 9.6 Map of Indo-European language family.

Renfrew, C. (1989). The origins of Indo-European languages. Scientific American 261, 4: 106–114. https://doi.org/10.1038/scientificamerican1089-106. Used with permission of Cambridge University Press.

- About 7000–5000 years ago, migrants moved south, to Armenia and Anatolia (present-day Turkey).
- About 5000, the same source population expanded into Eastern Europe, including the Balkans, and into central Europe.
- Migrants from the same population also ventured eastward, into South Siberia and Central/South Asia.

The evidence is genetic, archaeological, and linguistic. Just as with species, languages can be classified and clustered, and be organized into phylogenetic trees. It is possible to reconstruct a common ancestor of Indo-European languages. The Proto-Indo-European language had words for wheel and axle, implying that its speakers had wagons and/or chariots. This is evidence that the region of the Yamnaya culture was the location of both the emergence of Proto-Indo-European, and of the invention of wheeled transport. The Yamnaya who entered Europe 5000 years ago brought with them both language and technology.

Domestication of the dog

The dog (*Canis lupus familiaris*) was the first species that humans domesticated, before crops and other animals. The wild progenitor is the grey wolf, *Canis lupus*. Dogs and wolves share 99.95% of their genome sequences.

The time, place, and exact progenitor subspecies of domestic dogs are not known precisely. The wild ancestor is extinct, and there has been extensive re-interbreeding of wolves with domesticated dogs. Indeed, there is evidence that dog domestication was not a unique event, just as it is known that pigs were domesticated twice, in China and in Anatolia. Analysis of SNPs shows that both East Asian and Near Eastern wolf populations contribute DNA to modern dog breeds. Moreover, multiple domestications might be masked by patterns of gene replacement from interbreeding.

There are genomic signatures of two bottlenecks, one associated with the original domestication(s); and the second, quite recent, with the formation of the different modern breeds. A plausible and widely accepted scenario for dog domestication comprises three successive stages:

(a) Pre-domestication scavenging around sites of human encampments by wolves. Some people suggest that it was wolves themselves who took the initiative for domestication, not the humans.

(b) Domestication of dogs with increasingly intimate human–dog interactions.

(c) Diversification of dogs into breeds by artificial selection by humans, within the last 200 years. Dogs of different breeds show the widest phenotypic divergence of any species, including behavioural and physiological traits as well as size and conformation. Nevertheless, genomic differences typically appear in only a few loci. This is consistent with the idea that different breeds arose as 'sports', propagated for attractive features or even just for novelty.

Where and when? Chinese indigenous dogs have not undergone intensive diversifying selection. They show the greatest genetic variation, and their sequences are intermediate between wolves and breed dogs. Although bottlenecks can compromise inferences from genetic variability, patterns of genetic diversity and archaeological evidence both suggest that dogs were domesticated in South East Asia, about 33 000 years ago, and that they migrated out of Asia about 15 000 years ago. Interbreeding with wolves was long-term and extensive, muddying the genomic signals of domestication events. Spreading first to the Middle East, dogs arrived in Europe about 10 000 years ago. At about the same time, there was a back migration to China. These patterns do not correlate with contemporaneous human migrations.

Surveys of genome sequences of extant animals include wolves, primitive dogs from various locations, and modern dogs of different breeds. One palaeosequence is from a 35 000-year-old specimen of a Siberian wolf from the Taimyr peninsula, on the arctic coast of Siberia, north of Krasnoyarsk (and incidentally the most northern location on Earth that is not an island).

Figure 9.7 shows the patterns of relationship and divergence among wolves and dogs, including the Taimyr specimen. It emphasizes the long-term dog–wolf interbreeding. The archaic Taimyr wolf lineage has contributed between 1.4% and 27.3% of the ancestry of modern dog breeds from Siberia and Greenland. The initial domestication did not mark the end of dog–wolf introgression (hybridization including back-crossing—dogs mating with wolves).

What do these sequences reveal about the genetic changes associated with domestication?

There have been different motives for the domestication of different animals. For many mammals, a goal of domestication was to provide milk and meat. For cattle, horses, and dogs, the animals provided labour, including but not limited to enhancements in transportation. Dogs are primarily working and companion animals. However, all animal domestications require behavioural/psychological changes, to promote effective human–animal interaction and partnership.

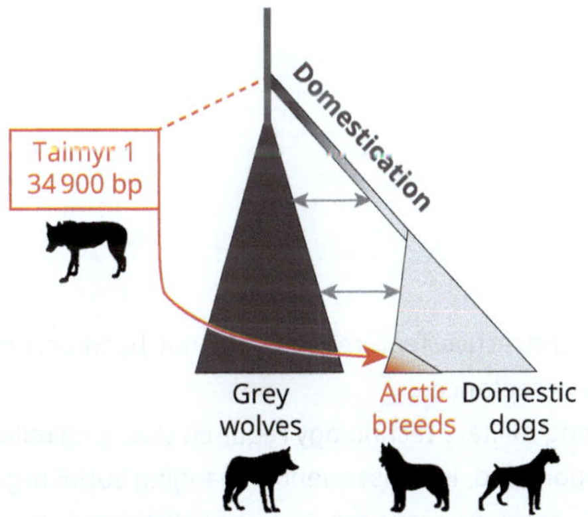

Figure 9.7 Relationships among progenitors of grey wolves and domesticated dogs. Note that there has been extensive dog–wolf interbreeding during (and before and after) domestication.

Used with permission of Elsevier Science & Technology Journals, from Skoglund, P., Ersmark, E., Palkopoulou, E., *et al.* (2015). Ancient wolf genome reveals an early divergence of domestic dog ancestors and admixture into high-latitude breeds. Current Biology 25, 11: 1515–1519. https://doi.org/10.1016/j.cub.2015.04.019. Permission conveyed through Copyright Clearance Centre, Inc.

With dogs, psychological adjustments were originally the major focus. (Subsequently, the development of different breeds also selected for size and appearance.) Domestication reduced fear of, and aggression towards, humans. To see how this worked at the genetic level one can (a) compare wolf and dog genomes, or (b) look at genes known to be associated with differential aggression in different dog breeds, and/or affecting aggressive behaviour in humans.

The epinephrine biosynthesis pathway was one target of domestication, affecting aggression. Genes implicated involve the synthesis, transport, and degradation of several neurotransmitters, including dopamine and noradrenaline. Genes affected include *TH* (encoding tyrosine hydroxylase), *DDC* (encoding L-DOPA decarboxylase), and *SLC6* (a neurotransmitter transporter). These play a role in behaviour, including but not limited to the flight-or-fight response.

Some proteins showing selection at early stages of dog domestication involve memory and learning, or nervous system development. These include GRIA1 (a neurotransmitter receptor), MBP (myelin basic protein), FGF13 (fibroblast growth factor 13), and SLC9A6 (sodium/hydrogen exchanger protein 6).

Some genes show significant allelic frequency differences between dogs and wolves. These include genes encoding monoamine oxidases (MOAO and MOAB), and tyrosine hydroxylase (TH). Genes implicated in behavioural differences among breeds encode dopamine 4 receptor (DRD4), dopamine transporter (SLC6A3), dopamine β-hydroxylase (DHB), and, again, MOAO, MAOB, and TH.

Not all the genetic differences between wolves and dogs involve behaviour. For instance, modern dogs have increased and variable copy numbers of AMY2B, the gene for starch-digesting pancreatic amylase, compared to wolves. (Most wolves have two copies, dogs anywhere from 4 to 30.) This may be a sign of adoption of a starch-rich diet, as a result of integration with humans. Some wolves also have multiple copies of AMY2B, but this may be the result of back-crossing (see Exercise 9.3).

Domestication of the horse

Of all the animals that humans have domesticated, the horse has had the greatest consequences for human technology and social organization. Like dogs, horses are both work and companion animals. They entered myth and fantasy—the Elgin marbles represent centaurs, Pegasus had wings. That the horse was the symbol of the city of Troy was essential to the success of the Greek ruse. More recently, countless movies record the importance of the horse to the development of the Americas in general and the West of what is now the US in particular.

Beginning in the Eurasian Neolithic, horses provided meat and milk, allowed development of the nomadic lifestyle, sped up the spread of goods and information—including languages, knowledge, and beliefs—and transformed warfare. To take full advantage of the horse's potential required development of crucial hardware, including the recursive bow, the wheel, and the stirrup (see Box 9.6).

Three successive stages of domestication involved (1) keeping horses as herd animals, (2) riding, and (3) harnessing horses to pull wagons and chariots.

(1) Archaeological evidence suggests that horse domestication, in the form of herding, began about 7000 years ago. A single human, perhaps assisted by a dog, can control a relatively large herd of animals. This is because the herd will have a natural leader, and control of the leader achieves control of the herd.[3] Horse domestication was relatively late, compared to sheep, goats, pigs, and cattle. Starting about 5200 years ago, shifts in climate producing colder winters (the Piora Oscillation) may have stimulated horse domestication: An advantage of horses is that they can forage for themselves over winter. Horses use their hooves to break crusted snow and ice to find food. Other animals require storage and provision of fodder.

[3] Rule 58 of UK Highway code (last updated October 2015), 'Animals being herded', contains the admonition: '... It is safer not to move animals after dark, but if you do, then wear reflective clothing and ensure that lights are carried (white at the front and red at the rear of the herd).' There is an implicit assumption that the animals will remain in order, else you'd have to keep transferring the lights.

BOX 9.6 Horses on the battlefield

The domestication of horses revolutionized warfare. Successive advances depended on development of technology as well as on the animals. Invention of the wheel led to the chariot, which for a time ruled the battlefield. Mounted archery was initially difficult, for an ordinary bow of adequate power must have an ungainly length. With invention of the recurve bow, shorter from tip to tip, riders on horseback—cavalry—could take advantage of greater mobility to overcome chariots. The stirrup was a later invention, perhaps 2000 years ago. Stirrups allow a knight to thrust a lance with the combined momentum of rider plus horse, without being propelled backwards off the mount. Historian Lynn White's suggestion that the stirrup was responsible for the feudal system in mediaeval Europe has many attractive features, but has not been accorded universal acceptance.

Changing military technology required changing military organization, and, in consequence, changing social organization. To form an army requires specialist warriors, who accept subservience to generals (see Homer's *Iliad*, among many sources). The heroic individual warrior became the depersonalized soldier, subject to orders. This was not only a social change but an ideological one. It is interesting that these changes in conduct and attitudes of humans, as a result of development of the application of horses, are in some ways analogous to the domestication of the horse itself. Conversely, to instil in a soldier the necessary aggression and even the willingness to kill, is a de-domestication.

(2) Humans first mounted on horseback about 5500–5700 years ago. Evidence for riding is the pattern of bit wear on the teeth of archaeological specimens. Riding probably started in the Pontic-Caspian steppes (the region from eastern Romania, through southern Moldova, Ukraine, Russia, and northwestern Kazakhstan). The first direct ancestors of modern domestic horses appeared about 4000 years ago.

(3) Invention of the wheel allowed horses to pull wagons, supporting nomadic lifestyles. Wagons permitted herders to migrate with their herds; the greater radius of residence enhancing interactions between populations, promoting the spread of languages and technology. Eurasia was transformed from a series of unconnected cultural pockets to a single interacting system—not necessarily an amicable one (then or now).

The wild progenitor of modern horses, *Equus ferus ferus,* is extinct. There is an extant wild horse, Przewalski's horse, *Equus ferus przewalskii.* Although Przewalski's is not the progenitor of modern horses, there is evidence that they interbred. The two subspecies are interfertile, despite the fact that they have different chromosome counts (modern horses have 32 chromosome pairs, Przewalski's have 33, hybrids have 32 chromosome pairs plus one unpaired chromosome).

Modern horses show great genetic diversity among females, but great homogeneity in males. That is, domestication involved relatively few stallions and many mares. This is attributable to the fact that mares but not stallions tend to be docile. Taming a stallion was presumably a difficult and rare event. And one stallion can impregnate many mares.

Comparative genomic studies have sequenced over 700 modern horses; 12 Przewalski's horses; and the remains of almost 300 ancient horses, from several possible sources of domestication including Iberia, Anatolia, Western Eurasia, and Central Asia. Two other ancient specimens sequenced were found in the Siberian permafrost, dated ~47000 years ago and ~16000 years ago; plus an ancient horse from a skeleton from North America, deposited approximately 700000 years ago. The sequences show current domesticated horses had their origin in populations from the Western European Steppes. Figure 9.8 shows the evolutionary tree.

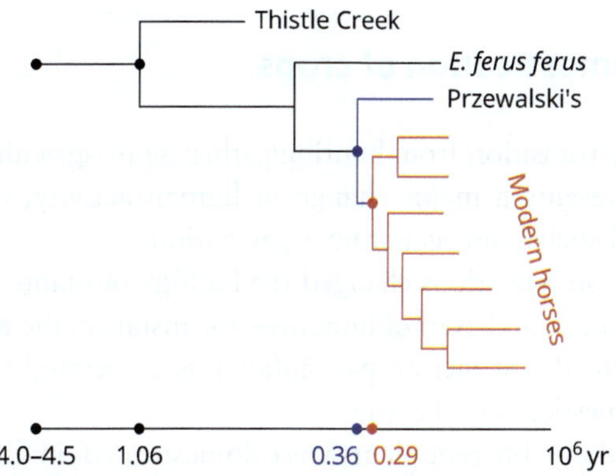

Figure 9.8 Evolutionary tree of extant horses and their extinct relatives.

The genomes allow identification of alleles present in domesticated horses but not in their progenitors.[4] The effects, in modern horses, of mutations in these genes reveals their functions. Sets of genes so identified (see Weblems 9.3 and 9.4) are involved in:

- Musculature (*ACTA1, C-SKI, MYBPC1, SGCD, VRK1,* and *TCTN1*). Defects in these genes involve myopathies or problems in locomotion, including difficulties in balance and coordination.
- Cardiac function (*ACAD8, BRAF,* and *CACNA1D*) and blood pressure regulation (*NR3C2, SCPEP1, WNK2,* and *CACNA1D*). These genes permit the increased energetic demands on horses in transportation (including chariotry), racing, and agriculture.
- Brain development (*NINJ1, NTM, MATN2, DCC, ASTN1, DLGAP1, ALK, NUMB, B3GALTL, GNPTAB, NIPBL, VDAC1, JPH3, VDAC1, GRID1,* and *CACNA1D*, yet again). These genes affect cognitive abilities, learning capacity, and docility.[5]

The very great interest in race-horse breeding has led to identification of genes that enhance performance on the track. It is interesting that the alleles associated with winning races were already present in wild horses. This is consistent with the idea that sports are a metaphor for real-life contests.

[4] Schubert, M., Jónsson, H., Chang, D., *et al.* (2014). Prehistoric genomes reveal the genetic foundation and cost of horse domestication. Proc. Natl Acad. Sci. USA 111: E5661–5669.

[5] In *Gulliver's Travels*, Swift describes the Houyhnhnms, a population of civilized horses: 'The behaviour of these animals was... orderly and rational... acute and judicious.'

Domestication of crops

The transition from hunting/gathering to agriculture represents a major change in human activity, diet, and social and economic organization.[6]

Domestications changed the biology of plants and animals, and even of humans—for instance, the ability to digest lactose past infancy is associated with domestication of cattle.

Many different plants were domesticated, in different regions around the world (Figure 9.9). Although these domestications were independent events, there are many common features:

- Characteristics that improve the product:
  - enlargement of fruit and/or seed
  - improved flavour and/or nutrition
- Characteristics that facilitate harvesting:
  - synchronization of ripening time
  - larger central stalks relative to side shoots—technically, increased apical dominance. For instance, contemporary maize has a central stalk, with the ears growing at the tips of short branches. The ancestral species, teosinte, was highly branched (see Figure 9.10). Apical dominance allows more plants per unit area tilled; it is analogous to building skyscrapers in cities.
  - seeds do not fall off the plant (called shattering). However, to facilitate harvesting the link of the seed to the plant should be relatively weak. The loss of seed dispersal can render the plant no longer viable in the wild.
- Tillering—shoots fill empty spaces between plants. This makes it unnecessary to plant seeds at specific intervals.
- Increased self-pollination.

These favourable properties are the result of genetic changes during domestication. Documented

[6] Not necessarily for the better. Jared Diamond has called the adoption of agriculture 'The Worst Mistake in the History of the Human Race'. He adduces in support of this contention the nutritional inferiority of diet, compared to hunter-gatherers; and social organization, including the problems associated with crowded living conditions, including but not limited to disease epidemics.

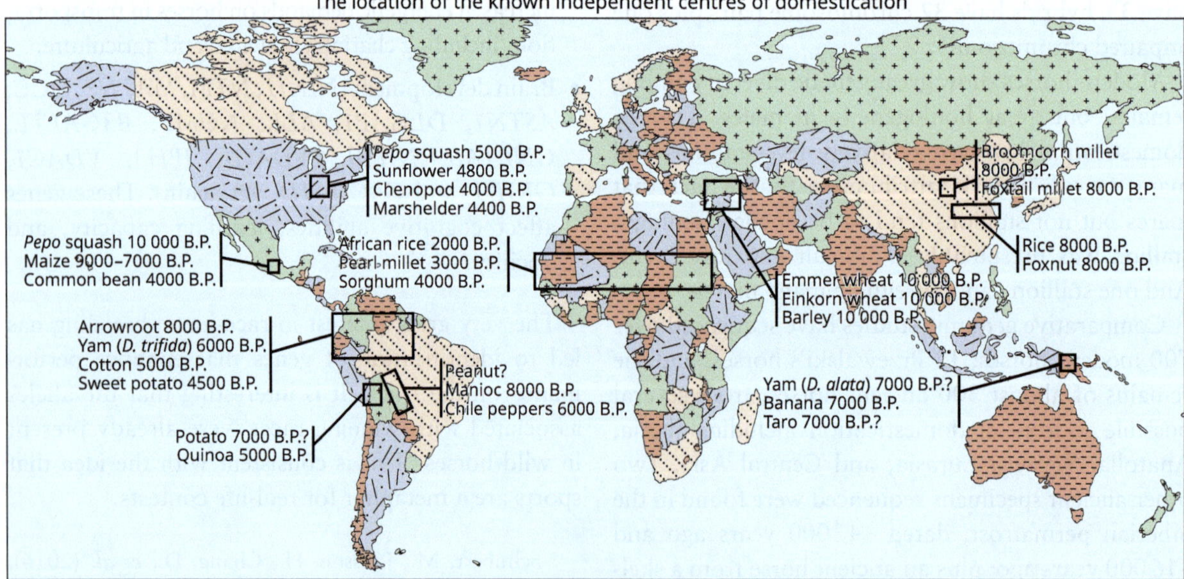

Figure 9.9 Populations in many regions of the world have domesticated plants. Dates shown are based on archaeological evidence, not DNA sequence analysis. (B.P. = before present.)

(a)

(b)

(c)

(d)

Figure 9.10 Comparison of modern maize with its teosinte progenitor (*Z. mays* ssp. *parviglumis*). (a) Teosinte grows many long, tasselled branches. (b) In modern maize, many short branches bear the ears at their tips. (c) The kernels of teosinte are encapsulated in a hard compartment. This picture shows both mature (left, dark) and immature (right) kernels. (d) Comparison of kernels of teosinte and modern maize.

Sources: (a) From US Department of Agriculture, Natural Resources Conservation Service, Plant Materials Program, Plant Release Photo Gallery. (b) Photo by David T. Webb, distributed by the Botanical Society of America. (c, d) Photographs by Hugh Iltis, with permission of Professor John Doebley.

types of changes include: amino-acid substitutions, deletion/truncations altering the functions of individual proteins, transposon insertion, regulatory changes, splice-site mutation, and gene duplications.

In general, domesticated plants show lower genetic diversity than their wild, progenitor species. Causes of loss of genetic diversity include (a) population bottlenecks and/or (b) selective sweeps (reduction of local genetic diversity produced by recent, strong positive selection).

Comparisons with genes that are selectively neutral, with respect to domestication phenotypes, reveal the relative importance of these two effects. Selectively neutral genes lose genetic diversity only through bottlenecks.

A very interesting question about the genetics of domestication is: to what extent were the genes selected for in domestication present in progenitor populations, and to what extent are they *de novo* mutations? Examples of both types are known. The large genetic variability in the progenitor population makes it harder to find a rare allele even if it was originally present.

KEY POINT

Studies of the genomes of crop plants have applications to agriculture, including the search for varieties that produce yields improved in quality and quantity, require less fertilizer and pesticides, and are resistant to disease. If genomes of wild progenitor species are also available, it is possible to study the genomics of domestication.

Maize (*Zea mays*)

The cultivation of maize (*Zea mays*) supported the pre-Columbian civilizations of Central and South America. Maize was brought to Europe in the fifteenth and sixteenth centuries and quickly spread around the world. Maize is now the world's third-largest crop plant, after rice and wheat. The annual harvest amounts to 10^{12} kg, raised on 2×10^6 km^2 of land. Maize is grown primarily as food, for humans and animals, but some is converted into ethanol for fuel.

In the late twentieth century, the intensive development of high-yielding crop varieties began. This 'green revolution', in addition to improving yields, bred maize for a higher content of lysine. Maize was formerly lysine-poor. As a result, people with diets based primarily on maize were traditionally lysine-deficient.

In the US and Canada, maize is called corn.

Maize is a domesticated form of a grass called teosinte. Several varieties of wild teosinte survive in Mexico. Analyses of isozyme and microsatellite diversity—'paternity tests for species'—have identified the progenitor of maize as a teosinte subspecies, *Z. mays* ssp. *parviglumis*, with contributions from a second wild relative *Z. mays* ssp. *mexicana*. A combination of archaeological and molecular evidence suggests that teosinte was domesticated in southern Mexico between 6000 and 9000 years ago. The earliest artefacts showing domesticated maize are cobs found in a cave in Oaxaca, Mexico, dated 6250 years ago.

Rafael Guzmán, an undergraduate student at the Universidad de Guadalajara, discovered the teosinte progenitor strain of maize during field work in southwestern Mexico in late 1978. Guzmán was stimulated by a challenge contained in a New Year's card from botanist Hugh Iltis. The importance of his discovery for maize science cannot be overestimated. Access to the progenitor strain makes possible detailed comparisons of genome sequences. Maize and teosinte are still interfertile, permitting the reintroduction of specific alleles. The teosinte that Guzmán discovered contains unique virus-resistant genes that have been bred into the maize used in agriculture.

Notable differences between teosinte and modern maize include:

- Teosinte has many long branches tipped by tassels. The tassels correspond to male flowers. Modern maize has a single main stalk, with the tassel at the top. The many short lateral branches bear the ears at their tips. The ears, containing the seeds, of course develop from female flowers. (Separate male and female flowers are a feature of varieties of teosintes and maize, not shared by other grasses.)
- The teosinte ear contains 5–12 kernels. Their structure adapts them for dispersal by passage through the digestive tracts of birds and mammals. Individual seeds grow inside an encasing **glume** hardened by silica and lignin (see Figure 9.10c). For human consumption, the kernels would have to be ground, or 'popped' by heating. At maturity, teosinte seeds separate spontaneously, another aid to dispersal. In contrast, an ear of modern maize has several hundred kernels, with no hard casing, and the kernels remain fixed to the ear (see Box 9.7).

These very large phenotypic differences between teosinte and maize appear in plants that have very similar genomes. The resolution of this paradox must emerge from study of the genomes, the proteins, and the expression patterns. How did the genome change upon domestication? How many genes were involved? Which genes? How have the changes in the genome produced the phenotypic effects?

KEY POINT

For maize we are fortunate to have both the modern varieties and the progenitor, teosinte.

The first clues to the extent of the genetic change came from classical genetics. G. Beadle—better known for his 'one-gene–one-enzyme' hypothesis—crossed a strain of maize with teosinte and examined

BOX 9.7 Glume and doom: modern maize could not survive in the wild

Loss of teosinte's hard and adhering seed casing in modern maize is fatal for its natural method of seed dispersal. If eaten by birds or other animals, the seeds would not pass through and be sown. They would be digested, as they are when we eat maize. As a result, modern maize could not survive in the wild. It depends on humans to plant seed each year.

There is a symmetry, literally a symbiosis: human populations—including Mesoamerican civilizations and the earliest pilgrim settlements in New England—have been dependent on maize for food, and maize is dependent on humans for sowing its seed. One could say—depending on one's point of view—that maize has domesticated humans. This is a characteristic of many but not all domestications.

the second generation (F2). He concluded that domestication involved about five major loci. The experiment was not intended to pinpoint the number of altered genes—Beadle wanted primarily to distinguish whether a large or a small number of loci were involved.

KEY POINT

Some people had believed that the phenotypic differences were so great as to preclude conversion of teosinte to maize by human selection. Beadle's point was that the genetic differences might be simpler than suspected.

The 2.3-Gb genome of maize was sequenced in 2009. With the availability of sequence information, it is possible to determine which genes in maize have been subject to selection. *Sites selected during domestication should show enhanced loss of genetic diversity*. There should be a signal from these sites, observable in comparisons of sequence diversity between maize and teosinte. This effect is amplified by the high natural variation in the progenitor population: in teosinte, *Z. mays* ssp. *parviglumis*, the nucleotide diversity at silent sites is as high as 2–3%. (Of course, these numbers may not accurately reflect the characteristics of a small founder population that gave rise to maize.) The overall loss in diversity upon domestication is estimated at ~25–30%, attributable to the 'domestication bottleneck'. The signal of selection must stand out, at a higher conservation level, above this background.

KEY POINT

What are the signatures of selection? A decrease in nucleotide diversity, increased linkage disequilibrium, and altered population frequencies of polymorphic nucleotides in a gene and linked regions.

The results of such sequence comparisons suggest that about 3%, or about 1200 genes, were targets of selection during maize domestication. At least 50 contribute to agronomically significant traits. Many of the genes involved in morphological changes are clustered around loci that may correspond to Beadle's observations. Others affect biochemical characters that Beadle did not investigate.

Some of the genes involved have been identified as transcription regulators.

- One gene that differs between teosinte and maize is *tb1* (tb = teosinte branched). The maize version represses lateral shoot development and converts tassels to ears (compare Figures 9.10a and 9.10b). Maize has ~30% of the diversity in teosinte populations in the protein-coding region of this gene—approximately the background level of diversity changes upon domestication—but only ~2% of diversity of teosinte in the region 5' to the gene. TB1 is a repressor, which binds to specific sites in promoters of cell-cycle genes. In this case, selection has been applied not to the coding region of the gene but to an upstream regulatory region. Indeed, the transcription level of *tb1* in maize is higher than that of teosinte. This appears to be the primary mechanism of action of the genomic difference: there has been selection for expression level, rather than for a changed amino-acid sequence of the protein expressed. For *tb1*, the alleles selected for domestication are present in wild teosinte.
- Another gene that differs between teosinte and maize is *tga1* (tga = teosinte glume architecture). This gene affects the structure of the glume, the surroundings of the seed. The teosinte allele produces the hard seed case. The maize allele reduces the glume to a soft membrane underneath the kernels. For *tga1*, the expression level is similar in teosinte and maize. However, there is a specific single nucleotide change in one exon, substituting a lysine in a teosinte protein with an asparagine in maize. The maize allele has not been found in wild teosintes.

Can we know whether these and other genetic differences appeared at the time of domestication? It has been possible to sequence DNA from cobs found in sites of a variety of ages, the oldest dated at 4400 years ago. The 4400-year-old cob shows modern-maize alleles for three genes tested, including *tb1*, showing that all three were present in the maize population at least that long ago. However, teosinte-specific alleles of some genes appear in 2000-year-old cobs from New Mexico, in the southwest US. Selection was incomplete then, at least in regions distant from the site of original cultivation.

Rice (*Oryza sativa*)

Rice is one of the world's most important crop plants. The 2022/2023 harvest produced 7.8×10^{11} kg. Rice forms the major component of agriculture in East Asia and India. It provides the staple diet of half the world's population. Rice accounts for about 20% of human calorie intake worldwide.

In Asia, the two major cultivars are: *Oryza sativa* ssp. *japonica* and *Oryza sativa* ssp. *indica*. The closest wild relative of O. *sativa* is the Asian wild grass *Oryza rufipogon*. O. *rufipogon* grows over a very wide geographic range, motivating many claimants for the site of the original domestication. Domesticated species show larger grains, and are non-shattering. (In this context, shattering means that the seeds fall off the plant as soon as they become ripe, to aid dispersal in the wild. For rice and many other crops, domestication involves suppression of shattering: The consequent retention of seeds on the plant facilitates harvesting.)

Do not confuse the wild progenitor *Oryza rufipogon* of domesticated rice *O. sativa* with the grain known as 'wild rice', *Zizania palustris* and certain other *Zizania* species, now grown primarily in the northern midwest US. These are not close relatives of *Oryza*.

Familiar from grocery shops are many different varieties of rice, used in different national cuisines. The characteristics of the major ones are:
Indica: long, slender grains; grown mainly in tropical areas
Japonica: rounder, shorter grains; temperate and tropical varieties
Minor varieties:
Aus: drought tolerant; Bangladesh and West Bengal
Aromatic: Iran, Pakistan, India, Nepal; including Basmati.

Genes changed from O. *rufipogon* to both subspecies of O. *sativa* include:

sh4—the mutation contributes to the suppression of shattering. *sh4* encodes a helix-turn-helix transcription factor controlling genes involved in cell wall degradation, part of the process of breaking the link between the seed and the plant.

rc—the mutation changes pericarp colour from dark (brownish-red) to white, by blocking proanthocyanidin synthesis (see Figure 9.11). Proanthocyanidins are also present in grapes, giving red wine its colour, and antioxidant effects. Because proanthocyanidins protect the plant against pathogens and predators, mutations that block their synthesis would be selected against in the wild. In O. *rufipogon, rc* also encodes a helix-loop-helix protein that is a regulator of gene expression. The observed mutations knock out the activity, blocking expression of proteins in the proanthocyanidins biosynthetic pathway.

Control of flowering time

Crop productivity depends in part on the duration of the growing season between sprouting and flowering. In rice, a major trigger for flowering is 'short-day'; that is, plants flower when day lengths fall below a threshold. (Flowering also responds to other environmental cues, including light quality, and availability of water and nutrients.) The farther from the equator, the earlier in the year the days grow shorter, threatening to truncate the rice growing season. To extend the regime of rice cultivation from tropical to temperate regions, it would be useful to delay flowering.

The control of flowering in rice is a very complex process. The gene *Hd3a* (= Heading data 3a; 'heading' is the immediate precursor of flowering) is one major focus of the control pathway for flowering. *Hd3a* is itself under control of circadian genes that sense day length.

An immediate precursor of Hd3a in the control pathway is Hd1, a zinc-finger-containing transcription regulator. Under short-day conditions, Hd1 enhances *Hd3a* expression, promoting flowering. Under 'long-day' conditions it reduces *Hd3a* expression, repressing flowering. Cultivars with mutant, non-functional Hd1 also have later flowering times.

Rice subspecies *japonica* and *indica* differ in their geography of cultivation. *Indica* is cultivated throughout tropical and subtropical regions of Asia. *Japonica* is grown in subtropical and temperate regions. Surveys show that *Hd1* alleles are sharply separated between temperate and subtropical populations of *japonica*. *Hd3a* expression in *japonica* is repressed north of 31° North, the latitude of Shanghai. Contributing to domestication was the selection of appropriate *Hd1* alleles, including but not limited to non-functional

(a)

(c)

(b)

Figure 9.11 Comparison of colours of kernels of (a) *O. rufipogon* and (b) *O. sativa*. (c) 'Brown rice' is non-dehulled kernels of *O. sativa*. Among the nutrients lost with the hull is vitamin B1 (thiamine). Although many people prefer the appearance, taste, and texture of white rice, brown rice is nutritionally superior. The deficiency disease beriberi is common in regions in which the typical diet contains white (dehulled) rather than brown rice, and there is no other dietary source of thiamine.

Used with permission of Oxford University Press, from Sweeney, M., & McCouch, S.R. (2007b). The complex history of the domestication of rice. Annals of Botany 100, 5: 951–957. https://doi.org/10.1093/aob/mcm12. Permission conveyed through Copyright Clearance Centre, Inc.

versions of Hd1, that allowed expansion of rice cultivation to more northernly latitudes.

How does Hd3a induce flowering? Downstream of *Hd3a* in the control pathway is *OsMADS15*, a homeotic gene, a general developmental trigger for flowering. Hd3a binds proteins 14-3-3 and OsFD1 to form a 'Florigen activation complex', which binds to DNA and activates transcription of *OsMADS15*.

History of rice domestication

What was the phylogenetic path of rice domestication in Asia? How many separate domestication events were there? Both subspecies, *japonica* and *indica*, are more closely related to different wild varieties than they are to each other. This suggests two independent domestications. The puzzling feature is that the *sh4* and *rc* genes share the *same* mutations in both subspecies (see Figure 9.12). How can these observations be reconciled? The current model is that O. *japonica* spread to India, and was hybridized with local varieties to produce O. *indica* (see Figure 9.13).

Where in China did domestication occur? A survey of geographic genome variation in wild grasses suggests the Pearl River valley.

When? A difficulty with this question is that cultivation might have preceded—by a very long time—the appearance of the alleles distinguishing the modern cultivars. Han dynasty records from 2000 years ago distinguish short-grained (*japonica*) and long-grained (*indica*) varieties. The oldest archaeological samples, from Baligang, Hunan province, are radiocarbon dated to 6300 years ago. The survey of

Rice sh4

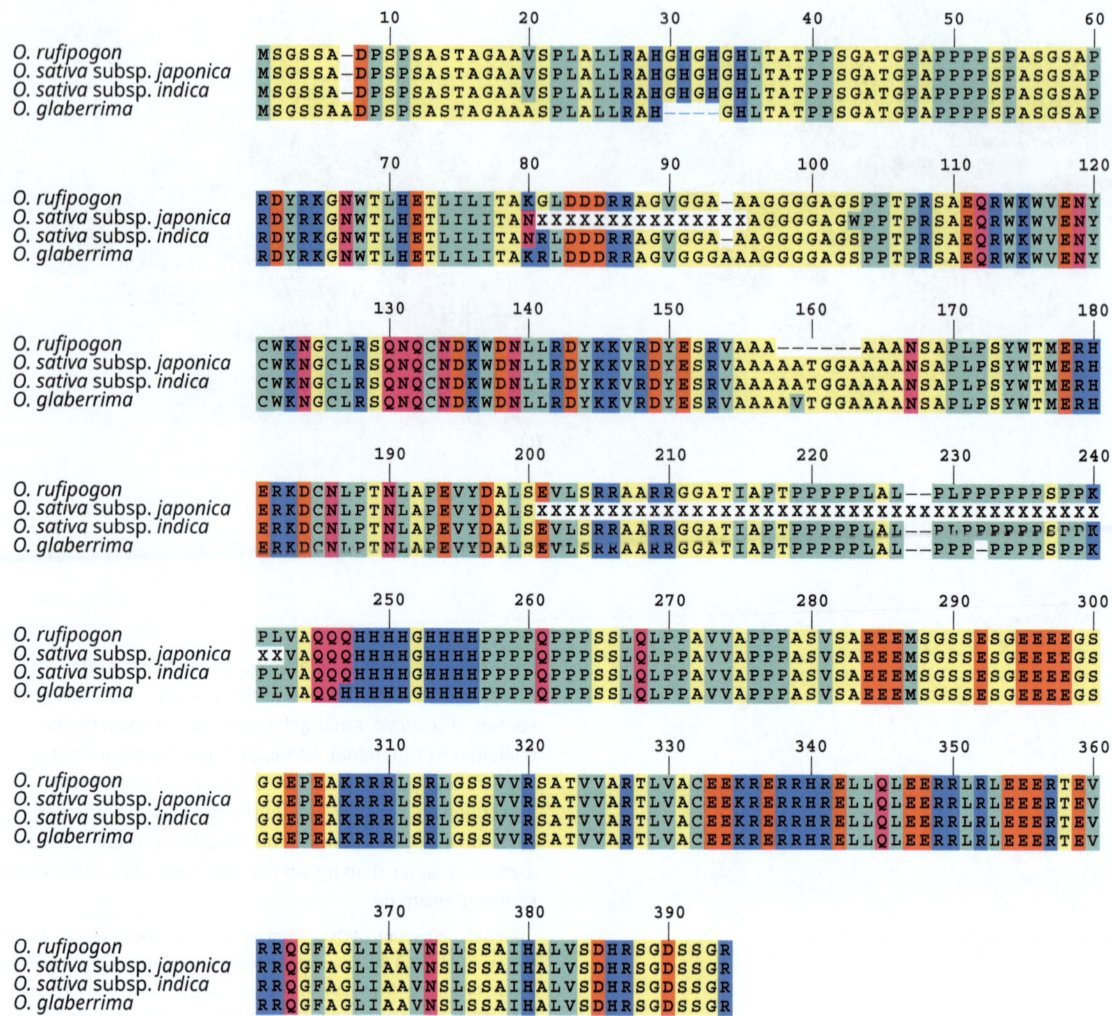

Figure 9.12 Sequence alignment of SH4 proteins in wild species *O. rufipogon,* cultivated species *O. sativa* sspp. *japonica* and *indica*, and *O. glaberrima*. The amino-acid sequences of the proteins from the two *O. sativa* subspecies are the same. Where is there a difference that distinguishes the *O. sativa* domesticates from the other two?

genome variation suggested that the original domestication occurred 8200–13 500 years ago.

Rice was brought to Europe by armies of Alexander the Great. The greatest eastern extent of his empire overlapped territory that is now India. Classic Greek texts mention rice. Rice also arrived in Europe from the Middle East, through trans-Mediterranean contacts.

An independent domestication of rice occurred in Africa, in what is now Mali, about 2000–3000 years ago, to produce *Oryza glaberrima*. Its progenitor is the wild grain *Oryza barthii*. *O. glaberrima* was cultivated in Africa, then brought to the Americas with the slave trade. It was cultivated in large areas of North, Central, and South America.

O. sativa was independently brought to the New World. Rice appears on a provision list from the Mayflower, the first ship bringing British colonists from Southampton to Massachusetts, US, in 1620. The Spanish introduced it to Central and South America, and the Portuguese to Brazil.

In both Africa and the Americas, *O. sativa* has largely replaced *O. glaberrima*. However, it continues to be cultivated as a minority crop, in part for traditional ritual or medicinal uses. *O. sativa* also has traditional ritual uses, notably at weddings.

Breeders have created many varieties of rice. Genetic engineering has created strains with enhanced nutritional values (see Box 9.8).

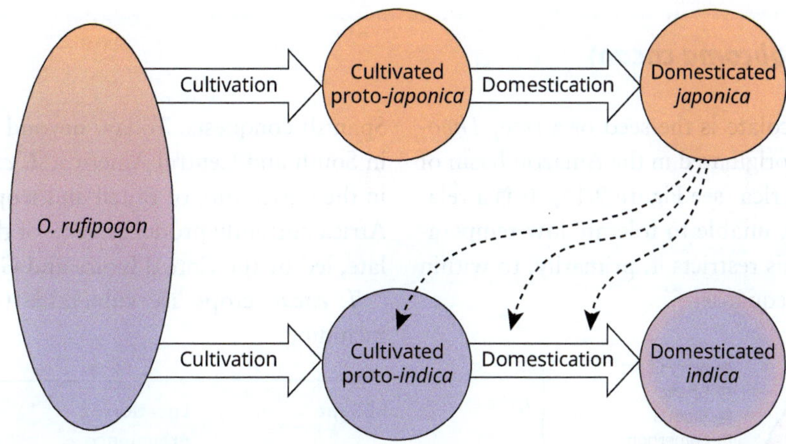

Figure 9.13 Suggested pathways of domestication to produce contemporary cultivars *Oryza sativa* ssp. *japonica* and *Oryza sativa* ssp. *indica* (circles in right column). The multiple colours assigned to the wild progenitor *O. rufipogon* indicate genetic diversity. The two-step process comprises first cultivation, a primarily passive use of a wild species without modification; followed by domestication, active modification of the wild species. Consistent with genome sequence data is the model that *japonica* was first domesticated from *O. rufipogon*, then exported to India where it was hybridized with local cultivars to create the domesticated *O. sativa* ssp. *indica*.

Gross, B.L., & Zhao, Z. (2014). Archaeological and genetic insights into the origins of domesticated rice. Proceedings of the National Academy of Sciences of the United States of America 111, 17: 6190–6197. https://doi.org/10.1073/pnas.1308942110

BOX 9.8

Golden rice

Vitamin A deficiency is an important public health problem, prevalent in regions where rice is an important dietary component. Golden rice is characterized macroscopically by yellow grains, and biochemically by high concentrations of β-carotene, precursor of vitamin A (see Figure 9.14). Golden rice was created by genetic engineering of rice by inserting two genes in the β-carotene synthesis pathway. These genes are under the control of an endosperm-specific promoter. The genes introduced are: phytoene synthase, from maize; and carotene desaturase from the soil bacterium *Erwinia uredovora*. It is estimated that an individual could achieve the minimum daily requirement of vitamin A by eating 75 g of golden rice per day. Whether cultivation of golden rice, like other genetically modified crops, will be allowed, is a matter of intense debate. It is permitted in some countries, forbidden in others.

Figure 9.14 The yellow colour of golden rice, compared to ordinary white rice, is the result of high concentrations of β-carotene, a precursor of vitamin A. Many people depending on rice for their nutrition lack adequate amounts of vitamin A. Symptoms of vitamin A deficiency include night blindness, dry skin, and frequent infections. Golden rice would help to alleviate this condition.

Courtesy Golden Rice Humanitarian Board, www.goldenrice.org. (See also: Wu, F., Wesseler, J., Zilberman, D., et al. (2021). Opinion: Allow Golden Rice to save lives. Proc. Natl. Acad. Sci. U.S.A. 118.

Chocolate (*Theobroma cacao*)

The source of chocolate is the seed of a tree, *Theobroma cacao*, that originated in the Amazon basin of western South America (see Figure 9.15). It is a relatively finicky plant, unable to tolerate low temperatures or aridity. This restricts it, primarily, to within 10° latitude of the equator:

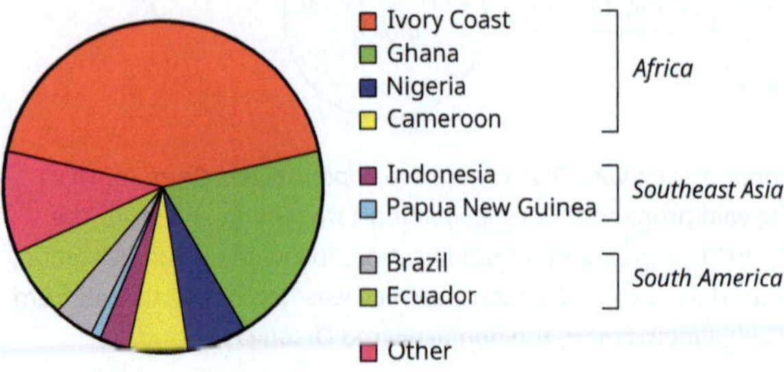

Chocolate drinks were important in the major central American Olmec, Maya, and Aztec cultures (see Box 9.9). It became popular in Europe after the Spanish conquests. Today, beyond its native habitat in South and Central America, *T. cacao* is important in the agriculture of Brazil and tropical Africa. West Africa currently produces 70% of the world's chocolate, led by the Côte d'Ivoire and Ghana.

T. cacao crops are vulnerable to several blights, including:

Blight	Location of prevalence	Causative organism
Witches' broom	South America	*Moniliophthora perniciosa*
Black pod rot	Central and West Africa	*Phytophthora megakarya*
	Global	*Phytophthora palmivora*
Vascular-streak dieback	Southeast Asia	*Ceratobasidium theobromae*

(a)

(b)

Figure 9.15 (a) *Theobroma cacao* tree showing fruit. (b) A ripe pod split open, showing the beans. To create chocolate from beans, the seeds are fermented and dried. Processing extracts cocoa butter (triacylglycerols) and cocoa powder (proteins and polysaccharides, plus small molecules such as flavonoids, terpenes, and theobromine).

Photo (a) by Paul Bolstad, University of Minnesota, Bugwood.org, (b) by Keith Weller, USDA Agricultural Research Service, Bugwood.org.

Some history of chocolate

In the Maya and Aztec civilizations in Central America, chocolate was a medicinal and ceremonial drink, and an expensive one.[7] Fermentation could produce an alcoholic beverage. The Aztecs prepared chocolate, not as a sweet as we know it, but mixed with chili pepper to form a spicy, bitter drink. In fact, the Aztec word xocolatl literally means xococ (bitter) + atl (water). A Maya vase in the collection of the Princeton University art museum shows a woman pouring chocolate from a height into a receptacle, to produce froth (http://artmuseum.princeton.edu/collections/objects/32221). Even more analogous to wine, in the Maya culture chocolate served sacramental, nutritious, and medicinal purposes. The beans also served as currency: a turkey cost 100 cacao beans.

European contact with cacao beans began in 1502 when Columbus captured the cargo of two Mayan trading canoes on his fourth voyage. The beans made little impact. Wider appreciation of chocolate awaited Cortez's return from his conquest of Mexico. In 1528, Cortez presented to Charles V a sample of cacao beans, and the tools and recipe for preparing them. In 1544, a group of Maya nobles accompanied Franciscan friars to the court of Spanish King Philip II and demonstrated the preparation.

In contrast to the spicy, bitter Mayan and Aztec beverages, the Spanish added sugar. This initiated the transition of chocolate from a sacramental and medicinal to a recreational substance. Chocolate beverages containing alcohol are now prepared by mixing chocolate and spirits.

Chocolate was introduced to Europe before coffee or tea were. It is widely believed that the marriage of Philip II's daughter Anne of Austria to Louis XIII of France in 1615 brought chocolate across the Pyrenees. Chocolate was the first exposure of Europeans to stimulant alkaloids such as caffeine. The taste for chocolate grew and spread, first as a drink and later in the solid, familiar 'bar' form. The first chocolate house in London opened in 1657. The popularity of these beverages derived originally from a belief in their medicinal value, as well as their flavour. Samuel Pepys's diary mentions drinking chocolate on several occasions, alluding to its curative properties:

> Waked in the morning with my head in a sad taking through the last night's drink, which I am very sorry for; so rose and went out with Mr. Creed to drink our morning draft, which he did give me in chocolate to settle my stomach. (24 April 1661)

The eighteenth-century physician and architect Sir Hans Sloane was the inventor of milk chocolate.

> Sloane is better known as the founding donor of the original collection of the British Museum. London's chic Sloane Square was named after him because his heirs owned the land on which it was developed. The design of the classic red British phone box, now an endangered species, was based on Sloane's tomb, which he himself designed.

The odour and taste of natural chocolate derives from unique flavonoids. Pharmacologically active compounds in chocolate include:

- theobromine (10% by weight of dark chocolate): heart stimulant, cough suppressor, vasodilator, lowering blood pressure leading to feelings of relaxation, diuretic. Pet lovers be warned: theobromine is toxic to dogs and cats.
- phenethylamine: a stimulant, similar in effect to amphetamines.

[7] '... chocolate drinks occupied the same niche as expensive French champagne does in our own [US] culture.' Coe, S.D., & Coe, M.D. (2007). *The History of Chocolate*, 2nd edn. London: Thames and Hudson, p. 61.

'Witches' broom' disease has hit Brazil's plantations very hard, but so far has not affected trees grown in Africa. A search for resistant varieties has motivated many expeditions to the headwaters of the Amazon. These expeditions also provided an opportunity to map out the geographic distribution of genetic variation in *T. cacao* trees. Areas of the highest variability are in the upper reaches of the Amazon, in what are now parts of Ecuador. This identifies the probable region of origin of the species.

Fourteen genetic clusters have been identified, extending the known varieties, which have been differentiated according to the appearance of the beans—as well as their flavour. Cultivars of *T. cacao* important in agriculture include: *criollo* (*T. cacao* ssp. *cacao*), the best tasting, and used for the finest products; *forestero* (*T. cacao* ssp. *sphaerocarpum*), inferior in taste but resistant to disease (90% of the world's cacao crop is *forestero*); and *trinitario*, a *criollo–forestero* hybrid, superior in taste to *forestero* and contributing to its hardiness.

KEY POINT

Unlike maize and rice, *Theobroma cacao* did not have to undergo a substantial genetic modification to domesticate it. Originally, the seeds from the native varieties were harvestable directly.

The *T. cacao* genome

T. cacao projects have sequenced the *criollo* and *forestero* varieties.

T. cacao is a diploid organism containing ten chromosomes. The assembly of the 420 Mbp *criollo* genome at 16.7× coverage includes 76% of the genome. Much of the remainder comprises repetitive regions. A linkage map allowed 67% of the 326 Mb sequenced to be anchored within the ten chromosomes.

28 798 protein-coding genes were identified, more than in *A. thaliana*. The average gene size was 3346 bp, with a mean of 5.03 introns/gene. Figure 9.16 shows the shared and unique genes from *T. cacao*.

Motivations for sequencing the *T. cacao* genome include the desire to understand:

- the determinants of resistance to witches' broom and other diseases
- the sources of the flavours in the chocolate produced
- the place of *T. cacao* in the general phylogeny of angiosperm plants.

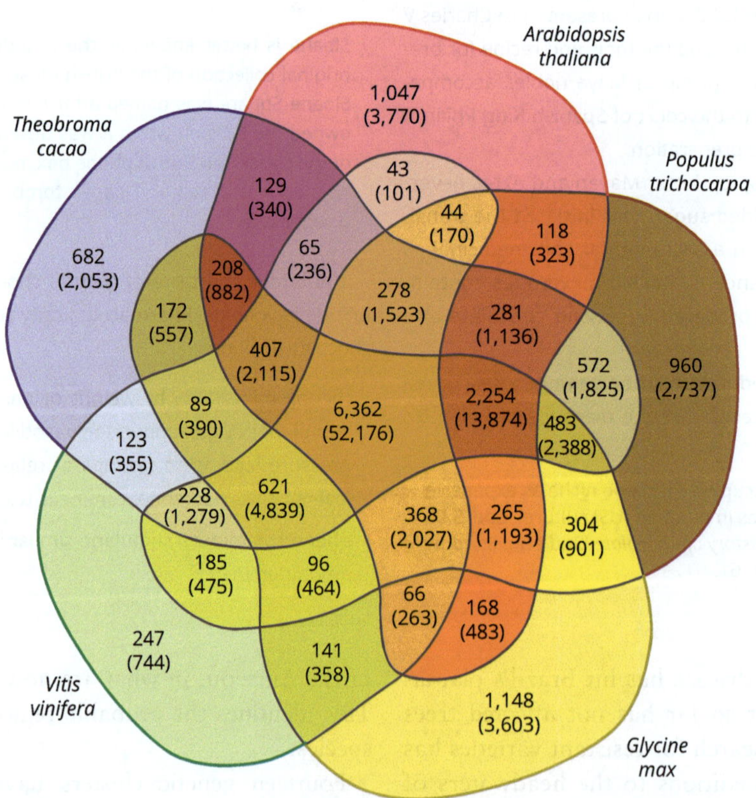

Figure 9.16 Numbers of shared and unique gene families and, in parentheses, numbers of genes from chocolate (*Theobroma cacao*), thale cress (*Arabidopsis thaliana*), black cottonwood (*Populus trichocarpa*), soya bean (*Glycine max*), and wine grape (*Vitis vinifera*).

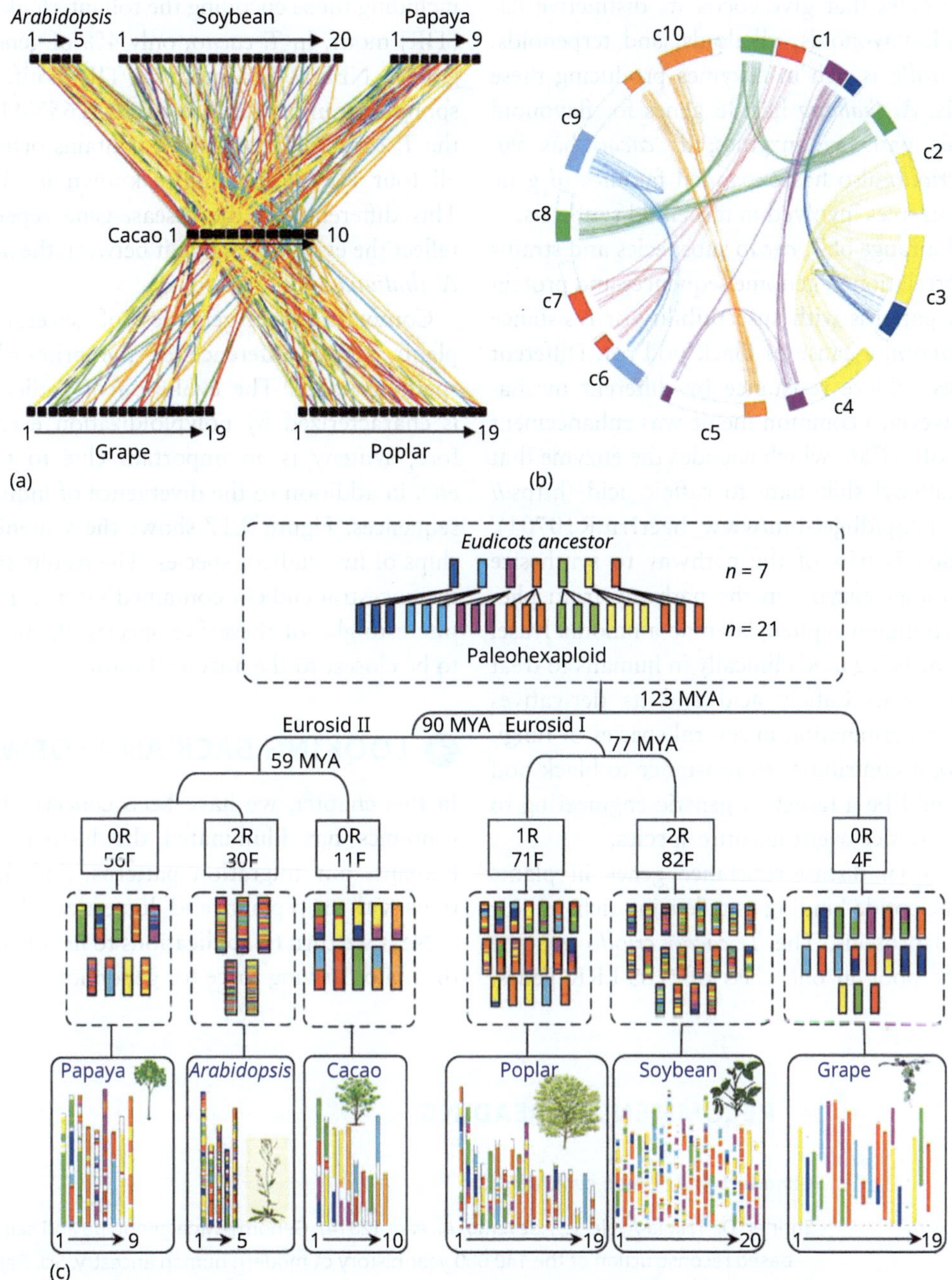

Figure 9.17 Syntenic relationships of eudicot genomes, comparing *T. cacao* with thale cress (*Arabidopsis thaliana*), soya bean (*Glycine max*), black cottonwood (*Populus trichocarpa*), papaya (*Carica papaya*), and wine grape (*Vitis vinifera*), and inferring the chromosome structure of the common ancestor. (a) Mapping of orthologues from *T. cacao* to other five genomes. For each species chromosomes are numbered consecutively. Different colours represent seven ancestral eudicot linkage groups. (b) Syntenic relationships *within* the *T. cacao* genome. There is evidence for ancestral triplication of the genome. For example, light green links regions of chromosomes 1, 2, and 8. (c) A model for the evolutionary history of the karyotype. The six species studies derived from an original eudicot ancestor with seven chromosomes. To produce the current karyotypes there were whole-genome duplications (R) and chromosome fusions (F). From comparisons of the chromosomes of the six species, it is possible to trace the path from the current chromosome structure of the six species (shown at the bottom), back to the common ancestor. The gene distribution among *T. cacao* chromosomes is the closest, of these six species, to the ancestral form.

The molecules that give cocoa its distinctive flavour include flavonoids, alkaloids, and terpenoids. *T. cacao criollo* is rich in enzymes producing these compounds. *A. thaliana* has 36 genes for flavonoid biosynthesis pathway enzymes; *T. cacao* has 96. *T. cacao criollo* also has expanded families of gene encoding enzymes involved in terpenoid synthesis.

Study of a range of *T. cacao* subspecies and strains allowed correlation of genome sequences and protein expression patterns with susceptibility or resistance to *P. palmivora,* a cause of black pod rot. Different populations achieve resistance by different mechanisms. However, a common theme was enhancement of activity of *TcCSE*, which encodes the enzyme that converts caffeoyl shikimate to caffeic acid (https://www.genome.jp/dbget-bin/www_bget?rn:R10701). This reaction is part of the pathway to synthesize lignin. (Another enzyme in the pathway from phenylalanine to lignin is phenylalanine-ammonia lyase, which is now being used clinically in humans to treat phenylketonuria.) Caffeic acid and its derivatives inhibit spore germination in several species of fungi. In *T. cacao*, it contributes to resistance to black pod rot, and could be a target of genetic engineering to enhance resistance over the entire species.

Other known disease-resistance genes in plants encode nucleotide-binding site/leucine-rich repeat (NBS-LRR) proteins. The *T. cacao criollo* genome is relatively poor in one class of NBS-LRR genes, including those encoding the toll interleukin receptor (TIR) motif. In *T. cacao*, only 4% of genes orthologous to NBS-LRR contain the TIR motif. The corresponding number for *A. thaliana* is 65%! In contrast, the *T. cacao criollo* genome contains orthologues of all four NPR1 subfamilies known in *A. thaliana.* This difference in the disease-gene repertoire may reflect the evolutionary split between the ancestors of *A. thaliana* and *T. cacao*.

Comparing the genomes of several flowering plants allows inference of the primordial angiosperm genome. The evolution of eudicot genomes is characterized by polyploidization events. Therefore, synteny is an important clue to the phylogeny, in addition to the divergence of individual gene sequences. Figure 9.17 shows the syntenic relationships of five eudicot species. The results suggest that the ancestral eudicot contained seven chromosomes. Interestingly, of these five species, *T. cacao* appears to be closest to the ancestral form.

LOOKING BACK AND FORWARD

In this chapter, we have been concerned with how genomics has illuminated the history of modern humans, our migration patterns, and our interactions with crop plants and domesticated animals.

Next we turn to applications to health and disease, the major driving force for genomics research.

RECOMMENDED READING

Genomics and human evolution

Shriner, D., Tekola-Ayele, F., Adeyemo, A., *et al.* (2014). Genome-wide genotype and sequence-based reconstruction of the 140,000 year history of modern human ancestry. Sci. Rep. 4: 6055.

Pääbo, S. (2014). *Neandertal Man: In Search of Lost Genomes*. New York: Basic Books.

Simonti, C.N., & Capra, J.A. (2015). The evolution of the human genome. Current Opinion in Genetics Development 35: 9–15.

Harris, E.E. (2015). *Ancestors in our Genome: The New Science of Human Evolution*. Oxford: Oxford University Press.

Sánches-Quinto, F., & Lalueza-Fox, C. (2016). Almost 20 years of Neandertal palaeogenomics: Adaptation, admixture, diversity, demography and extinction. Phil. Trans. Roy. Soc (London) B379: 21030374.

Rogers, A.R., Bohlender, R.J., & Huff, C.D. (2017). Early history of Neanderthals and Denisovans. Proc. Natl. Acad. Sci. USA 37: 9859–9863.

Cappellini, E., Prohaska, A., Racimo, F., *et al.* (2018). Ancient biomolecules and evolutionary inference. Annu. Rev. Biochem. 87: 1029–1060.

Wolf, A.B., & Akey, J.M. (2018). Outstanding questions in the study of archaic hominin admixture. PLoS Genet. 14: e1007349.

Dannemann, M., & Racimo, F. (2018). Something old, something borrowed: Admixture and adaptation in human evolution. Curr. Opin. Genet. Dev. 53: 1–8.

Liu, Y., Mao, X., Krause, J., *et al.* (2021). Insights into human history from the first decade of ancient human genomics. Science 373: 1479–1484.

Ahlquist, K.D., Bañuelos, M.M., Funk, A., *et al.* (2021). Our tangled family tree: New genomic methods offer insight into the legacy of archaic admixture. Genome Biol. Evol. 6: evab115.

Schaefer, N.K., Shapiro, B., & Green, R.E. (2021). An ancestral recombination graph of human, Neanderthal, and Denisovan genomes. Sci. Adv. 2021 16: eabc0776.

Shao, Y., Zhou, L., Li, F., *et al.* (2023). Phylogenomic analyses provide insights into primate evolution. Science 380: 913–924.

History of migrations

Kamusella, T. (2009). *The Politics of Language and Nationalism in Modern Central Europe*. Basingstoke: Palgrave, Macmillan.

Bellwood, P. (2013). *First Migrants: Ancient Migration in Global Perspective*. Chichester: Wiley Blackwell.

Novembre, J. (2015). Human evolution: Ancient DNA steps into the language debate. Nature 522: 164–165.

Non, A.L., & Thayer, Z.M. (2015). Epigenetics for anthropologists: An introduction to methods. Am. J. Hum. Biol. 27: 295–303.

Serrano, J.G., Ordõnez, A.C., & Fregel, R. (2021). Paleogenomics of the prehistory of Europe: Human migrations, domestication and disease. Ann. Hum. Biol. 48: 179–190.

Human migrations revealed by microbiology

Yamaoka, Y. (2009). *Helicobacter pylori* typing as a tool for tracking human migration. Clin. Microbiol. Infect. 15: 829–834.

Moodley, Y., & Linz, B. (2009). *Helicobacter pylori* sequences reflect past human migrations. Genome Dyn. 6: 62–74.

Renfrew, C. (2009). Anthropology. Where bacteria and languages concur. Science 323: 467–468.

Dominguez-Bello, M.G., & Blaser, M.J. (2011). The human microbiota as a marker for migrations of individuals and populations. Annu. Rev. Anthropol. 40: 451–474.

Mégraud, F., Lehours, P., & Vale, F.F. (2016). The history of *Helicobacter pylori*: From phylogeography to paleomicrobiology. Clin. Microbiol. Infect. 22: 922–927.

Waskito, L.A., & Yamaoka, Y. (2019). The story of *Helicobacter pylori:* Depicting human migrations from the phylogeography. Adv. Exp. Med. Biol. 1149: 1–16.

Databases of human mitochondrial sequences

Lott, M.T., Leipzig, J.N., Derbeneva, O., *et al.* (2013). mtDNA variation and analysis using MITOMAP and MITOMASTER. Curr. Protocols in Bioinformatics 44: 1.23–1-26.

Ingman, M., & Gyllensten, U. (2006). mtDB: Human Mitochondrial Genome Database, a resource for population genetics and medical sciences. Nucl. Acids Res. 34: D749–D751.

Lee, Y.S., Kim, W.Y., Ji, M., *et al.* (2009). MitoVariome: A variome database of human mitochondrial DNA. BMC Genomics 3: Suppl 3: S12.

Database of human Y-chromosome sequences

Tiirikka, T., & Moilanen, J.S. (2015). Human chromosome Y and haplogroups; introducing YDHS Database. Clin. Trans. Med. 4: 60.

Domestication of animals

Larson, G., & Bradley, D.G. (2014). How much is that in dog years? The advent of canine population genomics. PLOS Genetics 10: e1004093.

Jensen, P. (2014). Behavior genetics and the domestication of animals. Ann. Rev. Animal Biosciences 2: 85–104.

Jones, P., & McCune, S. (2016). Genetic components of companion animal behavior. In: L.S. Freund, S. McCune, L. Esposity, N.R. Gee, and P. McCardle (eds). *The Social Neuroscience of Human-Animal Interaction*. Washington: American Psychological Association.

Cagan, A., & Blass, T. (2016). Identification of genomic variants putatively targeted by selection during dog domestication. BMC Evolutionary Biology 16: 10.

Trench, C.P.C. (1970). *A History of Horsemanship*. Garden City, NY: Doubleday and Co., Inc.

Anthony, D.W. (2007). *The Horse, the Wheel, and Language: How Bronze-Age Riders from the Eurasian Steppes Shaped the Modern World*. Princeton, NJ: Princeton University Press.

Frantz, L.A.F., Bradley, D.G., Larson, G., *et al.* (2020). Animal domestication in the era of ancient genomics. Nat. Rev. Genet. 21: 449–460.

Orlando, L. (2020). The evolutionary and historical foundation of the modern horse: Lessons from ancient genomics. Annu. Rev. Genet. 54: 563–581.

Orlando, L. (2020). Ancient genomes reveal unexpected horse domestication and management dynamics. Bioessays 42: e1900164.

Librado, P., Khan, N., Fages, A., *et al.* (2021). The origins and spread of domestic horses from the Western Eurasian steppes. Nature 598: 634–640.

Domestication of crop plants

Doebley, J.F., Gaut, B.S., & Smith, B.D. (2006). The molecular genetics of crop domestication. Cell 127: 1309–1321.

Zeder, M.A. (2006). Central questions in the domestication of plants and animals. Evol. Anthropol. 15: 105–117.

Doebley, J. (2006). Plant science. Unfallen grains: How ancient farmers turned weeds into crops. Science 312: 1318–1319.

Staller, J., Tykot, R.H., & Benz, B. (2006). *Histories of Maize: Multidisciplinary Approaches to the Prehistory, Linguistics, Biogeography, Domestication, and Evolution of Maize*. London: Elsevier.

Murphy, D.J. (2007). *People, Plants, and Genes. The Story of Crops and Humanity*. Oxford: Oxford University Press.

Tang, H., Sezen, U., & Paterson, A.H. (2010). Domestication and plant genomes. Curr. Opin. Plant Biol. 13: 160–166.

Harris, S. (2016). *What Have Plants Ever Done for Us? Western Civilization in Fifty Plants*. Oxford: Bodleian.

Montagna, M.T., Diella, G., Triggiano, F., *et al.* (2019). Chocolate, "Food of the Gods": History, science, and human health. Int. J. Environ. Res. Public Health 16: 4960.

Díaz-Valderrama, J.R., Leiva-Espinoza, S.T., & Aime, M.C. (2020). The history of cacao and its diseases in the Americas. Phytopathology 110: 1604–1619.

Marelli, J.P., Guest, D.I., Bailey, B.A., *et al.* (2019). Chocolate under threat from old and new cacao diseases. Phytopathology 109: 1331–1343.

The contribution of genomics to understanding the history of agriculture

Armelagos, G.J., & Harper, K.J. (2005). Genomics at the origins of agriculture. Evol. Anthropol. 14: 68–77, 109–121.

Krug, A.S., Drummond, E.B.M, Van Tassel, D.L., *et al.* (2023). The next era of crop domestication starts now. Proc. Natl. Acad. Sci. USA 120: e2205769120.

EXERCISES, PROBLEMS, AND WEBLEMS

Exercises

Exercise 9.1 How many chromosomes is it likely that Neandertals and Denisovans had?

Exercise 9.2 In patrilocal social organization, a married couple lives with the kin group of the husband's parents; in matrilocal social organization, with the kin group of the wife's parents. A site at El Sidrón, in Spain, contained at least 12 Neandertal individuals. The remains are ~48 000 years old. Sequencing shows that three adult males have the same mitochondrial DNA haplotype. Three adult females and juveniles show two other, distinct lineages. Does this suggest patrilocal or matrilocal residence? Explain your reasoning.

Exercise 9.3 Assuming that increase in amylase copy number in both humans and dogs, over their progenitors, was a response to a starch-rich diet, why was the salivary amylase affected in humans and the pancreatic amylase in dogs?

Exercise 9.4 A 'gain-of-function' mutation is one in which the mutated protein shows a new activity, not possessed by the protein encoded by the unmutated progenitor gene. A 'loss-of-function' mutation produces a protein that lacks an activity of the protein encoded by the unmutated progenitor gene. The mutations in the *sh1* and *rc* genes in *O. rufipogon* to produce *O. sativa* are both recessive. Does this suggest that they are gain of function or loss-of-function mutations? Which would be easier for evolution to find, a gain-of-function or a loss-of-function mutation? Why?

Exercise 9.5 Does rice flower in response to short days, or long nights? What relatively simple experiment could you do to distinguish these?

Exercise 9.6 Chocolate (*Theobroma cacao*) is not shown on the map in Figure 9.9. Insert *T. cacao* on a photocopy of this figure.

Exercise 9.7 One of the papers on the recently sequenced *Theobroma cacao* genome states that the Belizean *criollo* genotype '... is suitable for a high-quality genome sequence assembly because it is highly homozygous as a result of the many generations of self-fertilization that occurred during the domestication process'. Why do these characteristics make this variety suitable for a high-quality genome sequence assembly?

Exercise 9.8 (a) How many genes in *Theobroma cacao* do not have homologues appearing in *Arabidopsis thaliana, Populus trichocarpa*, *Glycine max*, and *Vitis vinifera*? (b) How many gene families are shared by all four species? (c) How many gene families appear in *A. thaliana*, *P. trichocarpa*, *Glycine max*, and *V. vinifera*, but *not* in *T. cacao?*

Exercise 9.9 How many syntenic groups in the *T. cacao* genome show evidence for triplication of the genome? List the triplets of chromosomes linked by each group.

Problems

Problem 9.1 Chromosomal DNA from non-African humans contains ~1–4% Neandertal contributions, but Neandertal mitochondrial DNA does not appear in modern humans. What explanations are possible? How could you test them?

Problem 9.2 If humans first entered Alaska from Asia 20 000 years ago and archaeological remains are found in Monte Verde, Chile, dated 14 000 years ago, what mean annual rate of southward migration speed is implied? What two villages or cities in the vicinity of where you live are as far apart as this rate would imply for ten years of migration?

Problem 9.3 George Beadle crossed a strain of maize with teosinte and examined the second generation (F2). Approximately 1/500 plants was identical to the maize grandparent and approximately 1/500 was identical to the teosinte grandparent. Assuming Mendelian inheritance of *n* unlinked genes, such that the teosinte grandparent was homozygous for *n* teosinte alleles and the maize grandparent was homozygous for *n* maize alleles, estimate the value of *n* such that the fraction of F2 plants genetically identical to the grandparents at all *n* loci is approximately 1/500.

Problem 9.4 Numerous genera of ants cultivate fungi for food. Suggest some similarities and some differences between the domestication of fungi by ants and the domestication of agricultural crops by humans.

Problem 9.5 What accounts for the paucity of disease-resistant genes in *T. cacao?* It has been suggested that the genes encoding toll interleukin receptor motif proteins are older than the angiosperm–gymnosperm split, and that they have been lost in lineages including genes in *T. cacao*. (a) What alternative hypothesis is suggested by the observation that *T. cacao* is close to the ancestral eudicot? (b) How might these hypotheses be tested?

Problem 9.6 Waskito and Yamaoka have studied patterns of migration within Africa by following changes in *H. pylori* sequences (see Figure 9.18).

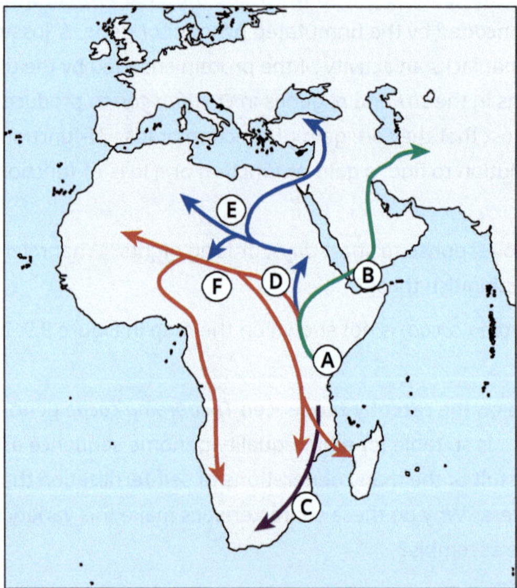

Figure 9.18 (A) The first colonizaton of *H. pylori* to modern humans, ~ 88 – 116 kya (kya = thousands of years ago). (B) First 'out of Africa' migration. (C) Dispersal to the South, characterized by the hpAfrica2 strain. (D) Divergence of hpAfrica1 and hpNEAfrica strains. (E) Expansion to the western part of Africa (~ 6 – 9 kya), followed by migration to Europe via the Levant. (F) The expansion of Bantu speakers, to Western Africa and the Eastern coast, then to Madagascar.

Used with permission of Springer Nature, from Waskito, L.A., & Yamaoka, Y. (2019). The story of *Helicobacter pylori*: Depicting human migrations from the phylogeography. In: S. Kamiya & S. Backert (eds), *Helicobacter pylori in Human Diseases. Advances in Experimental Medicine and Biology*, vol 1149. Cham: Springer. https://doi.org/10.1007/5584_2019_356 Permission conveyed through Copyright Clearance Centre, Inc.

Compare the results shown in this figure based on *H. pylori* strains with those based on human mitochondrial DNA in Figure 9.4. To what extent can you correlate *H. pylori* strains with mitochondrial haplotypes?

Weblems

Weblem 9.1 A region of human mitochondrial DNA sequence, positions 16 047–16 385, was determined to have the following mutations, relative to the Cambridge Reference Sequence (CRS):

Position (nt)							
Source	16 147	16 172	16 189	16 223	16 248	16 320	16 355
CRS	C	T	T	C	C	C	C
Unknown	A	C	C	T	T	T	T

To what mitochondrial haplogroup does the sequence belong?

Weblem 9.2 For the genes implicated in dog domestication, determine from OMIM™ and OMIA the effects of mutations.

Weblem 9.3 For the genes listed as contributing to the domestication of the horse, (a) What do these genes encode? (b) What are the functions of their products? (c) For which is it relatively clear how a change in the gene could affect the anatomy or physiology of the horse?

Weblem 9.4 (a) For the genes listed as contributing to the domestication of the horse, determine from Online Mendelian Inheritance in Man and Online Mendelian Inheritance in Animals (OMIM™ and OMIA) the consequences in humans and horses of mutations. (b) For at least one of the genes, recover the DNA sequences of the genes from humans, contemporary horses, and progenitors of domesticated horses. Convert to amino-acid sequences and align them.

Report the amino-acid sequence differences. Which are shared between humans and domesticated but not progenitor horses? For *CACNA1D*, for which the structure is known, submit the mutations to one of the sites that offer to predict the structural effect of a mutation, and discuss the results.

Weblem 9.5 What is the difference in structure between theobromine and caffeine?

Weblem 9.6 (a) Are there any other species in the genus *Theobroma?* (b) What other genus of plant is most closely related to *Theobroma?*

CHAPTER 10

The Impact of Genomics on Human Health and Disease

LEARNING GOALS

- *Recognize broad categories of causes of disease*, including infection by pathogens; and dysfunction of proteins or nucleic acids arising from mutations, including breakdowns in the control of cell proliferation that result in cancer.
- *Know the sources of diseases caused by abnormalities specifically in haemoglobin genes*, including sickle-cell disease and thalassaemias.
- *Recognize that phenylketonuria is a disease caused, in most cases, by a mutation in a specific gene*; it is an example of success in neonatal screening to abrogate development of symptoms formerly only through lifestyle adjustment, in this case strict lifetime dietary control; now also with drugs.
- *Know the causes and symptoms of Alzheimer's disease*, one of the most important threats to the well-being of older adults, and the genetic changes that enhance risks of its early onset.
- *Understand the variability of phenotypic responses to mutations. Penetrance* is the fraction of individuals carrying the allele for a condition who actually manifest it. Among those who carry the allele, *variable expressivity* is the relative severity of the condition.
- *Know the story of the discovery of the gene for cystic fibrosis*, and how such tasks are accomplished now with available techniques of next-generation sequencing.
- *Be familiar with the methods and results of Genome-Wide Association Studies (GWAS)*. GWAS can identify genes that are the primary causes of a condition, and also other genes that affect penetrance and variable expressivity.
- *Understand the human microbiome and its role in health and disease*, and how remediation can ameliorate problems arising from abnormal microbiome flora. Understand how the human microbiome fits into the general scheme of metagenomics, as discussed in Chapter 7.
- *Know the role of genomics in the aetiology, diagnosis, and treatment of cancer*. Recognize that cancer is a disease of genome instability. Many possible causes lead to a common effect, the breakdown of controls over cell proliferation.
- *Appreciate the success of cancer genome sequencing to identify particular mutations associated with different types of cancer*; and the application of this knowledge to more accurate and confident diagnosis, prognosis, and therapy.
- *Conclude that the illumination of disease by various genome-sequence-based investigations has made 'personalized medicine' an effective reality.*

Introduction

The primary motivation for support of the Human Genome Project was its potential for improving human health. Successful applications of the sequence data fall into broad categories:

- Identification of features of the genome associated with particular diseases. Comparisons of sequences from patients presenting with a condition, with sequences from healthy counterparts, have in many cases allowed pinpointing a genetic condition correlated with a clinical abnormality.
- Studying the non-human genome sequences of pathogens that cause human disease, to understand, for example, the mechanism of development of drug resistance, and to identify potential drug targets.
- Identifying changes in DNA sequences associated with cancer.
- Discovering the relationship between epigenetic changes and diseases.
- Recognizing the importance of the human microbiome in health and disease.

Translations of this knowledge into clinical practice include:

- Recognizing from the genome sequence an inevitability of development of a disease (for instance, Huntington's), or an enhanced risk of development of a disease. For instance, mutations in the *BRCA1, BRCA2*, or *PALB2* (partner and localizer of *BRCA2*) genes predispose individuals to breast and ovarian cancer.
- In some cases, responding to genetic predispositions to disease by recommendations of changes in lifestyle.
- Genetic counselling of potential parents at risk of bringing together in a child dangerous recessive genes.
- Using sequence information to choose effective targets for drug development.
- Tailoring treatment to the genetic background of an individual patient, including but not limited to pharmacogenomics—the choice of drug and dosage on the basis of a patient's genome sequence.
- Adjustments in the microbiome to restore it to a healthy condition.
- Gene editing, by CRISPR/Cas, to remedy a condition arising from a mutation has been successful. This will certainly become more common in clinical practice.
- In cancers, using comparisons of healthy and tumour cells from the same individual to aid precise diagnosis, more accurate prognosis, and choice of the most effective therapy.

Some diseases are associated with mutations in specific genes

Haemoglobinopathies—molecular diseases caused by abnormal haemoglobins

Sickle-cell disease

Linus Pauling and co-workers showed in 1949 that haemoglobin isolated from patients with sickle-cell disease differed in electric charge from normal haemoglobin. Patterns of inheritance showed that sickle-cell disease is a genetic disease. Pauling's discovery was therefore the first evidence that genes precisely control the structures of proteins. This preceded the first determination of the amino-acid sequence of a protein.

Sickle-cell disease was the first recognized molecular disease. Recall that haemoglobin contains four polypeptide chains, two α chains and two β chains (Figure 1.10). The sickle-cell mutation changes residue 6 of the β chain from a charged sidechain, glutamic acid, to a non-polar one, valine (β6Glu→Val). This creates a 'sticky patch' on the surface of the molecule. As a result, the mutant haemoglobin forms polymers within the erythrocyte, in the deoxy form typical of venous blood, where haemoglobin is not binding oxygen.

To flow through small capillaries, erythrocytes must be deformable, as their typical size, 7.8 μm

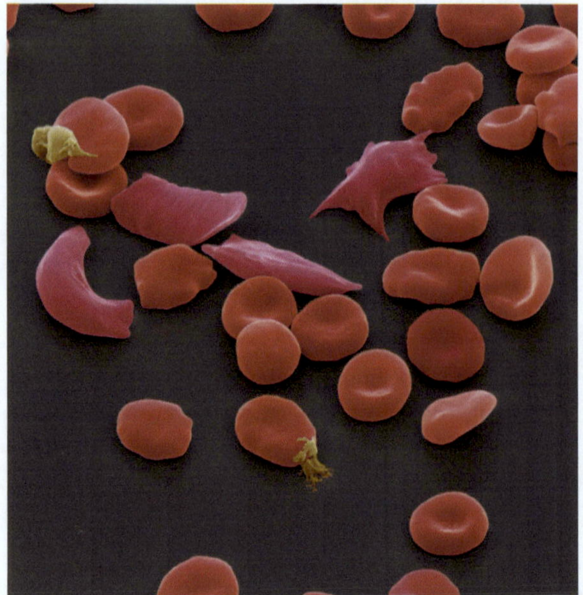

Figure 10.1 Coloured scanning electron micrograph of normal (round) and sickle-shaped erythrocytes. The shape of the cells is caused by aggregation of mutant haemoglobin in its deoxy state. The symptoms of the disease are caused not so much by the shape of the cells but by their rigidity: their inability to distort their shape in order to pass through the narrow capillaries. (M108/0493—EYE OF SCIENCE/SCIENCE PHOTO LIBRARY)

(in humans), is larger than the diameter of small capillaries. The formation of the polymers has a rigidifying effect on the erythrocyte, impeding its flow and blocking capillaries (Figure 10.1). In the traffic jam building up behind a plugged capillary, arriving red cells release their oxygen to surrounding hypoxic tissues, become deoxygenated, and thereby aggravate the problem. This produces pain—because of reduced oxygen supply resulting from capillary congestion—and anaemia and jaundice, consequences of the rapid breakdown of red blood cells.

Approximately 12 500–15 000 infants with sickle-cell disease are born in the UK each year. The most common SNP associated with sickle-cell disease changes the codon gag to gtg. It is possible to test specifically for this mutation. Alternatively, sequencing the entire region would pick up other mutations.

Thalassaemias

Thalassaemias are genetic diseases associated with defective or deleted haemoglobin genes: most people of European ancestry have four genes for the α-chain of normal adult haemoglobin, two alleles of each of the two tandem genes α_1 and α_2. Therefore, α-thalassaemias can present clinically in different degrees of severity, depending on how many genes encode normal α chains. Only deletions leaving fewer than two active genes present as symptomatic under normal conditions. Observed genetic defects include deletion of both genes (a process made more likely by the tandem gene arrangement and sequence repetition, which make crossing over more likely) and loss of chain termination leading to transcriptional 'read through', creating extended polypeptide chains, which are unstable.

β-thalassaemias are usually point mutations. These may be:

- missense mutations (amino-acid substitutions). The sickle-cell mutation is also an example of a missense mutation;
- nonsense mutations (changes from a triplet coding for an amino acid to a stop codon) leading to premature termination and a truncated protein;
- mutations in splice sites;
- mutations in regulatory regions;
- certain deletions, including the normal termination codon and the intergenic region between δ and β genes, creating $\delta-\beta$ fusion proteins.

Phenylketonuria

Phenylketonuria (PKU) is a genetic disease caused by deficiency in a metabolic enzyme, phenylalanine hydroxylase, the enzyme that converts phenylalanine to tyrosine (see Figure 10.2). If untreated, phenylalanine and metabolic byproducts accumulate in the blood, to toxic levels. The high levels of phenylalanine cause a variety of developmental defects, including mental retardation, microcephaly, and seizures. The disease cannot be cured, but symptoms can be avoided by lifestyle control: a phenylalanine-free diet, and, recently, specific drugs.

PKU is an example of how understanding the molecular mechanism of a disease can, for some conditions, help restore and preserve health through suggested changes in lifestyle and/or drug therapy.

PKU is an autosomal recessive trait, associated with mutations in the phenylalanine hydroxylase

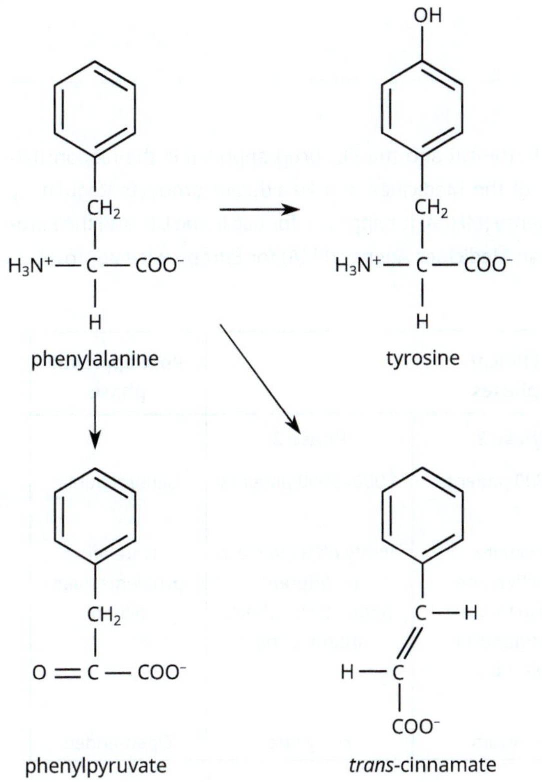

Figure 10.2 Phenylalanine (top left) and three metabolites. Top right: tyrosine, produced by the normal phenylalanine hydroxylase enzyme. Bottom left: phenylpyruvate, a degradation product formed in large quantities by people suffering from untreated phenylketonuria. Bottom right: *trans*-cinnamate, produced from phenylalanine by the enzyme phenylalanine ammonia-lyase. A derivative of this enzyme is now approved for clinical use in the US.

gene on the long arm of chromosome 12 (12q22–12q24.1). In the UK and US the prevalence is about 1 in 10 000 individuals; 1 out of 50 are carriers. PKU is the subject of neonatal screening in many countries.

A large number of known mutations are associated with PKU, appearing in all 13 exons of the gene and in flanking sequences (http://www.biopku.org/home/home.asp). Many but not all of them are SNPs. These include many that affect catalytic activity, somewhat fewer that affect regulation, and even fewer that affect the assembly of the tetrameric enzyme.

The classical test for PKU depended on the build-up of phenylalanine and its degradation products such as phenylpyruvate in neonatal blood and urine. (Phenylpyruvate is a ketone, hence the name of the disease.) From a blood sample taken from a neonate, mass spectrometry can measure abnormal concentrations of phenylalanine or tyrosine. It is also possible to detect mutations in the gene by sequencing, although should a novel mutation appear it might not be possible to conclude with confidence that the mutant protein is dysfunctional. Genomic sequencing could also detect carriers, allowing counselling of potential parents.

Formerly, management of PKU depended on enforcing a low-phenylalanine diet. This is not entirely satisfactory: compliance is a common problem, as low-phenylalanine foods are relatively unpalatable and do not provide complete nutrition (see Weblem 10.9). Artificial mixtures of amino acids—phenylalanine-free formulas—replace high-protein foods.

An alternative to a lifelong low-phenylalanine diet is enzyme replacement therapy. It is not satisfactory to administer functional phenylalanine hydroxylase itself, as is done with insulin for diabetes. This is because phenylalanine hydroxylase is a tetramer, requires a cofactor, and is subject to complex regulatory controls. Instead, it is possible to use phenylalanine ammonia-lyase, an enzyme found in plants, fungi, and bacteria, which converts phenylalanine to *trans*-cinnamic acid (see Figure 10.2). This enzyme is more stable, and does not require cofactors. The product, *trans*-cinnamic acid, degrades to hippuric acid, excreted in the urine. The problem of antigenicity is addressed by attachment of polyethylene glycol (PEG). As a drug, pegylated phenylalanine ammonia-lyase has been approved for clinical use in the US (see Box 10.1).

Many of the common mutants in phenylalanine hydroxylase could be targets for genome editing.

It is particularly tricky to manage PKU women in pregnancy. Remember that PKU is an autosomal *recessive* trait. A woman with PKU must be homozygous for defective phenylalanine hydroxylase (although the two alleles need not necessarily bear the same mutation). If such a woman becomes pregnant, it is likely that the foetus is only a carrier (unless the father is also a carrier). The problem is to control phenylalanine levels in the mother so as

BOX 10.1 Phases of clinical trials

Before approval for use in general clinical practice in the US, a drug or treatment must successfully navigate successive phases of clinical trials in humans:

In the UK and the EU, drug approval is the responsibility of the Medicines and Healthcare products Regulatory Agency (MHRA) for approval for use in the UK, and the European Medicines Agency (EMA) for Europe-wide approval.

	Pre-clinical phase	Clinical phases			Post-approval phase
		Phase 1	Phase 2	Phase 3	
Subjects	Laboratory and animal study	20–100 healthy volunteers	100–300 patients	1 000–3 000 patients	General public
Purpose	Assess safety, biological activity, and pharmacokinetics	Determine safety dosage, metabolism	Evaluate effectiveness relative to other treatments or placebo	Verify effectiveness in different populations, check interactions	Note any problems; take action
Typical duration	>1–2 years	3–4 years	4–5 years	6–8 years	Open-ended

⇑ File Application: Investigational New Drug

⇑ Approval to market

to provide adequate nutrition to the foetus, without subjecting the mother to toxic levels.

Alzheimer's disease

The symptoms of Alzheimer's disease are degradation of cognitive functions, characterized by: loss of train of thought, progressive memory problems, missing important appointments, etc. It is associated with the deposition in the brain of amyloid plaques and neurofibrillary tangles containing polymers of tau protein.

Alzheimer's disease is a very severe public-health problem, especially in view of increasing lifespans. The most common form is late-onset Alzheimer's, appearing in people over the age 65. Approximately 50% of people over age 85 suffer from it. Early-onset Alzheimer's, defined as first appearing at age <65, is rarer. Even rarer is familial Alzheimer's, involving <1% of cases, appearing at age 40–60.

The risk of late-onset Alzheimer's disease is correlated with SNPs in apolipoprotein E (ApoE). The basic function of this 317-residue protein is to remove cholesterol from the blood. The gene for ApoE is on chromosome 19. There are four common alleles, which differ by SNPs:

ApoE1 = rs429358(C) + rs7412(T) [minor variant]
ApoE2 = rs429358(T) + rs7412(T)
ApoE3 = rs429358(T) + rs7412(C) [~55%]
ApoE4 = rs429358(C) + rs7412(C)

The correlations with risk of Alzheimer's disease are:

At least one E4 allele → increased risk of Alzheimer's
At least one E2 allele → decreased risk of Alzheimer's

Identification of genes associated with inherited diseases

Some conditions show simple Mendelian genetics. These are attributable to a defect in a single gene. Sickle-cell disease, and phenylketonuria, are examples. In contrast, many other conditions, such as asthma, arise from a combination of many genes, plus environmental factors.

Methods for establishing genotype–phenotype correlations are clearly easier for the single-gene case. But even then, rarely is the gene-disease relationship simple. Consider a set of autosomal dominant traits. All phenotypic characters, including likelihood of developing particular diseases, differ in their penetrance—the fraction of individuals, carrying the allele for a condition, who actually manifest it. Among those who carry the allele and develop the condition, variable expressivity measures the severity of the condition. (Variable expressivity refers to the different degrees of manifesting a trait among people with same genetic condition.) In many cases, contributions of other genes affect the phenotypic consequences of the mutation.

For example, the symptoms of Marfan's syndrome include skeletal abnormalities such as long and very flexible fingers,[1] lengthening of the long bones, and scoliosis; and cardiovascular abnormalities including dissecting aneurysms. Some Marfan's sufferers are only mildly affected, and lead normal lives; others, more severely affected, have shortened lifespans—often as a result of the aneurysms.

Given a disease attributable to a defective protein:

- If we know the protein involved, we can pursue rational approaches to therapy. If the gene involves loss of function, it may be difficult to devise therapy to restore it, other than by editing the gene. In those cases, it is useful to look to other genes that modify penetrance and expressivity as targets for intervention. For some proteins it is possible to supply normal versions; well-known examples include insulin for diabetes and blood coagulation factors for haemophilia.

[1] The nineteenth-century violinist Niccolò Paganini exhibited these anatomical characters—and very good use he put them to! It has been suggested that he had Marfan's, or the related Ehlers–Danlos syndrome.

- If we know the gene involved, we can devise tests to identify sufferers or carriers.

In many cases, knowledge of the chromosomal location of the gene is unnecessary for either therapy or detection. (This is not true of diseases arising from chromosomal abnormalities.)

For instance, in the case of sickle-cell disease, we know the protein involved. The disease arises from a single point mutation in the β-chain of haemoglobin. We can proceed directly to protein-structure-based drug design. We need the DNA sequence for genetic testing and counselling, and, as has recently been achieved, curing the disease by gene editing.

- In contrast, if we know neither the protein nor the gene, we must somehow retrace the steps back to the gene from the phenotype. Before widespread whole-genome sequencing, the classical method for identifying a gene causing a condition showing simple Mendelian inheritance was based on studies linking the condition to mapped genes, and 'walking' along the chromosome, sequencing regions to extend a contig of known sequence, until a mutation was discovered. The process, called *positional cloning* or *reverse genetics*, involved a cascade from the gene map to the chromosome map to the DNA sequence.

For example, to find the gene associated with cystic fibrosis it was necessary to begin with the gene map, using linkage patterns of heredity in affected families to localize the affected gene to a region of a particular chromosome. It was then possible to sequence the DNA in that region to identify candidate genes, and finally to pinpoint and sequence the gene responsible (see Box 10.2).

The cystic fibrosis story begins with the implication from simple Mendelian inheritance that the disease involves only a single gene. In contrast, many diseases do not show simple inheritance. Even if only a single gene is involved, heredity may create only a predisposition, the clinical consequences of which depend on other genes, and on environmental factors, including lifestyle. In these cases the full human

BOX 10.2 Identification of the cystic fibrosis gene

Cystic fibrosis, a disease known to folklore since at least the Middle Ages and to science for about 500 years, is an inherited recessive autosomal condition. Its symptoms include intestinal obstruction, reduced fertility including anatomical abnormalities (especially in males); and recurrent clogging and infection of lungs—the primary cause of death now that there are effective treatments for the gastrointestinal symptoms. With improvements in treatment, the median expectancy has almost tripled from less than 20 years in 1980 to 53 years. Cystic fibrosis affects 1/2500 individuals in the American and European populations. Approximately 1/25 people of European ancestry carry a mutant gene, and 1/65 African–Americans.

Clinical observations provided the gene hunters with useful clues. It was known that the problem had to do with Cl^- transport in epithelial tissues. Folklore had long recognized that children with excessive salt in their sweat—tasteable when kissing an infant on the forehead—were short-lived. Modern physiological studies showed that epithelial tissues of cystic fibrosis patients cannot reabsorb chloride. When closing in on the gene, the type of protein implicated, and the expected distribution, among tissues, of its expression, were useful guides.

How was the gene found?

A search in family pedigrees for a linked marker localized the gene to chromosome 7, band q3. Other markers found were linked more tightly to the target gene, bracketing it within a region of approximately 500 kb. A region this long could well contain 100 genes.

A technique called chromosome jumping made the exploration of the region more efficient. A 300 kb region at the right distance from the markers was cloned. A succession of linked probes allowed extension of the sequenced portion. The sequence allowed identification of genes in the region.

Checking in animals for genes similar to the candidate genes turned up four likely possibilities. Checking these possibilities against cDNA libraries from sweat glands of cystic fibrosis patients and healthy controls identified one probe with the right tissue distribution for the expected expression pattern of the gene responsible for cystic fibrosis. One long coding segment had the right properties, and indeed corresponded to an exon of the cystic fibrosis gene.

Proof that the gene was correctly identified included:

- Seventy per cent of cystic fibrosis alleles have a certain deletion. It is not found in people who are neither sufferers nor likely to be carriers.
- Expression of the wild-type gene in cells isolated from patients restores normal Cl^- transport.
- Knockout of the homologous gene in mice produces the cystic fibrosis symptoms.
- The pattern of gene expression matches the organs in which it is expected.
- The protein encoded by the gene would be expected to contain a transmembrane domain, consistent with involvement in transport.

In 1989, the gene for cystic fibrosis was isolated and sequenced. This gene—called *CFTR* (cystic fibrosis transmembrane conductance regulator)—codes for a 1480 amino acid protein that normally forms a cyclic AMP-regulated epithelial Cl^- channel. The gene, comprising 24 exons, spans a 250-kb region. For 70% of mutant alleles, the mutation is a three base-pair deletion, deleting the residue 508Phe from the protein. This mutation is denoted del508. The effect of the deletion is defective translocation of the protein, which is degraded in the endoplasmic reticulum rather than transported to the cell membrane. Over 2000 different mutations are known; their effects vary widely in the severity of the resulting condition.

Knowing the gene had clinical consequences: An *in utero* test for cystic fibrosis is based on recovery of foetal DNA. A PCR primer is designed to give a 154 bp product from the normal allele and a 151 bp product from the del508 allele.

Clinicians have taken advantage of the fact that the affected tissues of the airways are easily accessible, to show success with gene therapy, using inhaled liposomes as the delivery system. Alternative approaches using viral carriers of the normal gene are under investigation. Genome editing using CRISPR/Cas is being developed.

genome sequence, and measurements of expression patterns, are essential to identify the genetic contributions to the phenotype. Some of the modifiers interact with the primary mutated gene, perhaps controlling its expression; others may interact with environmental contributors, such as certain genes that protect against the effects of smoking.

The protein that is defective in cystic fibrosis also acts as a receptor for uptake of *Salmonella typhi*, the pathogen that causes typhoid fever. Increased resistance to typhoid in heterozygotes—who do not develop cystic fibrosis but are carriers of the mutant gene—probably explains why the gene has not been eliminated from the population.

Genome-wide association studies (GWAS)

The goal of a genome-wide association study is to identify regions in the genome for which sequence differences correlate with particular phenotypes. In many cases, the target traits are: diseases caused by mutations in the genome sequence, or risk or mitigating factors associated with diseases, or clinical responses to drugs.

A common protocol is to select two sets of subjects, one a group of people that shows a phenotype of interest, for instance, enhanced susceptibility to a disease; and another group that does not: these are the controls, or normals, or, in the case of diseases, healthy people. Formerly, each subject in each group would be screened for the presence of a large number of SNPS, typically 500 000–2 000 000. A significant correlation of nucleotides at the SNP with the presence of the trait identifies the region as one of interest.

It follows that GWAS are most powerful if one, or no more than a few, loci are responsible for a trait (see Box 10.3). For traits that depend on combinations of small effects of genes at a large number of sites, each individual SNP may be struggling for statistical significance.

In actual practice, there is considerable bias in choice of subjects. The vast majority (91%) of complex trait GWAS studies have been performed in European-ancestry populations. Apart from East Asians (~5%), other ethnic groups are poorly represented in published GWAS studies.

Because even 2 000 000 positions is sparse in the whole genome, the hope is to hit a haplotype block that contains a gene involved in the trait (see Figure 10.4). Further sequencing of the region is necessary to pin down the exact sequence difference responsible for phenotype. As the technology advances, whole-genome sequencing will replace interrogations of SNPs only, in GWAS.

Regime of effectiveness of GWAS for identifying trait-associated genes

A disease or other phenotypic trait can depend on the genome with varying degrees of subtlety. How many genes are involved? What are the frequencies in the population of the mutant alleles? What is their penetrance (= the extent to which people with the trait-associated gene actually show the condition, and with what severity).

The simplest cases are those caused by a single gene, showing Mendelian inheritance, with complete penetrance. Phenylketonuria and Huntington's disease are examples. At the other extreme are **multifactorial** conditions, with many genes contributing small effects, some of which may be modifiers (just to confuse the issue), giving low penetrance.

For some of the Mendelian cases, it was possible to identify the gene by classical linkage analysis and 'chromosome walking'. Cystic fibrosis is an example (see Box 10.2). Conversely, a rare variant, with low penetrance, requires a very large sample size to pick up its effect with statistical significance.

Bush and Moore have provided a chart showing the relationship between the effects of allele frequency and penetrance (Figure 10.3). GWAS have been most effective in a diagonal band in this chart. There is some trade-off; that is, lower allele frequency requires higher penetrance. Progress in technique and extension to larger sample sizes will drive down the lower boundary of this diagonal band.

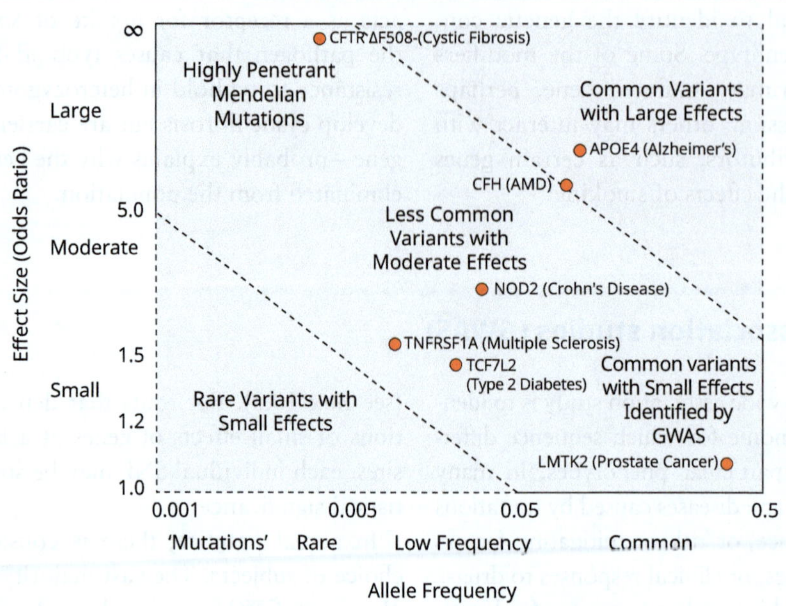

Figure 10.3 Factors governing the difficulty of identifying genes responsible for a disease or other phenotypic trait depend on the allele frequency in the population studied (*x*-axis, logarithmic scale) and the effect size, or penetrance (*y*-axis, also logarithmic scale). The odds ratio is the fraction of people with the *affected* gene who show the condition divided by the fraction of people with the *normal* gene who show the condition. The severity of the condition is not taken into account, except as a threshold for deciding which people are affected. (AMD = age-related macular degeneration)

Bush, W.S., & Moore, J.H. (2012). Chapter 11: Genome-wide association studies. PLoS Comput Biol 8, 12: e1002822. https://doi.org/10.1371/journal.pcbi.1002822

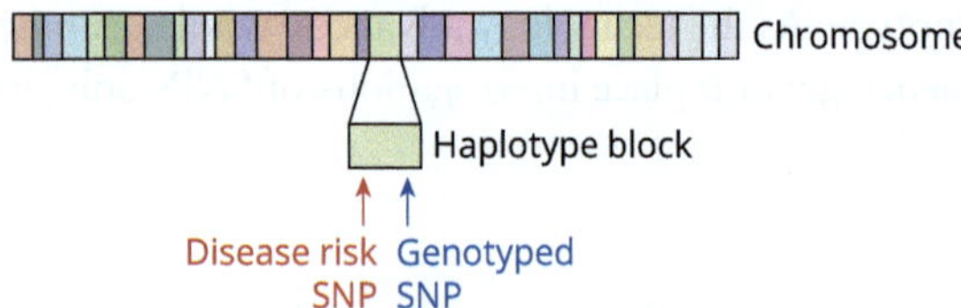

Figure 10.4 In typical GWAS the number of SNPs checked is a small fraction of the entire genome. Therefore, it is extremely unlikely that one of them will itself be responsible for a phenotype (in this example, disease risk). However, if the SNP checked and the position causing the phenotype are in the same haplotype block, an association between the SNP checked and the phenotype will be observed. Each SNP is a vicar of all positions in the same haplotype block.

Localizing an SNP to a haplotype block does not identify the actual site of the mutation associated with a trait. Indeed, many of the phenotype-associated SNPs identified by GWAS are in introns or intergenic regions. Readers must beware of a tendency by authors of papers in the field, uncritically to report the gene closest to the SNP as the relevant one. But it may well be that the actual site associated with the phenotype interacts with a gene distant on the same chromosome, or even on a different chromosome, affecting the trait indirectly.

Bush, W.S., & Moore, J.H. (2012). Chapter 11: Genome-wide association studies. PLoS Comput Biol 8, 12: e1002822. https://doi.org/10.1371/journal.pcbi.1002822

Intermediate between SNPs and whole-genome sequences is **whole-exome sequencing**. This is partial sequencing of a human genome, determining only the regions expressed as proteins. In many cases these include genes associated with disease, or other phenotypic characters. An SNP in a protein-coding region may change the function of the transcript by altering the amino-acid sequence of the protein, or affect expression level by altering mRNA stability or transcription-factor-binding affinity. However, many important disease traits are associated with non-coding sequences, such as regulatory regions. Technology has now advanced to the point that sequencing the entire genome is the only reasonable approach.

In 2009, whole-exome sequencing was sufficient to explain a mysterious disease of six-year-old Nicholas Volker (see Box 10.4).

Most GWAS projects report significant SNPs independently. But it is feasible, and useful, to look for common features among the identified genes. If, for instance, they affect parts of a common pathway, this provides valuable insight into the mechanism of the disease. What is more difficult is analysis of the

When all else fails read the instructions

Nicholas Volker, a six-year-old boy living in Wisconsin—in the midwest US—suffered from inflammatory bowel disease. The cause of his illness stumped physicians: Misdiagnoses led to ineffective treatments, including over 100 surgeries. Unable to understand the cause of his condition, physicians could do no more than keep him alive by treating his symptoms.

In 2009, a team at the Medical College of Wisconsin decided to do large-scale genomic sequencing. (Not as simple in 2009 as it would be today: whole-exome sequencing then cost US$75 000.) They found a novel mutation that impaired Nicolas' immune system: attacking his digestive tract, leading to inflammation and wounds recalcitrant to healing; and also impairing his ability to fight infections. There was a single point mutation in the *XIAP* gene, X-linked inhibitor of apoptosis, linked to immune deficiency. (In the protein encoded, the mutation produced the substitution Cys203→Tyr.) A stem cell transplant provided a normal *XIAP* gene. It worked! Nicholas is leading a much more nearly normal life.

data to consider logical combinations, such as: the trait is correlated with the presence of both SNP A and SNP B, or the trait is correlated with the presence of SNP A and SNP B provided that the position of SNP C is normal. There has been some progress along these lines, but the sheer number of combinations is daunting.

The European Bioinformatics Institute in collaboration with the US National Human Genome Research Institute maintains a catalogue of published GWAS studies (see Figure 10.5).

Genome-wide association studies of sickle-cell disease

The main genetic impairment in sickle-cell disease is an SNP in a single gene, that encodes the β–chain of haemoglobin. It shows Mendelian inheritance. However, other genes can modify the phenotype.

People with alleles for HbS—the version of the gene for the β–chain of haemoglobin that contains the mutation—do not show sickle-cell disease before an age of about six months. Before a programmed developmental switch, they synthesize foetal haemoglobin, HbF. Foetal haemoglobin does not polymerize in the deoxy state, nor can it even co-aggregate with HbS. As a result, individuals with the HbS allele who also have a condition in which foetal haemoglobin continues to be expressed have reduced severity of the disease, or may even be symptom-free.

GWAS studies searching for genetic signals associated with persistence of foetal haemoglobin synthesis have identified several loci, including the second intron of the gene *BCL11A* (SNP rs1427407) (see Figure 10.6). *BCL11A* is a C2H2 type zinc-finger protein, a major regulator of haemoglobin gene switching, and specifically regulation of foetal haemoglobin expression.

Experiments with mouse models of the disease confirm this. Complete homozygous germline knockout of *BCL11A* causes neonatal lethality. The gene has an essential role in development of the immune system. However, it is possible to create *conditional* knockouts that inactivate the gene only in adults. In a transgenic mouse model for sickle-cell disease, adult *BCL11A* knockout led to copious production of foetal haemoglobin, and a haematologically normal phenotype.

Conditional knockout of *BCL11A* is a proof-of-principle. A problem is that *BCL11A* controls other processes than foetal haemoglobin expression. People with natural deletions in and near *BCL11A* show neurological problems: autism spectrum disorder and developmental delay.

Is there any way to enhance the specificity of our intervention (see Box 10.5)? How about taking one step back in the control cascade? Expression of *BCL11A* depends on tissue-specific enhancers. The CRISPR/Cas system has made it possible to engineer knockouts in an ethryroid-tissue-specific enhancer of *BCL11A*. The result is a reduction of *BCL11A* expression in precursors of erythrocytes, leaving other tissues unaffected. Foetal haemoglobin expression levels in transgenic mice were enhanced, to an extent comparable to cells in which *BCL11A* itself was knocked out.

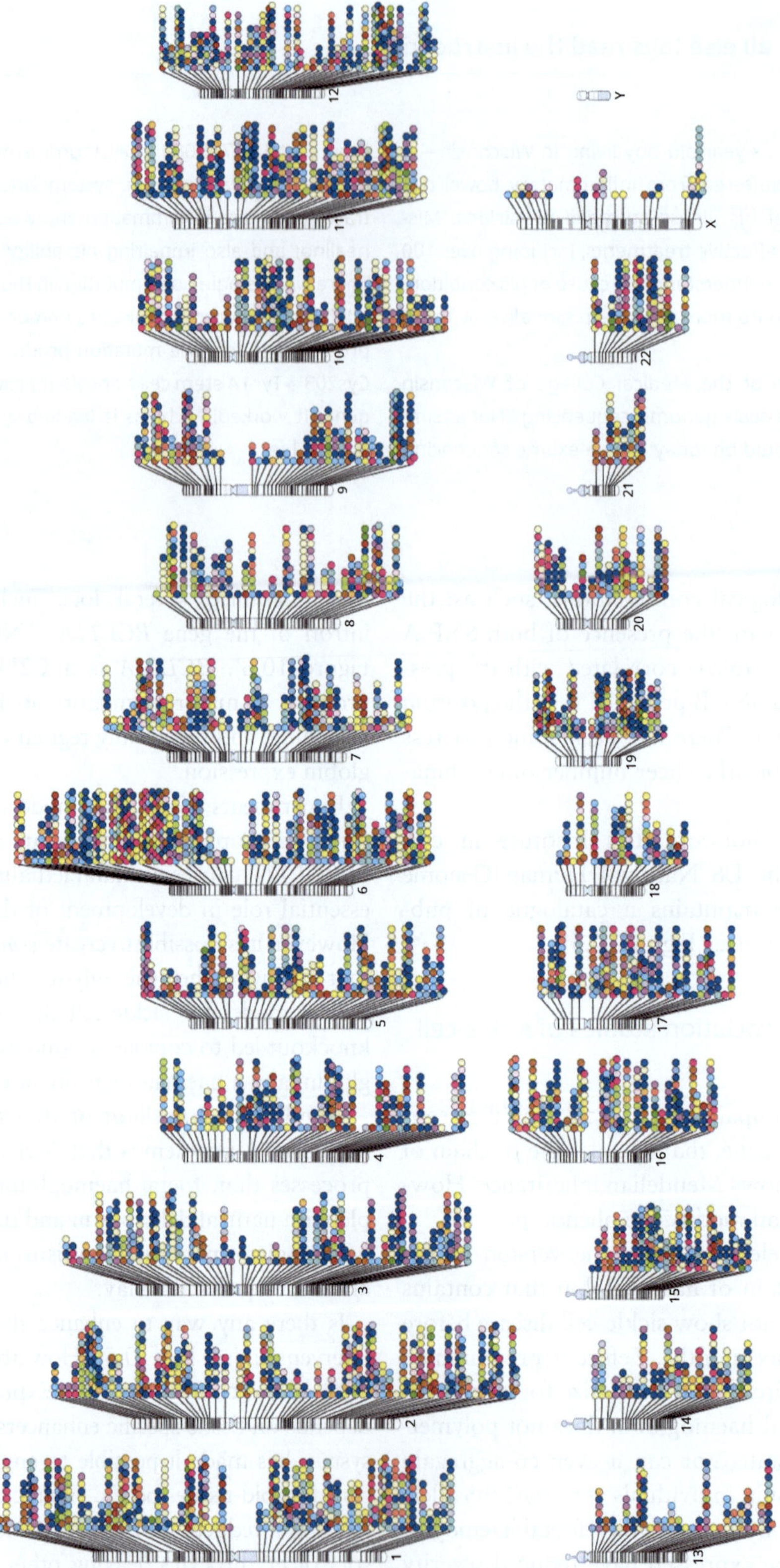

Figure 10.5 Chart of results of genome-wide association studies (GWAS) mapped onto human chromosomes. Colour codes refer to categories of traits. The site https://www.ebi.ac.uk/gwas/diagram contains an interactive version. (Reproduced courtesy of the European Bioinformatics Institute.)

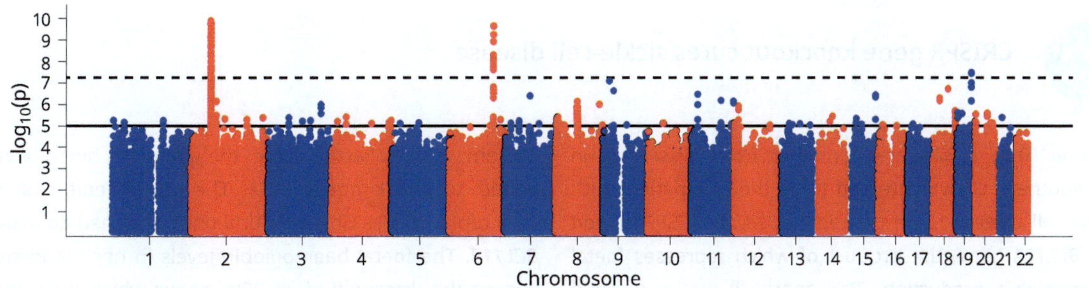

Figure 10.6 Results of genome-wide association study of foetal haemoglobin levels in a population from Tanzania. Sickle-cell disease is prevalent in Tanzania, with persistence of foetal haemoglobin synthesis a major ameliorating factor for the symptoms. The subjects included 1213 individuals homozygous for HbS and not on drug therapy. A panel of ~2.4×10^6 SNPs were tested for significant association with foetal haemoglobin levels.

This figure shows what is called a 'Manhattan plot', because of its putative resemblance to a city skyline (although New York, of which Manhattan is a part, is by no means the only city with skyscrapers). The *x*-axis shows the genomic distribution of the SNPs, in order of chromosomes. The *y*-axis reports the statistical significance of the association of particular regions with foetal haemoglobin levels. P is the probability of observing the result if the SNP were unrelated to HbF level. (A high negative value of P signifies a very small probability of observing the result.) The dashed horizontal line marks $P = 5 \times 10^{-8}$. The solid line marks $P = 10^{-5}$.

The peaks >8 correspond to *BCL11A* (SNP rs1427407), on chromosome 2q16; and *HMIP* (SNP rs9494145), on chromosome 6q23. (rs numbers identify SNPs.) *BCL11* is discussed in the text. *HMIP* lies within an enhancer of the gene encoding cMYB, a transcription factor regulating erythropoiesis and haematopoiesis. Studies of other populations, in which other relevant alleles are more prevalent, identify additional alleles correlated with foetal haemoglobin expression, including the β–globin locus itself.

Mtatiro, S.N., Singh, T., Rooks, H., *et al.* (2014). Genome wide association study of fetal hemoglobin in sickle cell anemia in Tanzania. PLoS ONE 9, 11: e111464. https://doi.org/10.1371/journal.pone.0111464 © 2014 Mtatiro et al.

Genetic diseases and possible approaches to intervention

Various combinations of enhancement and repression appear in the following diagram:

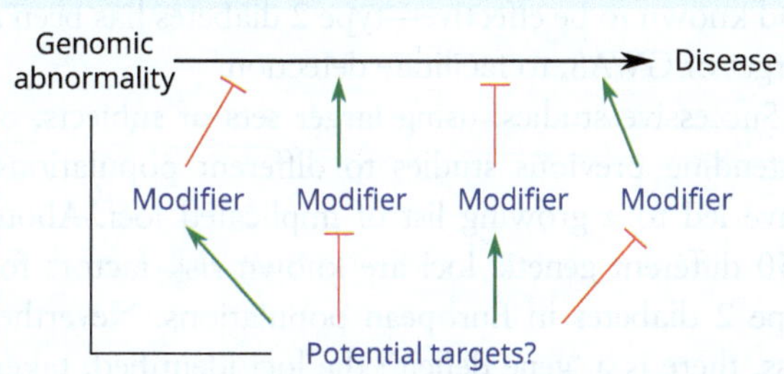

If the primary genomic abnormality causes the disease by creating *greater* than normal activity in a protein, and a modifier ameliorates the symptoms by *reduction* in activity associated with the primary abnormality (red 'tee', upper level), the goal of intervention would be to increase the activity of the modifier (green arrow, lower level).

If the primary genomic abnormality causes the disease by *reduction* in expression or activity of a protein, and a modifier ameliorates the symptoms by *increasing* the activity of the primary abnormality (green arrow, upper level), the goal of intervention would be to increase the activity of the modifier (green arrow, lower level).

Other combinations are possible.

Because the primary genomic abnormality may cause the disease by virtue of either increased or decreased activity relative to the normal level, the goal of intervention targeting the primary activity might be to reduce or increase the activity. Hence, in different cases, the link from potential targets to the primary genomic abnormality could end in either an arrow or a tee.

BOX 10.6 CRISPR gene knockout cures sickle-cell disease

Victoria Gray, a 32-year-old mother from Mississippi in the southern US, volunteered to be the first patient with sickle-cell disease to be treated by use of CRISPR to edit the *BCL11A* gene, the product of which represses foetal haemoglobin production. This approach was suggested by the observation that sickle-cell patients with a natural mutation in the *BCL11A* gene continue into adulthood to produce erythrocytes containing foetal haemoglobin, and do not show the typical sickle-cell symptoms.

Stem cells collected from the patient's blood were edited to knock out *BCL11A*. The CRISPR-edited stem cells replaced Mrs Gray's natural ones, that had an active *BCL11A*. The foetal haemoglobin levels in her blood surpassed the threshold of 20–30%, above which the symptoms disappear.

The results are impressive: the patient no longer suffers from sickle-cell disease. She is effectively normal. Cured.

The ideal would of course be to edit the mutant human haemoglobin β–chain gene to 'revert' the mutation. But gene knockout is easier than gene editing. Knockout of *BCL11A* by CRISPR/Cas has cured patients with sickle-cell disease (see Box 10.6).

Genome-wide association studies of type 2 diabetes

Diabetes is a metabolic disease in which absence or insufficiency of insulin production, or dysfunction of insulin receptors, produces elevated blood glucose levels. Untreated, this condition leads in the long term to many severe health problems.

Insulin is a hormone secreted by cells in the pancreas, that acts to promote transport of glucose from the bloodstream into cells. A specific protein, glucose transporter type 4 (GLUT4), mediates the entry of glucose into cells of skeletal and cardiac muscle, adipose tissue, and the central nervous system. Insulin causes GLUT4 to move from intracellular vesicles to the plasma membrane where it facilitates glucose entry into the cell.

Normally, our bodies respond to an elevation of blood glucose levels by release of insulin.

Type 1 diabetes is an autoimmune disease which destroys the insulin-producing cells in the pancreas. Injections of insulin can provide the protein. But this regimen lacks the control mechanism whereby a normal body reacts quickly to fluctuations in blood glucose levels.

Type 2 diabetes is more common, accounting for approximately 90% of the incidence of diabetes. Patients with type 2 diabetes can produce some insulin, but often not enough. In addition, many patients show insulin resistance, a dysfunction of the insulin receptor system.

Many lifestyle conditions contribute to type 2 diabetes, notably obesity and low physical activity. However, there is a hereditary component, estimated at 30–70%. Environmental factors grow in importance as the age of onset increases, suggesting a cumulative effect. Because the recommendations for lifestyle change for individuals at high risk are so obvious—and known to be effective—type 2 diabetes has been a target of GWAS, to facilitate detection.

Successive studies, using larger sets of subjects, or extending previous studies to different populations, have led to a growing list of implicated loci. About 150 different genetic loci are known risk factors for type 2 diabetes in European populations. Nevertheless, there is a 'gene deficit': the loci identified, taken together, account for only about 15% of the hereditary risk. Moreover, 80% of type-2-diabetes-associated SNPs are intronic or intergenic. This makes it difficult to interpret their effect at the molecular level.

Several genes, which confer enhanced risk of type 2 diabetes in mutant/normal heterozygotes, have drastic effects for mutant homozygotes. For instance, mutation in *WFS1* causes adult-onset type 2 diabetes in heterozygotes. Homozygotes for the mutation show the much more serious Wolfram syndrome. Wolfram syndrome comprises a combination including but not necessarily limited to diabetes, optic atrophy, and deafness, with juvenile onset.

The exact function of the protein wolframin, encoded by *WFS1*, is not known. It has been suggested that in the pancreas it helps to achieve the correct folding of the insulin precursor proinsulin. Wolframin is also involved in regulation of calcium transport. Lack of wolframin activity leads to dysfunction of the endoplasmic reticulum, triggering **apoptosis** (programmed cell death). Loss of cells in the pancreas and in the optic nerve could explain some of the symptoms.

What is the relationship between the functions of the SNPs identified as type-2-diabetes risks and the underlying molecular biology? There have been attempts to integrate the data into a biologically meaningful, and clinically applicable, picture. Clustering of risk genes according to function can point to certain pathways as foci of type-2-diabetes risk. These can suggest drug targets.

Some mutations are, unsurprisingly, directly related to insulin production or resistance. For instance, one locus associated with very significant risk is *TCF7L2*. This was discovered before application of GWAS techniques. A common SNP in an intron of this gene increases risk of type 2 diabetes with an odds-ratio of ~1.5. *TCF7L2* encodes a transcription factor expressed in many tissues. The mutation produces dysfunction of the insulin-producing cells, not insulin resistance. Indeed it is even possible to partition some of the type-2-diabetes-associated SNPs into one set that affects function of insulin-producing cells, and another that affects insulin resistance.

A connection discovered by GWAS is a link between melatonin receptor 1B and type-2-diabetes risk. Sequencing the coding exons of the protein in 7682 Europeans, 28% of them type-2 diabetes patients, confirmed that the patients showed significant amounts of mutation in this gene, relative to the control group.

This result suggests a possible approach to therapy. Melatonin receptor 1B is a G-protein-coupled receptor. It modulates circadian rhythms in response to daylight. Melatonin receptor agonists are already in clinical practice and prescribed for depression and sleep disorders. Some people feel that melatonin is helpful in managing type 2 diabetes, but this is still controversial.

Genome-wide association studies of schizophrenia

Schizophrenia is chronic and severely disabling neuropsychiatric illness, characterized by a number of features indicating a weakened link with reality. Symptoms may include delusions, impaired social relations, blunting of emotion and goal-directed motivation, problems with attention, and specific impairments of verbal working memory. A typical age of onset is between 16 and 30.

The disease is widespread, affecting approximately 1% of the UK population at some point in their lives. Symptoms may vary in severity. There is certainly a genetic component—schizophrenia tends to run in families—with heritability estimates ranging from 65–80%. However, development of the disease is linked to environment and life history also.

GWAS studies have identified over 100 loci associated with schizophrenia. The locations of the SNPs do not necessarily identify a specific gene, but merely a region of the genome. Indeed, many of the mutations are in non-coding regions—introns or intergenic regions—making it hard to identify the source of the signal and its function. It is likely that the mutations associated with schizophrenia affect gene expression patterns.

A recent study showed a very strong GWAS signal from the region of the major histocompatibility complex (MHC) (see Figure 10.7). This region encodes proteins involved in immune recognition of foreign proteins (see Chapter 1). Focusing in on this region identified a specific protein, complement component 4 (C4). Tandem genes encode two related proteins, C4A and C4B. Each can have long (L) and short (S) forms. C4AL and C4BL contain a retroviral insertion in intron 9 that lengthens the gene from 14 to 21 kb but does not affect the protein sequence. The region is variable, individuals differing in copy numbers of different forms of the gene.

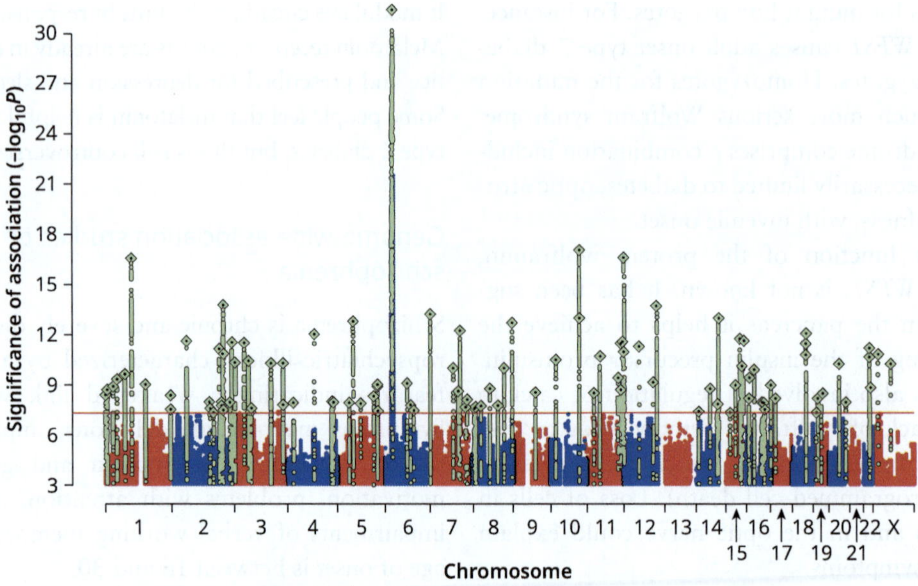

Figure 10.7 Manhattan plot of results of genome-wide association study showing loci correlated with schizophrenia. *x*-axis = chromosomal position, *y*-axis is statistical significance. The major 'hit', on chromosome 6, is in the region of the MHC complex.

Used with permission of Springer Nature BV, from Schizophrenia Working Group of the Psychiatric Genomics Consortium (2014). Biological insights from 108 schizophrenia-associated genetic loci. Nature 511: 421–427. https://doi.org/10.1038/nature13595. Permission conveyed through Copyright Clearance Centre, Inc.

The higher the level of expression of C4A, the greater the association with schizophrenia. One function of C4 is control of pruning of synapses during neural development. Developmental pruning involves elimination of synapses during childhood (see Figure 10.8). Starting *in utero*, the network of neurons grows more highly connected until the age of about two to three years. After that, there is normally a reduction in synaptic connections. Such pruning occurs between early childhood and the onset of puberty. The brain of a three-year-old child has approximately 1000 trillion synapses. After puberty, the number has decreased by about half.

C4 mediates synaptic pruning: C4 deposits another protein, C3, at synapses. This tags the synapses for pruning, carried out by microglial phagocytosis.

Schizophrenia patients tend to have higher expression of C4A than controls. The consequence is greater elimination of synapses. This is consistent with the observations that the typical age of onset is teenage years or early twenties, and that schizophrenics show a thinning of prefrontal layers.

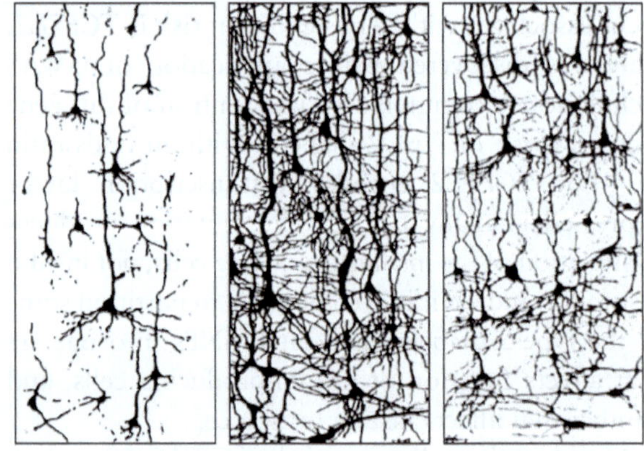

Figure 10.8 Changes in synapse density during development. Left: birth; centre: 6 years old; right: 14 years old. Synaptic pruning reduces synapse density between the ages of 6 and 14.

From: Shore, R. (1997). *Rethinking the Brain: New Insights into Early Development*. New York: Families and Work Institute; reproduced by courtesy of Dr H.T. Chugani.

Given the integration of C4-induced synaptic pruning with healthy aspects of neural development, the translation of this new understanding to the clinic may be quite challenging.

The human microbiome

It has long been common knowledge that our bodies are hosts to microorganisms. What has emerged recently is better definition of the extent and distribution of our microbial companions, and recognition of the role of the personal microbiome in health and disease.

Microbes are on us and in us. The human microbiome contains bacteria, archaea, viruses, and fungi. It is estimated that if a typical person were 'shrink-wrapped', the enclosed volume would contain approximately 30 trillion human cells and similar numbers of bacterial cells. (These are average values; there is substantial variability from person to person.)

The concept of infectious disease is that specific bacteria or viruses invade our bodies, temporarily (we hope), to be eliminated by the immune system and/or drug therapy. Classical clinical microbiology involved the study of cloned pathogens. Although undeniably valuable, the limitations of this approach are that (1) most microbiome organisms cannot be cultured under laboratory conditions, especially the anaerobes, and (2) one has given up on observing the interactions among the microbes, and between the microbes and the host. Interactions among microbes include small-molecule signalling, e.g. quorum sensing, and gene transfer: The microbiome has been likened to a 'chatroom': Interactions between microbes and host include modulation of patterns of host gene expression, and important roles in shaping the immune system.

Indeed, the microbiome is an integral part of our bodies, often providing useful functions rather than causing disease.

KEY POINT

Two of the most deeply ingrained ideas of biology are the organism and the species. We discussed the difficulties with the species concept in Chapter 1. Now it is perhaps time to call into question the idea of the organism, or individual, also. Human beings are not merely assemblies of human cells and tissues, but a combination of human and microbial cells. Aside from temporary infections by pathogens, the microbial communities are not transient features but a relatively stable community. The human microbiome can include bacteria, archaea, viruses, including bacteriophages, and fungi.

Modern techniques better define the microbiome by treating it as a problem in ecology and applying methods for metagenomics. One approach is to look at 16S ribosomal RNA sequences, as introduced by Woese to define three kingdoms of life (see Figure 1.14). The 16S rRNA contains conserved and variable regions. The conserved regions are targets of PCR primers that can amplify all the different molecules sharing the conserved sequence. The sequences of the variable regions provide species-specific signatures distinguishing different components of the microbiome. However, this method does not reliably determine relative abundances.

The field is naturally turning to next-generation whole-genome sequencing, and to proteomics and metabolomics of microbiomes. The microbiome counterpart of GWAS is MWAS: metagenome-wide association studies. For clinical applications, sequencing power will permit determination, not only of individual human genome sequences, but individual microbiome data, which can contribute to precision or personalized medicine. One looks ahead to a catalogue of microbiome fingerprints of different diseases.

Figure 10.9 shows a common distribution of components of a typical healthy human microbiome. Our bodies create large numbers of microenvironments, hospitable to different species of microbes. It is estimated that our intestinal microbiome contains about 150 times as many different genes as the human genome itself.

The microbiome is robust in the absence of disease, but also variable systematically during our lifetimes, on short-term and long-term timescales. Microbiome acquisition begins at birth. A stable microbiome profile is established by one year of age.

In the short term, microbiomes change with diet and other environmental influences, specifically including antibiotic therapy. Some changes are cyclical; for instance, the microbiome of the respiratory tract changes with the seasons, in locations with large temperature changes between summer and winter. Long-term changes may be related to changes in diet and lifestyle, others related to physiological transitions such as puberty and pregnancy, or to progressive changes as we age. An obvious example of age-dependent microbiome changes is a consequence of the eruption of teeth,

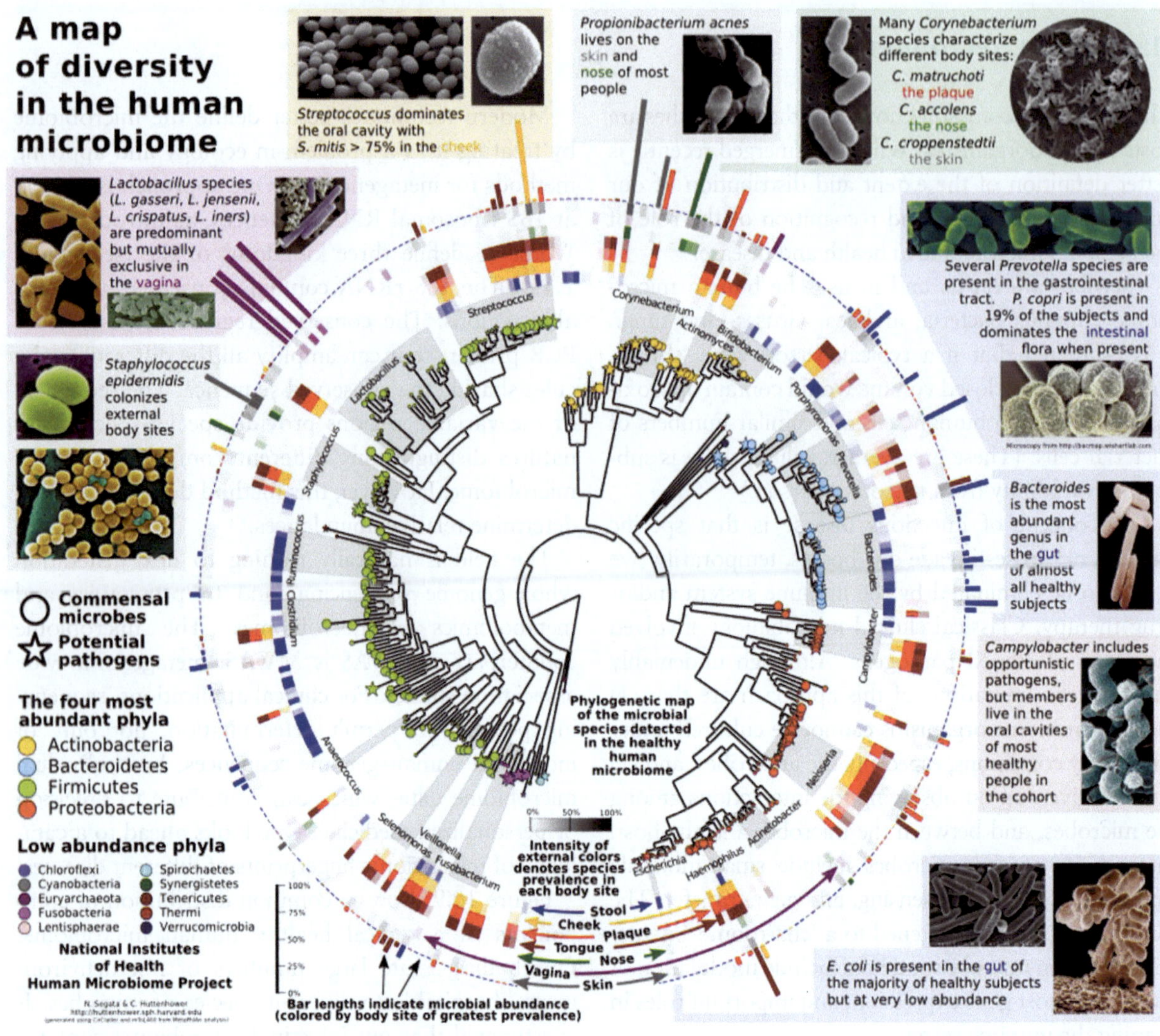

Figure 10.9 The diversity of the human microbiome. Arranged around the circle is a phylogenetic map of the most abundant organisms. Four phyla are prominent: Actinobacteria, Bacteroidetes, Firmicutes, and Proteobacteria. Concentric circles and colour coding indicate different locations of components of the microbiome, as do the bars around the outer ring, their lengths indicating abundances in the preferred body site of each species.

Used with permission of Elsevier Science & Technology Journals, from Morgan, X.C., Segata, N., & Huttenhower, C. (2013). Biodiversity and functional genomics in the human microbiome. Trends in Genetics 29, 1: 51–58. https://doi.org/10.1016/j.tig.2012.09.005. Permission conveyed through Copyright Clearance Centre, Inc. Data from the Human Microbiome Project and MetaPhlan; microscopy images from BacMap.

which creates novel microenvironments in the mouth and alters its bacterial flora. (Anton von Leewenhoek's original observation of bacteria was in scrapings of material from between his teeth.)

In addition to the cellular microbiome, we are host to many viruses. The human virome includes viruses that infect eukaryotic cells, and bacteriophages. Host genomes contain virus-derived elements—retroviral elements in human cells, and prophages in prokaryotes. In addition, there are active viral infections. Some are related to known diseases, but metagenomic sequencing suggests that most of the viruses, and their effects, are unknown.

In health our microbiomes are beneficial in several ways:

- They supplement dietary intake by synthesizing certain essential nutrients, including cobalamin (= vitamin B12) and folic acid.
- They metabolize undigestible compounds. It is well known that symbiotic bacteria in ruminants digest cellulose; less well known that microbes in the

human gut can depolymerize certain complex carbohydrates that human-encoded enzymes cannot, including fructooligosaccharides, a human-amylase-resistant component of starch, and even cellulose.

- They defend against pathogens, by competing for epithelial cell receptors, by synthesizing antimicrobial compounds, by modulating the immune response, and by making body fluids inhospitable to invaders by adjusting their pH. (On the other hand, bacterial acidification of saliva causes dental caries.)

A famous bit of anecdotal evidence for the importance of the microbiome for health involved a man suffering from an infection of the left ear only, that proved recalcitrant to treatment. His doctor tried antibiotics, vinegar washes, and antifungal agents, all without success. Then the patient showed up for a visit at which he and the doctor agreed that the infection had disappeared completely. What had happened? The patient admitted that he had transferred some ear wax from his good right ear to his affected left ear. The microbiome transplant did the trick!

Microbiome information can identify risk factors for disease, and in some cases suggest methods of treatment, including but not limited to change in lifestyle and diet. For example:

- Features of the gut microbiome are markers of, and contributors to, obesity. There is an association of microbiome composition in childhood and risk of obesity in later life. In experiments with mice, transplants of gut microbes from facultatively obese strains of mice (or from obese humans) to normal mice resulted in weight gain. Simultaneous transplant of microbes from lean mice ameliorated this effect. The mechanism seems to involve, at least partly, the contribution of the microbiome to the patient's ability to derive energy from food.
- The effects of the gut microbiome are not limited to gut-related conditions. Children at high risk of type 1 diabetes show a lower abundance of bacterial species that produce lactate and butyrate. In diabetic children, the abundance of Actinobacteria and Firmicutes was lower, the Firmicutes to Bacteroidetes ratio decreased, the abundance of Bacteroidetes significantly increased, relative to healthy children.
- Rheumatoid arthritis is a chronic inflammatory disorder that affects joints, causing painful swelling and, eventually, loss of bone, and deformity. It is an autoimmune disorder: the result of an attack by the immune system on the body's own tissues. Genome-wide association studies have identified some implicated genes, but in fact the heritability of the disease is small. In Europe, the concordance of the disease in monozygotic twins is less than 15% (comparable to the general population risk).

A predisposition to rheumatoid arthritis is associated with changes in the gut, dental, and salivary microbiomes. In fact there is a strong correlation of rheumatoid arthritis with periodontal disease. In comparison with controls, patients with rheumatoid arthritis showed depletion of *Haemophilus* species, and increased abundance of *Lactobacillus salivarius*. These differences were correlated with increased levels of serum autoantibodies.

- Acetaminophen (N-acetyl-*p*-aminophenol, Tylenol) is a common drug for pain and inflammation. The metabolism of this drug varies, according to the gut microbiome of the patient. Normally, acetaminophen is converted to harmless products by sulphation or glucuronosylation. A third, normally minor, pathway converts it to the toxic *N*-acetyl-*p*-benzoquinone imine (NAPQI). In some individuals gut bacteria generate *p*-cresol, which competes with acetaminophen for sulphation. This leads to increased production of *N*-acetyl-*p*-benzoquinone imine, risking liver damage.
- There is an association of consumption of red meat and atherosclerosis and cardiovascular disease. Microbes in the gut, including *Acinetobacter,* metabolize L-carnitine in meat to trimethylamine, converted to trimethylamine-N-oxide in the liver. Trimethylamine-N-oxide promotes cardiovascular problems as well as several other dangerous conditions. Dietary adjustments are the obvious suggestion. A vegetarian diet selects against carnitine-metabolizing gut flora. The compound 3,3-dimethyl-1-butanol reduces plasma trimethylamine-N-oxide levels. Foods containing 3,3-dimethyl-1-butanol include several associated with the 'Mediterranean diet': balsamic vinegars, red wines, and olive oils.

Data about microbiomes can have direct implications for diagnosis, prognosis, and treatment of disease. Crohn's disease is a type of inflammatory bowel disease. It can be focused in the distal ileum (large intestine) or the colon. The two forms differ in their response to antibiotics. The microbiome composition can distinguish them: reduced *Faecalibacterium prausnitzii* and enrichment in *E. coli* signal the ileal site. This provides a guide to treatment.

Treatment of abnormal microbiome composition

Although the human microbiome is variable, deviations from the common pattern are often associated with disease. Without pausing to distinguish cause and effect, it is reasonable to try to restore the pattern of healthy individuals. The human microbiome is malleable by diet, medical intervention, or hygienic practice. Perturbations of microbiomes are consequences of oral antibiotic therapy, and of the not uncommon obsession with cleanliness. Without denying the effectiveness of good hygiene as a defence against infectious disease, it may also militate against a robust microbiome.

Treatments aimed at 'healing' an atypical microbiome include:

- Dietary 'probiotics' such as those containing Lactobacilli.
- Prebiotics—compounds that encourage the growth of organisms deemed to be beneficial.
- Transplants (the ear-wax transfer was an example of this).

Cancer genomics

KEY POINT

Cancer is the leading cause of death in many developed countries.

Cancer is a disease of the genome. Under normal conditions, cell division is carefully regulated. In many tissues of adults, cell division occurs no more than necessary to make up for cell loss. In contrast, cells in tumours have escaped the controls over proliferation. One of the mechanisms of control is contact inhibition, the role of adjacent cells in inhibiting proliferation. A similar mechanism normally inhibits cell migration, keeping cells of an organ such as the liver contained in a coherent mass that is stably localized. Breakdown of this mechanism allows cells from a tumour to escape and metastasize, to colonize remote sites in the body with secondary tumours.

Cancer comes in many different guises, in different forms and in different tissues. This section contains examples of several different types of cancers, but let the reader recognize that the presentation is anecdotal, not encyclopaedic.

Cancer is a disease of genomic instability. The genomes of tumours and normal tissue from the same individual differ quite substantially. Not only do mutations initiate cancer, they continue to accumulate, creating a diverging population of cells. Tumour cells are thereby subject to selection, which can lead to drug resistance.

KEY POINT

Cancer is neither a unique nor a stationary target. We cannot expect a 'magic bullet'—in Ehrlich's sense. A more reasonable goal is individualized therapy. Even this is complicated by the observation of the high degree of genomic heterogeneity, not only among cancers of the same tissue, but in populations of cells of a tumour in a single individual.

Cancer results from accumulated mutations that break down the controls on cell growth. The source can be in three classes of genes: genes that regulate cell proliferation, genes required for repair of DNA damage, and genes that control apoptosis. The mutations can directly affect molecules that control proliferation or apoptosis, or that repair damaged DNA. Or they may affect controllers of activity or expression of these molecules.

Any breakdown in the mechanisms that reduce mutation is a threat to produce cancer-causing mutations. The protective mechanisms include DNA repair upon incorrect replication, or apoptosis to rid the system of cells that have accumulated too many mutations. That many of the mutations have indirect effects is not surprising, given the complexity of control and signalling pathways. Conversely, there are many potential targets for therapy.

A very large number of genes are associated with cancer. An **oncogene** is a gene with the potential to promote tumour development. Oncogenes often start out as **proto-oncogenes** with harmless sequences, until mutations activate them. Examples of oncogenes include:

- Ras is a GTPase involved in a signalling pathway controlling cell proliferation. Mutations in Ras are associated with adenocarcinomas of the pancreas and colon, thyroid tumours, and myeloid leukaemia.
- myc is a transcription factor that regulates expression of genes that induce cell proliferation. Mutations in myc are associated with malignant T-cell lymphomas and acute myeloid leukaemias, breast cancer, pancreatic cancer, retinoblastoma, and small cell lung cancer.

Some genes contributing to the problem come from distant regions of the biochemical map. *IDH1* encodes a metabolic enzyme that converts isocitrate to α-ketoglutarate, in the Krebs cycle. Mutations in *IDH1* are associated with brain cancer and leukaemia. Some mutants lose their normal enzymatic activity but instead convert isocitrate to D-2-hydroxyglutarate. This product is a potent inhibitor of methyltransferases involved in DNA methylation. The result is aberrant epigenetic modifications, which can change patterns of gene expression.

Recent research, empowered by the new powerful sequencing methods, has compared the genomes of tumour cells and normal tissues from the same patient (see Box 10.7). A very large number of genes have been implicated in carcinogenesis. Signalling molecules and transcription factors are very well represented among them.

Other mechanisms protect us against cancer. The *p53* gene encodes a **tumour-suppressor** protein, in effect the opposite of an oncogene. This protein acts in multiple ways: Upon recognition of DNA damage, it can activate DNA repair proteins. By halting the cell cycle at a 'checkpoint', it can give the DNA repair mechanisms time to do their job. And if the damage cannot be repaired, p53 can initiate apoptosis. p53 also inhibits expression of facilitators of metastasis.

Loss of DNA repair activity deprives us of its protective effect in restraining mutagenesis. In almost half of all human cancers, p53 is lost or mutated. Alternatively, it may be present as the wild type but inactivated, by ubiquitinylation and hydrolysis by proteasomes.

Large-scale cancer genome projects

- The cancer genome atlas http://cancergenome.nih.gov maintains a database of key genomic changes in 33 types of cancer, comprising over two petabytes of publicly available data.
- The International Cancer Genome Consortium https://dcc.icgc.org/ has the goal of assembling comprehensive catalogues of genomic abnormalities—including somatic mutations, abnormal gene expression patterns, and epigenetic modifications—from 50 different cancer types.
- The Cancer Genome Project https://www.sanger.ac.uk/group/cancer-genome-project/ is active in research as well as database assembly and dissemination. It maintains and distributes *COSMIC*, a database of somatic mutations curated from the published literature, systematic sequencing studies, and international consortia, such as the International Cancer Genome Consortium (ICGC) and The Cancer Genome Atlas (TCGA). https://cancer.sanger.ac.uk/cosmic
- GDSC (Genomics of Drug Sensitivity in Cancer) collects results on measurements of drug sensitivity for >1000 cancer cell lines and hundreds of potential therapeutic compounds, plus analysis of the relationship between drug sensitivity data and genomic, epigenomic, and transcriptomic features of the cell lines. http://www.sanger.ac.uk/science/tools/gdsc-genomics-drug-sensitivity-cancer

The proteins encoded by *BRCA1*, *BRCA2*, and *PALB2* are also involved in DNA repair. These genes encode long proteins unrelated in sequence and structure (see Table 10.1). Both BRCA proteins are required for chromosome stability, participating in mechanisms of repair of DNA double-strand breaks.

Mutations in these genes enhance the lifetime risk of breast and ovarian cancer:

In the general population:

~12% of women will develop breast cancer;

~1.4% of women will develop ovarian cancer.

Of women with a harmful mutation in *BRCA1* or *BRCA2*:

~60% will develop breast cancer;

~15–40% will develop ovarian cancer.

Many *BRCA1* mutations are known. (Many are not SNPs.) Their prevalence varies among populations, showing strong founder effects (see Table 10.2). Testing for mutations in these genes is now quite common (see Box 10.8).

Tumours display many types of genome changes. These include: SNPs; relatively short insertions or deletions; copy-number variations; and **aneuploidy**: alterations in structure or number of chromosomes, including whole-genome duplication. In some cases, there have been no detected differences between sequences in tumour cells and normal cells from the same patient. The difference can be at the epigenetic level.

A single SNP may be enough to trigger cancer. A specific mutation in the p21 protein appears in a large fraction of bladder tumours. p21 is a growth factor receptor, involved in transmitting from the cell surface to the nucleus a signal initiating cell division. The mutation Gly12→Val results in permanent activation of p21, constitutively promoting cell division.

Many other SNPs have been observed in cancers. Many are specific to cancers of particular organs. Tumours in children typically show 0.1–0.2 mutations per megabase of DNA sequence. Most adult solid tumours typically show one to two mutations per megabase, an order of magnitude higher. There are some cases of much higher mutation rates, attributable to repair mechanisms broken by protein dysfunction or epigenetic silencing of mismatch repair genes, or to mutation of the DNA polymerase ε which has proofreading and repair activity.

Determination of mutation patterns in cancers is clinically useful for confidence and precision in diagnosis, prognosis, and choice of treatment. Different patients with similar tumours may have different mutations. Different cells in a tumour from a single patient can also show sequence diversity. In cases where only one type of cell within a tumour is

Table 10.1 The genes *BRCA1*, *BRCA2*, *and PALB2*

Gene	Chromosome band	Gene length	Protein length (amino acids)	Number of exons
BRCA1	17q21.31	81 189	1 863	24 (22 coding, exon 11 very large)
BRCA2	13q13.1	85405	3418	27
PALB2	16p12.2	38 198	1 186	13

Table 10.2 Common *BRCA1* and *BRCA2* mutations

Population	Common *BRCA1* mutations	Common *BRCA2* mutations
Ashkenazi Jews	185delAG, 5382insC	6174delT
Iceland		999del5
Denmark	2594delC, 5208T→C	
Lithuania	4153delA, 5382insC, 61G→C	
China	589delCT, IVS7-27del10, 1081delG, 2371-2372delTG	3337C→T

sensitive to a particular drug, the killing of the sensitive cells and selection for the non-sensitive ones will result in development of resistance.

Mutations and cancer

Mutations are relevant to cancer research and treatment in several ways:

(1) Mutations detectable in the genome indicate propensity for development of cancers. Mutations in *BRCA1*, *BRCA2*, and *PALB2* as indicators for likelihood of breast and ovarian cancer development are probably the best known (see Box 10.8).

(2) Sequence analysis can predict disease progression and outcome.

(3) Sequence analysis can help choose optimal treatment.

(4) Tumour progression often involves mutations and divergence of cell lines.

For autosomal oncogenes, development of cancer may require inactivation of both copies. Consider, as an example, retinoblastoma, a rare childhood tumour of the eye. Approximately 30–40% of cases are familial; the rest sporadic. The familial form shows an autosomal dominant inheritance pattern. Clinical characteristics of familial retinoblastoma that distinguish it from the sporadic picture are early onset, and the appearance of multiple tumours, affecting both eyes.

BOX 10.8 Screening for mutations in breast cancer genes *BRCA1*, *BRCA2*, and *PALB2*

Breast cancer is a leading killer of women in the UK, western Europe, and the US, affecting over 13% of the population (100 times as many women as men). Testing for mutations in *BRCA1*, *BRCA2*, and *PALB2* can provide a risk-alerting system for breast and ovarian cancer. However, planning of population-wide screening programmes must take into consideration cost/benefit analysis. In many countries, medical care delivery policy is set by a national health service. (The US is a notable exception.) Governments must decide how to apportion resources, taking into account the cost of any procedure and the utility of the information it produces. Mutations in *BRCA1* and *BRCA2* are associated with only about 5–10% of breast cancers. Conversely, not all women with a mutation in either of these genes develop cancer, although over 50% do. The *BRCA1* and *BRCA2* genes are long, multi-exon sequences each >100 kb long, until recently making total resequencing of the genes a complicated procedure. Many mutations are known, spaced widely within the exons. Given that even complete resequencing of the genes would not provide a helpful prognosis in many cases, it has so far not been deemed useful, even with current technology, to screen the entire population fully for *BRCA1* and *BRCA2* mutations.

It is generally accepted that the logic of the decision changes for individuals suspected to be at high risk. This includes women who

- have a close relative known to have a *BRCA1*, *BRCA2*, or *PALB2* mutation;
- have close relatives who have been diagnosed with early-onset (age <50 years) breast or ovarian cancer; or
- have themselves been diagnosed with breast cancer and want to know their likelihood of developing ovarian cancer.

For these individuals, the likelihood of finding useful information certainly justifies genetic testing. On the basis of the results, some women, most famously actress Angelina Jolie, have opted for prophylactic surgery (see Chapter 2).

In the UK, screening is available on the NHS only for women deemed at high risk because of personal or family medical history. In the US, many insurance plans 'cover' testing only for women deemed at high risk. Private sequencing services also provide analysis for *BRCA1* and *BRCA2* mutations.

Rejecting the high-risk criteria, Mary-Claire King, discoverer of *BRCA1*, has argued for universal screening of genes implicated in increased breast cancer risk. She points out that approximately half the women carrying mutations in *BRCA1* or *BRCA2* have no family history of breast or ovarian cancer.

The cost of testing has dropped precipitously, partly through advances in technology, and partly as a result of a 2013 US Supreme Court decision that invalidated the patents on *BRCA1* and *BRCA2* (see Box 2.2). Soon the only question will be the benefit; costs will in most cases be a very diminished issue.

The 'two-hit hypothesis' interprets the relation between sporadic and familial forms of disease in terms of the need to mutate both copies of such genes. This hypothesis offers an explanation for the differential age of onset, and severity, of familial and sporadic retinoblastoma. The idea is that non-familial cases require inactivation of both copies of retinoblastoma gene, each of which was originally functional. Separate and independent mutations are necessary. In contrast, familial retinoblastoma affects a person who has inherited one defective and one functional copy of the gene. That is, the first hit is inherited; all that is needed is the second hit (see Figure 10.10).

Whole-genome sequencing association studies of breast cancer

Conventional genome-wide association studies (GWAS) using panels of SNPs are very powerful, but have limitations: The aggregate of SNPs is a very small fraction of the genome, and none may share a haplotype block with a gene of interest. Comprehensive exome sequencing is also powerful, but is blind to regions of the genome that do not encode proteins. The logical step, made possible now by technical advances, is to apply whole-genome sequencing to patients presenting a disease, and control groups.

Known genes linked to breast and ovarian cancer—including but not limited to *BRCA1*, *BRCA2*, and *PALB2*—help people to evaluate their risk of developing breast and ovarian cancer. However, the known genes explain less than a quarter of cases linked to a shared family history. Other genes must be involved. Conversely, many people with mutations in *BRCA1* do not develop cancer. Are there modifiers? Moreover, many people with mutations in *BRCA1* have no family history of breast or ovarian cancer. The mutations must have arisen in the patient's own genome, rather than being inherited.

Therefore, genome sequencing illuminates cancer in ways other than finding hereditary predispositions. Cancer is a disease of genome instability. The primary genomic lesion may remove protection against further, progressive mutation—for instance, BRCA1 is a DNA repair enzyme. It is essential to understand the full spectrum of mutations that characterize the development of the disease.

To this end, a large international group carried out whole-genome sequencing of 560 breast cancers and normal tissue, from 556 females and 4 males. Previous investigations provided data from an additional 776 individuals. The data revealed regions of the genome containing clusters of sites that differed between tumour and normal tissue, thereby identified as cancer-associated.

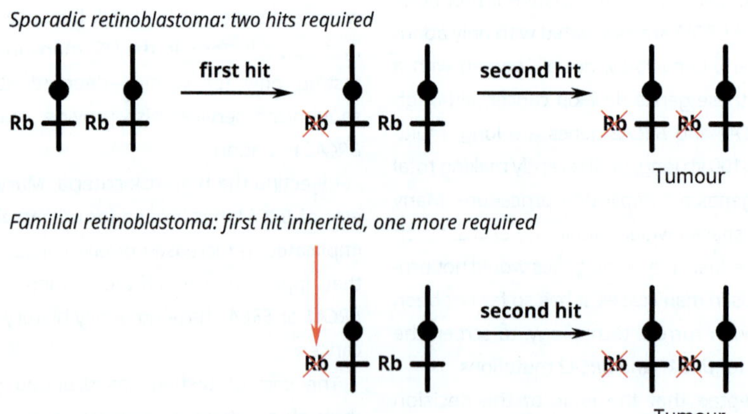

Figure 10.10 Explanation of the difference between sporadic and familial retinoblastoma, a rare cancer of the retina. According to the two-hit theory, two copies of a gene must be inactivated. In sporadic retinoblastoma, both copies are originally functional, and two separate, independent mutations are required to inactivate them. In familial retinoblastoma, one defective copy of the gene is inherited. Only a single mutation, in the other allele, is required to produce the disease.

Six new breast cancer-associated genes reported are:

Gene	Function
ATRIP	required for checkpoint signalling after DNA damage
MED23	subunit of the CRSP (cofactor required for SP1 activation) complex, required to activate transcription
FOXP1	transcription factor, regulation of transcription
MLLT4	fusion partner of acute lymphoblastic leukaemia (ALL-1) gene, in t(6;11)(q27;q23) translocation
XBP1	transcription factor that regulates expression of MHC class II gene; may enhance expression of T-cell leukaemia virus type 1 proteins
ZFP36L1	regulates response to growth factors

There is now a total of over 100 genes known to be associated with breast cancer; and almost 1000 genes associated with all types of human cancers.

An advantage of whole-genome sequencing is access to non-protein-coding regions. A statistically significant number of mutations identified, as breast cancer related, three promoter genes—*WDR74*, *TBX1D12*, and *PLEKHS1* and two genes for long non-coding RNAs: *NEAT1* and *WDR74*.

Clustering of different types of mutations yields signatures, that characterize different tumours within the dataset. There are 12 signatures involving sets of substitution mutations, 6 signatures involving rearrangements, and 2 signatures involving insertions/deletions. The aim is to be able to classify the tumour of any patient, and correlate signature with choice of effective treatment.

Copy-number alterations in cancer

One category of differences between genome sequences is copy-number variation—the loss or multiplication of regions. More than 30% of the human genome is contained in some region of varying copy number. The regions can vary in size from microscopic to megabases in size. In principle, complete chromosome duplication—as in the trisomy-21 associated with Down's syndrome—or even whole-genome duplication or monosomy—that is, chromosome loss, as in Turner's syndrome: females with only one X—are large-scale copy-number variations. As is even the difference between XY and XX male and female mammals.

Gene duplication contributes to protein evolution, facilitating the development of novel functions. (If a protein provides an essential function, duplication allows one copy to retain its function and the other to diverge.) However, duplications of oncogenes are a common cause of cancer. In some cases, the gene duplication pattern is specific to a particular type of cancer, providing useful cues to diagnosis and treatment.

Diffuse large B cell lymphoma is a cancer of B cells, the cells that produce antibodies. A recent study showed a signature pattern of copy-number variations in diffuse large B cell lymphoma, compared to non-haematologic cancers (see Figure 10.11). In some patients, a set of copy-number variations included genes active in apoptotic and cell cycle pathways. These sequences allowed prediction of success of conventional chemotherapy for the condition, and proposals of treatment strategy based on the individual patient's copy number alteration profile.

Chromosomal aberrations

The Philadelphia chromosome is an abnormal, shortened chromosome 22, arising from a translocation—an exchange of chromosomal segments—between chromosome 22 and chromosome 9. The breakpoints are at bands 9q34 and 22q11, and this translocation is denoted t(9;22)(q34;q11) (see Figures 10.12 and 10.13).

The disastrous effects of the Philadelphia translocation arise because both breakpoint sites are within genes. This results in genes for two chimaeric, or fusion, proteins, combining *ABL1* (Abelson murine leukaemia viral oncogene homologue 1), a tyrosine kinase, from chromosome 9; and *BCR*, the breakpoint cluster region gene, from chromosome 22. In normal cells, *ABL1* and *BCR* are separate. As a result of the translocation, chromosome 22 encodes a *BCR–ABL1* fusion and chromosome 9 encodes an *ABL1–BCR* fusion.

The *BCR–ABL1* gene on chromosome 22 encodes a fusion protein with tyrosine kinase activity insensitive to the normal regulation of *ABL1*. (The *ABL1–BCR* gene on chromosome 9 is apparently silent.)

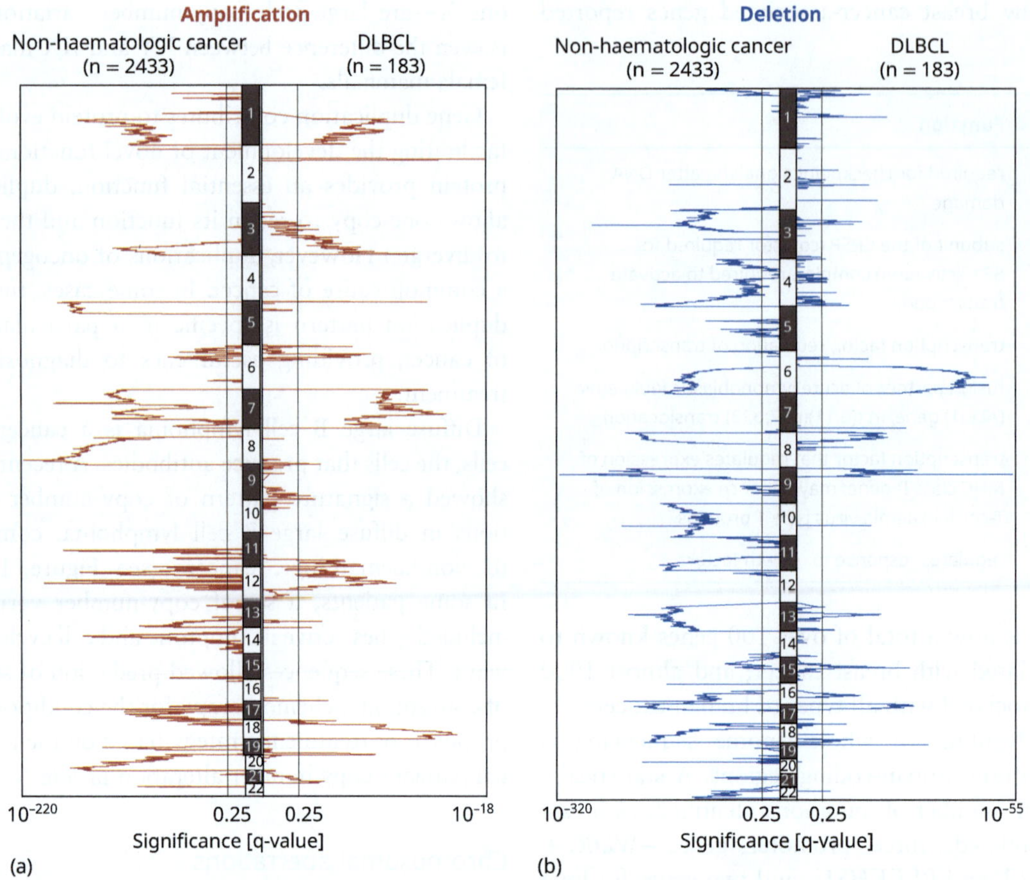

Figure 10.11 Locations of copy-number variants in primary diffuse large B cell lymphomas compared with non-haematologic cancers. Copy-number gain in red; copy-number loss in blue. x-axis: significance, expressed as false discovery rate q-value (a q-value of 0.25, marked by the black line, implies that 25% of significant tests will produce false positives; many of the peaks have much smaller false discovery rates), y-axis: chromosome position.

Used with permission of Elsevier Science & Technology Journals, from Monti, S., Chapuy, B., Takeyama, K., *et al.* (2012). Integrative analysis reveals an outcome-associated and targetable pattern of p53 and cell cycle deregulation in diffuse large B cell lymphoma. Cancer Cell 22, 3: 359–372. https://doi.org/10.1016/j.ccr.2012.07.014. Permission conveyed through Copyright Clearance Centre, Inc.

There is good evidence for the association with cancer of the product of this abnormal gene: (1) the fusion protein is expressed in the proliferating cells; and (2) a drug—imatinib mesylate (Gleevec®)—that inhibits the kinase activity is therapeutically effective.

Because the translocation produces unique DNA sequences at the site of the join, it is possible to design FISH (fluorescent *in situ* hybridization) probes that span the common breakpoints of both the *ABL1* and *BCR* genes. A red signal detects sequences on the *ABL1* side of the breakpoint; a green signal detects sequences on the *BCR* side of the breakpoint. Normal cells would show two green and two red signals, on different chromosomes. Diseased cells show overlapping red and green signals (appearing yellow) on the aberrant chromosomes, and single red and green signals for the normal homologues (see Figures 10.12 and 10.13).

Specific chromosomal aberrations can in some cases provide a reliable guide to prognosis (see Figure 10.14).

Moreover, knowing that the protein responsible for the cancer was a rogue tyrosine kinase suggests the appropriate target for drug design. The first inhibitor used for treatment of chronic myeloid leukaemia was imatinib (STI571, Gleevec®). A measure of its success is the increase in five-year survival rate for people with chronic myeloid leukaemia,

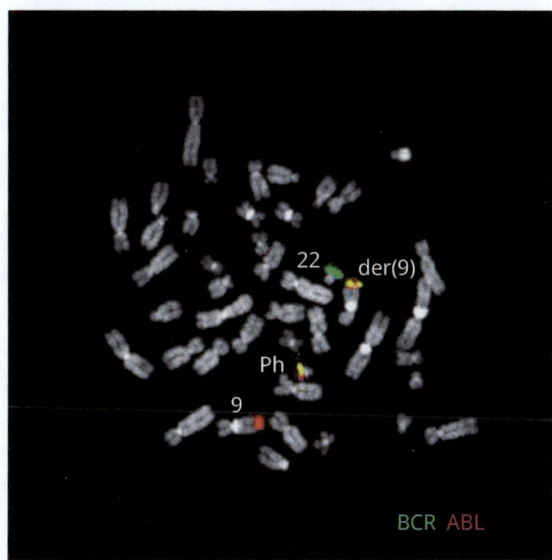

Figure 10.12 Detection by FISH probes of the Philadelphia (Ph) chromosome, associated with chronic myeloid leukaemia. It arises from a reciprocal exchange between chromosomes 9 and 22. The figure shows a representative image of a metaphase cell from a patient with chronic myeloid leukaemia analysed by FISH with fusion (red/green or yellow) signals marking the Ph and der(9) chromosomes (der stands for derivative: a **derivative chromosome** combines segments from two or more normal chromosomes). The normal 9 and 22 homologues are shown by a red and a green signal, respectively.

Figure courtesy of Dr E. Nacheva, Royal Free and University College Medical School, UK.

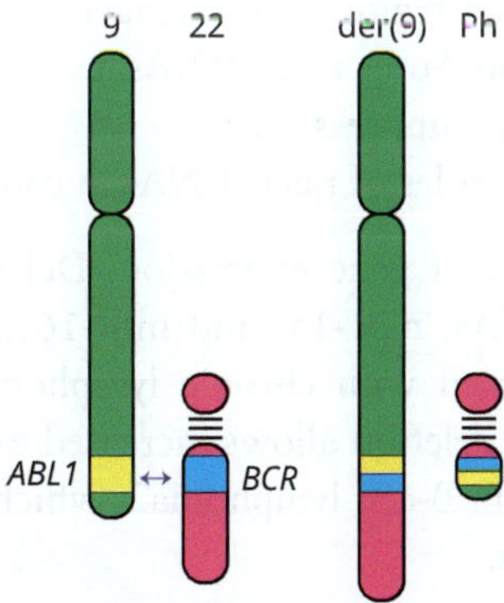

Figure 10.13 Transposition of material between chromosomes 9 and 22 forms the Philadelphia chromosome (Ph), associated with chronic myeloid leukaemia. Left: normal chromosome 9 (green) containing the *ABL1* gene (yellow) and normal chromosome 22 (magenta) containing the *BCR* gene (cyan). The arrow shows the positions of the breakpoints for the transposition. Right: the der(9) and Ph chromosomes, showing exchange of material at the bottom. Note that the site of translocation on each chromosome is within each gene.

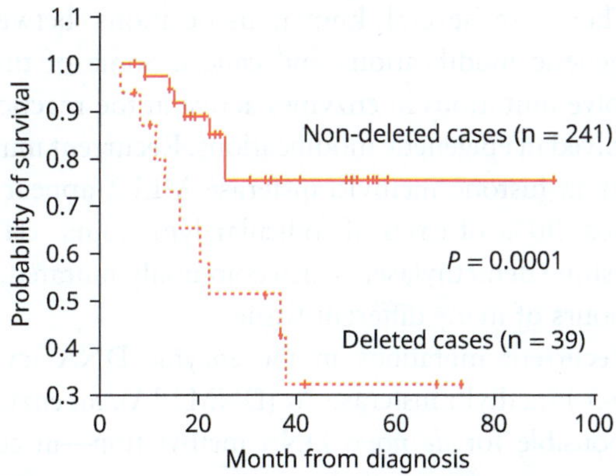

Figure 10.14 Comparison of survival rates in chronic myeloid leukaemia patients with and without deletions of 9q/22q sequences. Patients showing the deletions lived no longer than 40 months after diagnosis. Long-term survival of patients not showing the deletions was in most cases substantially longer.

From: Sinclair, P.B., Nacheva, E.P., Laversha, M., *et al.* (2000). Large deletions at the t(9;22) breakpoint are common and may identify a poor-prognosis subgroup of patients with chronic myeloid leukaemia. Blood 95: 738–744.

which nearly doubled from 31% in 1993 to 59% in 2003–2009. (The patients in the study reported in Figure 10.14 were not treated with imatinib, which received US approval in 2001, and UK approval in 2009.)

Epigenetics and cancer

Epigenetics is the modification of DNA and the protein components of nucleosomes, exerting control over gene expression. Examples include:

(1) Methylation of bases. Methylation of cytosine at the 5 position, in CpG islands, is the most common. It is a mechanism for turning gene expression off. Conversely, hypomethylation of a gene that is methylated normally, activates the gene.

Other base modifications include 6-O-methylation of guanine, and 5-hydroxymethylation of cytosine. Comparisons between malignant and normal tissues from the same individual show significant differences in DNA methylation.

(2) Modification of histones. Histones are the core proteins of nucleosomes (see Figure 1.3). Histone modification has the effect of exposing, or sequestering, regions of DNA, controlling the expression of genes in the affected regions.

There are several known associations between epigenetic modifications and cancer. Some of them involve mutations in enzymes active in the reactions involved in epigenetic modifications. Recurrent mutations in histone methyltransferase MLL2 appear in almost 90% of cases of follicular lymphoma. UTX, a histone demethylase, is also commonly mutated, in tumours of many different tissues.

Recurrent mutations in the enzyme DNA (cytosine-5-)-methyltransferase 3α (DNMT3A), an enzyme responsible for *de novo* DNA methylation—in contrast to copying methylation patterns upon DNA replication—appear in up to 25% of patients with acute myeloid leukaemia.

The discovery of epigenetic changes associated with cancer suggests potential targets for therapy.

Temozolamide is an alkylating agent that methylates guanine at the O6 or N7 position, especially at polyG tracts. It has profound effects on gene expression. Some cells can reverse this modification, by expressing the repair enzyme O-6-methylguanine-DNA-methyltransferase. Tumours in which this expression of this protein is epigenetically silenced are more vulnerable to the drug.

Temozolamide is the treatment of choice for glioblastoma. The survival duration of patients treated with temozolamide depends on their state of methylation of the promoter of O-6-methylguanine-DNA-methyltransferase. The group with suppressed expression had median survival times of 27.7 months; those with unsuppressed expression, 12.7 months. This implies that the epigenetic status of the repair enzyme would be a useful guide to therapy, and that inhibitors of O-6-methylguanine-DNA-methyltransferase such as O-6-benzylguanine, or RNA interference-mediated silencing, can be useful enhancers of temozolamide therapy. This is a subject of active research.

There have been several approaches to methylation of DNA at specific sequences. One idea is to split a methylase into two parts, altered so that their assembly into a functional unit is compromised. By fusing each part to one of the zinc finger proteins HS1 and HS2 that bind DNA at specific nucleotide sequences, the adjacent binding of the two fragments shifts the equilibrium towards assembly of an active methylase, producing a specific modification of a site on the DNA between the two zinc-finger binding sites (see Figure 10.15) (see also Exercise 10.6). The idea is generally similar to the two-hybrid screening method for discovering protein–protein interactions.

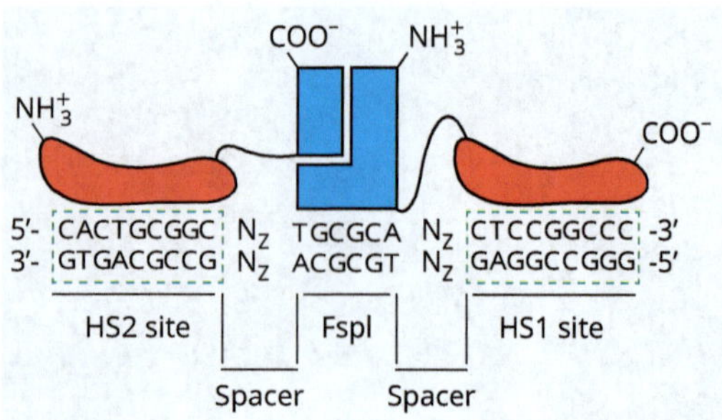

Figure 10.15 Schematic of system to methylate specific sites in DNA. Zinc finger proteins can be designed to bind to specific base sequences. Only if the two zinc finger moities (left and right) bind at the right relative positions will the fragments of the methylase (center) associate into an active form, and modify a base. FspI marks a restriction site. Methylation would block cleavage at this site by the restriction enzyme. This provided a test of successful specific methylation.

Dulla, K., Aguilà, M., Lane, A., *et al.* (2018). Splice-modulating oligonucleotide QR-110 restores CEP290 mRNA and function in human C.2991+1655A≫G LCA10 models. Molecular Therapy - Nucleic Acids 12: 730–740. https://doi.org/10.1016/j.omtn.2018.07.010

MicroRNAs and cancer

Many non-coding RNAs, exclusive of ribosomal and transfer RNAs, regulate gene expression. They function by binding to complementary sequences in the 3'-untranslated region of messenger RNAs, inhibiting translation. Some microRNAs are oncogenic and others tumour-suppressive.

Among the roles of microRNAs in cancer are:

(1) regulation of gene expression. Deletion of two microRNAs, miR-15[2] and miR-16, at 13q14.3, is associated with chronic lymphocytic leukaemia. The deletion allows increased expression of the protein B-cell lymphoma2, which suppresses apoptosis.

(2) control of cell differentiation. Reduction of activities in certain microRNAs prevents terminal differentiation, retaining a proliferative state.

Each tumour type has an individual miRNA signature pattern. This is different from normal tissue and

[2] microRNAs are numbered simply in order of discovery. All microRNAs discussed here are human. If necessary to indicate source species, the nomenclature allows extension to hsa-miR-15.

from other types of cancer. Measurement of these patterns allows (1) better diagnosis—a sharper classification of subtypes, (2) better prognosis, including monitoring of expression levels to warn of relapse or metastasis, (3) prediction of response to treatment, and (4) devising therapy directed at microRNAs themselves.

Several microRNAs suppress metastasis. For example, miR-335 targets the transcription factor SOX4 and extracellular matrix component tenascin C. Silencing of miR-335 by deletion or promoter hypermethylation is associated with metastatic breast cancer.

Therapies targeting microRNAs are an attractive goal. The molecules are typically short enough that synthesis is not onerous. However, simple replacement is problematic because of short half-life and difficulty of delivery to target tissues. Viral delivery systems have come to the rescue. Reduced miR-26a expression is associated with hepatocellular carcinoma. Adenovirus delivery of miR-26a suppressed tumourigenesis in a mouse model of this disease.

Another goal would be to silence microRNAs that promote the tumour development. MicroRNA miR-10b is a node in a control cascade associated with cancer. It exerts control via two downstream proteins (see Figure 10.16):

- Ras homolog gene family, member C (RHOC) is a small GTPase signalling protein. Its overexpression is associated with cell proliferation and malignancy.

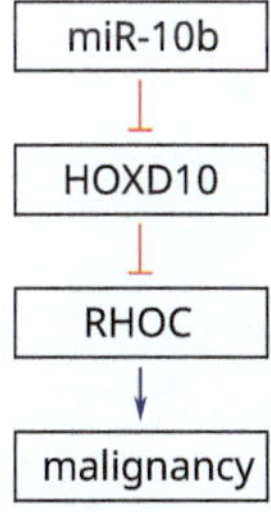

Figure 10.16 The logical relationship of microRNA miR-10b, Homeobox D10 (HOXD10), the ras homolog gene family, member C (RHOC), and malignancy.

Malignancy is associated with overexpression of RHOC. RHOC is downregulated by HOXD10 (lower red 'tee') which should have a protective effect. But HOXD10 is downregulated by miR-10b (upper red 'tee'). This weakens the protection, and can give rise to malignancy (blue arrow). Conclusion: Inhibit activity of miR-10b, to increase expression of HOXD10, which lowers the expression of RHOC, which lowers chance of development of malignancy.

- Homeobox D10 (HOXD10) protein is a transcription factor involved in differentiation and limb development. It downregulates RHOC.
- The microRNA miR-10b blocks translation of the mRNA encoding HOXD10, leading to increased expression of RHOC.

The implication is that blocking of the action of miR-10b would be a useful goal. An antisense oligonucleotide complementary to miR-10b, chemically modified to increase its lifetime, was shown to inhibit metastasis in a mouse model of breast cancer.

Cancer therapy based on microRNAs has many attractive features. Several drugs are in clinical trials, some in phase III.[3]

Immunotherapy for cancer

One difference between growing tumours and replicating pathogens is that the tumour cells are sufficiently non-foreign as to fail to elicit an adequate immune response.

Some classes of tumours show cell-surface signals that are accessible to therapeutic antibodies. The human epidermal growth factor receptor 2 (HER2) gene encodes a receptor that receives and processes signals that trigger cells to multiply. Approximately 15–25% of breast cancers overexpress this receptor. They thereby become oversensitive to the signals, and proliferate. The monoclonal antibody herceptin blocks the HER2 receptors, and interrupts the signal. (Unlike the very large variety of antibodies produced by our immune systems in response to a challenge, monoclonal antibodies are copies of a single protein, with desired binding affinity.) It can also trigger the immune system to destroy the cells to which it is bound.

But even if a human cannot raise antibodies to his or her own tumours, a rat can. Inject human tumour cells into a rat, harvest the antibodies raised, and use them for therapy back in the human. Then the problem is that the rat antibody is a protein foreign to the patient, and elicits an antigenic response. To get around this dilemma,

[3] See: Tian, Z., Liang, G., Cui, K., *et al.* (2021). Insight into the prospects for RNAi therapy of cancer. Front. Pharmacol. 12: 644718; Table I.

copy the binding site of the rat antibody into a human antibody. The result, a **humanized antibody**, contains the minimum required amount of the rat sequence, and is mostly a human protein (see Box 10.9 and Figure 10.17). This reduces the antigenic response, and produces a useful therapeutic agent.

Although the first humanized antibodies were based on antibodies from rodents, phage display is now the method of choice for producing specific high-affinity monoclonal antibodies.[4]

[4] See: Lesk, A.M. (2016). *Introduction to Protein Science: Architecture, Function and Genomics*, Chapter 7. Oxford: Oxford University Press.

BOX 10.9 Humanizing antibodies for therapy

The immune system does not easily recognize cancer cells as foreign. However, a rodent can raise antibodies against human cancers. These rodent antibodies can be effective in human therapy. A problem that limits their clinical utility is that the rodent antibody *is* recognized as a foreign protein by the patient's immune system, and elicits an antigenic response. Because the antigen-binding site resides in a small portion of the antibody—the complementarity-determining regions, CDRs—it is possible to engineer a 'humanized antibody' by grafting the rodent CDRs onto a human framework. Molecules so produced retain binding affinity with greatly reduced antigenicity.

Humanized antibodies have been effective in the treatment of a number of diseases, including cancers, multiple sclerosis, macular degeneration, rheumatoid arthritis, and some viral infections. They have had major impacts on the treatment of recalcitrant diseases and on the economics of the biotechnology industry.

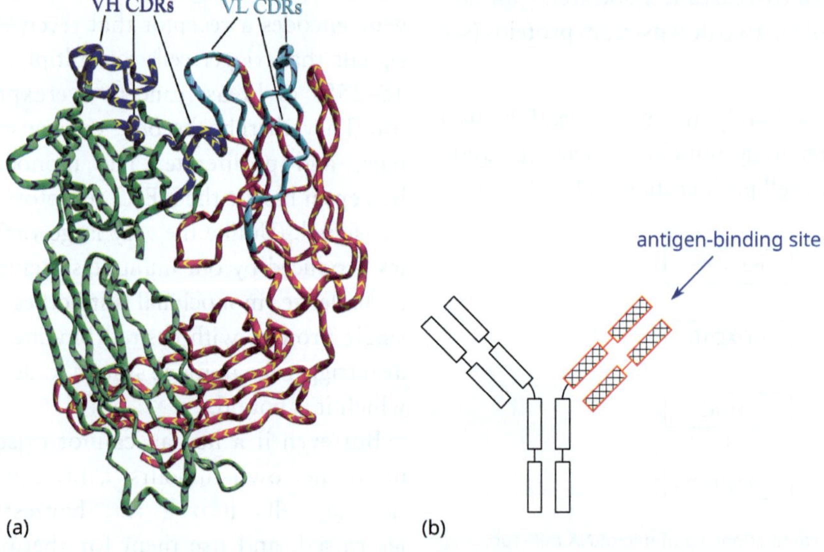

Figure 10.17 (a) The structure of the Fab fragment of CAMPATH-1H, the first humanized antibody, used in successful therapy for chronic lymphocytic leukaemia, cutaneous T-cell lymphoma, non-Hodgkin lymphoma, and even multiple sclerosis. The antigen-binding loops are coloured royal blue and cyan, the rest of the two chains green and crimson. Only the antigen-binding loops, a relatively small fraction of the entire structure, do not correspond to a human antibody sequence. This reduces the overall antigenicity of the molecule in human patients, relative to the rat antibody from which the antigen-binding loops were copied. The molecule used clinically is a complete Immunoglobulin G (IgG), which contains two copies of the structure shown here plus another fragment (the Fc) of comparable size but with no non-human amino acids at all [1CEI]. (b) Schematic diagram of domain structure of a complete IgG. An IgG contains two light chains, each with two domains, and two heavy chains, each with four domains. The portion shown in (a) corresponds to the four domains shown in patterned red, comprising one complete light chain and two domains of one heavy chain. The antigen-binding site is at the 'wing-tip'. There is a second, equivalent symmetrically related antigen-binding site, at the upper-left of this diagram.

LOOKING BACK AND FORWARD

In this chapter, we have focused on applications of genomics to human health. What has been the impact of genomics on medicine—in aspects of both prevention and treatment? Genomics illuminates disease in general, has applications to improved precision in diagnosis and prognosis, and in design and optimum usage of therapies. Applied to individual patients, genomics supports 'personalized medicine'.

We shall go on to traverse the central dogma, treating transcriptomics, proteomics, and metabolomics. Finally, systems biology has the task of integrating all these many separate streams of data and ideas.

RECOMMENDED READING

Applications of genomics to understanding and treatment of disease

Precone, V., Del Monaco, V., Esposito, M.V., *et al.* (2015). Cracking the code of human diseases using next-generation sequencing: Applications, challenges, and perspectives. Biomed. Res. Int. 161648.

Royer-Bertrand, B., & Rivolta, C. (2014). Whole genome sequencing as a means to assess pathogenic mutations in medical genetics and cancer. Cell Mol. Life Sci. 72: 1463–1471.

Nebert, D.W., Zhang, G., & Vesell, E.S. (2008). From human genetics and genomics to pharmacogenetics and pharmacogenomics: Past lessons, future directions. Drug Metab. Rev. 40: 187–224.

Mattick, J.S., Dziadek, M.A., Terrill, B.N., *et al.* (2014). The impact of genomics on the future of medicine and health. Med. J. Aust. 201: 17–20.

Price, A.L., Spencer, C.C., & Donnelly, P. (2015). Progress and promise in understanding the genetic basis of common diseases. Proc. Roy. Soc. (London) 282: 20151684.

Handley, S.A. (2016). The virome: A missing component of biological interaction networks in health and disease. Genome Med. 8: 32.

Braicu, C. (2021). Functional genomics in health and disease. Int. J. Mol. Sci. 22: 12944.

Pirmohamed, M. (2023). Pharmacogenomics: Current status and future perspectives. Nat. Rev. Genet. 24: 350–362.

Zhao, Q., Chen, Y., Huang, W., *et al.* (2023). Drug-microbiota interactions: An emerging priority for precision medicine. Signal Transduct. Target Ther. 8: 386.

Juan, D., Santpere, G., Kelley, J.L., *et al.* (2023). Current advances in primate genomics: Novel approaches for understanding evolution and disease. Nat. Rev. Genet. 24: 314–331.

Genome-wide association studies

Corvin, A., Craddock, N., & Sullivan, P.F. (2010). Genome-wide association studies: A primer. Psychol. Med. 40: 1063–1077.

Bush, W.S., & Moore, J.H. (2012). Chapter 11: Genome-wide association studies. PLoS Comput. Biol. 8: e1002822.

Edwards, S.L., Beesley, J., French, J.D., *et al.* (2013). Beyond GWASs: Illuminating the dark road from association to function. Amer. J. Human. Gen. 93: 779–797.

Appasani, K. (ed.) (2015). *Genome-Wide Association Studies: From Polymorphism to Personalized Medicine*. Cambridge: Cambridge University Press.

Barr, A.J. (2018). The biochemical basis of disease. Essays Biochem. 62: 619–642.

Neu, M.B., Bowling, K.M., & Cooper, G.M. (2019). Clinical utility of genomic sequencing. Curr. Opin. Pediatr. 31: 732–738.

Eisenstein, M. (2021). Fix the gene, cure the disease. Nature 596: S1–S4.

McGuire, A.L., Gabriel, S., Tishkoff, S.A., *et al.* (2020). The road ahead in genetics and genomics. Nat. Rev. Genet. 21: 581–596.

Claussnitzer, M., Cho, J.H., Collins, R., *et al.* (2020). A brief history of human disease genetics. Nature 577: 179–189.

Uffelmann, E., Huang, Q.Q., Munung, N.S., *et al.* (2021). Genome-wide association studies. Nature Reviews Methods Primers 1: 59.

He, J., & Gai, J. (2023). Genome-wide association studies (GWAS). Methods Mol. Biol. 2638: 123–146.

Fitipaldi, H., & Franks, P.W. (2023). Ethnic, gender and other sociodemographic biases in genome-wide association studies for the most burdensome non-communicable diseases: 2005–2022. Hum. Mol. Genet. 32: 520–532.

The human microbiome

Murray, P.R. (2013). The human microbiome project: Beginning and future status. Ann. Clin. Microbiol. 16: 162–167.

Morgan, X.C., Segata, N., & Huttenhower, C. (2013). Biodiversity and functional genomics in the human microbiome. Trends Genet. 29: 51–58.

Relman, D.A. (2015). The human microbiome and the future practice of medicine. J. Amer. Med. Assoc. 314: 1127–1128.

Lepage, P., Leclerc, M.C., Joossens, M., *et al.* (2013). A metagenomic insight into our gut's microbiome. Gut 62: 146–158.

Marchesi, J.R., Adams, D.H., Fava, F., *et al.* (2016). The gut microbiota and host health: A new clinical frontier. Gut 65: 330–339.

Jeyakumar, T., Beauchemin, N., & Gros, P. (2019). Impact of the microbiome on the human genome. Trends Parasitol. 35: 809–821.

Requena, T., & Velasco, M. (2021). The human microbiome in sickness and in health. Rev. Clin. Esp. (Barc.) 221: 233–240.

Manos, J. (2022). The human microbiome in disease and pathology. Acta Pathologica, Microbiologica, et Immunologica Scandinavica 130: 690–705.

Hayes, W., & Sahu, S. (2020). The human microbiome: History and future. J. Pharm. Pharm. Sci. 23: 404–411.

Das, B. (2022). An introduction to human microbiome. Prog. Mol. Biol. Transl. Sci. 191: 1–28.

Leonard, J.M., & Toro, D.D. (2023). Defining the microbiome components (Bacteria, Viruses, Fungi) and microbiome geodiversity. Surg. Infect. (Larchm't.) 24: 208–212.

Aggarwal, N., Kitano, S., Puah, G.R.Y., *et al.* (2023). Microbiome and human health: Current understanding, engineering, and enabling technologies. Chem. Rev. 123: 31–72.

Genomics and cancer

Dellair, G., & Arceci, R.J. (2014). Cancer genomics: Historical perspective and current challenges of cancer genomics. In: G. Dellaire, J.N. Berman, & R.J. Arceci (eds), *Cancer Genomics: From Bench to Personalized Medicine*, Chapter 1, pp. 3–10. London: Elsevier.

Chmielecki, J., & Meyerson, M. (2014). DNA sequencing of cancer: What have we learned? Ann. Rev. Med. 65: 63–79.

Wang, L., & Wheeler, D.A. (2014). Genomic sequencing for cancer diagnosis and therapy. Ann. Rev. Med. 65: 33–48.

Roos, A., & Byron, S.A. (2019). Genomics-enabled precision medicine for cancer. Cancer Treat. Res. 178: 137–169.

Cieslik, M., & Chinnaiyan, A.M. (2020). Global genomics project unravels cancer's complexity at unprecedented scale. Nature 578: 39–40.

Sakamoto, Y., Sereewattanawoot, S., & Suzuki, A. (2020). A new era of long-read sequencing for cancer genomics. J. Hum. Genet. 65: 3–10.

Ginsburg, O., Ashton-Prolla, P., Cantor, A., *et al.* (2021). The role of genomics in global cancer prevention. Nat. Rev. Clin. Oncol. 18: 116–128.

Menghi, F., & Liu, E.T. (2022). Functional genomics of complex cancer genomes. Nat. Commun. 7: 5908.

Waldmann, H. (2002). A personal history of the CAMPATH-1H antibody. Mol. Oncol. 19, Suppl. S3–S9.

EXERCISES, PROBLEMS, AND WEBLEMS

Exercises

Exercise 10.1 From Figure 10.9, (a) what is the most abundant phylum of bacteria in the mouth (or cheek)? (b) In what location on or in the body are *Haemophilus* species found?

Exercise 10.2 What are examples of different combinations of modifiers and targets such as shown in Box 10.5?

Exercise 10.3 Why, in principle, would GWAS studies using similar protocols provide less information if applied to African populations than to Caucasian populations?

Exercise 10.4 Suppose there are strong correlations in the occurrences of different human SNPs, because they fall into the same haplotype block. How might this information be used to simplify genome-wide association studies based on detection of SNPs by microarrays?

Exercise 10.5 Victoria Gray was cured of sickle-cell disease by using CRISPR to edit her stem cells to knock out the *BCL11A* gene, to enhance her production of foetal haemoglobin. Would any children she bore, subsequent to the treatment, be in danger of inheriting a gene for sickle-cell disease from her?

Exercise 10.6 On a diagram of the structure of a guanine–cytosine base pair indicate the result of methylating the guanine at the O6 or N7 position. Would either of these modifications affect its base-pairing capability?

Exercise 10.7 Suppose it were possible to replace the nuclease in the CRISPR/Cas system with a domain that would methylate DNA bases. How might this be applied in cancer chemotherapy?

Exercise 10.8 On a photocopy of Figure 10.17a, (a) draw lines separating the two domains of each chain, and (b) draw a circle around each of the antigen-binding loops.

Problems

Problem 10.1 Duchenne muscular dystrophy (DMD) is an X-linked inherited disease causing progressive muscle weakness. DMD sufferers usually lose the ability to walk by the age of 12, and life expectancy is no more than about 20–25 years. Becker muscular dystrophy (BMD) is a less severe condition. Both conditions are usually caused by deletions in the same gene, dystrophin. In DMD, there is complete absence of functional protein; in BMD, there is a truncated protein retaining some function. Some of the deletions in cases of BMD are longer than others that produce DMD. What is likely to distinguish the two classes of deletions causing these two conditions?

Problem 10.2 A rare mutation on the X chromosome is associated with T- and NK-cell lymphoproliferative disorder. To what extent would the two-hit theory apply to this disease?

Problem 10.3 What would be the effect of loss of function of *both* HOXD10 and RHOC on the likelihood of development of malignancy?

Problem 10.4 On a photocopy of Figure 10.11: (a) circle a region that shows a deletion in both DLBCL (diffuse large B-cell lymphoma) and non-haematological cancer; (b) circle a region that shows a deletion in DLBCL but not in non-haematological cancer; (c) circle a region that shows amplification in non-haematological cancer but not DLBCL; (d) circle a region that shows amplification in DLBCL but not in non-haematological cancer. (e) For each of the regions you chose, on which chromosome does it appear? (f) If you wanted to search for a target to develop therapy specific for DLBCL, which of these regions would you investigate? Explain your answer.

Weblems

Weblem 10.1 Using the Online Mendelian Inheritance in Man™ website (http://www.ncbi.nlm.nih.gov/omim), what are the clinical consequences of the mutations associated with (a) haemoglobin Barts? (b) haemoglobin Linkoping? How do these abnormal haemoglobins differ structurally from normal adult haemoglobin?

Weblem 10.2 Determine a feature of a human MHC haplotype that is associated with a relatively slow progression to AIDS after HIV infection.

Weblem 10.3 In chickens, what MHC haplotype determines resistance to Marek's disease virus?

Weblem 10.4 What are the most common mutations in *BRCA1* among Swedish women?

Weblem 10.5 The three most frequent mutations in *BRCA1* and *BRCA2* genes in Ashkenazi Jews are 185delAG and 5382insC in *BRCA1* and 6174delT in *BRCA2*. In what exons of these genes do these mutations appear?

Weblem 10.6 The mutation causing sickle-cell anaemia is a single base change A→T, causing the change Glu→Val at position 6 of the β chain of haemoglobin. The base change occurs in the sequence 5′-GTGAG-3′ (normal) → GTGTG (mutant). What restriction enzyme can be used to distinguish between these sequences, to detect carriers? What is the specificity of this enzyme?

Weblem 10.7 What gene appears between the genes for C4A and C4B in the human genome?

Weblem 10.8 What mutation is the most common cause of phenylketonuria (PKU)?

Weblem 10.9 Suggest menus, as close as possible to your normal meals, for breakfast, dinner, and supper, appropriate for a university student with PKU. Include recommended amounts.

Weblem 10.10 Find three examples of mutations in the CFTR gene (associated with cystic fibrosis) that produce reduced but not entirely absent chloride channel function. What are the clinical symptoms of these mutations?

Weblem 10.11 Find an example of a genetic disease that is: (a) autosomal dominant, (b) autosomal recessive (other than cystic fibrosis), (c) X-linked dominant, (d) X-linked recessive, (e) Y-linked, (f) the result of abnormal mitochondrial DNA, (g) the result of abnormal copy-number expansion, and (h) the result of a deletion of a region longer than 1kb.

Weblem 10.12 (a) Identify a state of the US in which newborn infants are routinely tested for tyrosinemia type III. (b) Identify a state of the US in which newborn infants are *not* routinely tested for tyrosinemia type III. (c) Identify a state of the US in which newborn infants are routinely tested for isobutyrylglycinuria. (d) Identify a state of the US in which newborn infants are *not* routinely tested for isobutyrylglycinuria. (e) What are the clinical consequences of failure to detect tyrosinemia type III or isobutyrylglycinuria?

Weblem 10.13 (a) What is the normal function of the protein that is defective in Menke disease? (b) Is there a homologue of this gene in the *A. thaliana* genome? (c) If so, what is the function of this gene in *A. thaliana*?

Weblem 10.14 Open a window on the EBI-NHGRI GWAS catalogue: https://www.ebi.ac.uk/gwas/diagram. Select Haematological measurement. Find SNPs that are associated with foetal haemoglobin measurement. Which SNP has the smallest p-value? In what chromosome band does it appear?

Weblem 10.15 The mutation responsible for Nicholas Volker's disease was a single amino-acid substitution Cys203→Tyr in the protein XIAP (X-linked inhibitor of apoptosis), linked to immune deficiency. The structure of this protein is known (wwPDB entry [4J3Y]. Draw a picture of this protein structure indicating the site of the mutation. Is it on the surface of the protein?

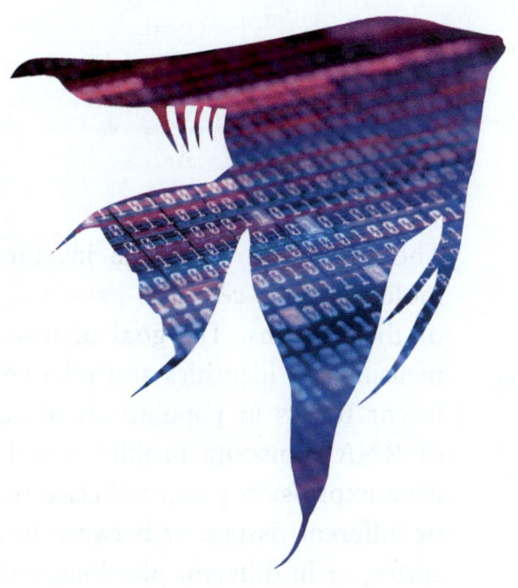

CHAPTER 11

Transcriptomics

LEARNING GOALS

- *To have a clear sense of the difference between the genome and the transcriptome.* The genome of an organism is static, at least in the short term. The transcriptome and the proteome are dynamic.
- *To understand the relationship between microarray technology and RNAseq.*
- *To know the principles underlying microarray technology.*
- *To know the principles underlying RNAseq methods.*
- *To be aware of the types of application which transcriptomics can support.*
- *To know what a gene expression table is and how it is derived* from a microarray or an RNAseq experiment.
- *To be able to distinguish gene-based and sample-based analysis* and how they support different types of application.
- *To understand how transcriptomics can be applied to the study of changing gene expression patterns* in different physiological states.
- *To understand the general features of the correlation of gene expression patterns* with development in *Drosophila melanogaster.*
- *To recognize how transcriptomics can be used to work out the genes and proteins responsible for specific features of a phenotype.*
- *To appreciate developments in metagenomics and metaproteomics*: the applications of methods from the laboratory to natural ecosystems.
- *To appreciate how transcriptomics can be used to study the molecular biology of higher mental processes in mammals*, including learning and the evolution of human language abilities.

Introduction

The transcriptome is the inventory of the RNA molecules in a cell. The proteome is the inventory of the proteins. The goal of transcriptomics is to measure the identities and relative amounts of different RNAs in populations of cells. Comparison of RNA transcript profiles reveals the changes in gene expression patterns between different species, or different tissues, or between healthy and disease states, or in different physiological states, or under different external conditions, or as a function of time or developmental stage.

Some, but by no means all, of transcribed RNAs are messenger RNAs (mRNAs). For these, there is correspondence between the transcriptome and the proteome. But the cellular RNA universe is much richer. More than 90% of the human genome is transcribed into RNA, but no more than 5% is protein coding.

Two approaches to determining the transcriptome are:

1. Microarrays have been in use since the late 1980s. A large set of oligodeoxynucleotides is affixed to a regular array of positions on a chip. These oligos bind fluorescently tagged RNA fragments from a sample. The fluorescent signal pattern indicates which sequences on the chip have complements within the sample. Other types of arrays can analyse short DNA molecules, or even proteins.
2. RNAseq is the sequencing of the RNA molecules in a sample, usually after converting them to cDNA. Although microarrays supported many classic experiments, RNAseq has superseded microarrays for measurements of differential gene expression (see Box 11.1). Microarrays are still sometimes used for genotyping, for instance, by the commercial sequencing project 23andMe, as a lower cost alternative to sequencing.

> **BOX 11.1 RNAseq has several significant advantages over microarrays, and is now the method of choice for most research:**
>
> - To construct a microarray, one must make a choice of what probe sequences to include. RNAseq will report whatever is there, with no prior commitment to any set of sequences. For instance, RNAseq but not microarrays will pick up novel mutations, including translocations, and post-transcriptional edits.
> - Sequencing methods give more precise measurements of RNA concentrations and have higher dynamic range.
> - If a population of cells includes both a host and a pathogen, the transcriptomes of both are simultaneously measurable. In contrast, without a specially designed chip, a microarray would most likely contain probes only from the host.

RNAseq—or whole transcriptome shotgun sequencing—measures RNA levels by applying high-throughput sequencing techniques. RNA isolated from cells is fragmented, reverse-transcribed to cDNA, and sequenced. Direct sequencing of RNA is also possible. Assembly is easiest by aligning with a reference genome.

> Methods for expression profiling include microarrays and whole transcriptome shotgun sequencing, or RNAseq. For most research applications, RNAseq has replaced microarrays.

Applications of transcriptomics

- *Investigating cellular states and processes.* Profiles of gene expression that change with cellular state or growth conditions can give clues to, for instance, the mechanism of sporulation, or to the change from aerobic to anaerobic metabolism. In higher organisms, variations in expression patterns among different tissues, or different physiological or developmental states, illuminate underlying biological processes.

- *Comparison of related species.* The very great similarity in genome sequence between humans and chimpanzees suggested that the profound phenotypic differences must arise at the level of regulation and patterns of protein and RNA expression, rather than in the few differences between the amino-acid sequences of the proteins themselves.
- *Diagnosis of genetic disease.* Testing for the presence of mutations can confirm the diagnosis of a suspected genetic disease. Not all mutations can be detected in genome sequences, for instance disease-correlated splice variants. Detection of carriers can help in counselling prospective parents.
- *Genetic warning signs.* Some diseases are not determined entirely and irrevocably by genotype, but the probability of their development is correlated with genes or their expression patterns. Genome sequencing and transcriptomic profiling can warn of enhanced risk.
- *Precise diagnosis of disease.* Different related types of leukaemia can be distinguished by signature patterns of gene expression. Knowing the exact type of the disease is important for prognosis and for selecting optimal treatment.
- *Target selection for drug design.* Proteins showing enhanced transcription in particular disease states might be candidates for attempts at pharmacological intervention.
- *Drug selection.* Gene expression patterns can be detected that govern responses to drugs, which in some patients render treatment ineffective and in others cause unusual or serious adverse reactions. (Pharmacotranscriptomics)
- *Suggestion of gene function.* A gene with an expression pattern similar to genes in a metabolic pathway is also likely to participate in the pathway.
- *Pathogen resistance.* Comparisons of genotypes or expression patterns between bacterial strains susceptible and resistant to an antibiotic point to the possible involvement of the RNAs or proteins in the mechanism of resistance.
- *Following temporal variations in protein expression.* This permits timing the course of (1) responses to pathogen infection, (2) responses to environmental change, (3) changes during the cell cycle, and (4) shifts in expression patterns during development.

Microarrays

For an integrated characterization of cellular activity, we want to determine what RNAs and proteins are present, in what amounts, and where in the cell.

Hybridization is an accurate and sensitive way to detect whether any particular nucleic-acid sequence is present. Microarrays achieve high-throughput analysis by running many hybridization experiments in parallel (see Box 11.2).

KEY POINT

The immobilized material on the chip is the *probe.* The sample tested is the *target.*

Expression patterns can help to identify genes that underlie diseases. Some diseases, such as cystic fibrosis, arise from mutations in single genes. For these, isolating a region by genetic mapping can help to pinpoint the lesion. Other diseases, such as asthma, depend on interactions among many genes, with environmental factors as complications. To understand the aetiology of multifactorial diseases requires the ability to determine and analyse expression patterns of many genes, which may be distributed around different chromosomes.

Microarrays are also used to screen for mutations and polymorphisms. Microarrays containing many sequence variants of a single gene can detect differences from a standard reference sequence.

The basic innovation of microarrays is parallel processing

Compare the following types of measurement:

- *'One-to-one'*. To detect whether one oligonucleotide has a particular known sequence, test whether it can hybridize to the oligonucleotide with the complementary sequence.
- *'Many-to-one'*. To detect the presence or absence of a query oligonucleotide in a mixture, spread the mixture out and test each component of the mixture for binding to the oligonucleotide complementary to the query. This is a **northern** or **Southern** blot.
- *'Many-to-many'*. To detect the presence or absence of *many* oligonucleotides in a mixture, synthesize a set of oligonucleotides, one complementary to each sequence of the query list, and test each component of the mixture for binding to each member of the set of complementary oligonucleotides. Microarrays provide an efficient, high-throughput way of carrying out these tests in parallel.

To achieve parallel hybridization analysis a large number of DNA oligomers are affixed to known locations on a rigid support, in a regular two-dimensional array. The mixture to be analysed is prepared with fluorescent tags to permit the detection of the hybrids. The array is exposed to the mixture. Some components of the mixture bind to some elements of the array. These elements now show the fluorescent tags. Because we know the sequence of the oligomeric probe in each spot in the array, measurement of the *positions* of the hybridized probes identifies their sequences. This identifies the components present in the sample (see Figure 11.1).

Such a DNA microarray is based on a small wafer of glass or nylon, in area typically 2 cm^2. Oligonucleotides are attached to the chip in a square array, at densities between 10000 and 250000 positions per cm^2. The spot size may be as small as ~150 μm in diameter. The grid is typically a few centimetres across. A yeast chip contains over 6000 oligonucleotides, covering all known genes of *Saccharomyces cerevisiae*. A DNA array, or DNA chip, may contain 400000 probe oligomers. Note that this is larger than the total number of genes, even in higher organisms (excluding immunoglobulin genes). However, the technique requires duplicates and controls, reducing the number of different genes that can be studied simultaneously. Nevertheless, it is possible to buy a single chip containing all known human genes (not all

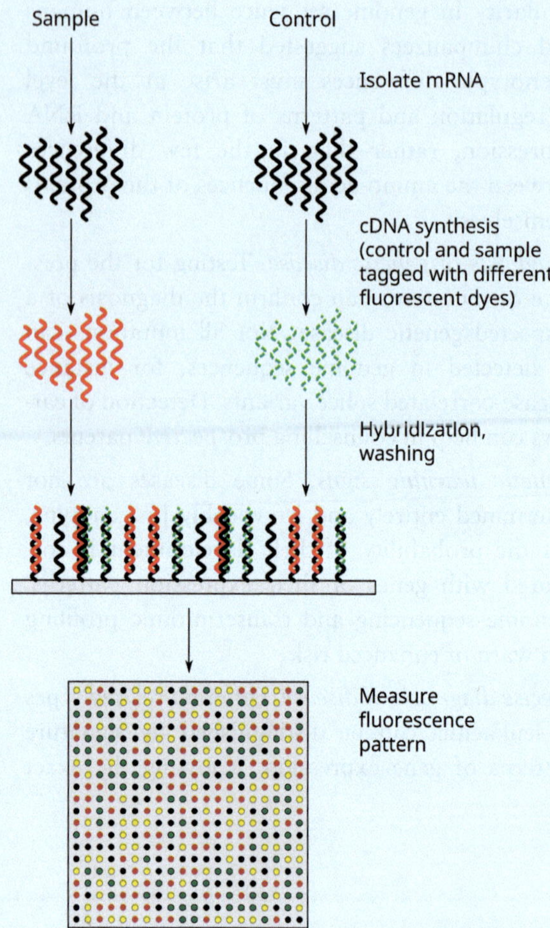

Figure 11.1 Schematic diagram of a microarray experiment. A sample to be tested is compared with a control of known properties. From each source, RNA is isolated and converted to cDNA, using reagents bearing a fluorescent tag, with different colours for the control and sample. After hybridizing to the microarrays and washing away unbound material, the bound target oligonucleotides appear at specific positions. A red spot indicates binding of oligonucleotides from the sample. A green spot indicates binding of oligonucleotides from the control. A yellow spot indicates binding of both. Each probe, represented here by a wavy black line affixed to the support, really contains *many copies of a single oligonucleotide*. Indeed, for accurate measurement, the concentration of the target must greatly exceed the concentration of the probe. If both red- and green-tagged targets are complementary to the oligonucleotide probe at one spot, both can bind to different probe molecules within the same spot.

immunoglobulin genes, of course). Also available is a set of 'tiling' chips that cover the entire human genome sequence.

A mixture is analysed by exposing it to the microarray under conditions that promote hybridization, then washing away any unbound oligonucleotides. To compare material from different sources, the samples are tagged with differently coloured fluorophores. Scanning the array collects the data in computer-readable form.

Different types of chip support different investigations.

- In an *expression chip*, the immobilized oligonucleotides are cDNA samples, typically 20–80 bp long, derived from mRNAs of known genes. This is by far the most common type of microarray. The target sample might contain RNAs from normal or diseased tissue, for comparison.

  Typically, one position on the chip contains an oligonucleotide with the exact sequence we want to test for and, as a control, another position contains a corresponding mismatched oligonucleotide, differing by one base near the centre of the sequence. These form a probe pair. To detect a single RNA, a chip may contain 16–20 probe pairs, spread over the RNA sequence.

- In *genomic hybridization*, one looks for gains or losses of genes, or changes in copy number. The probe sequences, fixed on the chip, are large pieces of genomic DNA from known chromosomal locations, typically 500–5000 bp long. The target mixtures contain genomic DNA from normal or disease states. For instance, some types of cancer arise from chromosome deletions, which can be identified by microarrays.

- *Mutation or polymorphism microarray analysis* is the search for patterns of single-nucleotide polymorphisms (SNPs). The oligonucleotides on the chip are selected from reference genomic data. They correspond to many known variants of individual genes.

- *Protein microarrays* are arrays of protein detectors—usually antibodies—that detect protein–protein interactions.

- *Tissue microarrays* collect and assemble microscopic samples of tissue. They permit comparative analysis of the molecular biology and immunohistochemistry of the samples.

Microarray data are semiquantitative

Microarrays are capable of comparing concentrations of target oligonucleotides. This allows investigation of responses to changed conditions. Unfortunately, the precision is low. Moreover, mRNA levels, detected by the array, do not always quantitatively reflect protein levels. Indeed, usually mRNAs are reverse transcribed into more stable cDNAs for microarray analysis; the yields in this step may also be non-uniform. Microarray data are, therefore, semiquantitative: although the distinction between presence and absence is possible, determination of relative levels of expression in a controlled experiment is more difficult, and measurement of absolute expression levels is beyond the capability of microarray techniques. A change in expression levels of a gene between two samples by a factor ≥ 1.5–2 is generally considered a significant difference.

RNAseq

The data measured in microarray experiments are also obtainable by high-throughput sequencing methods. A major advantage of sequencing is that it is 'hypothesis free': It is not necessary to design and select sets of probe oligos to appear on chips. What you sequence is what you get.

The goal of RNAseq is a comprehensive determination of the RNAs in a sample, or even in a single cell. Although all expressed proteins correspond to some messenger RNA, by no means all RNA encodes proteins. About 85% of the genome is transcribed, but only about 3% codes for proteins. Indeed, the transcriptome is an object of very high interest in itself; hence the very wide application of RNAseq. Also, RNAseq is suited perfectly for studying divergence of RNA viruses such as hepatitis C or HIV-1, during the course of the disease in an infected patient.

In addition, sequencing RNA is also a convenient way to get at the sequences of the expressed proteins. What can RNA sequencing tell us about a cell's proteins that whole genome sequencing cannot? Conversely, what can RNA sequencing *not* tell us about a cell's proteins? RNA sequencing can report on splice variants, and on the results of RNA editing. By measuring the amounts of each RNA fragment present in the original sample one can derive approximate transcription and expression profiles. Absolute quantitation, even of RNAs, is not possible, although RNAseq can identify up- and downregulated genes in controlled studies. The conversion step from RNA to cDNA introduces additional bias, and potential errors, into the results.

RNAseq can tell us only indirectly the amounts of different proteins in the cell, as these are imperfectly correlated with abundances of different mRNAs. RNAseq cannot tell us about quaternary structure of proteins, nor about post-translational modifications.

Figure 11.2 gives an outline of the procedure. Classically, RNAseq approached the determination of the transcriptome by using reverse transcriptase to convert to DNA all the RNA extracted from a sample and then fragmenting and sequencing the DNA. Now, it is possible to sequence RNA directly.

If a complete genome of the organism is available, mapping the sequences of the fragments onto the genome is an easy route to assembly. A pileup of fragments signifies an exon. Gaps in the mapping between the RNA and genome indicate introns: A junction read is a fragment that overlaps intron–exon boundaries. It will not map fully onto the reference genome (see Figure 11.3).

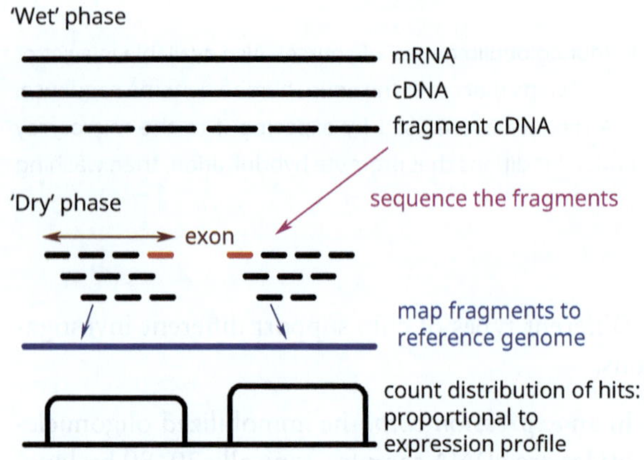

Figure 11.2 Components of the RNAseq measurement. First, the 'wet phase': RNA is collected, and reverse-transcribed. The cDNA is diced into fragments and sequenced. Next, the 'dry phase': map the reads onto a reference genome, if available, to assemble them. Take an inventory to give an approximate transcription profile. Identifying the genes that contributed to the RNA extracted identifies the proteins expressed. It is possible to extrapolate from transcription profiles of mRNAs to the distribution of expressed protein, but the inference is indirect and approximate.

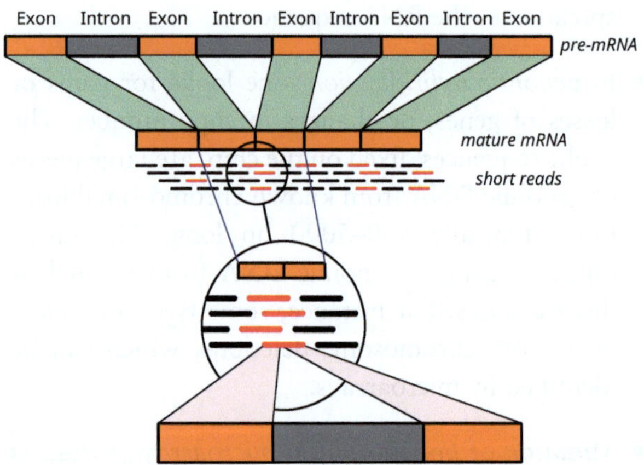

Figure 11.3 A 'junction read' is a fragment that straddles an intron–exon boundary. The sequence is continuous in the mature messenger RNA but fragmented in the genomic DNA. Orange segments represent exons; grey segments represent introns. Short reads coloured red are junction reads. They overlap the boundary between two exons in the mature mRNA, and therefore do not map back onto the pre-messenger RNA or the genomic DNA.

Several software systems address the assembly problem if a complete genome is not available.

What information can RNAseq reveal, that is not contained in the genome sequence itself?

- *Differential gene expression*: Comparison of transcriptomes between strains or species or individual cells, in different tissues, across different developmental stages, or disease conditions and treatments.
- *Variant detection and allele-specific expression*: Single-nucleotide polymorphisms (SNPs, differences in single, isolated nucleotide positions) may lead to allele-specific expression.
- *Small RNA profiling*: Small RNA species include microRNA (miRNA), small interfering RNA (siRNA), piwi-interacting RNA (piRNA), small nucleolar RNA (snoRNA), and small nuclear RNA (snRNA). These non-coding RNAs play roles in gene silencing and post-transcriptional regulation of gene expression. They are involved in development, cell proliferation and differentiation, and apoptosis.
- *Identification of alternative splicing patterns*: Multiple proteins arising from a single gene, by different choices of exons retained during messenger RNA maturation.
- *System biology*: The integration of gene regulation and expression patterns, including analysis of co-expression networks and of pathways.
- *Metatranscriptomics*: The totality of RNAs (protein-coding and non-coding) present in a natural environmental niche, including sites within the human body, or the soil, or the sea. Metatranscriptomics describes the abundance, diversity, and differential expression of genes expressed by different source organisms or cells, and the variability with environmental changes and with time—including periodic and non-periodic changes. Metatranscriptomics illuminates the structure, function, and mechanisms of interactions between components of the niche.

High-throughput RNA sequencing has identified many novel RNA-genome viruses. A recent study discovered a fivefold increase in known virus diversity over the population known previously.

RNAseq has the following advantages over microarrays:

(1) A major difference is that microarrays require specification in advance of what you will be looking for, in order to populate the chip. The array will not detect a target unprovided for on the chip; even though the goal of many experiments is to detect such variants. Cancers often produce novel genetic variants to which microarrays are blind. RNAseq reports what the sample contains, with no preconceptions or limitations.

(2) RNAseq has a greater dynamic range.

(3) RNAseq can be tuned to look for rare transcripts. RNAseq is better than microarrays at identifying novel, low-frequency RNAs associated with disease processes, including but not limited to splice variants.

Variable splicing in the retina

Variable splicing is the processing of the initial RNA transcript of a gene to produce multiple mature messenger RNAs that encode different proteins.

The retina, typical of neuronal tissues, exhibits highly ramified cellular differentiation, including very high levels of variable splicing. The panoply of alternative splicing events (see Figure 11.4) includes retinal-specific splicing factors, and tissue-specific exons. This demands regulation of exquisite precision, both spatially and temporally.

The complexity creates a challenge to molecular biology. Clearly, information about splice variants is not directly available from even a complete genome sequence. It must be measured at the RNA or protein level. Moreover, accurate and complete detection of splice variants requires long-read RNA sequencing. PacBio and Oxford Nanopore provide the required technology, able to sequence a typical mRNA in 'one go'.

A few genes are expressed only in the retina. Rhodopsin is an example. The Human Protein Atlas (https://www.proteinatlas.org/humanproteome/tissue/tissue+specific) reports, for the retina, tissue-enriched gene expression for 123 genes (not distinguishing

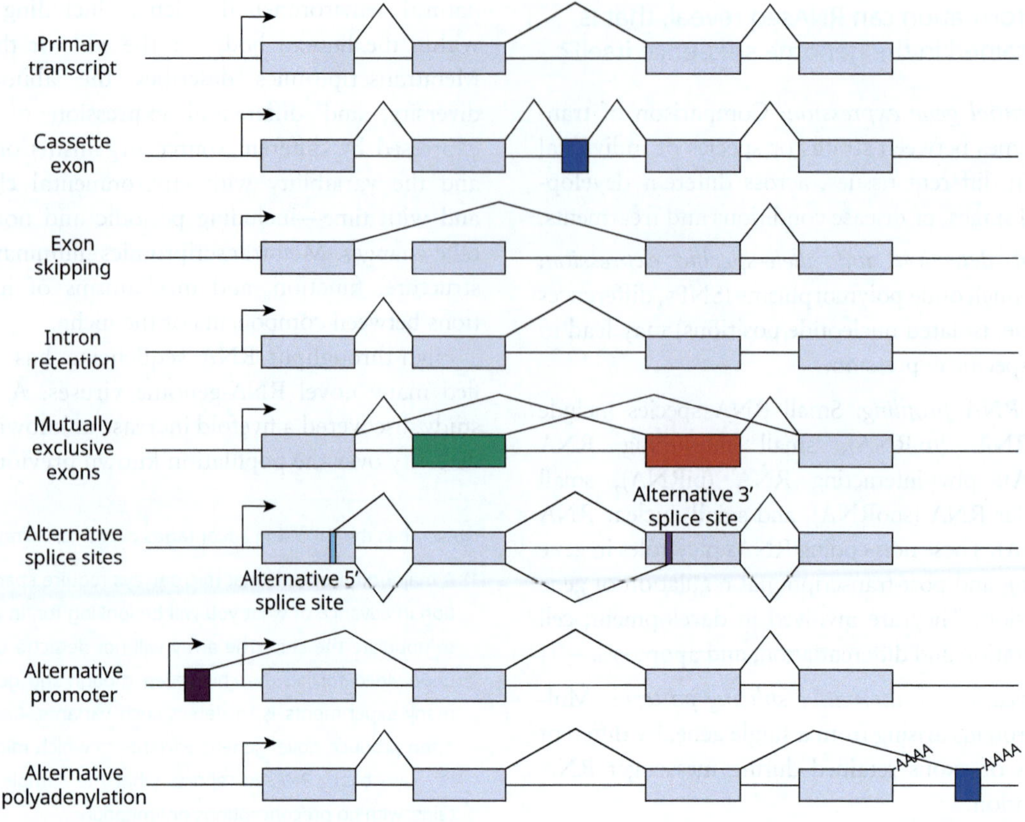

Figure 11.4 Types of splice variants observed in the retina. Boxes are exons, horizontal lines are introns, arrows are promoters, -AAAA indicates a polyadenylation site. Coloured boxes indicate exons included in different transcripts. The mature messenger RNA *excludes* the sections linked by triangles: For instance, the primary transcript includes all four exons but no introns; the mature mRNA corresponding to the intron retention line includes all four exons AND the intron between exon 3 and 4; the mature mRNAs corresponding to the mutually exclusive exons contain either exons 1, 2, and 4 or exons 1, 3, and 4.

Aísa-Marín, I., García-Arroyo, R., Mirra, S., *et al.* (2021). The Alter Retina: Alternative splicing of retinal genes in health and disease. International Journal of Molecular Sciences 22, 4: 1855. https://doi.org/10.3390/ijms22041855

splice variants). Tissue-enriched is defined as: >four-fold higher mRNA level in a particular tissue compared to any other tissue.

Approximately 300 genes involved in inherited retinal diseases are known (https://web.sph.uth.edu/RetNet/). Many of the associated mutations involve variable splicing. Over 85% of these genes display several protein isoforms; most (54%) between 2 and 5 coding transcripts.

Conversely, more easily than internal organs the retina is accessible to therapy, and the blood-retina barrier reduces potential side effects that might follow exchange of therapeutic agents with other organs, including triggering immune responses.

An example: Leber congenital amaurosis type 10 (LCA10) is an inherited retinal dystrophy. The most common mutation is an SNP in an intron of a mitochondrial gene for the protein CEP290: c.2991+1655A>G. (Nucleotide 2991 is the end of the exon (exon 26) preceding the mutation site, which is 1655 positions into the following intron.) This mutation creates a cryptic splice site leading to insertion into the mature mRNA of additional coding sequence (a pseudoexon), that produces a premature stop codon at the position of Cys998, precluding translation to a functional protein.

Antisense oligonucleotides designed to correct the splicing defect restored mRNA and protein function in mice and rabbits (see Figure 11.5a), and are in human clinical trials. CRISPR-Cas9 is also being used, to delete a short region around the cryptic splice site (Figure 11.5b). This eliminates the

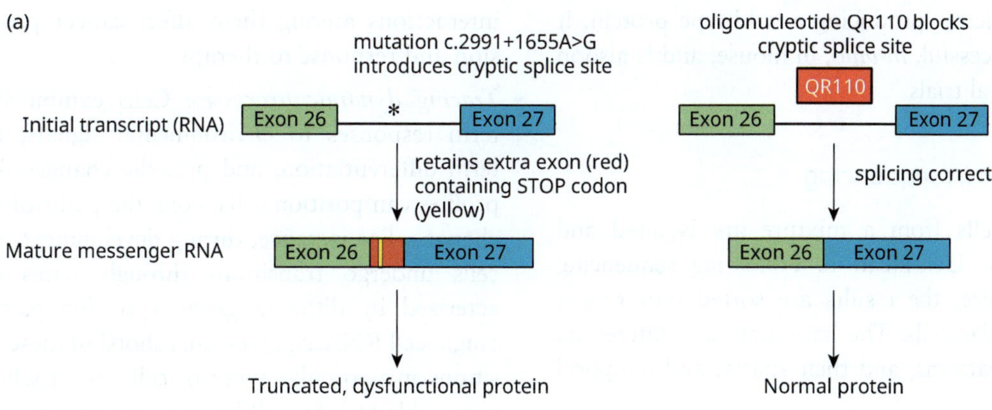

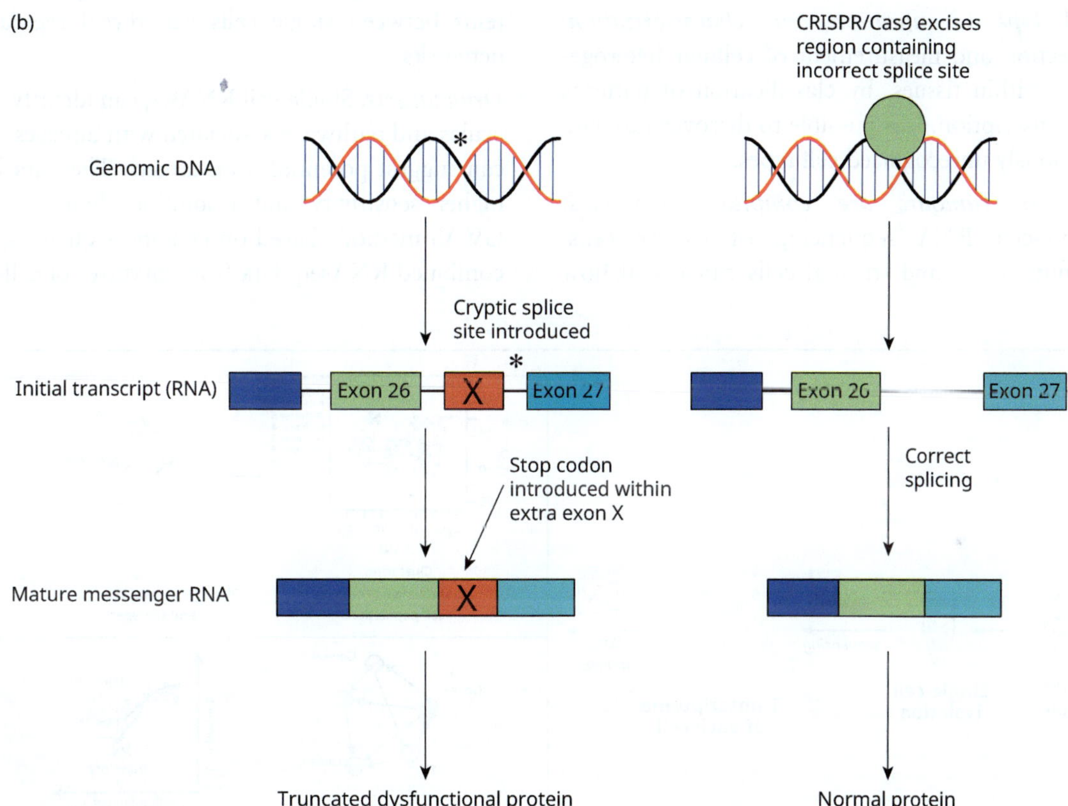

Figure 11.5 Leber congenital amaurosis type 10 (LCA10) arises from a mutation causing improper splicing in the CEntrosomal Protein CEP290.

(a) Improper splicing produces a mutant CEP290 leading to a dysfunctional photoreceptor. An antisense 17-residue oligonucleotide QR110 (sepofarsen) binds to the pseudo-exon region, restoring proper splicing, producing a wild-type CEP290 and a healthy photoreceptor. Sepofarsen = (2′-O-methyl-P-thio)(G-G-U-G-G-A-U-C-A-C-G-A-G-U-U-C-A); all bases are modified 2′-O-methyl derivatives, and all phosphates have a sulphur in place of one of the non-backbone oxygens. Sepofarsen is now in phase 3 clinical trials.

(b) CRISPR/Cas9 can excise a region surrounding the cryptic splice site introduced by the mutation, restoring proper splicing and a correct translated protein.

(a) Dulla, K., Aguilà, M., Lane, A., *et al.* (2018). Splice-modulating oligonucleotide QR-110 restores CEP290 mRNA and function in human C.2991+1655A≫G LCA10 models. Molecular Therapy - Nucleic Acids 12: 730–740. https://doi.org/10.1016/j.omtn.2018.07.010

incorrect splicing, producing a wild-type protein. It has been successful, *in vivo,* in mouse, and is also in human clinical trials.

Single-cell RNA sequencing

Individual cells from a mixture are isolated and barcoded for identification. Following sequencing of the mixture, the results are sorted into results from particular cells. The basic data are differential expression patterns, and their spatial and temporal distributions.

Applications of single-cell RNA sequencing include (see Figure 11.6):

- *Cell type identification and characterization:* Detection and measurement of cellular heterogeneity within tissues, by classification of patterns of transcription. It is possible to discover rare and previously unrecognized cell types.
- *Cancer: Tumours are complex ecosystems.* Single-cell RNA sequencing of cancer cells, immune cells, and stromal cells can reveal how interactions among them affect cancer progression and response to therapy.
- *Tracing dynamic processes:* Cells exhibit short-term responses to environmental signals, long-term differentiation, and periodic changes. RNA profiles can position cells along the paths of these changes. For instance, during development, some cells undergo transitions through states characterized by different gene expression patterns. Single-cell RNAseq gives 'snapshots' of these transitions in particular types of cells, from which it is possible to infer cell lineages and trajectories of differentiation.
- *Covariation patterns* among transcripts or proteins between single cells can reveal regulatory networks.
- *Drug targets:* Single-cell RNAseq can identify molecules and pathways associated with diseases. This can suggest potential drug targets. The data have higher sensitivity and resolution than standard GWAS methods based on genome sequences, and combined RNAseq data from mixtures of cells.

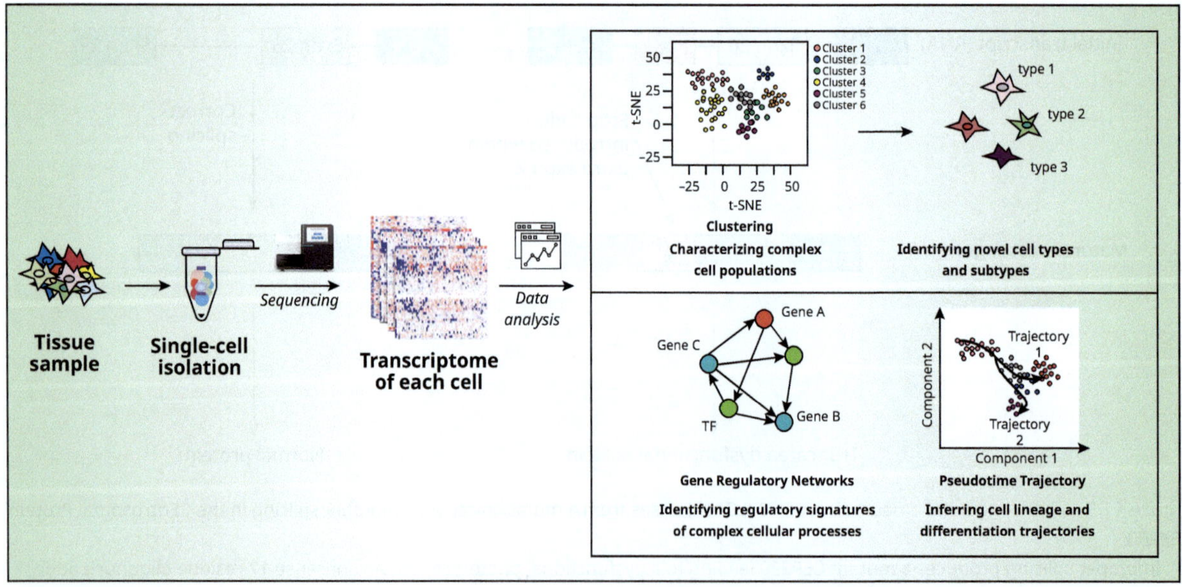

Figure 11.6 Technique and applications of single-cell RNA sequencing. Single cells are isolated from a tissue sample, and individual transcriptomes measured, revealing cellular heterogeneity. Clustering allows classification into different cell types. The data are of very high dimension, but there are methods for projecting onto variables most efficient at distinguishing classes (see Chapter 3). It is possible to infer gene regulatory networks, and trajectories of differentiation of individual types of cells.

Used with permission of Elsevier Science & Technology Journals, from Nayak, R., & Hasija, Y. (2021). A hitchhiker's guide to single-cell transcriptomics and data analysis pipelines. Genomics 113, 2: 606–619. https://doi.org/10.1016/j.ygeno.2021.01.007. Permission conveyed through Copyright Clearance Centre, Inc.

- *Drugs can perturb the transcriptome.* Single-cell RNAseq allows checking how treatment affects different types of cells. It can thereby identify new targets of approved drugs, warn of potential side effects that may have gone unnoticed in preclinical trials, and characterize development of resistance.
- *Metatranscriptomics* adds additional dimensions of complexity to the data. In single-cell RNA sequencing of human tissues, for instance, all cells have (very nearly) exactly the same genome sequence. This is not true of transcriptomes measured from a collection of cells in a natural ecosystem.

Analysis of differential gene expression data

A very common approach to analysis of RNAseq data refers back to the genome. The initial goal of data processing is a **gene expression table**. This is a matrix containing relative expression levels of genes. The rows of the matrix correspond to different genes, and the columns to different sources of material.

A typical experiment *compares* expression patterns in material from two sources—perhaps a control of known properties and a sample to be characterized. We may wish to compare organisms growing under different experimental conditions and/or physiological states, or DNA from different individuals or different tissues, or a series of developmental stages, or healthy and ill individuals.

Two general approaches to the analysis of a gene expression matrix involve (1) comparisons focused on the genes, i.e. comparing distributions of expression patterns of different genes by comparing *rows* in the expression matrix; or (2) comparisons focused on samples, i.e. comparing expression profiles of different samples by comparing *columns* of the expression matrix.

- *Comparisons focused on genes: how do gene expression patterns vary among the different samples?* Suppose a gene is known to be involved in a disease, or linked to a change in physiological state in response to changed conditions. Other genes co-expressed with the known gene may participate in related processes contributing to the disease or change in state. More generally, if two rows (two genes) of the gene expression matrix show similar expression patterns across the samples, this suggests a common pattern of regulation and some relationship between their functions, possibly including (but not limited to) a direct physical interaction.
- *Comparisons focused on samples: how do samples differ in their gene expression patterns?* A consistent set of differences among the samples may distinguish and characterize the classes from which the samples originate. If the samples are from different controlled sources (for instance, diseased and healthy animals), do samples from different groups show consistently different expression patterns? If so, given a novel sample, we could assign it to its proper class on the basis of its observed gene expression pattern.

How can we measure the similarity of different rows (genes) or columns (samples)? Each row or column of the expression matrix can be considered as a vector in a space of many dimensions. In the row vectors, or *gene vectors* (a row corresponds to a gene), each position refers to the same gene in different samples. The gene vector has as many elements as there are samples. In the column vectors, or *sample vectors* (a column corresponds to a sample), each entry refers to a different gene in a single sample. The sample vector has as many elements as there are genes reported. It is possible to calculate the 'angle' between different gene vectors, or between different sample vectors, to provide a measure of their similarities. The smaller the angle, the more similar the pattern.

The gene vectors and sample vectors correspond, separately, to points in spaces of many dimensions. The number of dimensions is either the number of genes or the number of samples. We may not be able to visualize easily points in a space with more than three dimensions, but all of our intuition about geometry works fine. For instance, it is natural to ask whether subsets of the points form natural clusters—points with high mutual similarity—characterizing either sets of genes or sets of samples. After finding clusters we can bring similar genes and samples together. This amounts to reordering the rows and columns of the gene expression matrix. The results are often displayed as a chart, coloured according to the difference in expression pattern. Figure 11.7 contains an example.

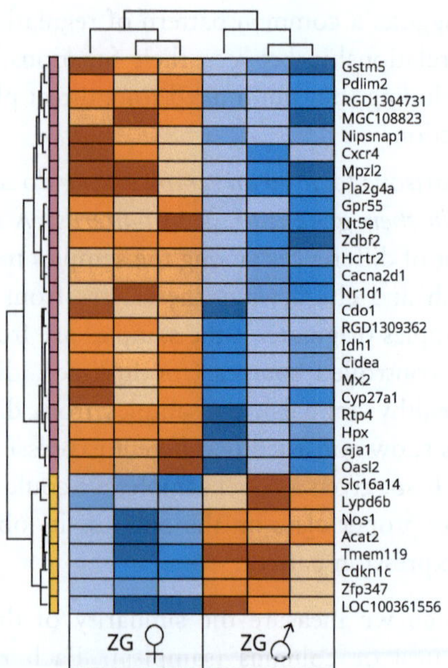

Figure 11.7 The adrenal glands of adult rats differ in size between males and females, a feature associated with functional differences in hormone secretion levels. Trejter and colleagues performed whole-transcriptome analysis of ~27 000 genes, on different tissues from within the adrenal glands. This figure shows the identification of genes differentially expressed in males and females, from the outer layer of the adrenal cortex, the zona glomerulosa. Blue corresponds to increased and orange to reduced relative expression. It was possible to pick out the relatively few differentially expressed genes from the full complement of ~27 000.

In the figure, the data are clustered based on differential expression level (gene-vector, according to rows), and *also* based on expression in male and female animals (sample-vector, according to columns). Note that the clusterings by gene and by male/female are independent: it would be possible to change the arrangement of the columns without altering the arrangement of the rows, and *vice versa*. The trees at the top and at the left indicate the similarities among the results, according to sample-vector and gene-vector, respectively. The sample-vector tree at the top separates the samples from male and female rats. The gene vector shows three classes of differentially expressed genes. Clusters 1 and 2 (nearer the top of the chart) involve genes responsible for ion transport. Genes in cluster 3 are involved in hormone response.

Used with permission of Spandidos Publications, from Trejter, M., Hochol, A., Tyczewska, M., *et al.* (2015). Sex-related gene expression profiles in the adrenal cortex in the mature rat: Microarray analysis with emphasis on genes involved in steroidogenesis. International Journal of Molecular Medicine 35: 702–714. https://doi.org/10.3892/ijmm.2015.2064

Depending on the origin of the samples, what is already known about them, and what we want to learn, data analysis can proceed in different directions.

1. The simplest case is a carefully controlled study, using two different sets of samples of *known characteristics*. For instance, the samples might be taken from bacteria grown in the presence or absence of a drug, from juvenile or adult fruit flies, or from healthy humans and patients with a disease. We can focus on the question: what differences in gene expression pattern characterize the two states? Can we design a classification rule such that, given another sample, we can assign it to its proper class? This would be useful in diagnosis of disease. Subject to the availability of adequate data, such an approach can be extended to systems of more than two classes.

In computer science, training such a classification algorithm is called 'supervised learning'. The expression pattern of each sample is given by a vector corresponding to a single column of the matrix. This corresponds to a point in a many-dimensional space—as many dimensions as there are genes. In favourable cases, the points may fall in separated regions of space. Then a scientist, or a computer program, will be able to draw a boundary between them. In other cases, separation of classes may be more difficult (see Chapter 5).

2. In a different experimental situation, we might not be able to *pre-assign* different samples to different categories. Instead, we hope to extract the classification of samples from the analysis. The goal is to cluster the data to *identify* classes of samples and then to investigate the differences among the genes that characterize them.

An intrinsic problem—and a severe one—in interpreting gene expression data is the fact that almost always the number of genes is much larger than the number of samples. We are trying to understand the relationship of one space of very many variables (the genes) to another one (the phenotype) from only a few measured points (the samples). The sparsity of the observations does not give us anywhere near adequate coverage. Statistical methods bear a heavy burden in the analysis to give us confidence in the significance of our conclusions.

Many clustering algorithms have been applied to transcriptomic data, including those that try to work out simultaneously *both* the number of clusters and the boundaries between them. All algorithms must face the difficulty arising from the sparsity of sampling. Sometimes it is possible to simplify the problem by identifying a small number of combinations of genes that account for a large portion of the variability. This is called reduction of dimensionality.

KEY POINT

Processing the data from a transcriptomic experiment produces a gene expression table, or matrix. The rows index the genes and the columns index the samples. We can either focus on the genes, and ask: how do patterns of expression of different genes vary among the different samples? Or we can focus on the samples, and ask: how do the samples differ in their gene expression patterns?

Transcriptome databases

Major institutions host several databases of transcriptome data:

- EBI: ExpressionAtlas https://www.ebi.ac.uk/gxa/home
- NCBI ENTREZ: Gene expression omnibus http://www.ncbi.nlm.nih.gov/geo/
- NIH GTEx https://commonfund.nih.gov/gtex

There are many others. Some are general, providing different 'front ends' to the major repositories, and others are specialized to organisms, tissues, or diseases. Study of expression profiles associated with cancer is a major industry (see Box 11.3). A number of databases specialize in cancer transcriptomes, linking the sequence data with clinical information:

• The Cancer Genome Atlas	http://cancergenome.nih.gov/
• International Cancer Genome Consortium, Cancer Genome Project	http://www.sanger.ac.uk/genetics/CGP/
• Therapeutically Applicable Research to Generate Effective Treatments	http://target.cancer.gov/

These are among the large-scale projects that curate, archive, and make available the mutations observed in cancers.

The alert reader will recognize that we have now discussed two approaches to cancer therapy: drugs targeting variant proteins produced by a cancer (Box 11.3), and the use of antibodies to enlist the body's immune system in the battle (in Chapter 10). It has not escaped anyone's notice that these approaches could well be synergistic and perhaps their combination could provide more effective treatments.

Expression patterns in different physiological states

A fundamental question in biology is how different components of cells smoothly integrate their activities.

Comparisons of gene expression patterns in alternative physiological states of an organism offer the possibility of extracting, from an entire genome, a subset of genes that underlie a particular life process. An example of shift in physiological state in microorganisms is diauxy. Diauxy, or double growth, is the switch in metabolic state of a microorganism when, having exhausted a preferred nutrient, it 'retools' itself for growth on an alternative. The organism may show a biphasic growth curve, with a lag period while the changed complement of proteins is synthesized.

A few examples of changes in gene expression pattern are understood in a fair amount of detail. These include:

1. The diauxic shift in yeast (the switch between aerobic and anaerobic metabolism).
2. The control of the lac operon in *E. coli*.
3. The switch between lytic and lysogenic states in bacteriophage λ.

We shall discuss these in Chapter 14.

BOX 11.3 Transcriptomics of cancer

Applied to cancer, DNA sequencing and RNA sequencing are complementary tools. Changes in the transcriptome are the effects of changes in the genome. Before a patient presents with cancer, DNA sequencing can identify risk factors. Detecting mutants in the *BRCA1*, *BRCA2*, and *PALB2* genes is a well-known example. The corresponding proteins normally enhance processes of DNA repair.

Once a cancer appears, the clinical challenge begins. Because so many things can go wrong, cancer is in fact a great variety of diseases. Understanding changes in the transcriptome and proteome is already an important aid to precise diagnosis of particular types of cancer, which can guide treatment. Sequence features are also a guide to the expected course of the disease. For example, NKp44 is a receptor expressed by natural killer cells involved in immune defence to acute myeloid leukaemia. The splice-variant profile of the expressed NKp44 gives clues to the prognosis. This is detectable by RNAseq but not by genome sequencing.

Beyond the individual patient, the challenge to the research establishment is to reveal fundamental biological principles that underlie the clinical consequences. High-throughput sequencing has become a basic avenue that, one hopes, will lead to this understanding, and that effective clinical applications will ensue.

Cancer is not one specific disease, but a panoply. Every patient has an individual set of molecular variants, although many features are typical and common. Indeed, the development of resistance to drugs that is often observed argues for genetic divergence and change in the cancer cells within the patient. RNAseq can identify genes that are overexpressed in cancer cells, relative to normal tissue from the patient, and can identify splice variants (see Figure 11.8). Transcriptomic lesions in cancer include point mutations, insertions/deletions, translocations, changes in splice variant, gene fusion, and changes in expression level. Knowing the set of lesions helps precisely classify the cancer and dictate treatment, use biomarkers to test the progress of treatment, and provide a guide to prognosis.

One typical genome change in cancer is gene fusion, which can activate normally quiescent oncogenes, often kinases. The fusions can arise through a variety of mechanisms, including translocations, insertion/deletion, and inversion. RNAseq can detect fusion events provided that the aberrant genes are transcribed. Knowing the molecular offender may provide approaches to targeted therapy. For example, in some types of lung cancer, a chromosomal inversion creates a fusion of two genes: *EML4* (*echinoderm microtubule-associated protein-like* 4) and *ALK* (*anaplastic lymphoma kinase*). The drug crizotinib specifically inhibits the kinase activity of the fusion protein.

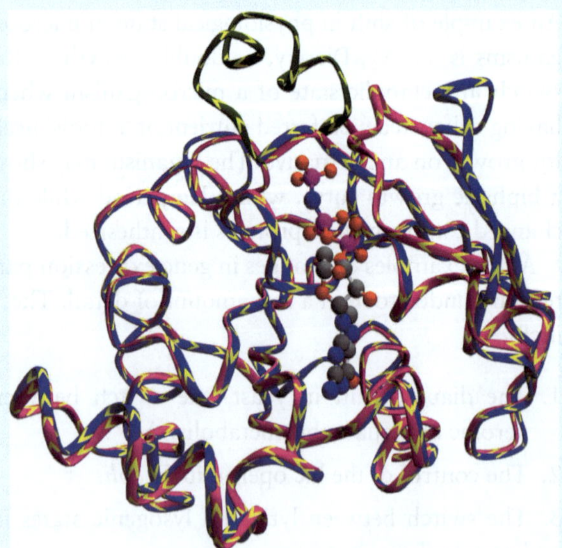

Figure 11.8 Rac1 is a small GTPase controlling signalling pathways. Rac1b is an alternatively spliced version associated with breast and colon cancer. Rac1b contains a 19-amino acid insert, corresponding to exon 3b from the gene. The picture shows the structure of Rac1 in blue, [1MH1] and Rac1b in magenta, with the insert in green [1RYH]. The insert protrudes above the top of the structure. The extra peptide blocks binding to Rho-GD1, a regulator of Rac activity. The effect is to maintain Rac1b in an activated form. The ligand is a GTP analogue, phosphoaminophosphonic acid-guanylate ester (that is, GTP with an NH replacing an O linking the two distal phosphates). Shown is only the ligand from the RAC1 structure, but it is virtually identical in conformation and position in both structures.

Jacques Monod discovered diauxy approximately 70 years ago, during his predoctoral work in Paris. He described his observations in his 1941 thesis. Resuming his research career after the war, he and his colleagues at the *Institut Pasteur* made their fundamental discoveries about the mechanism of gene regulation.

Expression pattern changes in development

Variation of expression patterns during the life cycle of Drosophila melanogaster

During their lifetime, insects undergo macroscopic changes in body plan that are more profound than any post-embryonic development in humans and other mammals, and even amphibians. This allows juvenile and adult flies to occupy different ecological niches. The major stages of a fly's life are embryonic, larval, pupal, and adult. Metamorphosis occurs during the pupal stage: flies spend their 'adolescence' sequestered within a pupa (to the envy of many a parent of a human teenager).

Fly development has been intensively studied at the molecular level. An impressive understanding has been achieved of the mechanism of translation of molecular signals into macroscopic anatomy. The genesis of specific organs—eyes, legs—has been carefully analysed. We have already encountered *HOX* genes and their relationship to body plan.

Arbeitman and co-workers examined changes in transcription patterns in *Drosophila melanogaster* during different stages of its life. When they took up the problem, it was known from earlier work that large-scale changes in gene expression occurred. Microarrays made possible a more systematic and thorough study.

cDNAs containing representatives of 4028 genes (about one-third of the total estimated number in *D. melanogaster*) revealed expression patterns for 66 selected time periods from embryo through to adulthood (see Figure 11.9). Expression levels were compared with pooled mRNA from all life stages, to represent a (weighted) average expression level. The interval between measurements varied from 1 hour (for embryos) up to several days (for adults) until a total age after fertilized egg of up to 40 days.

Most of the genes tested—3483 out of 4028, or 86%—changed expression levels significantly at *some*

Stage	Approximate interval between measurements
Embryo	1 hour
Larva	5 hours
Pupa	8 hours
Adult	3–4 days

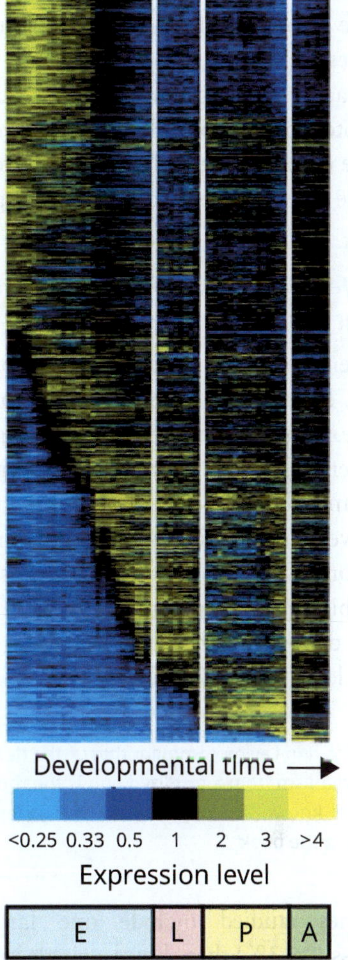

Figure 11.9 The figure shows that in the earliest stages of development, about half the genes have high expression levels and half have low expression levels. Some of the genes with initial high expression levels lose expression towards the end of the embryo (E) stage and remain low during the larval, pupal, and adult stages. Some of the genes with initial low expression levels change to high expression levels towards the end of the embryo (E) stage. Particularly interesting are genes that show an alternation of expression levels: initially low in embryo stage, high in larval stage, low again in pupal stage, and high again in adult stage.

Arbeitman, M.N., Furlong, E.E.M., Imam, F., *et al.* (2002). Gene expression during the life cycle of *Drosophila melanogaster*. Science 297, 5590: 2270–2275. https://doi.org/10.1126/science.1072152 Reprinted with permission from AAAS.

stage(s) of life. Of these, 3219 varied by a factor of >4 between their maximum and minimum values.

The data show that in *D. melanogaster*, as in other species, genes participating in a common process often exhibit parallel expression patterns and similar perturbations of these patterns in mutants. For instance, the expression pattern of the *eyes absent* mutant, which produces an eyeless or at least a reduced eye phenotype, forms a cluster, in the analysis of expression patterns, with 33 genes. Of these, 11 are already known to function in eye differentiation or phototransduction. The other 22 are likely to as well; the data provide at least hypotheses, and at best reliable clues, to the function of these genes.

Different life stages make different demands on different genes

Different genes exhibit different temporal patterns of expression. Most of the developmentally modulated genes are expressed in the embryonic stage, as the whole system is getting started. Genes expressed in the early embryo include transcription factors, proteins involved in signalling and signal transduction, cell-adhesion molecules, channel and transport proteins, and biosynthetic enzymes. A third of these are maternally deposited genes; many of these fall off in expression level within 6–7 hours.

> Stathopoulos and Levine comment that '... the genesis of a complex organism from a fertilized egg is the most elaborate process known in biology, and thereby depends on "every trick in the book" '.*

The genes studied include one large stage-specific class (36.3%) that shows a single major peak of expression. Some of these remain constitutively expressed (at lower levels) subsequently. Others show sharp peaks in expression level.

Another group of genes (40.3%) shows two peaks in expression. Two patterns are common: genes with their first onset of enhanced expression early in embryogenesis generally have their second at pupation, with elevated expression levels continuing into the pupal stage. Genes with their first onset of enhanced expression late in embryogenesis generally have their second at the late pupal stage, with elevated expression levels continuing into the adult stage. The remaining 23.4% of genes show multiple peaks in expression level.

The observation of similarities between expression patterns in embryonic stages and pupal stages, and between larval and adult stages, is intriguing. Certain analogies suggest themselves. In both embryonic and pupal stages, body structures are forming and physical activity is minimal. In contrast, larval and adult stages have a more active lifestyle but stasis of anatomical form, although larvae but not adults grow substantially in size. Consistent with the notion of a 'go back and get it right this time' aspect to metamorphosis is the occurrence of some dedifferentiation in the pupa.

Ideas of this sort can be examined in light of the nature of genes that show different lifelong expression patterns. For example, maximal expression levels of most metabolic genes occur during larval and adult stages. Another set of genes involved in larval and adult muscle development has a similar two-peak expression pattern. More precise analysis is possible and shows that steps in the regulatory hierarchy for muscle development show peaks at different times, with genes expressed later being downstream in the regulatory hierarchy. Similar time-of-onset sequences appear in both larvae and pupae. The two stages at which body plans are formed—embryo and pupa—re-utilize not only the same materials but also some of the same mechanisms.

KEY POINT

Measurement of expression patterns at different stages reveals which batteries of genes are active in development. Particularly striking, in *Drosophila*, is the alternation of time-of-onset of the expression patterns of some genes: embryo and pupa show similarities, and larva and adult.

Flower formation in roses

Roses have long been favourites of gardeners and lovers: for their appearance, for their scent, and, perhaps less commonly, for their purported medicinal

* Stathopoulos, A., & Levine, M. (2002). Whole-genome expression profiles identify gene batteries in *Drosophila*. Dev. Cell 3: 464–465.

qualities—rose hips (fruits) are rich in vitamin C. Breeders have responded by creating tens of thousands of varieties.

Roses have been known in Europe since antiquity and appear frequently in literature, famously in works by Dante and Shakespeare. Roses symbolized the two armies fighting for control of England at Bosworth in 1485.

Roses have been cultivated for 5000 years. Rose varieties brought to Europe from China at the end of the eighteenth century introduced the very desirable property of recurrent blooming throughout the season, rather than flowering only once a year. Some arrived in 1794 with Lord Macartney on his return from the famous embassy to the court of the Qianlong emperor. It took only a single gene change to achieve recurrent flowering. The Chinese strains also enlarged the palette to include yellow and scarlet.

Flower formation involves tissue differentiation and formation of an organ with specialized structure and function. Flowers are the reproductive organs of plants. Their beautiful appearance and scent attract insect pollinators. In addition to homeotic genes that generate the structure of the flower, novel metabolic pathways are activated to produce the small molecules responsible for the colours and scents. Colour and scent are properties of the petals, the focus of this section.

Plant developmental biologists distinguish six stages in the development of a rose flower (see Figure 11.10). In formation of petals, an initial stage of cell division may stop while the flower is less than half its final size. Subsequent development occurs by differential cell elongation.

Regulation of pathways that produce scents in roses

Compounds from roses are an important source of fragrances (see Table 11.1). Their production rises to a peak in the mature flower, in consequence of raised expression levels of the enzymes that synthesize them.

KEY POINT

Terpenes are compounds formed from isoprene units (2-methyl-1,3-butadiene). Sesquiterpenes are 15-carbon compounds formed from three isoprene units.

Figure 11.10 Three of the six stages in the development of a rose flower: stages 1 (left), 4 (centre), and 6 (right). At stage 1, petals are beginning to emerge. At stage 4, cells are actively elongated and show increasing pigment concentrations. Finally, at stage 6, the flower is fully open. This figure illustrates the cultivar Fragrant Cloud.

Table 11.1 Major classes of molecule in the scent of Fragrant Cloud roses

Compounds	Amount emitted (mg/flower per day)
Esters	61
Aromatic and aliphatic alcohols	37
Monoterpenes	18
Sesquiterpenes	10

In order to identify enzymes involved in rose scent production, Guterman and co-workers compared expression patterns in different stages of flower development.[1] They contrasted two tetraploid cultivars, called Fragrant Cloud and Golden Gate. Fragrant Cloud gives large, red flowers with a strong scent (this is the cultivar illustrated in Figure 11.10). Golden Gate has yellow flowers and a less distinct odour to humans. Two experiments might identify the proteins involved in scent production: (1) a comparison of mature flowers between Fragrant Cloud (highly scent producing) and Golden Gate (poorly scent producing); and (2) a comparison of Fragrant

[1] Guterman, I., Shalit, M., Menda, N., *et al.* (2002). Rose scent: Genomics approach to discovering novel floral fragrance-related genes. Plant Cell 14: 2325–2328.

Cloud flowers at an early developmental stage, before scent production ramps up, with the mature Fragrant Cloud flower, which is rich in scent.

The cDNA libraries created from both flowers in stage 4 contained 2139 unique sequences. Of these, 1288 were found only in Fragrant Cloud and 746 only in Golden Gate. Expression patterns were studied of 350 Fragrant Cloud genes associated with:

- primary and secondary metabolism;
- development;
- transcription;
- cell growth;
- cell biogenesis and organization;
- cell rescue;
- signal transduction; and
- unknown functions.

Taking, as a threshold of significance, a twofold difference in expression level, 77 genes (about one-fifth) had higher expression levels in Fragrant Cloud than in Golden Gate, and only three had higher levels in Golden Gate; the rest had similar expression levels in both varieties. Comparing Fragrant Cloud flowers in stage 1 and stage 4, 65 genes had higher expression levels in stage 4 and 14 had lower levels. Common to both sets were 40 genes that were more highly expressed in both Fragrant Cloud relative to Golden Gate and in Fragrant Cloud stage 4 relative to Fragrant Cloud stage 1 flower development (see Figure 11.11).

What are these 40 genes? Fifteen are involved in metabolism and seven appear to have roles in secondary metabolism, i.e. reactions outside of the core metabolic pathways responsible for normal growth, development, and reproduction. Two very strongly upregulated genes encode proteins similar in sequence to known enzymes: glutamate decarboxylase, and (+)-δ-cadinene synthase. The enzyme (+)-δ-cadinene synthase is involved in the synthesis of sesquiterpenes, one of the classes of floral scent molecule.

Cloning of the (+)-δ-cadinene synthase gene to produce recombinant enzyme permitted direct functional studies. It catalyses the reaction of farnesyl diphosphate to produce the sesquiterpene germacrene D, the major sesquiterpene component of the scent of Fragrant Cloud (see Figure 11.12).

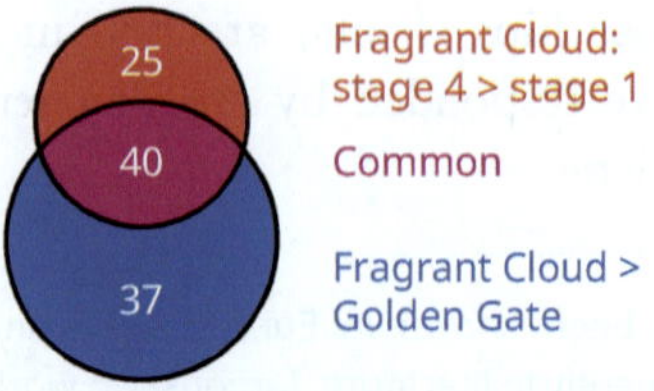

Figure 11.11 Two rose cultivars, Fragrant Cloud and Golden Gate, differ in their maximal odour production in stage 4 of flower development: Fragrant Cloud is rich in scent and Golden Gate is poor. Scent development in Fragrant Cloud is fully developed in stage 4 relative to the immature flowers in stage 1.

In comparisons of expression patterns, 25 genes were expressed more highly in stage 4 Fragrant Cloud rose flowers than in stage 4 Golden Gate rose flowers; 37 genes were expressed more highly in stage 4 Fragrant Cloud rose flowers than in stage 1 Fragrant Cloud rose flowers; and 40 were expressed more highly in stage 4 Fragrant Cloud flowers than in both stage 4 Golden Gate flowers and stage 1 Fragrant Cloud flowers.

These experiments can help to identify proteins preferentially expressed in stage 4 Fragrant Cloud flowers that are involved in scent biosynthesis.

OPP
Farnesyl diphosphate
Germacrene D

Figure 11.12 Conversion of precursor farnesyl diphosphate to germacrene D, a common scent molecule produced by rose petals. OPP = pyrophosphate.

Colour: the elusive blue rose

A walk through many neighbourhoods in temperate climates will reveal the many colours that roses can display, notably red, pink, peach, and yellow. Conspicuous by its absence is blue. It has not been possible to produce a blue rose by classical selective breeding. This is not for lack of trying: in 1840 the horticultural societies of Great Britain and Belgium offered a prize of 500 000 francs for one.

The major pigments of flower petals are anthocyanins: cyanidin, pelargonidin, and delphinidin. All are synthesized from a common precursor, dihydrokaempferol (see Figure 11.13). It is delphinidin that is blue, but roses lack the gene for the enzyme that produces dihydromyricetin from dihydrokaempferol (marked by a green * in the figure).

Scientists at the Australian Commonwealth Scientific and Industrial Research Organisation (CSIRO), in collaboration with a Melbourne company,

Dihydroquercetin ← Dihydrokaempferol → (* F3′,5′H) Dihydromyricetin

DFR ↓ DFR ↓ DFR ↓

Cyanidin 3-glucoside Pelargonidin 3-glucoside Delphinidin 3-glucoside

Figure 11.13 Simplified scheme showing structures and metabolic relationships of anthocyanin flower pigments. The difficulty in breeding a blue rose is that roses lack the gene that encodes the enzyme flavonoid 3′,5′-hydroxylase (F3′,5′H), which produces the precursor of the blue pigment delphinidin. In other plants, this enzyme acts at the point marked by a green *. The steps corresponding to the vertical arrows are all catalysed by a single enzyme, dihydroflavonol reductase (DFR).

Florigene, and the Suntory Corporation of Japan, have created a blue rose by genetic engineering. There were two challenges: to produce the blue pigment, and to turn off synthesis of red and orange pigments. To suppress the red and orange pigments, the target was the enzyme dihydroflavonol reductase (DFR). DFR modifies precursor molecules in all three branches of the pathway; a *DFR*⁻ plant would have white flowers, providing a 'clean slate' for production and display of blue pigments. However, *DFR* is also needed for synthesis of delphinidin.

It was necessary to manipulate the pathways with precision. An interfering RNA (RNAi) was engineered to knock out the rose *DFR*. Then, introduction of a *DFR* from iris, together with the gene for flavonoid 3′,5′-hydroxylase from pansy to supply the enzyme missing in roses, produced a plant with high levels of petal delphinidin and only small amounts of cyanidin. The rose and iris *DFR* genes are quite similar but have two important differences: (1) the natural specificity profile of the iris DFR produces delphinidin predominantly and (2) the RNAi can be sufficiently specific that it selectively inactivates the rose *DFR* and not the iris homologue.

The new rose is a shade of blue (see Figure 11.14). The reasons it is not a purer blue are: (1) the RNAi did not suppress completely the rose DFR expression (this could presumably be achieved using CRISPR/

Figure 11.14 Picture of delphinidin-rich rose flower, produced by methods similar to those described in the text. It is called 'Applause'. Photograph from Suntory Ltd.

Cas techniques) and (2) the pH of rose petals is relatively acidic compared with other flowers. (Indeed, anthocyanins can be used as pH indicators.)

Blue roses do not exist in nature, and attempts to breed them have been unsuccessful. Understanding of the underlying metabolic pathways of the pigments allows rational attempts at genetic engineering. It is interesting that the pH of the petals has an effect—the pigments are 'indicators'. If this effect helped to stimulate Sydney Brenner's fascination for molecular biology,[2] that may be its greatest significance for the field.

Expression patterns in learning and memory: long-term potentiation

Learning and memory involve changes in the structure and biochemistry of nerve cells. Nervous systems are dynamic networks, passing signals among cells. Synapses are the sensitive points at which neurons interact. As each neuron integrates inputs from several others, its output depends on the 'weighting' of its inputs—the distribution of the strengths of the input synapses. Increasing or reducing the strengths of individual synaptic connections modulates the dynamics of the network.

Learning and memory must involve some permanent structural change. The observation that memory survives periods of coma, during which neural activity ceases, proves this. The change in the network cannot, therefore, be purely a change in the dynamic state.

Long-term potentiation (LTP) is a neural phenomenon underlying learning and memory. LTP is a persistent increase in strength of a synaptic connection as a result of stimulation of the upstream cell. The original observation, first described by Bliss and Lomo in 1973, was that high-frequency stimulation of a synapse during a finite time interval produced a persistent subsequent enhancement of the postsynaptic response. It is believed that transient effects, lasting <1 hour, require modifications of pre-existing proteins at the synapse. Longer lasting effects involve protein synthesis and gene transcription, and resulting structural remodelling of synapses.

Learning must be regarded as a specialized form of development.

In addition to neural plasticity, learning also involves generation of new neurons. LTP stimulates both neurogenesis and enhanced survival of new cells.

Park and co-workers studied genes that change their expression pattern in response to LTP.[3] They studied cells from the mouse dentate gyrus, a structure within the hippocampus. The hippocampus is known to be involved in memory formation, especially spatial memory (see Box 11.4). Following high-frequency stimulation for four separated 1-second intervals, total RNA was extracted after several time intervals—30, 60, 90, and 120 minutes after stimulation—and then reverse transcribed into cDNA and hybridized onto Affymetrix GeneChip arrays. The chip reported 12 000 genes and ESTs. Samples of unstimulated tissue provided controls.

Park and colleagues found, within all time points, 1664 genes with statistically significant changed expression patterns. Of these, 39% were upregulated and 61% downregulated. The genes identified suggest that LTP produces changes in a variety of processes affecting cell morphology and affects interactions among cells and between cells and the extracellular matrix.

Specific functional assignment showed several categories of genes, shown in Table 11.2 (this table shows a composite containing genes identified as having changed expression at *any* of the time points; see also Figure 11.15). Most but not all of the categories contain examples of both up- and downregulated genes.

[2] Brenner, S. (2001). *My Life in Science*, pp. 5–6. London: BioMed Central Ltc.

[3] Park, C.S., Gong, R., Stuart, J., *et al.* (2006). Molecular network and chromosomal clustering of genes involved in synaptic plasticity in the hippocampus. J. Biol. Chem. 281: 30195–30211.

BOX 11.4

London taxi drivers: spatial memory and the hippocampus

London taxi drivers must have an exhaustive knowledge of the metropolitan geography, of optimal routes between points both famous and obscure, and of variations in traffic patterns. As portrayed in the famous 1979 film *The Knowledge*, drivers must pass a strict test to earn their licence before taking control of a classic black London cab. (This was before GPS![4]) Consistent with the involvement of the hippocampus in spatial memory, brain scans show that London taxi drivers have a larger hippocampus than a control group, and that the hippocampus enlarges with time spent behind the wheel.[5]

[4] Rosen, J. (2014). The Knowledge, London's legendary taxi-driver test, puts up a fight in the age of GPS. *The New York Times Style Magazine*, https://www.nytimes.com/2014/11/10/t-magazine/london-taxi-test-knowledge.html

[5] Maguire, E.A., Gadian, D.G., Johnsrude, I.S., *et al.* (2000). Navigation-related structural change in the hippocampi of taxi drivers. Proc. Natl. Acad. Sci. USA 97: 4398–4403; Maguire, E.A., Spiers, H.J., Good, C.D., *et al.* (2003). Navigation expertise and the human hippocampus: A structural brain imaging analysis. Hippocampus 13: 250–259.

Expression patterns can identify particular genes involved in neural plasticity. Many of the genes identified by changed expression patterns were already known to play roles in synaptogenesis, synapse differentiation, neurite outgrowth, and synaptic plasticity. Others were not previously known to be involved in LTP, but their altered expression pattern makes them candidates to be tested for their effects on LTP.

For example, transglutaminase is known to be expressed in neural tissue and appears in synapses. However, it was not known to be implicated in LTP. A connection was confirmed by showing that cystamine—a specific antagonist of transglutaminase—impairs LTP. (Cystamine is an inhibitor of transglutaminase and also causes disulphide exchange producing unfolding. Transglutaminase also helps to produce protein aggregates in Huntington's disease. Cystamine does ameliorate Huntington's disease in the mouse, but the mechanism is unclear. So far, clinical trials in humans have not produced a convincing effect.)

Some of the differentially expressed genes are coordinated into coherent pathways. These include enhanced expression of genes in the MAPK signalling

Table 11.2 Genes that change expression pattern in response to long-term potentiation (LTP)

Functional category	Upregulated	Downregulated
The extracellular matrix and its regulation	+	+
Membrane protein/cell surface/adhesion molecule	+	+
Neurosteroid hormone metabolism	–	+
Cytokine/growth factor/receptor	+	+
Other receptors/signalling	+	+
Ion channel	+	+
Transcription factor/regulation	+	+
Translation	+	+
Neurotransmitter receptor/neuromodulator	+	+
Regulation of cytoskeleton	+	+
Mitochondrial/energy production	+	+
Proteases/protease inhibitors	+	+
Immunoresponsive proteins/oxidative stress/neuroprotection/cell death	+	+
Myelin-related proteins	–	+
Chromatin structure	+	–

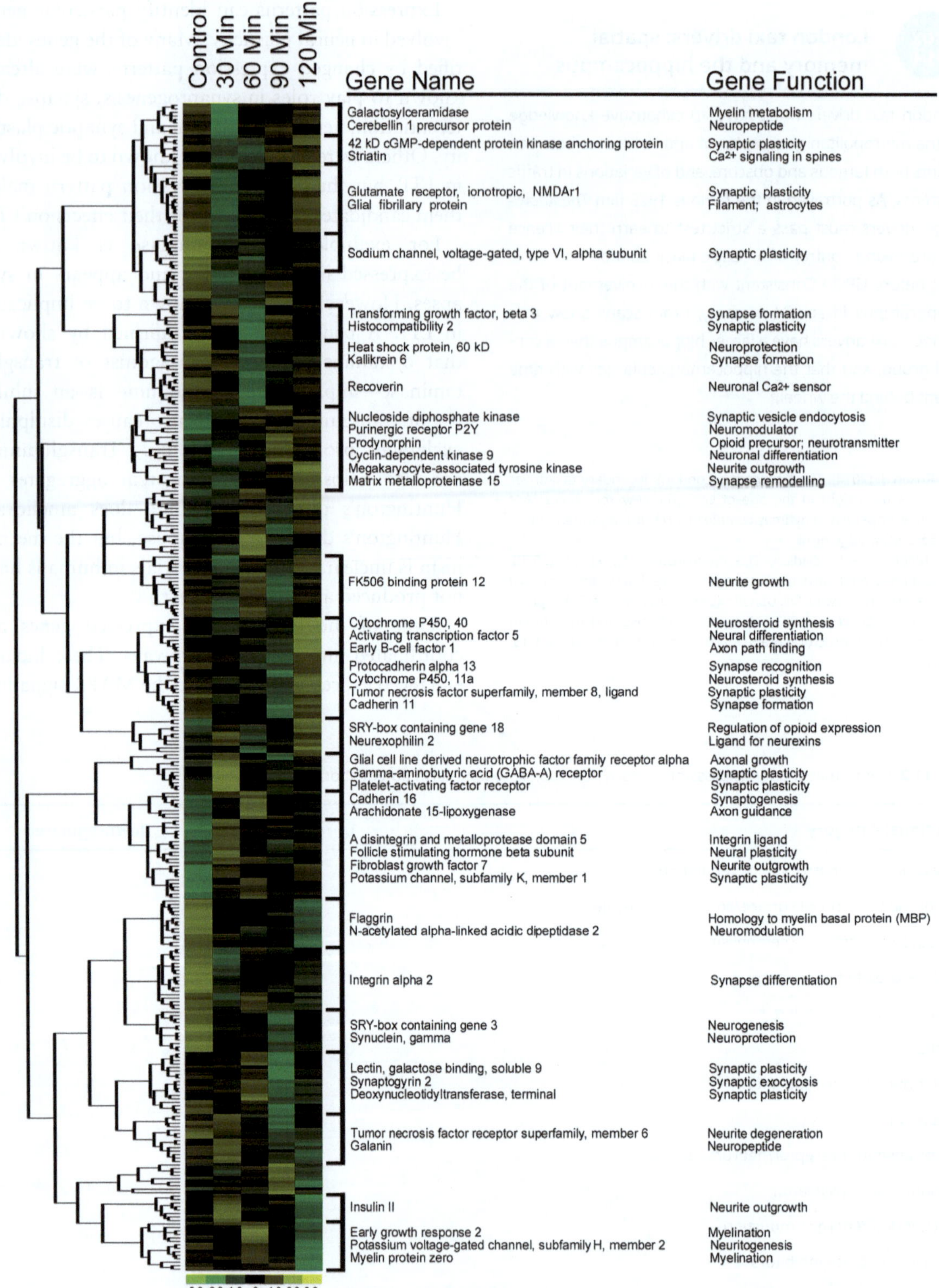

Figure 11.15 Clustering of genes differentially expressed after induction of LTP. The scale at the bottom indicates the expression level relative to the control: green = enhanced expression; yellow = reduced expression. Brackets indicate clusters of genes with known neural or synaptic functions.

Park, C.S., Gong, R., Stuart, J.M., *et al.* (2006). Molecular network and chromosomal clustering of genes involved in synaptic plasticity in the hippocampus. Journal of Biological Chemistry 281, 40: 30195–30211. https://doi.org/10.1074/jbc.m605876200

cascade (which was already known to be important in LTP) and the Wnt signalling pathway (which had not previously been connected with LTP).

Comparison of expression patterns at different times after LTP induction revealed the temporal expression patterns of different genes. An interesting observation is that many genes in the same general functional groups have similar temporal expression profiles. Conversely, different time points are associated with a 'schedule' of activity of genes with particular types of function.

- *Genes with common time profiles*. Genes involved in responses to external stimuli are upregulated at 30 and 60 minutes after LTP induction. These may be involved in interactions between pre- and post-synaptic components. Genes involved in signal transduction and transcription regulation provide less-clear time profiles. It is likely that their effects are indirect.
- *Events happening at particular times*. Many genes active at 30 minutes are involved in cell–cell interaction, synapse formation and remodelling, and neurite outgrowth. These represent a relatively early component of the response. Genes related to the cytoskeleton are downregulated at 30 minutes, but upregulated at 60, 90, and 120 minutes. These may be involved in structural changes at the synapse.

In this chapter, we shall examine the *effects* of the reconfiguration of the expression patterns, comparing the different complements of genes active in the two states. In Chapter 14, we shall return to the yeast diauxic shift to examine the control mechanisms that regulate the transcriptional reprogramming.

Conserved clusters of co-expressing genes

Mapping the loci of the differentially expressed genes showed that they are concentrated in specific chromosomal regions (tandem duplicates were removed). These clusters tended preferentially to contain genes with similar functions. Comparison of the distributions of homologues in the genomes of rats, humans, *Drosophila*, and *C. elegans* suggest that the clustering is conserved in evolution. The clustering may contribute to a mechanism for common regulation of expression, reminiscent of a bacterial operon.

Evolutionary changes in expression patterns

The very high similarity in genome sequence between humans and chimpanzees suggests that the evolutionary differences between such closely related species would not lie primarily in the relatively small changes in sequences of individual proteins, but in expression patterns. Nevertheless, amino-acid sequence changes in proteins may be significant, even though they may be small. One example is the *FOXP2* gene, discussed in Chapter 2. Another is a contributor to the control of overall cerebral cortical size (a crude but not entirely irrelevant feature of our mental evolution), the gene *ASPM* (abnormal spindle-like microcephaly associated), which has undergone positive selection in the lineage leading to humans.

Carefully controlled experiments show that there are significant differences between human and chimpanzee expression patterns that arose during evolution:

- *The set of genes that show different expression patterns between humans and chimpanzees is rich in transcription factors*. This is in accord with King and Wilson's hypothesis.
- *The differences in expression pattern are not uniform in different tissues*. As far as expression pattern is concerned, our hearts and livers have diverged from chimpanzees in expression pattern more than our brains, both in terms of the numbers of differentially expressed genes and the amount of the differences in transcription levels. (Would you have guessed this?) However, looking at the course of evolution using the macaque, for instance, as an outgroup, shows that the human brain is particularly rich in genes with *increased* expression levels relative to the chimpanzee, consistent with the distinct differences in cognitive abilities. In other tissues, there is a more even distribution of genes

expressed more highly in humans or more highly in the chimpanzee.

- *Changes in expression patterns tend to be lower in X and higher in Y chromosomes than in autosomes.* For brain tissue, the average human/chimpanzee ratio of expression level is about 1.51 for autosomes, 1.43 for the X chromosome, and 2.14 for the Y chromosome.
- *Duplicated genes tend to show a higher divergence in expression pattern than non-duplicated genes.* One possible consequence of gene duplication is divergence and specialization of function. This is consistent with the requirement for differential control of gene expression. (Recalling Chapter 1, vertebrate α– and β–globins are exceptions to this paradigm: it is necessary to *calibrate* their levels of expression. It is not yet clear what mechanism achieves this. Synthesis of different amounts of α– and β–globin causes thalassaemias.)
- *The differences in expression patterns are not uniform, even across different autosomes.* During the 6–7 million year period of divergence, there has been substantial chromosome rearrangement between humans and chimpanzees (see Figure 4.6). In cortical tissue, changes in expression patterns are larger among genes in rearranged chromosomes than syntenic ones (see Figure 11.16).

The comparative analysis of expression patterns in human and chimpanzee brains has been pursued to high resolution. Khaitovich and colleagues used an array containing ~10 000 human genes and arrays containing ~40 000 human transcripts to test several areas of human and chimpanzee brains[6] (see Table 11.3).

Differences between the species must be extracted from the variation among individuals. Within each species, typically a few hundred genes (out of ~10 000 tested) vary in expression in different brain regions.

There is relatively low variation within the cortex itself, but over 1000 genes show differences in expression pattern between the cerebellum and other regions.

It is surprising to observe a *similarity* of expression pattern in humans between Broca's area, associated with speech, and the homologous right-hemisphere area, which usually is not. (In some left-handed people, the right-hemisphere area is dominant for language.) This observation suggests that the achievement of language in humans did *not* depend on localized changes in transcription patterns.

[6] Khaitovich, P., Muetzel, B., She, X., *et al.* (2004). Regional patterns of gene expression in human and chimpanzee brains. Genome Res. 14: 1462–1473.

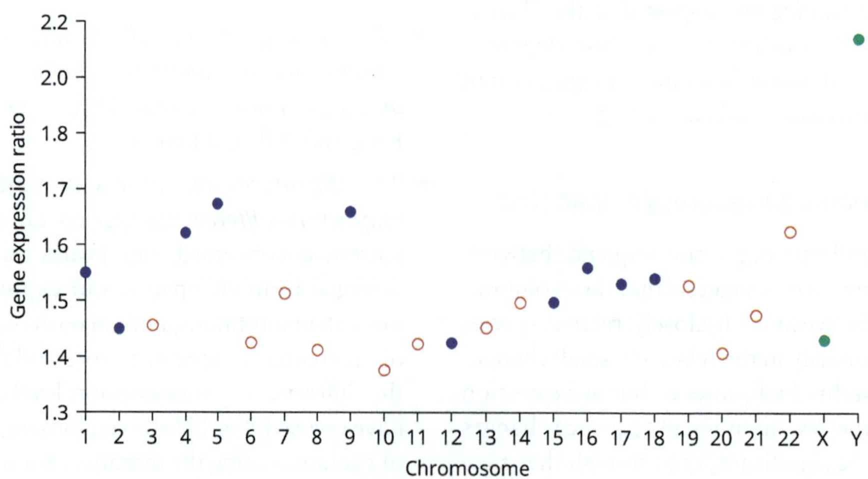

Figure 11.16 Average ratio of gene expression levels between human and chimp cortical tissue in collinear (hollow red) and rearranged (solid blue) chromosomes. X and Y chromosomes shown on the far right in green. For instance, the data point for chromosome 1, y = 1.55, means that for homologous genes on chromosome 1, on average the human expression level was 1.55× that of the chimpanzee.

Table 11.3 Areas of brain for which human/chimpanzee expression pattern differences were measured

Area of brain	Function in human	Function in chimpanzee if known to differ from human
Dorsolateral prefrontal cortex	Important for higher brain functions: working memory, conscious control of behaviour	
Anterior cingulate cortex	Autonomic functions: heart rate and blood pressure and cognitive functions such as reward anticipation, decision making, empathy, and emotion	
Broca's area	Mainly language, plus action	Gesture, especially control over orofacial action, including communicative acts
Central part of cerebellum	Coordinating complex movements such as walking	
Caudate nucleus	Regulation and organization of information sent to frontal lobes	
Pre-motor cortex	Sensory guidance of movement, activating proximal and trunk muscles	
Area in right hemisphere homologous to Broca's	Not entirely clear; some involvement in communication without syntactic contribution to language use	

Functional analysis showed that genes encoding proteins involved in signal transduction, cell–cell communication, differentiation, and development show a greater-than-random tendency to vary in expression between regions, within both species. Genes encoding proteins involved in protein synthesis and turnover tend to show conserved expression patterns.

Approximately 10% of the genes studied differ in expression pattern between humans and chimpanzees. Most of the differences appear in two or more regions of the brain. The cerebellum contains several genes showing species-specific differences in expression *not* shared by other regions. An analysis of functional categories of the genes showing enhanced or reduced interspecies expression differences does not reveal an enrichment in specific families.

If our goal is to understand at the molecular level the phenotypic differences between humans and chimpanzees involving higher mental functions such as cognition and language, our data must be very accurate and detailed, for as the traits grow more subtle, the molecular signal grows correspondingly fainter. At some point, it will be necessary to trace the origin of the changes in expression patterns to protein and genomic sequences. This does not contradict the King and Wilson hypothesis; after all, amino-acid sequence changes modulate the functions of regulatory proteins. We may also be required to address different levels of complexity: the different traits may depend on very complicated patterns of interactions, difficult to infer from properties of individual genes.

Applications of transcriptomics in medicine

Development of antibiotic resistance in bacteria

The growth in bacterial resistance to antibiotics has created a crisis in disease control.

One of the most powerful antibiotics available for use in humans is vancomycin, a 1.5 kDa glycopeptide antibiotic isolated from a soil bacterium in Borneo, *Amycolatopsis orientalis* (see Figure 11.17). Vancomycin was first used clinically in 1958 when infectious strains of staphylococci developed penicillin resistance (see Box 11.5). It became the antibiotic of choice for many infections and the drug of last resort for some.

Development of drug resistance by pathogenic microorganisms threatens to deprive us of the ability to control infectious disease. Widespread use of antibiotics, not only in human clinical medicine, but in raising animals, has aggravated the problem.

Vancomycin acts by interfering with cell wall synthesis. The cell wall in Gram-positive bacteria is a combination of polysaccharides and peptides. Linear polysaccharides, formed from alternating *N*-acetylglucosamine and *N*-acetylmuramic acid units, are cross-linked by short peptides: L-ala–D-gln–L-lys–D-ala–D-ala. Vancomycin acts by binding to the oligopeptide, preventing the cross-linking. Without a robust cell wall, bacteria cannot stand up to their internal osmotic pressure.

The development of resistant *Staphylococcus aureus* strains occurred gradually. Vancomycin-resistant enterococci appeared in 1977. Twenty years later, *S. aureus* developed resistance. The strains were already methicillin resistant. Vancomycin-resistant *S. aureus* (VRSA) strains appeared in 2002. They have been found in Europe and the US (see Box 11.5).

Resistance is measured by an increase in the minimum inhibitory concentration (MIC), which is related to the clinically effective dose (Table 11.4).

The **minimum inhibitory concentration (MIC)** is the lowest concentration of an antibiotic that inhibits growth of a strain of bacteria.

One contribution to the spread of vancomycin resistance may be the practice in Europe of widespread feeding of avoparcin (see Figure 11.17) to animals. Homologues of the resistance genes are present in the source organism for vancomycin, *A. orientalis*. The

Table 11.4 Effect of minimum inhibitory concentration (MIC) on resistance

Resistance of *S. aureus* strain	Minimum inhibitory concentration (MIC)	Year first appeared
Sensitive	1 µg/ml	–
Intermediate (VISA)	4–8 µg/ml	1997
Resistant (VRSA)	>16 µg/ml	2002

BOX 11.5 Development of vancomycin resistance—a chronology*

1941	First clinical use of penicillin G
1942	Appearance of penicillin-resistant *Staphylococcus aureus*
1950s	Multidrug-resistant *S. aureus* widespread
1956	Vancomycin described
1958	First clinical use of vancomycin
1960	First clinical use of methicillin
1961	Appearance of methicillin-resistant *S. aureus*
1960s	Spread of methicillin-resistant *S. aureus*
1970s	Methicillin-resistant *S. aureus* widespread
1988	Appearance of vancomycin-resistant enterococci
1992	Laboratory transfer of high-level vancomycin resistance from enterococci to methicillin-resistant *S. aureus*
1997	Appearance of vancomycin-intermediate *S. aureus* in clinical setting
2002	Appearance of vancomycin-resistant *S. aureus* in clinical setting
2006–2014	Increase of global prevalence from 2% to 5%
2015–2020	Increase of global prevalence to 7%

* Pfeltz, R.F., & Wilkinson, B.J. (2004). The escalating challenge of vancomycin resistance in *Staphylococcus aureus*. Curr. Drug Targets Infect. Disord. 4: 273–294.

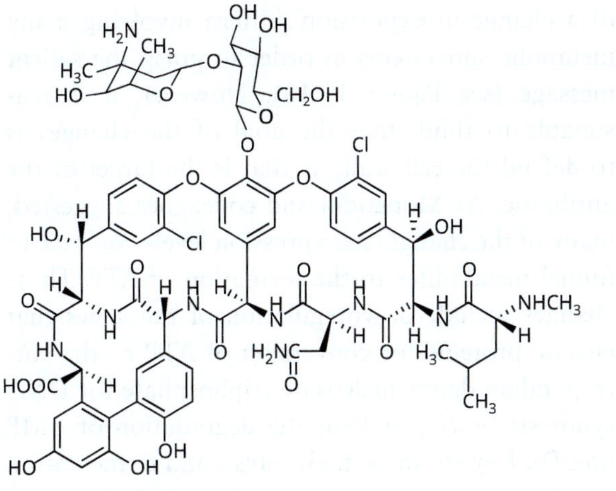

Vancomycin

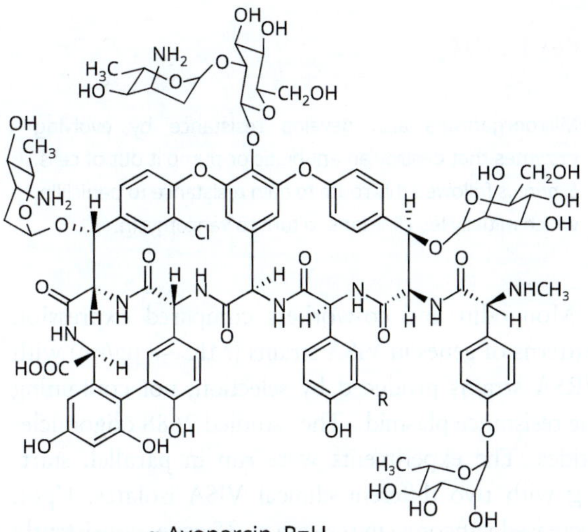

α-Avoparcin R=H
β-Avoparcin R=Cl

Figure 11.17 (left) Vancomycin, a glycopeptide antibiotic produced by the bacterium *A. orientalis.* (right) Antibiotics α– and β–avoparcin, related to vancomycin and produced by *Streptomyces candidus*.

Lu, K., Asano, R., & Davies, J. (2004). Antimicrobial resistance gene delivery in animal feeds. Emerging Infectious Diseases 10, 4: 679–683. https://doi.org/10.3201/eid1004.030506.

finding of bacterial DNA contamination in animal feed-grade avoparcin-containing sequences related to the resistance gene cluster strongly suggests that the use of avoparcin has led to gene transfer to bacteria which could be taken up by the animals. Eventually the genes found their way to bacteria that infect humans.

The toxicity of vancomycin in its early days was caused by impurities—the brown preparations were nicknamed 'Mississippi mud'. Since the mid-1980s, purification procedures and the safety of the preparations have improved.

S. aureus has achieved vancomycin resistance in two ways. These can be thought of as defence and attack. Both are effective, if success—for the bacterium—can be defined as attaining a level of resistance that survives doses of vancomycin that would be intolerably toxic to the patient.

VISA = vancomycin-intermediate *Staphylococcus aureus*
VRSA = vancomycin-resistant *Staphylococcus aureus*

Acting 'defensively', *S. aureus* achieves the *intermediate* stage of vancomycin resistance (VISA) by a number of structural changes, including reduced growth rate, reduction in cell wall cross-links, increased cell-wall thickness, and the appearance of D-glutamic acid instead of D-glutamine in the peptide. The genomic changes responsible for the VISA phenotype can be magnified by vancomycin challenge and selection to produce VRSA strains with MIC = 32 μg/ml.

The alternative, for the bacterium, is a 'counter-attack' on vancomycin, to 'pull its sting'. *S. aureus* has achieved high vancomycin resistance by picking up a specific plasmid from a resistant *Enterococcus*. The plasmid contains a cluster of genes, leading to changing the D-ala–D-ala at the C terminus of the cross-linking pentapeptide to D-ala–D-lactate (Table 11.5). The modified peptide can enter the cell wall but has a lower binding affinity for vancomycin by a factor of ~1000.

Table 11.5 Genes in vancomycin resistance cluster

Gene	Action of gene product
VanH	Reduces pyruvate to D-lactate
VanA	Esterifies D-ala-D-lactate
VanX	Hydrolyses D-ala-D-ala, leaving D-ala-D-lactate to build the cell wall
VanS	A kinase that senses vancomycin and activates *VanR*
VanR	Stimulates expression of vancomycin resistance genes In the absence of vancomycin, they are not expressed

KEY POINT

Microorganisms also develop resistance by evolving enzymes that destroy an antibiotic or pump it out of cells. *S. aureus* followed this route to gain resistance to penicillin, which initially led clinicians to turn to vancomycin.

Mongodin and co-workers compared expression patterns of genes in VISA strains (MIC ~8 μg/ml) with VRSA strains produced by selection, not containing the resistance plasmid.[7] They studied 2688 oligonucleotides. The experiments were run in parallel, starting with two different clinical VISA isolates. Upon increased vancomycin resistance, 35 genes consistently showed increased expression, some as high as 30-fold, and 16 consistently showed decreased expression.

Genes upregulated with increased vancomycin resistance are associated with the following:

- purine biosynthesis, which is a large component of the change in expression: 15 of the 35 upregulated genes involved purine biosynthesis or transport, and there was a mutation in the regulator of the purine biosynthesis operon;
- cell envelope synthesis, remodelling, and degradation;
- proteins involved in transport and binding of amino acids, peptides and amines, and nucleic acid components (including purines);
- synthesis of staphyloxanthin, an orange carotenoid that gives *S. aureus* its golden colour;
- folic acid synthesis; and
- unknown functions.

Genes downregulated with increased vancomycin resistance are associated with:

- energy metabolism;
- cell envelope biosynthesis;
- proteins involved in transport and binding of carbohydrates, organic alcohols, and acids;
- salvage of nucleic acid components;
- regulatory functions; and
- tetracycline resistance.

[7] Mongodin, E., Finan, J., Climo, M.W., *et al.* (2003). Microarray transcription analysis of clinical *Staphylococcus aureus* isolates resistant to vancomycin. J. Bacteriol. 185: 4638–4643.

It is not always easy to put together the details of a change in expression pattern involving many metabolic subsystems in order to grasp the salient message (see Figure 11.18). However, it is reasonable to think that the goal of the changes is to defend the cell wall, as that is the target of the antibiotic. As Mongodin and colleagues suggested, many of the changes in expression levels combine to funnel metabolites to the formation of ATP. These changes include downregulation of the genes that encode proteins for conversion of ATP to the corresponding deoxynucleoside triphosphate for DNA synthesis (*nrdD*) and for the degradation of AMP (*deoD*). Key enzymes in glycolysis and fermentation are downregulated, diverting glucose 6-phosphate through the pentose phosphate pathway to form the ribose component of ATP.

Synthesis of the thickened cell wall is a very energy-intensive process. The ratio of cell-wall volume to total cell volume increased by 41% in the vancomycin-resistant cells. Perhaps reduced cellular growth rate is a price that must be paid if a larger fraction of the cell's energy budget goes into cell-wall synthesis.

Drug resistance in pathogens is a crucial problem in contemporary medicine. Learning, in detail, how it comes about will be essential for developing ways to prevent it.

Fighting back

Several approaches to restoring the efficacy of vancomycin are being pursued:

- Inhibition of the products of any of the genes in the VAN cluster recovers vancomycin activity. For instance, hydroxyethylamines, phosphinate, and phosphonate transition-state analogues inhibit VanA. Other potential drugs inhibit VanX.
- Bacteriophages are an attractive source of clinically effective therapies, as they are microbe-specific and should not show side effects. AB-SA01 is a mixture of three complementary phages, active against a broad range of *S. aureus* strains. *In vitro,* it killed 95% of 205 multidrug-resistant clinical isolates, including methicillin-resistant and vancomycin-intermediate strains. Resistance did not develop: the bacteria would have to develop *simultaneous* resistance to all three phages. AB-SA01 is effective therapy for *S. aureus* infections in mice, rats, and rabbits. In

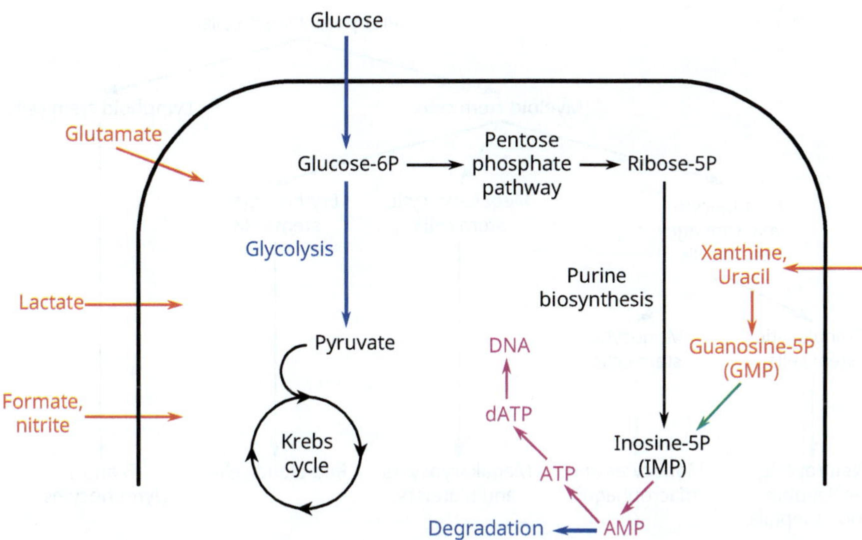

Figure 11.18 With enhancement of vancomycin resistance in the laboratory by selection after vancomycin challenge, expression patterns of genes associated with some processes are upregulated (thicker blue arrows) and others are downregulated (red arrows). A major target of upregulation is purine synthesis, aimed at enhanced production of ATP for energy requirements.

humans, its use is allowed for patients with serious or immediately life-threatening *S. aureus* infections for which no alternative treatment(s) are currently available. It is now in clinical trials.

Childhood leukaemias

Haemopoietic stem cells are the undifferentiated precursors of all types of blood cell. They mature by differentiating along one of two pathways (see Figure 11.19). B and T cells of the immune system have followed the lymphoid path; red blood cells have followed the myeloid path. An abnormal genetic transformation leading to unregulated proliferation of blood cells, at any stage of differentiation, gives rise to leukaemia. Leukaemias can be classified according to the type of cell that is proliferating.

Acute lymphoblastic leukaemia is the most common cancer in children, representing almost one-third of childhood cancers. The main treatment is chemotherapy, the success rate of which has greatly improved in the past quarter of a century. Nevertheless, conventional therapy is unsuccessful in about 25% of patients.

Measurements of gene expression patterns have permitted molecular classification of disease subtypes, and guide prognosis by correlations with response to chemotherapy, and with likelihoods of rapid or delayed recurrence or long-term survival.

- From the expression pattern of a group of 50 genes, it is possible to distinguish almost perfectly between lymphoblastic and myeloid leukaemias and, for lymphoblastic leukaemias, to distinguish B- and T-cell lineages. The results are calibrated against established methods based on flow cytometry. The variability in expression pattern is unusually high in acute lymphoblastic leukaemia relative to other types of cancer. It is thereby feasible to create a molecular taxonomy of childhood leukaemias.
- Expression patterns can predict the likelihood of a favourable outcome. This combines questions of the success of therapy, the likelihood of spontaneous relapse after remission, and the development of secondary tumours. A complicating factor of studies of this type in humans is the fact that the samples are taken from patients under a variety of treatments.

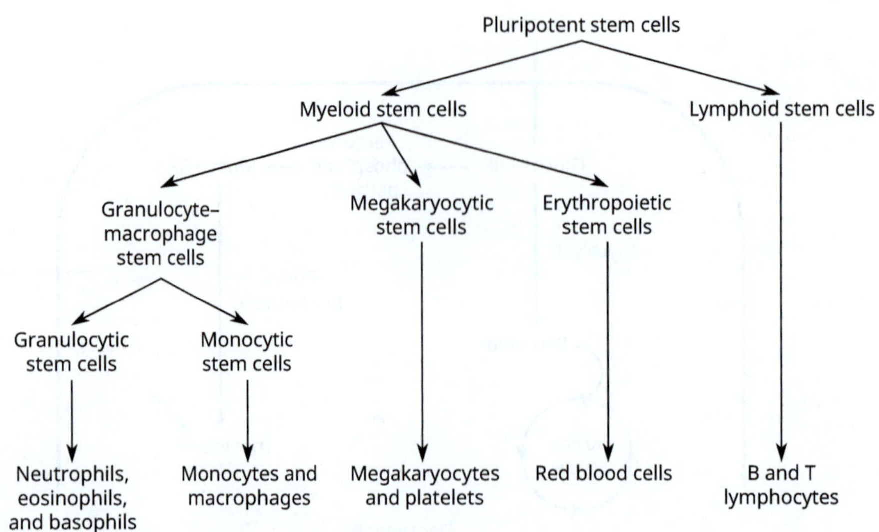

Figure 11.19 Haematopoiesis is the formation of new blood cells. Our blood contains many types of cell:

Cell type	Function
Neutrophils	Respond to bacterial infection
Eosinophils	Respond to allergens and to infections by parasites
Basophils	Respond to allergens
Monocytes and macrophages	Remove dead tissue and respond to infections by bacteria and fungi
Megakaryocytes	Precursors of platelets
Platelets	Involved in blood clotting
Red blood cells (erythrocytes)	Contain haemoglobin and transport O_2 and CO_2
B lymphocytes	Produce specific antibodies
T lymphocytes	Destroy infected cells and regulate immune responses

These cells arise by different developmental pathways from a common stem-cell precursor, which can potentially differentiate into any of the mature cell types, a property called **totipotency**. As maturation proceeds, cells first become **pluripotent** (able to mature into some but not all cell types) and then finally committed to a single ultimate form.

Normally, haematopoiesis produces, every day, approximately 175 billion red cells, 70 billion granulocytes (neutrophils, eosinophils, basophils), and 175 billion platelets. When we are challenged by infection, production can be stepped up by an order of magnitude.

Leukaemia is the uncontrolled proliferation of any of these types of blood cell, either mature forms or their precursors. In mammals, mature erythrocytes, being enucleate, cannot themselves proliferate. However, mutations can occur in *precursor* cells. For example, a mutation affecting the JAK2 signalling pathway in the stem-cell precursor can result in overproduction of erythrocytes, a disease called primary polycythaemia vera. (Secondary polycythaemia vera, also characterized by overproduction of red cells, is a response to lack of oxygen; possible causes include heavy smoking, emphysema, or moving without acclimation to a high altitude.)

Clinical experience shows that the time interval of first remission is a good predictor of long-term survival. A group of genes was found with an expression pattern correlated with length of remission; i.e., these genes are differentially expressed in patients with early and late relapse. Testing the expression levels of these genes can improve the precision of prognosis. Identifying the pathways in which the genes are involved can illuminate the underlying biology of the disease progression. Genes involved in cell proliferation and DNA repair were upregulated in the early-relapse group.

Development of a second cancer, not related to leukaemia in any obvious way, is a common and very serious complication of acute lymphoblastic leukaemia. Brain tumours are one of the most common secondary malignancies. Several genes have been identified, the expression patterns of which correlate with the risk of secondary brain tumours.

- *Expression patterns can predict effectiveness of treatment and guide the choice of therapy.* In a study using 14500 probe sets and samples from 173 patients, sets of 20–40 genes were identified that distinguish resistance and sensitivity to four different drugs: prednisolone, vincristine, asparaginase, and daunorubicin. These results, even taken as purely empirical correlations, have clinical utility in guiding treatment. Their interpretation at the genetic level reveals that the activities of the drugs involve some different as well as some common pathways. A set of 45 genes was found to be correlated with resistance to *all four* of the drugs. The majority of these genes involve transcription, DNA repair, cell-cycle maintenance, and nucleic acid metabolism.
- *Identification of specific genes involved in diseases can suggest targets for drug development.* For instance, one type of acute lymphoblastic leukaemia is associated with overexpression of the gene FLT3 for a receptor tyrosine kinase. Patients with mutations that produce constitutively active receptors have a poor prognosis. Several FLT3 inhibitors have been approved for clinical use in Europe and the US.

Thus, expression profiles can permit tailoring of drug therapy, both to the specific disease and subtype, and to the patient.

KEY POINT

Expression profiling can (a) permit precise diagnosis of the subtype of the disease, (b) predict the likely course of the disease, and (c) guide choice of therapy.

The Encyclopedia of DNA Elements (ENCODE)

The goal of the ENCODE (Encyclopedia of DNA Elements) project is to determine function across the entire human genome. At the start of the project, outside the protein-coding regions and others known to be involved in transcription control, most of the genome had no assigned function.

The ENCODE consortium comprises several hundred scientists worldwide. Plans for the project included consecutive phases. First, a pilot project, treating a selected ~1% of the human genome—about 30 Mba—served as a testbed for development of effective methods. Then the project was scaled up to the entire genome. Currently, the system is in production phase. modENCODE extends the project to *C. elegans* and *D. melanogaster*. modERN (model organism encyclopedia of regulatory networks) merges the worm and fly groups. MouseENCODE is another extension.

ENCODE has produced a genome-wide catalogue of transcripts, and identified their subcellular localization. The most surprising conclusion from ENCODE is how pervasive transcription is. RNA was detected that covers three-quarters of the genome. Cells are much richer in non-coding RNAs than suspected. Many of these arise from regions adjacent to protein-coding regions. Many of the transcripts show variable splicing patterns, enriching the RNAome still further.

What do these transcripts do?

Traditionally, DNA outside protein-coding regions was considered 'junk', but for no reason other than ignorance. Why then did the human genome retain so much sequence of unknown function? The pufferfish genome is 400 Mb long, roughly 1/8 the size of the human genome. This disparity suggests that much of the human genome might be dispensable. Sydney Brenner distinguished junk from garbage: 'Garbage you throw away, junk you keep.' Is the human's junk the pufferfish's garbage?

There has been extensive and heated debate over the attribution of function to transcripts without being able to assign specific functions to them. Part of the problem is that many transcripts for which unspecified function was claimed do not show conservation among mammals. This was the basis for many criticisms of the claims of the ENCODE consortium in its early publications. The rule, which does apply to proteins, that nature conserves important things but not unimportant things, does not necessarily apply to these transcripts. Sequence-structure-function relationships can be very different for RNAs and for proteins. The argument that transcribed regions are functional—rather than just 'junk RNA'—rests on the observation of differential transcription in cell-specific patterns, and dynamic regulation of transcription during development and tissue differentiation, and in disease.

Even many transcribed pseudogenes would, by these criteria, appear to be functional. Examples are known of pseudogenes, originally derived from protein-coding sequences, that are transcribed to form siRNAs, functional in regulating gene expression. In some cases, a genome contains a gene for a functional protein and a homologous pseudogene in reverse orientation. In this situation, transcription of the pseudogene produces an RNA that is approximately the reverse complement of the transcript of the proper gene. The two RNAs can pair, and ('. . . as two spent swimmers, that do cling together and choke their art'), suppress translation. It can go even farther: Dicer, a protein component of the RNAi pathway, can fragment the RNA double helix into siRNAs. Incorporation of these siRNAs into the RNA-induced silencing complex leads to degradation of the mRNA from the gene coding the active protein.

LOOKING BACK AND FORWARD

The preceding two chapters have dealt with human biology, historical and clinical aspects. Now we have changed focus, to a traversal of the Central Dogma: DNA makes RNA makes protein. This chapter treats RNA. The next treats proteins.

RECOMMENDED READING

Transcriptional regulation

Payne, J.L., & Wagner, A. (2015). Mechanisms of mutational robustness in transcriptional regulation. Frontiers in Genetics 6: 322.

Kumar, D., Bansal, G., Narang, A., *et al.* (2016). Integrating transcriptome and proteome profiling: Strategies and applications. Proteomics 16: 2533–2544.

Meyer, M.M. (2016). The role of mRNA structure in bacterial translational regulation. Wiley Interdiscip. Rev. RNA 8: 1.

Kuznetsova, T., & Stunnenberg, H.G. (2016). Dynamic chromatin organization: Role in development and disease. Int. J. Biochem. Cell Biol. 76: 119–122.

Lenstra, T.L., Rodriguez, J., Chen, H., *et al.* (2016). Transcription dynamics in living cells. Annu. Rev. Biophys. 45: 25–47.

Reilly, S.K., & Noonan, J.P. (2016). Evolution of gene regulation in humans. Annu. Rev. Genomics Hum. Genet. 17: 45–67.

Bashiardes, S., Zilberman-Schapira, G., & Elinav, E. (2016). Use of metatranscriptomics in microbiome research. Bioinform. Biol. Insights 10: 19–25.

Zoabi, Y., & Shomron, N. (2021). Processing and analysis of RNA-seq data from public resources. Methods Mol. Biol. 2243: 81–94.

Microarrays

Butte, A. (2002). The use and analysis of microarray data. Nat. Rev. Drug Discov. 1: 951–960.

Penkett, C.J., & Bähler, J. (2004). Getting the most from public microarray data. Eur. Pharm. Rev. 1: 8–17.

RNA sequencing

Wang, Z., Gerstein, M., & Snyder, M. (2009). RNA-Seq: A revolutionary tool for transcriptomics. Nat. Rev. Genet. 10: 57–63.

Ozsolak, F., & Milos, P.M. (2011). RNA sequencing: Advances, challenges and opportunities. Nature Reviews Genetics 12: 87–98.

Mutz, K.-O., Heilkenbrinker, A., Lönne, M., *et al*. (2013). Transcriptome analysis using next-generation sequencing. Curr. Opin. Biotechnol. 24: 22–30.

Han, Y., Gao, S., Muegge, K., *et al*. (2015). Advanced applications of RNA sequencing and challenges. Bioinform. Biol. Insights 9 (Suppl 1): 29–46.

Geraci, F., Saha, I., & Bianchini, M. (2020). Editorial: RNA-Seq analysis: Methods, applications and challenges. Front. Genet. 11: 220.

Pyatnitskiy, M.A., Arzumanian, V.A., Radko, S.P., *et al*. (2021). Oxford Nanopore MinION Direct RNA-Seq for Systems Biology. Biology (Basel) 10: 1131.

Gondane, A., & Itkonen, H.M. (2023). Revealing the history and mystery of RNA-Seq. Curr. Issues Mol. Biol. 45: 1860–1874.

Applications

GTEx Consortium (2020). The GTEx Consortium atlas of genetic regulatory effects across human tissues. Science 369: 1318–1330.

The problem of bacterial drug resistance and the development of novel antibiotics

Amábile-Cuevas, C.F. (2003). New antibiotics and new resistances. Am. Sci. 91: 138–149.

Projan, S.J., & Shlaes, D.M. (2004). Antibacterial drug discovery: Is it all downhill from here? Clin. Microbiol. Infect. Suppl. 4: 18–22.

Projan, S.J., Gill, D., Lu, Z., *et al*. (2004). Small molecules for small minds? The case for biologic pharmaceuticals. Expert Opin. Biol. Ther. 4: 1345–1350.

Thomson, C.J., Power, E., Ruebsamen-Waigmann, H., *et al*. (2004). Antibacterial research and development in the 21st century—an industry perspective of the challenges. Curr. Opin. Microbiol. 7: 445–450.

Overbye, K.M., & Barrett, J.F. (2005). Antibiotics: Where did we go wrong? Drug Discov. Today 10: 45–52.

Barrett, J.F. (2005). Can biotech deliver new antibiotics? Curr. Opin. Microbiol. 8: 498–503.

Talbot, G.H., Bradley, J., Edwards, J.E., *et al*. (2006). Bad bugs need drugs: An update on the development pipeline from the Antimicrobial Availability Task Force of the Infectious Diseases Society of America. Clin. Infect. Dis. 42: 657–668.

Strathdee, S., Patterson, T., & Barker, T. (2019). *The Perfect Predator: A Scientist's Race to Save Her Husband from a Deadly Superbug: A Memoir*. New York: Hachette Books.

Schooley, R.T., & Strathdee, S. (2020). Treat phage like living antibiotics. Nat. Microbiol. 5: 391–392.

Yilmax, N.K., & Schiffer, C.A. (2021). Introduction: Drug resistance. Chem. Rev. 121: 3235–3237.

Larsson, D.G.J., & Flach, C.F. (2022). Antibiotic resistance in the environment. Nat. Rev. Microbiol. 20: 257–269.

The ENCODE project

ENCODE Project Consortium (2011). A user's guide to the encyclopedia of DNA elements (ENCODE). PLoS Biol. 9: e1001046.

The ENCODE Project Consortium (2012). An integrated encyclopedia of DNA elements in the human genome. Nature 489: 57–74.

Mattick, J.S., & Dinger, M.E. (2013). The extent of functionality in the human genome. The HUGO Journal 7: 2.

RNA sequencing

Ketkar, S., Burrage, L.C., & Lee, B. (2023). RNA sequencing as a diagnostic tool. J. Amer. Med. Assoc. 329: 85–86.

Shangguan, Y., Li, C., Lin, H., *et al.* (2020). Application of single-cell RNA sequencing in embryonic development. Genomics 112: 4547–4551.

Thind, A.S., Monga, I., Thakur, P.K., *et al.* (2021). Demystifying emerging bulk RNA-Seq applications: The application and utility of bioinformatic methodology. Briefings in Bioinformatics 22: bbab259.

Bow, A. (2021). A streamlined approach to pathway analysis from RNA-Sequencing data. Methods Protoc. 4: 21.

Wang, S., Sun, S.T., Zhang, X.Y., *et al.* (2023). The evolution of single-cell RNA sequencing technology and application: Progress and perspectives. Int. J. Mol. Sci. 24: 2943.

Haque, A., Engel, J., Teichmann, S.A., *et al.* (2017). A practical guide to single-cell RNA-sequencing for biomedical research and clinical applications. Genome Med. 9: 75–87.

Aldridge, S., & Teichmann, S.A. (2020). Single cell transcriptomics comes of age. Nature Communications 11: 4307.

Vlasova-St. Louis, I. (2021). *Applications of RNA-Seq in Biology and Medicine*. Rijeka, Croatia: Intech-Open.

Chen, J.W., Shrestha, L., Green, G., *et al.* (2023). The hitchhikers' guide to RNA sequencing and functional analysis. Brief Bioinform. 24: bbac529.

Gondane, A., & Itkonen, H.M. (2023). Revealing the history and mystery of RNA-Seq. Curr. Issues Mol. Biol. 45: 1860–1874.

EXERCISES, PROBLEMS, AND WEBLEMS

Exercises

Exercise 11.1 In the third level (under Hybridization, washing), Figure 11.1 shows, schematically, a row of 14 probe oligomers, corresponding to the leftmost 14 elements in one of the rows of the 19 × 19 square array of fluorescent dots in the fourth level. Which row?

Exercise 11.2 In Figure 11.8, the cofactor analogue phosphoaminophosphonic acid-guanylate ester has an NH group replacing an O linking the two distal phosphates. On a copy of this figure, circle that nitrogen (the hydrogen is not shown).

Exercise 11.3 In Figure 11.4, on each line colour the first exon green, the second orange, the third peach and the fourth red. (Exons 2 and 3 will appear as they are coloured in the line Mutually exclusive exons.) For each line, draw the mature mRNA that results. For instance, the mature mRNA that corresponds to the primary transcript will appear:

Exercise 11.4 A professor of molecular biology wanted to design a microarray experiment to detect tRNA sequences. He suggested using the sequences of the anticodon stem–loop as target oligonucleotides (see Figure 1.8). A student pointed out that this was not likely to be successful. For what reason?

Exercise 11.5 Using RNAseq, you would like to map the RNA sequences determined onto the reference genome. Reads that cross exon boundaries in mature messenger RNA will not map to the genome because the genomic DNA but not the RNA contains intron sequences (see Figure 11.3). What relatively easy step could you take to create a pseudo-genome to which many of these 'junction reads' could be mapped? What problems would remain?

Exercise 11.6 Why would it be preferable to use RNAseq, rather than microarrays based on human probes, if the goal of an experiment is a quantitative comparison of human and chimpanzee mRNA levels in corresponding tissues?

Exercise 11.7 (a) From the data shown in Figure 11.7, is there any gene that distinguishes with perfect consistency between males and females? That is, is there any gene that is significantly upregulated in all three male subjects and significantly downregulated in all three female subjects (or *vice versa*)? (b) If the first subject (corresponding to the leftmost column) had been omitted from the study, what would be the answer to the question in part (a)? (c) What pairs of genes give the most consistent results (that is, agree best) across all six subjects?

Exercise 11.8 On a photocopy of Figure 11.9, indicate the position of a gene that is first upregulated and subsequently downregulated during the development of *Drosophila melanogaster*.

Exercise 11.9 Transcriptome profiling is the measurement of patterns of mRNA concentrations. However, 90% of the RNA in a cell is ribosomal. How could you apply the fact that mRNAs carry a 3′ poly A tail to avoid interference from the high background levels of ribosomal RNA?

Exercise 11.10 You wish to insert a human gene into *E. coli*, in order to produce large amounts of pure protein. The human gene has introns, and *E. coli* cannot splice them out. How could you use RT-PCR (here RT does NOT stand for Real-Time) to overcome this difficulty?

Exercise 11.11 Many disease-linked genetic variants are outside protein-coding regions. The condition must arise from defects in regulation rather than dysfunctional expressed proteins bearing mutations. Consider a mutant outside a protein-coding region that associated with a disease. Does it follow that the mutation must affect the function of an RNA transcript?

Exercise 11.12 The genome of the pufferfish (*Tetraodon nigroviridis*) is 385 Mb long, roughly 1/9 of the human genome. It has at least as many protein-coding genes as humans. One difference is the much smaller proportion of repetitive sequences in pufferfish relative to human. Nevertheless, estimate the amount of RNA transcribed from the human genome that could not correspond to any RNA transcribed in the pufferfish.

Problems

Problem 11.1 The average ratio of gene expression levels between human and chimpanzee is ~ 1.5 for collinear chromosomes and can be as high as ~ 1.6–1.7 for rearranged chromosomes (Figure 11.16). (a) Qualitatively describe the differences in banding pattern between human and chimpanzee for human chromosomes 5, 6, 9, and 10 (see Figure 6.12). (b) Are your results consistent with the data in Figure 11.16? (c) Which would you expect to have an average ratio of expression levels closer to 1.0, genes on human chromosome 3 and their gorilla homologues, or genes on human chromosome 8 and their gorilla homologues?

Problem 11.2 Describe the reasons for and against including antibiotics routinely in animal feed. Decide on a conclusion to be drawn from these arguments and formulate a paragraph of recommendations.

Problem 11.3 A recent study of 1737 patients treated with vancomycin for *S. aureus* infections concluded that many patients received less than an adequate dose to maintain serum concentrations above the MIC.[8] For patients infected by vancomycin-sensitive *S. aureus*, this characterized 7.9% of patients given continuous infusions and 19% of patients to whom the drug was administered by periodic intravenous injections. For patients infected by *S. aureus* strains with

[8] Kitzis, M.D., & Goldstein, F.W. (2006). Monitoring of vancomycin serum levels for the treatment of staphylococcal infections. Clin. Microbiol. Infect. 12: 92–95.

intermediate-level resistance to vancomycin, this characterized 79.1% of patients given continuous infusions and 87.8% of patients given periodic intravenous injections. What are the expected effects of this situation on (a) the individual patients involved and (b) the spread of vancomycin-resistant *S. aureus* strains?

Problem 11.4 You have a sample of mRNA which you convert to cDNA. From this material, what information can you derive from a microarray or direct RNA sequencing that would not be available from a DNA sequencing?

Weblems

Weblem 11.1 An EST for rose germacrene D synthase appears in dbEST (GenBank) as entry BQ105086. Blast this sequence against general sequence databases. What properties of rose germacrene synthase can you infer from the results?

Weblem 11.2 (a) What is the chromosome location of the *ASPM (abnormal spindle-like microcephaly associated)* gene? (b) What mutations are known? (c) What is their clinical effect?

Weblem 11.3 Search for articles on discovery of novel antibiotics and study a number of them to learn about the current situation. (Suggested search terms: future antibiotic development. Alternatively, consult articles cited as recommended reading in this chapter.) (a) Provide facts and figures that describe trends in funding for research by large pharmaceutical companies devoted to novel antibiotic development. (b) Assuming that these figures confirm that large pharmaceutical companies are curtailing the relevant research programmes, what are the reasons for this, in the face of high demand for novel antibiotics? (c) Can you recommend appropriate actions by governments, industry, and the academic community? Give credible reasons that justify the conclusion that if your actions were adopted they would accelerate the development and approval of novel antibiotics. What do you see as the most serious obstacles to acceptance of your ideas? How would you suggest overcoming them? (Warning: consider the possibility that the less ambitious your suggestions, the more credibly you can justify them.)

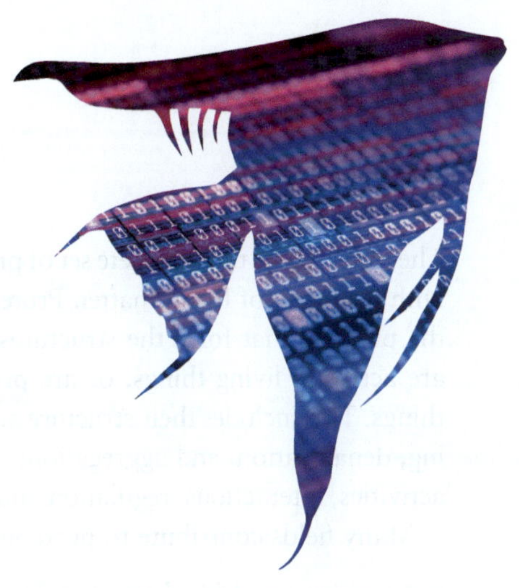

CHAPTER 12

Proteomics

LEARNING GOALS

- *Understand the fundamental chemical structure of proteins*: the mainchain and sidechains, types of sidechain, and common post-translational modifications.
- *Know how to describe protein conformation.*
- *Be able to distinguish primary, secondary, tertiary, and quaternary structures.*
- *Understand the use of polyacrylamide gel electrophoresis (PAGE) to separate proteins.*
- *Understand the technique and uses of mass spectrometry.*
- *Recognize the power of experimental methods for protein structure determination*: X-ray crystallography, NMR spectroscopy, and cryo-electron microscopy.
- *Appreciate the importance of the recent breakthrough in reliable prediction of protein structure* from amino-acid sequence, by AlphaFold2 software.
- *Understand the principles of classifications of protein-folding patterns.*
- *Recognize that many properties of proteins are invisible in genome sequences*, including but not limited to variable splicing, post-translational modifications, and formation of complexes of interacting proteins.
- *Know what underlies diseases of protein aggregation.*
- *Appreciate the goals and difficulties of structural genomics projects.*
- *Know about creation of useful artificial proteins*, by directed evolution, and by *a priori* design.

Introduction

The proteome is the complete set of proteins associated with a sample of living matter. Proteomics deals with the proteins that form the structures of living things, are active in living things, or are produced by living things. This includes their structure and function, folding, denaturation, and aggregation; their distribution, activities, interactions, regulation, and evolution.

Many fields contribute to proteomics:

- *Chemistry and biochemistry*. These include physical methods, such as spectroscopy, kinetics, and techniques of structure determination, and organic and biochemical methods for working out mechanisms of enzymatic catalysis. Techniques for separation and analysis of proteins have their sources in chemistry and molecular biology.
- *Molecular and cellular biology*. These disciplines help to coordinate our knowledge of individual proteins into an understanding of the biological context, and of how cells properly integrate protein activities.
- *Evolutionary biology*. Proteins evolve. Their evolutionary pathways explore variations in amino-acid sequences, protein structures, interactions and functions, and patterns of protein expression.
- *Structural genomics*. This activity applies advances in X-ray crystallography, nuclear magnetic resonance (NMR), and cryo-electron microscopy (cryo-EM), and predicted structures, to comprehensive, high-throughput delivery of coordinate sets of proteins contained in an integral source.
- *Protein structure prediction*. Bioinformatics coordinates its efforts with structural genomics to supplement experimental structure determinations by prediction of protein structures from amino-acid sequences. A recent breakthrough, the computer program AlphaFold2, has been successful at predicting protein structure from amino-acid sequence. The importance of this achievement cannot be overemphasized, given the great disparity between the very large number of known amino-acid sequences and the relatively few experimentally determined structures.
- *Bioinformatics* brings together the many data streams of genomics, expression patterns, and proteomics, to assemble databases and create links among them. This enables their coordinated application to problems of biology, clinical medicine, agriculture, and technology.

Proteins are where the action is (much of it, anyway)

- *Proteins have a great variety of functions*. There are structural proteins (our hair, fingernails, molecules of the cytoskeleton, epidermal keratin, viral coat proteins); catalytic proteins (enzymes); transport and storage proteins (haemoglobin, retinol-binding protein, ferritin); regulatory proteins (including hormones, many kinases and phosphatases, and proteins that control gene expression); and proteins of the immune system and the immunoglobulin superfamily (including antibodies and proteins involved in cell–cell recognition and signalling). How can proteins accomplish so many different things? By coming in a great variety of structures, each specialized to carry out different functions.
- *The amino-acid sequences of proteins dictate their three-dimensional structures and their trajectories of folding*. Under physiological conditions of solvent and temperature, most proteins fold spontaneously to an active native state (see Box 12.1). The amino-acid sequence of a protein must not only preferentially stabilize the native state, it must also contain a 'road map' telling the protein how to get there, starting from the many diverse conformations that comprise the unfolded state. This is called the folding pathway.
- *Advances in protein science have spawned the biotechnology industry*. It is now possible to identify and apply useful natural proteins, design and test modifications of known proteins, and to design novel ones with desired functions.

BOX 12.1

Reversible denaturation of proteins shows that amino-acid sequence determines protein structure

In the 1950s, a set of classic experiments by C.B. Anfinsen showed that proteins could exhibit *reversible* denaturation. Anfinsen used the small protein Ribonuclease A (142 amino acids), an enzyme that hydrolyses RNA. Anfinsen showed that exposure of a dilute solution of Ribonuclease A to guanidinium hydrochloride (which breaks hydrogen bonds between protein atoms) and β-mercaptoethanol (which breaks disulphide bridges) denatured the protein, and caused it to lose enzymatic activity. But upon removing the guanidinium hydrochloride and reoxidizing the disulphide bridges, the protein refolded to the native state, regaining enzymatic activity.

This result has been extended to many other proteins, establishing the *very* important principle: **protein structure is determined by amino-acid sequence**. This deserves extension of Crick's central dogma:

DNA → RNA → amino–acid sequences of proteins → protein structures → protein functions

Anfinsen's experiments established the general principle that protein denaturation is *reversible:* if a denatured protein is brought back to normal conditions of solvent and temperature it will recover its native structure and function. But as readers may well point out, if one boils an egg and subsequently cools it down, one does not recover a raw egg. Does this contradict Anfinsen's principle?

The difference is that the proteins in an egg are very concentrated. Heating denatures them, and the denatured proteins aggregate. It is the **aggregation**, not the denaturation, that is the irreversible step.

Old molecular biologist's joke: 'How might you reverse the boiling of an egg? Answer: Feed it to a chicken!'

In some cases, mutations destabilize proteins, sometimes to moderate extents only. α_1-antitrypsin is an elastase inhibitor that protects the lungs from neutrophil elastase. Dysfunction of α_1-antitrypsin raises the risk of emphysema.

The Z-mutant of α_1-antitrypsin, Glu342→Lys, is destabilized to the point that it can denature and aggregate in temperatures attainable in high fevers. Many readers will know that it is not uncommon for ill babies to 'spike' high fevers, with body temperatures above the denaturation temperature for Z-α_1-antitrypsin. Suppose an infant is homozygous for the Z mutant of α_1-antitrypsin. What clinical advice would you give parents, warning them of what to look out for and what action to take should their infant become ill?

Protein structure

The chemical structure of proteins

Chemically, protein molecules are long polymers typically containing several thousand atoms, composed of a uniform repetitive **backbone** (or **mainchain**) with a particular **sidechain** attached to each residue (see Figure 12.1 and Box 12.2). The residues are chosen from a set of **20 standard amino acids** (with very few exceptions). The amino-acid sequence of a protein states the order of the sidechains. The base sequence of the gene coding for each protein specifies the amino-acid sequence.

KEY POINT

A protein is a message written in a 20-letter alphabet.

The sidechains in proteins show a variety of physicochemical features: some are charged, some are uncharged but polar, and others are **hydrophobic**—the sidechains are bulky but not charged or polar.

Residue $i-1$ Residue i Residue $i+1$

S_{i-1} S_i S_{i+1} }Sidechains variable

··· N—Cα—C(=O)—N—Cα—C(=O)—N—Cα—C(=O)—··· }Mainchain constant

Figure 12.1 Proteins contain a mainchain of constant structure. Attached at regular intervals are sidechains, each chosen (with few exceptions) from the canonical set of 20 amino acids. Here S_{i-1}, S_i, and S_{i+1} represent successive sidechains. Different sequences of sidechains characterize different proteins. It is this sequence that gives each protein its individual structural and functional characteristics.

BOX 12.2

The amino acids

The amino acids shown both as chemical structures and in space-filling representation. For glycine, the entire amino acid is shown; for the others, only the sidechains. Atoms with positive charges are in blue, atoms with negative charges are in red.

(a)

Glycine	$H{-}C\alpha{-}H$, with $^{+}NH_3$ above and COO^- below $C\alpha$
Alanine	$-C\alpha-CH_3$
Serine	$-C\alpha-CH_2-OH$
Cysteine	$-C\alpha-CH_2-SH$
Threonine	$-C\alpha-CH(OH)-CH_3$
Proline	N−Cα (ring)
Valine	$-C\alpha-CH-(CH_3)_2$
Leucine	$-C\alpha-CH_2-CH-(CH_3)_2$
Isoleucine	$-C\alpha-CH(CH_3)-CH_2-CH_3$
Methonine	$-C\alpha-CH_2-CH_2-S-CH_3$
Phenylalanine	$-C\alpha-CH_2-$(benzene ring)
Tyrosine	$-C\alpha-CH_2-$(benzene ring)$-OH$
Aspartic acid	$-C\alpha-CH_2-COO^-$
Glutamic acid	$-C\alpha-CH_2-CH_2-COO^-$

(b)

Histidine	$-C\alpha-CH_2-C$ (ring: N^+=C, C−NH)
Asparagine	$-C\alpha-CH_2-CONH_2$
Glutamine	$-C\alpha-CH_2-CH_2-CONH_2$
Lysine	$-C\alpha-CH_2-CH_2-CH_2-CH_2-NH_3^+$
Arginine	$-C\alpha-CH_2-CH_2-CH_2-NH-C(NH_2)(=NH_2^+)$
Tryptophan	$-C\alpha-CH_2-$(indole ring, N−H)

(c)

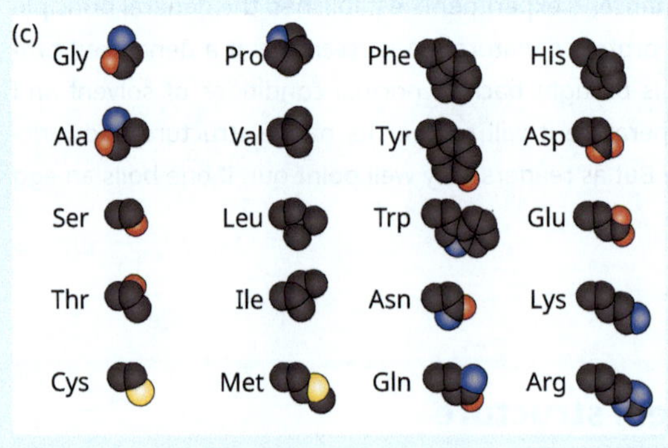

Different types of residue make different types of interactions, including:

- *Hydrogen bonding*. The hydrogen bond is an interaction between two polar atoms (oxygen or nitrogen; occasionally sulphur) mediated by a hydrogen atom. Several types of hydrogen bond are *extremely* important to biology:
  - Water is an extensively hydrogen-bonded liquid. This accounts for its physicochemical properties, for instance its high boiling point. The structure of water determines the solubility of different substances. Water is not merely the medium of most biochemical processes, it is an active participant in many.

- Hydrogen bonds between mainchain C=O and H–N groups stabilize the folding patterns of proteins. Often they form standard substructures, helices and sheets (see Box 12.4).
- Hydrogen bonds are about 20 times weaker than covalent chemical bonds. Hydrogen bonds in aqueous solution can be easily broken and reformed. The solvent water can form hydrogen bonds to polar groups in nucleic acids and proteins. This creates a competition between intramolecular hydrogen bonds and solute–solvent hydrogen bonds.
- In nucleic acids, hydrogen bonds mediate the complementarity between adenine and thymine, and between guanine and cytosine.

- *Hydrophobic interactions*. Hydrophobic residues have sidechains that are primarily hydrocarbon in nature. They have thermodynamically unfavourable interactions with water. Salad dressing is an everyday example of the hydrophobic effect: the thermodynamic unfavourability of dissolving oil in water causes a phase separation. In proteins, it is energetically favourable to bury hydrophobic sidechains in the interior of the structure, where they are not exposed to the solvent. This is a general feature of globular proteins.
- *Disulphide bridges*. In addition to the primary chemical bonds in the individual residues and the peptide bonds joining the residues into a polymer, cysteine residues in proteins, with sidechain $-CH_2SH$, can form disulphide bonds: $-CH_2S-SCH_2-$. Disulphide bonds contribute to the stability of native states. In order to denature proteins fully, it is necessary to break any disulphide bonds.

Different possible conformations of the backbone of a protein bring different types of residue into spatial proximity and expose some but not all residues to the solvent. Every conformation therefore has a different associated energy that depends on the distribution of favourable and unfavourable interactions. The native state of a soluble globular protein is the conformation that optimizes the set of interactions among the residues and between the residues and the solvent.

KEY POINT

Different types of residues make different types of interactions, including hydrogen bonds, hydrophobic interactions, and disulphide bridges. Formation of the native structure allows optimal formation of favourable inter-residue and residue–solvent interactions.

Non-ribosomal peptides

In addition to the standard mechanism of protein synthesis based on the Central Dogma, certain microbial cells have alternative mechanisms of direct peptide synthesis, not encoded by DNA sequences. In this respect, these peptides are simply metabolites.

Their claim on our attention derives from the great range and importance of their pharmacological and biological activities. Well-known examples include antibacterials (penicillin and vancomycin), anti-tumour agents (bleomycin), and immunosuppressants (cyclosporine), and there are many others.

These molecules are synthesized by specialized enzymes, non-ribosomal peptide synthetases (NRPS) (the NRPS ARE encoded in the microbial genome). Each NRPS incorporates a single amino acid into a growing peptide chain. The adduct is not necessarily one of the 20 canonical amino acids. NRPS can incorporate a very large variety of residues, including D-amino acids, β-amino acids, methylated, glycosylated, and phosphorylated residues, heterocycles, and even fatty acids. From this their products derive their great versatility of biological activity.

Conformation of the polypeptide chain

The conformation of a polypeptide chain can be described in terms of angles of internal rotation around the bonds in the mainchain (see Figure 12.2). The bonds between the N and Cα, and between the Cα and C, are single bonds. Internal rotation around these bonds is not restricted by the electronic structure of the bond, only by possible steric collisions in the conformations produced (see Box 12.3).

The entire conformation of the protein can be described by these angles of internal rotation. Each set of four successive atoms in the mainchain defines a dihedral angle. (The dihedral angle defined by four points is the angle between

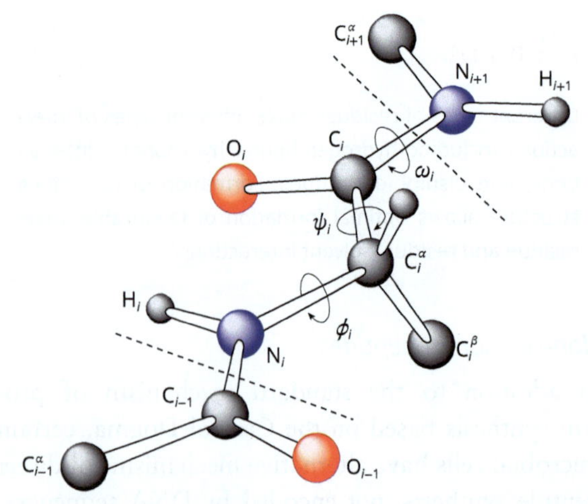

Figure 12.2 Conformational angles describing the folding of the polypeptide chain.

The mainchain of each residue (except the C-terminal residue) contains three chemical bonds: N-Cα, Cα-C, and the peptide bond C-N linking the residue to its successor. The conformation of the mainchain is described by the angles of rotation around these three bonds:

Rotation around:	N-Cα bond	Cα-C bond	Peptide bond (C-N)
Name of angle:	ϕ	ψ	ω

ω is restricted to be close to 180° (*trans*) or, infrequently, close to 0° (*cis*).

[Based on: https://iupac.qmul.ac.uk/misc/ppep3.html#fig5]

the plane defined by the first three and the plane defined by the last three. (If you open a door you are changing the dihedral angle between the plane of the door and the plane of the wall to which it is attached.) In each residue i (except for the N and C termini), the angle ϕ_i is the angle defined by atoms C(of residue $i-1$)–N–Cα–C, and the angle ψ_i is the angle defined by atoms N–Cα–C–N(of residue $i+1$). Then ω_i is the angle around the peptide bond itself, defined by the atoms Cα–C–N(of residue $i+1$)–Cα(of residue $i+1$).

The peptide bond has partial double-bond character and adopts two possible conformations: *trans* (by far the more common) and *cis* (rare). Values of angles ω are restricted to be close to 180° (*trans*) or 0° (*cis*).

Proline is an exception: the sidechain is linked back to the N of the mainchain to form a pyrrolidine ring. This restricts the mainchain conformation of proline residues. It disqualifies the N atom as a hydrogen-bond donor, for instance in helices or sheets. Also, the energy difference between *cis* and *trans* conformations is less for proline residues than for others. Most *cis* peptides in proteins appear before prolines.

The Sasisekharan–Ramakrishnan–Ramachandran diagram (Box 12.3) describes the *local* structure of a polypeptide chain. The great variety of protein structures derives from the *global* conformation; that is, the path in three-dimensional space that the chain traces out.

Protein folding patterns

Focusing on the backbone, in the native state the polypeptide chain follows a curve in space. The general spatial layout of this curve defines a **folding pattern**. Many proteins have similar folding patterns: there is great but not infinite variety. Native states are selected from a large but finite repertoire.

In describing protein structures, the Danish protein chemist K.U. Linderstrøm-Lang defined a hierarchy of levels of protein structure. The amino-acid sequence—roughly the set of chemical bonds—is called the **primary structure**. The assignment of helices and sheets—the hydrogen-bonding pattern of the mainchain—is called the **secondary structure**. The spatial disposition and interactions of the helices and sheets is called the **tertiary structure**. For proteins composed of more than one subunit, J.D. Bernal called the assembly of the monomers the **quaternary structure** (see Box 12.4 and Figure 12.4).

KEY POINT

Helices and sheets are recurrent structures, stabilized by mainchain hydrogen bonding, that appear in many protein structures.

Some proteins change their quaternary structure as part of a regulatory process. Cyclic AMP activates protein kinase A by a mechanism involving subunit dissociation. The resting, inactive form of protein kinase A is a tetramer of two catalytic subunits and two regulatory subunits. In this resting state, the regulatory subunits inhibit the activity of the catalytic subunits. Binding of cyclic AMP to protein kinase A

The Sasisekharan-Ramakrishnan-Ramachandran diagram

The mainchain conformation of each residue is determined primarily by the two angles ϕ and ψ, assuming the common *trans* conformation of the peptide bond, $\omega = 180°$.

For some combinations of ϕ and ψ, atoms would collide, a physical impossibility. V. Sasisekharan, C. Ramakrishnan, and G.N. Ramachandran first plotted the sterically allowed regions (see Figure 12.3). There are two main allowed regions, one around $\phi = -57°, \psi = -47°$ (denoted α_R) and the other around $\phi = -125°, \psi = +125°$ (denoted β) with a 'neck' between them. The mirror image of the α_R conformation, denoted α_L, is allowed for glycine residues only. (As glycine is achiral—identical to its mirror image—a Ramachandran plot specialized to glycine must be right-left symmetric. For non-glycine residues, collisions of the Cβ atom forbid the α_L conformation.)

The two major allowed conformations of the mainchain, α_R and β, correspond to the two major types of secondary structure: **α-helix** and **β-sheet** (see Box 12.4). The α-helix is right-handed, like the threads of an ordinary bolt. In the β region, the chain is nearly fully extended.

A graph showing the ϕ and ψ angles for the residues of a protein against the background of the allowed regions is a **Sasisekharan-Ramakrishnan-Ramachandran diagram**, often called a Ramachandran plot for short.

It is no coincidence that the same conformations that correspond to low-energy states of individual residues also permit the formation of structures with extensive mainchain hydrogen bonding. The two effects thereby cooperate to lower the energy of the native state.

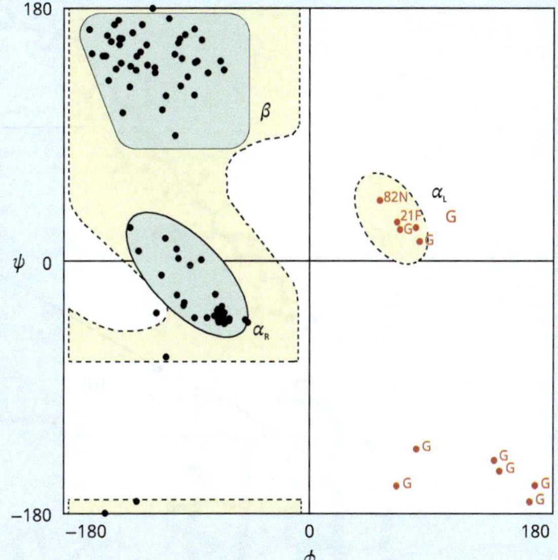

Figure 12.3 A Sasisekharan-Ramakrishnan-Ramachandran plot of bovine acylphosphatase [2ACY]. Sterically most-favourable regions are shown in green and sterically allowed regions in yellow. Most residues map to the β and α_L regions. Residues with $\phi > 0$, mostly glycines, appear in red.

Helices and sheets

Underlying the great variety of protein folding patterns are some recurrent structural themes. Helices and sheets are two conformations of the polypeptide chain that appear in many proteins. They satisfy the hydrogen bonding potential of the mainchain N-H and C=O groups, while keeping the mainchain in an unstrained conformation. They thereby solve certain structural problems faced by *all* globular proteins. They present general solutions—where in this context 'general' means 'compatible with all (or at least almost all) amino-acid sequences'.

Helices and sheets are like Lego® pieces, standard units of structure of which many proteins are built and which can be put together in different ways. Helices are formed from a single consecutive set of residues in the amino-acid sequence. They are therefore the structure of a contiguous region of the polypeptide chain, i.e. they form from a set of residues consecutive in the sequence. The mainchain hydrogen-bonding pattern of an α-helix, the most common type of helix, links the C=O group of residue *i* to the H-N group of residue *i* + 4 (see Figure 12.4, upper right).

Sheets form by lateral interactions of several independent sets of residues to create a hydrogen-bonded network that is often nearly flat, but sometimes cylindrical (forming a **β-barrel** structure). Unlike helices, sheets need not form from consecutive regions of the chain but may bring together sections of the chain separated widely in the sequence (Figure 12.4, upper right, below picture of helix).

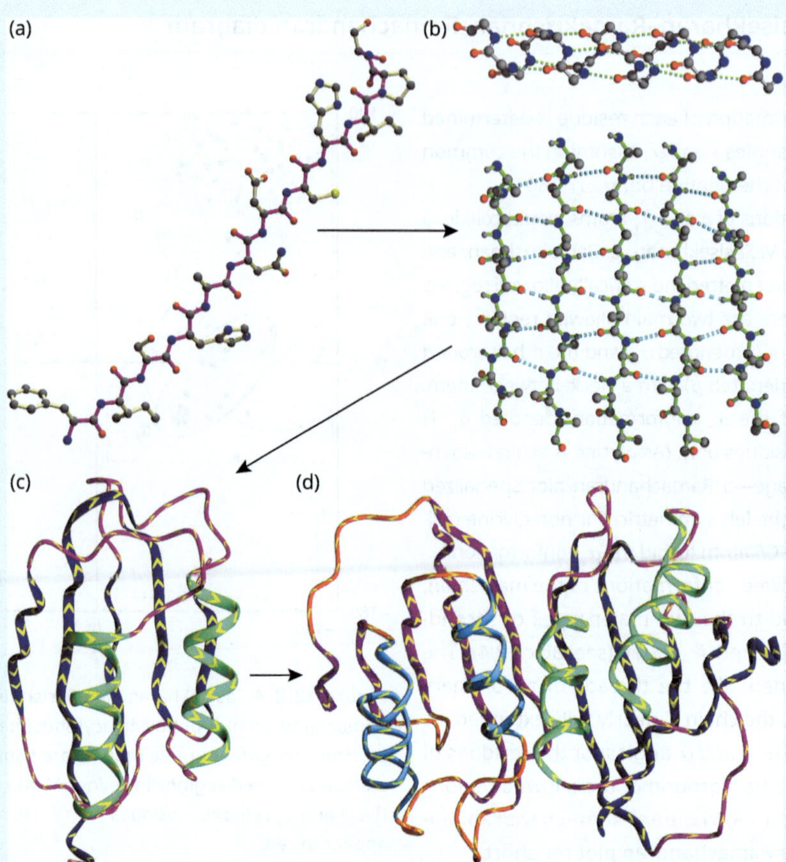

Figure 12.4 Underlying the great variety of protein folding patterns are a number of common structural features. For instance, α-helices and β-sheets are standard elements of the 'parts list' of many protein structures. α-helices and β-sheets were modelled by L. Pauling before their experimental observation. Pauling recognized that helices and sheets provide convenient ways for the residues to achieve comfortable steric relationships and satisfy the requirements for backbone hydrogen bonding in an (almost) sequence-independent manner.

This figure shows: (a) The primary structure in terms of a simple extended chain. (b) The standard secondary structures, the α-helix and β-sheet, with hydrogen bonds indicated by broken lines. (c) Tertiary structure is represented by a monomeric acylphosphatase from the cow, which contains two α-helices packed against a five-stranded β-sheet. (d) Quaternary structure: A dimeric acylphosphatase from *Pyrococcus horikoshii.* The five-stranded sheets in the monomers link together to form a ten-stranded sheet in the dimer.

dissociates the tetramer, releasing individual catalytic subunits in active form.

In some cases, evolution can merge proteins—changing quaternary to tertiary structure. For example, five separate enzymes in *Escherichia coli* that catalyse successive steps in the pathway of biosynthesis of aromatic amino acids correspond to five regions of a single protein in the fungus *Aspergillus nidulans*. (In rare cases, multiple proteins in eukaryotes correspond to a single protein in prokaryotes.)

KEY POINT

We describe protein folding patterns according to a hierarchy of primary, secondary, tertiary, and quaternary structures. See Box 12.4 and Figure 12.4.

Supersecondary structures

In addition to the common secondary structures, α-helices and β-sheets, many proteins show common patterns of interaction between sets of α-helices and strands of β-sheet that appear (a) consecutively in the sequence, and (b) close together in the three-dimensional structure. Supersecondary structures are local structures, formed by residues in a contiguous segment of the sequence, that contain two or more α-helices and/or strands of β-sheet, packed against each other to form an internal local compact unit. Common supersecondary structures include the α-helix hairpin, the β-hairpin, and the β-α-β unit (see Figure 12.5). Supersecondary structures appear, in the hierarchy of protein architecture, between

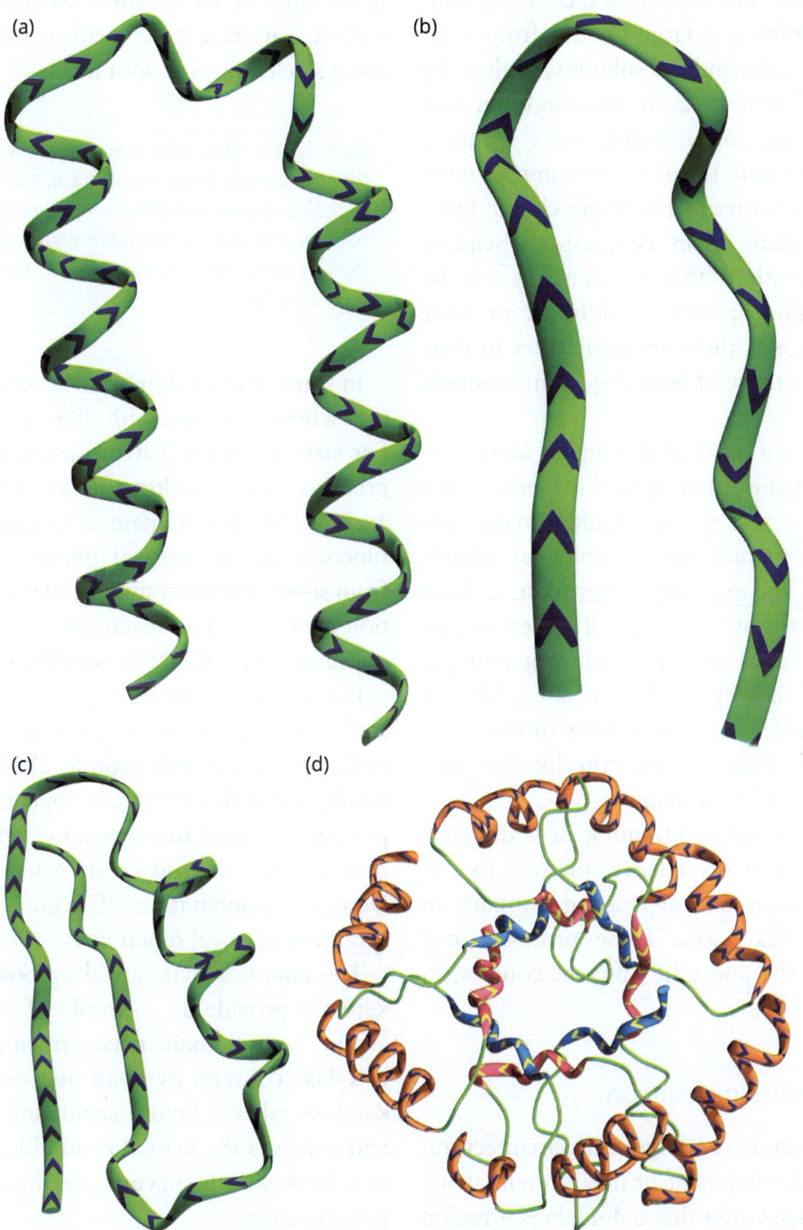

Figure 12.5 Common supersecondary structures: (a) α-helix hairpin, (b) β-hairpin, (c) a β-α-β unit. (d) A combination of individual supersecondary structures forms a complete structure. Spinach glycolate oxidase is an example of a very common β-barrel structure (a β-α barrel) formed from a succession of β-α-β units. The eight-stranded sheet is closed into a cylinder. Helices are shown in orange, with blue chevrons. Alternate strands in the β-barrel are shown in red and blue, with yellow chevrons. (Eight helices appear on the outside of the figure; eight strands forming a cylinder appear near the centre.)

secondary and tertiary structures. Such small structural patterns that recur in many proteins are also known as motifs.

Box 12.6 summarizes the terms used to describe protein structure.

Domains

One way that proteins have evolved increasing complexity is by assembling a large protein from a set of smaller quasi-independent subunits, either by forming stable oligomers, as in haemoglobin (see Figure 1.10a), or by concatenating units within a *single* polypeptide chain. **Domains** are compact units within the folding pattern of a single chain. Justifications for regarding them as quasi-independent include the observations that (1) domains can be recombined in different ways in different proteins and (2) in many cases, there are similarities in their folding patterns to those of homologous monomeric proteins.

Domains form the basis of the higher level protein structural organization typical of eukaryotic proteins. **Modular proteins** are multidomain proteins that often contain many copies of closely related domains. For example, fibronectin, a large extracellular protein involved in cell adhesion and migration, contains 29 domains including multiple tandem repeats of three types of domain, F1, F2, and F3 (see Figure 6.10). It is a linear array of the form: $(F1)_6(F2)_2(F1)_3(F3)_{15}(F1)_3$. Fibronectin domains also appear in other modular proteins.

To create new proteins, inventing new domains is an unusual event. It is far more common to create different combinations of existing domains in increasingly complex ways. These processes can occur independently, and take different courses, in different phyla.

Domain reassembly in evolution

Domain recombination is a large-scale, architectural approach to the development of new protein structures and functions. Given that a domain is a region in a protein that would have independent stability on its own, proteins are free to 'mix and match' different combinations of domains.

For example, several proteins active in blood coagulation arise as different combinations of a set of domains (see Figure 12.6). Each domain is a relatively small compact unit in its own right. The domain composition and the order of domains along the polypeptide chain differ among these proteins.

In many cases, variable splicing can contribute to domain recombination. Conversely, mutations in splice sites or termination codons, or deletions in mRNA, can have drastic effects on proteins, almost always deleterious if not fatal.

> Comparisons of protein sequences and structures confirm that the domain is an important unit of protein evolution. Domains appear in different proteins in different combinations. Thereby, from a relatively small roster of domain families, evolution can assemble a large number of complete proteins.

In some cases, a domain will retain the same function when associated with different partner domains. For instance, many dehydrogenases are multidomain proteins (some multimeric as well) that combine an NAD-binding domain of common structure and function with a range of partner catalytic domains from seven different families, that vary with the reaction catalysed. The reactions catalysed are coupled to reduction of NAD^+ or $NADP^+$.

In contrast, many examples are known in which a change in partners, or even a change in domain order along the polypeptide chain, can create or modify catalytic activity. It appears much easier for protein evolution to adapt an existing structure to a new function than to create a new folding pattern. Domain recombination offers great opportunities for evolution of novel functions.

The enzymes pyruvate decarboxylase and transketolase provide an example of functional change arising from domain reassortment. Pyruvate decarboxylase converts pyruvate to acetaldehyde. Transketolase takes a ketose sugar and an aldose sugar, and converts the ketose to an aldose and the aldose to a ketose. Both enzymes use the cofactor thiamine pyrophosphate.

Structurally, both pyruvate decarboxylase and transketolase contain three domains (see Figure 12.7). They share two of the three domains, but the

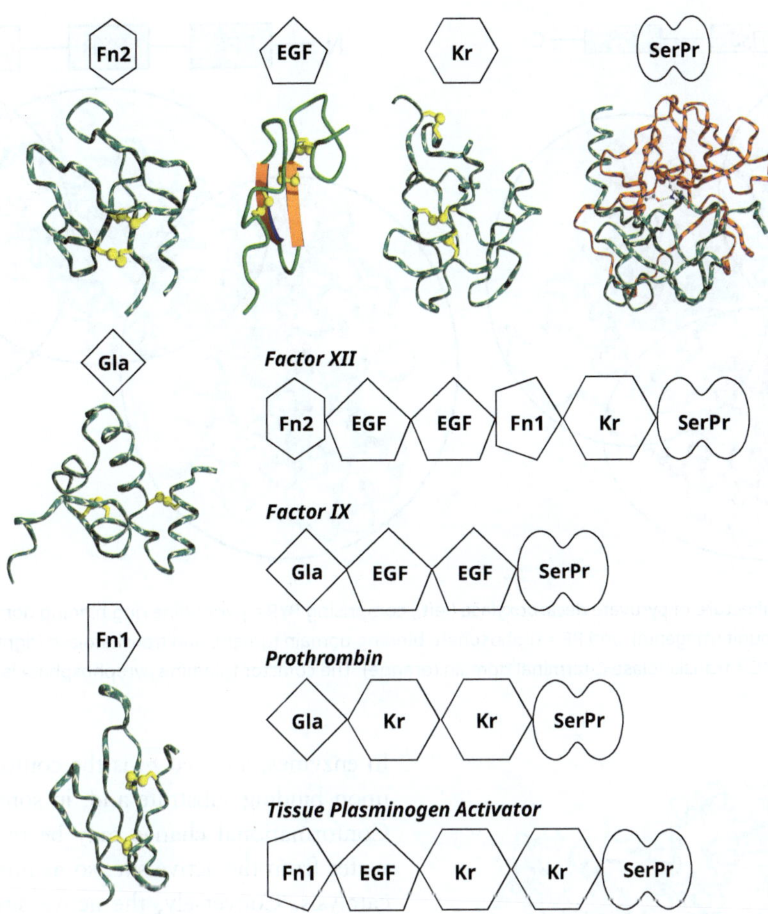

Figure 12.6 Several proteins involved in blood coagulation show structures based on the same domains, recombined in different ways. The figure illustrates the structures of six domains, and an icon symbolizing them: SerPr = serine protease, Kr = kringle, EGF = epidermal growth factor, Fn2 = fibronectin 2, Gla = γ-carboxyglutamic acid-rich domain, Fn1 = fibronectin 1. The figure shows choice and order of these domains in the proteins: Factor XII, Factor IX, Prothrombin, and Tissue Plasminogen Activator.

Note that all contain a serine protease domain at the C-terminus: Blood coagulation involves a cascade of proteolytic activations.

domains appear in different orders along the polypeptide chains. Nevertheless, the interface between the PYR and PP domains is preserved between the two structures (see Figure 12.8). The active site is formed from residues in these two domains.

> Proteins fairly freely 'mix and match', and reassemble domains, with a variety of functional consequences:
>
> - In some cases, such as the NAD-binding domains of dehydrogenases, the domain retains its function when associated with different partner domains.
> - In other cases, entirely novel function emerges.
> - In the case of pyruvate decarboxylase, substantial domain rearrangement within the sequence is consistent with retention of a cofactor-binding site in the structure.

Disorder in proteins

Crystal structures portray protein structures in fixed native conformations. There is no choice, really: Any mobile structures or substructures are invisible in the electron-density maps derivable from X-ray diffraction. This made for a very simple and attractive picture: The fully disordered denatured state *v.* the native state with a unique conformation. This was in accord with Fischer's authoritative 'lock-and-key' model of enzyme-substrate specificity; ignored for a century (by almost everyone) was the unquestioned fact that locks contain mechanisms that move when keys are inserted and turned.

And yet there was evidence that this picture was overdrawn. There was suspicion that crystal structures

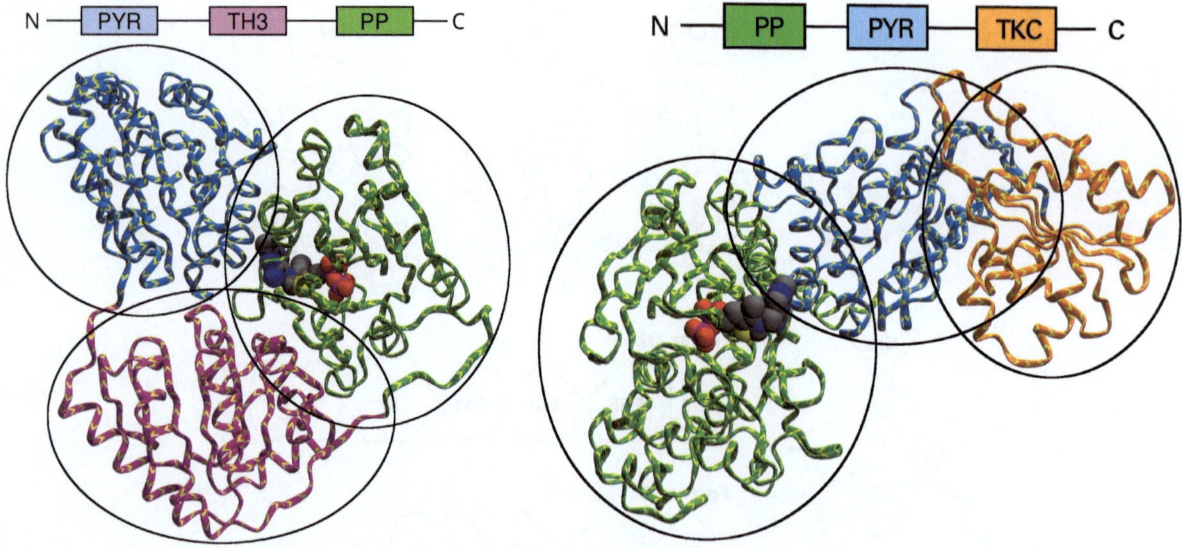

Figure 12.7 Domain architecture of pyruvate decarboxylase (left), comprising PYR = pyrimidine ring binding domain (blue), TH3 = transhydrogenase dIII subunit (magenta), and PP = diphosphate binding domain (green); and transketolase (right), comprising PP (green), PYR (blue), and TKC = transketolase C-terminal domain (orange). The cofactor thiamine pyrophosphate is shown in a shaded-sphere representation.

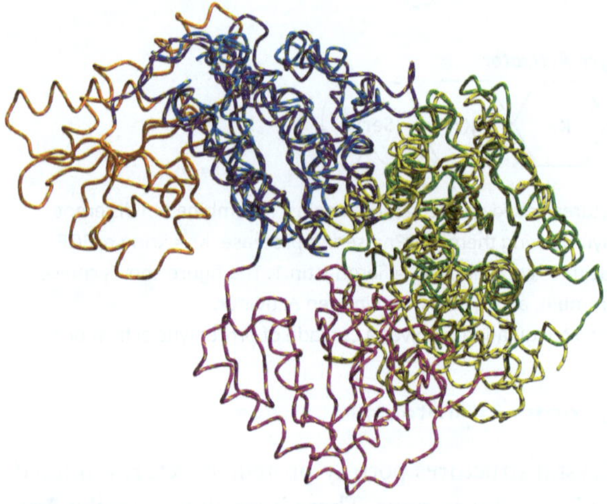

Figure 12.8 Pyruvate decarboxylase and transketolase, superposed on PYR and PP domains. Despite the difference in overall domain architecture in these two proteins, the geometric relationship between these two domains is preserved. The colours of the domains of pyruvate decarboxylase are the same as in the preceding figure: PYR blue, TH3 magenta, PP green. But in this figure the transketolase domains appear thus: PP yellow, PYR purple, TKC remains orange. The colours of the PP and PYR domains of transketolase have been changed in order to distinguish them from the superposed domains of pyruvate decarboxylase.

might be statues, capturing only one pose of a dynamic and active subject. Thus, crystal structures of proteins in different states of ligation gave different structures. Oxy- and deoxyhaemoglobin were the classic example.

In enzymes, induced fit is the conformational change upon binding substrate and, in some cases, cofactor. Conformational change may be required to exclude water from the active site, so as not to interfere with catalysis. Conversely, the active site must obviously remain solvent and solute accessible in the unligated state. Especially interesting examples of multiple conformers appear in icosahedral viruses. Most viruses assemble capsids from multiple copies of a unique coat protein. In many cases, the coat proteins have the same sequence but adopt different conformations.[1]

With the development of protein NMR spectroscopy, evidence mounted that proteins were more flexible than crystallography had suggested: (1) Measurements of genuine disorder, from NMR and other sources; and (2) discovery of more and more exceptions to the 'lock-and-key' model, in the form of molecules that showed less specificity in their binding and interactions.

In many cases, a portion—maybe almost all—of the protein is rigid, but some regions are disordered. Such regions can play a role in zymogen activation. Digestive enzymes such as trypsin and chymotrypsin are synthesized in an inactive form containing

[1] Why viruses do this is a very interesting story; see Lesk, A.M. (2015). *Introduction to Protein Science: Architecture, Function and Genetics*, Chapter 6. Oxford: Oxford University Press.

disordered regions. Cleavage of an N-terminal oligopeptide by enterokinase causes a conformational change in which interactions with the new N-terminus fix the active structure of the catalytic site.

The idea that has emerged is that of a *spectrum* of protein flexibility. At one end is the classic unique native state, with a rigid binding site and high specificity. The lock and key model applies. In the direction of greater flexibility, one encounters states in which ligation induces a fixed conformation from a more-disordered unligated state. (Zymogen activation and induced fit are examples.) This is compatible with ligand specificity, but also with conformational selection in which different ligands induce different bound-state conformations. The unligated state can be close to the bound conformation, but, alternatively, could be disordered (see Figure 12.9).

In most of these cases, the picture is of flexibility (and possible promiscuity) of the unbound state but rigidity (and possible multiple rigid conformations) of the ligated state (see Box 12.5).

Moving even further away from rigidity is the 'fuzzy complex' in which even the conformation of the ligated state is not rigid. Such a complex may pay a price—in terms of affinity—relative to a rigid complex, but functional reasons may require a molecule capable of less specific and weaker binding. An example is the complex between the cyclin-dependent kinase inhibitor Sic1 (stoichiometric inhibitor of Cdk1-Clb) and the SCF subunit of Cdc4 (Cdc4 = cell-division control protein-4; SCF = a ubiquitin ligase (E3) complex: the Skp1-Cdc53/CUL-F

BOX 12.5 The distinction between flexibility and disorder

Flexibility and disorder are related concepts, but it is useful to distinguish them. Regions in proteins can show two, or more, stable mainchain conformations. If this set of conformations is discrete and small, let us call such regions *flexible*. In other cases, regions in a protein may be entirely *disordered*—there appear to be no preferred conformations, at least no stable ones. They are just 'flapping in the breeze'.

A disordered region can become flexible; for instance, it may adopt different, fixed, conformations, when the protein binds two different ligands. But a flexible region may not necessarily be disordered, short of complete denaturation of the protein.

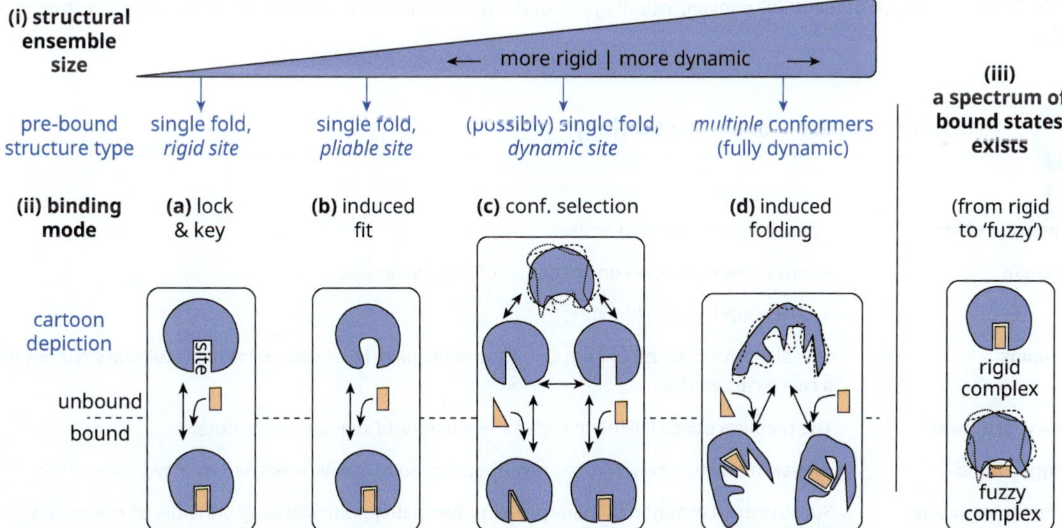

Figure 12.9 (i) The spectrum of protein native conformations, from the classical unique, rigid fold at left, through states of increasing flexibility and dynamics on right. The increase in ensemble size indicates that the greater the flexibility, the larger the number of individual microscopic conformations that the system can adopt. (ii) The corresponding nature of ligation, from the classical 'lock-and-key' model at left, to ligation-induced selection of conformational state at right, allowing for the possibility of reduced specificity of binding and ligand-dependent conformations. (iii) The alternatives of rigid ('lock + key') binding and 'fuzzy' binding, which gives up our idea of fixed complementarity between protein and ligand. Although this figure is drawn in terms of a large protein binding a relatively small ligand, similar considerations can apply to protein–protein interactions.

From: Mannige, R.V. (2014). Dynamic new world: refining our view of protein structure, function and evolution. Proteomes 2, 128–153.

Box receptor E3 complex). Sic1 is intrinsically disordered in both the unligated and ligated states.

In some cases, domains even change *structure* when paired with different partners. The Nuclear Coactivator Binding Domain (NCBD) of the creb-binding protein is disordered in the unligated state. It becomes ordered, but forms *different* structures in complex with the Activator for Thyroid Hormone, and with Retinoid Receptor domain of p160 and with Interferon Regulatory Factor 3 (see Figure 12.10). The structure induced depends on the ligand.

Finally, there are some proteins that never adopt a fixed native state at all. Casein, a milk protein that primarily serves to nourish babies, is an example. (It also inhibits the precipitation of calcium phosphate that is formally supersaturated in milk.) Casein neither needs nor possesses a typical native structure.

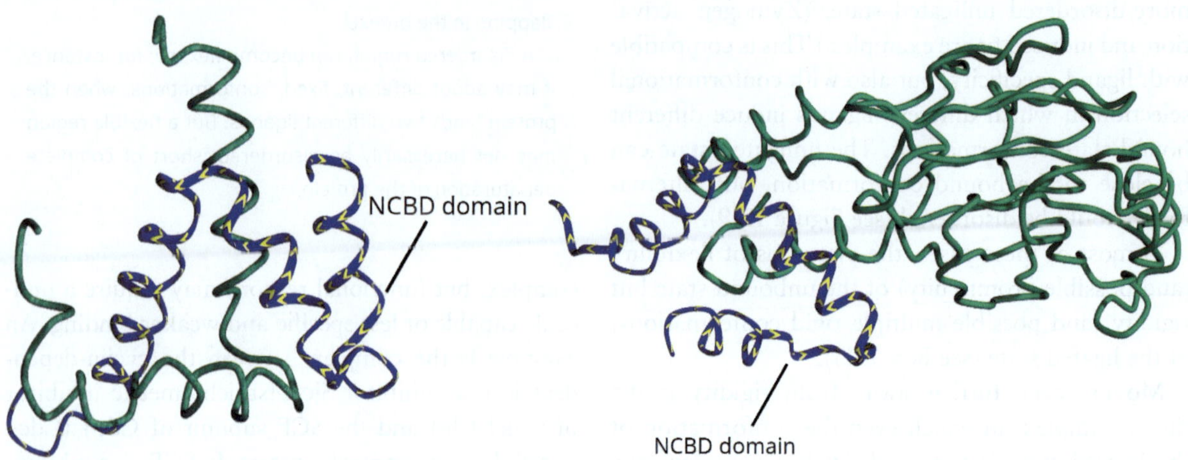

Figure 12.10 The NCBD (IbID) domain of the creb-binding protein (CBP) is disordered in the unligated state. It forms different structures in complex with (left) the ACTR domain of p160 and (right) with IRF3. In both pictures, the common NCBD domain is shown in blue. The C-terminal region of the NCBD domain, including two of the three helices, has approximately the same structure in both complexes.

BOX 12.6

Protein structure—basic vocabulary

Polypeptide chain	Linear polymer of amino acids.
Mainchain	Atoms of the repetitive concatenation of peptide groups . . . N-Cα-(C=O)N-Cα-(C=O) . . .
Sidechains	Sets of atoms attached to each Cα of the mainchain. Most sidechains in proteins are chosen from a canonical set of 20.
Primary structure	The chemical bonds linking atoms in the amino-acid sequence in a protein.
Hydrogen bond	A weak interaction between two neighbouring polar atoms, mediated by a hydrogen atom.
Secondary structure	Substructures common to many proteins, compatible with sterically unhindered mainchain conformations, free of interatomic collisions, and stabilized by hydrogen bonds between mainchain atoms. Secondary structures are compatible with all amino acids, except that a proline necessarily disrupts the hydrogen-bonding pattern.
α-Helix	Type of secondary structure in which the chain winds into a helix, with hydrogen bonds between residues separated by four positions in the sequence.
β-Sheet	Another type of secondary structure, in which sections of mainchain interact by lateral hydrogen bonding.

Folding pattern	Layout of the chain as a curve through space.
Tertiary structure	The spatial assembly of the helices and sheets, and the pattern of interactions between them. (Folding pattern and tertiary structure are nearly synonymous terms.)
Quaternary structure	The assembly of multisubunit proteins from two or more monomers.
Native state	The biologically active form of a protein, which is compact and low energy. Under suitable conditions, proteins form native states spontaneously.
Denaturant	A chemical that tends to disrupt the native state of a protein; for instance, urea.
Denatured state	Non-compact, structurally heterogeneous state formed by proteins under conditions of high temperature, or high concentrations of denaturant.
Post-translational modification	Chemical change in a protein after its creation by the normal protein-synthesizing machinery.
Disulphide bridge	Covalent sulphur–sulphur bond between two cysteine sidechains. A simple example of a post-translational modification.

Post-translational modifications

How much does genomics actually tell us about the proteome? Even if we could identify coding regions of genomes with complete accuracy, we would not know about:

- levels of transcription—or even absence of transcription;
- formation of different splice variants (in eukaryotes);
- mRNA editing—exclusive of splicing—before translation, which alters the amino-acid sequence;
- the nature and binding sites of ligands integral to the final structure; and
- post-translational modifications, the subject of this section.

The ribosome synthesizes proteins by using the genetic code to direct the incorporation of a sequence of amino acids chosen from the canonical 20. Selenomethionine and pyrrolysine are two natural rare extensions of the standard genetic code.

Nevertheless, the protein world is richer than the standard genetic code suggests. Many proteins contain ligands, such as metal ions or small organic molecules, as intrinsic and permanent parts of the structures. The nature of the binding between protein and ligand depends on the protein as well as the ligand. For instance, the haem group is bound covalently to cytochrome *c* but non-covalently to (almost all) globins. (Of course, proteins bind many molecules *transiently*. Enzyme–substrate complexes provide many examples.)

Post-translational modifications can take several forms (see also Box 12.7).

- Attaching various groups to sidechains, including but not limited to acetate, phosphate, lipids, and carbohydrates. Disulphide bridge formation is a related example. Some additions, such as sulphation, are permanent modifications; others, notably phosphorylations, are in many cases reversible, and serve as on-off 'toggles' regulating activity.
- Conversions, for instance deamidation of asparagine (or glutamine) to aspartic acid (glutamic acid), or deimination of arginine to citrulline.
- Removing peptides, either from a terminus or from the middle of the chain, and in a few cases even making cyclic permutations.
- Addition of other peptides or proteins, not always by extension of the mainchain, through peptide linkages.

BOX 12.7 Major types of post-translational modification

- Attachments of groups to termini and sidechains. Although acetylation of protein N-termini is not uncommon, most modifications involve sidechains. Many possible derivatives are observed:
  - Reversible phosphorylation of serine, threonine, or tyrosine sidechains is a very common means of regulating protein activity. However, *irreversible phosphorylation* of tau protein contributes to the development of Alzheimer's disease. The neurotoxicity of some organophosphorus compounds arises from their irreversible phosphorylation of acetylcholinesterase.
  - Attachment of sugars or oligosaccharides to proteins to make glycoproteins. In mammals, many glycoproteins appear on cell surfaces, to mediate cell–cell recognition and communication and immune system recognition. The difference between O, A, and B blood groups resides in the carbohydrate attached to serum glycoproteins and to (non-protein) glycolipids on cell surfaces. Lectins are carbohydrate recognition proteins that mediate cell–cell recognition and communication and sugar transport. In vertebrates, recognition of sugars on the surfaces of bacteria is a component of the immune response to infection. Many viruses, including influenza and HIV-1, gain entry to cells via cell-surface glycoprotein receptors.

    Deficiencies in turnover lead to glycoprotein storage diseases, involving accumulation of incompletely degraded glycoproteins or oligosaccharides. Examples include α– and β–mannosidosis and aspartylglucosaminuria. Tay–Sachs disease is a related condition, part of a larger family of diseases called the lysosomal storage diseases. Tay–Sachs disease arises from a mutation in the α subunit of the hexosaminidase A gene. In Tay–Sachs disease, the dysfunction of the mutant protein impedes the degradation of a ganglioside, rather than the degradation of a glycoprotein.
  - Addition of oligomers of the small protein ubiquitin to lysine residues targets proteins for degradation by the **proteasome**. Conversely, the methylation of lysine is believed to 'protect' proteins against ubiquitinylation and thereby against degradation.
- Post-translational modification by proteolytic cleavage implies that the protein is synthesized with extra amino acids beyond those required to form the native state. What is the purpose of the additional residues?
  - Some proteins are synthesized with N-terminal signal peptides that direct their transport to particular subcellular compartments or organelles, or mark them for secretion.
  - It would be dangerous to turn unregulated—rogue—proteases loose on the cells in which they are synthesized. Many proteases are synthesized in inactive forms and then activated by cleavage. (The proteins of the blood-coagulation cascade, Figure 12.6, are examples.)
  - Facilitation of folding. Insulin contains two polypeptide chains, one of 21 residues and the other of 30 residues. The precursor proinsulin is a single 81-residue polypeptide chain from which *excision of an internal peptide* produces the mature protein. Insulin contains one intrachain and two interchain disulphide bridges. Attempts to renature mature insulin—after unfolding and breaking the disulphide bridges—give poor yields. Many incorrectly paired disulphide bridges form. *In vivo*, the precursor proinsulin folds into a three-dimensional structure with the cysteines in proper relative positions to form the correct disulphide bridges. Excision of a central region by endopeptidases then produces the mature dimer. Unfolded proinsulin *can* spontaneously refold correctly.
- Most post-translational cleavage reactions are carried out by proteases. Alternatively, **inteins** are proteins that have a 'self-splicing' activity. They autocatalytically excise internal peptides and join the ends. (In contrast, peptide excision from proinsulin leaves two chains that are *not* joined by a peptide bond.)
- The lectin concanavalin A is synthesized in a precursor form that is a cyclic permutation of the final structure. Thus, during maturation of the protein, there is cleavage of an internal peptide bond and formation of a new peptide bond between the original N and C termini. For concanavalin A, the DNA sequence of the gene is not co-linear with the amino-acid sequence of the mature protein.

Why is there a common genetic code with 20 canonical amino acids?

Almost all organisms synthesize proteins containing a canonical set of 20 amino acids.

However, both nature and the laboratory show that 20 amino acids are not a fundamental limitation. Selenomethionine and pyrrolysine are natural exceptions. P. Schultz and co-workers have extended the genetic code by introducing modified tRNAs and synthetases into *E. coli*, yeast, and even mammalian cells in tissue culture.[2] The basic idea was to recode one or two STOP codons to specify an amino acid. Over 200 novel amino acids are now available to be introduced into proteins at specific sites. No more than two novel amino acids can be introduced into a protein, but these can be positioned anywhere in the sequence. Some of the novel amino acids show designed steric or electronic properties; others contain chromophores as fluorescent reporters or are susceptible to photocross-linking; there are glycosylated amino acids; iodine derivatives to facilitate X-ray structure determination; and sidechains containing other types of reactive group.

In understanding the contents and layout of the common genetic code, can we go beyond F. Crick's comment that the code is a 'frozen accident'?

There is now consensus that prokaryotes were and are engaged in widespread horizontal gene transfer (see Chapter 7). This suggests leaving aside the question of what the optimal genetic code should be—that it would be to the advantage of any participating species to conform to *some* standard, for that would give it access to everyone else's genes. Analogously, anyone can run any operating system on a computer that they want, but the obvious advantages of running the same system as many other people exert pressure to conform to *some* standard. Perhaps that is at least a partial explanation of why almost all species have the same genetic code.

It is true that if different species adopted different genetic codes, this might protect them against viruses jumping from other species. The idea has been applied to create a strain of *E. coli* with modified genetic codes, which do in fact resist viral infections and gene transfers.

But why not a code with many more than 20 amino acids? Certainly, one perfectly feasible way to introduce greater versatility into the components of proteins is by expanding the genetic code. However, keeping within the general framework of a triplet code, introducing more amino acids at the expense of the redundancy of the code threatens to reduce robustness to mutation. An alternative approach to greater versatility without this cost is to effect post-translational modifications of individual amino acids. Whether or not this reasoning is the correct explanation, post-translational modification is the choice that nature seems largely to have made.

Protein purification

The complete complement of a cell's proteins is a large and complex mixture of molecules. Metazoa contain tens of thousands of protein-encoding genes. Different splice variants multiply the number of possible proteins. Vertebrate immune systems generate billions of molecules by specialized techniques of combinatorial gene assembly.

To give some idea of the 'dynamic range' required of detection techniques, the protein inventory of a yeast cell varies from 1 copy per cell to 1 million copies per cell.

Suppose one starts with a heterogeneous 'soup' of proteins. Each step in a protocol to isolate a particular target protein will involve choosing some physical property, and separating the molecules according to the value of that property. This will bring together the target protein and other proteins from the mixture that share the value of the selected property. For example, selecting molecules on the basis of molecular volume will enrich the mixture in the target protein but retain also others that have similar sizes. Isolating the target

[2] See http://schultz.scripps.edu/research.html and Wang, L., & Schultz, P.G. (2004). Expanding the genetic code. Angew. Chem. Int. Ed. Engl. 44: 34–66.

Table 12.1 Techniques for purification of proteins from a mixture

Property	Experimental method
Solubility	ammonium sulphate precipitation
Size/Shape	size-exclusion chromatography
Charge	ion-exchange chromatography
Size/Isoelectric point	two-dimensional SDS polyacrylamide gel electrophoresis
Specific binding	affinity chromatography

will usually require several consecutive steps, using, at each step, a different physical property, and a different method of selection (Table 12.1).[3]

Isolation of a particular protein from a mixture is a multi-step process, each step segregating components of a mixture with a particular value of some physical property, for instance size or charge. After several steps the remaining sample will retain only proteins with specified values of many different physical properties. The goal is that the overall procedure produces a sample containing a unique protein.

Ammonium sulphate precipitation

Ammonium sulphate precipitation is a method for purifying proteins based on differences in their variation in solubility as a function of the concentration of dissolved salts.

Protein solubility requires that proteins do NOT interact strongly with one another: Protein–protein interactions can lead to aggregation and precipitation. Charges on proteins mediate interactions: Protein surfaces bristle with charges, and a negatively charged sidechain on the surface of one protein will attract a positively charged sidechain on the surface of another. In very dilute salt solution, this leads to aggregation and precipitation (see Figure 12.11a).

Conversely, proteins will be more soluble if they interact more strongly with solvent than with one another. In moderately concentrated salt solutions (Figure 12.11b), the ions from the salt cluster around the charged groups on the protein surface, and screen the electrostatic interactions. As a result, with increasing salt concentration, the solubility of the protein *increases,* called **salting in**.

However, salt ions *compete* with proteins for interaction with water. At high salt concentrations, the salt 'wins': the protein molecules dehydrate, aggregate, and precipitate, called **salting out** (Figure 12.11c).

Figure 12.12 shows the variation of solubility of a protein with the concentrations of different salts. (The **ionic strength** of a solution is measurement of salt concentration: $I = \frac{1}{2}\Sigma_i c_i z_i^2$, where c_i is the concentration of the ith type of ion, and z_i is the charge on the ion.)

In Figure 12.12, the curve corresponding to the salt commonly used in protein purification, ammonium sulphate—$(NH_4)_2SO_4$—is shown in red. Although the general picture is the same for different salts, different ions have different solvation energies and affect protein solubility in individual ways.

The application to protein purification implements these observations. If the salt concentration in a solution of a protein is increased, typically the protein will precipitate when the salt concentration reaches a threshold. The threshold concentration for precipitation differs between different proteins, making it the basis of a useful technique for protein purification. Traditionally, ammonium sulphate is the salt used.

Suppose your target protein will precipitate at an ammonium sulphate concentration of 3 M. Then the basic procedure is to raise the salt concentration in your solution to—perhaps—2.5 M. Then some proteins will have precipitated but the target has remained soluble. Remove the insoluble material, by centrifugation or filtration, and retain the supernatant (see Figure 12.13a). Then raise the salt concentration to a value *above* the threshold—perhaps 3.5 M (Figure 12.13b)—and again separate precipitate and supernatant. This time discard the supernatant (Figure 12.13c), and redissolve the precipitate (Figure 12.13d). This will produce a new solution enriched in the target protein.

[3] Occasionally, during a lecture, I have carried out the following 'experiment'. First, I ask all male students to step to the left side of the room, and females to step to the right side. Then I ask all students wearing clothing showing the logo of the university to step to the front of the room, and the others to move to the back of the room. Now there are four groups. By a succession of such steps it would be possible to separate out a single individual.

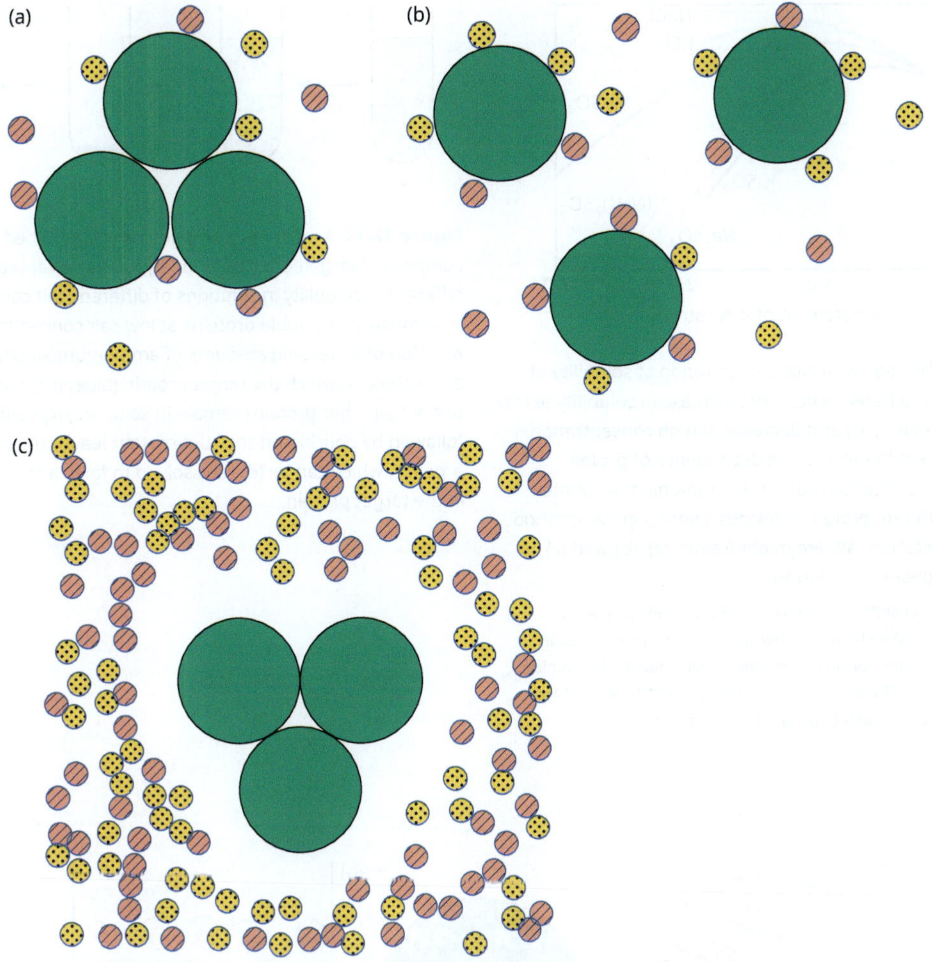

Figure 12.11 Protein solubility is determined by a competition between protein–protein interactions and protein–solvent interactions. In these figures, large green discs represent protein molecules, and small discs represent ions arising from dissociation of salt, perhaps NH_4^+ (red) or SO_4^{2-} (yellow). (a) Very dilute salt concentration. Insufficient concentration of salt ions to shield protein charges; protein molecules attract and aggregate. (b) Moderate salt concentration: salt ions screen charges, protein molecules soluble. (c) High salt concentration: water tied up by salt ions, protein molecules dehydrate, aggregate, and precipitate.

Size-exclusion chromatography

It is possible to create beads with cavities of uniform volume. Packing a chromatography column with these beads, and running a mixture through it, will result in a separation of molecules with volumes *smaller* than the cavity, from molecules with volumes *larger* than the cavity (see Figure 12.14a). Large molecules will not enter the crevices in the gel; they will pass unimpeded through the column. Small molecules will take 'detours' into the crevices, and their passage through the column will be retarded (Figure 12.14b). Collection of successive fractions from the column will allow separation of larger from smaller molecules.

Because inorganic salts are much smaller than proteins, size-exclusion chromatography is a very convenient way to 'desalt' a protein solution.

An analogy I have used in lectures: suppose a mixture of students—some drinkers and some non-drinkers—are walking down a street. Whenever the group passes a pub, the drinkers enter and have a beer; the non-drinkers just keep walking. The drinkers and non-drinkers will be separated.

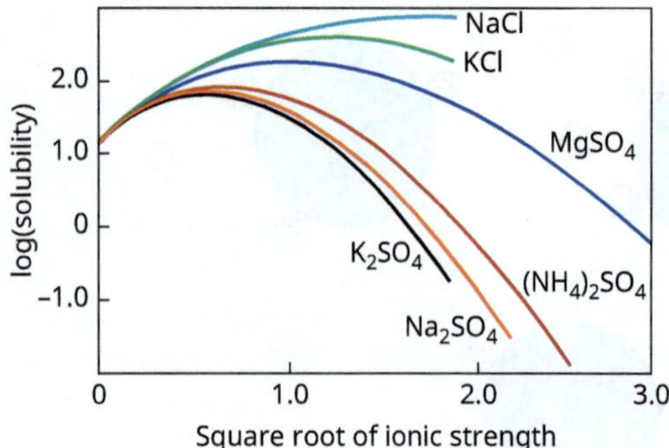

Figure 12.12 Variation with salt concentration of solubility of haemoglobin, for different salts. Note increase in solubility at low concentration ('salting in') and decrease at high concentration ('salting out'). The difference in the dependence of protein solubility in different concentrations of ammonium sulphate $(NH_4)_2SO_4$ for different proteins provides a very popular method for protein purification. (Where might frames (a), (b), and (c) of Figure 12.11 appear in this figure?)

Used with permission of Royal Society of Chemistry from Salis, A., & Ninham, B.W. (2014). Models and mechanisms of Hofmeister effects in electrolyte solutions, and colloid and protein systems revisited. Chemical Society Reviews 43, 21: 7358–7377. https://doi.org/10.1039/c4cs00144c. Permission conveyed through Copyright Clearance Centre, Inc.

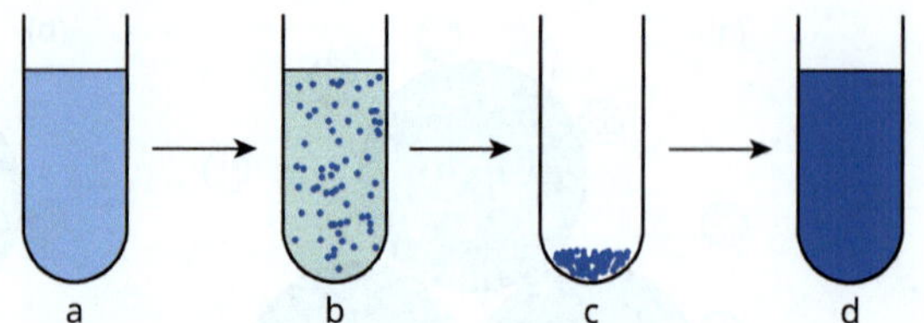

Figure 12.13 A mixture of proteins can be enriched in a component targeted for isolation by taking advantage of differential solubility in solutions of different salt concentration. (a) A mixture of soluble proteins at low salt concentration. (b) Addition of increasing amounts of ammonium sulphate reaches a threshold at which the target protein (blue) precipitates. (Some but not all other proteins remain in solution.) (c) centrifuging followed by pouring off the supernatant leaves the precipitate in a 'pellet', which can be (d) redissolved to form a solution enriched in the target protein.

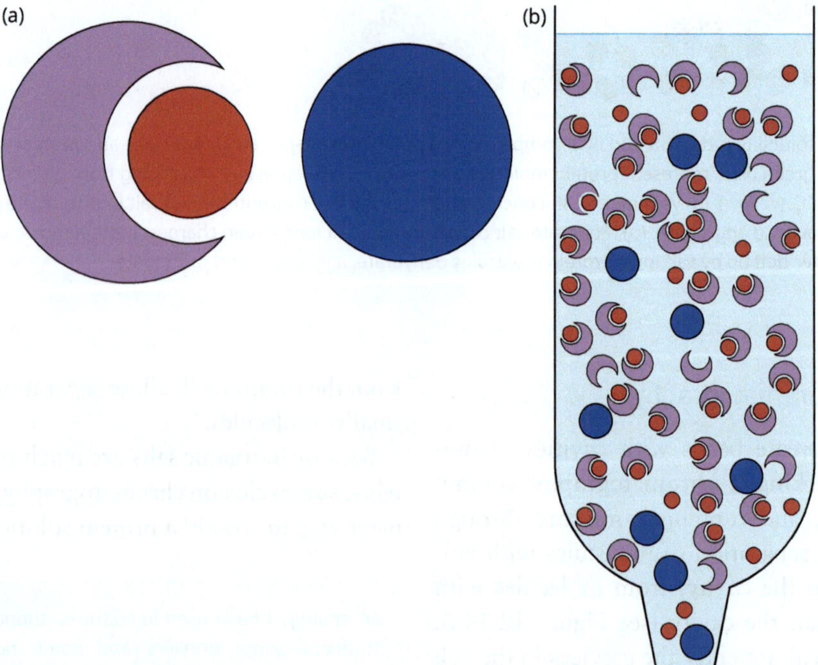

Figure 12.14 (a) A polysaccharide bead (purple) contains a crevice. Small molecules (red) can fit into the crevice; large molecules (blue) cannot. (b) If a column is packed with such beads, running a mixture of different-sized molecules through it will result in retardation of the passage of smaller molecules, and unimpeded passage of the large molecules. Therefore, the 'blue' molecules will emerge from the column before the 'red' molecules, achieving a separation on the basis of molecular size.

Ion-exchange chromatography

Different proteins have different charges. This depends on the amino-acid composition of surface-exposed residues. Is there a preponderance of negatively charged sidechains, or of positively charged sidechains? Pack a column with beads to which charged molecules are attached. If the beads are negatively charged, the passage through the column of positively charged proteins will be retarded. Conversely, if the beads are positively charged, passage of negatively charged proteins will be retarded. Either choice allows separation of molecules with different net charges.

Affinity chromatography

Suppose only one protein in a mixture binds to a particular ligand. Then a convenient way to purify the protein is to affix the ligand to a set of beads, pack a column with the beads, and run a mixture containing the target protein through the column. The target proteins will bind to the column—or at least be retarded in their passage through it. They are thus separated from other proteins that do not bind, and which will pass freely through the column. Subsequently, the bound proteins must be released from the column, and collected. This allows fairly clean separation of proteins with the particular binding specificity.

Examples include:

(a) The **His-tag** (also called 6xHis-tag) is one of the simplest and most widely used affinity purification tags, with six or more consecutive histidine residues. These residues readily coordinate with transition metal ions such as Ni^{2+} or Co^{2+} immobilized on beads or a resin for purification.

(b) Some proteins bind specific carbohydrates, useful as affinity tags.

(c) It is also possible to use an antibody specific for the target protein as the ligand in an affinity column.

Specific expression of a target protein

With the determination of DNA sequences, it has become possible to insert a gene encoding a protein of interest into a cell, and cause the cell to express the protein copiously. This produces a starting mixture already enriched in the target protein, from which it is more easily isolatable. (A natural example of this is the overproduction of specific antibodies by myeloma patients. These served as sources of purified 'monoclonal' antibodies for research, before hybridoma and genetic engineering techniques became available.)

An advantage of producing a specific human protein in *E. coli* is that there will be no other *human* proteins as impurities. This is important in synthesizing human growth hormone for therapeutic purposes. Formerly, the source of human growth hormone was extraction from pituitary glands of deceased humans. The problem was that some of the extracted material was contaminated with prions, leading to Creutzfeldt–Jacob disease in patients. Using human growth hormone synthesized in *E. coli* eliminated this danger.

On the other hand, a problem with synthesizing human proteins in bacteria is that the bacterial cells cannot perform post-translational modifications, notably **glycosylation.** Use of mammalian cells to produce the desired proteins can overcome this deficiency.

Polyacrylamide gel electrophoresis (PAGE)

In electrophoresis, an electric field exerts force on a molecule. The force is proportional to the molecule's total or net charge. In a vacuum, the corresponding acceleration would be inversely proportional to the mass. However, counteracting the acceleration from the electric field are retarding forces from the medium through which the proteins move. *Polyacrylamide gels* contain networks of tunnels, with a distribution of sizes. Smaller proteins can enter smaller tunnels as well as larger ones, and therefore move faster through the gel than larger proteins. Proteins with different mobilities move different distances during a run, spreading them out on the gel.

The mobility of a native protein depends on its mass and its shape. Higher mass tends to reduce mobility; more compact shape tends to increase it. In particular, the mobility of denatured proteins is lower than that of the corresponding native states. To achieve a separation that depends solely on molecular weight, denature the proteins. Common denaturing media include urea (which competes for hydrogen bonds), and the reducing agent dithiothreitol to break S–S bridges (and iodoacetamide to prevent their reformation).

Sodium dodecyl sulphate (SDS) is a negatively charged detergent that helps to denature proteins. Multiple detergent molecules bind all along the polypeptide chain. The result is a protein–detergent complex that has an extended shape, with a uniform charge density along its length.

Carrying out SDS-PAGE in one dimension spreads out a mixture of proteins or nucleic acids into bands. Running several samples on the same gel in parallel lanes is a familiar procedure if only from sequencing gels. The results of protein gels can be made visible ('developed') by staining with Coomassie Blue, or, if the samples are radioactively labelled, by autoradiography. Often markers of known molecular weight are run in a separate lane for calibration.

Two-dimensional polyacrylamide gel electrophoresis (2D-PAGE)

One-dimensional PAGE will not adequately separate a very complicated mixture of proteins. The bands in a lane on a gel will overlap, and contain mixtures of proteins with similar sizes. To achieve better resolution, a two-stage procedure first separates proteins according to charge; then an SDS-PAGE step, run in a direction 90° from the original direction, separates according to size.

The charge on a protein depends on the charged residues it contains, and the pH of the medium. At different values of pH, ionizable groups on proteins have different charges. For instance, a free histidine sidechain is uncharged below pH ~5, and positively charged above pH ~7. For any protein, there is a pH at which it has a net charge of 0. This is called its isoelectric point.

A protein at its isoelectric point will feel no force in an electric field. It will not migrate in electrophoresis. To separate proteins according to their isoelectric points, establish a pH gradient in a medium and apply an electrophoretic field. The proteins will migrate, changing their charge as they pass through regions of different pH, until they reach their isoelectric points; then they will stop. The result, called isoelectric focusing, spreads proteins out according to their charged sidechains.

After the proteins are spread out along a lane by isoelectric focusing, running PAGE at 90° spreads them out in two dimensions (see Figure 12.15). Spots of interest can be eluted and identified by mass spectrometry (see next section).

Two-dimensional PAGE (2D-PAGE) can give a detailed fingerprint of protein expression patterns, in different cells, under different conditions, or at different stages of development. Figure 12.16 shows the differences between three developmental stages of rose petals—see Figure 11.10.

KEY POINT

Polyacrylamide gel electrophoresis (PAGE) is a common method for protein separation. Proteins migrate through a gel with different mobilities depending on their mass, shape, and charge. Proteins separated by SDS-PAGE are denatured by the detergent SDS, creating a protein-detergent complex with a uniform layer of negative charge. Protein mobility in SDS-PAGE depends only on relative molecular mass.

In 2D-PAGE, the first step, isoelectric focusing, separates a mixture of proteins into components with the same isoelectric points. The second step, PAGE, further separates the mixture according to molecular weight. Each spot on the resulting gel contains a unique protein, or possibly several proteins, that have particular values of *both* isoelectric point and molecular weight.

Mass spectrometry

Mass spectrometry is a physical technique that characterizes molecules by measurement of the masses of their ions, or of ions formed from their fragments. Applications to molecular biology include:

- rapid identification of the components of a complex mixture of proteins;
- sequencing of proteins and nucleic acids;

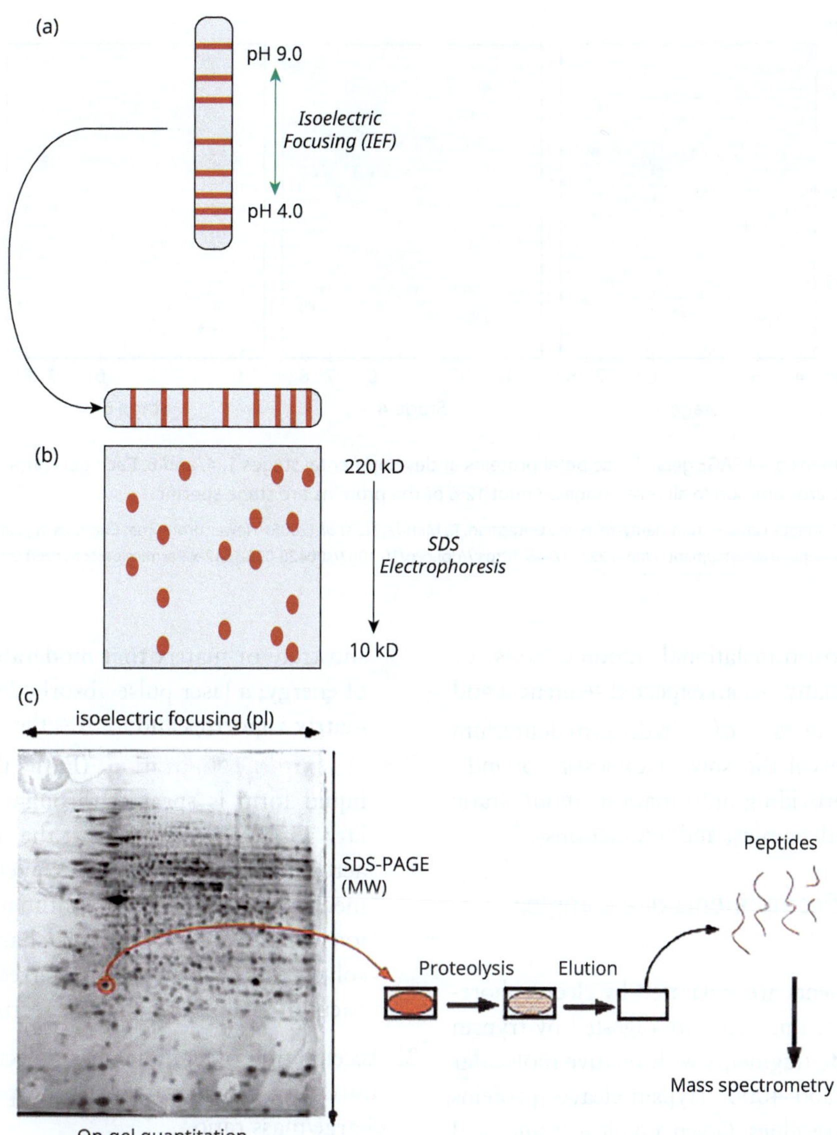

Figure 12.15 Two-dimensional gel electrophoresis is a two-step process. (a) First the proteins move, in *one* dimension, in an electric field. A pH gradient is established, such that different points in the path have different values of pH. Because proteins contain titratable groups, they will have different charges at different pH values of the medium. When charged they will migrate under the influence of the electric field. Some, initially bearing a positive charge, will migrate towards lower pH; others, initially bearing a negative charge, will migrate towards higher pH. But, for each protein, at some pH the net charge will be zero. That is called the *isoelectric point (pI)* of the protein. When each protein reaches the point in the gradient where the pH is equal to its isoelectric point, it will no longer move. (The electric field does not exert a force on uncharged molecules.) This is called **isoelectric focusing (IEF)**. The effect is to sort out proteins according to their isoelectric points. (This is related to the charge on the protein at pH7 but is not quite the same property.) However, there may be many proteins in the original mixture with the same or similar isoelectric points. The isoelectric focusing step will not separate them.

(b) Exposure of the sample to **SDS**, a detergent, unfolds the proteins, and coats them with a uniform negatively charged layer. The results of the first, isoelectric focusing, step are applied to an edge of a *two-dimensional* flat polyacrylamide gel. Proteins so treated migrate in electrophoresis according to their molecular weight. This separates into different components the proteins contained in each band arising in the first, IEF, step. The positions of the 'spots', and estimates of their concentrations, are available by staining the gel. ((a) and (b) are schematic.) (c) To identify the protein or proteins contained in any spot in the two-dimensional gel, it is possible to form tryptic peptides, and apply mass spectrometry.

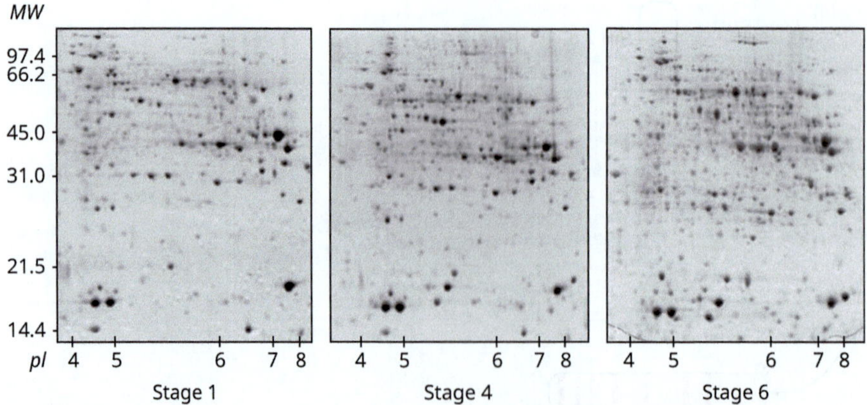

Figure 12.16 Two-dimensional PAGE gels of rose petal proteins at developmental stages 1, 4, and 6. Each gel contains over 600 proteins, of which 421 are common to all three stages. About 12% of the proteins are stage specific.

- analysis of post-translational modifications or substitutions relative to an expected sequence; and
- measuring extents of hydrogen–deuterium exchange to reveal the solvent exposure of individual sites (providing information about static conformation, dynamics, and interactions).

Identification of components of a complex mixture

First, the components are separated by electrophoresis, then the isolated proteins are digested by trypsin to produce peptide fragments with relative molecular masses of about 800–4000. Trypsin cleaves proteins after Lys and Arg residues. Given a typical amino-acid composition, a protein of 500 residues yields about 50 tryptic fragments. The mass spectrometer measures the masses of the fragments with very high accuracy (see Figure 12.17). The list of fragment masses, called the **peptide mass fingerprint**, characterizes the protein (see Figures 12.18 and 12.19). Searching a database of fragment masses identifies the unknown sample.

The operation of the spectrometer involves the following steps:

1. Production of the sample in an ionized form in the vapour phase. Proteins being fairly delicate objects, it has been challenging to vaporize and ionize them without damage. Two 'soft-ionization' methods that solve this problem are:
    - **matrix-assisted laser desorption ionization** (MALDI): the protein sample is mixed with a substrate or matrix that moderates the delivery of energy; a laser pulse absorbed initially by the matrix vaporizes and ionizes the protein;
    - **electrospray ionization** (ESI): the sample in liquid form is sprayed through a small capillary with an electric field at the tip to create an aerosol of highly charged droplets, which fragment upon evaporation, ultimately producing ions, which may be multiply charged, devoid of solvent; these ions are transferred into the high vacuum region of the mass spectrometer.
2. Acceleration of the ions in an electric field. Each ion emerges with a velocity proportional to its charge/mass ratio.
3. Passage of the ions into a field-free region, where they 'coast'.
4. Detection of the times of arrival of the ions. The time of flight (TOF) indicates the mass-to-charge ratio of the ions.
5. The result of the measurements is a trace showing the flux as a function of the mass-to-charge ratio of the ions detected (see Figure 12.20).

KEY POINT

Mass spectrometry is often used to characterize proteins isolated from mixtures. The peptide mass fingerprint—the list of fragment masses—is usually sufficient to identify a protein.

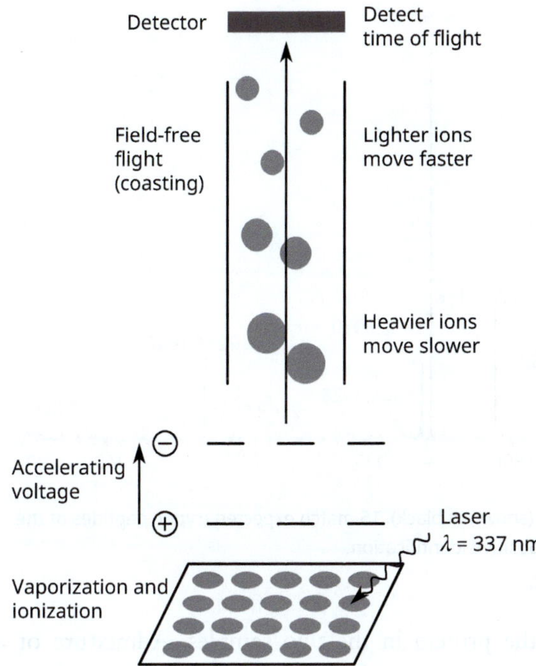

Figure 12.17 Schematic diagram of a mass spectrometry experiment.

Mass spectrometry is sensitive and fast. Peptide mass fingerprinting can identify proteins in subpicomole quantities. Measurement of fragment masses to better than 0.1 mass units is quite good enough to resolve isotopic mixtures. It is a high-throughput method, capable of processing 100 spots/day (although sample preparation time is longer). However, there are limitations. Only proteins of known sequence can be identified from peptide mass fingerprints, because only their predicted fragment masses are included in the databases. (As with other fingerprinting methods, it would be possible to show that two proteins from different samples are likely to be the same, even if no identification were possible.) Post-translational modifications interfere because they alter the masses of the fragments.

Used with permission of Elsevier Science & Technology Journals, from Edwards, J., Ruparel, H., & Ju, J. (2005). Mass-spectrometry DNA sequencing. Mutation Research/Fundamental and Molecular Mechanisms of Mutagenesis 573, 1–2: 3–12. https://doi.org/10.1016/j.mrfmmm.2004.07.021. Permission conveyed through Copyright Clearance Centre, Inc.

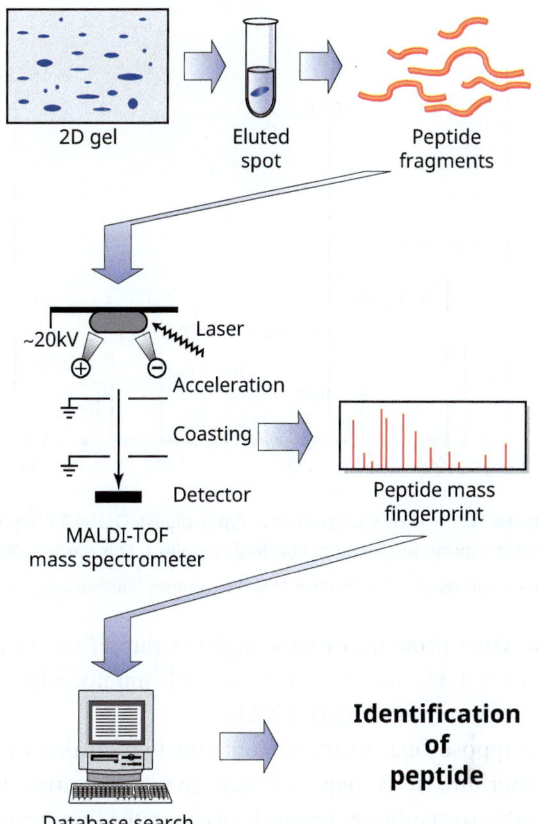

Figure 12.18 Identification of components of a mixture of proteins by elution of individual spots, digestion, and fingerprinting of the peptide fragments by MALDI–TOF (matrix-assisted laser desorption ionization–time of flight) mass spectrometry, followed by looking up the set of fragment masses in a database.

Protein sequencing by mass spectrometry

Fragmentation of a peptide produces a mixture of ions. Conditions under which cleavage occurs primarily at peptide bonds yield a series of ions differing by the masses of single amino acids. The y ions are a set of nested fragments containing the C terminus (see Figure 12.20a) (*b* ions are nested fragments containing the N terminus). The difference in mass between successive y ions is the mass of a single residue. The amino-acid sequence of the peptide is, therefore, deducible from analysis of the mass spectrum (see Figure 12.20b).

Two ambiguities remain: Leu and Ile have the same mass and cannot be distinguished, and Lys and Gln have almost the same mass and usually cannot be distinguished. Discrepancies from the masses of standard amino acids signal post-translational modifications. In practice, the sequence of about five to ten amino acids can be determined from a peptide of length <20–30 residues.

Quantitative analysis of relative abundance

Variation in yield in the preparation steps makes it difficult to use the intensity of the peaks in a mass spectrum to determine absolute abundances. An accurate method to determine *relative* abundances of

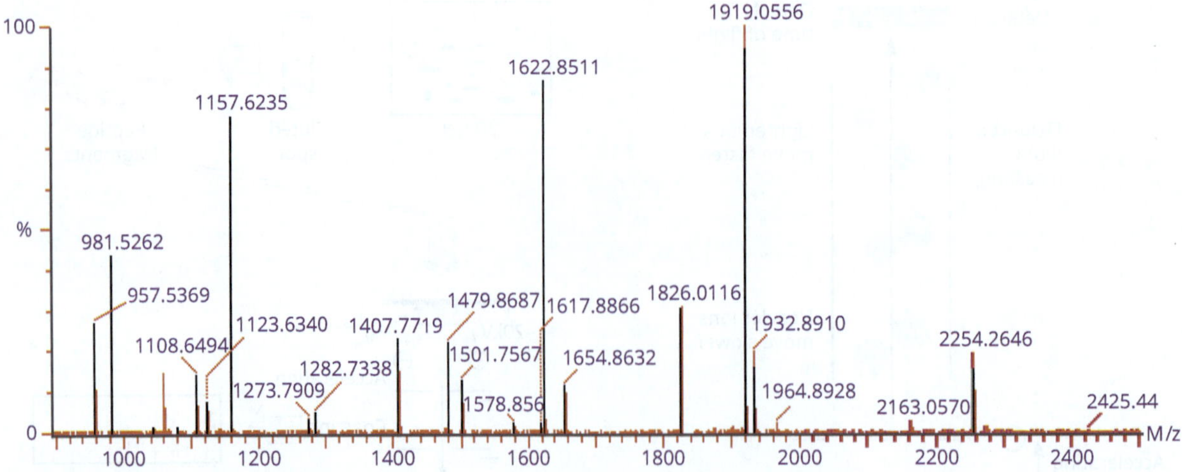

Figure 12.19 Mass spectrum of a tryptic digest. Of the 21 highest peaks (shown in black), 15 match expected tryptic peptides of the 39 kDa subunit of cow mitochondrial complex I. This easily suffices for a positive identification.

Figure courtesy of Dr I.M. Fearnley, MRC Dunn Human Nutrition Unit, Cambridge, UK.

the same protein in two samples is the SILAC technique (stable isotope labelling with amino acids in cell culture) (see Figure 12.21).

Suppose one wants to compare two strains of a bacterium. It is required that the two strains be unable to synthesize lysine. Cells from the two strains in question are grown in media containing unlabelled ('light') lysine and $^{13}C/^{15}N$-labelled ('heavy') lysine. Mix equal amounts of 'light' and 'heavy' cells. Then, cleavage by a lysine-specific endopeptidase produces fragments containing exactly one lysine residue. The masses of the corresponding fragments from 'light' and 'heavy' cells will differ by a constant amount (equal to the number of carbons and nitrogens in a lysine residue). The ratio of the peak heights from corresponding fragments gives the relative abundances of the protein in the two samples. Admixture of a known amount of selectively labelled recombinant proteins as a standard allows absolute quantitation.

Measuring deuterium exchange in proteins

If a protein is exposed to heavy water (D_2O), mobile hydrogen atoms will exchange with deuterium at rates dependent on the protein conformation. By exposing proteins to D_2O for variable amounts of time, mass spectrometry can give a conformational map of the protein. Applied to native proteins, the results give information about the structure. Applied to initially denatured proteins brought to renaturing conditions using pulses of exposure, the method can give information about intermediates in folding.

Experimental methods of protein determination

The first three-dimensional protein structures, of myoglobin and haemoglobin, were determined in the late 1950s by J.C. Kendrew, M.F. Perutz, and co-workers, using X-ray crystallography. For many years, X-ray crystallography was the only source of detailed macromolecular structures. A companion appeared in the 1980s, when K. Wüthrich, R.R. Ernst, and their co-workers developed methods for solving protein structures by nuclear magnetic resonance (NMR) spectroscopy. With recent advances in a third technique, cryo-electron microscopy (cryo-EM), it has been possible to determine structures of larger scale objects, including viral capsids, large protein complexes, and intact ribosomes. It is very exciting that cryo-EM techniques have improved to the point of being able to visualize individual atoms in large protein structures.

Taken together, these methods have given us over 200000 experimentally determined protein

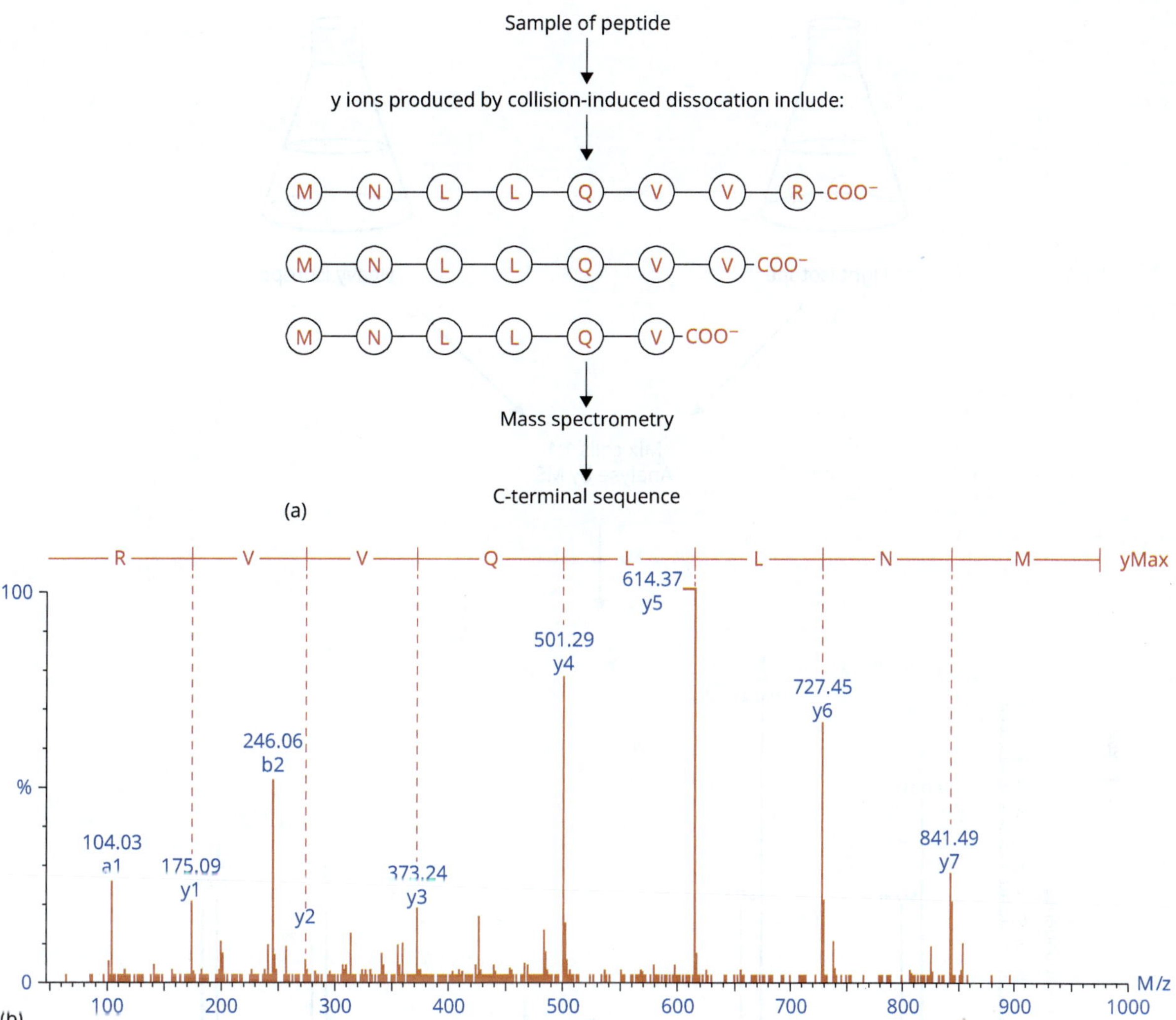

Figure 12.20 Peptide sequencing by mass spectrometry. Collision-induced dissociation produces a mixture of ions. (a) The mixture contains a series of ions, differing by the masses of successive amino acids in the sequence. The ions are not *produced* in sequence as suggested by this list, but the mass-spectral measurement automatically sorts them in order of their mass/charge ratio. (b) Mass spectrum of fragments suitable for C-terminal sequence determination. The greater stability of y ions over b ions in fragments produced from tryptic digests simplifies the interpretation of the spectrum. The mass differences between successive y-ion peaks are equal to the individual residue masses of successive amino acids in the sequence. Because y ions contain the C terminus, the y-ion peak of smallest mass contains the C-terminal residue, etc., and therefore the sequence comes out 'in reverse'. The leucines in this sequence could not be distinguished from isoleucine in this experiment.

From: Carroll, J., Fearnley, I.M., Shannon, R.J., *et al.* (2003). Analysis of the subunit composition of complex I from bovine heart mitochondria. Mol. Cell Proteomics 2: 117–126 (supplementary figure S1–38).

structures, archived and distributed by the World-wide Protein Data Bank (wwPDB).

X-ray crystallography of proteins

One evening late in April 1934, J.D. Bernal developed a photograph he had taken of the X-ray diffraction pattern of a crystal of pepsin, a proteolytic enzyme. Amazed, he recognized from this picture that someday a complete atomic structural model of a protein would be revealed:

> . . . The wet crystals gave individual X-ray reflections, which were rather blurred owing to the large size of the crystal unit cell, but which extended all over the films to spacings of about 2 Å. That night, Bernal, full of excitement, wandered around the streets of Cambridge, thinking

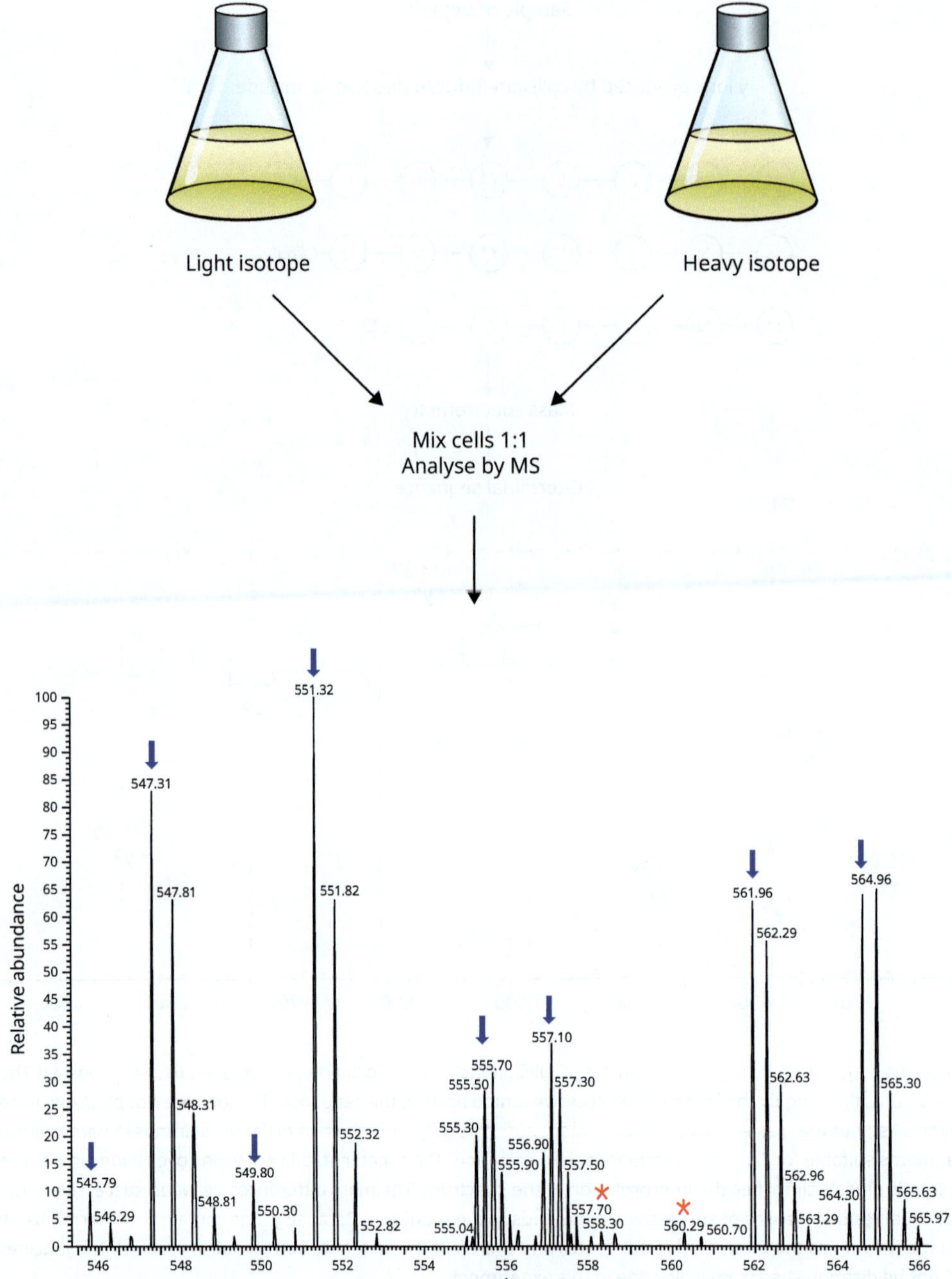

Figure 12.21 Stable isotope labelling with amino acids in cell culture (SILAC) technique to determine relative abundances of proteins in two different strains. Using cells that cannot synthesize lysine, one strain is grown in a medium containing unlabelled (^{12}C) lysine and the other is grown on a medium containing heavy lysine (^{13}C and ^{15}N). Mixing equal amounts of 'light' and 'heavy' cells, followed by digestion and analysis, gives a spectrum with pairs of peaks corresponding to the corresponding proteins. If the digestion is carried out with LysC, an endopeptidase that cleaves at lysine residues, each fragment will contain only one lysine, and the mass difference between corresponding peptides will be a constant. The relative heights of the corresponding peaks give the relative abundances of the protein in the two strains. Two peaks marked by red stars have no partners; they are likely to be contaminants.

of the future and of how much it might be possible to know about the structure of proteins if the photographs he had just taken could be interpreted in every detail.[4]

It took 25 years.

What then did Bernal see that was so exciting? The X-ray diffraction pattern appeared as a set of spots on a film. According to the interpretation discovered by W.L. Bragg, the larger the angle of deflection of a spot, the more detailed the information it signifies (Bragg's law). A photograph containing only a few spots, petering out in intensity at angles not far from the incident beam, would signify that the molecules in different unit cells are similar only to *low resolution*—perhaps they are 'blobs' of similar general shape, but different in detail. When Bernal saw spots 'which extended all over the films to spacings of about 2 Å', he realized that the proteins in different unit cells were identical in details not much larger than individual atomic spacings at least.

X-ray crystallography had determined the structures of simple inorganic compounds before the First World War. William Bragg, coinventor of X-ray crystal structure determination—with his father, William Henry Bragg—years later became Cavendish Professor of Physics at Cambridge in 1938. Bernal had just left Cambridge for Birkbeck College, London. Bragg's commitment to the technique he had invented, plus recognition of the importance of the goal, elicited his continued and unstinting support for the efforts of Max Perutz to extend X-ray structure determination to haemoglobin.

The X-ray structure determination of a protein begins with its isolation, purification, and crystallization. If a suitable crystal is placed in an X-ray beam, diffraction will be observed, arising from the regular microscopic arrangement of the molecules in the crystal. The pattern can be recorded on film (see Figure 12.22), or in digital form by detectors.

Protein structure determination by X-ray crystallography is now a mature technique. The equipment and technique of data collection, including synchrotrons, and software for data reduction and structure solution, are now integrated into an effective high-throughput technology. The rate-limiting steps are now in the preparation—expression and purification, and getting good crystals.

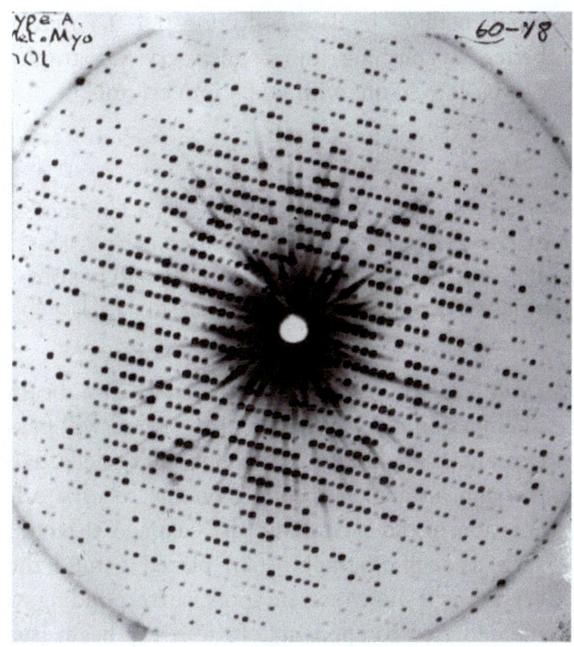

Figure 12.22 Photograph of X-ray diffraction of myoglobin, from work by J.C. Kendrew and his collaborators. Notice, as Bernal did in his 1934 picture, how the spots extend 'all over the film'. The distribution of the reflections in a diffraction pattern gives clues to the size and symmetry of the unit cell, and of the relative orientations of the subunits that it contains.

Reproduced courtesy of the MRC Laboratory of Molecular Biology.

The steps in a protein structure determination by X-ray crystallography are:

1. ***Purify the protein.*** Classically, biochemists isolated proteins from natural samples. Typical sources include beef heart or liver, or suspensions of large quantities of microorganisms. Blood samples, or placentas, or post-mortem material, provide access to human proteins.

   An alternative to isolating proteins from their natural sources is to insert a gene into a microorganism and overexpress the proteins in the host.

2. ***Crystallize the protein.*** The goal is a crystal showing high-quality internal order, and of adequate size. For data collection on a synchrotron, a crystal 0.1 mm in diameter is acceptable. For

[4] Hodgkin, D.C., & Riley, D.P. (1968). Some ancient history of protein X-ray analysis. In: A. Rich and N. Davidson (eds), *Structural Chemistry and Molecular Biology*, pp. 16–28. San Francisco: W.H. Freeman & Co. The quotation appears on p. 15.

other methods of data collection 0.3–0.5 mm is better. By pushing things, microcrystallography can be carried out with ~0.04 mm crystals.

KEY POINT

Purity of a sample is important for crystal formation. Whereas in classical organic chemistry one crystallizes a molecule in order to purify it, in working with proteins one purifies a molecule in order to be able to crystallize it.

3. ***Data collection.*** If a collimated X-ray beam impinges on a crystal, most of the beam will pass straight through. A small amount will be deflected, through an angle depending on the relative orientation of the crystal and the beam. A detector, such as a photographic film placed behind the crystal, will show discrete spots, corresponding to the deflected components of the X-ray beam (see Figure 12.22). The distribution of intensities in the pattern of spots is the experimental data. Today most datasets are collected at synchrotrons.
4. ***Solving the structure.*** The electron density in the molecule is related to the diffraction pattern by a pair of relationships called the Fourier transform. In principle, given the data, we should be able to compute the electron density in the crystal. Peaks in the electron density correspond to positions of atoms. Picking the peaks in the electron density gives us the coordinates. We must figure out which atoms go with which peaks, but knowing the primary structure of the protein this should not be too difficult.

Now the bad news. Each spot in the diffraction pattern corresponds to a structure factor F. But even in the simplest case, where the unit cell of the crystal has a centre of symmetry, the experiment determines the magnitude of the structure factor but not its sign. That is, for each structure factor F we can measure the absolute value $|F|$ but we don't know whether the actual structure factor is $+|F|$ or $-|F|$. For a film containing 1000 spots in a diffraction pattern, there are 2^{1000} sign combinations! If unit cell does not have a centre of symmetry, the information deficit is worse, *much worse*. This is the notorious *phase problem of X-ray crystallography*.

The phase problem is simpler for small-molecule crystallography, and indeed the structures of simple salts and minerals were solved very early on. It is much harder for proteins. However, there are now quite a few approaches that have proved successful.

Interpretation of the electron density: model and improvement

Given the experimental structure factor magnitudes, and even a reasonable estimate of phases, it is possible to calculate a map of the electron density in a unit cell of the crystal. With sufficient resolution, and sufficiently accurate phases, the map contains peaks that correspond—at least approximately—to positions of atoms. It is necessary to interpret the electron density in terms of a molecular structure.

For small molecules, the classic method of interpreting the pattern of peaks in the electron-density map used a 'needle and thread' approach: crystallographers would lay out x and y coordinates of the peaks on a sheet of graph paper pasted onto a styrofoam base. They would insert knitting needles into the styrofoam: for each peak at coordinates (x, y, z), they would push a needle into the styrofoam, at position (x, y), leaving a length z *protruding*. Tying woollen threads between peaks nearby in space would create a three-dimensional model of the molecular structure.

Interpretation of the first resolved electron-density maps of proteins involved an extension of this approach. To interpret electron-density maps of myoglobin, J.C. Kendrew designed brass wire molecular models of amino acids, and assembled them on a set of vertical rods, at a scale of 5 cm/Å (Figure 12.23). To display the electron-density map, contour maps of serial sections—like topographic maps—were drawn on plexiglass plates, and stacked. To make it possible to superpose the model and electron density visually, F.M. Richards introduced the use of a half-silvered mirror placed at 45° between the model and the electron-density map.

Computer graphics replaced all this. The program FRODO, by T. Alwyn Jones, achieved a near monopoly in the protein crystallographic community. After many years, Jones's program 'O' superseded FRODO (Figure 12.24). Such programs make use of interactive computer display devices that show selected portions of the electron density, and superpose a model of part or all of the protein structure. Devices such as dials allow manipulation of the model to adjust the fit to the experiment.

Figure 12.23 John C. Kendrew (left) and Max F. Perutz examining a model of sperm-whale myoglobin. (Note that models of several proteins require separate allocations of materials, and of space to house them; another advantage of computer graphics.)

Reproduced courtesy of the MRC Laboratory of Molecular Biology.

Figure 12.24 T. Alwyn Jones, author of FRODO and O, at a display showing part of an electron-density map, part of a protein model, and menus for manipulating them. A set of four dials stands at the left of the screen. Dials are suitable for interactive control of overall orientation of a molecule, or for internal rotation around selected bonds.

Photo courtesy frozentime.se.

The endgame—refinement

Once the atoms are positioned at close to their proper positions, it is possible to optimize the fit of experimental structure factors to those computed from the model by refinement. One is comparing the experimental observed structure factor magnitudes $|F_o|$ with the *magnitudes* of the structure factors computed from the model $|F_c|$. The usual function to be minimized includes the **R-factor** (*R* for residual or reliability):

$$R = \frac{\sum ||F_o| - |F_c||}{\sum |F_c|}$$

and also contains terms expressing the conformational energy.

Advances in computational methods have improved the power of refinement methods, to the point of blurring the distinction between the model-building and refinement steps of structure determination.

KEY POINT

As a rule of thumb, one expects the final R-factor of the structure to be about 1/10 the resolution of the data. That is, if you collect data on a protein to 2 Å the expected R-factor should be ~ 20%.

How accurate are the structures?

Most genome scientists are consumers rather than providers of crystal structures. It is essential to appreciate what details of the coordinates can be trusted.

Crystal-structure determinations are at the mercy of the degree of order in different parts of the molecule. (Order is the extent to which different unit cells of the crystal are *exact* copies of one another.)

The degree of order governs the available **resolution** of the experimental data. Resolution measures the fineness of the details that can be distinguished in the electron-density map. (**Atomic resolution** means that individual atoms can be distinguished.) Resolution is an index of potential quality of an X-ray structure determination. The resolution of an X-ray structure determination is a measure of how much data was collected. The more data, the greater the ratio of number of observations to the number of atomic coordinates to be determined, and in principle the more precise the results. Resolution is expressed in Å, with a *lower* number signifying a higher resolution (Figure 12.25).

It was the recognition that the pepsin crystal diffracted to high resolution that so excited Bernal in 1934.

KEY POINT

Accuracy and precision of X-ray structure determinations of proteins depend primarily on (1) quality of crystals (= degree of order), (2) care in analysing data, and (3) resolution of data collection.

Nuclear magnetic resonance spectroscopy in structural biology

Nuclear magnetic resonance (NMR) spectra measure the transitions between energy levels of the magnetic nuclei in atoms placed in a strong homogeneous external magnetic field. These energy levels are sensitive to the chemical environment of the atom. They are therefore very rich in information about chemical structure and dynamics.

Like X-ray diffraction, NMR spectroscopy is a technique invented by physicists, but taken over by chemists. NMR was first observed in 1945–1946, independently by E.M. Purcell, R.V. Pound, and H.C. Torrey; and F. Bloch, W. Hansen, and M. Packard. It was one of a very large number of important discoveries made by scientists returning to basic research after participating in projects related to support of the military during the Second World War.

NMR was first applied to studies of peptides and proteins in the 1970s. The first protein structure determination by NMR was that of the 57-residue bull seminal protease inhibitor, by K. Wüthrich and collaborators, in 1984. To prove the method so as to convince even the most hardened sceptics, the groups of Wüthrich in Zürich and R. Huber in Munich undertook independent structure determinations of the α-amylase inhibitor tendamistat, by NMR spectroscopy and X-ray crystallography, respectively. The structures were virtually identical. Subsequently, NMR spectroscopy took its place in structural biology alongside X-ray crystallography. There are currently over 14 000 NMR entries in the wwPDB, including proteins and nucleic acids.

As methods of macromolecular structure determination, X-ray crystallography and NMR spectroscopy each have advantages and disadvantages. The main advantage of NMR is that it is not necessary to produce crystals; this is sometimes a severe impediment to X-ray structure determination.

A crystallographer once boasted in a public lecture that, in the near future, when a new strain of flu emerged, we could crystallize a surface protein, solve the structure, computationally design a drug to bind it, and nip any epidemic in the bud. A member of the audience suggested that in that event flu would evolve so that its surface proteins would resist crystallization.

NMR also gives us a window into protein dynamics, on timescales of about 10^{-9}–10^{-6} s and slower than 10^{-3} s.

Proteins in solution are not subject to the constraints of crystal packing, and are expected to flounce around somewhat. Indeed, typically the result of an NMR structure determination is a set of similar but not identical models (often ~15–20 of them), all of which are comparably consistent with the combination of experimental data and general stereochemical restraints. This may be a manifestation of dynamics, or incompleteness in the experimental data that leave the structure underdetermined.

NMR has found a large and varied set of applications, not only in physics, chemistry, and molecular biology, but in medicine (magnetic resonance imaging), geology (lowering NMR spectrometers into bore holes to determine the composition and mobility of fluids in underground reservoirs), and materials science (chemical structure determination, imaging of structure, dynamics). NMR has important applications in drug development, for example as a method of screening protein–ligand interactions. The NMR–MOUSE is a portable spectrometer useful for subjects that cannot be brought into the laboratory, for example paintings in art museums, or rare books.

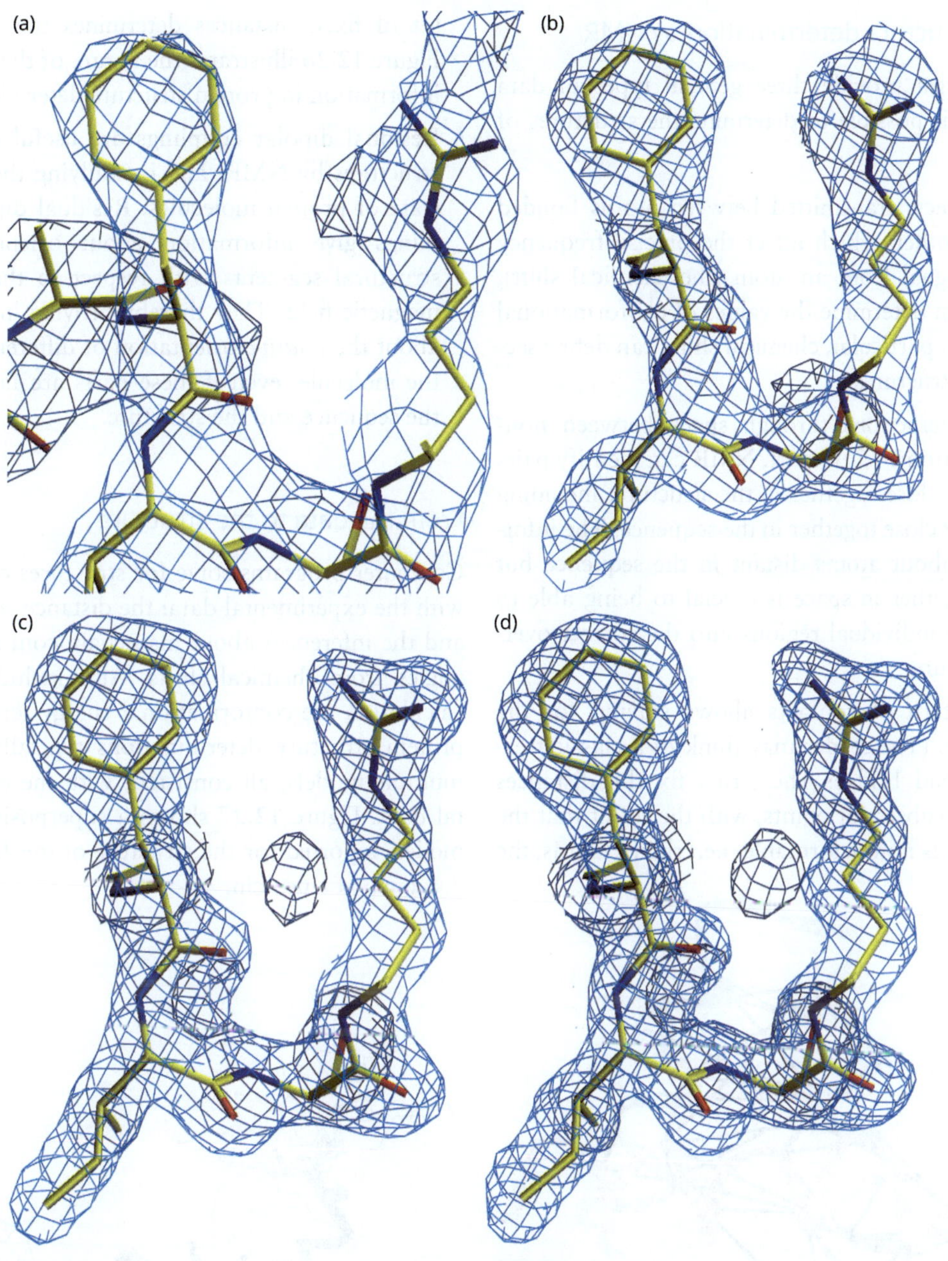

Figure 12.25 Electron-density maps and model fitting at different resolutions. (a) 4.0 Å. It is possible to trace the general course of the bonding but not to position individual atoms reliably. The 'lollipop' of the ring of the phenylalanine sidechain is identifiable. The arginine and the carboxyl group with which it is interacting are blurred together. The guanidinium group of the arginine is particularly ill-represented. (b) 2.8 Å. It is possible to place the mainchain and sidechains from their overall sizes and shapes. With a resolution of 2.8 Å or better, it is generally possible to build a model into the electron density. (Traditionally, growing crystals that would diffract to 2.8 Å or better was the protein crystallographer's goal.) (c) 2.3 Å. One can see better definition of some features than in the 2.8 Å map. The position of the guanidinium group of the arginine is clearer. The orientations of the carbonyl oxygens are better defined. (d) 1.8 Å. Model building into this map is a piece of cake! The zig-zag of the arginine sidechain is clear, the carbonyl oxygens are quite well defined. Even the lower electron density at the centre of the phenylalanine ring is beginning to appear. Notice that phenylalanine and arginine sidechains have very distinctive sizes and shapes, and can be recognized even at relatively low resolution. To decide whether a region of density is valine, isoleucine, or leucine would require quite high definition.

Protein structure determination by NMR

NMR spectra provide three general types of data that make it possible to determine the structures of proteins.

- From effects transmitted between atoms bonded to each other, which affect the precise frequency of the signal from an atom (the chemical shift), NMR can determine the values of conformational angles. In particular, chemical shifts can define secondary structures.
- From interactions through space between non-bonded atoms <5 Å apart, NMR can identify pairs of atoms close together in the structure, including those *not* close together in the sequence. The information about atoms distant in the sequence but close together in space is crucial to being able to assemble individual regions into the correct overall structure.

  The pattern of contacts allows solving for the structure. (The reader may think of a classic trestle railroad bridge: The struts fix the distances between subsets of points, with the result that the structure is non-deformable (i.e., rigid); that is, the set of fixed distances determines the structure.) Figure 12.26 illustrates the nature of the measured information in protein structure determination.
- Residual dipolar couplings are useful in solving structures by NMR, and in studying the dynamic state of protein molecules. Residual dipolar couplings give information about orientations of structural segments with respect to the external magnetic field. They thereby provide information about the *relative* orientation of different parts of the molecule, even if these parts are far apart in the sequence and the structure.

From the data to the structure

Computer programs solve for structures compatible with the experimental data: the distance constraints, and the inferences about mainchain conformational angles from chemical shifts. Also included in the calculation are conformational energy terms. NMR protein structure determinations typically produce multiple models, all consistent with the experimental data. Figure 12.27 shows a superposition of 19 models reported for the structure of the *Drosophila* antennapedia protein.

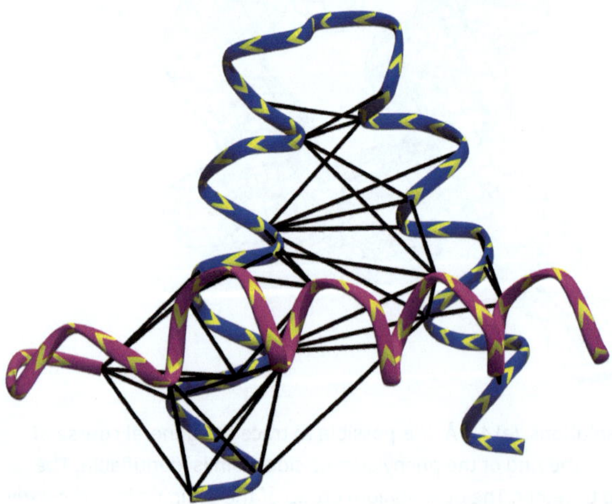

Figure 12.26 A simplified picture indicating experimental information used by NMR spectroscopists to determine protein structures. This picture shows residues 20–40 and 72–103 from *Borrelia burgdorferi* outer surface lipoprotein. The entire protein is 122 residues long. Black lines indicate experimentally determined pairs of residues close in space in the structure. (Other measured contacts between pairs of residues of these regions within five residues of each other in the sequence are not shown. Although this figure shows links between Cα atoms the experiment actually determines distances between protons.)

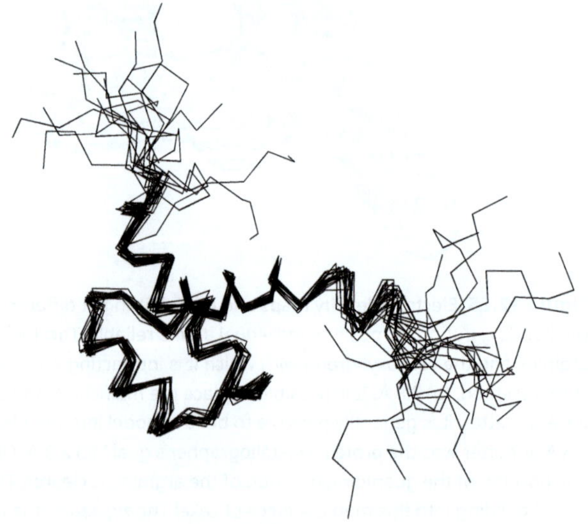

Figure 12.27 Models of *Drosophila* antennapedia protein, determined by NMR spectroscopy [1HOM]. The core of the molecule, comprising the interacting regions of secondary structure, is consistent among the models, but the regions at the N- and C-termini appear variable.

Low-temperature electron microscopy (cryo-EM)

Electron microscopy (EM) is the third experimental technique of protein structure determination. Recent breakthroughs have promoted it from a method that could produce only low-resolution structures—'blobs'—to the new successes that produce structures at atomic resolution (see Figure 12.28). Electron microscopy of specimens at liquid nitrogen temperatures has revealed structures of assemblies in the range $M_r = 500\,000$ to 4×10^8, 100–150 Å in diameter, such as the hepatitis B core shell or the clathrin coat.

Protein structure determinations by cryo-EM have achieved atomic resolution. Figure 12.28 shows high-resolution EM maps and fitted atomic models.

In cryo-EM a sample under native conditions is fast-frozen. The solvent is not given time to form the regular hexagonal equilibrium crystal form of ice that we see in snowflakes, but rather remains in a snapshot of the liquid state.

Shooting a beam of electrons through a molecule produces an image that integrates the electron density in the sample. Compare such an EM image with an ordinary photograph, such as a 'selfie' that you might take with your phone: that shows only the *surface* of the subject. EM images are penetrative—as if your 'selfie' contained not only your face, but the bones of your skull, your brain, and the ponytail (if present) behind your head. It would still be a two-dimensional image, but containing full three-dimensional information.

How can we derive a true three-dimensional structure from such images? Suppose we scatter many copies of a molecule, or complex, on a grid on the stage of the microscope. We can observe many images of the same object, but in different orientations. The challenge is to determine the orientations of the different images. But when this is achieved, we will be able to combine the images—different projections of the same object, in different orientations—into a three-dimensional structure.[5]

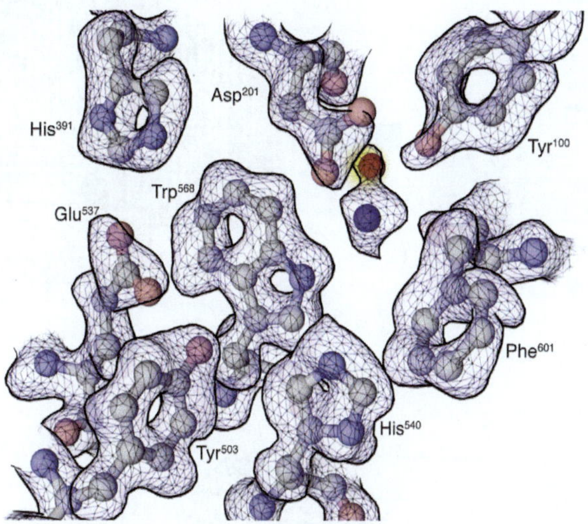

Figure 12.28 Density map and atomic model for selected active site residues in β-galactosidase, determined by cryo-electron microscopy.

Reprinted from Bartesaghi, A., Aguerrebere, C., Falconieri, V., *et al.* (2018). Atomic Resolution Cryo-EM structure of B-Galactosidase. Structure 26, 6: 848–856.e3. https://doi.org/10.1016/j.str.2018.04.004, with permission from Elsevier.

Figure 12.29 shows the overall course of a cryo-EM structure determination.

> The three main experimental methods for determining protein structure are: X-ray crystallography, NMR spectroscopy, and cryo-electron microscopy. They are experimental in the sense that structure determination depends on physical measurements; however, all these methods are crucially dependent on the contributions of computer programs for data analysis.

Protein structure prediction

Proteins fold spontaneously into unique three-dimensional native conformations. Thus, nature has an algorithm for predicting protein structure from amino-acid sequence. Some attempts to understand this algorithm are based solely on general physical principles; others are 'knowledge based' on observations of known amino-acid sequences and protein structures. A proof of our understanding

[5] In the table game Foosball different copies of the same structures appear, on a lattice, in different orientations. See: https://www.123rf.com/photo_82959806_people-playing-enjoying-foosball-table-soccer-game-recreation-leisure.html

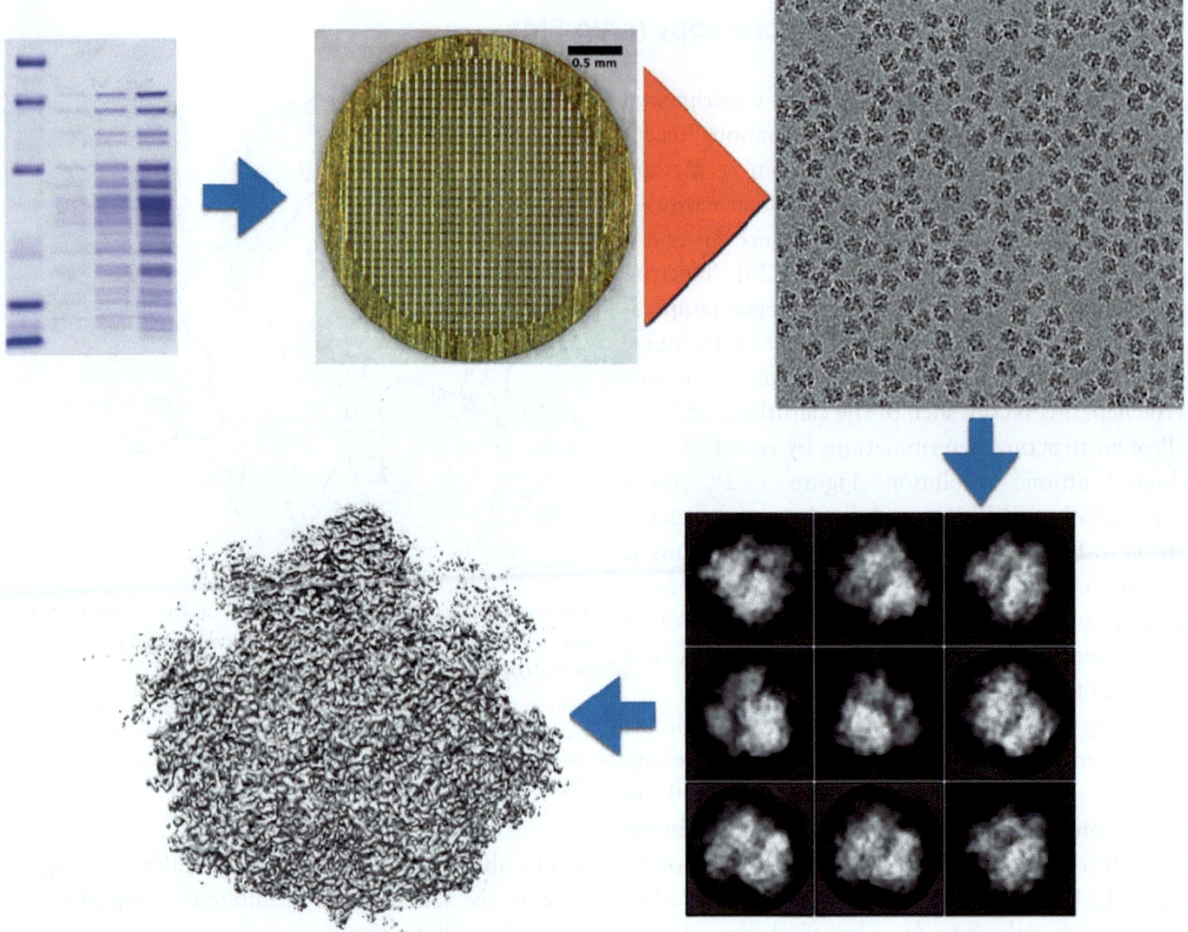

Figure 12.29 Steps in a cryo-EM protein structure determination. An isolated protein is presented on a grid, producing an image containing many views of the structure. These are isolated (bottom row, right) and programs determine the orientation. The information from the combination of orientations is integrated into a structure.

Reproduced with permission of Dr C Russo and the MRC Laboratory of Molecular Biology.

would be the ability to reproduce the algorithm in a computer program that could predict protein structure from amino-acid sequence. This has been the 'holy grail' of computational structural biology for 50 years. The achievement of AlphaFold2 represents a major breakthrough.

The Critical Assessment of Structure Prediction (CASP) programmes provide 'blind' tests of the state of this art (see Box 12.8).

Most attempts to predict protein structure from basic physical principles alone try to reproduce the interatomic interactions in proteins, to define a numerical energy associated with any conformation. Computationally, the problem of protein structure prediction then becomes a task of finding the global minimum of the conformational energy function over all possible backbone and sidechain conformations. This approach has proved difficult, partly because of the imprecision of the energy function and partly because the minimization algorithms tend to get trapped in local minima.

The alternative is empirical or 'knowledge-based' techniques. These are currently the most successful methods. Two approaches are good enough to have practical application to academic and clinical research:

- ***Homology modelling.*** Suppose a target protein of known amino-acid sequence but unknown structure is related to one or more proteins of known structure. Then we expect that much of the structure of the target protein will resemble that of the

BOX 12.8 Critical Assessment of Structure Prediction—CASP

The **Critical Assessment of Structure Prediction (CASP)** programme organizes blind tests of **protein structure predictions**. Crystallographers, NMR spectroscopists, and cryo-electron microscopists make public the amino-acid sequences of the proteins they are investigating, and agree to keep the experimental result confidential until predictors have had a chance to submit their models. After the deadline for deposit of predictions, the experimental structures are released and the predictions evaluated.

CASP runs on a two-year cycle. Early in each active year, the target sequences are published, and predictions collected prior to release of the experimental structures. At the end of the year a gala meeting brings together the predictors and evaluators to discuss the current results and to assess progress. An album of papers appears as a supplement to the journal *Proteins: Structure, Function and Bioinformatics*.

Protein structure prediction is a classic challenge in computational biology. For many years—as recorded in successive CASP programmes—progress took place in fits and starts. Recently, however, AlphaFold2 has effectively solved the problem.

known proteins (see Figure 5.13). The related proteins of known structure can therefore serve as a basis for a model of the target protein.

- ***a priori* structure prediction** models an unknown structure, without making *explicit* use of any other known structure. (However, many algorithms make use of *general* properties of known structures, such as the Sasisekharan–Ramakrishnan–Ramachandran plot.)

Homology modelling

In homology modelling, one predicts the structure of a protein from the known structures of related proteins. The task is to predict the *changes* in structure between the known structures used in the modelling, and the actual structure of the target. (One might consider *a priori* prediction 'from scratch' as the *integral* form of the protein structure prediction problem, and homology modelling as the *differential form*.)

Figure 12.30 shows the superposition of hen egg white lysozyme, and baboon α-lactalbumin. Suppose you knew the amino-acid sequence and three-dimensional structure of hen egg white lysozyme, but only the amino-acid sequence of baboon α-lactalbumin. Proposing the chain tracing of lysozyme, as shown in the figure, as a model for the structure of α-lactalbumin would give a pretty good answer.

To go beyond this, you would have to improve the prediction of the regions in which there were structural differences between lysozyme and α-lactalbumin. (Of course you would not know where in the structure these differences appear, as we are assuming that you do not know the α-lactalbumin structure.)

Modern homology modelling methods are quite sophisticated and successful. They make use not only of a single similar structure, but of a whole family of similar structures. The challenge is to identify the most useful features of each source structure, and to combine them into the best possible prediction. The program I-TASSER, by Y. Zhang and colleagues of the University of Michigan, US, has consistently provided superb results (see Figure 12.31).

Mature software for homology modelling is available:

I-TASSER	Y. Zhang, University of Michigan, USA	https://zhanggroup.org/I-TASSER/
SWISS-MODEL	T. Schwede, M.C. Peitsch, and N. Guex, now at the Geneva Biomedical Research Institute	http://www.expasy.org/swissmodel/SWISS-MODEL.html
MODELLER	A. Sali, University of California at San Francisco, USA	https://salilab.org/modeller/

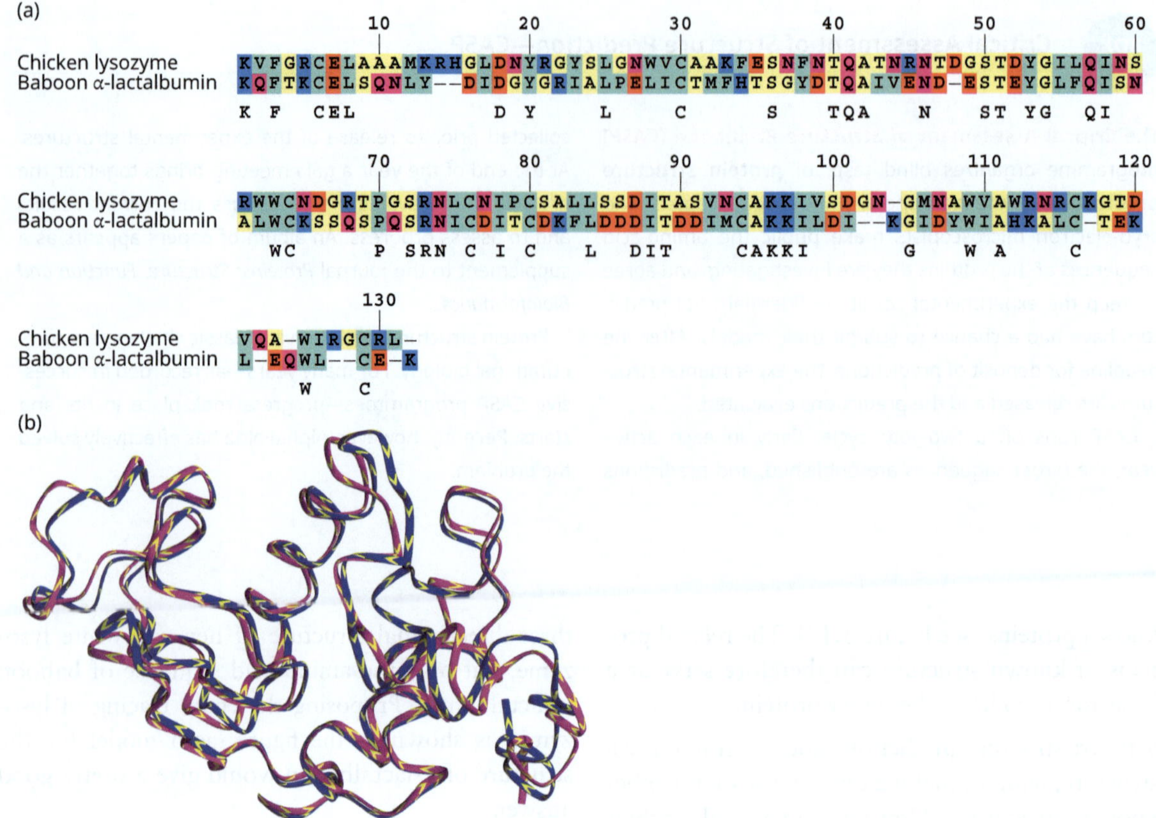

Figure 12.30 (a) Sequence alignment of hen egg white lysozyme and baboon α-lactalbumin. (b) Superposition of structures of hen egg white lysozyme (blue) and baboon α-lactalbumin (magenta). There are only 37% identical residues in an alignment of the amino-acid sequences, but the structures are nevertheless quite similar.

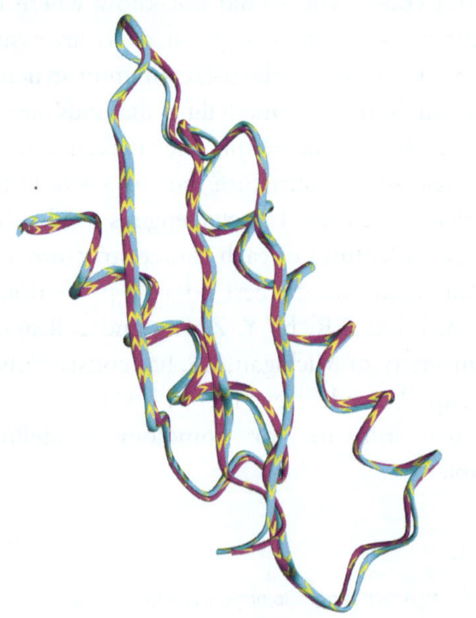

Figure 12.31 Prediction by I-TASSER of target T1006 from CASP13, magnetosome protein MamM CTD from *Desulfamplus magnetovallimortis BW-1*.

MODBASE (http://salilab.org/modbase) and 3DCrunch (http://swissmodel.expasy.org/SM_3DCrunch.html) collect homology models of proteins of known sequence.

A priori structure prediction

Given that the amino-acid sequence contains all information required to specify the native structure of a globular protein, it should be possible to write a computer program that predicts the three-dimensional structures of proteins from amino-acid sequences. Over the last several years, substantial progress was made by a group at the University of Washington. But their work has been overtaken by the program AlphaFold2, by Google subsidiary DeepMind. At CASP14, in 2020, AlphaFold2 achieved spectacular success in predicting accurate and reliable three-dimensional structures (see Figure 12.32). The approach made use of techniques

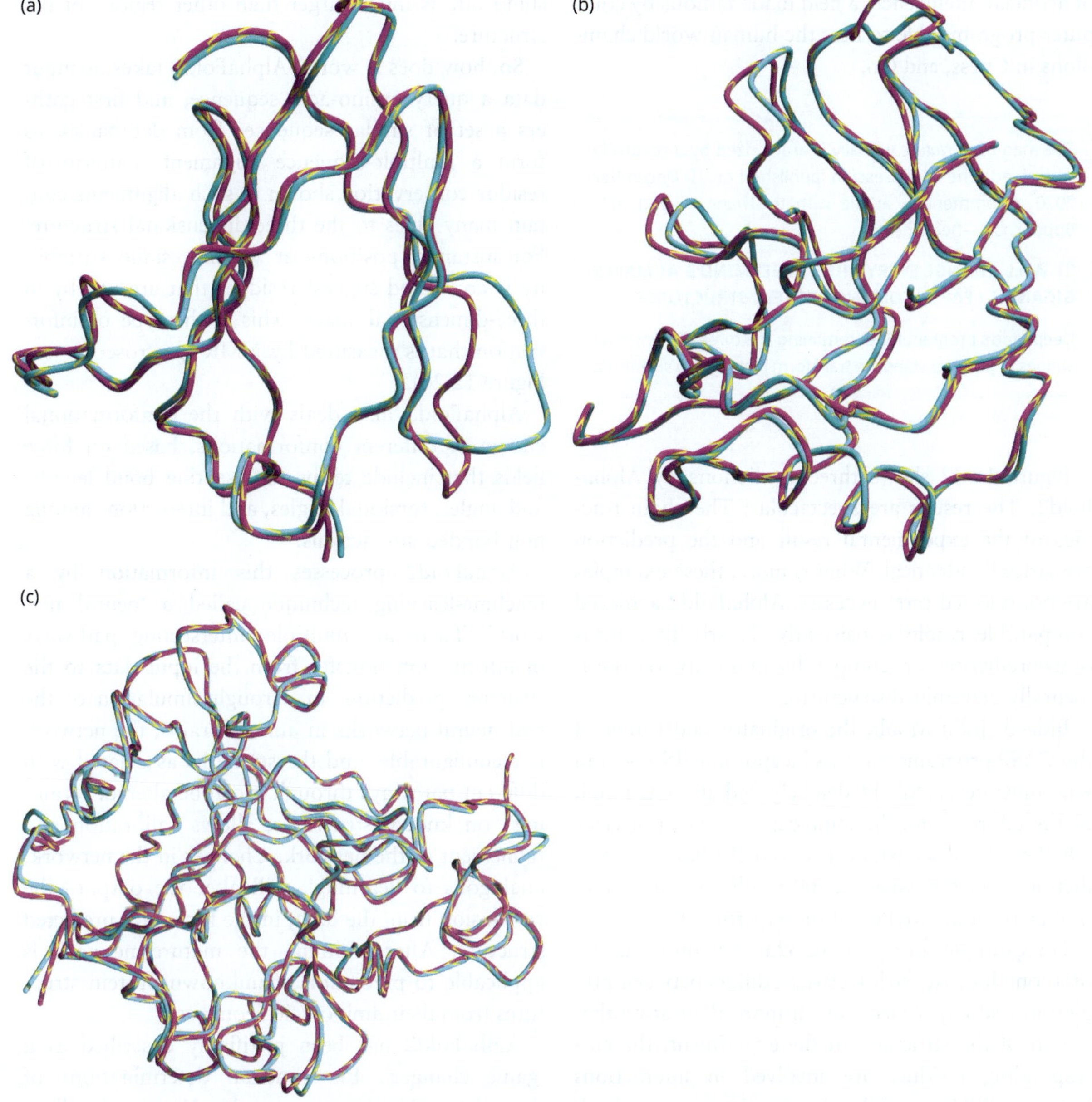

Figure 12.32 Predictions by AlphaFold2: (a) Domain 1 of target T1038 from CASP14, tomato spotted wilt orthotospovirus glycoprotein N ectodomain (Trypsin treated), 76 residues. The Cα atoms of 70 residues—excluding the loop at lower right—fit with an r.m.s.d. of 0.62 Å.

(b) Domain 1 of target T1049 from CASP 14, major virulence-associated fimbrial protein, MrpH, of the bacterial urinary tract pathogen *Proteus mirabilis*, 134 residues. 124/134 Cα atoms fit to an r.m.s.d. of 0.67 Å.

(c) Human carbonic anhydrase 2, 259 residues. 257/259 Cα atoms fit to an r.m.s.d. of 0.44 Å.

In each picture the experimental structure is shown in cyan and the predicted structure in magenta.

of artificial intelligence, a field made famous by computer programs able to beat the human world champions in Chess, and Go.

The magazine *Nature*, usually characterized by a relatively restrained tone of expression, published on 10 December 2020 a commentary article with the banner—and fully upper case!—headline:

'IT WILL CHANGE EVERYTHING': DEEPMIND'S AI MAKES GIGANTIC LEAP IN SOLVING PROTEIN STRUCTURES

DeepMind's program for determining the three-dimensional shapes of proteins stands to transform biology, say scientists.

Figure 12.32 shows three predictions by AlphaFold2. The results are spectacular: The chain tracings of the experimental result and the prediction are virtually identical. What is more, these examples are not selected rare successes; AlphaFold2 achieved comparable results consistently. Nearly two-thirds of its predictions are comparable in quality to experimentally determined structures.

Indeed, John Moult, the originator and leader of the CASP programs since its inception in 1994—and who must deservedly be dead chuffed at the triumph of the effort—said: 'In some cases, it was not clear whether the discrepancy between AlphaFold's predictions and the experimental results was a prediction error or an artefact of the experiment.'

For example, in Figure 12.32a, the conformation of a long loop at the lower right differs between prediction and experiment, much more than any other region of the structure. In the experiment, the corresponding residues are involved in interactions between different molecules in the crystal. Such crystal-packing contacts are known, in many cases, to perturb the conformations of the regions involved, in the resulting experimental structure. It may well be, in this case, that the predicted structure is a more accurate statement of the solution conformation than the X-ray crystal structure. How could you decide whether this is true? One clue is available from the AlphaFold2 calculation itself: In addition to the structure prediction, AlphaFold reports the expected reliability of different parts of each predicted protein structure, based on an internal confidence measure. For this prediction, the error estimates for the residues in this loop are above average, but do not stand out as much larger than other regions of the structure.

So, how does it work? AlphaFold2 takes as input data a query amino-acid sequence, and first gathers a set of similar sequences from databanks, to form a multiple-sequence alignment. Patterns of residue conservation shown in such alignments contain many clues to the three-dimensional structure. For instance, positions at which residue variability is correlated suggest residues that are nearby in three-dimensional space. This is the type of information that is measured by NMR spectroscopy (see Figure 12.26).

AlphaFold2 also deals with the conformational energy of different conformations, based on force fields that include terms representing bond lengths and angles, torsional angles, and interactions among non-bonded interactions.

AlphaFold2 processes this information by a machine-learning technique called a 'neural network'. There are multiple, intersecting pathways of information transfer from the input data to the structure prediction. In a rough simulation of the real neural networks in animal brains, the network is reconfigurable, and the strengths associated with different pathways through it can be altered. 'Training', on known structures, allows calibration and refinement of the network. Changes in the network, analogous to 'learning', will alter the output—the conclusion from the data, in the form of a predicted structure. After training, the mature network is applicable to prediction of unknown protein structures from their amino-acid sequences.

AlphaFold2 has been justifiably described as a 'game changer'. Experimental determinations of three-dimensional structure—by X-ray crystallography, NMR, or cryo-EM—remain labour-intensive. The number of known amino-acid sequences, derived from high-throughput DNA sequencing, has greatly exceeded the number of experimentally determined structures—the Worldwide Protein Data Bank contains slightly over 200 000 experimentally determined entries. AlphaFold2 models have unlocked the structural information inherent in the sequences: the AlphaFold2 database contains 2×10^8 entries (https://alphafold.ebi.ac.uk/), and the software is available.

There is consensus that the success of AlphaFold2 is one of the major scientific achievements of the century. News of the result has spread beyond the specialist literature; see, for example:

https://www.theguardian.com/technology/2020/nov/30/deepmind-ai-cracks-50-year-old-problem-of-biology-research

https://www.economist.com/science-and-technology/2020/11/30/how-do-proteins-fold

and even https://www.youtube.com/watch?v=ZHfumZVPjVA

Classifications of protein structures

The large and varied corpus of protein structures presents a challenge in how to classify them. There is consensus that the proper objects to compare are domains.

Ernest Rutherford famously said: 'All science is either physics or stamp collecting.' I reply: 'The study of protein structure combines the best elements of both'.

Several websites offer hierarchical classifications of the entire Protein Data Bank, according to the folding patterns of the proteins. These include:

- SCOP: Structural Classification of Proteins
  http://scop.mrc-lmb.cam.ac.uk/scop/
- CATH: Class/Architecture/Topology/Homologous superfamily
  http://www.cathdb.info
- DALI: based on extraction of similar structures from distance matrices
  http://ekhidna2.biocenter.helsinki.fi/dali/
- ECOD: Evolutionary Classification of Protein Domains
  http://prodata.swmed.edu/ecod/
- CE: a database of structural alignments
  https://www.sdsc.edu/pub/envision/v16.3/ce.html#DATABASE

These sites describe projects *derived* from the primary archival databases of macromolecular coordinates. They are useful general entry points to protein structural data.

KEY POINT

It is in the tertiary structure of domains that proteins show their individuality and variety. Classifying proteins according to their tertiary structure indicates evolutionary relationships (or, at the very least, interesting structural similarities) between proteins that might have diverged so far that the relationship is not detectable by comparing their amino-acid sequences.

Structural Classification of Proteins (SCOP)

SCOP, by A.G. Murzin, L. Lo Conte, B.G. Ailey, S.E. Brenner, T.J.P. Hubbard, and C. Chothia, organizes protein structures in a hierarchy according to evolutionary origin and structural similarity. At the lowest level of the hierarchy are individual domains. SCOP groups sets of domains into **families** of homologues, for which the similarities in structure and sequence (and sometimes function) imply a common evolutionary origin (see Figure 12.33). Families, containing proteins of similar structure and function but for which the evidence for evolutionary relationship is suggestive but not compelling, form **superfamilies**. Superfamilies that share a common folding topology, for at least a large central portion of the structure, are grouped as **folds**. Finally, each fold group falls into one of the general **classes**. The major classes in SCOP are α, β, $\alpha + \beta$, α / β, and miscellaneous 'small proteins', many of which have little secondary structure and have structures stabilized by disulphide bridges or ligands. Box 12.9 shows the SCOP classification of *E. coli* CheY (see Figure 12.34).

KEY POINT

SCOP (Structural Classification of Proteins) offers facilities for searching on keywords to identify structures, navigation up and down the hierarchy, generation of pictures, access to the annotation records in the PDB entries, and links to related databases.

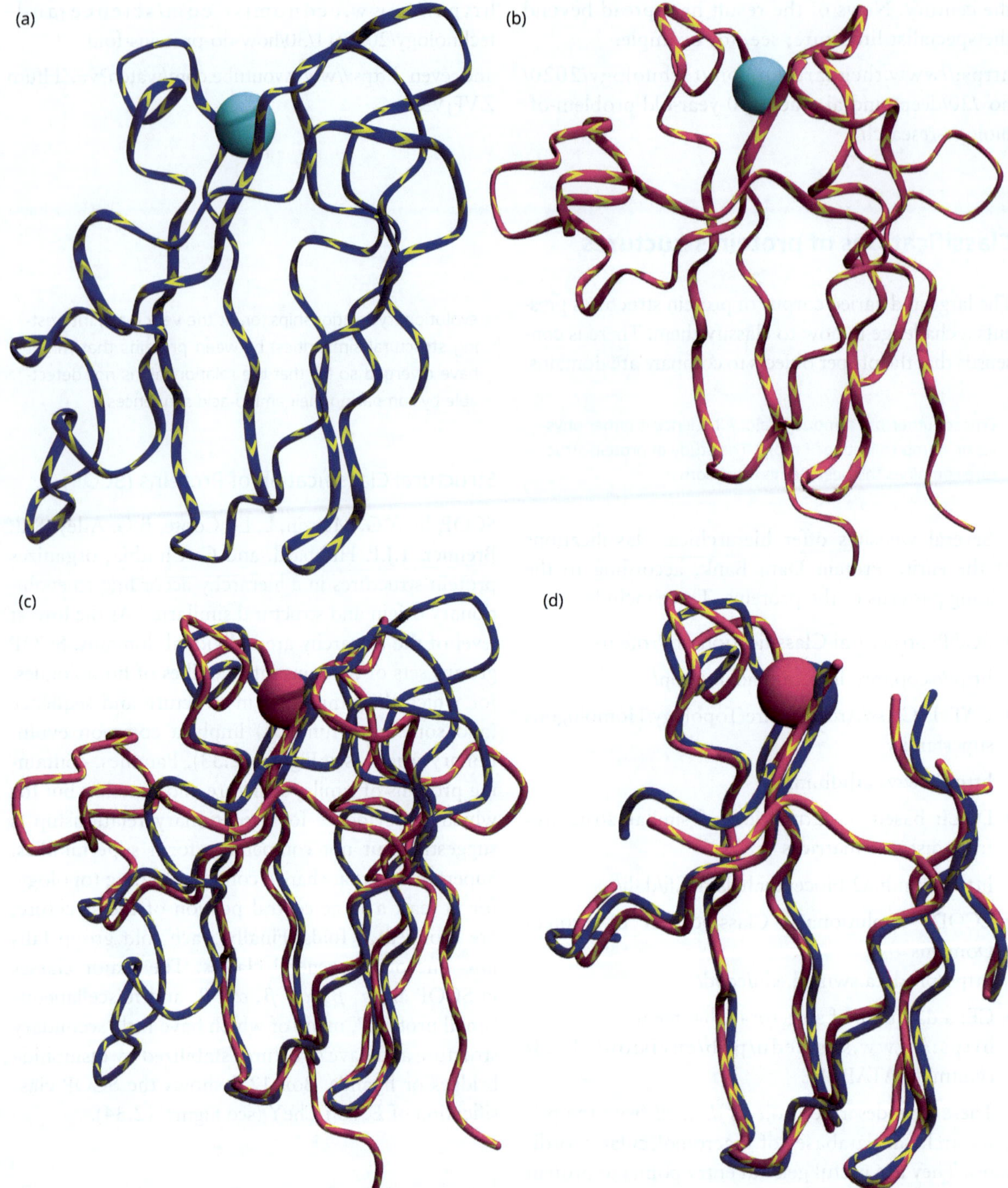

Figure 12.33 Two homologous proteins that share the same general folding pattern, but differ in detail. Circles represent copper ions. (a) Spinach plastocyanin [1AG6], (b) cucumber stellacyanin [1JER]. Superposition showing (c) the entire structures and (d) only the well-fitting core (plastocyanin, blue; stellacyanin, magenta). The main secondary structural elements of these proteins are two β-sheets packed face-to-face. It is seen in the superposition that several strands of β-sheet are conserved but displaced, and that the helix at the right of the cucumber stellacyanin structure has no counterpart in the spinach plastocyanin structure. Even the (relatively) well-fitting core shows the conservation of folding topology but nevertheless reveals considerable distortion.

BOX 12.9 SCOP classification of CheY protein of *Escherichia coli*

1. *Root*: SCOP.
2. *Class*: Alpha and beta proteins (α / β)

   Mainly parallel β-sheets (β-α-β units).
3. *Fold*: Flavodoxin-like.

   Three layers, α / $\beta\alpha$; parallel β-sheet of five strands, order 21345.
4. *Superfamily*: CheY-like.
5. *Family*: CheY-related.
6. *Protein*: CheY protein.
7. *Species*: *Escherichia coli*.

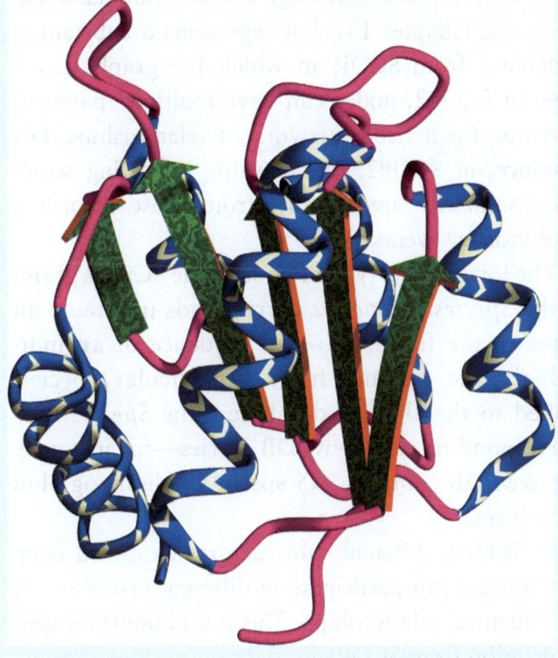

Figure 12.34 *E. coli* CheY protein. CheY is a bacterial signal-transduction protein involved in regulation of flagellar dynamics in chemotaxis. Activation of a receptor causes phosphorylation of CheY. Phosphorylated CheY interacts with flagellum protein FliM to induce tumbling [3CHY].

To locate a protein of interest in SCOP, the user can traverse the structural hierarchy, or search via keywords, such as protein name, PDB code, function (including Enzyme Commission number), and name of fold (for instance, barrel). For each structure, SCOP provides textual information, pictures, and links to other databases. An extended version, SCOPe, is available from the group of Steven Brenner at the University of California at Berkeley, in the US: https://scop.berkeley.edu/

SCOP2

There is consensus that SCOP has made a major contribution to our understanding of protein structure and evolution. Testimony includes the plaudits of its user community and the fidelity of its imitators. SCOP2 (scop2.mrc-lmb.cam.ac.uk) is a redesigned successor to SCOP. In presenting SCOP2, Andreeva *et al.* write:[6]

'... it has become clear that relationships between proteins are more complex than anticipated and that protein evolutionary pathways do not always conform to the same rules. The vast amount of structural information also provided new insights into the mechanisms underlying molecular recognition and evolution of protein structure.

. . .

'... Therefore, we endeavor to develop a more advanced framework for presentation of protein relationships, a new classification scheme that can be adapted to any particular case and evolutionary scenario.'

[6] Andreeva, A., Howorth, D., Chothia, C., *et al.* (2014). SCOP2 prototype: A new approach to protein structure mining. Nucl. Acid Res. 42: D310–D314.

SCOP2 is a Directed Acyclic Graph, in which the nodes represent relationships between selected regions of proteins. This form of graph is similar to those of The Gene Ontology Consortium classifications (see Chapter 13), but represents a substantial difference from SCOP, in which the graph was a tree. In SCOP2, nodes can have multiple 'parents', allowing for a richer network of relationships. For instance, in SCOP2, relationships indicating structural similarity are separate from those recording evolutionary events.

The lowest level type of node in the SCOP2 graph is the *Species*. A *Species* corresponds in almost all cases to the full amino-acid sequence of an individual gene product from a particular species, linked to the UniProt database. One *Species* may correspond to many wwPDB entries—for instance, the wwPDB contains 345 sperm whale myoglobin structures.

In SCOP2, different substructures extracted from any *Species* can participate in different evolutionary or structural relationships. This is an important generalization from SCOP.

In the SCOP2 graph, some edges connect *Species* nodes with other relationship nodes. The edges select regions of the *Species* that participate in, for example, relationships linking regions from different *Species* showing a common *Fold*. Thus the protein corresponding to one *Species* may contain several structural domains assigned to different *Folds*. A *Family* region can comprise several of these structural domains, provided that they occur in the same arrangement in all members of the *Family*. In addition, each of these structural domains may participate individually, or in combinations with others, in different *Superfamily* relationships. The substructure corresponding to a *Family* is typically longer than the substructures that correspond to *Superfamilies (Why?)*

Other edges in the graph link general categories of relationships with the more specific ones that they contain. For instance, *Evolutionary relationships* contains: *Hyperfamily, Superfamily, Family, Protein*, and *Species*. (A hyperfamily in SCOP2 is a region—perhaps fragmentary—shared by different superfamilies.)

The description of SCOP2 appearing in current protocols contains a number of scripts for sessions, with the authors holding your hand as you go through them.[7] Readers should work through these examples.

CATH

CATH is an acronym for the four levels in its hierarchical classification scheme (1) Class (C-level), (2) Architecture (A-level), (3) Topology (Fold or T-level), (4) Homologous Superfamily (H-level). CATH subclusters Homologous Superfamilies into distinct Functional Families (FunFams), based on measured or predicted similarity of function.

The CATH database and website were established in 1994; since then, CATH has been maintained regularly and kept up to date. To focus on the best-determined structures, CATH retains only PDB entries that are: experimentally determined protein structures, with (i) length >40 amino acid residues, (ii) resolution (of crystal structures) ≤4 Å, and (iii) >70% of sidechains resolved. CATH has integrated AlphaFold2 predictions into its database.

CATH is very rich in information-retrieval tools, including queries for:

- *Three-dimensional Structure:* Submit a sequence and ask whether it is possible to identify its fold.
- *Protein Evolution:* How did a selected family evolve?
- *Protein Function:* CATH Functional Families (FunFams) allows checking whether a submitted sequence belongs to a set of domains that share a function.
- *Conserved Sites:* Check a submitted sequence for conserved sites that are likely implicated in function.

[7] Andreeva, A., Howorth, D., Chothia, C., *et al.* (2015). Investigating protein structure and evolution with SCOP2. Curr. Protoc. Bioinform. 49: 1.26.1–1.26.21.

Protein complexes and aggregates

Within cells, life is organized and regulated by a set of protein–protein and protein–nucleic acid interactions.

Interacting proteins and nucleic acids span a range of structures and activities:

- simple dimers or oligomers in which the monomers appear to function independently;
- oligomers with functional 'cross-talk', including ligand-induced dimerization of receptors and allosteric proteins such as haemoglobin (Figure 1.10), phosphofructokinase, and asparate carbamoyltransferase;
- large fibrous proteins such as **collagen**, keratin, and;
- non-fibrous structural aggregates such as **viral capsids**;
- large aggregates with dynamic properties such as ATP synthase, pyruvate dehydrogenase, the GroEL–GroES chaperonin, and the proteasome;
- protein–nucleic acid complexes, including ribosomes, nucleosomes, transcription regulation complexes, splicing and repair particles, and viruses;
- many proteins, whether monomeric or oligomeric, which *function* by interacting with other proteins. These include all enzymes with protein substrates and many antibodies, inhibitors, and regulatory proteins.

Protein aggregation diseases

Protein interactions are frequently associated with disease, caused by misfolded or mutant proteins that are prone to aggregation. **Amyloidoses** are diseases characterized by extracellular **fibrillar deposits.** Alzheimer's and Huntington's diseases are also associated with protein aggregation. Many aggregates contain proteins in a common crossed-β-sheet structure, different from their native state. There are a variety of causes of protein aggregation, including overproduction of a protein, destabilizing mutations, and inadequate clearance in renal failure (see Box 12.10).

Alzheimer's disease

Alzheimer's disease is a neurodegenerative disease common in older adults. It is associated with two types of deposits:

a. dense insoluble extracellular protein deposits, called **senile plaques.** These contain the Aβ fragment (the N-terminal 40–43 residues) of a cell surface receptor in neurons, the β-protein precursor (βPP) or **amyloid precursor protein (APP).**

b. **neurofibrillary tangles,** twisted fibres inside neurons, containing microtubule-associated protein tau. There is some evidence that amyloid deposits promote tangle formation.

Abnormalities in tau appear in other neurodegenerative diseases, the **tauopathies.**

Prion diseases—spongiform encephalopathies

Prion diseases are a set of neurodegenerative conditions of animals and humans, associated with deposition of protein aggregates in the brain, and a characteristic sponge-like appearance of the brains of affected individuals seen in postmortem investigation; hence the name, **spongiform encephalopathies.**

Prion diseases are unusual among protein deposition diseases in that they are transmissible; and unusual among transmissible diseases in that the infectious agent is a protein. Moreover—another unusual feature—some prion diseases are hereditary, for example, familial **Creutzfeld–Jacob disease (CJD).** (Distinguish between a disease transmitted from mother to baby perinatally by passage of an

BOX 12.10

Diseases associated with protein aggregates

Disease	Aggregating protein	Comment
Sickle-cell anaemia	Deoxyhaemoglobin-S	Mutation creates hydrophobic patch on surface
Classical amyloidoses	Immunoglobulin light chains, transthyretin, many others	Extracellular fibrillar deposits
Emphysema associated with α1-antitrypsin	Mutant *Z*-antitrypsin	Destabilization of structure facilitates aggregation
Huntington's	Altered huntingtin	One of several polyglutamine-repeat diseases
Parkinson's	α-Synuclein	Found in Lewy bodies
Alzheimer's	$A\beta, \tau$ and $A\beta$ = 40–42 residue fragment	
Spongiform encephalopathies	Prion proteins	Infectious, despite containing no nucleic acid

infectious agent—as in many AIDS cases—and a truly hereditary disease depending on parental genotype.) All hereditary human prion diseases involve mutations in the same gene, one that encodes the protein found in the aggregates deposited in the brains of sufferers of *both* hereditary and infectious prion diseases.

The prion protein can exist in two forms: the normal PrP^{C} and the dangerous PrP^{Sc}. PrP^{Sc} but not PrP^{C} can (a) form aggregates, (b) catalyse the conversion of additional PrP^{C} to PrP^{Sc} within the brain of an individual person or animal, and (c) infect other individuals, by various routes including ingestion of nervous tissue from an affected animal (or person, in the case of kuru).

KEY POINT

- PrP^{C} = prion protein-*Cellular;*
- PrP^{Sc} = prion protein-*Scrapie.*

Prion disease presents widespread health problems for humans and animals. In 2001, a serious epidemic of bovine spongiform encephalopathy (colloquially, 'mad-cow disease') devastated the UK countryside. There was an apparent association with the appearance of human cases of **variant Creutzfeld–Jacob disease (vCJD)**. In the hereditary disease familial CJD, symptoms began to appear in people aged 55–75. Variant CJD affected people in their twenties. It is hypothesized that these outbreaks were associated with transmission of prion protein infections across species barriers: sheep to cows for BSE, and cows to humans for vCJD.

Prion proteins form a family of homologous proteins in many species of animals, and also in yeast, but apparently not in *C. elegans* or *Drosophila*. Normal human prion protein is synthesized as a 253-residue polypeptide. This comprises: an N-terminal signal peptide, followed by a domain containing ~5 tandem repeats of the octapeptide PHGGGWGQ (in mammals), a conserved 140-residue domain, and a C-terminal hydrophobic domain. The signal domain and the C-terminal domain are cleaved off, and the protein is anchored to the extracellular side of the cell membrane of neurons by a GPI (glycosylphosphatidylinositol) group bound to the C-terminal residue Ser231 of the mature protein.

The normal role of PrP^{C} is not clear. Mice in which PrP^{C} has been knocked out develop normally for a time, and eventually die of apparently unrelated developmental defects. In fact, PrP^{C}-knockout mice are not susceptible to infection with PrP^{Sc}, an observation important in proving the mechanism of the disease.

The mechanism by which PrP^{Sc} catalyses the transformation of additional PrP^{C} to PrP^{Sc} is also

Table 12.2 Some diseases associated with prion proteins

Disease	Species affected	Symptoms
scrapie	sheep	hypersensitivity, unusual gait, tremor
bovine spongiform encephalopathy, or 'mad-cow disease'	cow	similar to scrapie
kuru	human	loss of coordination, dementia
Creutzfeld–Jacob disease (CJD)	human, age 55–75	impaired vision and motor control, dementia
variant CJD	human, age 20–30	psychiatric and sensory anomalies preceding dementia
Gerstmann–Straüssler–Scheinker syndrome	human	dementia
fatal familial insomnia	human	sleep disorder

not clear. Inherited prion diseases are associated with mutants, presumably increasing the tendency for conformational mobility (Table 12.2). A related question concerns the kinetics of the process—what governs the rate of accumulation of aggregates that causes many prion diseases to appear only among older adults?

KEY POINT

Many diseases arise from formation of protein aggregates, including: sickle-cell anaemia, amyloidoses, Alzheimer's disease, Huntington's disease, familial and variant Creutzfeld–Jacob disease, and prion diseases. Most of these are genetic, some are infectious.

Properties of protein–protein complexes

Stoichiometry—what is the composition of the complex?

Protein complexes vary widely in the numbers and variety of molecules they contain. Some contain only a few proteins; others are very large. For example, pyruvate dehydrogenase contains hundreds of subunits and some viral capsids contain thousands.

Some prokaryotic proteins containing identical subunits are homologous to eukaryotic proteins containing related but non-identical subunits, arising by gene duplication and divergence. The proteasome is an example (see Box 12.11). Some viruses achieve diversity *without* duplication, by combining proteins with the same sequence but different conformations.

Evolution of the proteasome

The archaeal proteasome contains 14 identical α subunits and 14 identical β subunits. They are arranged in four stacked rings, $\alpha_7 - \beta_7 - \beta_7 - \alpha_7$. All β subunits have protease activity. The core of the eukaryotic proteasome also contains the $\alpha_7 - \beta_7 - \beta_7 - \alpha_7$ stacked ring structure, but each ring contains seven diverged and non-identical subunits. Thus, the eukaryotic proteasome contains seven homologous but non-identical α subunits and seven homologous but non-identical β subunits. Only three of the eukaryotic β subunits have protease activity. The eukaryotic proteasome also contains large regulatory subunits in addition to the α-β rings, which select ubiquitinylated proteins for degradation.

Affinity—how stable is the complex?

The measure of the **affinity** of a complex is the **dissociation constant,** K_D, the equilibrium constant for the *reverse* of the binding reaction:

$$\text{Protein–Ligand} = \text{Protein} + \text{Ligand} \qquad K_D = \frac{[P][L]}{PL}$$

where [P], [L], and [PL] denote the numerical values of the concentrations of protein (P), ligand (L), and protein–ligand complex (PL), respectively, expressed in mol litre^{-1}. The lower the K_D, the tighter the binding. K_D corresponds to the concentration of free ligand at which half the proteins bind ligand and half are free: [P] = [PL].

KEY POINT

The Michaelis constant of an enzyme is the dissociation constant of the enzyme–substrate complex.

Dissociation constants of protein–ligand complexes span a wide range, as shown in Table 12.3.

Structural studies have elucidated several important features of the interactions between soluble proteins, that contribute to affinity.

- *What holds the proteins together?* Burial of hydrophobic sidechains, **hydrogen bonds** and **salt bridges**, and **Van der Waals forces**. A typical protein–protein **interface** might involve 22 residues and 90 atoms, of which 20% would be mainchain atoms, and an occasional water molecule (Box 12.12). Burial of 1 Å^2 contributes ~100 J mol^{-1} to stability. There is, on average, one intramolecular hydrogen bond per 170 Å^2 of interface area. The average value of the surface area buried in binary protein complexes is ~1600 Å^2. The minimum **buried surface area** for stability of a protein–protein complex is ~1000 Å^2.

Table 12.3 Dissociation constants of some protein–ligand complexes

Biological context	Ligand	Typical K_D
Allosteric activator	Monovalent ion	10^{-4}–10^{-2}
Co-enzyme binding	NAD, for instance	10^{-7}–10^{-4}
Antigen–antibody complexes	Various	10^{-4}–10^{-16}
Thrombin inhibitor	Hirudin	5×10^{-14}
Trypsin inhibitor	Bovine pancreatic trypsin inhibitor	10^{-14}
Streptavidin	Biotin	10^{-15}

- *Do proteins change conformation in complexing?* In some cases, the interaction energy has to 'pay for' the conformational change and the interface tends to be correspondingly larger. Complexes that involve conformational changes generally bury >2000 Å^2.
- *What determines specificity?* Complementarity of the occluding surfaces, in shape, hydrogen-bonding potential, and charge distribution. Prediction of protein complexes from the structures of the partners is the **docking** problem. Reliable solution of this problem, together with progress in structural genomics, would permit *in silico* screening of proteomes for interacting partners.

How are complexes organized in three dimensions?

When two proteins form a complex, each leaves a 'footprint' on the surface of the other, defining the portion of the surface involved in the interaction. If two proteins interact using the *same* surface on both, the complex is *closed*. If two proteins interact through *different* surfaces, the complex is *open*. The significance is that a closed complex does not allow additional proteins to bind with the same interaction. An open complex, in which the surface of potential interaction is not occluded, can grow by accretion of additional subunits. Thus, open but not closed complexes are compatible with the formation of aggregates by continued addition of monomers making the same interaction.

A protein–protein interface: phage M13 gene III protein and *E. coli* TolA

During infection of *E. coli* by phage M13, a complex forms between the N-terminal domain of the minor coat gene 3 protein of the phage and the C-terminal domain of a receptor protein in the bacterial cell membrane, TolA (see Figure 12.35).

The complex is stabilized by burial of 1765 Å² of surface area, by combination of β-sheets from both proteins to form an extended β-sheet (see Figure 12.4) and by several linkages of sidechains by hydrogen bonds and salt bridges. The area buried in the complex is divided almost evenly between the two partners.

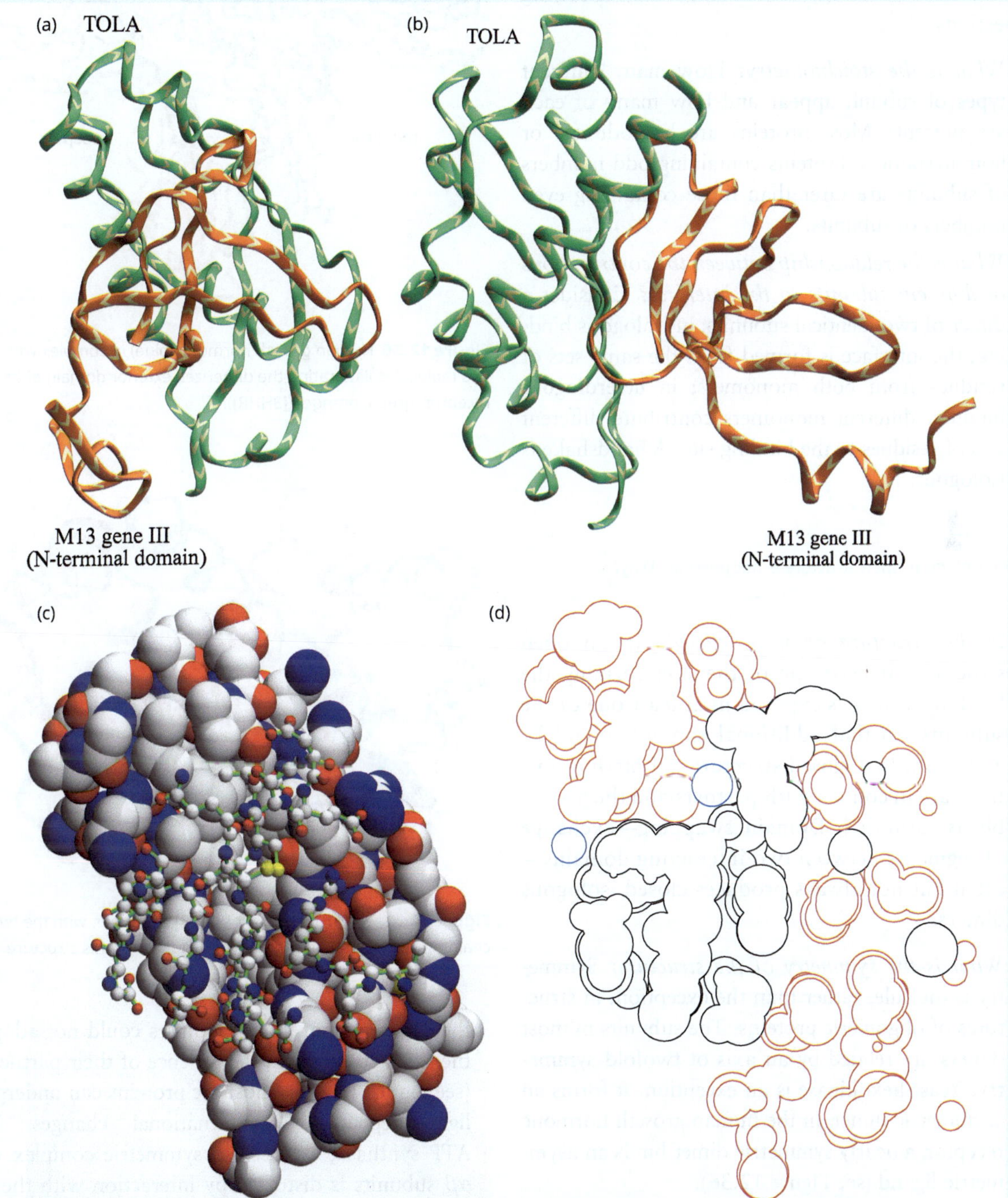

Figure 12.35 The interface between phage M13 gene III protein (N-terminal domain), orange, and *E. coli* protein TolA (C-terminal domain), green [1TOL]. (a, b) Folding patterns and relative orientation of domains, viewed approximately (a, c) perpendicular and (b) parallel to the interface. Note the β-sheet formed from strands contributed by both partners. (d) Slice through the interface, with TolA shown in black, gene III protein in red, and water molecules in blue. It is possible that another water molecule sits next to the one inside the structure.

Multisubunit proteins

An important class of protein–protein complexes is oligomeric or multisubunit proteins. We appeal to structural biology to address the following questions.

- *What is the stoichiometry?* How many different types of subunit appear and how many of each are present? Most proteins are homodimers or homotetramers. Proteins containing odd numbers of subunits are rarer than those containing even numbers of subunits.
- *What is the relationship between the contributions of different subunits to the interface?* Consider a dimer of two identical subunits: in **isologous** binding, the interface is formed from the same sets of residues from both monomers; in **heterologous** binding, different monomers contribute different sets of residues to the binding site. A handshake is isologous.

An isologous open structure is not possible. Why?

- *Is the structure open or closed?* In an open structure, at least one of the sites forming the binding surface is exposed in at least one of the subunits, so that additional subunits could be added on. In a closed structure, all binding surfaces are in contact with partners and the assembly is saturated. **Domain swapping**—exchange of segments between two interacting domains—often but not always produces closed isologous dimers.
- *What is the symmetry of the structure?* Symmetry is the rule, rather than the exception, in structures of oligomeric proteins. The subunits in most dimers are related by an axis of twofold symmetry. Yeast hexokinase is an exception. It forms an asymmetric dimer. In the human growth hormone receptor, a nearly symmetric dimer binds an asymmetric ligand (see Figure 12.36).
- *Do any of the subunits undergo conformational changes on assembly?* Often we don't know. In cases of extensively interlocked interfaces, such as the Trp repressor, the monomers could not adopt the same structure in the absence of their partners (see Figure 12.37). **Allosteric** proteins can undergo ligand-dependent conformational changes. In ATP synthase, a threefold symmetric complex of $\alpha\beta$ subunits is distorted by interaction with the γ subunit.
- *What is its classification within the 'periodic table' of protein complexes reported by Ahnert*

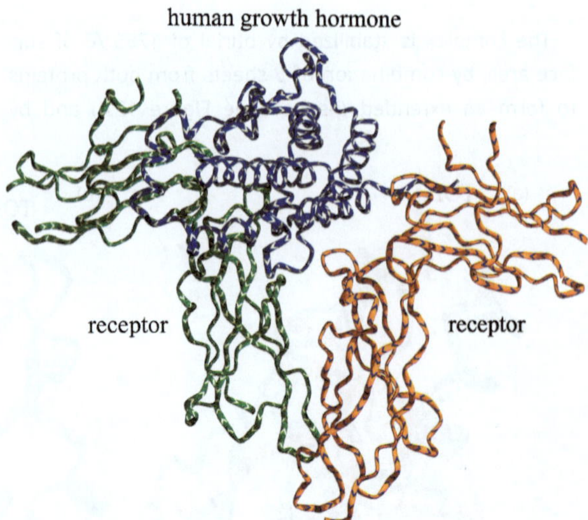

Figure 12.36 Human growth hormone (blue) in complex with two molecules illustrating the dimerized exterior domain of its receptor (green, orange) ([3HHR]).

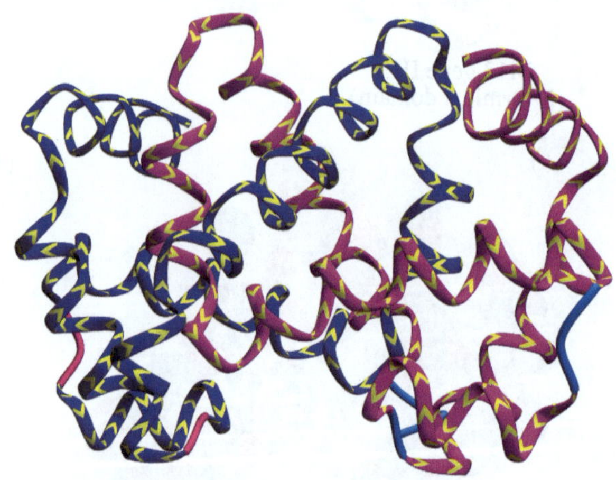

Figure 12.37 The Trp repressor [3WRP] is a dimer, with the two chains intertwined. A monomer could not adopt this structure.

et al.?[8] They derived the table from considerations of combinations of elementary assembly steps and patterns of intersubunit interfaces. Each complex is characterized by (a) the point-group symmetry (columns of the table) and (b) the number of subunits (rows).

Many proteins change conformation as part of the mechanism of their function

The fundamental principle is that proteins fold to unique native structures. However, the mechanism of action of many proteins requires flexibility, or conformational change, during the active cycle. Most protein conformational changes are responses to binding.

- An enzyme may change structure after binding a substrate and/or a cofactor.
- Conformational changes arising from interactions with one or more other proteins, or nucleic acids, are a common component of regulatory mechanisms.
- Some proteins are microscopic motors, interconverting chemical and mechanical energy; for instance, ATP synthase.
- Some **serpins** (serine protease inhibitors) are synthesized in a metastable active state and convert spontaneously to an inactive native state. This gives them a limited lifetime of activity, under tighter control than normal turnover processes. (Serpins also undergo an analogous structural change when they are cleaved by proteases, as part of their mechanism of inhibition.)

KEY POINT

Many proteins are microscopic machines, with internal parts moving in precise ways to support their function.

Experimental structure determinations almost always give us the end states of a conformational transition, but not the intermediates. **Molecular dynamics**—the computational simulation of the motion of proteins—has now advanced enough to be able to calculate detailed trajectories of conformational change.

Conformational change during enzymatic catalysis

The Michaelis–Menten model of enzyme catalysis is:

1. An enzyme (E) reversibly binds a substrate (S) to form a **Michaelis complex** (ES).
2. The ES complex changes to a transition state ($ES^{\ddagger}$), the peak of the thermodynamic barrier between substrate and product.
3. The transition state converts to the enzyme–product complex (EP).
4. The EP complex breaks down to release product (P) and re-form the original enzyme.

This simple model suggests the following dependence of Initial reaction velocity (v_0) on Substrate concentration [S]:

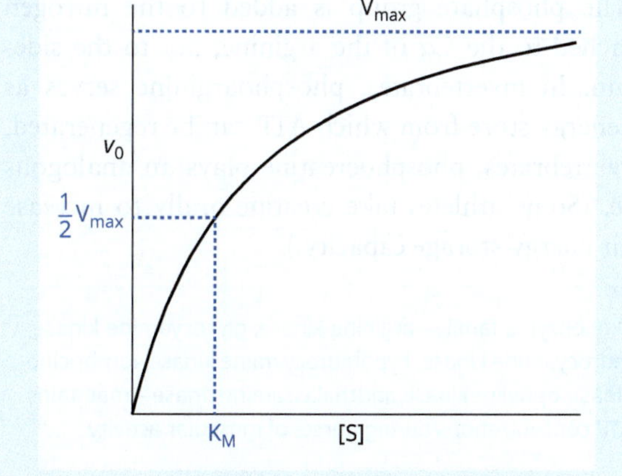

[8] Ahnert, S.E., Marsh, J.A., Hernández, H., *et al.* (2015). Principles of assembly reveal a periodic table of protein complexes. Science 350: aaa2245.

With increasing substrate concentration, the velocity approaches an asymptotic maximum velocity V_{max}. The substrate concentration at which the velocity is $\frac{1}{2}V_{max}$ is K_M. The two parameters V_{max} and K_M characterize the enzyme and the conditions of the measurements. K_M is also the dissociation constant of the enzyme–substrate complex: $K_M = \frac{[E][S]}{[ES]}$.

The equation of the curve is:

$$v_0 = \frac{V_{max}[S]}{K_M + [S]} \quad (v_0 = \text{velocity, [S] = substrate concentration})$$

Many enzymes have different structures in the unligated state, in the Michaelis complex, in the transition state, and/or in the enzyme–product complex.

Many binding sites occur in clefts between protein domains. Binding often induces conformational changes involving reorientation of domains, to close the structure around the ligand. Some of these changes can be described as a 'hinge motion', in which the two domains remain individually rigid but change their relative orientation by means of structural changes in only a few residues in the regions linking the domains. Hinge motion in myosin is responsible for the impulse in muscle contraction.

By their nature, transition states are reactive and difficult to trap long enough for structure determination. Possible solutions include enzymes binding transition-state analogues or inhibitors, or lowering the temperature to slow down the reaction.

Arginine kinase

Arginine kinase catalyses the reaction:

L-arginine + ATP = N-phospho-L-arginine + ADP

The phosphate group is added to the nitrogen attached to the Cα of the arginine, not to the side-chain. In invertebrates, phosphoarginine serves as an energy store from which ATP can be regenerated. In vertebrates, phosphocreatine plays an analogous role. (Some athletes take creatine orally to increase their energy storage capacity.)

One enzyme family—arginine kinase, glycocyamine kinase, taurocyamine kinase, hypotaurocyamine kinase, lombricine kinase, opheline kinase, and thalassamine kinase—maintains ATP concentrations during bursts of muscular activity.

The structure of arginine kinase from the horseshoe crab (*Limulus polyphemus*) differs between the unligated state and the state binding ADP, arginine, and nitrate, a simulation of the transition state in which the nitrate mimics the phosphate group being transferred (see Figure 12.38).

Closure of interdomain clefts also occurs in transport proteins. Transport proteins act as carriers for their ligands, without catalysing reactions.

Ribose-binding protein is one of many periplasmic proteins in bacteria that are involved in chemotaxis and transport (see Figure 12.39). These proteins scavenge for nutrients in the cell's environment by

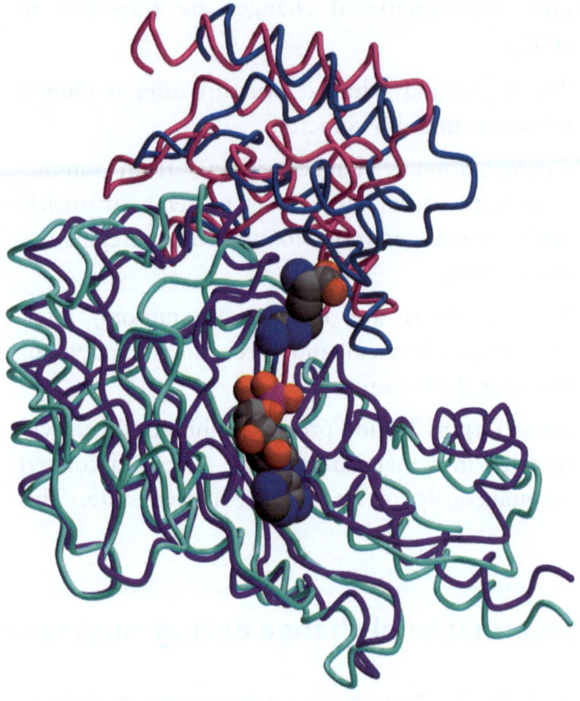

Figure 12.38 Superposition of two conformational states of horseshoe crab arginine kinase [1M15], [1M80]. The unligated state is shown in pink (upper region of figure) and purple (lower region of figure), the ligated state in blue (upper region of figure) and light green (lower region of figure). The reader will see that the domains shown in the lower region of the figure, in purple and light green, superpose well in the ligated and unligated states. In contrast, the domains shown at the top, in pink and blue, show a shift in position between the two states. The ligands, arginine and ADP, appear in the ligated structure only. There are steric clashes between the ligands and the unligated structure in its position in the picture.

The nature of the conformational change has reminded many people of the Venus flytrap. Regions of the structure have come together around the ligands. The motion of the small domain at the top of the picture is primarily a 'hinge' motion—the mobile domain moves almost rigidly around an axis through the interdomain interface. The axis is approximately perpendicular to the page.

The parts of the structure at the lower right also deform. This protein is showing 'induced fit'—in response to ligation.

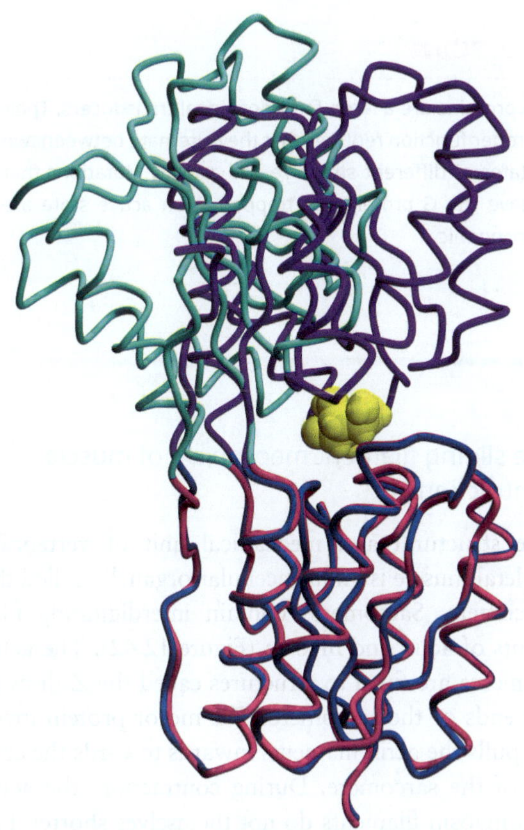

Figure 12.39 Superposition of two structures of ribose-binding protein [2DRI], [1URP]. The unligated structure is shown in pink (lower region of figure) and light green (upper region of figure); the ligated structure is shown in blue (lower region of figure) and purple (upper region of figure). The ribose, in yellow, appears only in the ligated structure. The reader will see that the domains shown in the lower region of the figure, in pink and blue, superpose well in the ligated and unligated states. In contrast, the domains shown at the top, in purple and green, show a shift in position between the two states.

Like arginine kinase (Figure 12.38), this is a 'hinge' motion: the individual domains remain nearly rigid. The conformational change is achieved by rotations about bonds in only a few residues in the hinge region itself.

coupling ligation to interaction with transporters or chemotaxis receptors in the inner membrane. Ribose-binding protein, like many other members of this family, undergoes conformational changes upon ligation such that domains of the protein close around the ligand. The structural changes increase the protein–ligand interactions. They also create a new surface recognized by transport complexes.

Regulation of G protein activity

GTP-binding proteins (or G proteins) are an important class of signal transducer. One of them, p21 Ras, is a molecular switch in pathways controlling cell growth and differentiation (see Figure 12.40). Ras has two conformational states, which differ in the structure of a local mobile region. The resting, inactive state binds GDP. Membrane-bound G-protein-coupled receptors (see Chapter 6) trigger a GDP–GTP exchange transition, associated with a conformational change (see Figure 12.41). Activated Ras binds Raf-1, a serine/threonine kinase. The

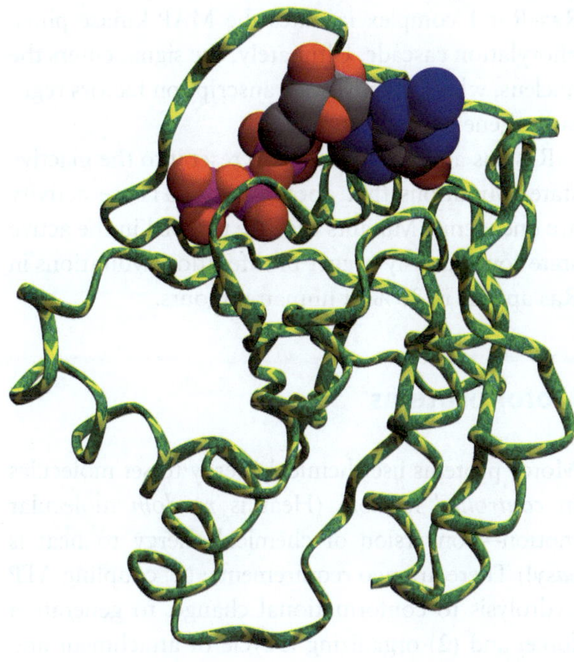

Figure 12.40 p21 Ras bound to GTP. Although an active GTPase, the system was stabilized for crystal-structure analysis by cooling to 100 K [1QRA].

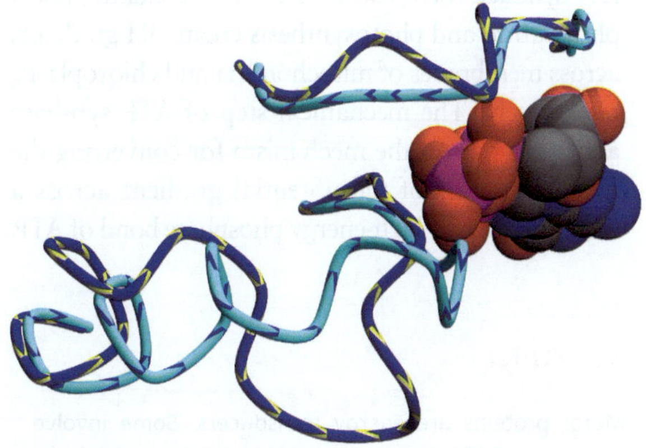

Figure 12.41 The conformational change in p21 Ras from the inactive GDP-binding conformation to the active GTP-binding conformation primarily involves two regions (shown here in light and dark blue) that form a patch on the molecular surface [1QRA], [1Q21].

Ras–Raf-1 complex initiates the MAP kinase phosphorylation cascade. Ultimately, the signal enters the nucleus, where it activates transcription factors regulating gene expression.

Ras has a GTPase activity to reset it to the inactive state. Mutations that abrogate the GTPase activity are oncogenic. Mutants that are trapped in the active state continuously trigger proliferation. Mutations in Ras appear in 30% of human tumours.

KEY POINT

G proteins are a large family of signal transducers. Their proper function requires that they alternate between two states of different structure and activity. Mutations that leave the G protein Ras trapped in an active state are oncogenic.

Motor proteins

Motor proteins use chemical energy to set molecules in *controlled* motion. (Heat is *random* molecular motion—conversion of chemical energy to heat is *easy*!) There are two requirements: (1) coupling ATP hydrolysis to conformational change, to generate a force; and (2) organizing a cycle of attachment and detachment to a mechanical substrate, to allow the force to generate movement.

Some motor proteins propel themselves—and their cargo—by exerting force against a stationary object, such as a cytoskeletal filament. Others remain stationary and propel movable objects.

- *Myosins* interact with actin during muscle contraction.
- *Kinesins* and *dyneins* interact with microtubules, mediating organelle transport, chromosome separation in mitosis, and movements of cilia and flagella.

Myosins, kinesins, and dyneins are primarily *linear motors*. In contrast, ATP synthase is a *rotary motor*.

- ATP synthase rotates during its action. Oxidative phosphorylation and photosynthesis create pH gradients across membranes of mitochondria and chloroplasts, respectively. The mechanical step of ATP synthase activity is part of the mechanism for converting the osmotic energy of the potential gradient across a membrane to the high-energy phosphate bond of ATP.

KEY POINT

Motor proteins are energy transducers. Some involve conversion of chemical energy—via ATP hydrolysis—to mechanical energy. ATP synthase converts chemiosmotic energy to chemical energy by ATP formation, with a rotary motor as part of its mechanism.

The sliding filament mechanism of muscle contraction

The structural and mechanical unit of vertebrate skeletal muscle is an intracellular organelle called the **sarcomere**. Sarcomeres contain interdigitating filaments of actin and myosin (Figure 12.42). The actin filaments are fixed to structures called the Z-discs at the ends of the sarcomere. The motor protein myosin pulls the actin filaments inwards towards the centre of the sarcomere. During contraction, the actin and myosin filaments do not themselves shorten but slide past one another, shortening the sarcomere by

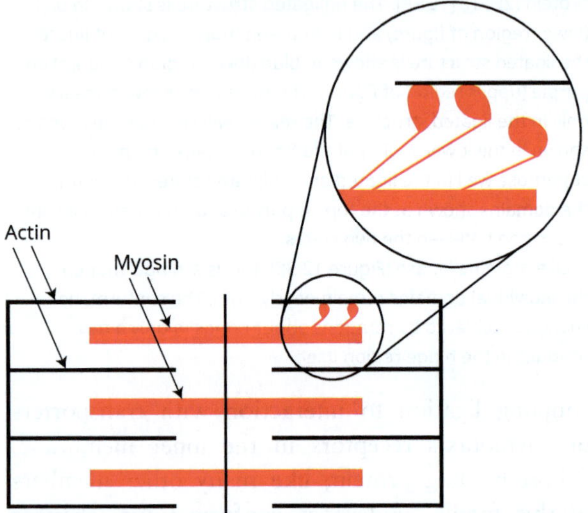

Figure 12.42 Schematic diagram of a sarcomere. Thick myosin filaments overlap thin actin filaments. In the main diagram, it is cursorily indicated that multiple myosin molecules from thick filaments interact with adjacent thin filaments. In fact, each thick filament contains several hundred myosin molecules. The inset shows different stages of the power stroke. From left to right: attachment, conformational change propelling the thin filament inwards by ~10 nm, detachment (followed by recovery of original conformation of the myosin head.)

increasing the region of overlap. Think of the shortening of a bicycle pump during its compression stroke.

A large muscle may contain ~ $10^4 - 10^5$ sarcomeres, laid end to end. Each sarcomere has a resting length of ~ 2.5 μm and can contract by ~ 0.3 μm. Therefore, the entire muscle can contract by about 1–2 cm.

Individual myosin molecules are large fibrous proteins of relative molecular mass ~ 5×10^5. They contain a fibrous section ~ 1.6 – 1.7 μm long, and a globular head. Each thick filament contains ~ 200 – 300 myosin molecules. The mechanical coupling between actin and myosin occurs through the myosin head, as shown in Figure 12.42.

During the power stroke, the myosin head undergoes a cycle of attachment–detachment and conformational change. From left to right in the inset in Figure 12.42, attachment of the myosin head is followed by conformational change that propels the actin towards the centre of the sarcomere. Detachment is followed by restoration of the initial conformation of the myosin. The myosin heads are like oars that 'row' the actin filaments towards the centre of the sarcomere. The displacement of the actin is ~10 nm per myosin molecule per cycle. Hydrolysis of one molecule of ATP during each cycle of each myosin molecule provides the energy.

Structures of fragments of myosin containing the globular head have defined the mechanism of the conformational change (see Figure 12.43).

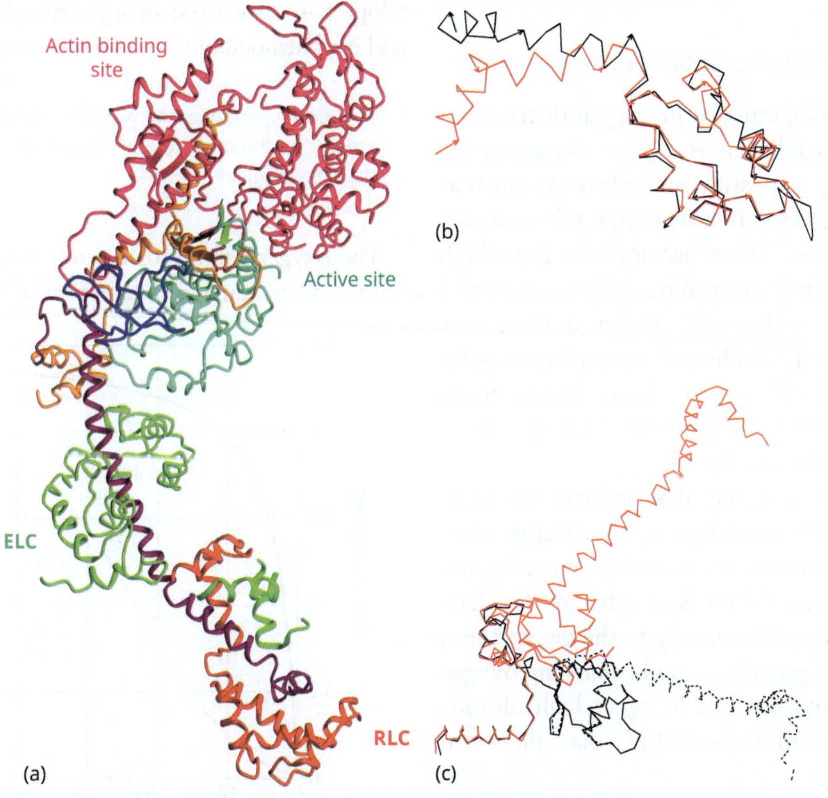

Figure 12.43 The contraction of muscle is a transformation of chemical energy to mechanical energy. It is carried out at the molecular level by a hinge motion in myosin, while myosin is attached to an actin filament. The cycle of *attach to actin–change conformation–release from actin* in a large number of individual myosin molecules creates a macroscopic force within the muscle fibre. (a) The structure of myosin subfragment 1 from chicken. The active site binds and hydrolyses ATP. ELC and RLC are the essential and regulatory light chains [2MYS]. (b) Hinge motion in myosin. Comparison of parts of chicken myosin open form [2MYS] (no nucleotide bound) and closed form binding the ATP analogue ADP-ALF_4^- [1BR2]. This shows the segments of the structure that surround the hinge region. (c) Model of the swinging of the long helical region in myosin as a result of the hinge motion. The dashed line shows a model of the position that the complete long helix would occupy in the closed form [2MYS] and [1BR1]. This conformational change is coupled to hydrolysis of ATP. It takes place while myosin is bound to actin, providing the power stroke for muscle contraction. In the context of the assembly and mechanism of function of a muscle filament, it is arguable that one should regard the helix as fixed and the head as swinging. However, this would not show the magnitude of the conformational change as dramatically.

Allosteric regulation of protein function

Allostery is modulation of the activity of a protein at one site by structural changes caused by binding a molecule at a distant site.

Allosteric proteins show 'action at a distance': ligand binding at one site affects activity at another. An impulse at the first site must transmit a conformational change affecting the second. In contrast, GTP-activated p21 Ras (Figures 12.40 and 12.41) shows ligand-induced regulation of activity, but the structural change is adjacent to the ligand. It is more challenging to explain the properties of haemoglobin, in which the shortest distance between binding sites is over 20 Å (2 nm).

Cooperativity in haemoglobin oxygen binding

To play its physiological role in oxygen distribution effectively, haemoglobin must capture oxygen in the lungs as efficiently as possible and release as much as possible to other tissues. To achieve this 'take from the rich, give to the poor' effect, haemoglobin has a high oxygen affinity at high oxygen partial pressure (pO_2) and a low affinity at low pO_2. Haemoglobin shows positive **cooperativity:** binding of oxygen *increases* the affinity for additional oxygen (Figure 12.44). Some proteins show negative cooperativity: binding *reduces* the affinity for additional ligand.

In contrast to a ligand that induces cooperative binding of the *same* ligand, an **effector** alters the activity of a protein towards a *different* ligand. Bisphosphoglycerate (BPG) is an effector for haemoglobin. It binds preferentially to the deoxy form, decreasing the oxygen affinity and enhancing oxygen release. People and animals living at high altitude have higher concentrations of BPG than their sea-level relatives.

KEY POINT

Allosteric proteins deviate from the Michaelis–Menten curve in ligand binding or, in the cases of allosteric enzymes, in reaction velocity as a function of substrate concentration. The cooperativity is achieved by ligation-induced conformational change.

A mammalian foetus depends on its mother for oxygen. It is perhaps surprising that the oxygen affinity of isolated human foetal haemoglobin is *lower* than that of adult haemoglobin. However, foetal haemoglobin has a lower affinity for the effector BPG. This difference in the interaction with the effector gives foetal haemoglobin a higher oxygen affinity than the maternal haemoglobin.

Allosteric changes in haemoglobin

The vertebrate haemoglobin molecule is a tetramer containing two identical α chains (α_1 and α_2) and two identical β chains (β_1 and β_2) (see Figure 1.10a). It can adopt two structures: deoxyhaemoglobin (unligated) and oxyhaemoglobin (four oxygen molecules bound).

The difference in colour between arterial and venous blood reveals the different state of the iron in ligated and unligated haemoglobin.

The oxygen affinity of the oxy form of haemoglobin is similar in magnitude to that of isolated α and β

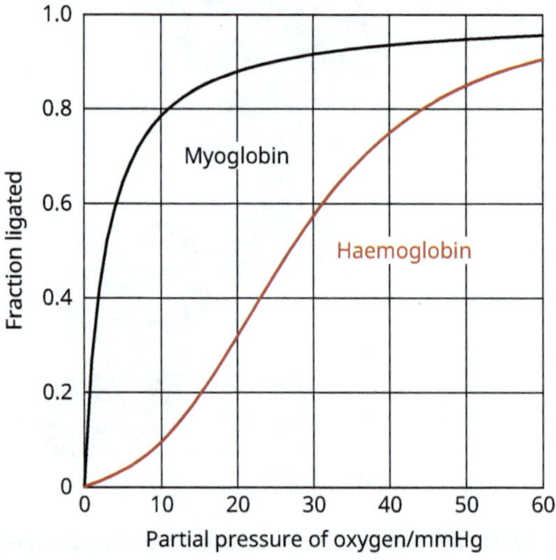

Figure 12.44 Oxygen-dissociation curves for myoglobin and haemoglobin. Myoglobin shows a simple equilibrium, with a binding constant independent of oxygen concentration. Haemoglobin shows positive cooperativity, the binding constant for the first oxygen being several orders of magnitude smaller than the binding constant for the fourth oxygen. The units for partial pressure are traditional in the literature about this topic: 760 mmHg = 1 atmosphere = 101 325 Pa.

subunits and to that of myoglobin, a monomeric globin. The oxygen affinity of the deoxy form is much less: the ratio of binding constants for the first and fourth oxygens is 1:150–300, depending on conditions. Therefore, it is the deoxy form that is special, as it has had its oxygen affinity 'artificially' reduced. J. Monod, J. Wyman, and J.-P. Changeux proposed that the reduced oxygen affinity of the subunits in the deoxy state of haemoglobin arises from structural constraints that hold the subunits in a **'tense' (T) state**, internally inhibited, whereas the oxy form is in a **'relaxed' (R) state**, as free to bind oxygen as the isolated monomer.

A general model for cooperativity is that the subunits of a protein are in equilibrium between the T and R forms. This model rationalizes the properties of haemoglobin.

1. At low partial pressures of oxygen, all of the subunits of haemoglobin are in the T form and are unligated (i.e., not binding oxygen). The binding constant for oxygen is low, because binding to the T state is inhibited.
2. At high partial pressures of oxygen, all of the subunits of haemoglobin are in the R form and each subunit binds an oxygen. The binding constant for oxygen is high, because binding of oxygen to the R state is unconstrained.
3. In the erythrocyte, haemoglobin is an equilibrium mixture of deoxy and oxy forms; the concentration of partially ligated forms is tiny. Binding of between two and three oxygen molecules shifts the subunits *concertedly* from all being T state to all being R state.
4. Effector molecules such as BPG modify oxygen affinity by shifting the T = R equilibrium, by preferentially stabilizing one of the two forms.

The interpretation of this scheme in structural terms was one of the early triumphs of protein crystallography. The structures of haemoglobin in different states of ligation have been studied with intense interest, because of their physiological and medical importance and because they were thought to offer a paradigm of the mechanism of allosteric change. The two crucial questions to ask of the haemoglobin structures are:

1. What is the mechanism by which the oxygen affinity of the deoxy form is reduced?
2. How is the equilibrium between low and high-affinity states altered by oxygen binding and release?

Comparison of the oxy and deoxy structures has defined the changes in tertiary structures of individual subunits, and in the quaternary structure. The allosteric change involves an interplay between changes in tertiary and quaternary structure (see Figure 12.45).

- The details of the quaternary structure—the relative geometry of the subunits and the interactions at their interfaces—is determined by the way the subunits fit together.
- The fit of the subunits depends on the shapes of their surfaces.
- The tertiary structural changes alter the shapes of the surfaces of the subunits, changing the way they fit together.

The haemoglobin tetramer can be thought of as a pair of dimers: $\alpha_1\beta_1$ and $\alpha_2\beta_2$. The allosteric change involves a rotation of 15° of the $\alpha_1\beta_1$ dimer with respect to the $\alpha_2\beta_2$ around an axis approximately perpendicular to their interface. (The motion is like that of a pair of shears with α_1 and α_2 as the blades and β_1 and β_2 as the handles.)

The impulse for the conformational change in haemoglobin is a change in the size of the iron ion. In the unligated state, the iron is in an electronic state with too large a radius to fit between the central nitrogen atoms of the porphyrin rings of the haem group. It lies out of the haem plane, and in fact the haem is domed in the direction of the iron, reflecting the attraction between the nitrogens and the iron. When oxygen binds, the electronic state of the iron changes. Its radius becomes smaller, and it can fit in the haem plane in the gap between the pyrrole nitrogens. This motion of the iron pulls on the proximal histidine. The force is transmitted through the subunit as far as its surface, altering its tertiary structure, and changing the shape of its surface.

The change in the shape of the surface of interaction between subunits destabilizes the deoxy quaternary structure. Formation of the oxy quaternary structure transmits the tertiary structural changes to other subunits.

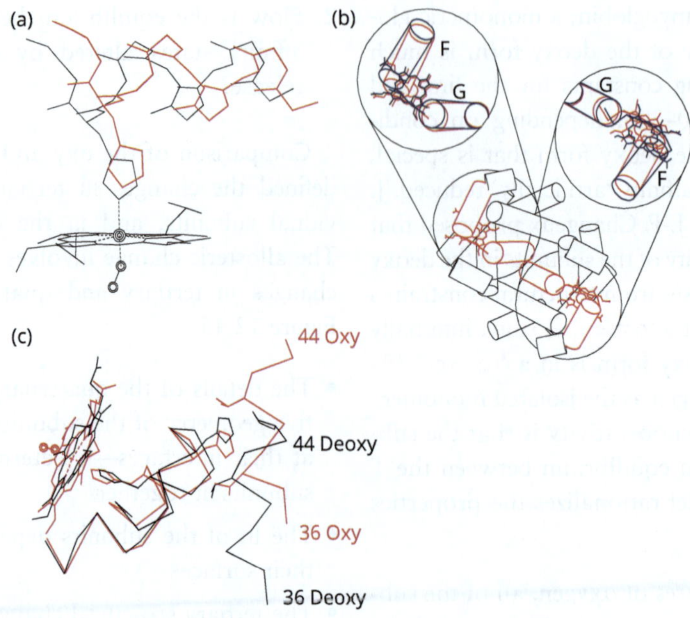

Figure 12.45 Some important structural differences between oxy- and deoxyhaemoglobin [1HHO], [2HHB]. (a) Changes at the haem group in human haemoglobin result in a change in state of ligation. This figure shows the F helix, proximal histidine, and haem group of the β chain in the oxy (black) and deoxy (red) forms; only the oxy haem is shown. Pulling the iron into the plane of the haem group exerts a force on the histidine, causing a shift in the position of the F helix. This shift in turn creates forces transmitted to the rest of the subunit, causing the changes in tertiary structure. For this picture, the structures were superposed on the haem group. (b) The $\alpha_1\beta_1$ dimer in oxy (red) and deoxy (black, in blown-up regions only) forms. In the blown-up regions, only the F helix, FG corner, and haem group are shown. The oxy and deoxy $\alpha_1\beta_1$ dimers have been superposed on their interface; in this frame of reference, there is a small shift in the haem groups and a shift and conformational change in the FG corners. (c) Alternative packing of α_1 and β_2 subunits in oxyhaemoglobin (red) and deoxyhaemoglobin (black). The oxy and deoxy structures have been superposed on the F and G helices of the α_1 monomer. Although for the purposes of this illustration we have regarded the α_1 subunit as fixed and the β_2 subunit as mobile, only the relative motion is significant.

In more detail:

- Starting from the deoxy structure, ligation of oxygen creates strain at the haem group, arising from a change in the position of the iron and the histidine sidechain linked to it.
- To relieve this strain, there are shifts in the F helix and the FG corner (the region of the chain between the F and G helices).
- To accommodate these shifts, a set of tertiary structural changes alter the overall shape of the $\alpha_1\beta_1$ and $\alpha_2\beta_2$ dimers, notably the shifting of the relative positions of the FG corners.
- In consequence, the deoxy quaternary structure is destabilized because the dimers no longer fit together properly (having changed their shape).
- Adopting the alternative quaternary structure requires the tertiary structural changes to take place even in subunits not yet liganded. As a result of the quaternary structural change, these unligated subunits have been brought to a state of enhanced oxygen affinity.

It is important to emphasize that this is a sequence of steps in a logical process and not a description of a temporal pathway of a conformational change. An analogy: Consider an Olympic vaulter. Suppose you have a picture of the athlete poised, still, before commencing the run-up, and another picture of the athlete again poised, still, after landing the dismount. Comparison of these two pictures gives no clue to the complex series of motions that took place between them. For haemoglobin M. Karplus and co-workers have simulated a detailed trajectory between oxy and deoxy states by molecular dynamics.

Protein evolution at the lab bench

So far, we have discussed protein evolution in nature. Creation of novel, artificial proteins in the laboratory has succeeded. There are two approaches: (1) Directed evolution and (2) Computational protein design.

Directed evolution

One strand of Darwin's thinking that led to the theory of evolution was the observation that farmers could improve the quality of livestock by selective breeding. He drew an analogy between this *artificial selection* and the idea of *natural selection* that he was proposing as the process of evolution. We now recognize that evolution by natural selection takes place at the molecular level. Why not artificial selection also?

Natural proteins do many things, but not everything we would like them to. For applications in technology, it would be useful to have proteins that would:

- have activities unknown in nature;
- show activity towards unnatural substrates, or altered specificity profiles;
- be more robust than natural proteins, retaining their activity at higher temperature or in organic solvents; and
- show different regulatory responses, enhanced expression, or reduced turnover.

Scientists have used **directed evolution**—or artificial selection—to generate molecules with novel properties starting from natural proteins.

Evolution requires the generation of variants and differential propagation of those with favourable features. Molecular biologists dealing with microbial evolution have advantages over the farmers that Darwin observed. We can generate large numbers of variants artificially. Screening and selection can, in many cases, be done efficiently, by stringent growth conditions, and there are virtually no limits on the size of the 'flock' or 'litter'. Darwin might well have been envious. He wrote:

> . . . as variations manifestly useful or pleasing to man appear only occasionally, the chance of their appearance will be much increased by a large number of individuals being kept. Hence, number is of the highest importance for success. On this principle Marshall formerly remarked, with respect to the sheep of parts of Yorkshire, 'as they generally belong to poor people, and are mostly in small lots, they never can be improved.'
>
> –*The Origin of Species*, Chapter 1.

The procedure of **directed evolution** comprises these steps:

1. Create variant genes by mutagenesis or genetic recombination.
2. Create a library of variants by transfecting the genes into individual bacterial cells.
3. Grow colonies from the cells and screen for desirable properties.
4. Isolate the genes from the selected colonies and use them as input to step 1 of the next cycle.

Strategies for generating variants include (a) single and multiple amino acid substitutions, (b) recombination, and (c) formation of chimaeric molecules by mixing and matching segments from several homologous proteins. Each method has its advantages and disadvantages. The smaller the change in sequence, the more likely that the result will be functional. Yet, multiple substitutions or recombinations give a greater chance of generating novel features. The choice depends in part on the nature of the goal. For instance, it is easier to lose a function than to gain one. (Why would you want to *lose* a function? Removal of product inhibition to enhance throughput in an enzymatically catalysed process is an example.)

Directed evolution of subtilisin E

Subtilisins are a family of bacterial proteolytic enzymes. Subtilisin E, from the mesophilic bacterium *Bacillus subtilis*, is a 275-residue monomer. It becomes inactive within minutes at 65°C. Directed evolution has produced interesting variants, with features including enhanced thermal stability and activity in organic solvents.

Enhancement of thermal stability by directed evolution

Thermitase, a subtilisin homologue from *Thermoactinomyces vulgaris*, remains stable up to 80°C. The existence of thermitase is reassuring, because it shows that the evolution of subtilisin to a thermostable protein is possible. However, subtilisin E and thermitase differ in 157 amino acid residues. Do we have to go this far? Are all of the changes essential for thermostability, or has there been considerable neutral drift as well?

A thermostable variant of subtilisin E, produced by directed evolution, differs from the wild type by only eight residue substitutions. The variant is identical to thermitase in its temperature of optimum activity: 76°C (17°C higher than the original molecule) and stability at 83°C (a 200-fold increase relative to the wild type).

The procedure involved successive rounds of generation of variants, and screening and selection of those showing favourable properties. The formation of mutations, via error-prone PCR to produce an average of two to three base changes per gene, was alternated with *in vitro* recombination to find the best combinations of substitutions at individual sites (see Figure 12.46). At each step, several thousand clones were screened for activity and thermostability.

The optimal variant differed from the wild type at eight positions: N188S, S161C, P14L, N76D, G166R, N181D, S194P, and N218S. Figure 12.47 shows their distribution in the structure. Most of the substitutions are far from the active site, which is not surprising as the wild type and variant do not differ in function. Most of the sites of substitution are in loops between regions of secondary structure. These regions are the most variable in the natural evolution of the subtilisin family. Two of the substitutions are in α-helices, including P14L. P14L has a certain logic: proline tends to destabilize an α-helix because it costs a hydrogen bond. Only two of the substitutions produce the amino acids that appear in those positions in thermitase.

> Directed evolution of subtilisin has succeeded in producing unnatural 'homologues' active (a) at high temperatures and (b) in organic solvents.

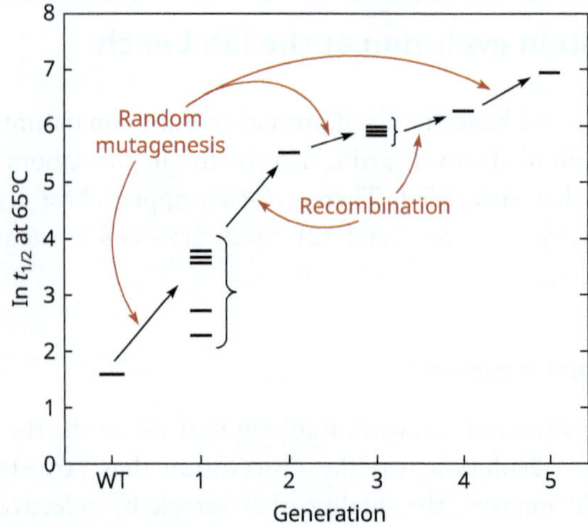

Figure 12.46 Directed evolution of a thermostable subtilisin. ($t_{1/2}$ measured in minutes.) The starting wild type (WT) was subtilisin E from the mesophilic bacterium *B. subtilis*. Steps of random mutagenesis were alternated with recombination. At each step, screening for improved properties and artificial selection chose candidates for the next round.

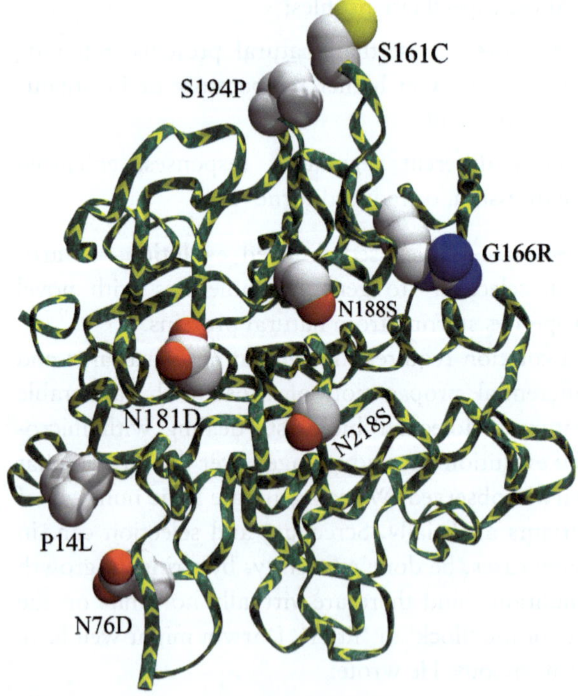

Figure 12.47 The sites of mutation in *B. subtilis* subtilisin E that produced a thermostable variant by directed evolution. The sidechains shown are those of the final product.

Designed enzymes

We know that mutations can change the structure and function of proteins, in some cases in radical ways. We observe this in natural protein evolution and can achieve it by artificial selection (see preceding section). We can do a reasonable job at predicting the structural changes arising from mutations. Putting these together suggests that we should be able to design proteins with altered or even novel functions *in silicio*.

Consider three related problems:

1. *Structure prediction*: Presented with an amino-acid sequence, predict the structure into which it folds.
2. *'Inverse folding problem'*: Presented with a mainchain conformation, devise an amino-acid sequence that will fold into a structure with that mainchain conformation.
3. De novo *protein design*: Devise an amino-acid sequence that will fold up into a structure that has some desired property, such as a novel enzymatic activity.

Extension of methods for protein structure prediction has achieved solutions of these problems. For example, enzymes designed computationally catalyse the retro-aldol reaction, a Kemp-elimination reaction, and the Diels–Alder condensation (see Figure 12.48). As with directed evolution, it has been possible to create enzymes with catalytic activities unknown in living cells. David Baker was awarded the 2024 Nobel Prize in Chemistry for protein design.

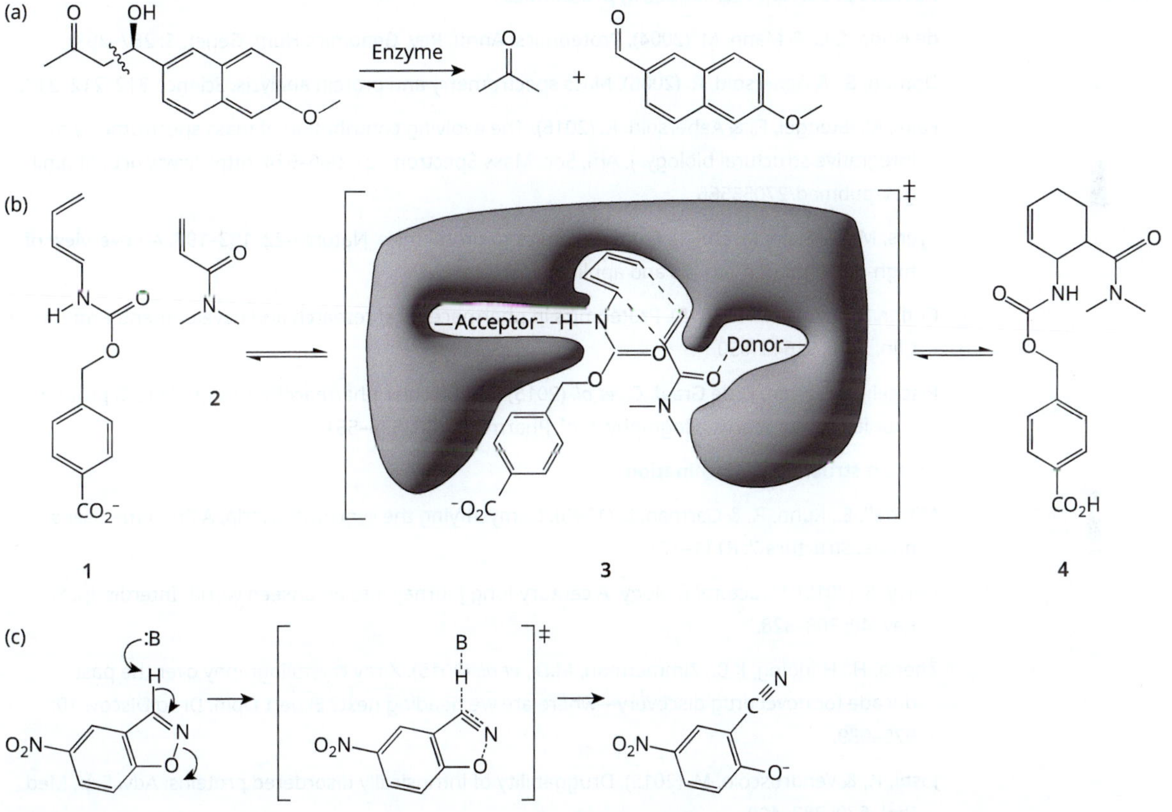

Figure 12.48 Reactions catalysed by computationally designed enzymes: (a) retroaldol reaction, (b) Diels–Alder condensation, (c) Kemp elimination reaction—here B is a general base, provided in the enzyme by a glutamate sidechain.

Panel b: Used with permission of American Association for the Advancement of Science, from Siegel, J.B., Zanghellini, A., Lovick, H.M., *et al.* (2010). Computational design of an enzyme catalyst for a stereoselective bimolecular Diels-Alder reaction. Science 329, 5989: 309–313. https://doi.org/10.1126/science.1190239. Permission conveyed through Copyright Clearance Centre, Inc.

Panel c: Used with permission of Proceedings of the National Academy of Sciences, from Privett, H.K., Kiss, G., Lee, T.M., *et al.* (2012). Iterative approach to computational enzyme design. Proceedings of the National Academy of Sciences of the United States of America 109, 10: 3790–3795. https://doi.org/10.1073/pnas.1118082108.

LOOKING BACK AND FORWARD

We have now followed the central dogma through transcription, discussing control over expression, to the results of translation, the structures of proteins. The major triumph of molecular biology has been to relate protein structure to protein function. This, together with the integration of function, is the subject of the next chapter.

RECOMMENDED READING

Two fine general references

Branden, C.-I., & Tooze, J. (1999). *Introduction to Protein Structure*, 2nd edn. New York: Garland.

Liljas, A., Liljas, L., Piskur, J., *et al.* (2009). *Textbook of Structural Biology*. Singapore: World Scientific.

Historical review

Giegé, R. (2013). A historical perspective on protein crystallization from 1840 to the present day. FEBS Journal 280: 6456–6497.

Reviews of current techniques in proteomics

de Hoog, C.L., & Mann, M. (2004). Proteomics. Annu. Rev. Genomics Hum. Genet. 5: 267–293.

Domon, B., & Aebersold, R. (2006). Mass spectrometry and protein analysis. Science 312: 212–217.

Faini, M., Stengel, F., & Aebersold, R. (2016). The evolving contribution of mass spectrometry to integrative structural biology. J. Am. Soc. Mass Spectrom. 27: 966–974. http://www.ncbi.nlm.nih.gov/pubmed/27056566

Tyers, M., & Mann, M. (2003). From genomics to proteomics. Nature 422: 193–197. An overview of high-throughput methods and applications.

Cutler, P., & Voshol, H. (2015). Proteomics in pharmaceutical research and development. Proteomics Clin. Appl. 9: 643–650.

Piscitelli, C.L, Kean, J., de Graaf, C., *et al.* (2015). A molecular pharmacologist's guide to G protein-coupled receptor crystallography. Mol. Pharmacol. 88: 536–551.

Protein structure determination

Mitchell, E., Kuhn, P., & Garman, E. (1999). Demystifying the synchrotron trip: A first time user's guide. Structure 7: R111–121.

Curry, S. (2015). Structural Biology: A century-long journey into an unseen world. Interdiscip. Sci. Rev. 40: 308–328.

Zheng, H., Handing, K.B., Zimmerman, M.D., *et al.* (2015). X-ray crystallography over the past decade for novel drug discovery—where are we heading next? Expert Opin. Drug Discov. 10: 975–989.

Joshi, P., & Vendruscolo, M. (2015). Druggability of intrinsically disordered proteins. Adv. Exp. Med. Biol. 870: 383–400.

Grabowski, M., Niedzialkowska, E., Zimmerman, M.D., *et al.* (2016). The impact of structural genomics: The first quindecennial. J. Struct. Funct. Genomics 17: 1–16.

Beis, K., & Evans, G. (eds.) (2018). *Protein Crystallography: Challenges and Practical Solutions*. London: Royal Society of Chemistry.

Serrano, P., Dutta, S.K., Proudfoot, A., *et al.* (2016). NMR in structural genomics to increase structural coverage of the protein universe. FEBS J. 283: 3870–3881.

Wlodawer, A. (2021). Protein crystallography: Alive and well. FEBS J. 288: 5786–5787. (Introduction to an entire issue collecting relevant articles.)

State of the art in protein structure prediction

Tramontano, A. (2006). *Protein Structure Prediction: Concepts and Applications*. Weinheim: Wiley-VCH.

Baker, D. (2014). Protein folding, structure prediction and design. Biochem. Soc. Trans. 42: 225–229.

Webb, B., & Sali, A. (2016). Comparative protein structure modeling using MODELLER. Curr. Protoc. Bioinformatics 54: 5.6.1–5.6.37.

Andreeva, A. (2016). Lessons from making the Structural Classification of Proteins (SCOP) and their implications for protein structure modelling. Biochem. Soc. Trans. 44: 937–943.

Zheng, W., Zhang, C., Bell, E.W., *et al.* (2019). I-TASSER gateway: A protein structure and function prediction server powered by XSEDE. Future Gener. Comput. Syst. 99: 73–85.

Jumper, J., Evans, R., Pritzel, A., *et al.* (2021). Highly accurate protein structure prediction with AlphaFold. Nature 596: 583–589.

Varadi, M., Anyango, S., Deshpande, M., *et al.* (2022). AlphaFold Protein Structure Database: Massively expanding the structural coverage of protein-sequence space with high-accuracy models. Nucl. Acids Res. 50: D439–D444.

David, A., Islam, S., Tankhilevich, E., *et al.* (2022). The AlphaFold database of protein structures: A biologist's guide. J. Mol. Biol. 434: 167336.

Lee, C., Su, B.-H., & Tseng, Y.J. (2022). Comparative studies of AlphaFold, RoseTTAFold and Modeller: A case study involving the use of G-protein-coupled receptors. Briefings in Bioinformatics 23: bbac308.

Porta-Pardo, E., Ruiz-Serra, V., Valentini, S., *et al.* (2022). The structural coverage of the human proteome before and after AlphaFold. PLoS Comput. Biol. 18: e1009818.

Terwilliger, T.C., Liebschner, D., Croll, T.I., *et al.* (2024). AlphaFold predictions are valuable hypotheses and accelerate but do not replace experimental structure determination. Nat. Methods 21: 110–116.

Bordin, N., Dallago, C., Heinzinger, M., *et al.* (2023). Novel machine learning approaches revolutionize protein knowledge. Trends Biochem. Sci. 48: 345–359.

Rajapaksa, S., Konagurthu, A.A., & Lesk, A.M. (2023). Sequence and structure alignments in post-AlphaFold era. Curr. Opin. Struct. Biol. 79: 102539.

Structural classifications of protein domains

Kolodny, R., Pereyaslavets, L., Samson, A.O., *et al.* (2013). On the universe of protein folds. Annu. Rev. Biophys. 42: 559–582.

Murzin, A.G., Brenner, S.E., Hubbard, T., *et al.* (1995). SCOP: A structural classification of proteins database for the investigation of sequences and structures. J. Mol. Biol. 247: 536–540.

Andreeva, A., Howorth, D., Chothia, C., *et al.* (2014). SCOP2 prototype: A new approach to protein structure mining. Nucleic Acids Research 42: D310–D314.

Andreeva, A., Howorth, D., Chothia, C., *et al.* (2018). Investigating protein structure and evolution with SCOP2. Current Protocols in Bioinformatics 49: 1–26.

Andreeva, A., Kulesha, E., Gough, J., *et al.* (2020). The SCOP database in 2020: Expanded classification of representative family and superfamily domains of known protein structures. Nucleic Acids Research 48: D376–D382.

Holm, L. (2020). Using Dali for protein structure comparison. Methods Mol. Biol. 2112: 29–42.

Sillitoe, I., Dawson, N., Thornton, J., *et al.* (2015). The history of the CATH structural classification of protein domains. Biochimie 119: 209–217.

Waman, V.P., Blundell, T.L., Buchan, D.W.A., *et al.* (2020). The Genome3D consortium for structural annotations of selected model organisms. Methods Mol. Biol. 2165: 27–67.

Bordin, N., Sillitoe, I., Lees, J.G., *et al.* (2021). Tracing evolution through protein structures: Nature captured in a few thousand folds. Front. Mol. Biosci. 8: 668184.

Sillitoe, I., Bordin, N., Dawson, N., *et al.* (2021). CATH: Increased structural coverage of functional space. Nucl. Acids Res. 49: D266–D273.

Sykes, J., Holland, B.R., & Charleston, M.A. (2022). A review of visualisations of protein fold networks and their relationship with sequence and function. Biol. Rev. Camb. Philos. Soc. 98: 243–262.

Protein complexes

Russell, R.B., Alber, F., Aloy, P., *et al.* (2004). A structural perspective on protein–protein interactions. Curr. Opin. Struct. Biol. 14: 313–324.

Sali, A., & Chiu, W. (2005). Macromolecular assemblies highlighted. Structure 13: 339–341.

Marsh, J.S., & Teichmann, S.A. (2015). Structure, dynamics, assembly, and evolution of protein complexes. Ann. Rev. Biochem. 84: 551–575.

Holguin-Cruz, J.A., Foster, L.J., & Gsponer, J. (2022). Where protein structure and cell diversity meet. Trends Cell Biol. 32: 996–1007.

Directed evolution and protein design

Arnold, F.H. (2008). The race for new biofuels. Eng. Sci. 71: 12–19.

Brustad, E.M., & Arnold, F.H. (2011). Optimizing non-natural protein function with directed evolution. Curr. Opin. Chem. Biol. 15: 201–210.

Arnold, F.H. (2015). The nature of chemical innovation: New enzymes by evolution. Quart. Rev. Biophys. 48: 404–410.

Renata, H., Wang, Z.J., & Arnold, F.H. (2015). Expanding the enzyme universe: Accessing non-natural reactions by mechanism-guided directed evolution. Angew. Chem. Int. Ed. Engl. 54: 3351–3367.

Korendovych, I.V., & DeGrado, W.F. (2020). *De novo* protein design, a retrospective. Quart. Rev. Biophys. 53: e3.

Tong, C.L., Lee, K.H., & Seelig, B. (2021). *De novo* proteins from random sequences through *in vitro* evolution. Curr. Opin. Struct. Biol. 68: 129–134.

Arnold, F.H. (2019). Innovation by evolution: Bringing new chemistry to life (Nobel Lecture). Angew. Chem. Int. Ed. Engl. 58: 14420–14426.

Derat, E., & Kamerlin, S.C.L. (2022). Computational advances in protein engineering and enzyme design. J. Phys. Chem. B. 126: 2449–2451. (Introductory paper to a full special issue on the subject.)

Mariz, B.P., Carvalho, S., Batalha, I.L., *et al.* (2021). Artificial enzymes bringing together computational design and directed evolution. Org. Biomol. Chem. 19: 1915–1925.

A companion volume to this one, treating the topics of this chapter in more detail:

Lesk, A.M. (2010). *Introduction to Protein Science/Architecture, Function and Genomics*. Oxford: Oxford University Press.

Non-ribosomal peptides

Marahiel, M.A. (2009). Working outside the protein-synthesis rules: Insights into non-ribosomal peptide synthesis. J. Pept. Sci. 15: 799–807.

Dincere, S., Takci, H.A.M., & Ozdenefe, M.S. (2022). Nonribosomal peptide synthesis. In *Molecular Cloning*. London: IntechOpen, Ltd.

EXERCISES, PROBLEMS, AND WEBLEMS

Exercises

Exercise 12.1 For each of the following amino acids, say whether they are hydrophobic, polar, positively charged at pH 7, or negatively charged at pH 7: (a) leucine; (b) aspartic acid; (c) glutamine; (d) phenylalanine; (e) lysine.

Exercise 12.2 (a) Identify a hydrophobic amino acid that is more bulky than alanine but less bulky than leucine. (b) Identify two amino acids that have almost the same size and shape (differing only in that one has a methyl group and the other has a hydroxyl group).

Exercise 12.3 On a photocopy of Figure 12.2, indicate the bond, a rotation around which would correspond to the conformational angle φ_{i-1}.

Exercise 12.4 Would it be possible, by rotation around bonds shown in Figure 12.2, to convert residue *i* from the L to the D conformation?

Exercise 12.5 Estimate the values of φ and ψ that correspond to the α_L conformation in Figure 12.3.

Exercise 12.6 Figure 12.2 shows the *trans* conformation of the polypeptide chain. (a) If the angle labelled ω_i in that figure is changed from $\omega = 180°$ *trans* to $\omega = 0°$ *cis*, keeping the positions of all atoms in residues *i*–1 and *i* fixed, what atom would occupy the position currently occupied by H_{i+1}? (b) In the structure shown in Figure 12.2, $\varphi_i = 180°$. The only unlabelled atom is the hydrogen connected to C_i^{α}. Assuming a rotation that keeps the positions of N_i, H_i and the atoms of residue $i-1$ fixed, estimate the value of φ_i that would place this unlabelled hydrogen atom at the position that C_i^{β} occupies in Figure 12.2.

Exercise 12.7 Describe and compare the nature of the accelerating and retarding forces on the molecules in mass spectrometry and SDS–PAGE.

Exercise 12.8 Hen egg white lysozyme has a relative molecular mass of about 14300. If mass spectroscopy can measure mass to within 0.01%, could the following be confidently distinguished from the unmodified protein: (a) N-terminal acetylation, (b) phosphorylation of a single serine residue, (c) a single Lys→Gln substitution?

Exercise 12.9 Cells are grown on a medium containing heavy lysine (^{13}C and ^{15}N). *All* lysine carbons are ^{13}C and both lysine nitrogens are ^{15}N. If a fragment of the protein contains a single lysine, what is the mass difference between the fragment derived from the protein from the cells grown on heavy lysine and the corresponding one derived from cells grown on unlabelled lysine?

Exercise 12.10 On photocopies of Figure 12.20b, indicate the positions of the peaks if the sequence were: (a) MNLVQVR, (b) GNLQVVR, (c) MNLQVVG.

Exercise 12.11 (a) What is the sequence of the fragment y6 in Fig. 12.20b? (b) To which peak in Fig. 12.20b does the fragment LQVVR correspond?

Exercise 12.12 Oligonucleotide samples may vary by the binding of a Na^+ or K^+ ion to a phosphate, instead of a proton. (a) What is the difference in mass between an oligonucleotide binding a proton or a Na^+ ion at a single site? (b) What base change has the closest mass difference to the H^+–Na^+ mass difference? (c) Would measuring mass to within 1 D be sufficient accuracy to distinguish this base change from the binding of a Na^+ ion instead of a proton, at a single site? (d) In a mass spectrum of an oligonucleotide, what is the difference in mass between an oligonucleotide with a proton or a Mg^{2+} ion at a single site? (e) What base change has the closest mass difference to the H^+–Mg^{2+} mass difference? (f) Would measuring mass to within 1 D be sufficient accuracy to distinguish this base change from the binding of a Mg^{2+} ion instead of a proton, at a single site?

Exercise 12.13 In a typical protein–protein interface of area 1700 Å^2 (a) how many intermolecular hydrogen bonds would you expect to be formed? (b) How many fixed water molecules would you expect to find in the interface? (c) If the entire buried area were hydrophobic, what contribution to the free energy of stabilization would you estimate it to make?

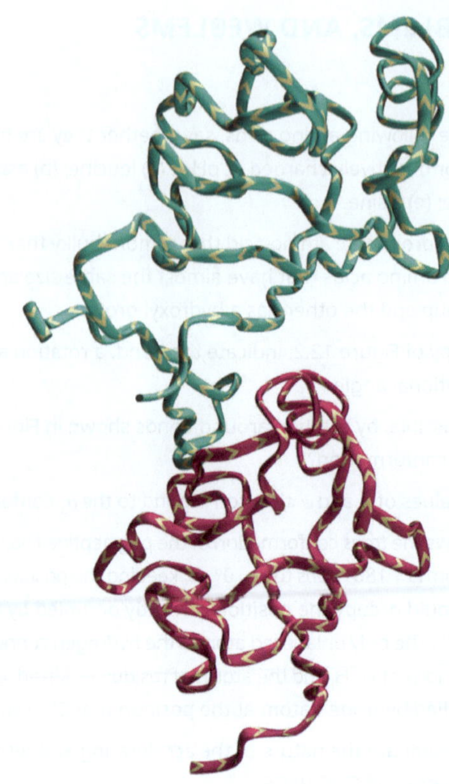

Figure 12.49 Interaction between PDZ domains in syntrophin (green) and neuronal nitric oxide synthase (magenta) [1qav].

Exercise 12.14 What are the symmetries of the structure of the haemoglobin molecule? (See Figure 1.10a.) If the α and β subunits were identical, what additional symmetries would there be?

Exercise 12.15 In the dimer between syntrophin and neuronal nitric oxide synthase (see Figure 12.49), (a) is the dimer structure open or closed? (b) What secondary structure element is shared between the two domains?

Problems

Problem 12.1 For dissociation of a complex involving a simple equilibrium: AB = A + B, the equilibrium constant is equal to the ratio of forward and reverse rate constants: $K_D = k_{off} / k_{on}$. For avidin–biotin, $K_D = 10^{-15}$. Suppose k_{on} were as fast as the diffusion limit, ~ 10^{-9} M s^{-1}. (a) What is the value of k_{off}? (b) What would be the half-life of the avidin–biotin complex? (c) Suppose k_{on} for avidin–biotin were 10^{-7} M s^{-1}. What would be the half-life of the complex?

Problem 12.2 Human haemoglobin α–subunits have helices A, B, C, E, F′, F, G, and H. Human haemoglobin β–subunits have helices A, B, C, D, E, F′, F, G, and H. The sites of interaction of the histidine residues with the iron in the haem group are in the F and G helices. In Figure 1.10a in what colours do the α–subunits appear and in what colours do the β–subunits appear?

Problem 12.3 P. Schultz has posed the question: would an extended genetic code—perhaps one in which one of the rarely used stop codons coded for a novel amino acid with a somewhat unusual size, shape, or charge distribution—be 'better' than the normal one? The question could be taken to apply to either natural or artificial extensions of the code. How would you design experiments to answer this question? What precautions would you consider necessary?

Problem 12.4 As a general rule, vertebrates use creatine as a phosphogen and invertebrates use arginine. Figure 12.50 shows the sequence alignment of creatine kinases (CK) from rabbit and chicken, and arginine kinases (AK) from sea cucumber, horseshoe crab, and abalone. The numbers of identical residues in pairs of sequences in this alignment are:

	Rabbit CK	Chicken CK	Sea cucumber AK	Horseshoe crab AK	Giant abalone AK
Rabbit CK	378	252	226	147	132
Chicken CK	252	381	219	129	129
Sea cucumber AK	226	219	370	154	137
Horseshoe crab AK	147	129	154	357	191
Giant abalone AK	132	129	137	191	358

(a) Does sea cucumber arginine kinase appear to be more related to vertebrate creatine kinases or other invertebrate arginine kinases? (b) On a photocopy of Figure 12.50, circle (at least two) regions, each at least four residues long, in which sea cucumber arginine kinase resembles vertebrate creatine kinases more closely than it resembles other invertebrate arginine kinases, and circle (at least two) regions, each at least four residues long, in which sea cucumber arginine kinase resembles other invertebrate arginine kinases more closely than it resembles vertebrate creatine kinases. (c) Can you identify any residues that might conceivably be responsible for the difference in substrate specificity between arginine and creatine? (d) Outline how you could test the hypothesis you presented as your answer to part (c), using only computational and not wet laboratory methods. (e) Which is more likely: that arginine kinase activity evolved once and that no protein in any ancestor of sea cucumber had creatine kinase activity, or that sea cucumber arginine kinase evolved from a precursor it shared with present vertebrate creatine kinases? Explain your reasoning.

Weblems

Weblem 12.1 Choose one of the following glycoprotein storage diseases: (1) aspartylglucosaminuria, (2) α-mannosidosis, (3) β–mannosidosis, (4) Sandhoff-Jatzkewitz disease, or (5) sialidosis. (a) What is the inheritance pattern of this disease? (b) What is the incidence of this disease in the UK; i.e., what fraction of the population is affected? (c) What is the biochemical defect that causes this disease?

Weblem 12.2 Some athletes take creatine in order to build up their ability to store energy as phosphocreatine. Athletes find that creatine certainly improves performance in 'burst' events—all-out effort for up to 10 seconds—but its efficacy for 'endurance events' is debated. Examples of burst events include the 100 metre dash, a point in tennis, or a play in football. In all of these cases, there is a pause between energy bursts to allow replenishment of phosphocreatine by oxidative metabolism. This takes ~ 30–60 seconds. (The International Tennis Federation rules allow no more than 20 seconds between points.) (a) How does the US Food and Drug Administration classify creatine? (b) Is creatine banned by the Olympics or major commercial sports enterprises? (c) Why would it be difficult to enforce a ban on creatine supplements?

Weblem 12.3 The bacterium *Pseudomonas fluorescens* and the fungus *Curvularia inaequalis* each possesses a chloroperoxidase, an enzyme that catalyses halogenation reactions. Do these enzymes have the same folding pattern?

Weblem 12.4 The normal human androgen receptor is a transcription regulator that mediates the response to androgens during development. Dysfunction of this protein produces *androgen-insensitivity syndrome,* with devastating consequences for sexual development in males.

The mutation 492G→S produces complete androgen-insensitivity syndrome in humans. (The residue number corresponds to the sequence in UniProtKB entry P10275, ANDR_HUMAN.)

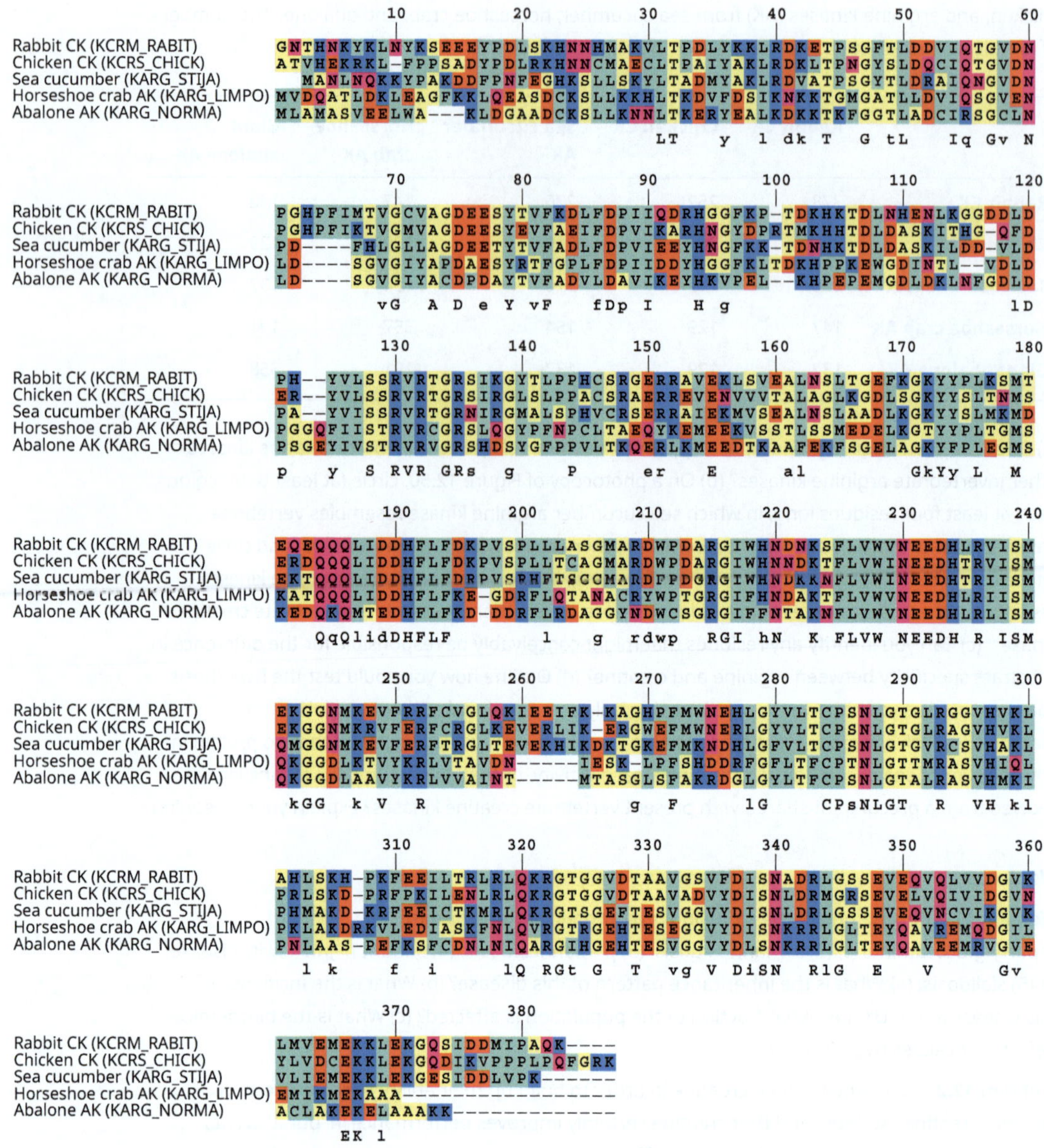

Figure 12.50 Alignment of creatine kinase (CK) from rabbit and chicken, and arginine kinase (AK) from sea cucumber, horseshoe crab, and abalone.

(a) Does a serine at this position appear in the normal, functioning androgen receptors of any other species? (b) If so, how would you go about exploring why the serine is compatible with normal activity of the protein in another species? Design a research project in which to explore some of the possibilities.

Weblem 12.5 wwPDB entry 4K8Z contains a DNA polymerase in binary complex with dsDNA, from *Thermococcus kodakarensis* (the same archaeon that was treated in Chapter 7, 'Comparative genomics of hyperthermophilic archaea: *Thermococcus kodakarensis* and *Pyrococci*'). By examining the structure, and, potentially, modelling in a 3′ hydroxyl, suggest mutations, or at least positions of mutations, that might allow this enzyme to accept an RNA template.

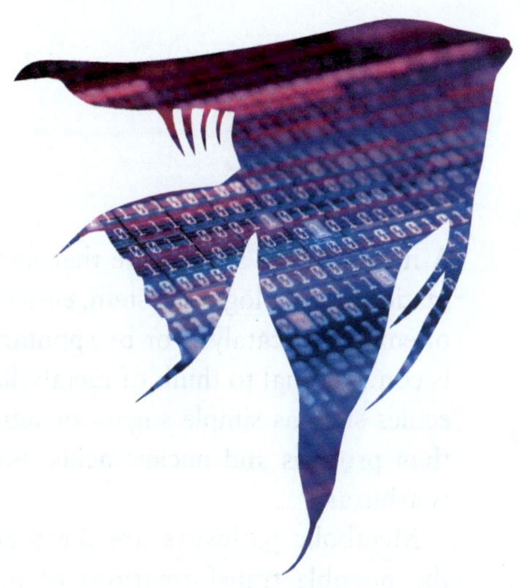

CHAPTER 13

Metabolomics

LEARNING GOALS

- *Recognize that metabolic networks of any organism correspond to graphs in which metabolites are the nodes*, and reactions connecting them are the edges. Enzymes label the edges corresponding to reactions which they catalyse.
- *Know the defining principles of the Enzyme Commission and the Gene Ontology Consortium classifications* of the functions of biological molecules. In what ways are they similar? In what ways do they differ?
- *Appreciate the importance of accurate annotation of enzyme function in databases*. Recognize that transfer of annotation among homologous proteins is by far the easiest way to proceed, but in the absence of experimental confirmation it is not always trustworthy.
- *Understand the structure, dynamics, and evolution of metabolic networks*.
- *Be familiar with databases of metabolic networks*.
- *Recognize some of the similarities and differences in metabolic pathways* among different species.
- *Know how it may be possible to reconstruct (perhaps only partially) the metabolic pathways of an organism* from its genome and proteome.
- *Understand the generalization of the idea of alignment to metabolic pathways*.
- *Understand the physicochemical basis of enzymatic catalysis*, and the quantities needed to characterize their kinetics (see Chapter 12). Such information is necessary if we are to consider modelling flows through metabolic networks.
- *Appreciate the challenges in modelling the traffic patterns in metabolic networks*.

Introduction

A metabolite is a molecule that undergoes transformation in a biological system, either under the action of enzymatic catalysis or by spontaneous reaction. It is conventional to think of metabolites as small molecules such as simple sugars or amino acids, rather than proteins and nucleic acids, but the distinction is arbitrary.

Metabolic pathways are the road maps defining the possible transformations of metabolites. They form a network, represented as a graph, usually with the metabolites as nodes, and reactions connecting them as edges. The enzymes that catalyse each reaction label the edge. Irreversible reactions correspond to directed edges.

To compile a metabolic network, we need to know the possible reactions that can occur, and we need to know the catalytic activities of all the enzymes. These sets of data are complementary.

Generations of biochemists have charted metabolic pathways. These are fairly comprehensive for the best-studied organisms, including *E. coli*, yeast, mouse, and human. A sizable fraction of the pathways are common to many life forms. In many cases, the enzymes that catalyse corresponding reactions are homologous over a broad range of species. Indeed, this provides the most direct route to establishing the metabolic pathway network of a less-well-studied organism. Working out the individual reactions using classical methods such as following radioactive tracers remains very labour-intensive. It is much easier to sequence the genome, infer the amino-acid sequences of the enzymes, and look for sequences similar to enzymes of known function. Then assemble the metabolic networks from the assignable enzymatic functions. When this works, it is golden. The problem is, that it often fails.

KEY POINT

The goal of metabolomics is a comprehensive and synoptic view of metabolism. This depends on the details of the activity of individual enzymes and the flow patterns of metabolites. The goal is to integrate them.

Classification and assignment of protein function

The basic infrastructure of metabolomics depends on knowing the functions of enzymes. Two classifications of enzymes have attempted to impose order on this information.

The Enzyme Commission

The first detailed classification of protein functions was that of the Enzyme Commission (EC). In 1955, the General Assembly of the International Union of Biochemistry (IUB), in consultation with the International Union of Pure and Applied Chemistry (IUPAC), established an International Commission on Enzymes, to systematize nomenclature. The Enzyme Commission published its classification scheme, first on paper and now on the web: http://www.chem.qmul.ac.uk/iubmb/enzyme/.

EC numbers (looking suspiciously like IP numbers) contain four numeric fields, corresponding to a four-level hierarchy. For example, EC 1.1.1.1 corresponds to the reaction:

$$\text{an alcohol} + NAD^{+} = \text{the corresponding aldehyde or ketone} + NADH + H^{+}$$

Several reactions, involving different alcohols, would share this number (whether or not the same enzyme catalysed these reactions). However, the same dehydrogenation of one of these alcohols using the alternative cofactor NADP would not. It would be assigned EC 1.1.1.2.

The first field in an EC number indicates to which of the seven main divisions (classes) the enzyme belongs:

Class 1. Oxidoreductases	catalyse oxidation–reduction reactions
Class 2. Transferases	transfer of a functional group from one molecule to another
Class 3. Hydrolases	split a substrate by hydrolysis
Class 4. Lyases	non-hydrolytic addition or removal of groups from substrates
Class 5. Isomerases	intramolecular rearrangement
Class 6. Ligases	join two molecules together
Class 7. Translocases	control movement of molecules across membranes, or their separation within membranes.

The significance of the second and third fields depends on the class:

- For *Oxidoreductases,* the second number describes the substrate and the third number the acceptor.
- For *Transferases,* the second number describes the class of moiety transferred, and the third number describes either more specifically what they transfer or in some cases the acceptor.
- For *Hydrolases,* the second number signifies the kind of bond cleaved (e.g., an ester bond) and the third number the molecular context (e.g., a carboxylic ester or a thiolester). (Proteinases are treated slightly differently, with the third number indicating the mechanism: serine proteinases, thiol proteinases, and acid proteinases are classified separately.)
- For *Lyases,* the second number signifies the kind of bond formed (e.g., C–C or C–O), and the third number the specific molecular context.
- For *Isomerases,* the second number indicates the type of reaction and the third number the specific class of reaction.
- For *Ligases,* the second number indicates the type of bond formed and the third number the type of molecule in which it appears. For example, EC 6.1 for C–O bonds (enzymes acylating tRNA), EC 6.2 for C–S bonds (acyl–CoA derivatives), etc.
- For *Translocases,* the second number indicates the types of ion or molecule translocated, and the third number the reaction that provides the driving force (where relevant).

The fourth number gives the specific enzymatic activity.

The Gene Ontology™ Consortium protein function classification

In 1999, Michael Ashburner and many co-workers faced the problem of annotating the soon-to-be completed *Drosophila melanogaster* genome sequence. As a classification of function, the EC classification was unsatisfactory, if only because it was limited to enzymes. Ashburner organized the **Gene Ontology™ Consortium** to produce a standardized scheme for describing function.

KEY POINT

An ontology is a formal set of well-defined terms with well-defined interrelationships; that is, a dictionary and rules of syntax.

The Gene Ontology™ Consortium (http://www.geneontology.org) has produced a systematic classification of gene function, in the form of a dictionary of terms, and their relationships. Organizing concepts of the Gene Ontology project include three categories:

- *Molecular function*: a function associated with what an individual protein or RNA molecule does in itself; either a general description such as *enzyme*, or a specific one such as *alcohol dehydrogenase* (specifying a catalytic activity, not a protein). This is function from the biochemist's point of view.
- *Biological process*: a component of the activities of a living system, mediated by a protein or RNA, possibly in concert with other proteins or RNA molecules; either a general term such as *signal transduction*, or a particular one such as *cyclic AMP synthesis*. This is function from the cell's point of view.

Because many processes are dependent on location, Gene Ontology (GO) also tracks:

- *Cellular component*: the assignment of site of activity or partners; this can be a general term such as *nucleus* or a specific one such as *ribosome*.

Figure 13.1 shows an example of the GO classification.

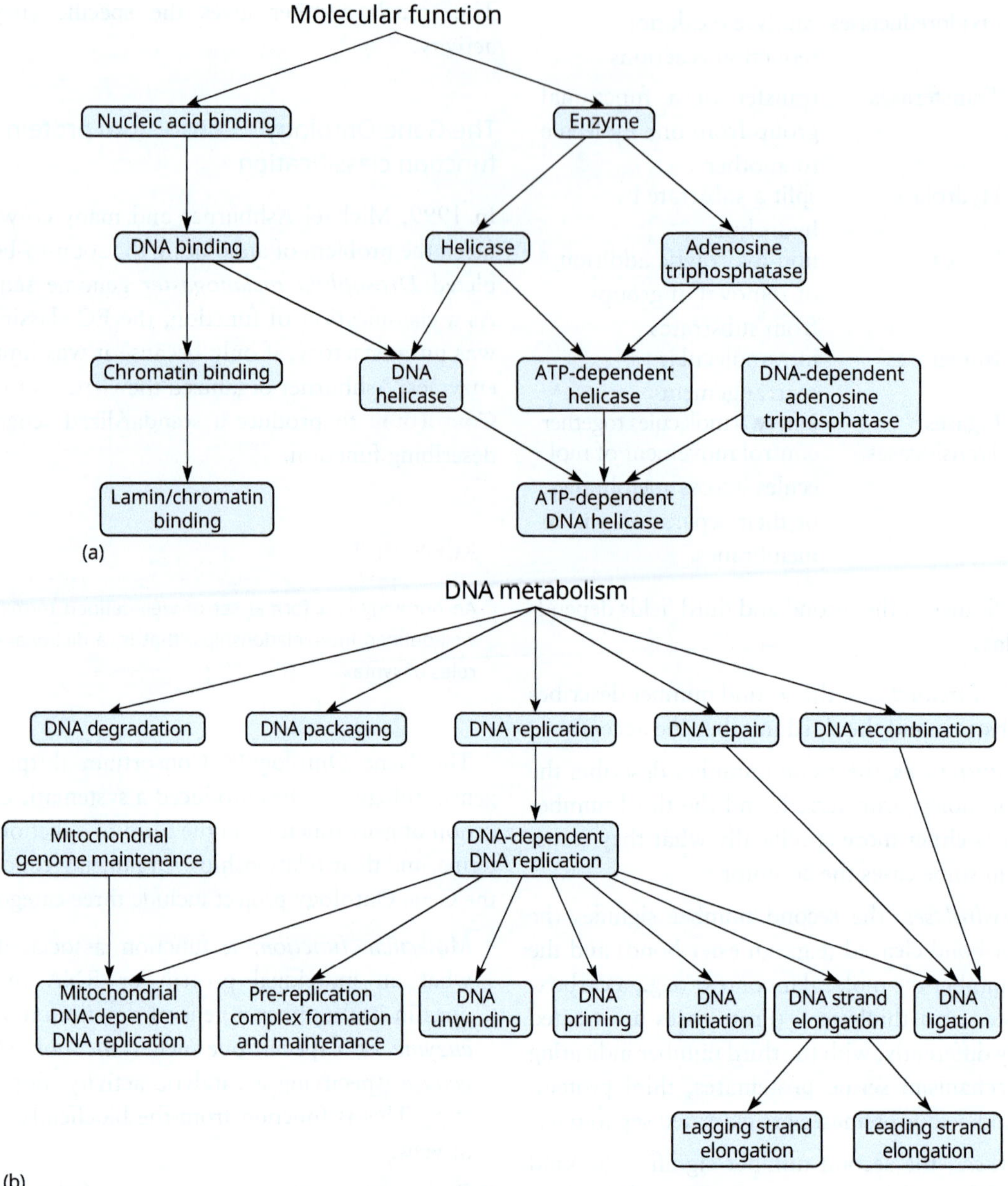

Figure 13.1 Selected portions of the three categories of Gene Ontology, showing classifications of functions of proteins that interact with DNA.

1. *Biological process*: DNA metabolism.
2. *Molecular function*: including general DNA binding by proteins, and enzymatic manipulations of DNA.
3. *Cellular component*: different locations within the cell, or molecular contexts.

These pictures illustrate the general structure of the Gene Ontology classification. Each term describing a function is a node in a graph. Each node has one or more parents and may have one or more descendants: arrows indicate direct ancestor–descendant relationships. A path in the graph is a succession of nodes, each node the parent of the next. Nodes can have 'grandparents', and more remote ancestors.

Unlike the EC hierarchy, the Gene Ontology graphs are not *trees* in the technical sense, because there can be more than one path from an ancestor to a descendant. For example, there are two paths in Figure 13.1a from *enzyme* to *ATP-dependent helicase.* Along one path *helicase* is the intermediate node. Along the other path *adenosine triphosphatase* is the intermediate node.

Although the nodes are shown on discrete levels to clarify the structure of the graph, all the nodes on any given level do not necessarily have a common degree of significance; unlike family, genus, and species levels in the Linnaean taxonomic tree, the ranks in military, industrial, academic, etc., organizations; or, indeed, the EC hierarchy. GO terms could not have such a common degree of significance, given that there can be multiple paths, *containing different numbers of edges,* between different nodes.

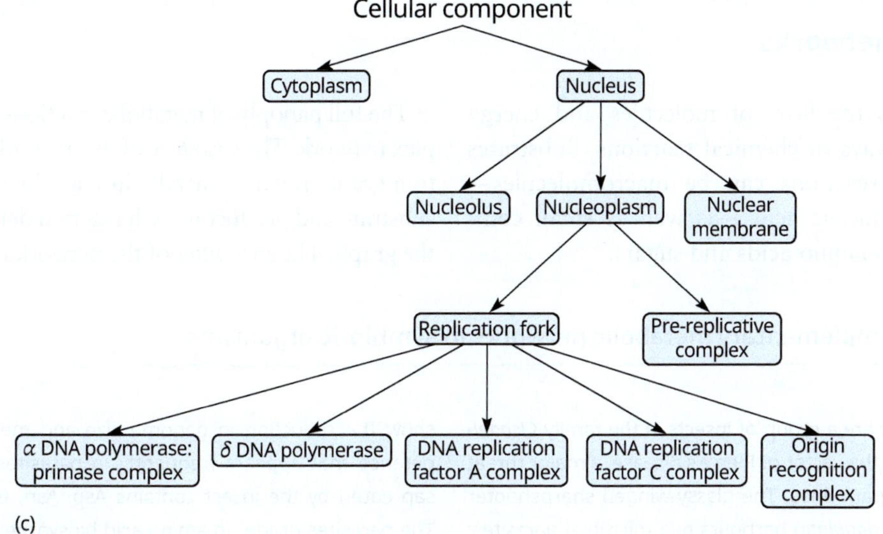

Figure 13.1 *(continued)*

Neither the EC nor the GO classification is an assignment of function to individual proteins. The EC emphasized that: *'It is perhaps worth noting, as it has been a matter of long-standing confusion, that enzyme nomenclature is primarily a matter of naming reactions catalysed, not the structures of the proteins that catalyse them'.* (See http://www.chem.qmul.ac.uk/iubmb/nomenclature/.) Assigning EC or GO numbers to proteins is a separate task. Such assignments appear in protein databases such as UniProtKB.

Comparison of Enzyme Commission and Gene Ontology classifications

Enzyme Commission identifiers form a strict four-level hierarchy, or tree. For example, isopentenyl-diphosphate Δ-isomerase is assigned EC number 5.3.3.2. The initial 5 specifies the most general category 5 = isomerases, 5.3 comprises intramolecular isomerases, 5.3.3 those enzymes that transpose C=C bonds, and the full identifier 5.3.3.2 specifies the particular reaction. In the molecular function ontology, GO assigns the identifier 0004452 to isopentenyl-diphosphate Δ-isomerase. (The numerical GO identifiers themselves have no interpretable significance.)

Figure 13.2 compares the EC and GO classifications of isopentenyl-diphosphate Δ-isomerase. The figure shows a path to GO:0004452 from the root node of the molecular function graph, GO:0003674. In this case, there are four intervening nodes, progressively more specific categories as we move down the figure. Note that the GO description of this enzyme as an oxidoreductase is inconsistent with the EC classification, in which a committed choice between oxidoreductase and isomerase must be made at the highest level of the EC hierarchy.

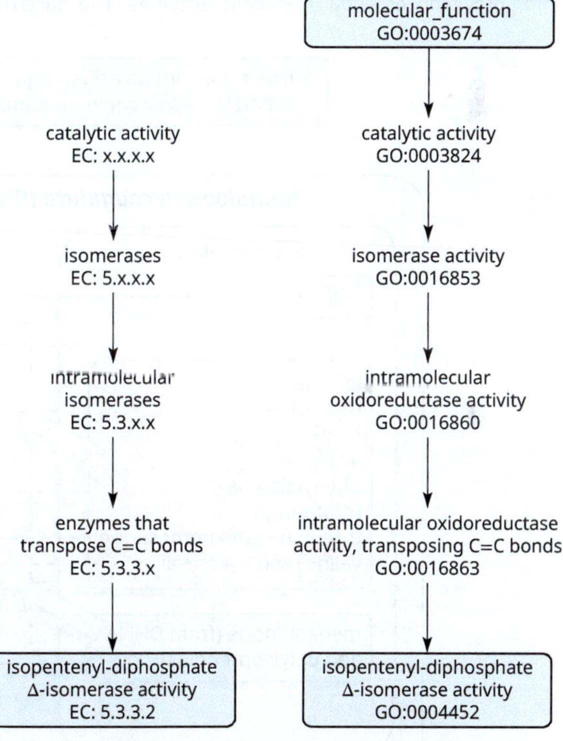

Figure 13.2 Comparison of Enzyme Commission and Gene Ontology classifications of isopentenyl-diphosphate Δ–isomerase.

Metabolic networks

Metabolism is the flow of molecules and energy through pathways of chemical reactions. Substrates of metabolic reactions can be macromolecules—proteins and nucleic acids—as well as small compounds such as amino acids and sugars.

The full panoply of metabolic reactions forms a complex network. The *structure* of the network corresponds to a graph in which metabolites are the nodes and the substrate and product of each reaction define an edge in the graph. The *dynamics* of the network depend on the

BOX 13.1 Complementary metabolic networks in symbiotic organisms

Sharpshooters are a group of insects in the family *Cicadellidae*. They are the vector of Pierce's disease, a major threat to California grapevines. The glassy-winged sharpshooter (*Homalodisca coagulata*) harbours two microbial parasites: *Sulcia muelleri,* a member of the Bacteroidetes, and *Baumannia cicadellinicola,* of the γ-proteobacteria. The three members of this tripartite symbiosis share responsibility for providing essential metabolic enzymes. The bacteria show the reduction in genome size and metabolic competence that appears frequently in parasites. The xylem sap eaten by the insect contains Asp, Asn, Glu, and Gln. The parasites divide up amino acid biosynthesis and other pathways: *Sulcia* can synthesize Leu, Ile, Lys, Arg, Phe, Trp, Thr, and Val. *Baumannia* can synthesize Met and His, and contributes to the synthesis of vitamins and cofactors (see Figure 13.3).

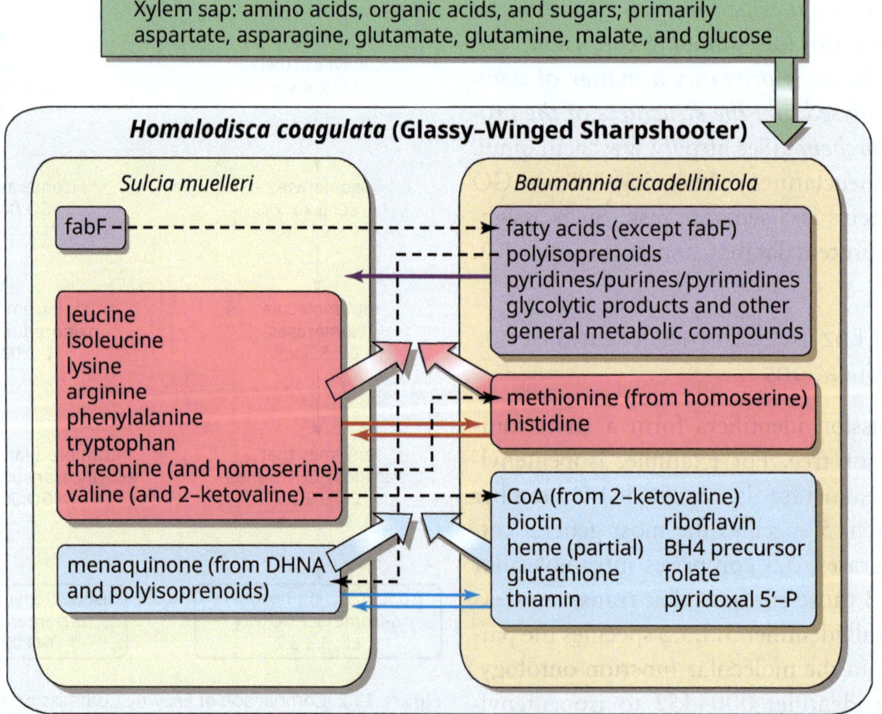

Figure 13.3 Complementarity in metabolic competence between coparasites of the glassy-winged sharpshooter (*Homalodisca coagulata*), *Sulcia muelleri* and *Baumannia cicadellinicola*. Large coloured arrows: compounds needed by the host that the bacterial symbionts produce. Small coloured arrows (and dashed arrows): compounds (believed to be) shared between the bacterial symbionts. Red: compounds, processes, or genes involved in biosynthesis of essential amino acids. Light blue: vitamin or cofactor biosynthesis. Purple: other metabolic functions.

McCutcheon, J.P., & Moran, N.A. (2007). Parallel genomic evolution and metabolic interdependence in an ancient symbiosis. Proceedings of the National Academy of Sciences of the United States of America 104, 49: 19392–19397. https://doi.org/10.1073/pnas.0708855104.

flow capacities of all of the individual links, analogous to traffic patterns on the streets of a city.

Some patterns within the metabolic network are linear pathways. Others form closed loops, such as the tricarboxylic acid (Krebs) cycle. Many pathways are highly branched and interlock densely. However, metabolic networks also contain recognizable clusters or blocks, for instance catabolic and anabolic reactions. There is a relatively high density of internal connections within clusters and relatively few connections between them.

In some cases, metabolic networks span different species in an ecological community. An example of a species interaction that could not be detected by trying to isolate and culture individual strains is the complementation in competence appearing in the symbiosis of insects that are host to multiple microorganisms (see Box 13.1).

Databases of metabolic pathways

Biochemists have learned a lot about different enzymes in different species. Approximately a seventh of the sequences in the UniProt database are enzymes (over 250 million sequences in all). They come from ~1.3 million species. They represent over 370 000 different enzymatic activities.

Databases organize this information, collecting it within a coherent and logical structure, with links to other databases that provide different data selections and different modes of organization. EcoCyc deals with *E. coli*. It is the model for—and linked with—numerous parallel databases, with uniform web interfaces, treating other organisms. BioCyc is the 'umbrella' collection. KEGG, the Kyoto Encyclopedia of Genes and Genomes, contains information from multiple organisms. Reactome collects pathways and provides tools for analysis and visualization (see Table 13.1).

Pathbank has a longer list of databases related to metabolomics: https://pathbank.org/others#metabolic.

EcoCyc

EcoCyc is a database representing our knowledge of the biology of *E. coli*, strain K-12 MG1655. It contains:

- *the genome*: the complete sequence, and for each gene its position and function if known;
- *transcription regulation*: operons, promoters, and transcription factors and their binding sites;
- *metabolism*: the pathways, including details of the enzymology of individual steps; for each enzyme the reaction, activators, inhibitors, and subunit structure;
- *membrane transporters*: transport proteins and their cargo; and
- *links to other databases*: protein and nucleic-acid sequence data, literature references, and comparisons to other *E. coli* strains.

Methionine synthesis in Escherichia coli

A tiny subset of the *E. coli* metabolic network is the pathway for synthesis of methionine from aspartate (see Figure 13.4).

To appreciate the logic of the system from diagrams such as Figure 13.4, keep in mind that both the reaction sequence and the control cascades are embedded in much larger networks:

- The first step, phosphorylation of L-aspartate, is common to the biosynthesis of methionine, lysine, and threonine. *E. coli* contains three aspartate kinases, encoded by three separate genes, each

Table 13.1 Some databases of metabolic pathways

Database	Home page
EcoCyc	http://ecocyc.org
BioCyc	http://www.biocyc.org
KEGG	http://www.genome.jp/kegg/
Human Metabolome Database	https://hmdb.ca/
BRENDA	https://www.brenda-enzymes.org/
PathBank	https://pathbank.org/
Reactome	https://reactome.org/
BiGG	http://bigg.ucsd.edu/

specific for one of the end-product amino acids. They catalyse the same reaction but are subject to separate regulation.

- The third step, conversion of L-aspartate semialdehyde to L-homoserine, is common to the methionine and threonine synthesis pathways. Two homoserine dehydrogenases are separately encoded. Regulation of expression of the aspartate kinases and homoserine dehydrogenases suffices to control all three pathways.
- The piece of the regulatory network is also extracted from a more complex tapestry. For example, CRP (catabolite repressor protein) regulates more than 200 genes!
- Methionine is converted to *S*-adenosylmethionine, a common participant in methyl group transfers. *S*-Adenosylmethionine activates the Met repressor (encoded by *metJ*) (see Figure 13.4b). This is a more complicated form of feedback. In classic feedback inhibition, a product interacts directly with an enzyme that produces one of its precursors. In this case, the product interacts with a repressor, which reduces the *expression* of enzymes that produce its precursors.

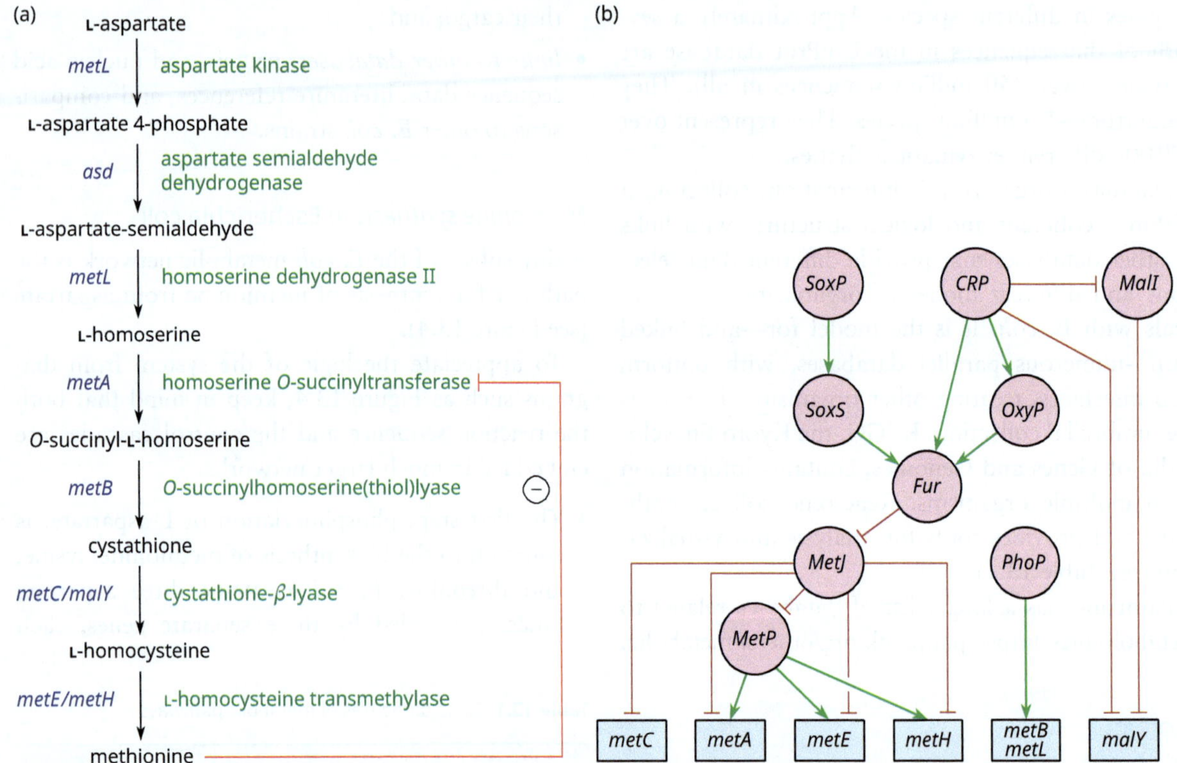

Figure 13.4 (a) The synthesis of methionine from aspartate in *E. coli* is a seven-step pathway through a linear sequence of intermediates (black). Different enzymes (green) catalyse different steps. They are encoded by the genes shown in blue italics. *metC* and *malY* encode alternative cystathione-*β*-lyases. The final step, conversion of L-homocysteine to L-methionine, is also catalysed by two different L-homocysteine transmethylases, encoded by two genes, *metE* and *metH*.

One mechanism of control is at the protein level: there is 'feedback inhibition' by the product, methionine, which inhibits homoserine *O*-succinyltransferase. This is shown by the red line; note that it connects the product, methionine, to an *enzyme*, not to a gene. (b) Expression of the genes in the L-aspartate → L-methionine pathway (blue rectangles) is subject to regulation. Circles contain the genes for the transcription factors that control expression. Regulated genes appear in the rectangles. Molecules that tend to *enhance* transcription are connected to their targets by green arrows. Molecules that tend to *repress* transcription are connected to their targets by red lines ending in a 'T'. In addition to the links shown here, most of the regulatory proteins feed back on themselves; in most cases, the self-regulatory signal is a repression.

Control is exerted on every protein of the pathway, from a variety of points of initiation.

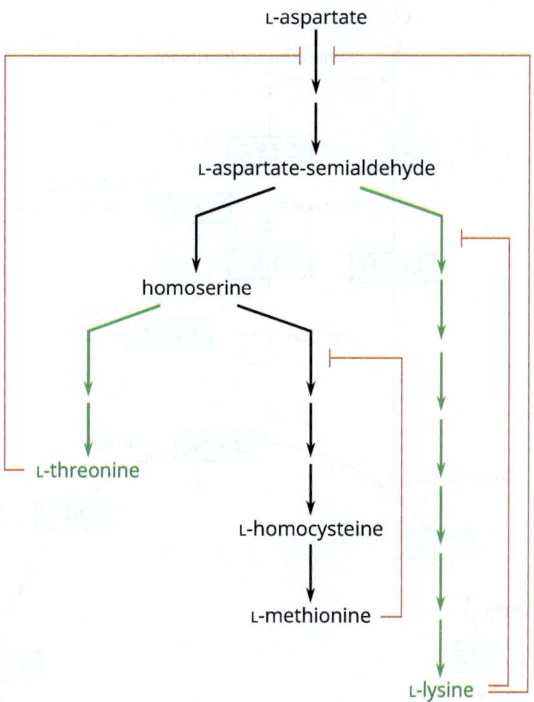

Figure 13.5 The pathway of amino acid biosynthesis from aspartate branches after aspartate semialdehyde. In this figure, the black sequence corresponds to the previous example, and the green pathways are the immediate context. The aspartate → methionine sequence (Figure 13.4a) is a subnetwork of the network shown here. Each amino acid plays a regulatory role, exerting feedback inhibition over its own synthesis, without affecting the others. It looks as if threonine and lysine both individually inhibit the first step of the synthesis of all three products, but this step is catalysed by *three* separate aspartate kinases, allowing specialized regulation.

In the EcoCyc web page that contains the information corresponding to this figure, the items are active. Links to other internal pages expand information about metabolites, cofactors, enzymes, genes, and regulators. It is possible to 'zoom' in or out by controlling the level of detail. For instance, asking for less detail than the contents of Figure 13.4a would first eliminate the information about the genes and enzymes and then reduce the pathway to an outline showing only critical intermediates:

⊖
L-aspartate →→→ homoserine →→→ L-homocysteine → L-methionine

> Readers are urged to explore the EcoCyc website on their own, either deliberately or serendipitously, or guided by weblems in this chapter or at the Online Resource Centre.

It is also possible to explore in other dimensions. The methionine synthesis pathway is embedded in larger networks. One of these involves synthesis of the amino acids lysine and threonine in addition to methionine, all starting with aspartate (see Figure 13.5).

The Kyoto Encyclopedia of Genes and Genomes (KEGG)

The Kyoto Encyclopedia of Genes and Genomes (KEGG) is an extremely comprehensive battery of databases for molecular biology and genomics. One of its special strengths is an integration of metabolic and genomic information. KEGG contains pathway maps, which describe potential networks of molecular

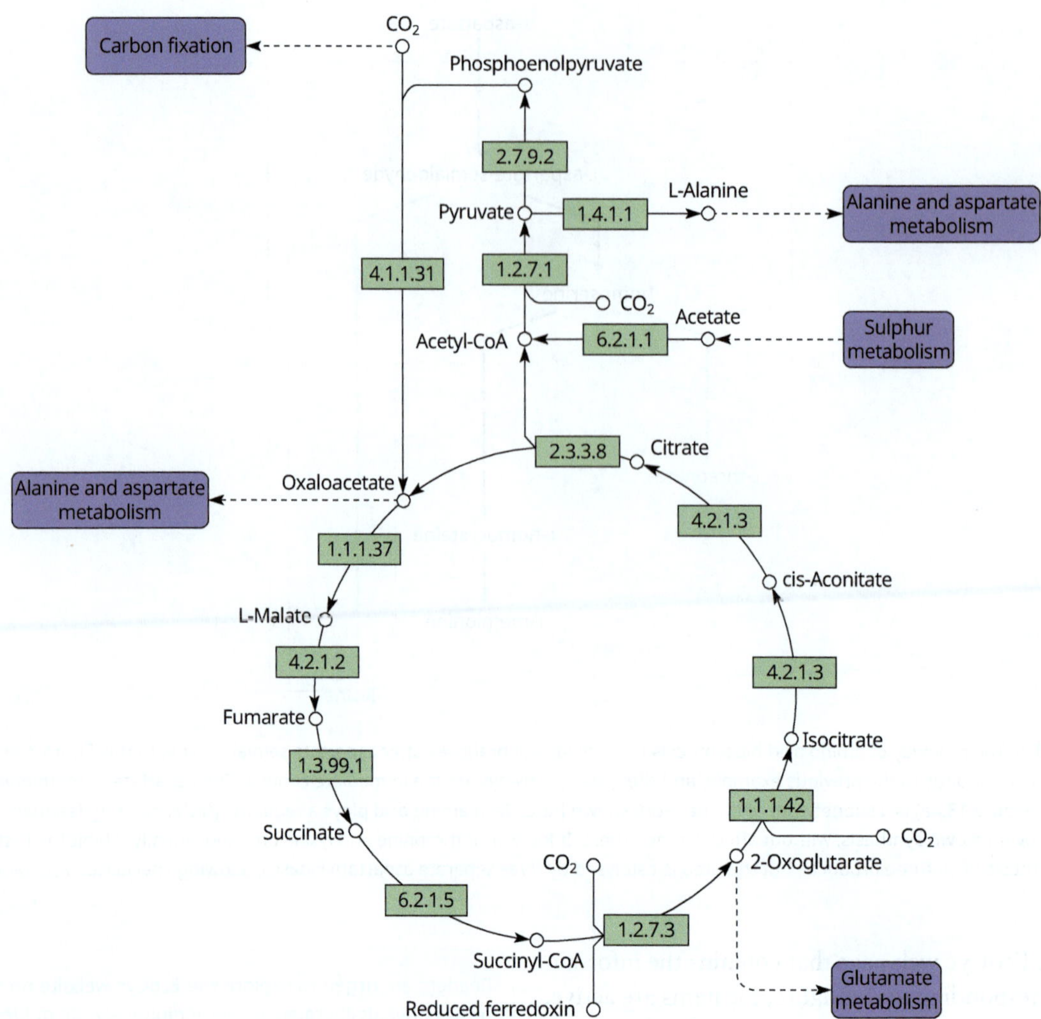

Figure 13.6 Metabolic pathway map from The Kyoto Encyclopedia of Genes and Genomes (KEGG). This figure shows the reductive carboxylate cycle, and its links to other metabolic processes. The numbers in rectangular boxes are EC numbers identifying the reactions at each step.

Kanehisa, M., Furumichi, M., Sato, Y., *et al.* (2022). KEGG for taxonomy-based analysis of pathways and genomes. Nucleic Acids Research 51, D1: D587–D592. https://doi.org/10.1093/nar/gkac963. Used with permission.

activities, both metabolic and regulatory. Figure 13.6 shows a pathway from KEGG, the reductive carboxylate cycle in photosynthetic bacteria. This pathway is basically the Krebs cycle, run backwards.

KEY POINT

Several databases present information about metabolic pathways. By linking the steps in metabolic pathways to individual genes and proteins, it is possible to reconstruct pathways in particular species, and to compare metabolic pathways in different species.

KEGG derives its power from the very dense network of links among these categories of information, and additional links to many other databases to which the system maintains access. Two examples of the kinds of questions that can be treated with KEGG are:

1. It has been suggested that simple metabolic pathways evolve into more complex ones by gene duplication and subsequent divergence, to create enzymes catalysing successive steps in a reaction sequence. Searching the pathway catalogue for

sets of enzymes that share a folding pattern will reveal clusters of linked paralogues.

2. KEGG can take the set of known enzymes from some organism and check whether they can be integrated into established metabolic pathways. A gap in a pathway suggests a missing enzyme or an unexpected alternative pathway.

KEY POINT

Several databases assemble biochemical reactions into metabolic pathways. Individual steps are linked to Enzyme Commission and Gene Ontology Consortium classifications of function, and to individual proteins that catalyse the reactions. These databases are useful in organizing the assignment of function to proteins identified in newly sequenced genomes.

The Human Metabolome Database

The Human Metabolome Database (HMDB) http://www.hmdb.ca contains information about small molecule metabolites found in the human body. HMDB includes four additional databases:

- *DrugBank*: information on ~1600 drugs and drug metabolites
- *T3DB*: information on toxins and environmental pollutants
- *SMPDB*: pathway diagrams for human metabolic and disease pathways
- *FooDB*: information on food components and additives

It supports a wide variety of query searches, and offers facilities for data visualization. There are copious links to KEGG, PubChem, MetaCyc, ChEBI, PDB, UniProt, and GenBank.

Evolution and phylogeny of metabolic pathways

Most organisms share common metabolic pathways. But there are many individual variations.

Some organisms have metabolic competence completely absent from others. Plants but not humans have enzymes for reactions involved in photosynthesis and cell-wall formation.

Some organisms achieve the same overall metabolic transformation but use alternative pathways; that is, different sets of intermediates. For instance, classical glycolysis—the Embden–Meyerhof pathway—and the Entner–Doudoroff pathway are alternative routes from glucose to pyruvate, in which there are alternative successions of reactions (Figure 13.7).

Often, organisms will share many steps in a metabolic transformation but some will extend or truncate the pathway. Most mammals can synthesize vitamin C. Humans cannot, and we must include it in our diet. Many parasites have dispensed with substantial biosynthetic competence.

In some cases, a particular species or strain may show a variant metabolic pathway. For instance, the normal Krebs (or tricarboxylic acid) cycle, memorized by generations of biochemistry students, includes the conversion of 2-oxoglutarate (a.k.a. α–ketoglutarate) to succinyl-CoA. Cyanobacteria, however, lack the enzyme 2-oxoglutarate dehydrogenase. Instead, they convert 2-oxoglutarate to succinate via succinic semialdehyde (Figure 13.8).

Greater differences in metabolic pathway emerge in studying more-distantly-related species.

B. Siebers and P. Schönheit have studied the metabolic pathways of carbohydrate metabolism in archaea. In the initial conversion of glucose to pyruvate, they observed a number of differences in the pathway, from either the standard Embden–Meyerhof glycolytic pathway, or the Entner–Doudoroff alternative (Figure 13.7).

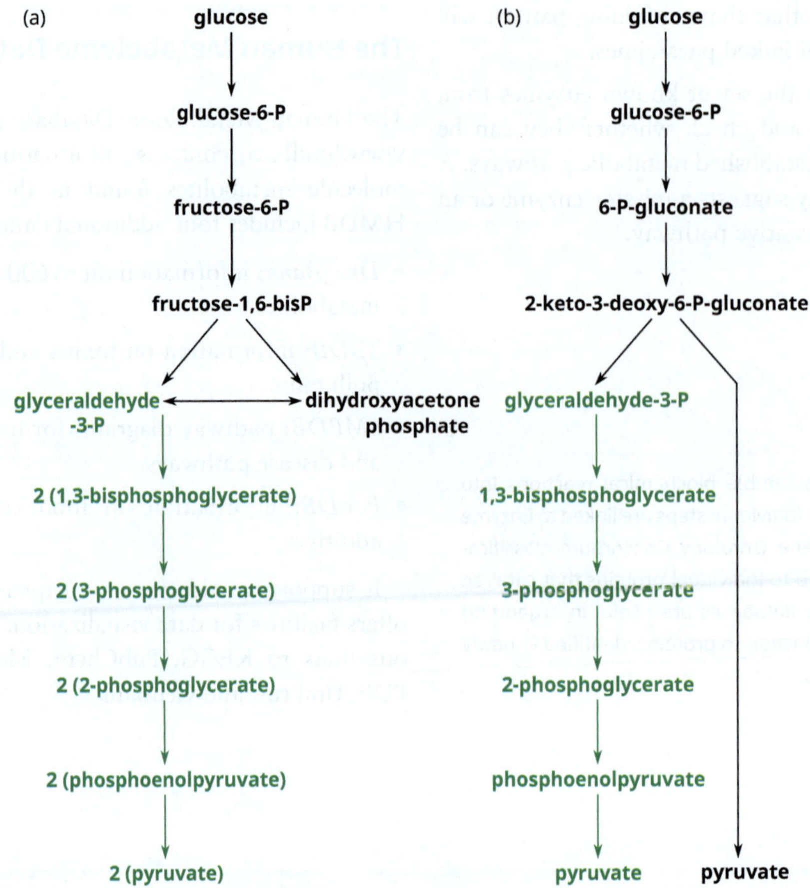

Figure 13.7 (a) Embden–Meyerhof glycolytic pathway, (b) Entner–Doudoroff pathway. Note that the enzymatic conversion of glyceraldehyde-3-phosphate to pyruvate is the same in both pathways (green branch).

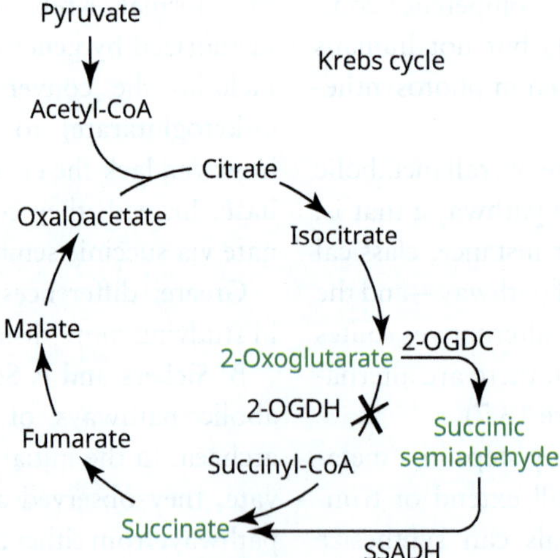

Figure 13.8 Cyanobacterial succinic semialdehyde shunt. 2-OGDH, 2-oxoglutarate dehydrogenase; 2-OGDC, 2-oxoglutarate decarboxylase; SSADH, succinic semialdehyde dehydrogenase.

Zhang, S., & Bryant, D.A. (2011). The tricarboxylic acid cycle in cyanobacteria. Science 334, 6062: 1551–1553. https://doi.org/10.1126/science.1210858. Reprinted with permission from AAAS.

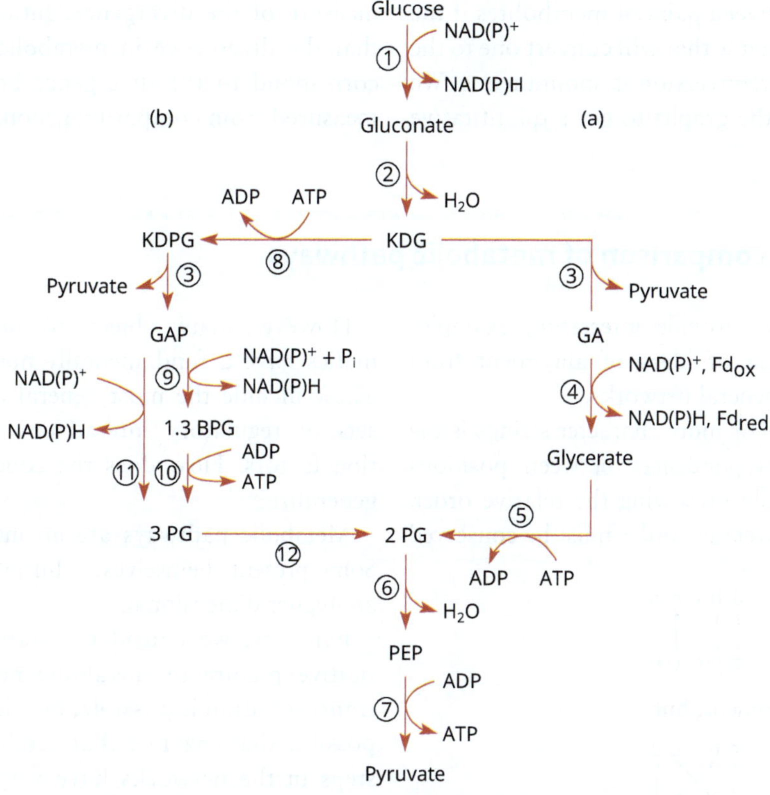

Figure 13.9 Modifications of the Entner–Doudoroff (ED) pathway in archaea. (a) The non-phosphorylative ED pathway in *Thermoplasma acidophilum*. (b) The semi-phosphorylative ED pathway in halophilic archaea. A branched ED (combining (a) and (b)) appears in *S. solfataricus* and *T. tenax*. Abbreviations: 1.3 BPG, 1,3-bisphosphoglycerate; Fdox and Fdred, oxidized and reduced ferredoxin; GA, glyceraldehyde; GAP, glyceraldehyde-3-phosphate; KDG, 2-keto-3-deoxygluconate; KDPG, 2-keto-3-deoxy-6-phosphogluconate; PEP, phosphoenolpyruvate; 2 PG, 2-phosphoglycerate; 3 PG, 3-phosphoglycerate. Enzymes are numbered as follows: 1, glucose dehydrogenase; 2, gluconate dehydratase; 3, KD(P)G aldolase; 4, glyceraldehyde dehydrogenase (proposed for *T. acidophilum*), glyceraldehyde:ferredoxin oxidoreductase (proposed for *T. tenax*) or glyceraldehyde oxidoreductase (proposed for *S. acidocaldarius*); 5, glycerate kinase; 6, enolase; 7, pyruvate kinase; 8, KDG kinase; 9, GAPDH; 10, phosphoglycerate kinase; 11, GAPN; 12, phosphoglycerate mutase.

Pyrococcus furiosus, *Thermococcus celer*, *Archaeoglobus fulgidus* strain 7324, *Desulfurococcus amylolyticus*, and *Pyrobaculum aerophilum* use a modified Embden–Meyerhof pathway. *Sulfolobos solfataricus* and *Haloarcula marismortui* use a modified Entner–Doudoroff pathway (Figure 13.9). *Thermoproteus tenax* uses both.

In addition to the differences in the sequence of metabolites, the enzymes that catalyse even the same reactions in archaea are almost always *not* homologues of bacterial or eukaryotic ones. Many of them use different cofactors. Bacterial and eukaryotic phosphofructokinases (that convert fructose-6-phosphate to fructose-1,6-bisphosphate) use ATP as the phosphoryl donor. The archaeal enzymes that catalyse this reaction can use ATP, ADP, or even inorganic pyrophosphate. In addition, some of the familiar enzymes from bacteria and eukaryotes are under allosteric control. The control relationships are not retained in the corresponding archaeal enzymes.

We can represent the metabolic networks of different species as graphs. The nodes are metabolites.

There are edges between pairs of metabolites if the organism has an enzyme that will convert one to the other, or if the interconversion is spontaneous. We can then compare the graphs to get a quantitative measure of the divergence. Intuitively, we expect that the divergence in metabolic network should correspond to the divergence between species as measured from comparing genome sequences.

Alignment and comparison of metabolic pathways

Metabolic pathways provide interesting examples of the generalization of ideas of alignment from sequences to more general networks.

Alignment of two or more character strings is the assignment of correspondences between positions in the strings, usually preserving the relative order. The constraint that relative order must be conserved means that:

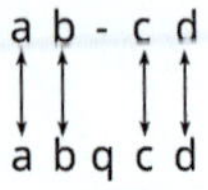

```
a b - c d
↕ ↕   ↕ ↕
a b q c d
```

is an allowable alignment, but

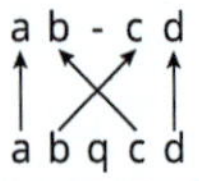

```
a b - c d
↕  ╲╱  ↕
↕  ╱╲  ↕
a b q c d
```

is not. The concept of alignment, including the relative-order constraint, carries over fairly directly to protein structures, because of the linear chemistry of the polypeptide chain: A structural alignment is still a correspondence between the amino-acid *sequences*, despite an appeal to three-dimensional data to determine it. (An exception would be the case of two multidomain proteins composed of homologous domains in different order.)

> **KEY POINT**
>
> Of particular interest for comparative genomics are facilities to compare pathways among different organisms. Alignment and comparison of pathways can expose how pathways have diverged between species. But even if the pathways are the same, in some cases the enzymes are non-homologous.

However, many objects of interest in bioinformatics have a fundamentally non-linear structure. These include the most general networks, such as sets of regulatory interactions among transcription factors. How does the concept of alignment generalize?

Metabolic pathways are an interesting example. Some present themselves as linear sequences, others are higher dimensional.

For now, we consider a static and non-quantitative picture of metabolic networks. Either a transformation is possible, or it is not. It is entirely possible that enzymes that catalyse corresponding steps in the networks have very different kinetic constants in two species, or are subject to different kinds of regulation. In this case, the dynamic patterns of traffic through the networks might be quite different, even if the topologies of the networks are the same. (That is, if the graphs are isomorphic.)

Comparing linear metabolic pathways

Many linear metabolic pathways are extractable from general metabolic networks. In principle, alignment of linear metabolic pathways is directly analogous to alignment of any other sequences. It is the extension to alignment of non-linear metabolic pathways that takes us out of our comfort zone. For example, an alignment of the pathways on the left and right branches of Figure 13.9 would be (for symbols see caption to Figure 13.9):

Glucose	Gluconate	KDG	KDPG	GAP	1,3BPG	3PG	–	–	2PG	PEP	Pyruvate
Glucose	Gluconate	KDG	–	–	–	–	GA	Glycerate	2PG	PEP	Pyruvate

An interesting generalization involves aligning metabolic pathways with genome sequences.

In its simplest form, a metabolic pathway is a sequence of metabolites. Associated with each step, in each organism, is an enzyme. Associated with each enzyme is a gene. In some cases, for example the tryptophan synthesis pathway in *E. coli,* the genes for successive steps of the pathway are collinear in the genome with the steps of the pathway (see Figure 13.10). Alignment methods can detect this. (In this case, we are aligning a metabolic pathway with elements of the genome sequence.) Such relationships are useful for reconstructing metabolic pathways (see Box 13.2).

Reconstruction of metabolic networks

Pathway comparison can be useful for annotation of genomes. It is often possible to assign function to proteins on the basis of similarity to sequences of proteins of known function in other organisms. However, sometimes there are several weak similarities to other proteins and it is unclear which is the true

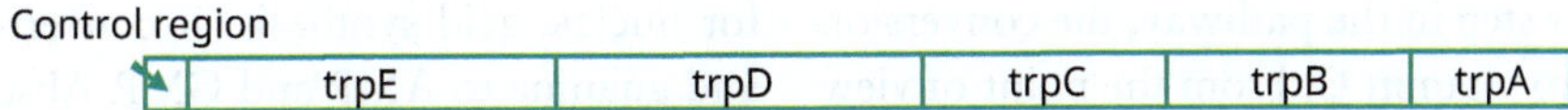

Figure 13.10 The trp operon in *E. coli* begins with a control region containing promoter, operator, and leader sequences. Five structural genes encode proteins that catalyse successive steps in the synthesis of the amino acid tryptophan from its precursor chor5smate:

chorismate → anthranilate → phosphoribosyl-anthranilate → indoleglycerolphosphate → indole → tryptophan
(1) (2) (3) (4) (5)

Reaction step (1): trpE and trpD encode two components of anthranilate synthase. This tetrameric enzyme, comprising two copies of each subunit, catalyses the conversion of chorismate to anthranilate. Reaction step (2): The protein encoded by trpD also catalyses the subsequent phosphoribosylation of anthranilate. Reaction step (3): trpC encodes another bifunctional enzyme, phosphoribosylanthranilate isomerase—indoleglycerolphosphate synthase. It converts phosphoribosyl anthranilate to indoleglycerolphosphate, through the intermediate, carboxyphenylaminodeoxyribulose phosphate. Reaction steps (4) and (5): trpB and trpA encode the β and α subunits, respectively, of a third bifunctional enzyme, tryptophan synthase (an $\alpha_2\beta_2$ tetramer). A tunnel within the structure of this enzyme delivers, without release to the solvent, the intermediate produced by the α subunit—indoleglycerolphosphate to indole—to the active site of the β subunit—which converts indole to tryptophan.

A separate gene, trpR, not closely linked to this operon, codes for the trp repressor. The repressor can bind to the operator sequence in the DNA (within the control region) only when binding tryptophan. Binding of repressor blocks access of RNA polymerase to the promoter, turning the pathway off when tryptophan is abundant. Further control of transcription in response to tryptophan levels is exerted by the attenuator element in the mRNA, within the leader sequence. The attenuator region (a) contains two tandem trp codons and (b) can adopt alternative secondary structures, one of which terminates transcription. Levels of tryptophan govern levels of trp-tRNAs, which govern the rate of progress of the tandem trp codons through the ribosome. Stalling on the ribosome at the tandem trp codons in response to low tryptophan levels reduces the formation of the mRNA secondary structure that terminates transcription.

BOX 13.2 Identification of *M. jannaschii* shikimate kinase

If an enzyme needed for a pathway cannot be identified, even by weak sequence similarity, it may be that the organism has evolved a non-homologous enzyme for the task. For example, the archaeon *Methanococcus jannaschii* has a pathway for biosynthesis of chorismate from 3-dehydroquinate. Enzymes for most of the steps have homologues in bacteria and/or eukaryotes. However, shikimate kinase was not identifiable from sequence similarity. *M. jannaschii* must have *some* protein with this function. How can it be found?

Although in bacteria, genes consecutive in pathways are often consecutive in operons in the genome (see Figure 13.10), this is not true of *M. jannaschii*. However, the genes for successive steps of the chorismate biosynthesis pathway *are* clustered and consecutive in another archaeon, *Aeropyrum pernix*. It was possible thereby to propose a gene for a shikimate kinase in *A. pernix* and to identify a homologue of that gene in *M. jannaschii*.

Experiments confirmed the prediction that the *M. jannaschii* gene thus identified (*MJ1440*) encoded a shikimate kinase. It has no sequence similarity to bacterial or eukaryotic shikimate kinases. A protein from a different family has been recruited for the archaeal pathway.

homologue. Conversely, sometimes an organism has a metabolic pathway but no annotated enzyme for an essential step. Confronting the unannotated proteins with the unassigned functions can sometimes identify the protein that fills the gap in the pathway.

Some differences in pathways between organisms are common knowledge. A vitamin is by definition *not* the product of a metabolic pathway. Humans and other primates require a diet containing vitamin C, because we cannot synthesize it. Most animals can synthesize vitamin C. All those that cannot lack the enzyme L-gulano-γ-lactone oxidase, the enzyme that catalyses the last step in the pathway, the conversion of L-gluconate to vitamin C. From the point of view of alignment of metabolic sequences, the pathway in humans is truncated, relative to that of animals such as the mouse that are competent to synthesize vitamin C. In primates, there is a deletion of a large component of the gene for L-gulano-γ-lactone oxidase.

Similar considerations apply to catabolic pathways. The end product of purine degradation—the form excreted—differs among animals in different phyla (Figure 13.11). Organisms with more water available in their immediate surroundings use more of the reactions. Most mammals degrade purines to allantoin, produced from uric acid by urate oxidase. Primates (and Dalmatian dogs) lack functional urate oxidase, and consequently excrete its substrate, uric acid.

The much lower solubility of uric acid relative to allantoin creates clinical problems in humans, including kidney stones and gout. The drug allopurinol inhibits xanthine oxidase, the enzyme that converts hypoxanthine → xanthine → uric acid. The precursors, hypoxanthine and xanthine, are more soluble than uric acid, and are cleared much faster by the kidneys. Moreover, in a mixture of hypoxanthine, xanthine, and uric acid each solute has *independent* solubility. Therefore, formation of a precipitate is less likely from a mixed solution of hypoxanthine, xanthine, and uric acid, than from a solution of the same total concentration of uric acid alone.

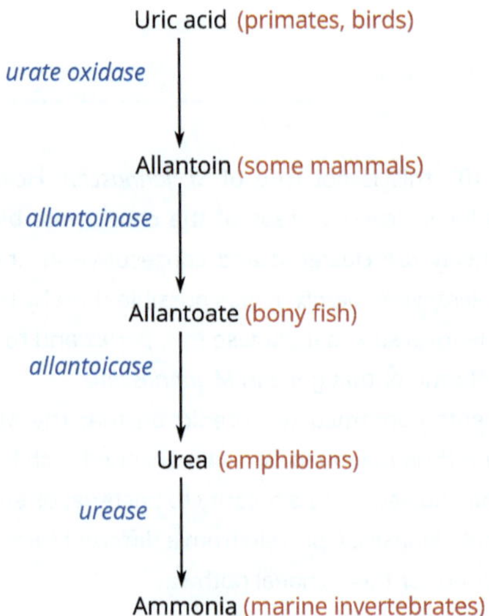

Figure 13.11 Succession of reactions to produce excreted forms of end products of nitrogen metabolism.

The enzyme hypoxanthine-guanine phosphoribosyltransferase (HGPRT) recovers degraded purines for nucleic acid synthesis. It converts hypoxanthine and guanine to AMP and GMP. Absence of HGPRT activity causes a buildup of uric acid, associated with the Lesch–Nyhan syndrome, an inherited metabolic disease. Gout and kidney stones are common symptoms, together with mental retardation and behavioural syndromes including uncontrollable lip and finger biting. (The Lesch–Nyhan syndrome was the first unambiguous correlation of a biochemical defect with a psychological abnormality.)

Comparing non-linear metabolic pathways: the pentose phosphate pathway and the Calvin—Benson cycle

The pentose phosphate pathway and the Calvin–Benson cycle in photosynthesis are two metabolic pathways involving transformations of sugars.

Metabolism of glucose can proceed through glycolysis and the Krebs cycle, to couple glucose oxidation to production of ATP. The pentose phosphate pathway is an alternative, that produces NADPH and ribose-5-phosphate. A cell that needs reducing power or ribose-5-phosphate for nucleic acid synthesis will divert some of its glucose metabolism through the pentose phosphate pathway. Several intermediates in the pentose phosphate pathway can be shuttled back into glycolysis.

The Calvin–Benson cycle is the route of CO_2 fixation in photosynthesis. The enzyme ribulose-1,5-bisphosphate carboxylase (RUBISCO) couples CO_2 to ribulose-1,5-bisphosphate to form an intermediate that breaks down spontaneously to two molecules of glyceraldehyde-3-phosphate. Of every six molecules of glyceralde-3-phosphate produced, five are used to reconstitute three molecules of ribulose-1,5-bisphosphate, and the sixth harvested for energy. (Five 3-carbon molecules ↔ three 5-carbon molecules.)

The pentose phosphate pathway and the Calvin–Benson cycle share many intermediates. Several intermediates link each pathway with 'mainstream' glycolysis.

A. Sillero, V.A. Selivanov, and M. Cascante presented three-dimensional diagrams of these two metabolic subnetworks, that bring out the similarities more clearly than standard two-dimensional textbook presentations do (see Figure 13.12). This may be considered an alignment of two multidimensional objects. In general, the alignment of graphs is an important and difficult problem that arises in many different contexts, not limited to genomics.

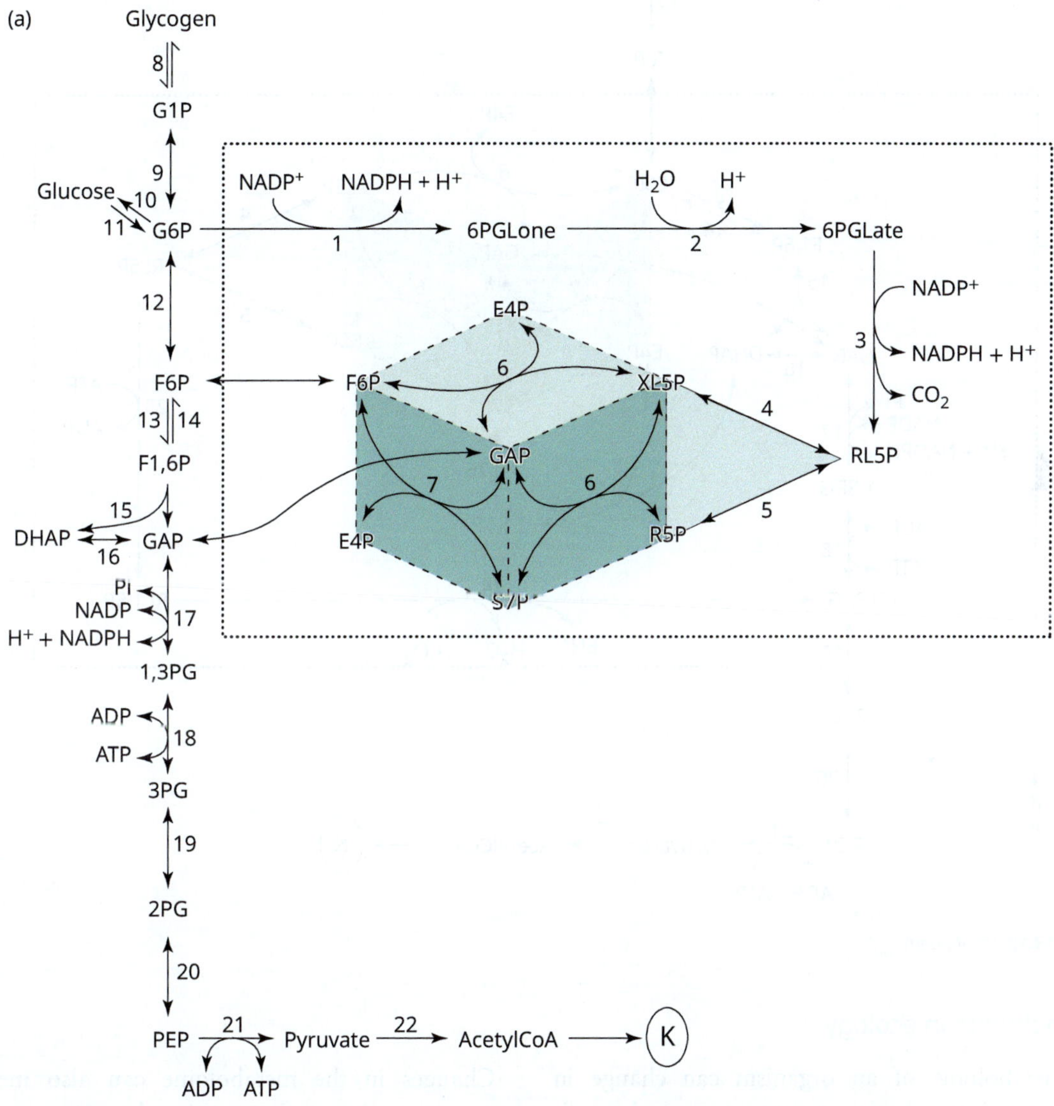

Figure 13.12 (a) The pentose phosphate cycle. (b) The Calvin–Benson cycle. In both panels (a) and (b), numbers in the figure correspond to the following enzymes: 1, glucose-6-P dehydrogenase; 2, gluconolactonase; 3, 6-P-gluconate dehydrogenase; 4, ribulose-5-P 3-epimerase; 5, ribulose-5-P-isomerase; 6, transketolase; 7, transaldolase; 8, enzymes acting in the interconversion glucose 1-P 7 glycogen; 9, phosphoglucomutase; 10, glucose-6-phosphatase; 11, hexokinase; 12, phosphoglucose isomerase; 13, 6-phosphofructokinase; 14, fructose-1,6-bisphosphatase; 15, aldolase; 16, triosephosphate isomerase; 17, glyceraldehyde-3-P dehydrogenase; 18, phosphoglycerate kinase; 19, phosphoglycerate mutase; 20, enolase; 21, pyruvate kinase; 22, pyruvate dehydrogenase. K represents the Krebs cycle.

In part (b): 23, phosphoribulose kinase; 24, rubisco; 25, transaldolase, 26, sedoheptulose-1,7-bisphosphatase.

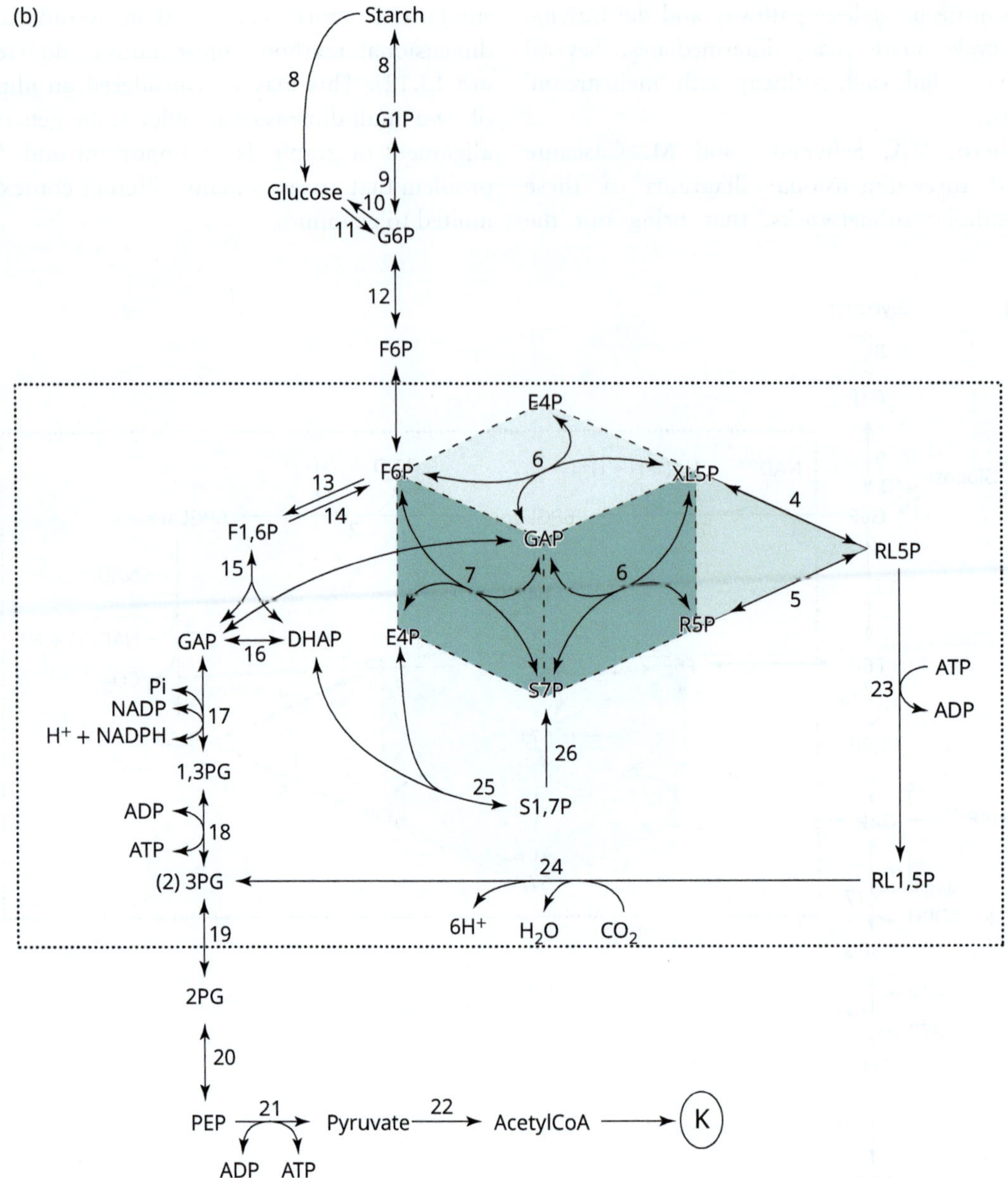

Figure 13.12 *(continued)*

Metabolomics in ecology

The metabolome of an organism can change in response to internal programmes or external conditions. Internal programmes include developmental stages and tissue differentiation. Changes in external conditions include the switch from aerobic to anaerobic metabolism, availability of different nutrients—for instance lactose or glucose in *E. coli*—or circadian rhythms. There is coordination between changes in the transcriptome and the metabolome.

Changes in the metabolome can also mediate responses to interactions with other organisms. In some cases, this involves synthesis of defence molecules. There is potential application to pest control.

A study of *A. thaliana* revealed its response to grazing by above-ground aphids (*Brevcoryne brassicae*) and below-ground nematodes (*Heterodera schachtii*) (see Figure 13.13). In the shoot, above ground, there is a change in the composition of

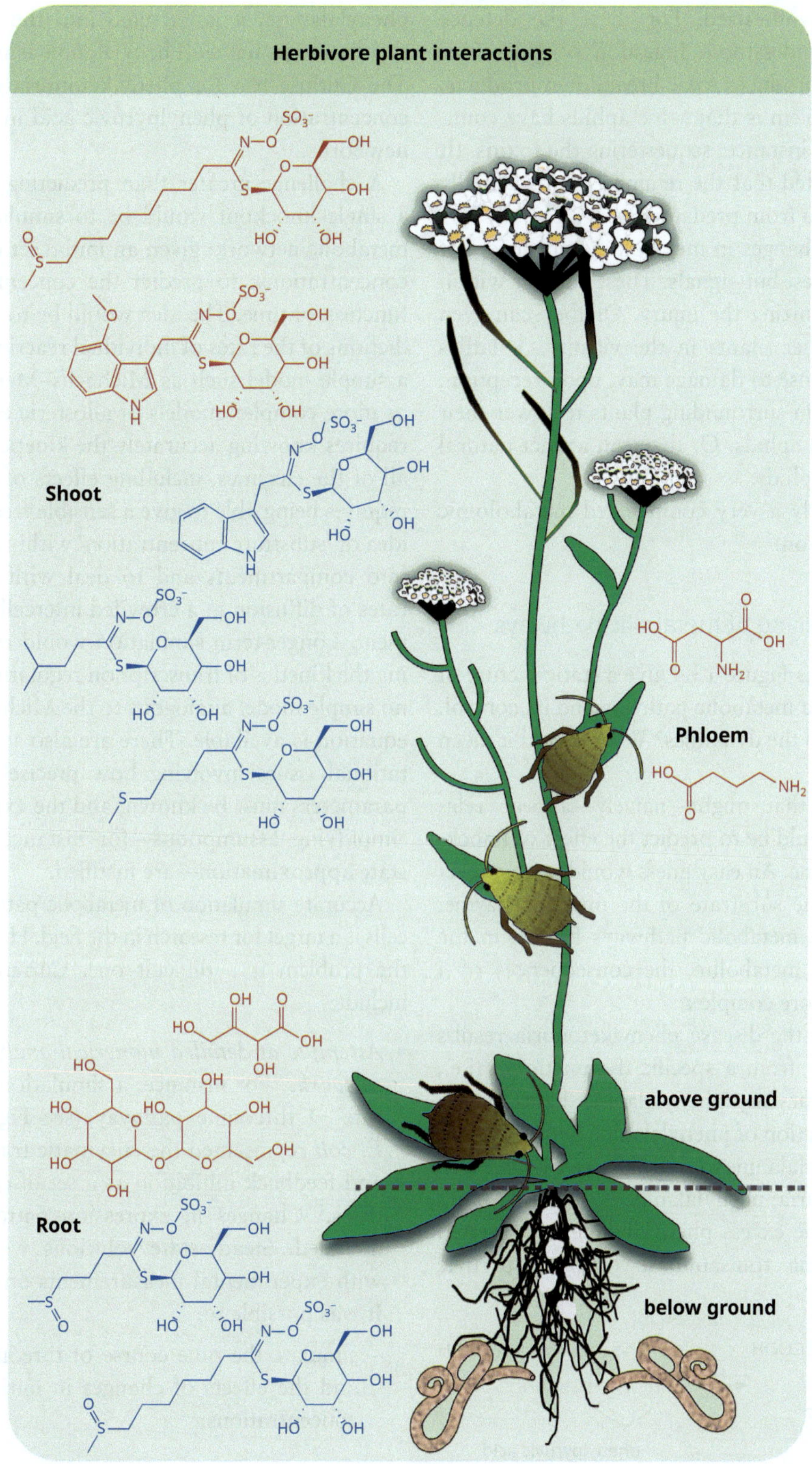

Figure 13.13 The metabolomic responses of *Arabidopsis thaliana* to above-ground aphid grazing and below-ground nematodes include regulation of aliphatic glucosinolates (Gls), amino acids, and sugars. Selected up- (red) and down-regulated (blue) metabolites are depicted for the shoot, the phloem, and the roots.

Kuhlisch, C., & Pohnert, G. (2015). Metabolomics in chemical ecology. Natural Product Reports 32, 7: 937–955. https://doi.org/10.1039/c5np00003c.

glucosinolates synthesized. For these the defence mechanism is understood. Ingestion of glucosinolates by aphids produces toxic breakdown products. The plant's problem is that some aphids have counter-defences, for instance, sequestering the toxins. (It has been suggested that the retained toxins actually protect the aphid from predators.)

Some of the changes in metabolome produce, not defence molecules, but signals. These can act within the plant, recognizing the injury. Or they can even act to warn other plants in the vicinity. Volatiles emitted in response to damage may, upon reception, induce changes in surrounding plants to lower their attractiveness to aphids. Or they can attract natural enemies of the aphids.

There is clearly a very complicated metabolomic arms race going on!

Dynamic modelling of metabolic pathways

Diagrams such as Figure 13.4 give a static picture of the structure of a metabolic pathway and its control.

Can we model the dynamics? What would it mean to do so?

A challenge that might—naively—appear relatively simple would be to predict the effect of knocking out an enzyme. An easy guess would be to expect a build-up of the substrate of the missing enzyme. However, if the metabolic pathways branch in the vicinity of that metabolite, the consequences of a knockout are more complex.

For example, the disease phenylketonuria results most commonly from a specific dysfunctional (i.e., knocked-out) enzyme, phenylalanine hydroxylase. The normal function of phenylalanine hydroxylase is to convert phenylalanine to tyrosine (see Chapter 10). In phenylketonuria, phenylalanine does indeed build up. However, the excess phenylalanine is converted by phenylalanine transaminase to phenylpyruvic acid:

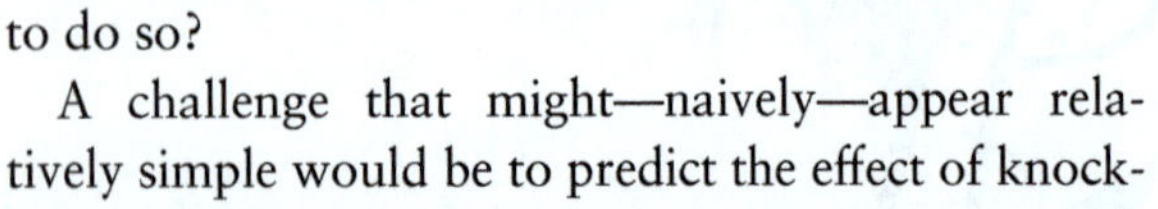

COOH
NH_2
phenylalanine
⇌
COOH
O
phenylpyruvic acid

Both compounds accumulate. As phenylpyruvic acid is less readily absorbed by the kidneys than phenylalanine, it is excreted into the urine, giving the disease its name. (Phenylalanine is not a ketone.) The Guthrie test for phenylketonuria measures the concentration of phenylpyruvic acid in the blood of newborns.

A challenge greater than predicting the effect of a single knockout would be to simulate the entire metabolic network: given an initial set of metabolite concentrations, to predict the concentrations as a function of time. The idea would be to combine predictions of the rates of individual reactions, assuming a simple model such as Michaelis–Menten kinetics, or more complex models of allosteric enzymes. This requires knowing accurately the kinetic constants of all of the enzymes, including effects of inhibitors. It requires being able to give a sensible treatment of the idea of 'substrate concentration' within a cell divided into compartments and to deal with questions of rates of diffusion in a crowded intercellular environment. Longer-term simulation would require knowing the kinetics of transcription regulation, for which no simple model analogous to the Michaelis–Menten equation is available. There are also serious computational issues involving how precisely the kinetic parameters must be known, and the extent to which simplifying assumptions—for instance, the steady-state approximation—are justified.

Accurate simulation of metabolic patterns of entire cells is a target for research in the field. However, clearly the problem is a difficult one. Current approaches include:

- *Attempts at detailed numerical analysis of simple networks*. For instance, a simulation of the aspartate → threonine pathway (see Figure 13.5) in *E. coli* represented the enzymatic transformations and feedback inhibition as a set of coupled equations.[1] Changes in expression pattern were not included. Steady-state solutions were compared with experimental measurements on cell extracts. It was possible to:
  - simulate the time course of threonine synthesis and the effects of changes in initial metabolite concentrations;

[1] Chassagnole, C., Raïs, B., Quentin, E., *et al.* (2001). An integrated study of threonine pathway enzyme kinetics in *Escherichia coli*. Biochem. J. 356: 415–423.

- predict the steady-state concentrations of intermediates;
- predict the effects of changes in concentrations of individual enzymes on overall throughput, expressed as **flux-control coefficients**; such data can help to guide development of microbial factories for increased yield of particular products;
- for different steps, distinguish whether the substrates and products are approximately at equilibrium.

KEY POINT

The flux-control coefficient is the percentage change in flux divided by the percentage change in amount of enzyme. It is not a property of the enzyme, but a property of a reaction within a metabolic network. A flux-control coefficient equal to 1 would correspond to a rate-limiting step.

- *Focusing not on individual enzymes but on potential sets of flow rates.* Represent the metabolic network as a graph. Metabolites are the nodes. Edges correspond to reactions: an edge connects two compounds if there is a reaction, or possibly several reactions, that interconvert them. The goal is to predict the flow rate through each edge. Recently the models have been generalized to include regulation of expression. There are general constraints on the set of flow rates:
  - under steady-state conditions, the fluxes through each node must add up to zero; i.e., for each compound, the amount that is synthesized or supplied externally must equal the amount used up or secreted;
  - the flux-control coefficients of all of the reactions contributing to a single flux must add up to 1;
  - the flux through any edge is limited by the values of the Michaelis–Menten parameter V_{max} for all enzymes contributing to the edge; and
  - the thermodynamic properties of each reaction determine whether or not the reaction is reversible: this is a property of the substrate and product of the reaction, not of the enzyme; the flux of an irreversible reaction must be ≥0.

It is interesting to see whether the space of possible metabolic states is connected or broken up into separated regimes.

In general, many possible flow patterns, or metabolic states, are consistent with the constraints. To determine a single metabolic state to compare with experiments, it is possible to select from the feasible states the one that is optimal for ATP production or for growth rate.

A variety of observable quantities are predictable.

- The effects of changes of medium or gene knockouts: which enzymes are essential for growth on different carbon sources?
- What are limiting factors in growth?
- What are maximal theoretical yields of ATP, or assimilation of carbon, etc.?
- What are the fluxes through individual pathways? This is difficult but not impossible to measure.
- What are the flux-control coefficients of different enzymes?
- For optimal growth, how much oxygen and carbon source are taken up?

Such models have been constructed for several organisms, including prokaryotes and eukaryotes. Predictions have generally achieved good agreement with experiments.

LOOKING BACK AND FORWARD

Most of this book is about nucleic acids and proteins—the stuff of the central dogma. This chapter treated material somewhat off the mainstream: interconversions of small molecules. This might superficially be thought to be a turning away from molecular biology, towards biochemistry. But it is naive to think that these subjects can be so easily disentangled in science and in the cell. Genomes are on stage for the whole scene.

Despite the changing subject matter, the theme of the final chapters is turning progressively to integration. The next chapter treats systems biology. Systems biology is not distinguished by a special subject matter; it encompasses everything we have discussed. Rather, systems biology is an *attitude*. It rejects attempts to simplify. It embraces complexity.

RECOMMENDED READING

General discussions of metabolomics

Fiehn, O. (2002). Metabolomics—the link between genotypes and phenotypes. Plant Mol. Biol. 48: 155–171.

Papin, J.A., Price, N.D., Wiback, S.J., *et al.* (2003). Metabolic pathways in the post-genome era. Trends Biochem. Sci. 28: 250–258.

Patti, G.J., Yanes, O., & Siuzdak, G. (2012). Innovation: Metabolomics: The apogee of the omics trilogy. Nat. Rev. Mol. Cell Biol. 13: 263–269.

Zhang, A., Sun, H., Xu, H., *et al.* (2013). Cell metabolomics. OMICS 17: 495–501.

Ramautar, R., Berger, R., van der Greef, J., *et al.* (2013). Human metabolomics: Strategies to understand biology. Curr. Opin. Chem. Biol. 17: 841–846.

Prosser, G.A., Larrouy-Maumus, G., & de Carvalho, L.P. (2014). Metabolomic strategies for the identification of new enzyme functions and metabolic pathways. EMBO Rep. 15: 657–659.

Rolfsson, Ó., & Pálsson, B.Ö. (2015). Decoding the jargon of bottom-up metabolic systems biology. BioEssays 37: 588–591.

Jacob, M., Lopata, A.L., Dasouki, M., *et al.* (2017). Metabolomics toward personalized medicine. Mass Spectrom. Rev. 38: 221–238.

Telenti, A. (2018). Integrating metabolomics with genomics. Pharmacogenomics 19: 1377–1381.

Kanehisa, M., Furumichi, M., Sato, Y., *et al.* (2021). KEGG: Integrating viruses and cellular organisms. Nucleic Acids Res. 49: D545–D551.

Guo, S., Zhang, C., & Le, A. (2021). The limitless applications of single-cell metabolomics. Curr. Opin. Biotechnol. 71: 115–122.

Medema, M.H., de Rond, T., & Moore, B.S. (2021). Mining genomes to illuminate the specialized chemistry of life. Nat. Rev. Genet. 22: 553–571.

Moco, S., & Buescher, J.M. (2023). Metabolomics: Going deeper, going broader, going further. Methods Mol. Biol. 2554: 155–178.

Santorelli, L., Caterino, M., & Costanzo, M. (2023). Proteomics and metabolomics in biomedicine. Int. J. Mol. Sci. 24: 16913.

Ford, L., Mitchell, M., Wulff, J., *et al.* (2022). Clinical metabolomics for inborn errors of metabolism. Adv. Clin. Chem. 107: 79–138.

Metabolomics of prokaryotes

Aldridge, B.B., & Rhee, K.Y. (2014). Microbial metabolomics: Innovation, application, insight. Curr. Opin. Microbiol. 19: 90–96.

Bundalovic-Torma, C., & Parkinson, J. (2015). Comparative genomics and evolutionary modularity of prokaryotes. Adv. Exp. Med. Biol. 883: 77–96.

Aguiar-Pulido, V., Huang, W., Suarez-Ulloa, V., *et al.* (2016). Metagenomics, metatranscriptomics, and metabolomics approaches for microbiome analysis. Evol. Bioinform. Online 12(Suppl 1): 5–16.

Heinken, A., Hertel, J., Acharya, G., *et al.* (2023). Genome-scale metabolic reconstruction of 7,302 human microorganisms for personalized medicine. Nat. Biotechnol. 41: 1320–1331.

Description of concepts used for analysis of metabolic networks

Lacroix, V., Cottret, L., Thébault, P., *et al.* (2008). An introduction to metabolic networks and their structural analysis. IEEE/ACM Trans. Comput. Biol. Bioinform. 5: 594–617.

Papale, F., Saget, J., & Bapteste, È. (2020). Networks consolidate the core concepts of evolution by natural selection. Trends Microbiol. 28: 254–265.

Metabolic network reconstruction

Feist, A.M., Herrgard, M.J., Thiele, I., *et al.* (2009). Reconstruction of biochemical networks in microorganisms. Nat. Rev. Microb. 7: 129–143.

Ozden, F., Siper, M.C., Acarsoy, N., *et al.* (2021). DORMAN: Database of Reconstructed MetAbolic Networks. IEEE/ACM Trans. Comput. Biol. Bioinform. 18: 1474–1480.

Tang, L. (2023). Microbial metabolic reconstruction. Nat. Methods. 20: 348.

Evolution of metabolic networks

Wagner, A. (2013). Metabolic networks & their evolution. In W. Dubitzky, O. Wolkenhauer, H. Yokota, & K.-H. Cho (eds), *Encyclopedia of Systems Biology*, pp. 1256–1259. New York: Springer.

Ramon, C., & Stelling, J. (2023). Functional comparison of metabolic networks across species. Nat. Commun. 14: 1699.

Modelling of network traffic

Chalancon, G., Kruse, K., & Babu, M.M. (2013). Metabolic networks, structure and dynamics. In: W. Dubitzky, O. Wolkenhauer, K.-H. Cho, & H. Yokota (eds), *Encyclopedia of Systems Biology*, pp. 1263–1267. New York: Springer.

Weaver, D.S., Keseler, I.M., Mackie, A., *et al.* (2014). A genome-scale metabolic flux model of Escherichla coli K-12 derived from the EcoCyc database. BMC Syst. Biol. 8: 79.

Basler, G., Nikoloski, Z., Larhlimi, A., *et al.* (2016). Control of fluxes in metabolic networks. Genome Res. 26: 956–968.

Orth, J.D., Thiele, I., & Pálsson, B.Ö. (2020). What is flux balance analysis? Nature Biotech. 28: 245–248.

Peres, S., & Fromion, V. (2020). Thermodynamic approaches in flux analysis. Methods Mol. Biol. 2088: 359–367.

Nicholson, J.K., Jia, W., Lasky-Su, J.A., *et al.* (2022). Metabolic modeling in health and disease. J. Proteome Res. 21: 559.

Hrovatin, K., Fischer, D.S., & Theis, F.J. (2022). Toward modeling metabolic state from single-cell transcriptomics. Mol. Metab. 57: 101396.

Metabolomics databases

Go, E.P. (2010). Database resources in metabolomics: an overview. J. Neuroimmune. Pharmacol. 5: 18–30.

Bansal, P., Morgat, A., Axelsen, K.B., *et al.* (2022). Rhea, the reaction knowledgebase in 2022. Nucl. Acids Res. 50(D1): D693–D700.

Karp, P.D., Billington, R., Holland, T.A., *et al.* (2015). Computational metabolomics operations at BioCyc.org. Metabolites 5: 291–310.

Caspi, R., Billington, R., Ferrer, L., *et al.* (2016). The MetaCyc database of metabolic pathways and enzymes and the BioCyc collection of pathway/genome databases. Nucl. Acids Res. 44 (Database issue): 471–480.

Paley, S., & Karp, P.D. (2021). The BioCyc metabolic network explorer. BMC Bioinformatics 21: 208.

Wishart, D.S., Guo, A., Oler, E., *et al.* (2022). HMDB 5.0: the Human Metabolome Database for 2022. Nucleic Acids Res. 50: D622–D631.

Karp, P.D., Paley, S., Caspi, R., *et al.* (2023). The EcoCyc database (2023). EcoSal Plus. 11, 1.

Kanehisa, M., Furumichi, M., Sato, Y., *et al.* (2023). KEGG for taxonomy-based analysis of pathways and genomes. Nucleic Acids Res. 51: D587–D592.

A 'niche' approach, but one of which the reader should be aware

Patumcharoenpol, P., Doungpan, N., Meechai, A., *et al.* (2016). An integrated text mining framework for metabolic interaction network reconstruction. PeerJ. 4: e1811.

EXERCISES, PROBLEMS, AND WEBLEMS

Exercises

Exercise 13.1 From Figure 13.1a, write the nodes appearing in four pathways from enzyme to ATP-dependent DNA helicase, including the starting and ending nodes enzyme and ATP-dependent DNA helicase.

Exercise 13.2 In Figure 13.1b, identify a node such that there are paths of *different* lengths to that node from the root, DNA metabolic process. Report two paths of different lengths, including the starting and ending nodes DNA metabolic process and the node you identified.

Exercise 13.3 From Figure 13.2, form an alignment of the nodes in the pathways to isopentenyl-diphosphate Δ-isomerase in the EC and Gene Ontology classifications.

Exercise 13.4 What would be the major class in the EC classification of the enzymes (a) hexokinase (catalysing glucose + ATP = glucose-6-phosphate + ADP) (b) ribulose-bisphosphate carboxylase (catalysing D-ribulose-1,5-bisphosphate + CO_2 + H_2O = 2(3-phosph-D-glycerate) + $2H^+$) (c) glutamine synthetase (catalysing ATP + L-glutamate + NH_3 = ADP + phosphate + L-glutamine). (Try to do this without treating it as a weblem.)

Exercise 13.5 From Figure 13.4b, (a) what would be the effect of increased expression of *metJ* on the expression of *metC*? (b) What would be the effect of increased expression of *OxyP* on the expression of *metJ*? (c) What would be the effect of increased expression of *metJ* on the expression of *metH?*

Exercise 13.6 Why would most animals not suffer from gout?

Problems

Problem 13.1 In one species, only enzyme A catalyses conversion S → P, the rate-limiting step of a reaction pathway. What is the flux-control coefficient of enzyme A? (b) In a related species, distinct but similar enzymes A and B both catalyse the S → P reaction. The kinetic characteristics of A and B are identical: S → P is still the rate-limiting step of the pathway. What is the flux-control coefficient of A?

Problem 13.2 Write detailed structures for all of the metabolites that appear in Figure 13.4a.

Problem 13.3 The network of metabolic pathways must obey constraints of thermodynamics and physical-organic chemistry. Meléndez-Hevia and colleagues suggested the principle that metabolic pathways are optimized, subject to the constraints, for the minimum number of steps. The non-oxidative phase of the pentose phosphate pathway converts six five-carbon sugars to five six-carbon sugars:

$$6 \text{ ribulose-5-phosphate} \rightarrow 5 \text{ glucose-6-phosphate}$$

A simplified model of a pathway for this conversion is a series of steps, each of which is either:

- transfer of a two-carbon unit from one sugar to another (a transketolase reaction); or
- transfer of a three-carbon unit from one sugar to another (a transaldolase or aldolase reaction).

Represent each sugar only by a number of carbon atoms. Starting with five five-carbon sugars, one possible initial step would be a transketolase step converting two five-carbon sugars to a three-carbon sugar and a seven-carbon sugar. Assume that all intermediates must have *at least* three carbon atoms.

Create a tableau with the following initial and final states (an initial transketolase (TK) step is also shown):

Step	number of carbons in sugar molecules					
0	5	5	5	5	5	5
	↘ TK ↙					
	↙ ↘					
1	3	7	5	5	5	5
		. . .				
N	6	6	6	6	6	0

Copy and fill in the tableau to find the shortest route from the top (step 0, six five-carbon sugars) to the bottom (five six-carbon sugars). Identify the intermediates created. Compare this with the observed metabolic pathway.

Weblems

Weblem 13.1 In the methionine biosynthetic pathway (see Figure 13.4), the product, methionine, inhibits an enzyme in the middle of the pathway, homoserine *O*-succinyltransferase. (a) The accumulation of which intermediate might this inhibition be expected to cause? (b) In what other pathways is this intermediate involved that might use it up?

Weblem 13.2 Figure 13.5 shows the amino acid biosynthesis leading from aspartate to methionine, threonine, and lysine. Look up, and report, on a photocopy of this figure, the names of the omitted intermediates at the unlabelled positions between consecutive arrows.

Weblem 13.3 The genes encoded by *metC* and *malY* in *E. coli* convert cystathione to L-homocysteine. Each has another function in addition. What are these other functions?

Weblem 13.4 Compare the pathways for biosynthesis of chorismate in *E. coli*, *M. jannaschii*, and *Aeropyrum pernix*. What is the earliest common intermediate in these pathways? What are its precursors in the three species?

Weblem 13.5 Define the following terms: (a) interactome; (b) signalome. (c) This is more difficult: can you think of, and define, a reasonable 'ome' that has not yet been proposed?

Weblem 13.6 Compare the methionine biosynthesis pathway from aspartate to methionine in *E. coli* and yeast. (a) Are there any differences in the series of intermediates? (b) Are the enzymes that catalyse similar transformations homologous? Show alignments of the amino-acid sequences where possible. Use EcoCyc for *E. coli* [http://ecocyc.org] and the *Saccharomyces* Genome Database [http://www.yeastgenome.org] for yeast, searching in each case for 'methionine biosynthesis'.

Weblem 13.7 Find the page in EcoCyc corresponding to Figure 13.4a. Choose different levels of detail and list what types of information are presented at each level.

Weblem 13.8 An enzyme with a function related to peptidylglycine monooxygenase (EC 1.14.17.3), and linked to it in the ENZYME DB, is 1-aminocyclopropane-1-carboxylate oxidase (EC 1.14.17.4). (a) What is the lowest common ancestor of the two reactions in the EC classification? (b) What is the lowest common ancestor of the two reactions in the Gene Ontology molecular function classification? (c) Are these two enzymes closely related in the Gene Ontology classification?

Weblem 13.9 From the UniProt entry, what identifiers does Gene Ontology associate with *E. coli* asparate aminotransferase, in the molecular function category? Arrange them in a directed acyclic graph, indicating the parent–child relationships between these identifiers.

Weblem 13.10 According to EcoCyc, what reactions can orotidine 5′-monophosphate undergo in *E. coli?* What enzymes catalyse these reactions? What genes encode these enzymes?

Weblem 13.11 Figure 13.6 shows the reductive carboxylate cycle, and the EC numbers of the enzymes that catalyse the individual steps. Find the corresponding information for the tricarboxylic acid cycle (or Krebs cycle), and for the glyoxylate cycle. Do an alignment of the metabolites participating in these cycles displaying the EC numbers of the enzymes that correspond to different reactions in different cycles. Report what is common to pairs or to all three of these pathways, in terms of (a) metabolites, (b) links between metabolites, corresponding to reactions, (c) enzymes that catalyse the reactions.

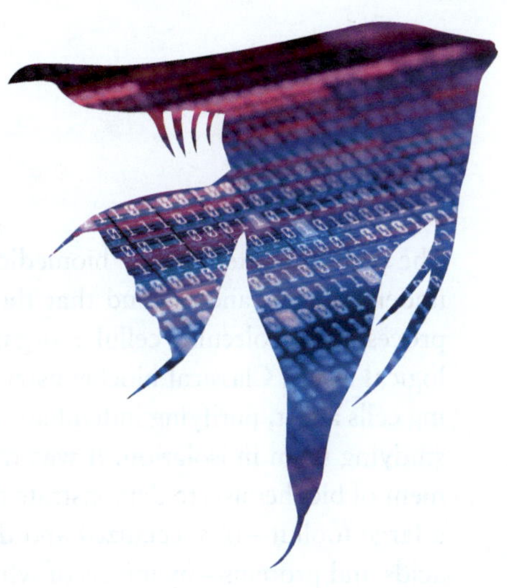

CHAPTER 14

Systems Biology

LEARNING GOALS

- *Gain a sense of the discipline of systems biology* as an integrative approach to all the 'omics disciplines.
- *Understand the idea of networks* and their representations as graphs.
- *Distinguish between static and dynamic aspects* of biological networks.
- *Know various types of regulatory mechanisms*, and the possible states of a network of activities that they can produce.
- *Appreciate the ideas of stability and robustness* and the mechanisms by which living systems—from cells to ecosystems—achieve them—or fail to.
- *Understand the structure, dynamics, and evolution of metabolic networks.*
- *Know the different ways of experimentally determining protein–protein and protein–nucleic acid interactions.*
- *Recognizing the great structural variety* of protein–DNA interactions.
- *Appreciate the details of some simple examples in which the mechanism of change in gene expression pattern has been worked out.*
- *Know the basic features of the regulatory network of E. coli*, including the classic example of the lac operon.
- *Appreciate the adaptability of regulatory networks*, in terms of the yeast regulatory network as an example.

Introduction

The goal of contemporary biomedical science is the understanding, and beyond that the control, of life processes at molecular, cellular, organismal, and ecological levels. Classical biochemistry focused on taking cells apart, purifying individual components, and studying them in isolation. It was the great achievement of biochemists to demonstrate that cells contain a large toolkit—of specialized and dedicated nucleic acids and proteins—by means of which they individually catalyse a large set of chemical reactions.

Molecular biology undertook to put things back together. It illuminated the regulatory networks that integrate the metabolic activities, and the expression patterns of the genes that underlie them. It was the great achievement of molecular biology to demonstrate the power of living things to effect the *controlled* manipulation of *matter*, *energy*, *and information*.

Systems biology is the study of the components of life, not in the lab, but 'on the job'—in the living context of their activities. It focuses on the coordination of the activities of the components; it seeks to establish connections, and to describe the physical and logical relationships that underlie these connections.

Several data streams feed the systems biology enterprise. Proteomics is the study of the distribution and interactions of proteins in time and space in a cell, organism, and ecosystem. What are the proteins in a sample? What are their functions? What are their abundances? What are their stabilities and turnover rates? With what partners do they physically and/or functionally interact? What are the control mechanisms that organize the various activities? Many features of the proteome are consequences of gene expression patterns. In addition, many non-coding RNAs play other important roles in cellular activities, alongside the proteins.

Although almost all cells of our bodies initially contain the same genome sequence, implementation of this genomic information varies in space and time. Differentiated cells form tissues. Programmed successions of processes characterize developmental stages. Transcriptomes, proteomes, and metabolomes respond to changes in the environment. The general underlying mechanism is the differential transcription and expression of genes, under the control of regulatory networks, involving specialized proteins and RNAs. Even viruses and bacteria show control networks; well-studied examples include the lytic–lysogenic switch in phage λ, and the control of gene expression in the lac operon of *E. coli*.

Systems biology tries to merge all these data into an integrated picture of the structure, dynamics, logistics, and ultimately the logic of living things. A systems biologist will combine study of the proteins and RNAs in a cell, their genes, the molecules that control their expression or their activities once expressed, and the set of other proteins with which they interact. A systems biologist will assemble into a metabolic network the chemical reactions catalysed by the enzymes of a cell, and assemble into control networks the mechanisms that regulate their expression and activities. Larger-scale ecological networks describe interspecies interactions; for instance, predator–prey relationships, including the spread of infectious disease in human, animal, and plant populations.

Regulatory mechanisms

Proteins, RNAs, and even genes are like social beings, and life depends on their interactions. Because individual molecules have specialized functions, control mechanisms are required to integrate their activities. The right amount of the right molecule must function in the right place at the right time. Failure of control mechanisms can lead to disease and even death.

> **KEY POINT**
>
> Systems biology focuses on the integration and control of gene, RNA, and protein activity.

Under unchanging environmental conditions, an organism's biochemical systems must be stable.

Under changing conditions, the system must be robust, accommodating both unchallenging and stressful perturbations. In some cases, processes must have their rates altered, or even be switched on and off. This regulation includes short-term adjustments, for instance in the stages of the cell cycle, circadian rhythms, or responses to external stimuli such as changes in the composition or levels of nutrients or oxygen. Longer-term regulatory activities include control over developmental stages during the entire lifetime of an organism.

Metabolism is the flow of molecules and energy through pathways of chemical reactions. The array of metabolic reactions forms complex traffic patterns. Some patterns are linear pathways, such as the multistep synthesis of tryptophan from chorismate. Others form closed loops, such as the tricarboxylic acid (Krebs) cycle. Moreover, the pathways interlock densely. The structure of the totality of metabolic pathways—its connectivity or topology—and its activity patterns can be analysed in terms of a mathematical apparatus dealing with graphs and flows and throughputs.

To control metabolic flow patterns, *regulatory pathways* connect genes, proteins, and metabolite concentrations. The structure and dynamics of the regulatory pathways are different from those of the metabolic pathways. Corresponding to the succession of enzymatic transformations in metabolism, a regulatory pathway is an assembly of signalling cascades.

Two parallel networks: physical and logical

In cells, two interaction networks operate in parallel: (1) a *physical network* of protein–protein and protein–nucleic acid complexes and (2) a *logical network* of control cascades. A logical network organizes regulatory interactions between processes, mediated of course by intermolecular interactions, though not necessarily by direct contacts.

Metabolic pathways partake of both: many but not all metabolic pathways involve physical protein–protein interactions and are regulated by logical interactions.

Examples of purely physical interactions include the assembly of oligomeric proteins such as haemoglobin, and photosynthetic reaction centres (complexes of proteins and cofactors that convert light to chemical energy), ATP synthase, assemblies of collagen in connective tissue, and the ribosome. Examples of logical interactions, *not* mediated entirely by direct physical interaction between proteins, include feedback loops in which the increase in concentration of a product of a metabolic pathway inhibits an enzyme catalysing one of the early steps in the pathway, or the secretion of a small molecule as a signal to other cells ('fire and forget' mode). (See Box 14.1 for an example of intercell communication.) In these cases, the logical interaction is transmitted by a diffusing small molecule. Many other examples make this a very common theme in the regulation of gene expression. A transcription factor, binding to DNA, may never interact physically with the proteins the expression of which it controls.

The allosteric change in the haemoglobin is an example of simultaneous physical and logical interaction: the subunits of haemoglobin respond to changes in oxygen levels by a conformational change that alters oxygen affinity. Another example is the transmission of a signal from the surface of a cell across the membrane to the interior by dimerization of a receptor. This can be the initial trigger of a process that ultimately affects gene expression. Not all links of this process need to involve protein–protein interactions. Some may be mediated by diffusion of small molecules such as cyclic AMP.

Even though certain protein–protein and protein–nucleic acid complexes participate in both physical and logical networks, the two networks remain distinct, and it is useful to keep the distinction in mind—*especially* when considering proteins that participate in both.

KEY POINT

Cells contain both physical and logical networks. Some interactions are common to both.

BOX 14.1 Cell–cell communication in microorganisms: quorum sensing

Control mechanisms *not* involving direct protein–protein interactions mediate intercellular signalling in microorganisms. *Vibrio fischeri* is a marine bacterium that can adopt alternative physiological states in which bioluminescence is active or inactive (literally a 'light switch'). The organism can live free in sea water, or colonize the light organs of certain species of fish or squid. It is bioluminescent only when growing within the animal.

The bacteria respond to the local density of cells, a form of communication called **quorum sensing**. In *V. fischeri*, quorum sensing is mediated by secretion and detection of a small signalling molecule, N-(3-oxohexanoyl)-homoserine lactone. Related species use other N-acyl homoserine lactones, generically abbreviated AHL. AHL can diffuse freely out of the cells in which it is synthesized. Within the light organs, culture densities can reach 10^{10}–10^{11} cells/mL, and the AHL concentration can exceed the threshold of about 5–10 nM for triggering the physiological switch.

Bacterial genes *LuxI* and *LuxR* govern the regulation. The product of *LuxI* is involved in the synthesis of AHL. The *LuxR* gene product contains a membrane-bound domain that detects the AHL signal, and a transcriptional activator domain. *LuxR* activates an operon that includes (1) genes for synthesis of luciferase (the enzyme responsible for the bioluminescence) and (2) *LuxI*, expression of which synthesizes additional AHL, amplifying the signal and sharpening the transition. This is the *opposite* of feedback inhibition.

The host also senses the bacteria: the light organs of squid grown in sterile salt water do not develop properly. This appears to be a reaction to the intensity of luminescence, rather than to the concentration of AHL. For the animal, the luminescence contributes to camouflage: disguise from predators inhabiting lower depths, by blending with illumination from the sky. The masking of shadows is a natural form of 'make-up'.

In quorum sensing, intercellular communication, via secretion and sensing of chemical signals, controls integrative behaviour of the system.

(The bioluminescence also regularly surprises diners in seafood restaurants, who jump to the conclusion that their glowing dinner is of extraterrestrial origin. However, most bioluminescent bacteria are harmless, although some strains of the related *Vibrio* species, *V. cholerae*, the causative agent of cholera, are weakly bioluminescent. In fact, the virulence of *V. cholerae* is also under the control of quorum sensing, by a related mechanism.)

Networks and graphs

In the abstract, networks have the form of graphs (see Box 14.2).

An example that will be familiar to many readers is the map of the London Underground.[1] The stations are nodes of a graph, and the tracks joining stations are the edges. Exercises 14.4 and 14.5, and Problem 14.2 also make use of this map, and maps of the tube (subway) systems of other cities. (Box 14.3 contains other examples of graphs.) The standard modern London Underground map shows the *topology* of the network; it does not quantitatively represent the geography of the area. An early map, from 1925, did maintain geographic accuracy.[2]

[1] http://www.transportforlondon.gov.uk/tfl/tube_map.shtml or http://www.afn.org/~alplatt/tube.html.

[2] http://www.clarksbury.com/cdl/maps.html; and see https://www.whatdotheyknow.com/request/224813/response/560395/attach/3/London\%20Connections\%20Map.pdf. This was possible when the system was simpler than it is now. Some of the maps now posted in the Paris Métro are fairly accurate geographically. *Considered as networks, a geographically accurate map and a simplified map with the same topology correspond to the same graph.*

BOX 14.2 The idea of a graph

- Mathematically, a graph consists of a set of vertices and a set of edges.
- Each edge links a pair of vertices.
- In a *directed graph*, the edges are *ordered* pairs of vertices.
- In a *labelled graph*, there is a value associated with each edge. In a road map, each edge might be labelled by the quality of the road (A road, B road, unpaved track); or the distance between vertices, even if the drawing is not geographically accurate. A directed graph is a special case of a labelled graph: consider the arrowheads as labels.

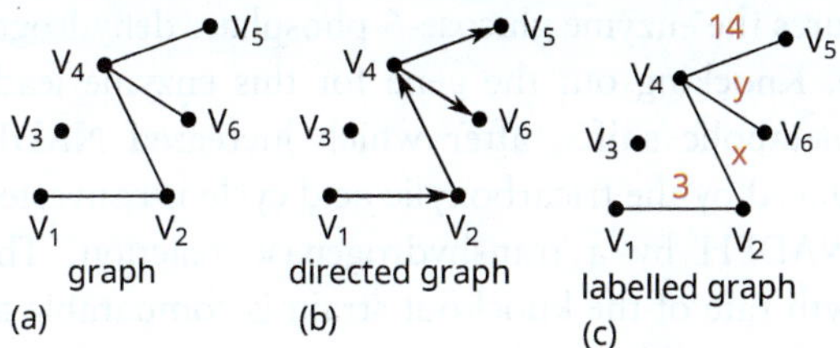

An undirected unlabelled graph specifies the connectivity of a network but not the distances between vertices (the topology but not the geometry, as in the modern London Underground map). Labels on the edges may indicate distances, but are unrestricted. For example, some phylogenetic trees indicate only the topology of the ancestry. Others indicate quantitatively the amount of divergence between species. Many phylogenetic trees are drawn with the lengths of the branches indicating the time since the last common ancestor. This is a pictorial device for labelling the edges.

Some graphs do not correspond to physical structures, and in any event edge labels need not reflect geometry in the usual sense. For example, the links in a network of metabolic pathways might be labelled to reflect flow patterns.

KEY POINT

Graphs are abstract representations of networks. They show the connectivity of the network. Labelled graphs can show physical distances between nodes or other properties of edges such as throughput capacity.

The London Underground map shows a static picture of the network and its connectivity. It does not show traffic patterns and the mechanisms that control them. These are essential aspects of biological networks—as well as of the London Underground.

BOX 14.3 Examples of graphs

- Sets of people who have met each other
- Electricity distribution systems
- Phylogenetic trees
- Metabolic pathways
- Chemical bonding patterns in molecules
- Citation patterns in the scientific literature
- The World Wide Web

The London Underground network is fully connected, in that there is a path between any two stations. Many questions familiar to commuters are shared in the analysis of biological networks; for example: What are the paths connecting Station A and Station B? Regarding different lines as subnetworks, how easy is it to transfer from one to another; that is, what is the nature of the patterns of connectivity? In case of failure of one or more links, is the network robust—does it remain fully connected?

Robustness and redundancy

Biological systems need to be robust, both for survival of individuals under stress and for the plasticity required for evolution. In yeast, for example, single gene knockouts of over 80% of the ~6200 open reading frames are survivable injuries.

In principle, networks can achieve robustness through redundancy. The most direct mechanism is simple **substitutional redundancy**: if two proteins are each capable of doing a job, knock out one and the other takes over. In the London Underground, this would correspond to a second line running over the same route. For instance, when the Circle Line is not running, passengers travelling between Paddington and King's Cross stations can travel by the District or the Hammersmith & City lines running on the same tracks.

In cells, some genes have closely related homologues resulting from gene duplication, and some of these contribute to substitutional redundancy. For example, in establishing mouse models for diabetes it was observed that mice and rats (but not humans) have two similar but non-allelic insulin genes. However, substitutional redundancy requires equivalence not only of function but of control of expression. This is the case for mouse insulin production: knocking out either insulin gene leads to compensatory increased expression of the other, resulting in a normal phenotype. We saw in Box 1.15 the mechanism of dosage compensation required because most mammalian females have twice the number of X-chromosome genes as males.

Equivalent expression patterns are more probable among duplicated genes than among unrelated ones. For an example of non-equivalent expression patterns, *E. coli* contains two fructose-1,6-bisphosphate aldolases. One, expressed only in the presence of special nutrients, is non-essential under normal growth conditions. However, the other is essential. In this case, functional redundancy does *not* provide robustness. These two enzymes are probably homologous, but they are distant relatives, not the product of a recent gene duplication. One is a member of a family of fructose-1,6-bisphosphate aldolases typical of bacteria and eukarya, and the other is a member of another family that occurs in archaea. *E. coli* is unusual in containing both.

An alternative mechanism of network robustness is **distributed redundancy**: the same effect achieved through different routes. In normal *E. coli*, approximately two-thirds of the NADPH produced in metabolism arises via the pentose phosphate shunt, which requires the enzyme glucose-6-phosphate dehydrogenase. Knocking out the gene for this enzyme leads to metabolic shifts, after which increased NADH produced by the tricarboxylic acid cycle is converted to NADPH by a transhydrogenase reaction. The growth rate of the knockout strain is comparable to that of the wild type.

KEY POINT

Biological networks need to be robust. The basic source of robustness is redundancy. Substitutional redundancy is roughly equivalent to multiple copies. (Think of 'back-ups'.) Distributed redundancy is roughly equivalent to alternative or even multiple providers of an essential function.

Connectivity in networks

If V_A and V_Z are vertices in a graph, a *path* from V_A to V_Z is a series of vertices: V_A, V_B, V_C, . . .V_Z, such that an edge in the graph connects each successive pair of vertices. The number of vertices in the chain, counting the starting and ending nodes, is called the *length* of the path. For instance, in the graph in Box 14.2 there is a path of length 4 from V_1 to V_5. A *cycle* is a path of length >2 in a non-directed graph for which the initial and final endpoints are the same, but in which no intermediate link is repeated.

A graph that contains a path between any two vertices is called **connected**. Alternatively, a graph may split into several **connected components**. The graph in Box 14.2 contains two connected components, one containing five vertices and one containing only one vertex. (In the extreme, a graph could contain many vertices but no edges at all.) It is often useful to determine the **shortest path** between any two nodes, and to characterize a network by the distribution of path lengths. The phrase 'six degrees of separation'—the title of a play by John Guare—refers to the assertion (attributed originally to Marconi) that if the people in the world are vertices of a graph and the graph contains an edge whenever two people know each other, then the graph is connected, and there is a path between any two vertices with length ≤6.

A **tree** is a special form of graph. A tree is a connected graph containing only one path between each pair of vertices. A hierarchy is a tree: Linnaean taxonomy is an example. Note that some family trees are not trees in the mathematical sense; examples are plentiful in the royal families of Europe. A tree cannot contain a cycle: If it did, there would be two paths from the initial point (= the final point) to each intermediate point. In the graph in Box 14.2, the subgraph consisting of vertices V_1, V_2, V_4, V_5, and V_6 is a tree. Adding an edge from V_1 to V_5 would create an alternative path from V_1 to V_5, and produce the cycle $V_1 \rightarrow V_2 \rightarrow V_4 \rightarrow V_5 \rightarrow V_1$; the graph would no longer be a tree.

The Enzyme Commission classification of protein function is a tree; the Gene Ontology classification is not.

The **density of connections** = the mean number of edges per vertex. It characterizes the structure of a graph. A fully connected graph of N vertices has $N-1$ connections per vertex; a graph with no edges has 0. Nervous systems of higher animals achieve their power not only by containing large numbers of neurons but also by high connectivities.

In some systems there are limits on numbers of connections. For many human societies, in the graph in which individuals are the vertices and edges link people married to each other, each node has connectivity 0 or 1. In neutral hydrocarbons, the graphs in which carbon and hydrogen atoms are the vertices and edges link atoms bonded to each other, each node has ≤4 connections. In other networks, connectivities follow observable regularities (see Box 14.5). For instance, the World Wide Web can be considered as a directed graph. Individual documents are the nodes, and hyperlinks are the edges. It is observed that the distribution of incoming and outgoing links follow power laws: $P(k)$ = probability of k edges is proportional to k^{-q} where $q = 2.1$ for incoming links and $q = 2.45$ for outgoing links.

KEY POINT

Topological features of networks that govern their general behaviour include:

- the density of connections;
- the topology of the connections—for instance, a tree is a graph with no cycles;
- the distribution of connections—for instance, small-world networks have a few nodes with large numbers of connections, called 'hubs' (see Box 14.4).

The density of connections is very important in defining the properties of a network. For instance, the interactions that spread disease among humans and/or animals form a network. Whether a disease will cause an epidemic depends not only on the ease of transmission in any particular interaction, but on the density of connections. As the density of connections—the rate of interactions—increases, the system can exhibit a *qualitative* change in behaviour,

BOX 14.4 'Small-world' networks

Many observed networks, including biological networks, the World Wide Web, and electric power distribution grids, have the characteristics of high clustering and short path lengths. They include relatively few nodes with very large numbers of connections, called **hubs**, and many that contain few connections. These combine to produce short path lengths between all nodes. From this feature they are called '**small-world networks**'. Such networks tend to be fairly robust—staying connected after failure of random nodes. Failure of a hub would be disastrous but is unlikely, because there are so few of them.

Many networks, notably the World Wide Web, are continuously adding nodes. The connectivity distribution tends to remain fairly constant as the network grows. These are called '**scale-free**' networks.

analogous to a phase change in physical chemistry, from a situation in which the disease remains under control to an epidemic spreading through an entire population. The classic approach of 'quarantine'—isolating people for 40 days—works by cutting down the degree of connectivity of the disease transmission network. A carrier who shows no symptoms—'Typhoid Mary'[3] was a classic case—serves as a hub of the transmission network.

Two historical epidemics associated with wars demonstrate the distinction between topology and geometry in network connectivity.

- In the early years of the Peloponnesian War, Athens suffered a severe epidemic. (From Thucydides' detailed description of the symptoms, the disease was probably bubonic plague.) A factor contributing to its transmission was the crowding of people into the city from the surrounding countryside, out of fear of greater vulnerability to military invasion.
- After the First World War, an epidemic of influenza killed over 50 million people, more than died in the war itself. Long-distance travel by soldiers returning from the war helped spread the disease. Any epidemic needs an infectious agent and a high density of routes of transmission.

These examples show that the controlling factor is the density of the *connections* and not the density of the people.

Dynamics, stability, and robustness

An unlabelled, undirected graph describes the *static* structure of the topology of a network. For molecular interaction networks, this may be an adequate description of many of the physical interactions.

For some networks, such as metabolic pathways or patterns of traffic in cities, the *dynamics* of the system depends on the transmission capacities of the individual links. These capacities can be indicated as labels of the edges of the graph. This allows modelling of patterns of flow through the network. Examples include route planning, in travel or deliveries. Note that the shortest path may well not give optimal throughput. In many cities, taxi drivers are exquisitely sensitive—and insensitively garrulous—about optimal traffic paths. In molecular biology, metabolic pathways and signal transduction cascades are networks that lend themselves to pathway and flow analysis.

Although much is known about the mechanisms of individual elements of control and signalling pathways, understanding their integration is a subject of current research. For instance, the idea that healthy cells and organisms are in stable states is certainly no more than an approximation (and in most cases a gross idealization). The description of the actual dynamic state of the metabolic and regulatory networks is a very delicate problem. Understanding *how* cells achieve even an apparent approximation to stability is also quite tricky. It is likely that great redundancy of control processes lies at its basis. Regulation is based on the resultant of many individual control mechanisms—here a short feedback loop, there a multistep cascade. Somehow, the independent actions of all the individual signals combine to achieve an overall, integrated result. It is like the operation of the 'invisible hand' that, according to Adam Smith, coordinates individual behaviour into the regulation of national economies.

Several types of dynamic states of a network are possible (see Box 14.5):

- equilibrium;
- steady state;
- states that vary periodically;
- unfolding of developmental programmes;
- chaotic states;
- runaway or divergence;
- shutdown.

[3] Mary Mallon (1869–1938) presented the following unfortunate combination of features: (1) she was infected with typhoid, (2) she did not show symptoms, and (3) she worked for many families as a cook.

KEY POINT

In addition to their static topology, networks have several possible dynamic states. Some of these are stable and others are unstable.

States of a network of processes

- At *equilibrium* one or more forward and reverse processes occur at compensating rates, to leave the amounts of different substances unchanging:

$$A \rightleftharpoons B$$

Chemical equilibria are generally self-adjusting upon changes in conditions, or upon addition of reactants or products.

- A steady state will exist if the total rate of processes that produce a substance is the same as the total rate of processes that consume it. For instance, the two-step conversion

$$A \rightarrow B \rightarrow C$$

could maintain the amount of B constant, provided that the rate of production of B (the process $A \rightarrow B$) is the same as the rate of its consumption (the process $B \rightarrow C$). The net effect would be to convert A to C.

A cyclic process could maintain a steady state in all its components:

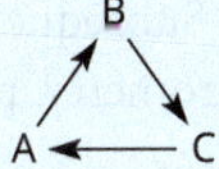

A steady state in such a cyclic process with all reactions proceeding in one direction at the same rate is very different from an equilibrium state. Nevertheless, in some cases it is still true that altering external conditions produces a shift to another, neighbouring, steady state.

- *States that vary periodically* appear in the regulation of the cell cycle, circadian rhythms, and seasonal changes such as annual patterns of breeding in animals and flowering in plants. Circadian and seasonal cycles have their origins in the regular progressions of the day and year, but have evolved a certain degree of internalization.

- Many equilibrium and some steady-state conditions are *stable*, in the sense that concentrations of most metabolites are changing slowly if at all, and the system is robust to small changes in external conditions. The alternative is a **chaotic state**, in which small changes in conditions can cause very large responses. Weather is a chaotic system: the meteorologist Lorenz asked, 'Does the flap of a butterfly's wings in Brazil set off a tornado in Texas?' In a carefully regulated system, chaos is usually well worth avoiding, and it is likely that life has evolved to damp down the responses to the kinds of fluctuations that might give rise to it. Chaotic dynamics does sometimes produce approximations to stable states—these are called **strange attractors**. Understanding stability in living dynamical systems subject to changing environmental stimuli is an extremely important topic, but beyond the scope of this book.

- *Unfolding of developmental programmes* occurs over the course of the lifetime of the cell or organism. Many developmental events are relatively independent of external conditions, and are controlled primarily by programmed regulation of gene expression patterns.

- *Runaway* or *divergence*. Breakdown in control over cellular proliferation leads to unconstrained growth in cancer.

- *Shutdown* is part of the picture. Apoptosis is the programmed death of a cell, as part of normal developmental processes, or in response to damage that could threaten the organism, such as DNA strand breaks. Breakdown of mechanisms of apoptosis—for instance, mutations in protein p53—is an important cause of cancer.

Protein complexes and aggregates

The physical basis of our understanding of how life within a cell is organized and regulated is the set of protein–protein and protein–nucleic acid interactions. We have already encountered protein complexes:

- Simple dimers or oligomers in which the monomers appear to function independently.
- Oligomers in which the monomers interact functionally, for instance ligand-induced dimerization of receptors, and allosteric proteins such as haemoglobin.
- Large fibrous proteins such as actin or keratin.
- Non-fibrous structural aggregates such as viral capsids.

- Protein–nucleic acid complexes, including ribosomes, nucleosomes, splicing and repair particles, and viruses. Of particular interest in systems biology are protein–nucleic acid complexes involved in the regulation of transcription.
- Many proteins, whether monomeric or oligomeric, *function* by interacting with other proteins. These include all enzymes with protein substrates, and many antibodies, inhibitors, and regulatory proteins.

The systems biology approach to protein interactions aims at a systematic and comprehensive identification of protein interactions. Most experiments reveal only pairwise interactions. The challenges are to integrate pairwise interactions into a network and then to study the structure and dynamics of the system.

Protein interaction networks

The units from which interaction networks are assembled are:

- for physical networks, a protein–protein or protein–nucleic acid complex;
- for logical networks, a dynamic connection in which the activity of a process is affected by a change in external conditions, or by the activity of another process.

Many techniques detect physical interactions directly:

- *X-ray and NMR structure determinations* can not only identify the components of the complex, but reveal how they interact, and whether conformational changes occur upon binding.
- *Two-hybrid screening systems*: Transcriptional activators such as Gal4 contain a DNA-binding domain and an activation domain. Suppose these two domains are separated, and one test protein is fused to the DNA-binding domain and a second test protein is fused to the activation domain. Then a reporter gene will be transcribed only if the components of the activator are brought together by formation of a complex between two test proteins (Figure 14.1). High-throughput methods allow parallel screening of a 'bait' protein for interaction with a large number of potential 'prey' proteins.

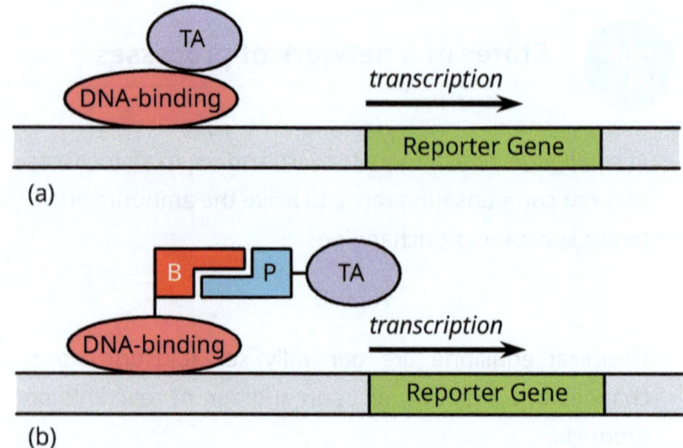

Figure 14.1 (a) A transcription activator contains two domains, a DNA-binding domain, and a transcription-activator domain (TA). Together they induce expression of a reporter gene. *lacZ*, encoding β-galactosidase, is a common choice of reporter gene because chromogenic substrates make β-galactosidase easy to detect. (b) Transcription proceeds if the DNA-binding domain and transcription-activator domain are separated in different proteins that can form a complex. A 'bait' protein B is fused to the DNA-binding domain and a 'prey' protein P is fused to the transcription-activation domain. Formation of a complex between bait and prey brings together the DNA-binding and TA domains, inducing transcription.

- *Chemical cross-linking* fixes complexes so that they can be isolated. Subsequent proteolytic digestion and mass spectrometry permit identification of the components.
- *Coimmunoprecipitation*. An antibody raised to a 'bait' protein binds the bait together with any other 'prey' proteins that interact with it.

  The interacting proteins can be purified and analysed, for instance by western blotting or mass spectrometry.
- *Chromatin immunoprecipitation* identifies DNA sequences that bind proteins (Figure 14.2).
- *Phage display*. Genes for a large number of proteins are individually fused to the gene for a phage coat protein to create a population of phage, each of which carries copies of one of the extra proteins exposed on its surface. Affinity purification against an immobilized 'bait' protein selects phage displaying potential 'prey' proteins. DNA extracted from the interacting phages reveals the amino-acid sequences of these proteins (see Figure 14.3).
- *Surface plasmon resonance* analyses the reflection of light from a gold surface to which a protein

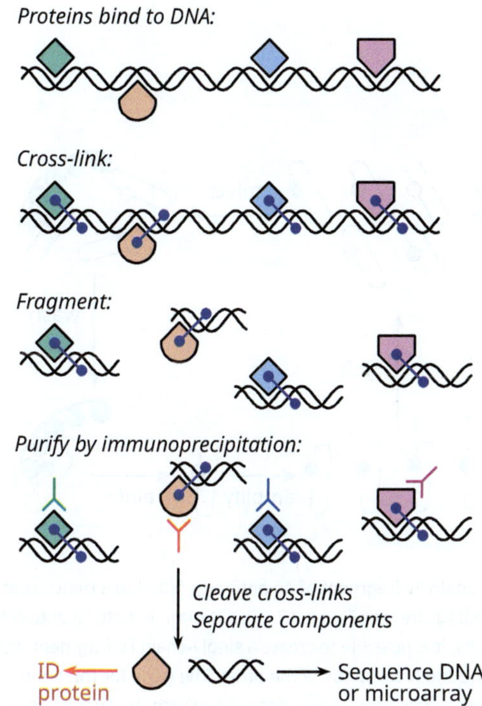

Figure 14.2 Chromatin immunoprecipitation. Treatment with formaldehyde cross-links proteins and DNA, fixing the complexes that exist within a cell. After isolation of chromatin, breaking the DNA into small fragments allows separation of proteins by binding to specific antibodies, carrying the DNA sequences along with them. Reversal of the cross-link followed by sequencing of the DNA identifies the specific DNA sequence to which each protein binds. To identify multiple sites in the genome to which the protein binds, the DNA fragments can be analysed using a microarray. To avoid the requirement for antibodies specific for each protein to be tested, the proteins can be fused to a standard epitope or to a sequence that can be biotinylated, taking advantage of the very high biotin-streptavidin affinity.

has been attached. The signal changes if a ligand binds to the immobilized protein. (The method detects localized changes in the refractive index of the medium adjacent to the gold surface. This is related to the mass being immobilized.)

- *Fluorescence resonance energy transfer*. If two proteins are tagged by different chromophores, transfer of excitation energy can be observed over distances up to about 60 Å.
- *Tandem affinity purification (TAP)* allows probing cells *in vivo* for partners that bind to a selected 'bait' protein. Fusion of the bait protein to *two* affinity tags, separated by a cleavage site, permits high efficiency in extraction of complexes. The method uses two successive, or tandem, affinity purification steps separated by a cleavage step to expose the second tag. The cleavage steps are specific and require only mild conditions in order to leave the bait–prey complex intact (see Figure 14.4).

  The fusion proteins contain an individual bait protein extended at its N or C terminus by a calmodulin-binding peptide, a TEV protease cleavage site (TEV is a cysteine protease from tobacco etch virus), and protein A (which binds to high affinity to an available antibody). The double tag, and the two-step purification, gives superior performance relative to a single-tag technique, in yield of low-concentration complexes.

  In vivo, the expressed tagged bait protein binds to a set of prey proteins. Figure 14.4 shows the purification protocol to recover the complexes from cell extracts.

  Determination of a protein interaction network requires measurements of many bait proteins. Because all bait proteins carry the same fusion sequences, the purification and cleavage steps in TAP are unique. There is no need to design different purification protocols for different bait proteins.

Other methods provide complementary information:

- *Domain recombination networks*. Many eukaryotic proteins contain multiple domains. A feature of eukaryotic evolution is that a domain may appear in different proteins with different partners. In some cases, proteins in a bacterial operon catalysing successive steps in a metabolic pathway are fused into a single multidomain protein in eukarya. The domains of the eukaryotic protein are individually homologous to the separate bacterial proteins.

  It is possible to create a network by defining an interaction between two protein domains whenever homologues of the two domains appear in the same protein. This is evidence for some functional link between the domains, even in species where the domains appear in separate proteins.
- *Coexpression patterns*. Clustering of RNAseq data identifies proteins with common expression patterns. They may have the same tissue distribution,

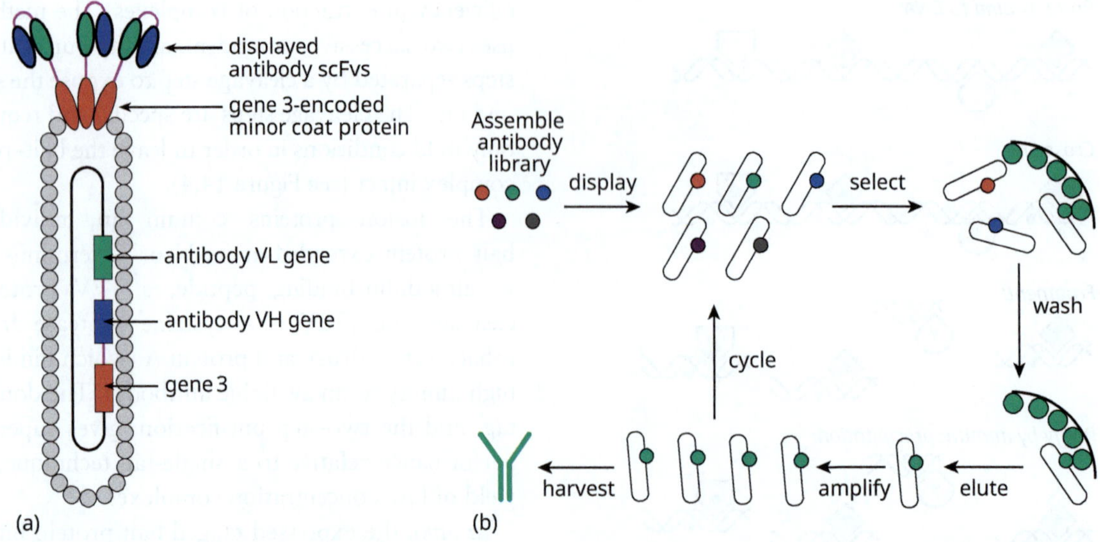

Figure 14.3 (a) Schematic diagram of a filamentous phage displaying a single-chain Fv fragment of an antibody, fused to a minor coat protein, on its surface. The Fv fragment of an antibody contains the antigen-binding site (see Figure 10.17). Although in natural antibodies the two domains constituting the Fv fragment are on different polypeptide chains, it is possible to create a single-chain Fv fragment (scFv) by introducing a linking oligopeptide between the C-terminus of one domain and the N-terminus of the other. The gene for the antibody fragment is linked to the gene for a minor coat protein. Expression of the tandem genes produces a coat protein with the Fv fragment attached. It appears on the surface of the phage. (b) The processes of preparation, selection, and harvesting in phage display. First, create a large population of phage each of which has a different antibody fused to a coat protein. This variety is represented by different colours. Exposure of the population of viruses to antigens fixed to a support, followed by washing, retains tight-binding antibodies. Several cycles of enrichment are typically necessary. Once the affinity has reached a target value, select the phages that are binding to the target with high affinity. Note the importance of being able to select for proteins but also having access to the genes that encode them.

or be up- or downregulated in parallel in different physiological states. This is also suggestive evidence that they share some functional link. But this must be verified. In the response of *M. tuberculosis* to the drug isoniazid, genes for the fatty acid synthesis complex are coordinately upregulated. They are on an operon-like gene cluster, and in fact these proteins do form a physical complex. On the other hand, alkyl hydroperoxidase (AHPC) is also upregulated in response to isoniazid. AHPC acts to relieve oxidative stress. There is no evidence that it physically interacts with the fatty acid synthesis complex, or that it mediates a metabolic transformation coupled to fatty acid synthesis. It is a second, independent, component of the response to isoniazid.

- *Phylogenetic distribution patterns*. The *phylogenetic profile* of a protein is the set of organisms in which it and its homologues appear. Proteins in a common structural complex or pathway are functionally linked and expected to coevolve. Therefore, proteins that share a phylogenetic profile are likely to have a functional link, or at least to have a common subcellular origin. There need be no sequence or structural similarity between the proteins that share a phylogenetic distribution pattern. A welcome feature of this method is that it derives information about the function of a protein from its relationship to *non-homologous* proteins.

Each of these methods provides a basis for a protein interaction network. The networks formed from each set of interactions are different, although they overlap, to a greater or lesser extent. They give different views of the kinds of relationships between proteins that exist in cells. It is possible to form a more comprehensive network by combining different types of interactions. For instance, the DIP database is a curated collection of experimentally determined protein–protein interactions, http://dip.doe-mbi.ucla.edu/. It contains data about 81923 interactions between 28850 proteins from 834 organisms. (For others, see Box 14.6.)

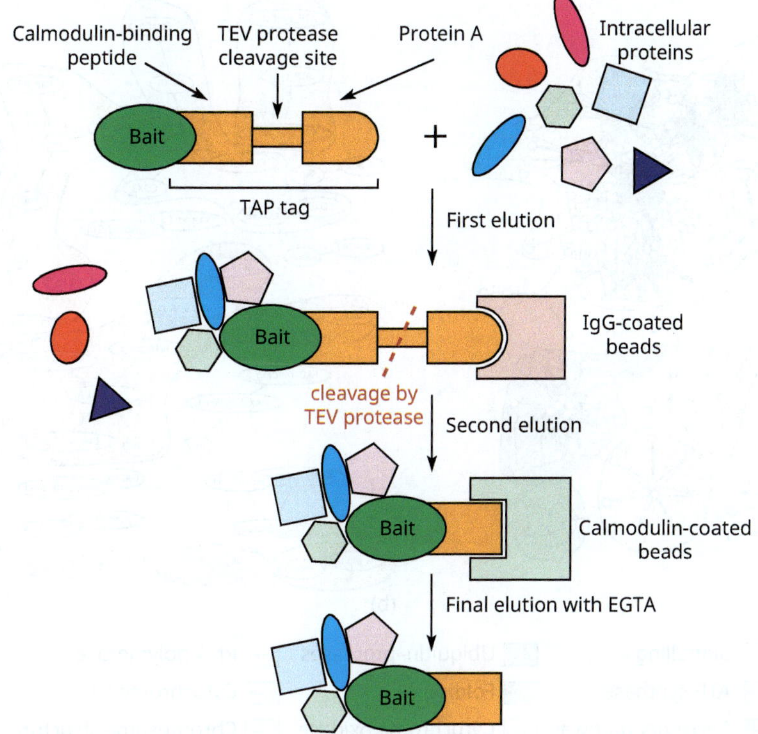

Figure 14.4 The construct and protocol in the tandem affinity purification (TAP) method for purification of complexes with a selected 'bait' protein. The fusion protein containing the bait and the two tags separated by the TEV cleavage site bind *in vivo* to proteins in the cell. A first affinity purification step binds the bait protein to a column containing IgG-coated beads that bind specifically to the first tag, protein A. After thorough washing, cleavage by TEV protease releases the bound complexes, and exposes the second tag. A second affinity purification step binds the bait protein to a column containing calmodulin-coated beads, which bind specifically to the second tag, the calmodulin-binding peptide. After washing, elution with the chelating agent ethylene glycol tetraacetic acid (EGTA) releases the purified complexes.

BOX 14.6

Web resource: interaction databases

Intact: An open source molecular interaction database	http://www.ebi.ac.uk/intact/
DIP: Database of Interacting Proteins	http://dip.doe-mbi.ucla.edu/
MIPS Comprehensive Yeast Genome Database	http://mips.gsf.de/
MIPS Mammalian Protein–Protein Interaction Database	http://mips.helmholtz-muenchen.de/proj/ppi
BIND: Biomolecular Interaction Network Database	http://www.bind.ca/
MINT: Molecular Interactions Database	http://cbm.bio.uniroma2.it/mint/
HPID: The Human Protein Interaction Database	http://www.hpid.org/
HPRD: Human Protein Reference Database	https://www.hprd.org/
IID: Integrated Interactions Database	http://iid.ophid.utoronto.ca/
BioGRID: Database of Protein, Genetic and Chemical Interactions	https://thebiogrid.org/
STRING: Protein–Protein Interaction Networks	https://string-db.org/

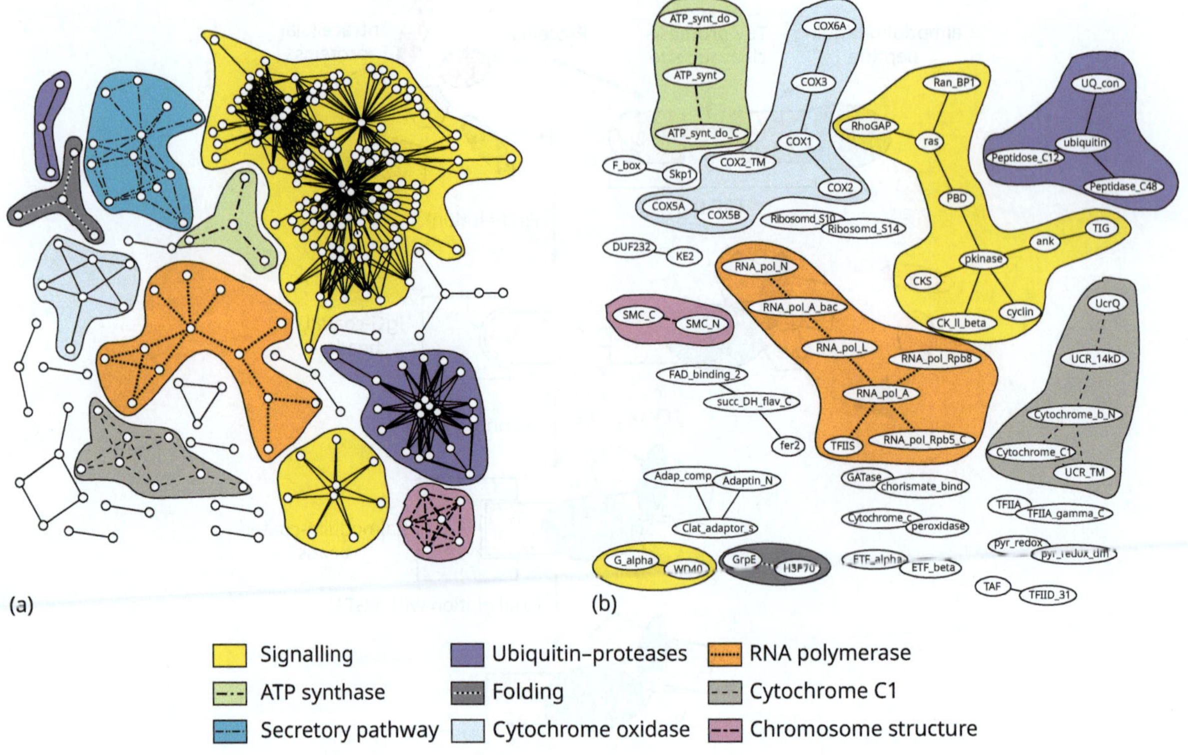

Figure 14.5 Portion of the interaction network of yeast proteins. (a) The interactions of individual proteins. (b) The interactions within a subnetwork based on representations of different protein families, in different functional categories, linked in (a). This figure is based on structural data and modelling. Each relationship implies a physical interaction between the proteins. Some of the interactions involve stable complexes (for instance, RNA polymerase II); others involve transient complexes.

Aloy, P., & Russell, R.B. (2005). Structure-based systems biology: A zoom lens for the cell. FEBS Letters 579, 8: 1854–1858. https://doi.org/10.1016/j.febslet.2005.02.014. Reprinted with permission of Elsevier.

Figure 14.5 shows a portion of an interaction network of yeast proteins, based on sets of proteins that have been found together in solved structures.

Protein–DNA interactions

In this section, we recognize that DNA is made of atoms, not character strings. 'One writes poems not with ideas, but with words.'—Mallarmé

DNA–protein complexes mediate several types of processes

These include:

- replication, including repair and recombination;
- transcription;
- regulation of gene expression;
- DNA packaging, including nucleosomes and viral capsids;
- the CRISPR bacterial defence system and its application in higher organisms.

Enzymes that replicate and transcribe DNA sequences are of necessity complex structures. Compare their operation with that of a typical metabolic enzyme that catalyses a single reaction of a small substrate and releases the product. In contrast, a DNA polymerase must carry out multiple manipulations of its substrate, some chemical and others mechanical. Addition of a nucleotide is a chemical step. Successive additions of nucleotides are mechanically linked, as the product is not released but translocated.

In adding different successive nucleotides, a DNA polymerase must evince a versatile but nevertheless exact specificity. It must be able to catalyse the addition of four different bases, and must select the proper one each time. This is different from promiscuous specificity, shown, for instance, by an exonuclease that can accept *any* nucleotide as a substrate at any step. Moreover, errors are unacceptable, as they would create mutations. In addition to the specificity (aimed at getting it right the first time), the enzyme does proofreading and error correction (just in case it didn't). To accomplish this, the enzyme has two active sites, one for addition of a base and an exonuclease site to remove an incorrectly incorporated one.

Other types of DNA–protein interactions require different degrees of DNA-sequence specificity. Differences in DNA-sequence specificity impose different requirements on the structures of the proteins that participate in different processes (see Box 14.7).

KEY POINT

There are a great variety of protein–nucleic acid complexes. Some are permanent, others are transient. Functions include, but are not limited to, catalysis of replication, transcription, translation, and regulation of transcription.

Structural themes in protein–DNA binding and sequence recognition

What does a protein looking at a stretch of DNA in the standard B conformation see (see Figure 4.11)? What could it hope to grab hold of? Prominent general features are the sugar-phosphate backbone, including charged phosphates suitable for salt bridges, and potential hydrogen-bond partners in the sugar hydroxyl groups. Contact with the bases is accessible through the major and minor grooves, although unless the DNA is distorted the bases are visible only 'edge-on'. Hydrogen-bonding patterns between bases in the grooves and particular amino acids account for some of the DNA-sequence specificity in binding (Figure 14.6). However, many protein–DNA hydrogen bonds are mediated by intervening water molecules, an effect that tends to *reduce* the specificity.

The idea that an α-helix has the right size and shape to fit into the major groove of DNA was noted in the 1950s. The structures of the first protein–DNA complexes confirmed this prediction. It became the paradigm for protein–DNA interactions. Indeed, when a student solving the structure of the Met repressor–DNA complex told his supervisor that in the

DNA-binding proteins show varying degrees of DNA-sequence specificity

- Some DNA-binding proteins are relatively non-specific with respect to nucleotide sequence, including DNA replication enzymes and histones.
- Some, for instance the restriction endonuclease EcoRV, *bind* to DNA with low specificity, but cleave only at GATATC. This combination permits a mechanism of finding the target sequence by initial non-specific binding followed by diffusion in one dimension along the DNA.
- Some recognize only specific nucleotide sequences. For example, the restriction endonuclease EcoRI binds specifically to GAATTC sequences with almost absolute specificity. It is a homodimer that recognizes palindromic sequences.
- Some DNA-binding proteins recognize consensus sequences. For example, the phage Mu transposase and repressor proteins bind 11 base-pair sequences of the form CTTTT[A/T]PyNPu[A/T]A[A/T] (where [A/T] = A or T, Py = either pyrimidine = C or T, Pu = either purine = A or G, and N = any of the four bases).
- Some recognize nucleotide sequences indirectly, via modulations of local DNA structure. For example, the TATA box-binding protein takes advantage of the greater flexibility of AT-rich sequences to form complexes in which the DNA is very strongly bent. The distinction between sequence specificity achieved through direct interaction with bases or through recognition of local structure has been termed 'digital versus analogue readout'.
- Some recognize general structural features of DNA, such as mismatched bases or supercoiling.
- Some DNA-binding proteins form an initial complex with high DNA-sequence specificity, followed by recruitment of other proteins of low specificity, to enhance overall binding affinity.

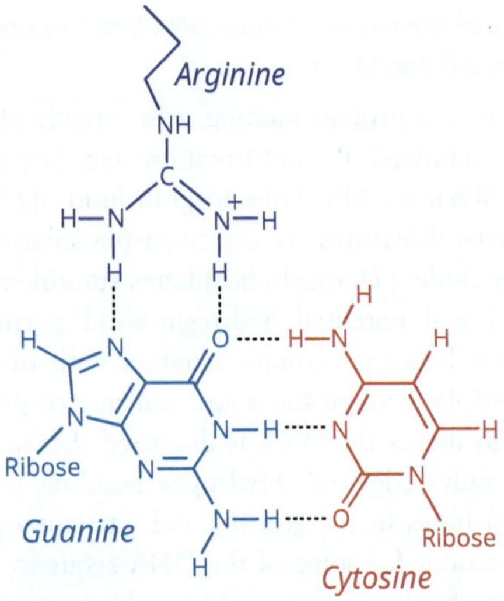

Figure 14.6 Specific hydrogen-bonding pattern between an arginine sidechain of a protein and a guanine of DNA.

electron-density map he was interpreting, a β-sheet appeared to bind in the major groove, he was advised, with patience strongly tinged with condescension, to go back and look for the helix.

We now recognize great structural variety in DNA–protein interactions. A few themes include:

- **Helix-turn-helix domains**. These appear in prokaryotic proteins that regulate gene expression, eukaryotic homeodomains involved in developmental control, and histones that package DNA in chromosomes.
- **Zinc fingers**, including eukaryotic transcription factors, and steroid and hormone receptors.
- Proteins with β-sheets that interact with DNA, for instance the gene-regulatory proteins, Met and Arc repressors, and the TATA-box binding protein.
- **Leucine zippers** that act as eukaryotic transcriptional regulators.
- The **high-mobility group** in eukaryotes and the prokaryotic protein HU, which bind sequences non-specifically and bend DNA.
- Enzymes that interact with DNA, involved in replication, translation, repair, and uncoiling. Some are relatively small; others are large multi-protein complexes. They show many different types of folding patterns. Many distort the DNA structure in order to get access to the bases that are the target of their activity. This category includes the nucleases associated with CRISPR.
- Viral capsid proteins form compact shells enclosing nucleic acid.

These examples are an anecdotal list, not a classification.

RNA-binding proteins have a separate variety. Some resemble DNA-binding proteins. Others bind to RNA molecules of defined structure; for instance, proteins of the spliceosome, responsible for excising introns from eukaryotic pre-mRNA; and enzymes that interact with tRNA, including but not limited to aminoacyl tRNA synthases, and the ribosome itself.

Examples of protein–DNA interactions

Bacteriophage T7 DNA polymerase

Bacteriophage T7 is a virus infecting many bacteria, including *E. coli*. It has its own DNA polymerase, which acts in conjunction with other proteins, including enzymes that unwind and prime the DNA, and that bind the unwound single-stranded DNA. The phage also recruits *E. coli* thioredoxin to keep the polymerase-template complex intact—in some cases, permitting replication of the entire 39937-base-pair T7 genome in one go.

S. Doublié, S. Tabor, A.M. Long, C.C. Richardson, and T. Ellenberger solved the structure of the T7 DNA polymerase, in complex with a primer-template double helix, and a nucleoside triphosphate about to be added to the growing strand. (To trap the structure, the 3' end of the primer strand is a dideoxynucleotide, to block the reaction.)

As part of its mechanism, the polymerase undergoes a conformational change from an open state,

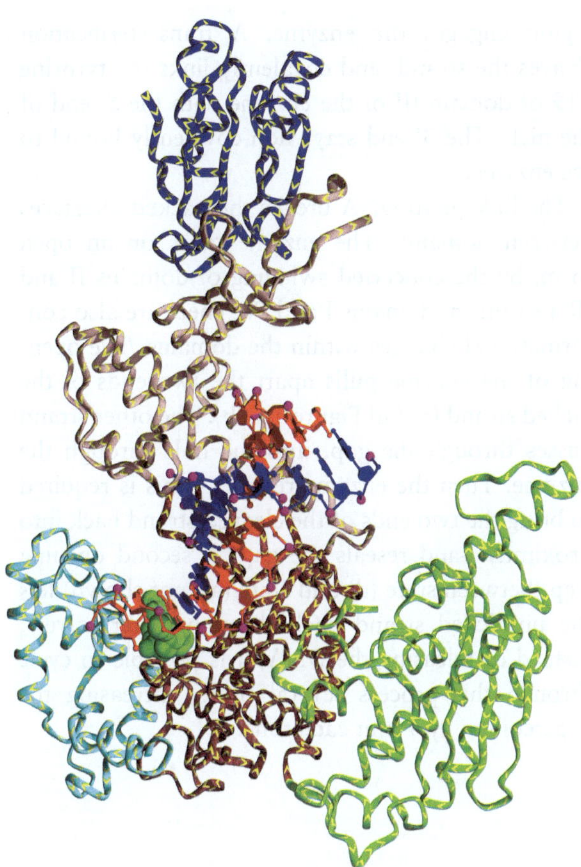

Figure 14.7 T7 DNA polymerase, binding primer-template double helix, with nucleoside triphosphate (green spheres) about to be incorporated. Green ribbon = exonuclease domain; blue, at top, thioredoxin.

which permits translocation of the primer and template into position for the next catalytic cycle, and a closed, catalytically active form.

Figure 14.7 shows T7 DNA polymerase in the act of adding a residue to the primer. The individual reaction step uses up one deoxynucleoside triphosphate, adds a nucleotide to the growing strand, and releases pyrophosphate.

The enzyme takes advantage of the stereochemical compatibility of different Watson–Crick base pairs to enhance fidelity of replication. A hydrophobic pocket that closely surrounds the nascent base pair selects against mismatches, in part by size discrimination and in part by an energetic penalty against unsatisfied hydrogen bonds in a structure sequestered away from water.

In case of incorporation errors, the speed of synthesis slows down, and the DNA shifts from the polymerase site to the exonuclease site to cleave off the incorrect base.

Some protein–DNA interactions require breaking the DNA chain

The equilibrium structure of the B form of DNA is the iconic double helix with ten base pairs per turn. If the two ends of a stretch of linearly extended double-helical DNA were grasped and twisted, the DNA would become either overwound or underwound. If the two ends were then linked to form a circular DNA molecule, the twisting could not be relieved by simple conformational changes.

Supercoiling is a measure of this twisting. Each full turn of unwinding counts as one negative unit of supercoiling. Each full turn of tightening, or overwinding, counts as one positive unit of supercoiling. The equilibrium form, with zero supercoiling, is relaxed.

To relieve the local strain associated with overwinding, a piece of circular DNA may form a superhelix. A similar phenomenon is familiar in the supercoiling of the helical wires attached to telephone receivers.

Supercoiling of DNA occurs during many natural processes, including DNA replication and transcription, segregation of daughter chromosomes, and recombination and repair. Even the packing of DNA into cells presents problems that must be addressed by higher-order structures; the DNA must condense, but yet be available to the replication and transcription apparatus.

DNA topoisomerases are enzymes that manipulate the higher-order levels of DNA structure by changing the supercoiling. In the process, they act as enzymes to catalyse the breakage and reformation of chemical bonds. However, the overall effect of their action is not a permanent change in the primary chemical bonding, but in the supercoiling of the DNA. They operate by temporarily breaking either one or both strands, passing a single strand or both strands through the break, and then rejoining the cleaved strands. This is topologically equivalent to local winding or unwinding. There are several families of topoisomerases, which differ in structure and activity.

E. coli topoisomerase III is type IA. Type IA enzymes act only on underwound—negatively supercoiled—DNA, to reduce the underwinding and bring the DNA closer to the relaxed state. Related enzymes appear in prokarya, archaea, and eukarya. Absence of functional type IA enzymes is lethal in *E. coli* and *Saccharomyces pombe*; *S. cerevisiae* survives but with substantial loss of robustness.

E. coli topoisomerase III has been the subject of structural studies by A. Mondragòn and co-workers to elucidate its mechanism of action. It has a four-domain structure, with interfaces between domains I and III and between domains III and IV (Figure 14.8). The active site is between domains I and III. A hole through the structure, 27 Å in diameter, is large enough to accommodate single- or double-stranded DNA. It is lined with positively charged sidechains to stabilize such a complex.

In the initial state, the enzyme is in a closed form (Figure 14.9a). The DNA must be in an underwound form, with a region in which the strands have separated. This can occur as a local mechanism to relieve the strain of underwinding. The single-stranded region engages the enzyme. A transesterification cleaves the strand, and covalently links the tyrosine 319 of domain III of the enzyme with the 5' end of the nick. The 3' end stays non-covalently bound to the enzyme.

The linkage to DNA breaks the packed interfaces between domains. The enzyme takes on an open form, by the concerted swinging of domains II and III away from domains I and IV. There are also conformational changes within the domains. The opening of the enzyme pulls apart the two ends of the nicked strand (red in Figure 14.9b). The other strand passes through the gap, into the hole through the enzyme. Then the enzyme recloses—this is required to bring the two ends of the cleaved strand back into proximity—and reseals the gap. A second opening step (between state (d) and state (a), not shown) lets the un-nicked strand back through. Alternatively, instead of releasing the DNA, it is possible to cycle through this process several times, increasing the supercoiling by 1 unit each time.

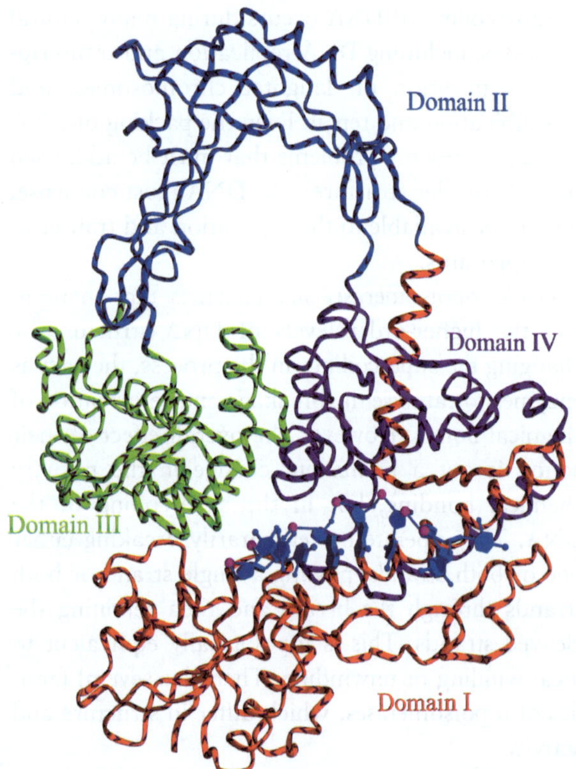

Figure 14.8 *E. coli* topoisomerase III binding short single strand of DNA. [1I7d].

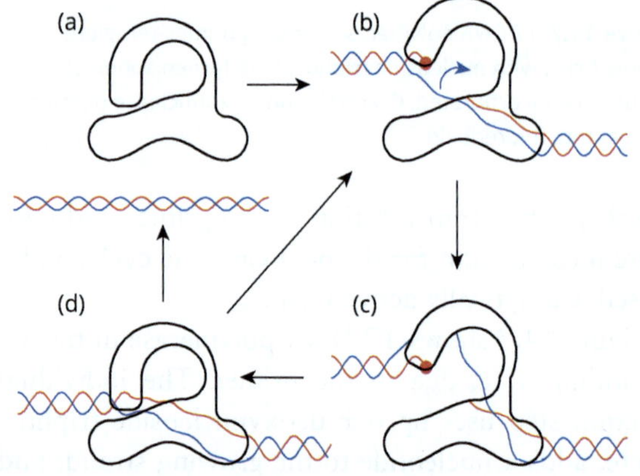

Figure 14.9 Steps in the action of type-IA topoisomerases. (a) Enzyme in closed form, plus DNA, shown here in relaxed form. DNA and enzyme unengaged. Enzyme–DNA interaction requires underwound DNA with available single-stranded region. (b) DNA strands locally separated. Enzyme has nicked one strand and changed its own conformation to open form. The 5' end of the nicked strand is covalently attached to tyrosine 319 (red dot). The 3' end is non-covalently attached to the enzyme. (c) The uncleaved strand has moved through the gap. (d) The enzyme has reclosed and reformed the strand. This increases the supercoiling by 1 unit. At this point the intact double strand may be released (d → a), or the cycle can repeat (d → b → c → d), increasing the supercoiling by 1 unit each time around.

Some idea of the structure and mechanism is observable in Rodin's statue, Le Penseur. The subject is seated with his left elbow leaning on his knee. If he were to sit up, raising his arm, a strand could enter the gap formed.

Some protein–DNA complexes that regulate gene transcription

Proteins involved in regulation of gene expression show great structural variety, reflecting the complexity and diversity of the network of transcription control mechanisms.

λ cro

Bacteriophage λ is a virus containing a double-stranded DNA genome of 48 502 base pairs. A λ phage infecting an *E. coli* cell chooses—depending on which genes are active—between lysis or lysogeny. λ can replicate, and *lyse* the cell, releasing ~100 progeny. Alternatively, it can integrate its DNA into the host genome. The phage in such a *lysogenized* cell is dormant, and can be released by stimuli that switch from the lysogenic to the lytic state.

λ cro is a transcription regulator involved in the lytic–lysogenic switch. It binds to DNA as a symmetrical dimer (Figure 14.10). Its target sequence is approximately palindromic.

The bases to which the protein makes contact are shown in red. The protein interacts with both strands. The DNA is slightly bent.

C T A T C A C C G C A A G G G A T A A
G A T A G T G G C G T T C C C T A T T

λ cro is an example of the helix-turn-helix structural motif. Following along the chain in Figure 14.10a, the first secondary structure is a helix, followed by two more helices that frame the motif. The second of these two helices (the third helix in the molecule)—called the ***recognition helix***—lies in the major groove and makes extensive contacts with the DNA (Figure 14.10b). The hairpin that follows is involved in dimerization of the protein. It interacts with the corresponding hairpin of the other monomer to form a four-stranded β-sheet. A long C-terminal tail wraps around the DNA, following the minor groove.

The eukaryotic homeodomain antennapedia

Homeodomains are highly conserved eukaryotic proteins, active in control of animal development. They regulate homeotic genes; that is, genes that specify the locations of body parts. Antennapedia is a *Drosophila* protein responsible for initiating leg development. (We saw the structure of free antennapedia in Figure 12.27.) The earliest mutations found in antennapedia produced ectopic (= out of place) legs at the positions of, and instead of, antennae. Loss-of-function mutations convert legs into antennae.

The structure of the antennapedia–DNA complex (Figure 14.11a) resembles, in some respects, prokaryotic helix-turn-helix proteins such as λ cro. However, the tail that wraps around into the minor groove is N-terminal to the helix-turn-helix motif in antennapedia, instead of C-terminal as in λ cro.

A comparison of the protein–DNA hydrogen bonds in the antennapedia complex (Figure 14.11b) with that of λ cro shows that in the antennapedia complex a greater fraction of DNA–protein hydrogen bonds involve phosphate oxygens rather than bases.

Leucine zippers as transcriptional regulators

Leucine zippers form part of another type of dimeric transcriptional regulator. The jun protein forms a homodimer consisting of an N-terminal domain

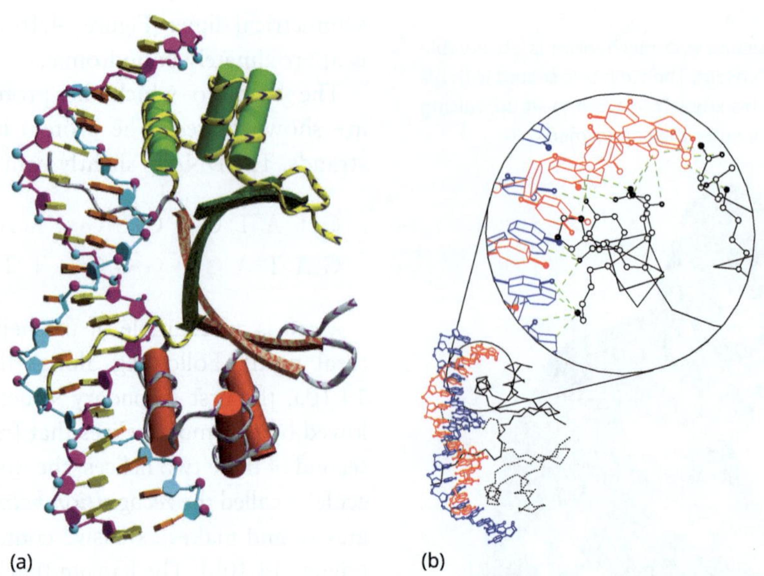

Figure 14.10 Phage λ cro repressor–DNA complex. (a) A dimer binds an approximately palindromic sequence in DNA. (b) Pattern of DNA–protein hydrogen bonds.

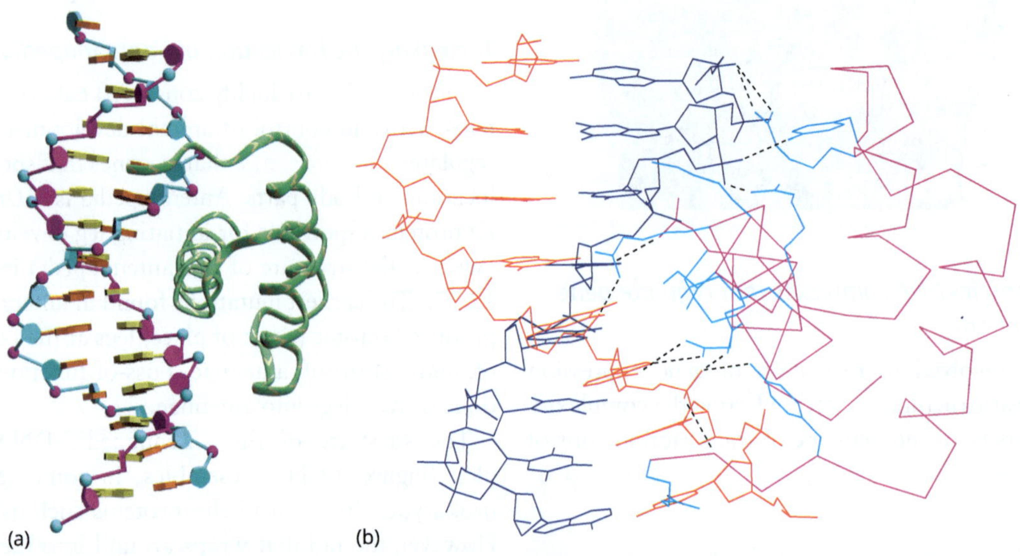

Figure 14.11 (a) Antennapedia homeodomain-DNA complex [9ANT]. (b) Details of protein–DNA interactions in antennapedia complex.

with many positively charged sidechains that bind to DNA, and a C-terminal leucine zipper domain involved in dimerization.

The proteins grip the DNA as if they were picking it up with chopsticks. The α-helices bind in major grooves on opposite sides of double helix (Figure 14.12). This structure shares with λ cro, and many other DNA-binding proteins, the *symmetry* of the complex, which mimics the dyad symmetry of the DNA double helix. This requires, on the part of the protein, formation of symmetrical dimers, and on the part of the DNA, an exactly or approximately palindromic target sequence. For jun dimers the target sequence is ATGACGTCAT.

Jun can dimerize not only with itself but with other related proteins, notably fos. Different dimers have different DNA-sequence specificities and different affinities, affording subtle patterns of control.

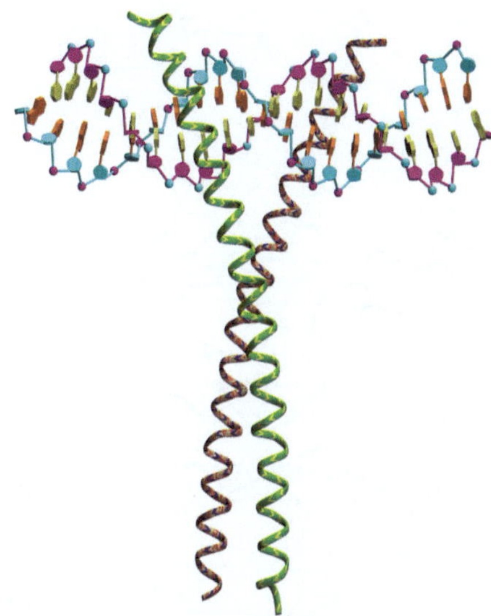

Figure 14.12 Bzip (basic DNA-binding domain)–leucine zipper transcriptional regulator jun homodimer binding to DNA [1JNM].

Zinc fingers

Zinc fingers are small modules found in eukaryotic transcription regulators. Each finger recognizes a triplet of bases in DNA. Tandem arrays of fingers recognize an extended region (Figure 14.13a). Understanding the relationship between the amino-acid sequences of Zn fingers and the DNA sequences they bind would permit design of gene-specific repressors. One fairly obvious target would require only one prototype finger: If one had a Zn finger that bound preferentially to the trinucleotide CAG, a polymer of such a domain might afford an approach to treating Huntington's disease, and other trinucleotide-repeat diseases such as Friedreich's ataxia and myotonic dystrophy. In fact this approach works. Another established technique is the fusing of Zn-finger domains to nucleases, to create artificial restriction enzymes that target specific genes.

Three types of zinc fingers differ in structure. In class 1, one Zn^{2+} ion binds two Cys and two His residues to stabilize the packing of a β-hairpin against an α-helix (Figure 14.13b). Residues in the helix interact with three consecutive nucleotides. (In class 2 zinc fingers, one Zn^{2+} ion binds four Cys sidechains. In class 3, two Zn^{2+} ions bind six cysteines.)

The E. coli *Met repressor*

Like many other DNA-binding proteins, the Met repressor binds as a symmetrical dimer. In the complex, each monomer contributes one strand of two-stranded β-sheet, which sits in the major groove (the student was right!), with sidechains making hydrogen bonds to bases (Figure 14.14).

The TATA-box binding protein

A **TATA box** is a sequence (consensus TATA[A/T]A[A/T]) upstream of the transcriptional start site of bacterial genes. Recognition of this sequence by the TATA-box binding protein (Figure 14.15) initiates the formation of the basal transcription complex, a large multiprotein particle. This is an example of initial binding of a protein to DNA followed by recruitment of other proteins to form an active complex. The most obvious feature of the complex is the very strong bending and unwinding induced in the DNA.

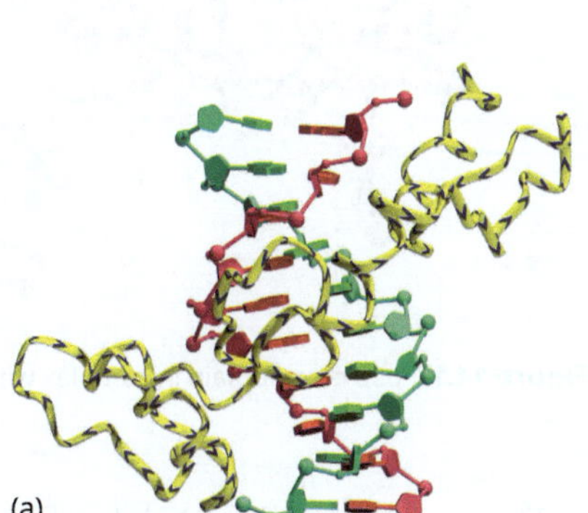

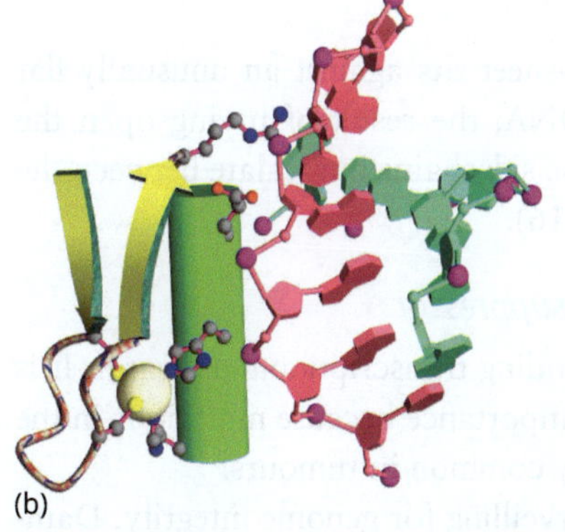

Figure 14.13 (a) Zif268, a tandem three-finger structure binding the sequence GCGTGGGCG [1AAY]. (b) Structure of a single module of a class 1 (Cys_2-His_2) zinc finger, taken from the Zif268–DNA complex. Each finger interacts with three consecutive bases. Three positions along the α-helix, non-consecutive in the amino-acid sequence, contain primary determinants of the DNA-sequence specificity. One of them, a histidine, binds to both the Zn^{2+} and a phosphate of the DNA. The other three sidechains that bind the Zn^{2+} are also shown.

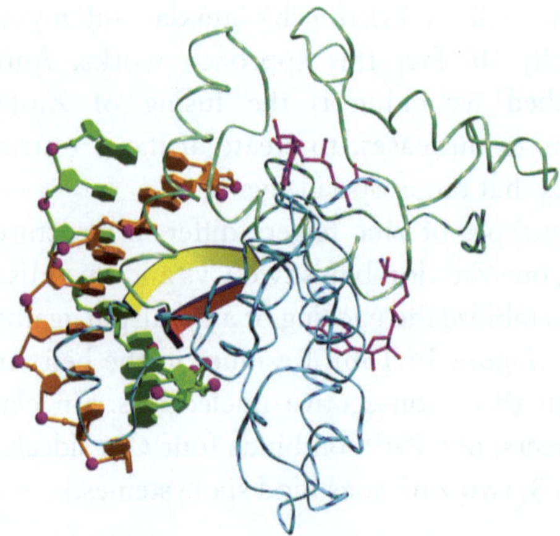

Figure 14.14 *E. coli* Met repressor–DNA complex [1CMA].

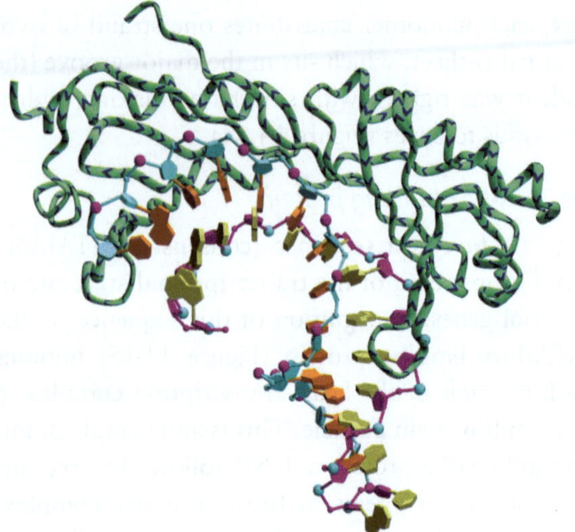

Figure 14.15 TATA-box binding protein YTBP [1YTB].

A long curved β-sheet sits against an unusually flat surface on the DNA, the result of prying open the minor groove. Phe sidechains intercalate between the bases (Figure 14.16).

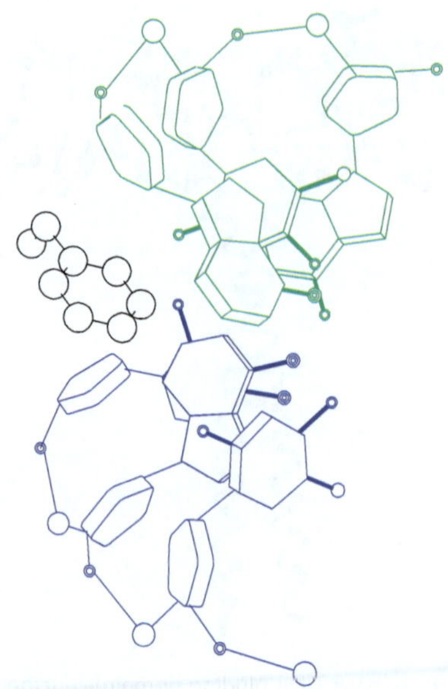

Figure 14.16 Intercalation of a Phe sidechain between bases in the distorted double helix in TATA-box binding protein–DNA complex [1YTB].

p53 is a tumour suppressor

p53 is a DNA-binding transcriptional activator. It is of great clinical importance because mutations in the p53 gene are very common in tumours.

p53 acts by surveilling for genome integrity. Damage to DNA induces enhanced expression of p53, which stalls cell-cycle progression. This gives time for DNA repair; if repair is unsuccessful, the 'fail-safe' mechanism is apoptosis.

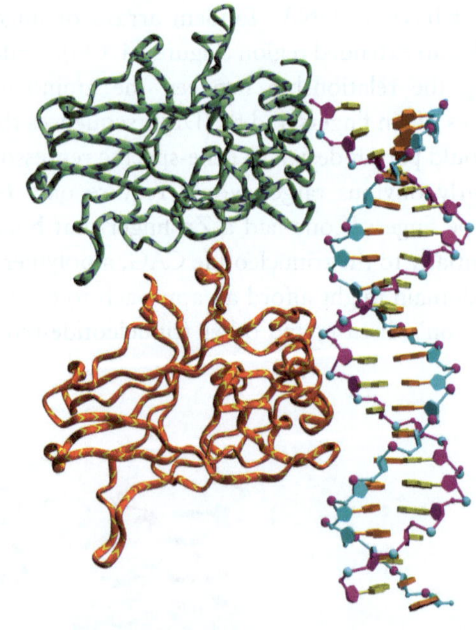

Figure 14.17 p53 core domain in complex with DNA [1TSR].

The structure of the DNA-binding subunit of p53 shows a double-β-sheet fold (Figure 14.17). A helix sits in the major groove, and sidechains from loops connecting strands of the β-sheet insert into the minor groove (Figure 14.18).

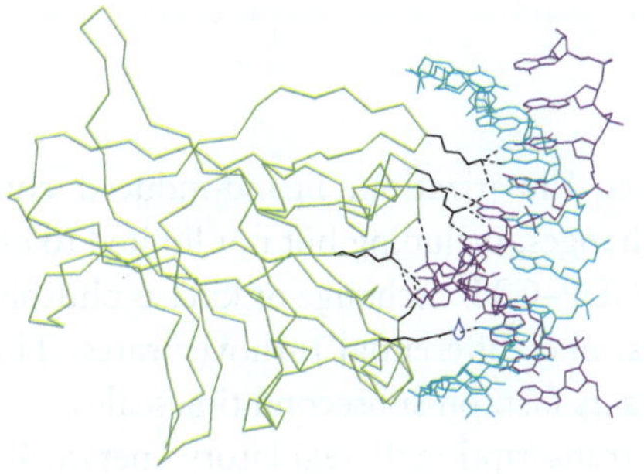

Figure 14.18 Details of protein–DNA interaction in p53 core domain–DNA complex 1TSR. One protein–DNA contact is water mediated.

The NAC domain is present in a large family of transcription regulators in plants, controlling key developmental processes, including regulation of flower and leaf development, secondary cell wall thickening, protein and lipid metabolic pathways, leaf senescence, fruit development, and seed germination (Figure 14.19).

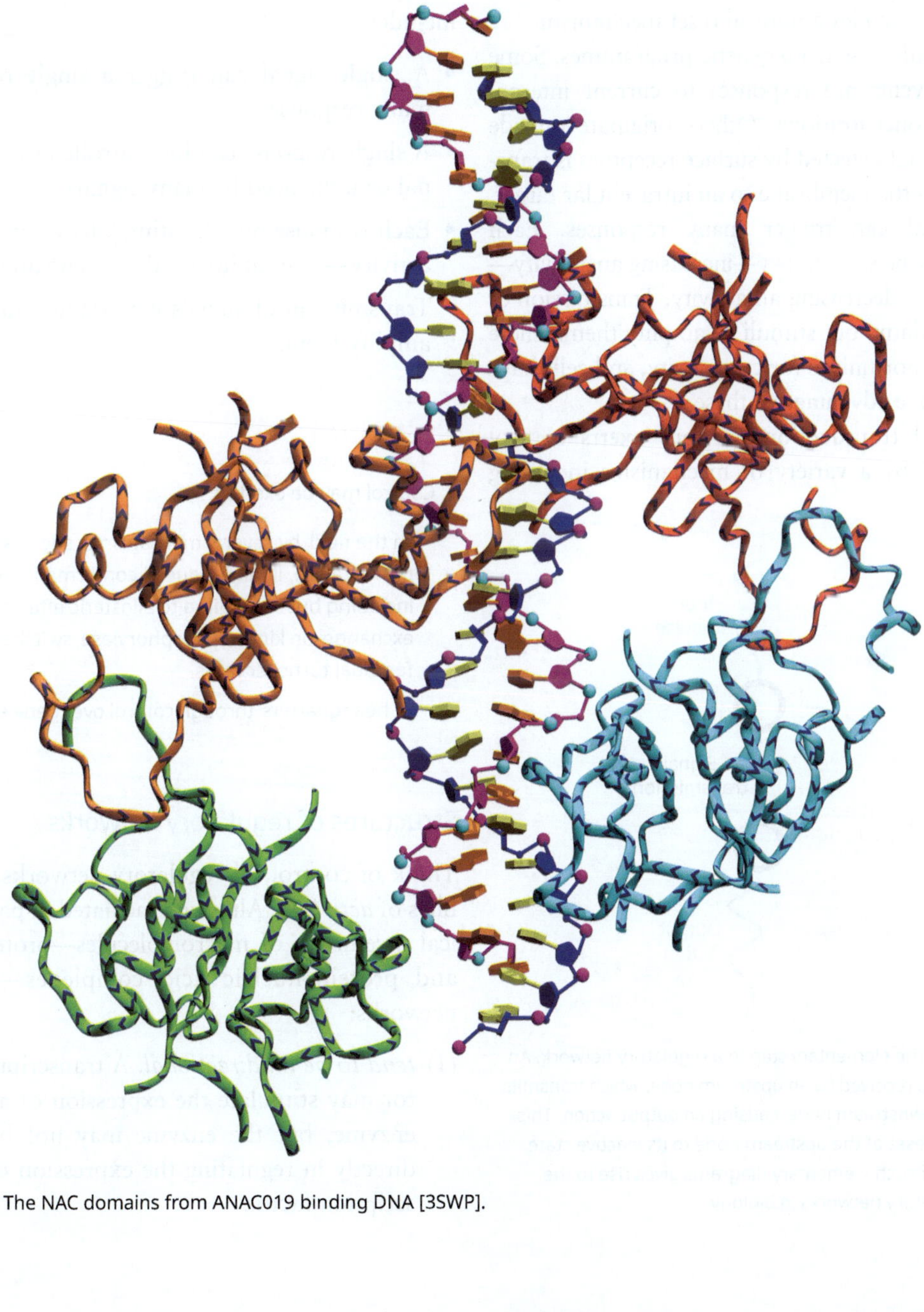

Figure 14.19 The NAC domains from ANAC019 binding DNA [3SWP].

Regulatory networks

Regulatory networks pervade living processes. Control interactions are organized into linear signal transduction cascades and reticulated into control networks.

Any individual regulatory action requires (1) a stimulus, (2) transmission of a signal to a target, (3) a response, and (4) a 'reset' mechanism to restore the resting state (Figure 14.20). Many regulatory actions are mediated by protein–protein complexes. Transient complexes are common in regulation, as dissociation provides a natural reset mechanism.

Some stimuli arise from genetic programmes. Some regulatory events are responses to current internal metabolite concentrations. Others originate outside the cell: a signal detected by surface receptors is transmitted across the membrane to an intracellular target.

One signal can trigger many responses. Each response may be **stimulatory**—increasing an activity—or **inhibitory**—decreasing an activity. Transmission of signals may damp out stimuli or amplify them. There are ample opportunities for complexity, and cells have taken extensive advantage of these.

The signal transduction network exerts control 'in the field' by a variety of mechanisms including inhibitors, dimerization, ligand-induced conformational changes, including but not limited to allosteric effects, GDP–GTP exchange or kinase-phosphorylase switches, and differential turnover rates. This component acts fast, on subsecond timescales.

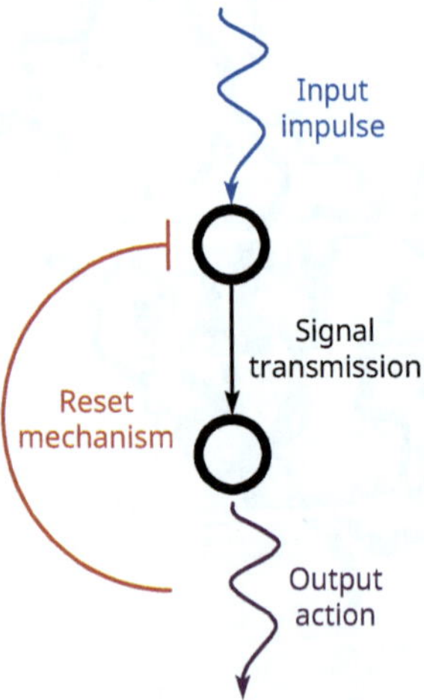

Figure 14.20 The elementary step in a regulatory network. An input impulse is received by an upstream node, which transmits a signal to a downstream node, causing an output action. This is followed by reset of the upstream node to its inactive state. Combination of such elementary diagrams gives rise to the complex regulatory networks in biology.

The transcriptional regulatory network exerts control 'at headquarters' through control over gene expression. This component is slower, acting on a timescale of minutes.

General characteristics of all control pathways include:

- A single signal can trigger a single response or many responses.
- A single response can be controlled by a single signal or influenced by many signals.
- Each response may be stimulatory—increasing an activity—or inhibitory—decreasing an activity.
- Transmission of signals may damp out stimuli or amplify them.

KEY POINT

Control may be exerted:

- 'in the field' by several mechanisms, such as inhibitors, dimerization, ligand-induced conformational changes, including but not limited to allosteric effects, GDP–GTP exchange or kinase-phosphorylase switches, and differential turnover rates.
- 'at headquarters' through control over gene expression.

Structures of regulatory networks

Think of control, or regulatory networks, as assemblies of *activities*. Although mediated in part by physical assemblies of macromolecules—protein–protein and protein–nucleic acid complexes—regulatory networks:

(1) *tend to be unidirectional*. A transcription activator may stimulate the expression of a metabolic enzyme, but the enzyme may not be involved directly in regulating the expression of the transcription factor.

(2) *have a logical dimension*. It is not enough to describe the connectivity of a regulatory network. Any regulatory action may stimulate or repress the activity of its target. If two interactions combine to activate a target, activation may require *both* stimuli (logical 'and') or *either* stimulus may suffice (logical 'or').

(3) *produce dynamic patterns*. Signals may produce combinations of effects with specified time courses. Cell-cycle regulation is a classic example.

The structure of a regulatory network can be described by a graph in which edges indicate steps in pathways of control. Regulatory networks are directed graphs: the influence of vertex A on vertex B is expressed by a directed edge connecting A and B. An edge directed from vertex A to vertex B is called *an outgoing connection* from A and *an incoming connection* to B. Conventionally, an arrow indicates a stimulatory interaction, and a T-symbol indicates an inhibitory interaction. An edge connecting a vertex to itself indicates autoregulation. A double-headed arrow indicates reciprocal stimulation of two nodes; note that this is *not* the same as an undirected edge.

A → B stimulatory interaction | A ⊣ B inhibitory interaction | A (self-loop) auto-regulatory interaction | A ↔ B reciprocal interaction

In a complex network, many 'upstream nodes' may govern the activity of nodes connected to them. In addition to the topological complexity of the pattern of connectivity in the network, there is the question of the logic applied to the set of inputs to the target node.

Structural biology of regulatory networks

Many molecules involved in regulation are multidomain proteins. Each domain in a multidomain protein is relatively free to interact with other molecules. An *interaction domain* is a part of a protein that confers specificity in ligation of a partner. Regulatory proteins contain a limited number of types of interaction domains which have diverged to form large families with different individual specificities. For instance, the human genome contains 115 SH2 domains and 253 SH3 domains. (*S*rc-*H*omology domains SH2 and SH3 are named for their homologies to domains of the src family of cytoplasmic tyrosine kinases.) Many individual interaction domains even bind to different partners as they participate in successive steps of a control cascade. Initial interactions may also trigger recruitment of additional proteins to form large regulatory complexes.

Figure 14.21 shows two types of interaction domain complexes with ligands, binding peptides (which may be attached to proteins), and protein–protein complexes.

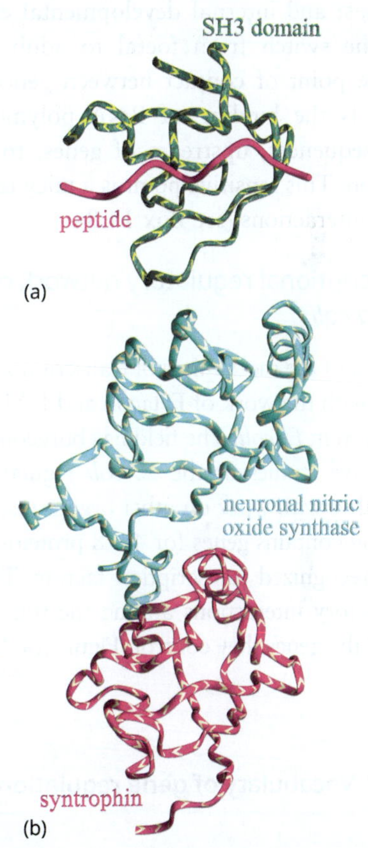

Figure 14.21 Some types of interactions involved in regulatory signalling. (a) Binding of a peptide (magenta) by an SH3 domain [1CKA]. SH3 domains are common constituents of regulatory proteins. Functions of SH3 domains include signal transduction, protein and vesicle trafficking, cytoskeletal organization, cell polarization, and organelle biosynthesis. (b) Domain–domain interaction: PDZ domains in syntrophin (magenta) and neuronal nitric oxide synthase (cyan) [1QAV].

Many interaction domains are sensitive to the state of post-translational modification of their ligands, for instance binding preferentially to states of a ligand in which specific tyrosines, serines, or threonines are phosphorylated. These and other post-translational modifications can function as switches, turning on or interrupting/resetting a signalling cascade.

KEY POINT

Regulatory networks:

- tend to be unidirectional;
- have a logical as well as a physical component;
- produce dynamic patterns.

Gene regulation

Cells regulate the expression patterns of their genes. They sense internal cues to maintain metabolic stability; external cues to respond to changes in the surroundings; and internal developmental cues, for instance, the switch from foetal to adult haemoglobin. The point of contact between genome and expression is the binding of RNA polymerase to promoter sequences, upstream of genes, to initiate transcription. This sensitive point is a juicy target for regulatory interactions (see Box 14.8).

The transcriptional regulatory network of *Escherichia coli*

Investigation of the mechanism of transcription regulation began with the work of F. Jacob and J. Monod on the lac operon in *E. coli*. The field has burgeoned, with comprehensive studies of the *E. coli* regulatory network, together with work on other organisms, notably yeast. *E. coli* contains genes for 4398 proteins, 167 of which are recognized transcription factors. There are 2369 regulatory interactions among the transcription factors and the genes they control (Figure 14.22).

There are many fewer known regulatory interactions than genes. Many genes in *E. coli* are organized into operons, under coordinated control—one regulatory interaction controlling many genes. *E. coli* is estimated to contain ~2700 operons. Conclusion: More interactions are still to be discovered.

This network has been the subject of many investigations. Some focus on questions about the *static* topology. Some of these address the local structure of the network, identifying the common types of small subgraphs or the motifs. The fork, scatter pattern, and feed-forward loop are motifs in the regulatory networks of *E. coli* and other organisms (see Box 14.9). The network contains a large number of autoregulatory connections. An autoregulatory activator amplifies responses; an autoregulatory repressor damps them out. In the *E. coli* regulatory network, links between different transcription factors are primarily activating, and autoregulatory interactions are often repressive. A high density of repressive autoregulatory interactions increases what might be thought of as the viscosity of the medium in which

BOX 14.8

Vocabulary of gene regulation

- **Operator**: a control region associated with a gene.
- **Promoter**: a region upstream of a gene, site of RNA polymerase binding to initiate transcription.
- **Repressor**: a DNA-binding protein that blocks transcription.
- **Operon**: a set of tandem genes in bacteria, usually catalysing consecutive steps in a metabolic pathway, under coordinated transcriptional control.
- **Cis-regulatory region**: a segment of DNA that regulates expression of genes on the same DNA molecule. The lac repressor binding site is a *cis*-regulator of the adjacent protein-coding genes *lacZ*, *lacY*, *lacA*.
- **Transcription start site**: position in the gene that corresponds to the first residue in the mRNA.
- **Constitutive mutant**: a mutant defective in repression of a gene, which in consequence is expressed continuously.

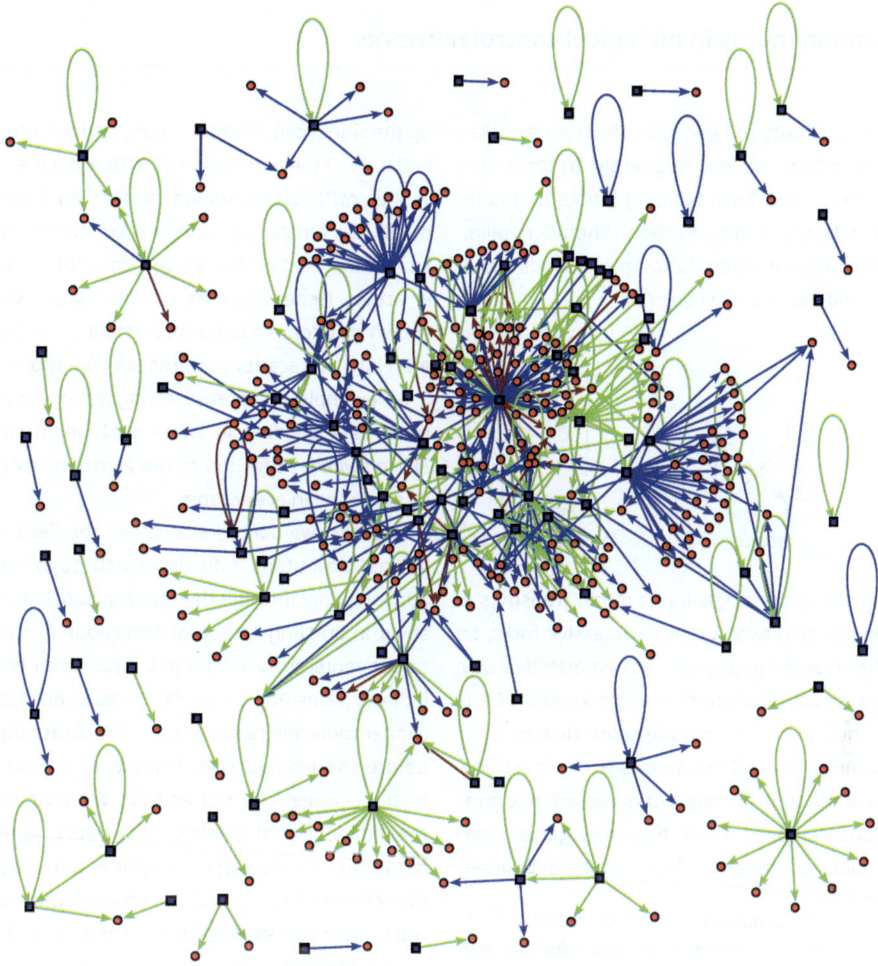

Figure 14.22 The *E. coli* transcriptional regulatory network represented as a directed graph. Colour coding of nodes: transcription factors are shown as blue squares; regulated operons are shown as red circles. Colour coding of links: blue, activators; green, repressors; brown, indeterminate.

Used with permission of Springer Nature, from Dobrin, R., Beg, Q.K., Barabási, A.L., *et al.* (2004). Aggregation of topological motifs in the *Escherichia coli* transcriptional regulatory network. BMC Bioinformatics 5: 10. https://doi.org/10.1186/1471-2105-5-10. https://doi.org/10.1002/bmb.2006.494034042627. Permission conveyed through Copyright Clearance Centre, Inc.

the network is active. The 'one-two punch', or feed-forward loop, motif can also act to filter out random fluctuations, preventing the propagation of noise (see Figure 14.23).

Combinations of the elementary motifs form modules within the network. These clusters of nodes are often dedicated to control of expression of genes with related physiological functions, such as a group of proteins responding to oxidative stress, or a group involved in aromatic amino acid biosynthesis. Such sets of genes need not be linked as a single operon.

KEY POINT

Regulatory networks are directed graphs. Some simple motifs, or common small subgraphs, form the lowest level of network structure. Networks can reprogram themselves within an organism and evolve between species.

Other analyses address the large-scale structure of the *E. coli* transcriptional regulatory network. The distribution of degrees of the nodes, that is, the histogram of the number of edges meeting at a node,

BOX 14.9 Common motifs in biological control networks

Within the high complexity of typical regulatory networks, certain common patterns appear frequently. In the architecture of networks, these form building blocks that contribute to higher levels of organization. Shen-Orr, Milo, Mangan, and Alon[4] have described examples including the *fork*, the *scatter*, and the *'one-two punch'*:

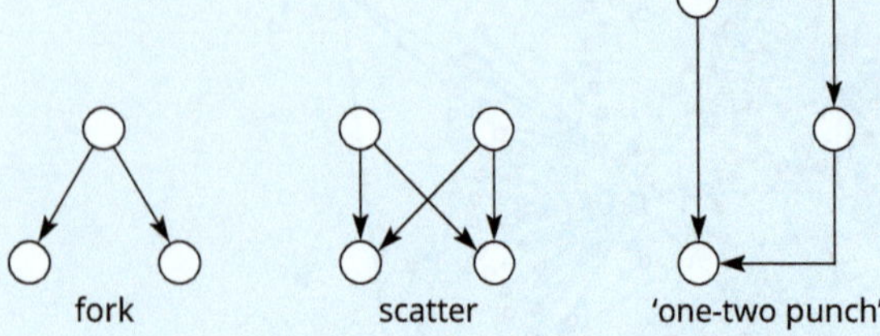

The *fork*, also called the single-input motif, transmits a single incoming signal to two outputs. Successive forks, or forks with higher branching degrees, are an effective way to activate large sets of genes from a single impulse. Generalizations of the binary fork include more downstream genes under common control (more tines to the fork), and autoregulation of the control node. Forks can achieve general mobilization. Moreover, if the regulatory genes have different thresholds for activation, the dynamics of building up the signal can produce a temporal pattern of successive initiation of the expression of different genes.

[4] Shen-Orr, S.S., Milo, R., Mangan, S., *et al.* (2002). Network motifs in the transcriptional regulation network of *Escherichia coli*. Nat. Genet. 31: 64–68.

The *scatter* configuration, also called the multiple-input motif, can function as a logical 'or' operation: both downstream targets become active if *either* of the input impulses is active. Generalizations of the square scatter pattern shown may contain different numbers of nodes on both layers. Note that scatter patterns are superpositions of forks.

An example of a scatter configuration, or multiple-input motif, is a building in which each room contains a fire-alarm switch; activation of the alarm in any room triggers its sounding in every room.

The *'one-two punch'*, also called the 'feed-forward loop', affects the output both directly through the vertical link and indirectly through the intermediate link. This motif can show interesting temporal behaviour if activation of the target requires simultaneous input from both direct and indirect paths (logical 'and'). Because buildup of the messenger molecule requires time, the direct signal will arrive before the indirect one. Therefore, a short pulsed input to the complex will not activate the output—by the time the indirect signal builds up, the direct signal is no longer active. The system can thereby filter out transient stimuli in noisy inputs (Figure 14.23). Conversely, the active state of the system can shut down quickly upon withdrawal of the external trigger.

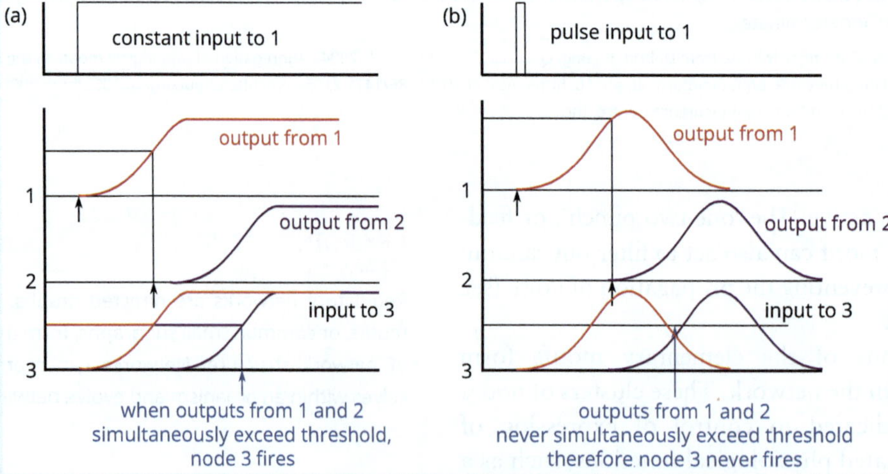

Figure 14.23 A 'one-two punch', or feed-forward loop, equipped with suitable AND logic at the downstream node, can filter out transient noise. (a) Constant input. (b) Pulse input.

follows a power law: number of nodes with k edges $\propto 10^{-\beta k}$, with $\beta = 0.8$. The scale-free topology means that some nodes have many connections, and form the 'hubs' of the network.

Is there substantial feedback from downstream nodes? (A social analogy: is the network hierarchical or democratic? That is, does your boss listen to you, or just give orders? Ring Lardner's classic sentence, 'Shut up, he explained', emphasizes an absence, or dysfunction, of receptors for feedback signals from lower levels back to higher ones.) Although the *E. coli* transcriptional regulatory network contains many autoregulatory interactions, it does not contain larger cycles; that is, paths in which gene A regulates gene B regulates gene C regulates . . . regulates gene A. Such cycles can lead to instabilities in the dynamics of a network.

Correlation of the topology of segments of the network with function suggests that short pathways, feed-forward loops, and repressive autoregulatory interactions are involved in control of metabolic functions, such as a switch to alternative nutrients. This type of network topology is adapted to maintenance of homeostasis. Long hierarchical cascades, and activating autoregulatory interactions, regulate developmental processes, such as biofilm formation and flagellar development involved in mobility and chemotaxis.

The dynamic properties of the network are also of interest. Even a constant network can produce different outputs from different inputs. Moreover, even within an organism, networks can change their structure in response to changes in conditions, as in the lac operon.

This can affect even some of the hubs of the network, the points at which changes have the most far-reaching effects. This has been examined most closely in yeast.

Similarities and differences in the regulatory interactions in related organisms illuminate how their networks evolve. The evolutionary retention of transcription factors is smaller than that of target genes. Even transcription factors that serve as hubs are not more highly conserved. Different organisms are relatively free to explore different regulatory pathways, even to regulate orthologous genes. This may well be the 'other side of the coin' of the redundancy in the networks that provides robustness.

There is evidence that larger changes in regulatory networks reflect changes in lifestyle: organisms with similar lifestyle—several species of soil bacteria, genuses *Bacillus, Corynebacterium*, and *Mycobacterium*—conserve regulatory interactions, as do intracellular parasites *Mycoplasma, rickettsiae*, and *chlamidiae*. Conversely, comparing organisms that adopt different lifestyles, individual transcription factors can be deleted.

'The road not taken . . .' some real-world decision making

The genetic switch of bacteriophage λ

Phage λ can adopt an active or passive lifestyle, effected by alternative gene expression profiles. It is the simplest form of life that makes a decision.

λ is a virus that infects *E. coli* (see Figure 14.24). The mature virion contains an icosahedral head that encapsulates the viral DNA; and a tail that recognizes and attaches to the host, and functions as a syringe to inject the viral DNA. The genome is a single molecule of double-stranded DNA 58402 bp long, containing 50 genes, organized into seven operons.

After attaching to an *E. coli* cell and injecting its DNA, the phage may follow either of two paths:

- In the *lytic* state: replication and intracellular reconstitution of daughter phage particles is followed, in about 45 minutes, by rupture of the host cell and release of the ~100 progeny. The expression patterns of several distinct sets of genes are under control of a developmental programme during this process.
- In the *lysogenic* state, the phage DNA becomes integrated into the bacterial genome, to form a prophage. Only one phage gene is expressed: the cI protein, which acts as a repressor to inhibit the expression of phage genes responsible for initiating viral multiplication, thereby maintaining the lysogenic state.

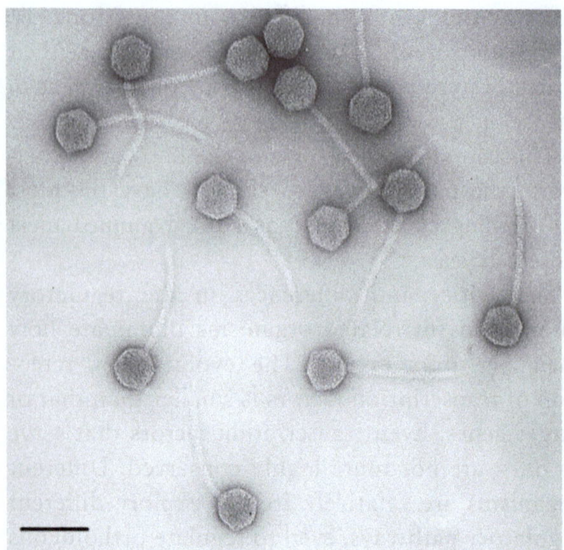

Figure 14.24 Bacteriophage λ. Bar at lower left indicates 100 nm.

From: ICTVdB—The Universal Virus Database, version 4. http://www.ictvdb.org (Picture courtesy Prof. R.B. Inman, University of Wisconsin.)

The strategy of the virus is to take advantage of a thriving host population to reproduce lytically, but to adopt lysogeny to get through 'lean' periods.

Given a healthy host population, the lytic state perpetuates itself as progeny viruses infect additional bacterial cells.

The lysogenic state of the phage is stable under normal conditions. The viral DNA, integrated into the bacterial genome, replicates with the bacterial DNA. This creates a population of infected bacteria. Although the virus does not reproduce completely to form intact progeny phage, the viral DNA *is* replicated, as a passenger in the dividing cells.

Sleeping Beauty can however be awakened, by damage, such as UV radiation, that threatens the host cell. The virus resumes active replication, in order to escape conditions that endanger its host. Of course, this *ensures* that the host cell will not survive.

What are the characteristics of the switch that must be implemented by DNA–protein interactions?

- The states of the switch must be mutually incompatible. Each state must repress the other.
- Under *constant conditions*, each state must be self-maintaining. In other words, not only is the choice of one or the other commitment enforced; once selected, the chosen state persists until conditions change.
- In response to *changing conditions*, it must be possible to move from one state to the other. We expect to find a simple trigger that leads to a complex cascade of consequences.

To implement this logic, the system has the following variables at its disposal:

- *DNA sequences*: in particular the sequences at sites of promoters and operators (see Box 14.10).
- *Local flexibility in DNA structure*: the ability of the DNA to form loops, bridged by interacting proteins bound to sites distant in the DNA sequence.
- *Protein–DNA interactions*: relative affinities of different proteins for different sites, including RNA polymerase and regulatory proteins.
- *Interactions among DNA–protein complexes*:
  - positive cooperativity: enhancement of binding by stabilizing protein–protein interactions on the DNA.

BOX 14.10

Sites of protein–DNA interactions in transcription control

A *promoter* is a site on DNA—typically ~60 bp in prokaryotes—near the beginning of a gene. It binds RNA polymerase, required for initiation of transcription. RNA polymerase is a bacterial enzyme. However, part of the developmental programme of lytic phage λ takes place through the modification of the bacterial polymerase by viral proteins, to alter its response to termination signals. The result is to extend the region of transcription of viral genes in successive stages of the lytic cycle.

A *operator* is a site on DNA that binds regulatory proteins. A repressor, or negative regulator, blocks the site where RNA polymerase binds to the operator, preventing transcription. Conversely, a positive regulator interacts with RNA polymerase to enhance its binding affinity to a promoter.

- negative cooperativity (or anticooperativity)—especially the blocking, by binding of one protein, of the binding site of another.

Which proteins will bind DNA, and where, depends on the:

- Intrinsic affinity of different sites for different proteins (a DNA-binding protein will choose the available site to which it has the highest affinity).
- Cooperativity of protein binding.
- Competition of proteins for sites.

Availability of a site may be denied by occlusion, caused by binding of another protein at or near the site. The blocking of RNA polymerase binding by repressor is an example. Conversely, favourable interaction with another protein bound at a neighbouring site may enhance affinity (cooperative binding). These effects may involve interactions among regulatory proteins or of regulatory proteins with RNA polymerase.

The materials

1. Proteins
   - *RNA polymerase*, the enzyme that transcribes DNA into RNA. RNA polymerase binds to available promoter sites.
   - *cro*, a transcription regulator that inhibits synthesis of cI.
   - *cI*, or *repressor*, a transcription regulator that inhibits expression of cro, and regulates its own expression.
2. Sites on the phage DNA
   - The accessibility of two adjacent promoters control the transcription of cI and cro.
   - Three operator sites, one within each promoter and a third overlapping both, that are binding sites for cro and cI. Figure 14.25 shows (a) the layout of promoter and operator sites on the DNA, and (b), (c) the two mutually exclusive states in which cro or cI are expressed.

The operation of the system depends on the relative values of the affinities of different operator sites for different proteins and for different combinations of proteins.

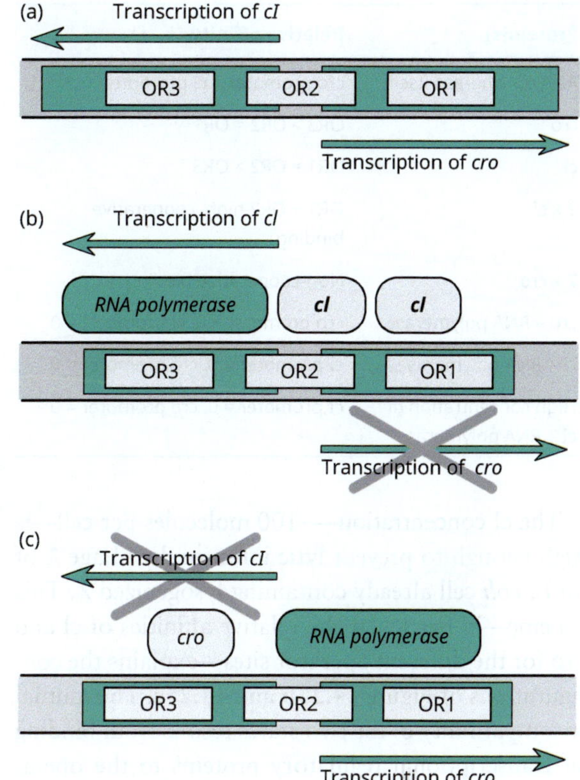

Figure 14.25 OR1, OR2, and OR3 are operator sites—binding sites for regulatory proteins. Each is about 15–20 base pairs long, or roughly two turns of DNA. OR1 overlaps the *cro* promoter, OR3 overlaps the *cI* promoter, and OR2 overlaps both. (a) Region of phage λ genome containing promoters for *cro* and *cI*. (b) In the lysogenic state, *cro* is off and *cI* (encoding repressor) is expressed. (c) In the lytic state *cro* is expressed and *cI* is off.

The relative affinities of *cro* and *cI* for the operator sites:

Protein	Relative affinity	Effect
cro	OR3 > OR1 = OR2	Binding of cro to OR3 prevents cI synthesis
cI	OR1, OR2 > OR3	Binding of 2 × cI to OR1/OR2 prevents cro synthesis

create alternative states:

State	cro	cI
Lytic	on	off
Lysogenic	off	on

Protein(s)	Relative affinity
RNA polymerase itself	cro promoter > cI promoter
cro	OR3 > OR2 = OR1
cI	OR1 = OR2 > OR3
2 × cI	OR1 + OR2 high, cooperative binding
2 × cro	Non-cooperative binding
cro + RNA polymerase	cro promoter > 0, cI promoter = 0
cI + RNA polymerase	cI promoter > 0, cro promoter = 0
High concentration of cI + RNA polymerase	cI promoter = 0, cro promoter = 0

The cI concentration—~100 molecules per cell—is high enough to prevent lytic infection by phage λ of an *E. coli* cell already containing lysogenized λ. This scheme—in particular the relative affinities of cI and cro for the different operator sites—explains the configurations of Figure 14.25b and 14.25c. The mutual incompatibility of the two states results from binding of transcriptional regulatory proteins to the operators, repressing one of the two genes and enhancing expression of the other. Binding of cI, preferentially and cooperatively to OR1 and OR2, turns off transcription of *cro* and, through favourable interaction with RNA polymerase on the DNA, stimulates transcription of *cI*. At higher concentrations of cI, after titration of the OR1 and OR2 sites, cI will bind to OR3, turning off *cI* transcription. This acts to regulate the ambient cI concentration.

The accompanying diagram shows the logical relationships between these two components. High concentrations of cro inhibit its further expression. The combination of both stimulatory and repressive links from cI to itself signifies the regulation of cI concentration: the autostimulatory link is active at low cI concentration and the autorepressive link is active at high cI concentration. The phage protein cII also activates expression of cI but using a different promoter.

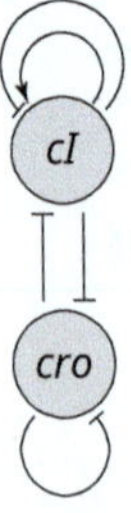

cI binds as a tetramer to OR1/OR2. There is an additional set of promoters and operators OL1, OL2, and OL3 ~2.3 kb from OR1, OR2, and OR3. A tetramer of cI, bound to OR1/OR2 and another tetramer of cI bound to OL1/OL2 can form an octamer, enhancing the affinity for DNA. To do this the DNA must loop around to allow apposition of the two tetramers.

How to 'throw' the switch

- To change from the lysogenic to the lytic state: UV irradiation or other hindrance to DNA replication causes bacterial protein RecA to cleave cI. This frees the OR1/OR2 sites to bind RNA polymerase, to express *cro*. Cro has its highest affinity for the OR3 site, turning off synthesis of cI. As the concentration of cro builds up, it binds also to OR1 and OR2, turning off its own expression. Expression of cro also initiates a cascade of events that effect the transition to the lytic state.
- The switch from lytic to lysogenic state can occur only upon infection, of necessity by a phage that has emerged from a lytic event. The phage may either remain lytic (the default) or become lysogenic.

The choice appears to be determined primarily by the concentration of a phage protein cII. cII activates transcription of:

- cI, the repressor (but via a different operator than that shown in Figure 14.25);
- int, a protein required for integration of phage DNA into bacterial genome;
- an antisense RNA that prevents the viral modification of bacterial RNA polymerase. This shuts down the lytic programme.

The transition to lysogeny requires buildup of concentration of cII. The concentrations of cII and cIII appear to depend primarily on the dose of viral DNA. About 1% of cells infected by one virus become lysogenic. About 50% of cells infected simultaneously by two or more viruses become lysogenic.

A high multiplicity of infection implies a low ratio of bacteria to phage. For the phage to remain lytic under these conditions would threaten to deplete the population of bacteria, to the point where progeny phase could not find hosts to infect. Similar considerations rationalize the greater frequency of lysogenation if the host cell is starved.

A bacterial protease HflB destroys cII, and a viral protein cIII inhibits HflB. Readers may wonder why the cell would synthesize a molecule that promotes lysis, and why the phage would synthesize one that promotes lysogeny. Both the bacterium and the phage appear to be acting to *reduce* the number of their own immediate progeny. However, there may be long-term benefits for the populations as a whole. In a population of lysogenized bacteria, occasionally a cell spontaneously goes lytic. The progeny phage does not damage the bacterial population—remember that lysogeny confers 'immunity' to phage infection—but may protect the bacterial population against competition with foreign invading susceptible bacteria. Sociobiologists may see this as an example of altruism.

Regulation of the lactose operon in *E. coli*

The lactose operon of *E. coli* contains structural genes and regulatory regions (see Figure 14.26a). Three regions of the operon encode proteins (see Table 14.1). *lacZ*, *lacY*, and *lacA* are cotranscribed into a single mRNA and separately translated.

The *lacI* gene, encoding the lac repressor protein, lies upstream of the operator. It is constitutively expressed.

The control regions of the lactose operon function as a switch, turning on and off transcription of the protein-encoding genes. Control is exerted through binding of CAP protein (catabolite activation protein) to the CAP site, and lac repressor protein to the operator site. Cyclic AMP (cAMP, produced in higher quantities if glucose is absent) and lactose analogues bind to CAP and lac repressor proteins, respectively, to control their binding affinities:

- Binding lactose induces a conformational change in the repressor, from a tightly binding form to a weakly binding one. *Lac repressor will bind promoter only in the absence of lactose.*

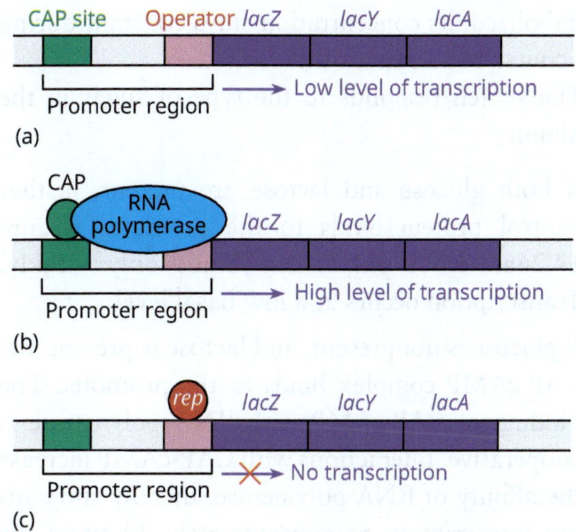

Figure 14.26 States of the lactose operon. (a) The promoter region contains regulatory sites upstream of protein-encoding genes *lacZ*, *lacY*, *lacA*. (b) Binding of CAP to its upstream site within the promoter enhances the binding affinity of RNA polymerase, turning transcription on. (c) Binding of repressor (rep) blocks binding of RNA polymerase, turning transcription off. Two additional subsidiary repressor binding sites are not shown. (Figure 14.26c is a simplification: lac repressor actually binds to three sites. The cover of the 1 March 1996 issue of *Science* magazine shows a model of the lac repressor, binding to two sites on a ~100-bp section of DNA, plus the CAP protein. The DNA is bent into a loop.)

- The presence of glucose reduces the concentration of cyclic AMP, causing a conformational change in CAP to a weakly binding form. *CAP will bind promoter only in the absence of glucose.*

The actual molecule that binds to repressor to reduce its affinity for its site on DNA is *allolactose*. Allolactose is an isomer of lactose, produced from lactose by β-galactosidase. Alternative lactose analogues also stimulate transcription. One that is useful in the laboratory is isopropylthiogalactopyranoside (IPTG), for two reasons: (1) IPTG enters the cell even if the *lacY*-encoded transporter is dysfunctional or not expressed and (2) because IPTG is not

Table 14.1 Genes in the lactose operon of *E. coli*

Gene	Enzyme	Function
lacZ	β-galactosidase	hydrolyses lactose → glucose + galactose, *and* isomerizes lactose → allolactose
lacY	β-galactoside permease	pumps lactose into cell
lacA	β-galactoside transacetylase	unknown, possibly detoxification?

metabolized, its concentration stays constant during the course of an experiment.

The switch responds to the type of sugar in the medium:

- If both glucose and lactose are present, neither control protein binds to the promoter (Figure 14.26a). RNA polymerase binds only weakly. Transcription occurs at a low basal level.
- If glucose is not present, and lactose *is* present, the CAP-cAMP complex binds to the promoter. The binding of CAP-cAMP with RNA polymerase is cooperative. Interactions with CAP-cAMP increase the affinity of RNA polymerase, thereby stimulating transcription to approximately 40 times the basal level (Figure 14.26b).
- If lactose is not present, the repressor binds to the operator site. This blocks RNA polymerase and turns off transcription (Figure 14.26c).

Transcription of the lac operon depends on a competition between glucose and lactose. In summary (– means 'absence of'; + means 'presence of'):

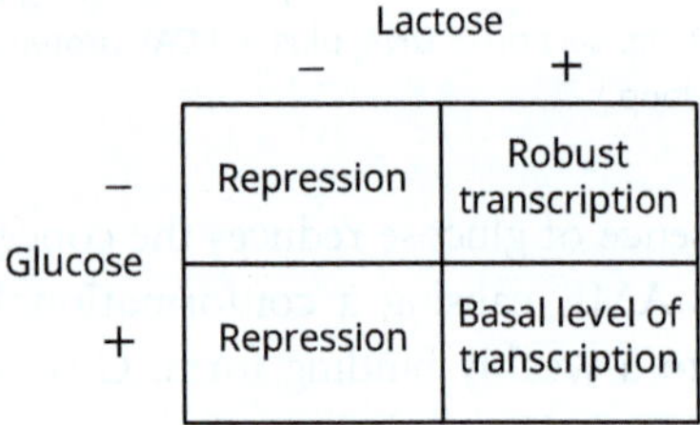

		Lactose –	Lactose +
Glucose	–	Repression	Robust transcription
Glucose	+	Repression	Basal level of transcription

The operator site is between the CAP site and the origin of transcription. As a result, *lactose absence 'trumps' glucose absence*. That is, in the absence of lactose, the binding of repressor stops transcription whether or not glucose is absent.

The effect is to express the proteins of the lac operon only if there is only lactose in the medium. The bacteria are saying, in effect, 'We prefer to grow on glucose. If glucose is there, we don't want high expression levels of the genes for lactose transport and metabolism, *even if* lactose is present. Only if lactose is present and glucose is not present, express the genes that transport and cleave lactose.'

The lactose operon switch is an example of a 'fire-and-forget' mechanism. Once the messenger RNA is synthesized, what happens on the DNA does not affect

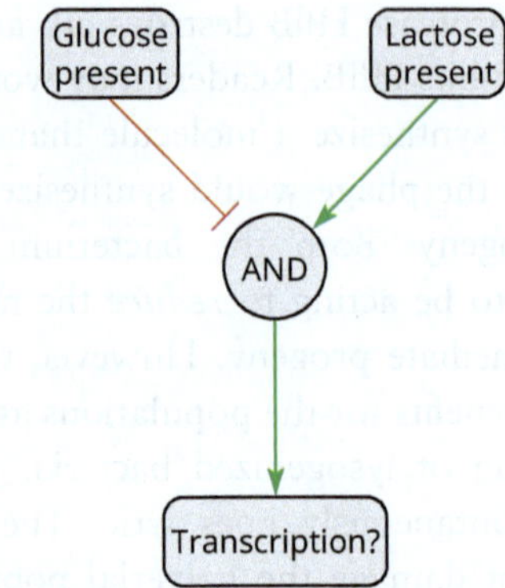

Figure 14.27 Logical diagram of lac operon. Green arrows show positive regulation. The red 'tee' shows negative regulation. The circle containing AND will pass through a positive signal only if glucose is NOT present (red 'tee') AND lactose IS present (green arrow). For many regulatory circuits, we know many of the inputs to a node but do not know the logic.

it. However, the mRNA for the protein-coding genes of the lac operon (*lacZ*, *lacY*, *lacA*) has a half-life of ~3 minutes. In the absence of continuous or repeated induction, synthesis of the lactose-metabolizing enzymes will cease within minutes. This resets the switch.

Figure 14.27 represents the lac operon control logic as a network. This diagram is almost equivalent to the table showing the response to the presence and absence of glucose.

KEY POINT

The lactose operon in *E. coli* encodes several proteins involved in the uptake and utilization of lactose. Control mechanisms activate this operon for transcription when the medium contains lactose but not glucose.

The diauxic shift in *Saccharomyces cerevisiae*

The diauxic shift in yeast is the transition from fermentative to oxidative metabolism upon exhaustion of glucose as an energy source.

Yeast is capable of adapting its metabolism to a variety of environmental conditions. In the presence of glucose, *S. cerevisiae* will—even in the presence of oxygen—preferentially use the Ebden–Meyerhof fermentative pathway, reducing glucose to ethanol.

Exhaustion of available glucose produces the 'diauxic shift', to oxidative metabolism, sending the products of fermentation through the Krebs (tricarboxylic acid) cycle and mitochondrial oxidative phosphorylation (see Figure 14.28).

The shift is not merely a redirection of metabolic flux through alternative pathways using pre-existing proteins, but involves a substantial change in expression pattern. Many genes are involved, not only enzymes in the awakened metabolic pathways. Another consequence of switching to respiratory metabolism is the danger of oxidative damage, and the oxidative stress response requires enhanced expression of many other genes.

The cells are effectively sensing the level of glucose. As glucose itself acts as a repressor of expression of many genes, depletion of glucose releases this repression—'turning on' a variety of genes.

An important component of the mechanism by which high glucose levels lead to repression is the state of phosphorylation of a transcriptional regulatory protein called Mig1. The phosphorylated form of Mig1 stays in the cytoplasm. On dephosphorylation it enters the nucleus and binds to promoters of various genes, repressing their expression. The activity of kinase Snf1-Snf4, which phosphorylates Mig1, depends on the AMP/ATP concentration ratio in the cell. Growth in high levels of glucose generates high levels of ATP via glycolysis, keeping the Snf1-Snf4 kinase inactive. This leaves Mig1 in its active, dephosphorylated form, repressing glucose-sensitive genes.

In a seminal paper in 1997, J.L. DeRisi, V.R. Iyer, and P.O. Brown reported on the changes in gene expression in the diauxic shift in *S. cerevisiae*. They observed that the patterns of gene expression were stable during exponential growth in a glucose-rich medium. This justifies considering the initial anaerobically growing population as being in a *state* that can be characterized by a set of expression levels of genes. (If, alternatively, fluctuations in expression levels were large compared to the differences between anaerobic and aerobic regimes, it would not be possible to analyse the diauxic shift in terms of expression patterns. In fact, if there are fluctuations, they average out over the population.)

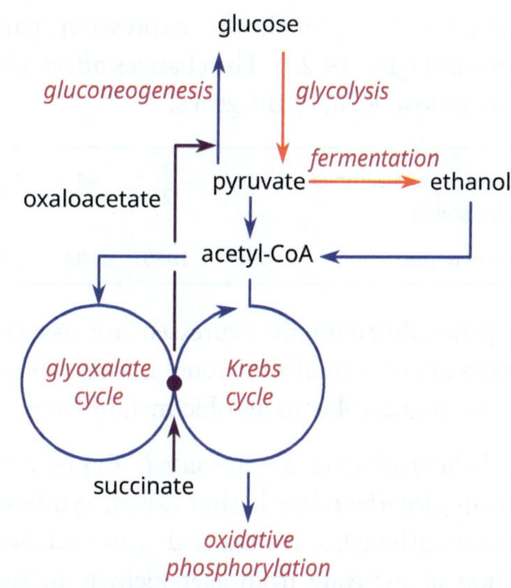

Figure 14.28 Some metabolic pathways in yeast affected by the diauxic shift. In the presence of ample glucose, yeast will adopt an anaerobic metabolic regime, converting glucose to ethanol via glycolysis and fermentation (red). On running out of glucose it will shift to an aerobic metabolic state in which ethanol is converted to CO_2 and H_2O, via the Krebs cycle and oxidative phosphorylation (blue). For the alternative pathways of energy release, pyruvate is the branch compound. However, the shift to a utilization of a different energy source creates two concomitant problems: (1) The oxidation of ethanol does not provide precursors for essential biosynthetic pathways. Oxidation of ethanol converts all carbon to CO_2. Some must be retained and converted to 3- and 4-carbon compounds, and even glucose, via the glyoxylate cycle and gluconeogenesis. Acetyl-CoA is the branch compound for this shift—it enters *both* the Krebs cycle and the glyoxylate cycle. The glyoxylate cycle shares several intermediates with the Krebs cycle, but, to conserve carbon, leaves out the decarboxylations that produce CO_2. Succinate (black dot), one of the metabolites common to the Krebs and glyoxylate cycles, is converted to oxaloacetate in mitochondria. It then feeds into numerous biosynthetic pathways (black arrow). (2) In addition to the metabolic pathways shown in this figure, yeast must also activate pathways of defence against oxidative stress. The danger is from potential chemical attack by ***reactive oxygen species***, produced by partial reduction of O_2: the superoxide radical O_2^-, hydrogen peroxide H_2O_2, and the hydroxyl radical $OH^{\bullet}$. In its defence, yeast increases expression of genes involved in detoxification of reactive oxygen species. The mechanism of induction of these genes depends on the transcription factor Yap1p. Increase in the concentration of reactive oxygen species leads to formation of disulphide bridges in Yap1p. This causes a conformational change that masks a nuclear export signal. The result is redistribution of the transcription factor to the nucleus, the site of its activity.

As glucose is depleted, the expression pattern changes (see Figure 14.29). The changes affect a large fraction, almost 30%, of the genes:

Expression ratio in aerobic/anaerobic states	>2	$<\frac{1}{2}$	>4	$<\frac{1}{4}$
Number of genes	710	1030	183	203

The genes differentially expressed are associated with proteins in several functional classes. On converting from anaerobic to aerobic metabolism:

- *Metabolic pathways are rerouted.* Expression of pyruvate decarboxylase is shut down; synthesis of pyruvate carboxylase is enhanced. This switches the reaction of pyruvate from acetaldehyde to oxaloacetate. Enhanced synthesis of genes for fructose-1,6-bisphosphatase and phosphoenolpyruvate carboxylase changes the direction of two steps in glycolysis. Expression is enhanced for genes encoding the enzymes that carry out the new Krebs and glyoxylate cycle reactions, and oxidative phosphorylation (blue arrows in Figure 14.28).
- *Genes related to protein synthesis show a decrease in expression level.* These include genes for ribosomal proteins, tRNA synthetases, and initiation and elongation factors. An exception is that genes encoding *mitochondrial* ribosomal molecules have generally enhanced expression.
- *Genes related to a number of biosynthetic pathways show reduced expression.* These include genes encoding enzymes of amino acid and nucleotide metabolism.
- *Genes involved in defence against oxidative stress from reactive oxygen species show enhanced expression.* Proteins encoded include catalases, peroxidases, superoxide dismutases, and glutathione S-transferases.

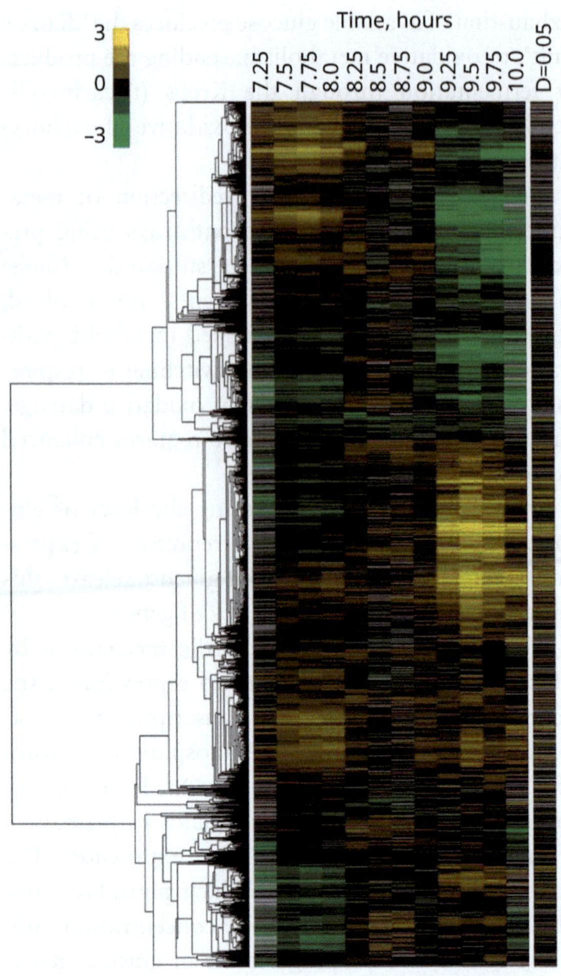

Figure 14.29 Expression patterns of yeast growing on glucose, then undergoing diauxic shift. Column headings indicate time in hours after exhaustion of glucose in the samples undergoing diauxic shift.

The genetic regulatory network of *Saccharomyces cerevisiae*

The detailed analysis of the diauxic shift in yeast shows the effects of the regulatory process. Let us now consider the agents that achieve these results.

A study of transcription regulation in yeast by N.M. Luscombe, M.M. Babu, H. Yu, M. Snyder, S.A. Teichmann, and M. Gerstein treated a network containing 3562 genes, corresponding to approximately half the known proteome of *S. cerevisiae*. The genes included 142 that encode transcription regulators, and 3420 that encode target genes exclusive of transcription regulators. There are 7074 known regulatory interactions among these genes, including effects of regulators on one another, and of regulators on non-regulatory targets.

KEY POINT

The yeast regulatory network comprises:

- regulatory genes: the products of regulatory genes affect the transcription of other genes and
- target genes: the products of target genes do not regulate other genes. The network is dense; each target gene feels the influence of a large number of regulatory genes.

Figure 14.30 is a cartoon-like sketch of a fragment of such a network, indicating rather loosely some of its general features. Nodes are divided into *transcriptional regulators*, shown as circles, and *target genes*, shown as squares. Target genes are distinguished by having no output connections. There is extensive interregulation among the transcription factors, to a much higher density of interconnections than can intelligibly be shown in this diagram. Think of a seething broth of transcription factors, within the shaded area, sending out signals to target genes. The shaded area indicates only the *logical* clustering of the transcriptional regulators. There is no suggestion about physical localization; indeed, transcriptional regulators interact with DNA, and almost never interact physically with the proteins the expression of which they control.

Each transcriptional regulator directly influences approximately 50 genes on average, although, as with other 'small-world' networks following power-law distributions of connectivities, the distribution is very skewed—some 'hubs' have very many output connections, but most nodes have very few. A few of the interregulatory connections between transcription factors are shown in black. In about 10% of the cases, two neighbours of the same transcription factor interact with each other. A pathway from one regulator (filled black circle) to one ultimate receptor (filled black square), through five intermediate nodes, is shown in black. The intermediate nodes are other transcriptional regulators, connected both

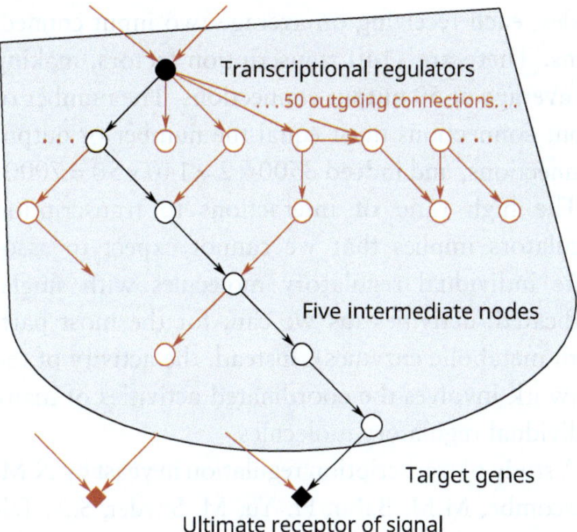

Figure 14.30 Simplified sketch illustrating some features of an 'average' segment of the pathways in the yeast interaction network. Transcriptional regulators appear as circles. Target genes appear as squares. A transcriptional regulator typically has direct influence over about 50 genes, indicated by multiple connections from the filled black circle to the circles on the line below it. Roughly one in ten of the neighbours of any node is connected to another neighbour, indicated by the horizontal arrow on the second row. The ultimate receptor of the signal lies at the end of a pathway typically containing about five intermediate nodes (shown in black). This ultimate target gene receives on average about two inputs. This diagram shows only a small fragment of a network that is in fact quite dense. The diagram also suggests that there is no 'feedback' in the form of cycles; in fact cycles are present but they are rare.

within the path drawn in black and off this path. Even the transcription factor shown as the origin of the path receives input connections. Although it is possible to identify target genes from the absence of outgoing connections, it is more difficult to identify ultimate initiators of signal cascades.

The ultimate receptor is a target gene that receives regulatory input but itself has no output links. This target is expected to receive (on average) a second control input. The black target node receives input via a black arrow, along the selected path, and via a red arrow, suggesting the second input. Of course the second input may arrive via a path that shares common nodes with the black path, including other routes from the filled black circle.

The dense forest of additional pathways, from which this fragment is extracted, is not shown. A 'back-of-the-envelope' calculation: there are ~3500 nodes, each receiving on average two input connections. There are ~140 transcription factors, making an average of 50 output connections. The number of input connections must equal the number of output connections, and indeed $3500 \times 2 = 140 \times 50 = 7000$.

The high ratio of interactions to transcription regulators implies that we cannot expect to associate individual regulatory molecules with single, dedicated, activities (as we can, for the most part, with metabolic enzymes). Instead, the activity of the network involves the coordinated activities of many individual regulatory molecules.

A study of transcription regulation in yeast by N.M. Luscombe, M.M. Babu, H. Yu, M. Snyder, S.A. Teichmann, and M. Gerstein treated a network containing 3562 genes, corresponding to approximately half the known proteome of *S. cerevisiae*. They compared the connectivities of the network in five different physiological states: cell cycle, sporulation, diauxic shift (the change from anaerobic fermentative metabolism to aerobic respiration as glucose is exhausted), DNA damage, and stress response. There is extensive reconfiguration, involving changing interaction patterns of the same transcription regulators.

Analysis of the overall network architecture reveals that:

- The distribution of incoming connections to target genes has a mean value of 2.1, and is distributed exponentially. Most target genes receive direct input from about two transcriptional regulators. The probability that a gene is controlled by k transcription regulators, $k = 1, 2, \ldots$, is proportional to $e^{-\alpha k}$, with $\alpha = 0.8$.
- The distribution of outgoing connections has a mean value of 49.8, and obeys a power law. The probability that a given transcriptional regulator controls k genes is proportional to $k^{-\beta}$, with $\beta = 0.6$. Power-law behaviour characterizes topologies in which a few nodes—the 'hubs'—have many connections while many nodes have few. In regulatory networks, hubs tend to be fairly far upstream, forming important foci of regulation with far-reaching control.
- The average number of intermediate nodes in a minimal path between a transcriptional regulator and a target gene is 4.7. The maximal number of intermediate nodes in a path between two nodes is 12.
- The clustering coefficient of a node is a measure of the degree of local connectivity within a network. If all neighbours of a node are connected to one another, the clustering coefficient of the node = 1. If no pairs of neighbours of a node are connected to each other, the clustering coefficient of the node = 0. The mean clustering coefficient, averaged over all nodes, is a measure of the overall density of the network. For the yeast transcriptional regulatory network, the mean clustering coefficient is 0.11.

Adaptability of the yeast regulatory network

The yeast regulatory network achieves versatility and responsiveness by reconfiguring its activities. This is seen by comparing the changes in the activities of networks controlling yeast gene expression patterns in different physiological regimes: cell cycle, sporulation, diauxic shift (the change from anaerobic fermentative metabolism to aerobic respiration as glucose is exhausted), DNA damage, and stress response. Cell cycling and sporulation involve the unfolding of endogenous gene expression programmes; the others are responses to environmental changes.

Different physiological states are characterized by both similarities and differences in gene expression patterns, and by the components of the regulatory network that are active. There is considerable shift in expression of target genes. About a quarter of the target genes are specialized to individual physiological states. That is, of the total of 3420 target genes, the expression of almost half (1514) do not show major changes in the different states. Of the 1906 that show altered expression levels in different states, almost half of them (803) are specialized to a single physiological state.

In contrast, different states show much more overlap in the usage of transcriptional regulators. For instance, for cell-cycle control, 280 target genes (8%) are differentially regulated by 70 (49%) of the transcription regulators. Clearly, there is a much greater degree of specialization in the target genes. In general, half the transcription factors are active in at least three out of the five physiological regimes. However, in contrast with the high overlap of usage of the transcriptional regulators (the nodes), the overlap of the activities within the network (the connections) is relatively low. Different components of the interaction network organize the different gene expression patterns in different states.

Whereas different physiological states are characterized by substitutions of different sets of synthesized proteins, the regulatory network uses much of the same structure but reconfigures the pattern of activity. Think of the transcription factors as 'hardware' and the connections as reprogrammable 'software'. The molecules do not change but the interactions do: in different states, many transcription regulators change most, or a substantial part, of their interactions. In particular, the set of transcription regulators that form the hubs of the network—those with many outgoing nodes that form foci of control—is not a constant feature of the system. Some hubs are common to all states, but others step forward to take control in different physiological regimes. The result of the reconfiguration of activity is that over half of the regulatory interactions are *unique* to the different states.

The effect of the changes in the active interaction patterns is to alter the topological characteristics of the network in different states. For instance, under panic conditions—DNA damage and stress—the average number of genes under control of individual transcriptional regulators increases, the average minimal path length between regulator and target decreases, and the clustering becomes less dense (that is, there is less interregulation among transcription factors). This can be understood in terms of a need for fast and general mobilization—the equivalent of shouting 'Go! Go! Go!' over the radio. Normal circumstances—cell-cycle control for instance—allow for a more dignified and precise regulatory state, which permits finer control over the temporal course of expression patterns. In cell-cycle control and sporulation, there is a much denser interregulation among transcription factors, and longer minimal path lengths between transcriptional regulators and target genes.

Different physiological states also differ in their usage of the common motifs—fork, scatter, and 'one-two punch' (see Box 14.9). Scatter motifs are more used in conditions of stress, diauxic shift, and DNA damage. They are appropriate to the need for quick action. Requirements for buildup of intermediates would delay the response. Conversely, the 'one-two punch' motif is more common in cell-cycle control. This is consistent with the need for a signal from one stage to be stabilized before the cell enters the next stage.

Much of evolution proceeds towards greater specialization. The human eye is a classic example. It is an intricate and fine-tuned structure, features that were once adduced as evidence *against* Darwin's theory. Many evolutionary pathways show a trade-off between specialized adaptation and generalized adaptability.

Regulatory networks are an exception. The reconfigurability of regulatory networks allows them to respond robustly to changes in conditions by creating many different structures, specialized to the conditions that elicit them. Evolution has produced structures that are both specialized *and* versatile.

Relationships among networks

How can we compare networks? To compare sequences and structures, the basic tool was alignment. This was made simple by the *linear* nature of the amino-acid sequences, or the order of (x_i, y_i, z_i) coordinates of the Cα atoms of residues in the polypeptide chain. Networks are in principle higher dimensional. This makes the alignment problem extremely difficult, computationally, and one that has not been adequately solved. We have already seen examples in metabolic networks (see Figure 13.12). (Even deciding whether two graphs are isomorphic is difficult for graphs containing more than five vertices.)

In some cases, we are given an alignment. For example, the yeast regulatory networks in the five different physiological states contain different sets of 'edges' between members of the same set of proteins. To extract common subnetworks, we can take the intersections of the edges of the networks in the different states, either pairwise comparisons, or all five at once. Other examples in which different networks are formed from the same sets of labelled nodes include metabolic networks (nodes labelled by metabolites) or protein interaction networks (nodes labelled by proteins, or sets of homologous proteins).

Alternatively, we can look locally, at the interactions of individual nodes in the different states. Define a *1-neighbourhood* of any node in a network as the subgraph of the network consisting of the selected node as the central node, all parent nodes from the network connected directly—that is, by a single edge—to the central node, all child nodes connected directly to the central node, plus any additional edges from the overall network that connect any pair of nodes from the set connected directly to the central node. (Technically: the vertex-induced subgraph involving the selected node, its parents, and its children.) Clusters of nodes containing pathways of length >1 to a central node would naturally define higher-order neighbourhoods.

We shall describe the similarities and differences in the structures and topologies of the 1-neighbourhoods of a few individual nodes in different states of the yeast regulatory network:

- Figure 14.31 shows the comparison of the 1-neighbourhoods of node YIR023W in the stress response and DNA damage networks. The central node, YIR023W, is shown in orange in both frames. Black arrows show edges connecting the central node to parent or to child nodes. Green arrows show edges connecting neighbours of the central node, but not the central node itself. For instance, the green arrow in the DNA damage panel connects a parent of the central node to a child of the central node. If multiple parent nodes are present (not in this case) they are shown in alphabetical order clockwise. If multiple child nodes are present they are shown in alphabetical order counterclockwise.

Nodes shown in purple are common to the 1-neighbourhoods of YIR023W in both networks. Each of these 1-neighbourhoods of YIR023W contains a single, common parent node, YKL112W. This is a relatively simple example, in which the 1-neighbourhood of the central node in the stress response network is *almost* a subset of the 1-neighbourhood in the DNA damage network, except for the child node YPR145W which does not appear in the DNA damage network. Incidentally YPR145W is a target gene; that is, in no network is there an edge for which YPR145W is the parent.

To what extent are the basic motifs (see Box 14.9)—MIM, SIM, and FFL—visible in the 1-neighbourhood? Any node with more than one child node exhibits one or more Single-Input Motifs (SIM), or forks (or, alternatively, a single fork with multiple tines). Every SIM in the network will appear in the 1-neighbourhood of the parent node of the SIM. The green arrow, connecting a parent node of YIR023W to a child node in the DNA damage network indicates a Feed-Forward Loop (FFL). Every FFL will appear in the set of 1-neighbourhoods, with the central node of the 1-neighbourhood corresponding to the intermediate node of the FFL, as in the case of the 1-neighbourhood of YIR023W in the DNA damage network. A Multiple-Input Motif (MIM) *might* occur within a 1-neighbourhood, provided that the 1-neighbourhood contains more than one parent node. But not all MIMs in the network will appear in the 1-neighbourhood of some node.

- Figure 14.31 showed a relatively simple example. Figure 14.32 shows the 1-neighbourhoods of YDR207C in the cell cycle and sporulation networks.

The 1-neighbourhood of YDR207C in the cell cycle network contains edges from three parent and to four child nodes. There are in addition two edges directly between a parent node and a child node (shown in green). The 1-neighbourhood of YDR207C in the cell cycle network does contain a Multiple Input Motif. It also contains two Feed-Forward Loops.

In contrast, the 1-neighbourhood of YDR207C in the sporulation network is among the most complicated observed in these networks. Not only is it very rich in edges to child nodes, but there are 16 edges directly connecting a parent of the central node YDR207C to a child, and 7 edges connecting two child nodes. In this case also, the cell cycle 1-neighbourhood is almost a subset of the sporulation neighbourhood: they have the same parent nodes, and share three child nodes (of the four in the cell cycle network). The two edges between the parent node YOL004W and the two child nodes YGL073W and YGR044XC are also common to both networks. However, overall, one can see that in the sporulation

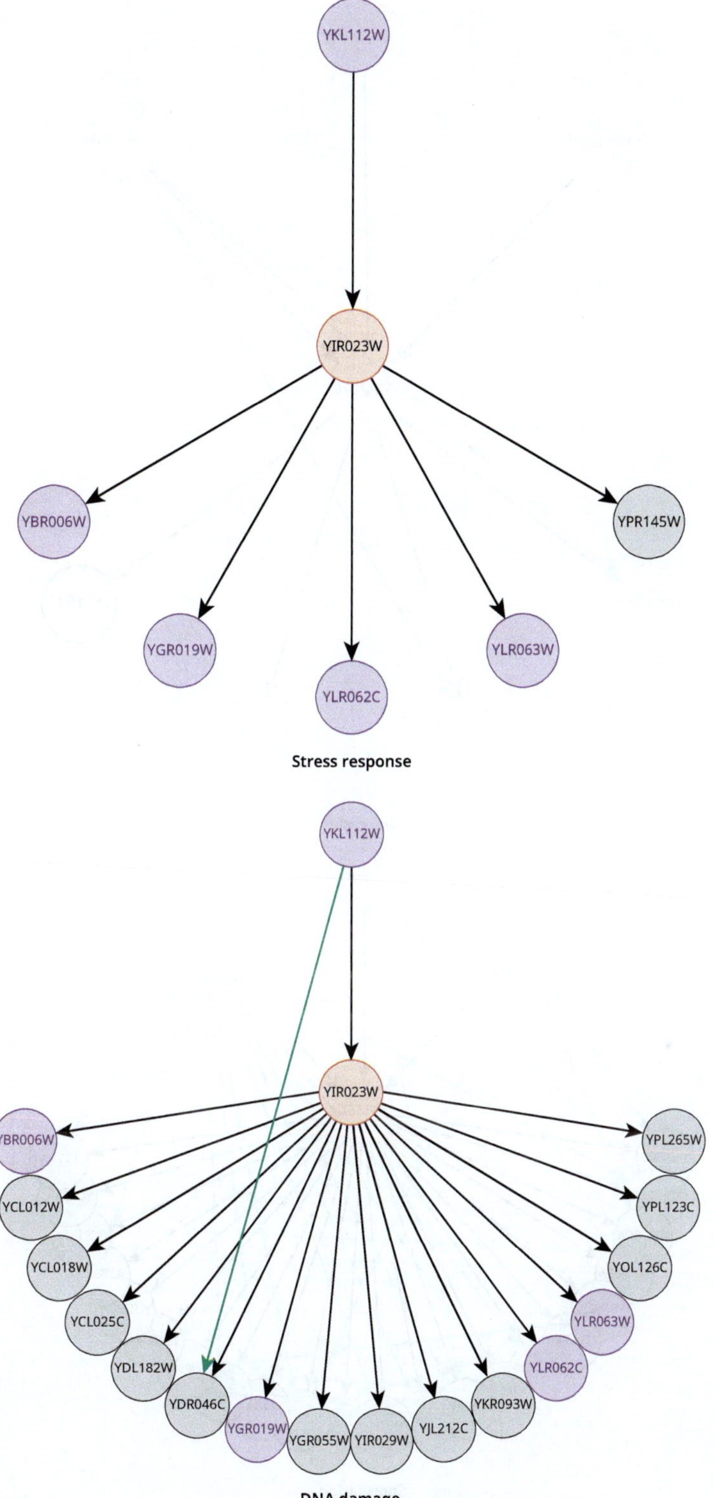

Figure 14.31 The 1-neighbourhoods of the selected central node YIR023W in two networks: stress response and DNA damage. In each of these networks the central node has only one parent neighbour. In the stress response network, YIR023W has four child connections. In the DNA damage network, YIR023W has 16 child connections, four of which are shared edges in stress response and DNA damage networks.

Used with permission of Oxford University Press, from Lesk, A.M., & Konagurthu, A.S. (2020). Neighbourhoods in the yeast regulatory network in different physiological states. Bioinformatics 37, 4: 551–558. https://doi.org/10.1093/bioinformatics/btaa831. Permission conveyed through Copyright Clearance Centre, Inc.

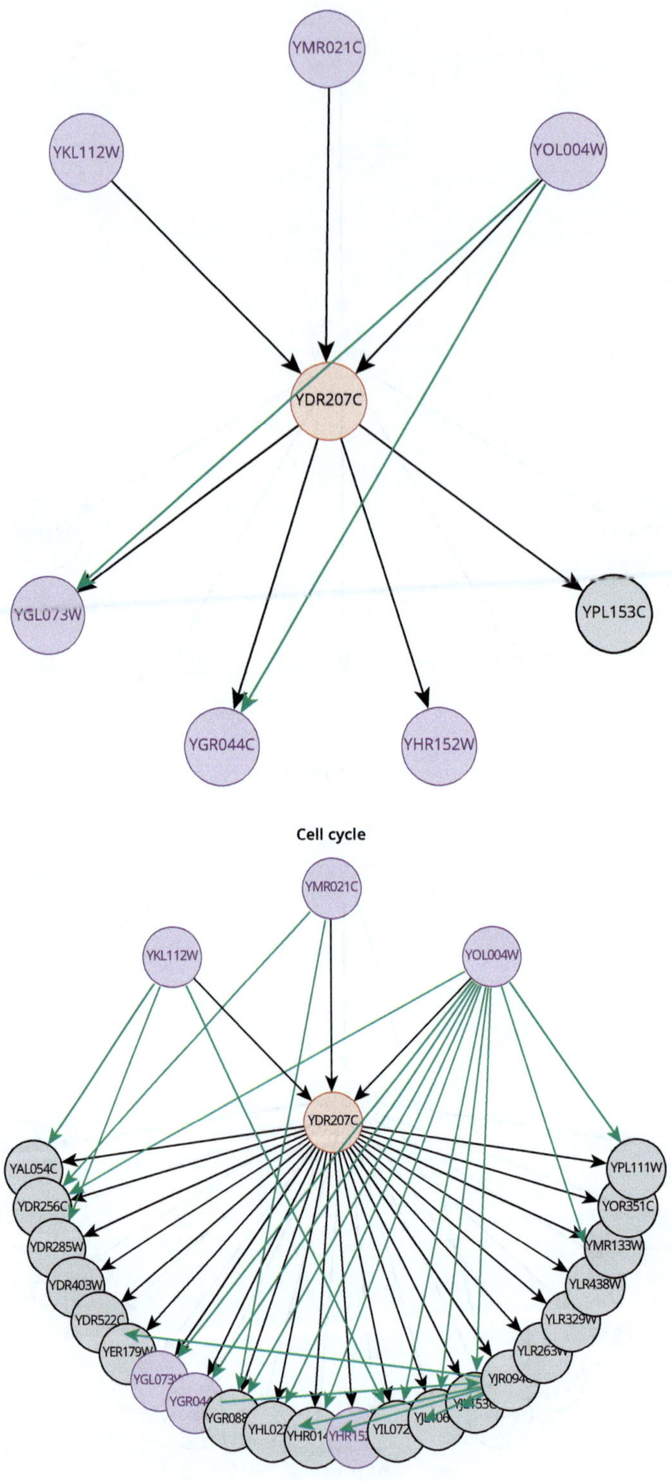

Figure 14.32 The 1-neighbourhoods of node YDR207C in the cell cycle and sporulation networks. The cell cycle 1-neighbourhood is of moderate complexity, containing edges from three parent and to four child nodes. The sporulation 1-neighbourhood is more complex, containing edges from the same three parent nodes, but edges to 22 child nodes; and many additional 'green' edges.

network YDR207C interactions are significantly more complex than the more moderate demands on the cell cycle network. Of course in the sporulation network there are many Multiple Input Motifs and Feed-Forward Loops.

- YGL209W illustrates a somewhat more complicated set of relationships between 1-neighbourhoods in different networks (see Figure 14.33). (YGL209W is one of the parents of YDR146C; see Figure 14.33.)

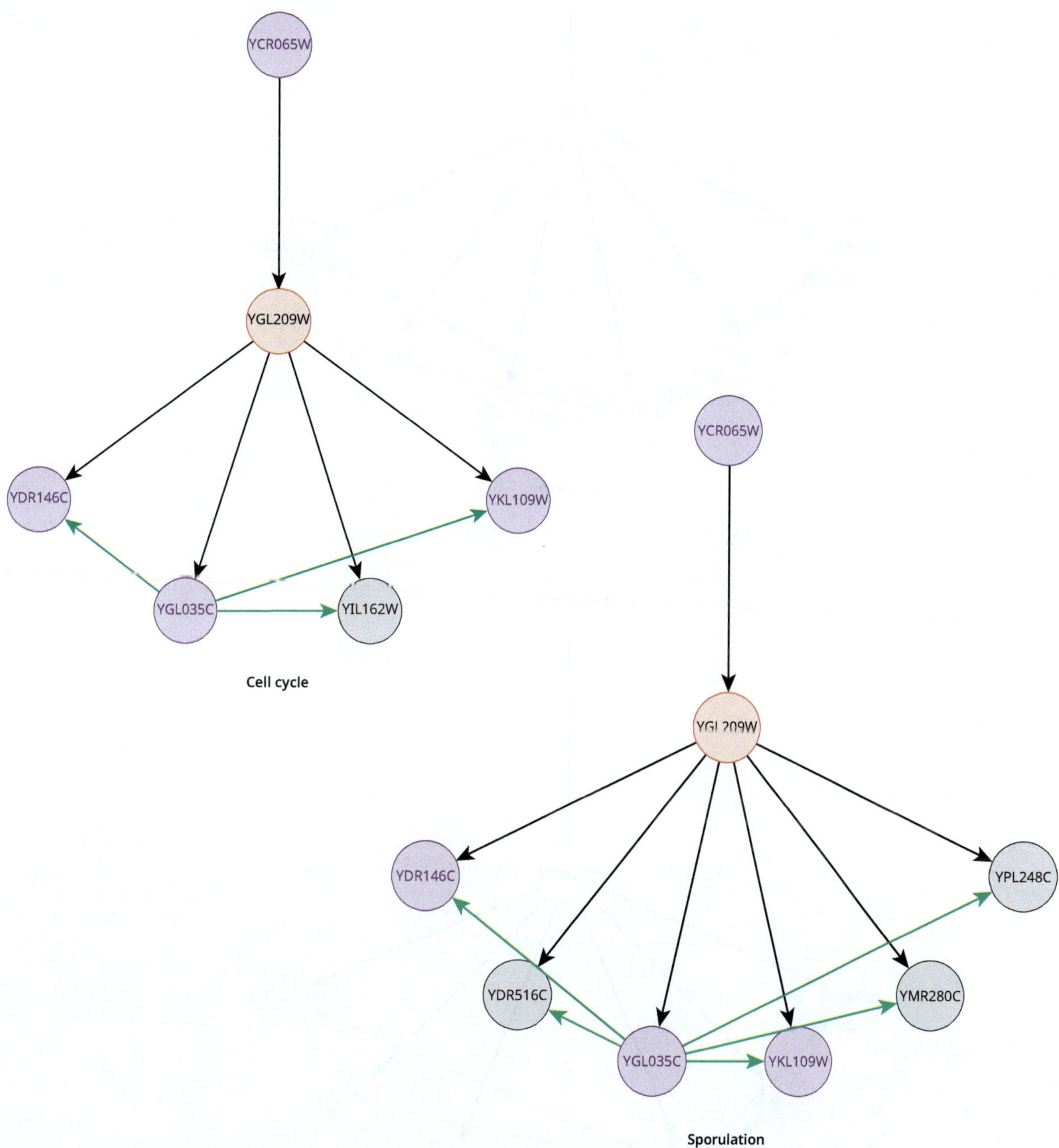

Figure 14.33 1-neighbourhoods of YGL209W. (a) Comparison of cell cycle and sporulation networks.

Used with permission of Oxford University Press, from Lesk, A.M., & Konagurthu, A.S. (2020). Neighbourhoods in the yeast regulatory network in different physiological states. Bioinformatics 37, 4: 551–558. https://doi.org/10.1093/bioinformatics/btaa831. Permission conveyed through Copyright Clearance Centre, Inc.

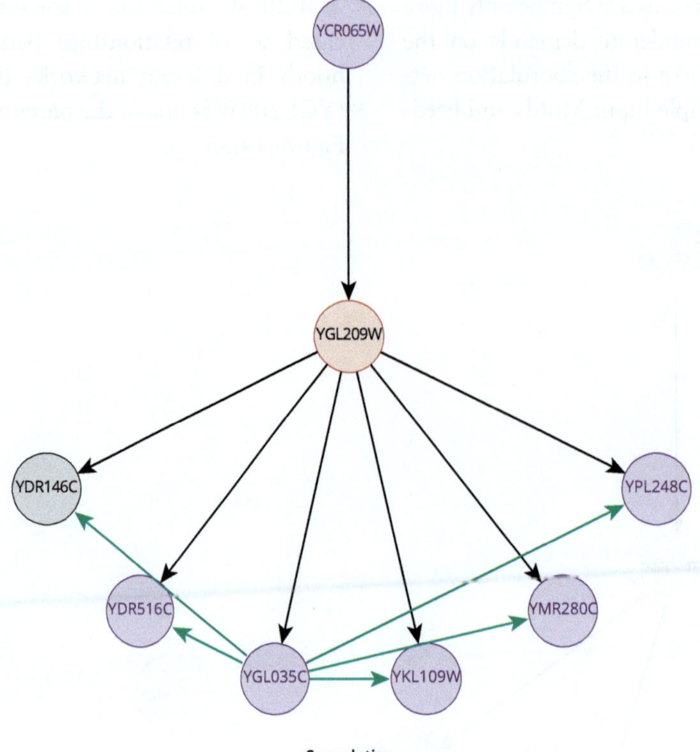

Sporulation

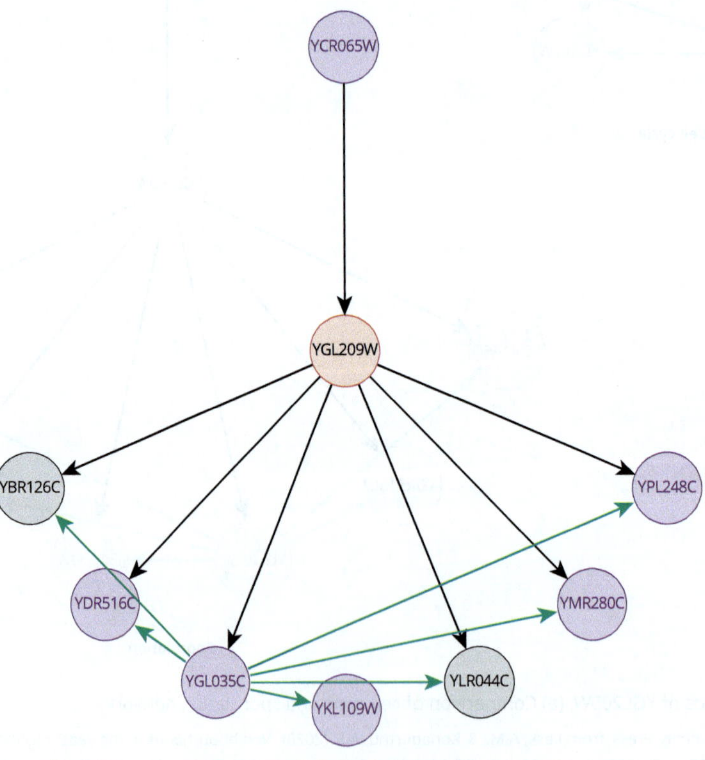

Stress response

Figure 14.33 (b) Comparison of sporulation and stress response networks. (The sporulation 1-neighbourhood is repeated because the colour patterns are different.)

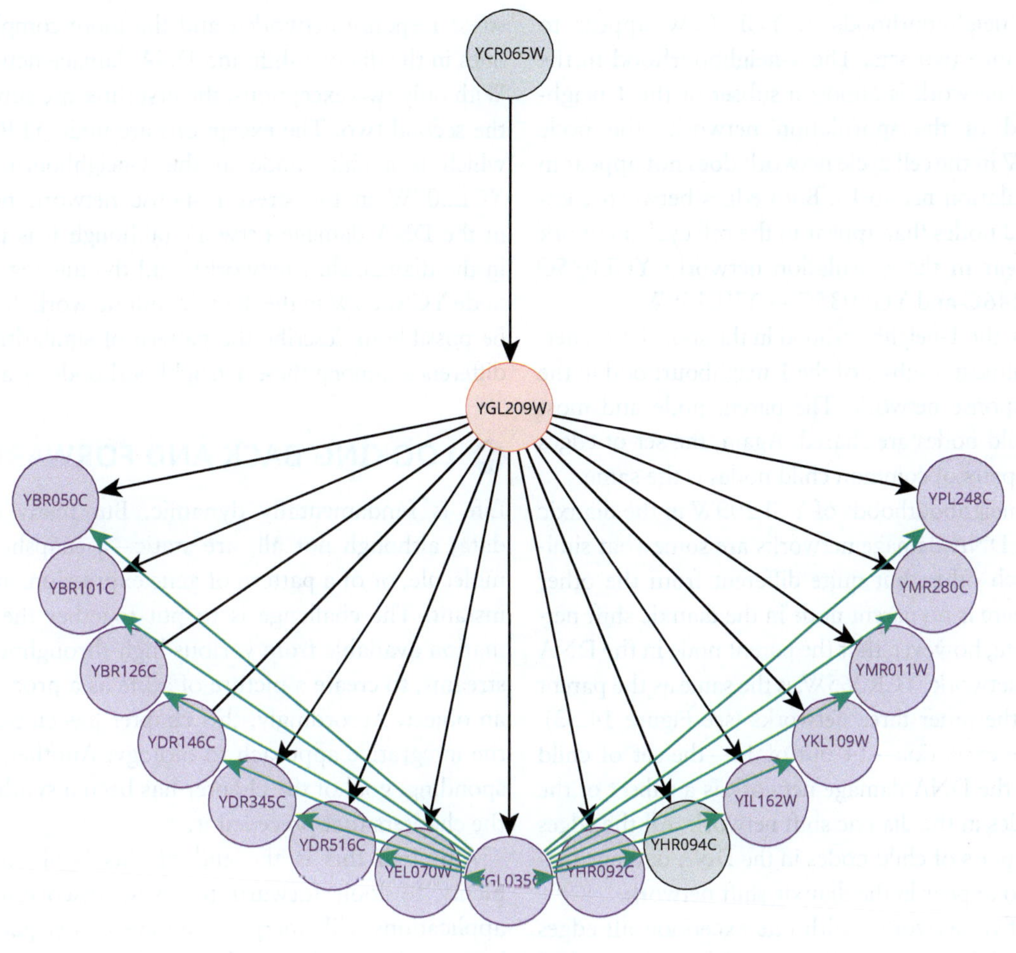

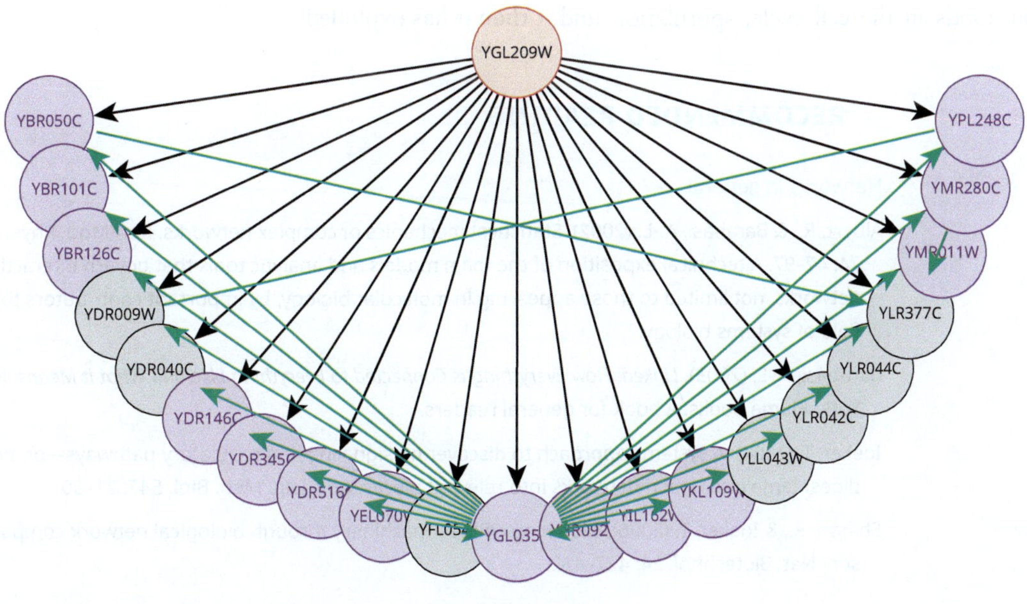

Figure 14.33 (c) Comparison of DNA-damage and diauxic-shift networks.

The 1-neighbourhoods of YGL209W appear to separate into two sets. The 1-neighbourhood in the cell cycle network is almost a subset of the 1-neighbourhood of the sporulation network (the node YIL162W in the cell cycle network does not appear in the sporulation network). Both edges between common child nodes that appear in the cell cycle network also appear in the sporulation network: YGL035C → YDR146C and YGL035C → YKL109W.

In turn, the 1-neighbourhood in the sporulation network is almost a subset of the 1-neighbourhood in the stress response network: The parent node and most of the child nodes are shared. Again, the set of edges between pairs of common child nodes is the same.

The 1-neighbourhoods of YGL209W in the diauxic shift and DNA damage networks are somewhat similar to each other, but quite different from the other three. There is no parent node in the diauxic shift network; note, however, that the parent node in the DNA damage network, YCR065W, is the same as the parent node in the other three networks (see Figure 14.33). With one exception—14 out of 15—the set of child nodes in the DNA damage network is a subset of the child nodes in the diauxic shift network. All the edges between pairs of child nodes in the DNA damage network also appear in the diauxic shift network.

In all five networks, with one exception all edges between child nodes of YGL209W have YGL035C as the parent node.

What is the relationship between the simpler 1-neighbourhoods in the cell cycle, sporulation and stress response networks, and the more complicated ones in the diauxic shift and DNA damage networks? With only two exceptions, the first three are subsets of the second two. The exceptions are node YLR044C, which is a child node in the 1-neighbourhood of YGL209W in the stress response network but not in the DNA damage network (although it is present in the diauxic shift network), and the missing parent node YCR065W in the diauxic shift network. It would be possible to describe the pattern of similarities and differences among these 1-neighbourhoods by a tree.

LOOKING BACK AND FORWARD

Life is fundamentally dynamic. But many of the data, although not all, are static—a snapshot of a molecule, or of a pattern of gene expression, at some instant. The challenge is to put together the information available from various high-throughput data streams, to create a picture of a life as a process, not an object. Accordingly, this chapter has emphasized the integrative approach to biology. Another, corresponding, goal of the chapter has been a synthesis of the chapters that preceded it.

Although this is the end of this book, there is plenty to look forward to. New discoveries and applications will emerge, at an ever-faster pace. The hope is that readers will appreciate them not only with intellectual satisfaction, but with a sense of wonder. Science has discovered many more miracles than it has exploded.

RECOMMENDED READING

Networks in general

Albert, R., & Barabási, A.-L. (2002). Statistical mechanics of complex networks. Rev. Mod. Phys. 74: 47–97. A technical exposition of the main models and analytic tools that govern interaction networks, not limited to those appearing in molecular biology, by important contributors to the field of systems biology.

Barabási, A.-L. (2003). *Linked: How Everything is Connected to Everything Else and What it Means.* New York: Plume Books. A book for general readers.

Ideker, T. (2004). A systems approach to discovering signaling and regulatory pathways—or, how to digest large interaction networks into relevant pieces. Adv. Exp. Med. Biol. 547: 21–30.

Sharan, R., & Ideker, T. (2006). Modeling cellular machinery through biological network comparison. Nat. Biotechnol. 24: 427–433.

Zupanic, A., Bernstein, H.C., & Heiland, I. (2020). Systems biology: Current status and challenges. Cell Mol. Life Sci. 77: 379–380. (Introduction to an entire issue containing articles on this topic.)

Milano, M., Agapito, G., & Cannataro, M. (2022). Challenges and limitations of biological network analysis. BioTech (Basel) 11: 24.

Vasina, M. (2022). Networks and systems biology in bioinformatics. J. Res. Dev. 10: 199.

Koutrouli, M., Karatzas, E., Paez-Espino, D., *et al.* (2023). A guide to conquer the biological network era using graph theory. Front. Bioeng. Biotechnol. 8: 34. (Corrigendum in Front. Bioeng. Biotechnol. 11: 1182500.)

Systems biology of important biological networks

Babu, M.M., Luscombe, N.M., Aravind, L., *et al.* (2004). Structure and evolution of transcriptional regulatory networks. Curr. Opin. Struct. Biol. 14: 283–291. A review article including information about how networks evolve.

Bechhoefer, J. (2005). Feedback for physicists: A tutorial essay on control. Rev. Mod. Phys. 77: 783–836. Description of the theoretical background of network function.

Azeloglu, E.U., & Iyengar, R. (2015). Signaling networks: Information flow, computation, and decision making. Cold Spring Harb. Perspect. Biol. 7: a005934.

Palsson, B.O. (2015). *Systems Biology: Constraint-based Reconstruction and Analysis*. Cambridge: Cambridge University Press.

Charitou, T., Bryan, K., & Lynn, D.J. (2016). Using biological networks to integrate, visualize and analyze genomics data. Genet. Sel. Evol. 48: 27.

Emmert-Streib, F., Dehmer, M., & Shi, Y. (2016). Fifty years of graph matching, network alignment and network comparison. Information Sciences 346–347: 180–197.

Milano, M., Agapito, G., & Cannataro, M. (2022). Challenges and limitations of biological network analysis. BioTech (Basel) 11: 24.

Evolution of networks

Wagner, A. (2005). *Robustness and Evolvability in Living Systems*. Princeton, NJ: Princeton University Press. (*Anything* by Wagner is worth reading.)

Ferrada, E., & Wagner, A. (2010). Evolutionary innovations and the organization of protein functions in genotype space. PLoS ONE 5: e14172.

Rodrigues, J., & Wagner, A. (2009). Evolutionary plasticity and innovations in complex metabolic reaction networks. PloS Computational Biology 5: e1000613.

Wagner, A. (2009). Networks in molecular evolution. In: R.A. Meyers (ed.), pp. 5655–5667. *Encyclopedia of Complexity and System Science*. Heidelberg: Springer.

Payne, J.L., Moore, J.H., & Wagner, A. (2014). Robustness, evolvability, and the logic of genetic regulation. Artificial Life 20: 111–126.

Payne, J.L., & Wagner, A. (2015). Function does not follow form in gene regulatory circuits. Sci. Reports 5: 13015.

Thompson, D., Regev, A., & Roy, S. (2015). Comparative analysis of gene regulatory networks: From network reconstruction to evolution. Annu. Rev. Cell Dev. Biol. 31: 399–428.

Aziz, M.F., & Caetano-Anollés, G. (2021). Evolution of networks of protein domain organization. Sci. Rep. 8: 12075.

EXERCISES, PROBLEMS, AND WEBLEMS

Exercises

Exercise 14.1 For which of the methods for determining interacting proteins ('Protein interaction networks') (a) must one of the proteins be purified and (b) must both of the proteins be purified?

Exercise 14.2 In the undirected, unlabelled graph in Box 14.2, (a) name two vertices such that if you add an edge between them at least one vertex has exactly four neighbours. (Note that two edges may cross without making a new vertex at their point of intersection.) (b) Name two vertices such that if you add an edge between them to the graph, the connected subgraph remains a tree. (c) Name two vertices (neither of them V_1) such that if you add an edge between them to the graph, the connected subgraph does not remain a tree. (d) Name two vertices such that if you add an edge between them to the original graph, there are alternative paths, of lengths 3 and 4, between V_1 and V_5, with no vertices repeated. (In determining the length of a path, you must count the initial and final vertices. A path of length 3 between V_1 and V_5 contains one intermediate vertex.) (e) Name two vertices such that if you add an edge between them to the original graph, there is exactly one path between V_1 and V_5, with no vertices repeated, and it has length 4.

Exercise 14.3 Of the examples of graphs in Box 14.3, (a) Which are directed graphs? (b) Which are labelled graphs? (c) In each example, what is the set of nodes? (d) In each example, what is the set of edges?

Exercise 14.4 In the London Underground: (a) What is the shortest path between Moorgate and Embankment stations? Note that, considered as a graph, the shortest path between two nodes is the path with the fewest intervening nodes, not the path that would take the minimal time or fewest interchanges, or be closest geographically. (b) What is the shortest cycle containing Kings Cross, Holborn, and Oxford Circus stations? (c) The clustering coefficient of a node in a graph is defined as follows. Suppose the node has k neighbours. The total possible connections between the neighbours is $k(k-1)/2$. The clustering coefficient is the observed number of neighbours divided by this maximum potential number of neighbours. If the neighbours of a station are the other stations that can be reached without passing through any intervening stations, what is the clustering coefficient of the Oxford Circus station? (If necessary, see this article on the BBC website. http://www.bbc.co.uk/london/travel/downloads/tubem.map.html.)

Exercise 14.5 In the London Underground: (a) what is the maximum path length between any two stations? That is, for which two stations does the shortest trip between them involve the maximum number of intervening stops? (b) If the District Line were not active, what stations, if any, would be inaccessible by underground? (c) If the Jubilee line were not active, what stations, if any, would be inaccessible by underground?

Exercise 14.6 On a photocopy of the three common network control motifs (Box 14.9) (a) indicate which nodes are controlled by only *one* upstream node; (b) indicate which node exerts control over only *one* downstream node.

Exercise 14.7 On a photocopy of the simplified fragment of the yeast regulatory network (Figure 14.30) indicate examples of the network control motifs (a) fork and (b) 'one-two punch'. (c) Add one arrow to create a scatter motif.

Exercise 14.8 In the overall yeast transcriptional regulatory network, the number of incoming connections to target genes follows an exponential distribution, that is, the probability that a gene is controlled by k transcriptional regulators is proportional to $e^{\beta k}$ with $\beta = 0.8$, $k = 1,2,\ldots$. What is the ratio of the number of target genes receiving four input connections to the number receiving two input connections?

Exercise 14.9 The binding site for λ cro ('Some protein–DNA complexes that regulate gene transcription') is an approximate palindrome, i.e., the two strands contain approximately the same sequence in reverse order. (a) On a copy of the binding site, indicate which six residues best fit the palindrome pattern. (b) A palindromic binding site can interact with a dimeric protein by presenting surfaces of similar structure to both protein subunits. How far apart are the six-residue regions identified in part (a)? How do you rationalize that distance in terms of features of the structure of DNA?

Exercise 14.10 Redraw Figure 14.27 with the top boxes containing glucose absent and lactose absent, instead of glucose present and lactose present.

Exercise 14.11 Match mutant to phenotype:

Dysfunctional gene	Resulting phenotypes
1. *lacI*	(a) Operon expressed, no lactose uptake
2. Mutation in repressor binding site	(b) Operon expressed, no glucose or galactose produced
3. Mutation in RNA polymerase binding site	(c) No expression
4. *lacZ*	(d) Constitutive expression of operon because no repression is possible
5. *lacY*	

Exercise 14.12 In Figure 14.33, (a) identify three nodes that form a fork motif; (b) identify four nodes that form a scatter, or multiple-input motif; and (c) identify three nodes that form a feed-forward loop, or one-two punch motif.

Problems

Problem 14.1 In the map of the London Underground, what is the distribution of numbers of neighbours of vertices?

Problem 14.2 Analyse the map of the London Underground by counting the number of connections made from each station in zone 1 (the central portion). Count connections *to* stations inside and outside zone 1 as long as they originate within zone 1. Count only one connection if two stations are connected by more than one line; in other words, for each station, the question is: how many other stations can be reached without passing through any intermediate stops? (a) What is the maximum number of connections of any station? (b) For each integer k from 1 to this maximum number, how many stations have k connections? (c) Plot these data on a log–log plot. Does the relationship appear reasonably linear? (d) If so, fit a straight line to the log–log plot and determine the exponent. Results of network analysis of this sort are more significant if the data cover several orders of magnitude, but this is not possible for this example.

Problem 14.3 In the overall yeast transcriptional regulatory network, the number of incoming connections to nodes follows an exponential distribution. That is, the probability P_k that a gene is controlled by k transcription regulators is given by $P_k = Ce^{-\beta k}$, $k = 1, 2, \ldots$, with $\beta = 0.8$. (a) Determine the constant of proportionality C in terms of β, by summing the series $\sum_{k=1}^{\infty} Ce^{-\beta k} = 1$.

(b) If $\beta = 0.8$, what is the maximum value of k for which at least 1% of the nodes would be expected to have at least k incoming connections? (c) If $\beta = 0.8$, plot the expected histogram for $1 \le k \le 7$. (d) Determine the mean value of k in terms of β. Hint: In the solution of (a) you expressed $\sum_{k=1}^{\infty} e^{-\beta k}$ as a function $f(\beta)$. Differentiate this relationship with respect to β to produce the equation: $\sum_{k=1}^{\infty} ke^{-\beta k} = f'(\beta)$. Then the mean value of k is given by $-f'(\beta)/f(\beta)$. (e) What is the mean value $\langle k \rangle$

corresponding to $\beta = 0.8$? (f) What is the median value of k? This is the value m such that half the nodes have $\leq m$ incoming connections, and half the nodes have $\geq m$ incoming connections. Find m in terms of β. Hint: if $\Sigma_{k=1}^{\infty} Ce^{-\beta k} = 1$ then $\Sigma_{k=m+1}^{\infty} Ce^{-\beta k} = \frac{1}{2}$, but $\Sigma_{k=m+1}^{\infty} Ce^{-\beta k} = e^{\beta m} \Sigma_{k=1}^{\infty} Ce^{-\beta k}$. In general, this approach will provide a non-integral estimate of m; just round this result to the nearest integer. (g) If $\beta = 0.8$, what is the median value m? How does it compare with the average value $\langle k \rangle$? Are the two values approximately equal?

Problem 14.4 Indicate how to connect a selection of the three common network control motifs so that a single input node can influence three output nodes.

Problem 14.5 (a) Citrate synthase and (b) pyruvate decarboxylase are two of the enzymes that change expression level in the diauxic shift in yeast. On a photocopy of Figure 14.28, mark the approximate positions of the reactions that they catalyse.

Weblems

Weblem 14.1 Find the sequences and structures of the two classes of *E. coli* fructose-1,6-bisphosphate aldolases. (For one of them there is at time of writing no experimental structure and you will have to use the AlphaFold2 prediction.) Align the sequences using (a) sequence-based techniques and (b) structure alignment. Describe the similarity of these two proteins. Does the evidence support their being homologues?

Weblem 14.2 Figure 14.19 showed the structure of a plant-specific transcription regulator NAC domain, binding DNA [3SWP]. Two other families of transcription factors are the GCM (glial cells missing) and WKRY (from a conserved sequence motif). wwPDB entry [1ODH] is an example of the GCM family and [2LEX] is an example of the WKRY family. Compare the three structures—[3SWP], [1ODH], and [2LEX]—by aligning the sequences, doing structural superpositions, looking for common ligand-binding sites, and comparing the nature of their interactions with DNA. Write a report stating the evidence in support of the idea that these three domains are homologous, and stating any reservations to this conclusion suggested by your analysis.

Weblem 14.3 Find one or more examples of a set of proteins that are multidomain proteins in prokaryotes but for which the domains are individually homologous to separate proteins in eukaryotes. (The opposite: multidomain in eukaryote, individual protein in prokaryote, is much more common.)

EPILOGUE

The past decade has already seen major achievements in genomics, in particular, high-throughput and accurate whole-genome sequencing. The human genome—the jewel in the crown—was completed in 2022. And yet, the field is still in an anticipatory stage. We can be confident that:

- Methods of sequence determination will continue to increase in power. The explosion in the number of complete sequences of many different human beings and of other organisms will accelerate. It is likely that within the lifetimes of many readers genomic sequencing will be near-universal.
- Tools for analysis will also show progress. Better algorithms, making effective use of the very large data sets, will produce more reliable inferences. Methods based on artificial intelligence will yield breakthrough results in many areas, as they have in protein structure prediction. Simulations will model larger-scale processes, towards the goal of a complete cell *in silicio*.
- Applications to medicine will become more successful and more numerous, and fluently make the transition from research laboratory to clinic. Personalized medicine, based on genome sequences and other molecular data, will transform the response to disease.
- We will achieve a more profound understanding of what life is and how it works.

As a result, we shall gain very, very powerful control over living systems. Many applications will emerge, in clinical, agricultural, and technological fields. Some of these are relatively simple extrapolations from what has already been achieved. Others seem more like the stuff of science fiction: a tightly coupled silicon–life interface; the *in vivo* deployment of nanoparticles sensing and interacting with our biochemical states; designer organisms—involving correcting mutations or even 'cosmetic genome engineering'. Nevertheless, they represent natural developments of the current state of the art.

Understanding the genome may ultimately release us from its constraints.

GLOSSARY

3D Crunch collection of homology models of proteins of known sequence

α_1-antitrypsin a serine protease inhibitor protecting the lungs against damage from elastase

α-helix one common hydrogen-bonded structure of a local region of the backbone of a protein

active site specific location on a protein that is the focus of biological activity; for instance, a crevice in the surface of an enzyme that binds substrate specifically and juxtaposes catalytic residues to the substrate molecule or molecules and cofactors

acylphosphatase a small enzyme containing α-helices and β-sheets

adaptor hypothesis the conjecture that there must be molecules that mediate information transfer between messenger RNA and the ribosome; discovery of tRNA proved the truth of the conjecture

aerobic metabolism in the presence of oxygen

affinity measure of the tenacity with which two or more molecules stick together

aggregate a grouping of objects; protein aggregates are often insoluble, arise under unnatural conditions, and are associated with diseases

algorithm a computational procedure for a calculation or other analysis

alignment the assignment of residue–residue correspondences in protein or nucleic-acid sequences

aliphatic having the properties of non-aromatic hydrocarbons

allele one possible sequence at any genetic locus

allopolyploid an individual with cells containing chromosomes derived from different species

allosteric a change in structure of one part of a protein affecting activity at another

AlphaFold2 computer program that successfully predicts protein three-dimensional structure from amino-acid sequence, a major breakthrough

Alzheimer's disease a neurodegenerative disease, primarily of the elderly, associated with deposition of protein aggregates in the brain

amino acid one of the constituents of proteins; proteins are polymers of amino acids

amplicon a piece of DNA or RNA of which it is intended to make many copies, usually by PCR

amyloidosis one of a class of diseases characterized by extracellular deposits of protein aggregates

amyloid precursor protein (APP) a component of senile plaques, protein deposits in Alzheimer's disease

anaerobic metabolism in the absence of oxygen

ancestry-informative markers polymorphisms for which allelic frequencies reflect the geographic origin of the population to which the subject belongs

aneuploidy an abnormal number of chromosomes, one or more chromosomes either extra or missing

antennapedia a mutation in a gene in *Drosophila* causing developmental defects

antibody one of a large set of proteins capable of defending the vertebrate body by specific binding to a foreign substance

anticodon a triplet of bases within tRNA that binds on the ribosome to a complementary triplet of bases in messenger RNA

antigen a foreign molecule that elicits production of antibodies that bind it specifically

apoptosis programmed cell death

archaea one of the basic divisions of life; includes, but is not limited to, numerous simple organisms adapted to unusual environments such as high temperatures

aromatic a molecule containing a set of conjugated double bonds, for instance benzene

aspartate carbamoyltransferase an allosteric protein catalysing the first step in pyrimidine biosynthesis starting from aspartate

assembly merging of sequenced fragments of DNA to form longer contiguous stretch of sequence

atomic resolution a very detailed X-ray structure determination, in which the experimental data unambiguously show the positions of individual atoms

autoimmune disease disease resulting from a body's immune system's reaction to non-foreign proteins

autoradiography exposure of photographic film to distribution of labelled molecules on a gel allowing visualization of their positions

β-barrel a type of assembly of regions in a protein in which strands of β-sheet wrap around into a closed cylindrical structure

β-hairpin two antiparallel strands of a β-sheet connected by a short loop

β-sheet a secondary structure in which strands interact laterally by mainchain hydrogen bonding

backbone the repetitive part of protein structures; a linear chain of amino acids linked by peptide bonds

bacteria one of the basic divisions of prokaryotic life; distributed ubiquitously in nature, responsible for many infectious diseases

bacteriophage virus that infects bacteria

bacteriorhodopsin a molecule related to G-protein-coupled receptors that functions as a light-driven proton pump

Barr body product of degeneration and silencing of one X chromosome in cells of eutherian mammalian females

biological taxonomy classification and naming of groups of organisms on the basis of similarities in shared characteristics

binding change mechanism a description of the mechanism of action of ATPase in which conformational changes cyclically alter the binding properties of different subunits

BioCyc a set of databases of metabolic pathways in different organisms

BioMagResBank a database of macromolecular structures determined by NMR, and associated information

BLAST an approximate but powerful method of searching large databases for sequences similar to a probe sequence

BLOSUM a substitution matrix giving a statistical description of the likelihood of single-site mutations in proteins

bovine spongiform encephalopathy a prion disease affecting cattle

Bragg's law the relationship between scattering angle θ and interplanar spacing within a crystal, d: $n\lambda = 2d \sin \theta$

buried surface area the surface area of atoms in a protein inaccessible to water molecules

Cambridge Crystallographic Data Centre a database of small-molecule crystal structures

capsid assembly of molecules forming the outer shell of a virus

CASP blind test of attempts to predict the structures of proteins

CATH a database of protein structure classification

central dogma 'DNA makes RNA makes protein' (Francis Crick, 1957)

chaotic state a dynamical state in which small changes in initial conditions may produce large changes in subsequent behaviour

chaperone a protein that catalyses protein folding by unfolding misfolded proteins

ChIP-seq combination of chromatin immunoprecipitation (ChIP) with DNA sequencing to identify binding sites of proteins to DNA

CID collision-induced dissociation, a technique for forming ions from a vaporized protein

circadian rhythm natural periodicities in biological activity, deriving from the 24-hour day

CLUSTAL-Omega a program for multiple sequence alignment

codon triplet of nucleotides in the genome that is translated into one amino acid, or stop signal

cofactor a non-amino acid molecule that forms part of a protein structure and assists function

coiled-coil structure in which two or more α-helices wind around each other to form a supercoil

collagen a fibrous protein forming structural material in our bodies

combined paternity index resultant of individual paternity indices from different loci

competitive inhibitor a molecule that interferes with the activity of an enzyme by binding to the active site

conformation spatial disposition of atoms in a molecule

conformational disease a disease arising from misfolding of a protein

connected component a subset of nodes of a graph such that there is a path in the graph between any two nodes

connected graph a graph containing a path between any two nodes

contig the result of merging the sequences of fragments to form a long connected region

Coomassie Blue a stain used for visualizing the positions of proteins on a gel

cooperativity deviation from independence of two or more binding sites on a protein

coverage average number of times each base appears in sequencing fragments

Creutzfeldt–Jakob disease a neurodegenerative disease associated with prion protein

CRISPR/Cas a defence mechanism against viral infection in prokaryotes, developed as a powerful laboratory tool for genome editing in eukaryotes including humans, with clinical applications

cryo-electron microscopy (cryoEM) structure determination by low-temperature electron microscopy

crystal structure atomic positions determined from the diffraction of X-rays by a crystal

DALI distance alignment: a program to align the amino-acid sequences of proteins based on their structures

denaturation loss of native protein structure

density of connections relative number of edges and nodes in a graph

derivative chromosome a structurally rearranged chromosome, the result of swapping material either within a single chromosome or between two or more chromosomes

diauxy transition between fermentative to oxidative metabolism

dihedral angle a quantitative description of the conformation of a molecule in terms of rotations around a bond

directed evolution selection of protein molecules with desired properties by laboratory rather than natural methods

directed graph a graph in which each edge is a 'one-way' street

dissociation constant a measure of the affinity of two molecules

distributed redundancy alternative sources of an activity, rather than duplication of a single source

disulphide bridge a post-translational modification in which two cysteine residues, which may be far apart in the sequence, come together in the structure and form an S—S bond

divergence progressive loss of similarity during evolution

docking prediction of the mode of association of two proteins from knowledge of the separate structures

domain a compact subunit of a protein structure

domain swapping a mechanism of dimer formation in which two proteins exchange regions

dotplot a matrix diagram that gives a quick overview of the similar features of two sequences

dynamic programming the basis of an algorithm for determining the optimal alignment of two sequences

EcoCyc a database of metabolic pathways of *Escherichia coli*

ectopic out of place

effector ligand that controls the state and activity of an allosteric protein

Ehlers–Danlos syndrome an inborn error of collagen formation

electrospray ionization (ESI) a method for vaporizing proteins for mass spectrometry

EMBOSS a collection of sequence-alignment tools

endoreplication replication of the genome in the absence of cell division, leading to polyploidy

ENTREZ a comprehensive database of the National Center for Biotechnology Information (NCBI)

Enzyme Commission (EC) group charged with developing a classification of enzyme function

epigenetic factors modifications of bases in genomes that do not affect DNA sequence but control gene expression

erythrocyte red blood cell

eukaryotes major grouping of life forms, those that have a nucleus

E-value estimate of number of sequences that would be retrieved from a database of random sequences the same size as a database probed

exon portion of a gene that can be translated into protein

expressed sequence tag (EST) short sequence of cDNA reverse transcribed from the terminal region of an mRNA

FAIR (Findability, Accessibility, Interoperability, and Reuse) general principles for scientific data management and presentation

familial Creutzfeld–Jacob disease (CJD) inherited version of Creutzfeld–Jacob disease, usually affecting people 55–75 years of age

fibre diffraction X-ray diffraction of proteins with regular structure in only one dimension

fibrillar deposits intracellular inclusions of fibrous protein aggregates

fibrous proteins proteins that form extended one-dimensional structures; tend to be insoluble and to have structural or mechanical functions

fluorescent in-situ hybridization (FISH) localization of a DNA sequence in a micrograph of chromosomes by binding a fluorescently labelled complementary sequence

flux-control coefficient differential change in the rate of a step of a process by a differential increase in the amount of enzyme catalysing that step

fold recognition the assignment of a protein of known amino-acid sequence but unknown structure to a member of a given library of structures, or concluding that it corresponds to no member of the library

folding pathway the mechanism of the transition from the denatured to the native state of a protein

folding pattern the topology of the three-dimensional structure of a protein

Fourier transform a mathematical relationship, in X-ray crystallography relating the measured data and the electron density, and in NMR relating the time dependence of a signal to its component frequencies

G protein a guanine-nucleotide-binding protein, involved in signal transduction cascades

G-protein-coupled receptor (GPCR) a protein that transmits external signals across a cell membrane, the signal received inside the cell by a G-protein

gap weighting parameter specifying the relative importance of substitutions and insertions/deletions in calculating optimal sequence alignments

gene a bearer of hereditary information

gene drive enhancing the propagation of particular genes through a population by altering the probability of transmitting specific alleles to offspring

gene duplication formation of a second copy of a gene or a portion or all of a genome, after which the copies can diverge in sequence, structure, and/or function, or be deleted

gene expression table the relative amounts of expression of different genes by different subjects

Gene Ontology a catalogue of protein functions, according to: molecular function, cellular component, and biological process

genetic drift evolutionary change not driven by selection

genome browser a website presenting genome sequences and associated information

genome-wide association studies (GWAS) identification of genes related to a trait—often, a disease—by correlating sequences with phenotypes

genomic imprinting epigenetic phenomenon in which the expression level of a gene depends on the parent of origin

genomics determination, analysis, and comparison of the complete genetic material of organisms of different species

Gibbs free energy measure of spontaneity and equilibrium under conditions of constant temperature and pressure

glycoprotein protein with attached sugars as post-translational modifications

graph mathematical structure containing nodes and edges providing useful models of networks

haemophilia disease involving defects in the mechanism of blood clotting

Hamming distance measure of difference between two character strings of the same length by counting numbers of non-identical characters

haplotype a group of closely linked genes that tend to be inherited as a block

helix-turn-helix domain a structural motif common in DNA-binding proteins

heterologous binding a protein–protein complex in which different monomers contribute different sets of residues to the site of interaction

high mobility group chromosomal proteins that regulate transcription

homeotic gene a gene responsible for determination of body plan

homologues two genes, proteins, or other biological objects related by evolution

homology modelling prediction of the structure of a protein from the structures of one or more related proteins

hub a node with many edges in a graph

humanized antibody an engineered antibody with CDRs from an antibody from a different species grafted into a human antibody framework

Huntington's disease a hereditary, adult-onset, polyglutamine-repeat disease characterized by movement disorder and dementia

hydrogen bonds weak bonds between polar atoms, mediated by hydrogen atoms

hydrophobic preferring an organic environment to an aqueous one

hydrophobic effect the unfavourable interaction of water with non-polar solutes

hyperthermophiles organisms that can thrive at very high temperatures

induced fit conformational change in a protein to enhance binding affinity

industrial melanism the takeover of a dark-colour allele, for camouflage against a background darkened by industrial pollution

information retrieval selection of items from a database by specification of desired features, for instance keywords

inhibitory causing a diminishment of an activity

insertion sequence mobile genetic elements in prokaryotes related to eukaryotic transposons

intein a protein that autocatalyses the removal of an internal peptide

interaction networks sets of interactions between objects, such as a group of proteins

interface the region of occlusion of two or more proteins in a multimer or aggregate

introns regions of a genome that are transcribed but never normally appear in a mature messenger RNA to be translated into a protein

isoelectric focusing separation of proteins with different isoelectric points by electrophoresis in a pH gradient

isoelectric point the pH at which the net charge on a protein is zero

isologous binding a protein–protein complex in which different monomers contribute the same residues to the site of interaction

junction read a sequence derived from a boundary between two exons in a mature messenger RNA, which cannot be mapped directly to the genome because the intron between them is absent

karyotype the number and appearance of the chromosomes in a eukaryotic cell

KEGG Kyoto Encyclopedia of Genes and Genomes, database dealing with genomes, biological pathways, diseases, drugs, and chemical substances

keratins fibrous proteins that are major constituents of hair, wool, and feathers

lac operon region of *E. coli* genome, the expression of which is responsive to presence of lactose and glucose in medium

Lesch–Nyhan syndrome a genetic disorder in the *HPRT1* gene leading to excess uric acid; effects include self-injury

Levenstein distance measure of difference between two character strings not necessarily of the same length by counting the minimum number of edit operations (substitution, insertion, deletion) required to transform one string into the other

life history integrated total experience, including physical and psychological environment

linkage the appearance of two genes on the same chromosome, leading to co-inheritance of alleles

linkage disequilibrium non-random association of alleles at different loci

local matching alignment of restricted subsequences

long-term potentiation increase in strength of synaptic connections between neurons, produced by frequent activation

lysis destruction of a cell, for example by bursting out of the products of a viral infection

lysogenic-lytic switch in bacteriophage λ change from quiescent integration into host genome, to active replication and bursting the host cell

lysogeny incorporation of bacteriophage genome into that of the host cell, the virus thereby entering a quiescent state

lysosomal storage disease rare genetic disorder resulting in intracellular accumulation of undigested molecules

mad-cow disease bovine spongiform encephalopathy, a prion disease

mainchain the common backbone of the polypeptide chain in proteins

major histocompatibility complex a family of proteins that present peptides to mediate the immune response

Marfan syndrome an inborn error of collagen formation

mass spectrometry identification or partial sequencing of proteins by measurements of masses of ions into which they are cleaved

matrix-assisted laser desorption ionization (MALDI) a 'soft' method for vaporizing proteins

MEROPS a database of proteinases

messenger RNA transcript of protein-coding gene which after processing is translated by a ribosome into amino-acid sequence of a protein

metagenomics comprehensive sequencing of organisms in an environmental sample

MHC protein protein of the immune system governing an individual's sensitivity to different antigens

Michaelis–Menten model general although simplified model of enzymatic catalysis, leading to useful description of dependence of initial enzyme-catalysed reaction rate on substrate concentration

Michaelis complex initial interaction between enzyme and substrate, before reaction occurs

microarray device to measure gene transcription patterns, or genotypes, by hybridization to a fixed set of oligonucleotides

microbiome collection of all microbes living on and in the body of a higher organism

mmHg unit of pressure: 760 mmHg = 1 atm = 101 325 Pa

MODBASE database of homology models of proteins

MODELLER a program to predict the structure of a protein knowing the structures of one or more close relatives

modular protein a protein containing a concatenation of regions, each of which folds into a compact unit (domain)

molecular dynamics computational simulation of the motion of atoms in a protein or other molecule

multifactorial disease a disease that is not transmitted by simple Mendelian inheritance

multiple sequence alignment determination of the correspondences among residues in more than two homologous sequences

NAD-binding domain a common folding pattern containing six β-α-β units, appearing in the NAD-binding domains of dehydrogenases

native state the compact, active conformation of a protein

neurofibrillary tangle aggregates of hyperphosphorylated tau protein, markers of Alzheimer's and other diseases

neutral evolution evolutionary change not driven by selection

northern blot detection, by hybridization probes, of RNA molecules that have been separated by electrophoresis

nuclear magnetic resonance spectroscopy measurement of structural information from transitions between nuclear spin energy states

nucleic acid structure databank (NDB) a database archiving, curating, and distributing nucleic acid structures and related information

Online Mendelian Inheritance in Man (OMIM™) a database linking diseases with sites in the human genome

open access free online availability of academic information, including but not limited to articles in scientific journals

open reading frame (ORF) a region of a genome that putatively codes for a protein

orthologues related proteins in different species

paired-end read sequencing of regions from both ends of a DNA fragment

palindrome a character string that is identical with its reverse, for instance 'MAXISTAYAWAYATSIXAM'; for nucleic acids, a sequence complementary to its reversal

paralogues related proteins within the same species

Parkinson's disease neurodegenerative disease common in the elderly, characterized by effects on movement; caused by reduction in dopamine levels in the brain as a result of neuronal death

paternity index strength of inference of paternity depending on the frequency of a VNTR length in the population

penetrance fraction of individuals bearing a gene for a trait who actually show the expected phenotype

peptide bond the covalent bond linking successive amino acids in a protein

peptide mass fingerprint a set of masses of fragments, measured by mass spectrometry, useful in identifying proteins

pharmacogenomics tailoring of drug treatment to the genotype of a patient

phase problem the loss of information essential for structure determination in measuring X-ray diffraction patterns

phenylketonuria (PKU) a genetic defect in which the enzyme that converts phenylalanine to tyrosine is dysfunctional, leading to accumulation of toxic amounts of phenylalanine and metabolic byproducts

phred score measure of the expected error rate of a DNA sequencing method

phylogeny ancestor–descendant relationships

phylogenetic tree set of ancestor–descendant relationships among a set of species or proteins, diagrammed as a tree structure

plasmid a small extra segment of DNA, often circular, external to the normal chromosome

plastids intracellular organelles in plants, including but not limited to chloroplasts

polyacrylamide gel electrophoresis (PAGE) separation of molecules according to mobility on a polymer sheet

polymerase chain reaction technique for amplifying selected DNA sequence, even if the target sequence is a very minor component of a sample

polypeptide chain the linkage of amino acids in proteins

polyploid containing more than one copy of a set of chromosomes

polyteny large chromosomes containing many attached chromatids, formed by repeated DNA replication

porphyrin ring heterocyclic organic ring system containing four pyrrole groups; the haem group found in globins, cytochrome c, and certain other proteins contain an iron-bound porphyrin

positive selection increase in allele frequency in a population as a result of selective pressure

post-translational modification chemical change in a protein after synthesis by the ribosome

primary structure the covalent bonds of a protein

prion protein that, if converted to a dangerous form, can cause disease

prion diseases infectious diseases associated with prions, unusual in that the transmissible element contains only protein and no nucleic acid

profile the distribution of different bases, or amino acids, at successive positions in a sequence

prokaryote a relatively simple form of life in which cells lack a nucleus

proteasome a large protein aggregate responsible for selective proteolysis

protein folding occurs spontaneously directed solely by amino-acid sequence

Protein Identification Resource (PIR) the protein sequence database at the National Biomedical Research Foundation, Georgetown University, US

proteome the set of proteins synthesized by a cell

proteomics measurement of the spatial and temporal distribution of protein expression patterns and interactions

PSI-BLAST program for a fast and powerful search of a sequence database for entries similar to a probe sequence, based on iterated compilation of a profile as a search object

PUBMED a database of the scientific literature

purifying selection removal of deleterious alleles as a result of selection

quaternary structure the assembly of a multichain protein

quorum sensing communication between microorganisms by secretion and mutual detection of a 'message' compound

read length the number of bases from a region of DNA determined together during sequencing

reading frame one of the ways to 'parse' a region of DNA into successive triplets, which may potentially code for protein

regulatory networks networks of control interactions; for instance, networks that control gene transcription

relaxed (R) state a state of haemoglobin in which oxygen affinity is not perturbed

renaturation reformation of the native state of a protein after denaturation

resolution measure of the amount of data collected in a protein structure determination by X-ray crystallography, limiting the ultimate quality of the structure determination, and usually based on the degree of identity of different unit cells of a crystal

restriction endonuclease an enzyme that cuts DNA at a specific nucleotide sequence

restriction fragment DNA segment resulting from cleavage by a restriction enzyme

restriction fragment length polymorphism (RFLP) individual variation in lengths of restriction fragments, useful in personal identification

retinoblastoma a form of cancer of the eye, the propensity for which is associated with a known oncogene

retrovirus a virus that can insert its genetic material into the host genome

R-factor a measure of the agreement between a molecular model and the X-ray diffraction data on which it is based

Rh incompatibility reaction of immune system of a mother to a particular foreign antigen present in a foetus

ribosome the large ribonucleoprotein molecular machine that synthesizes proteins at the direction of messenger RNA sequences

ribozyme an RNA molecule with catalytic activity

RNA editing modification of base sequence of messenger RNA between transcription from DNA and translation by ribosome

RNAseq sequencing of RNA molecules in a cell, either directly or by reverse transcription and sequencing of the resulting DNA

root-mean-square deviation measure of structural similarity—the average distance between corresponding atoms after optimal superposition

ROSETTA program for the prediction of protein structure from an amino-acid sequence

salt bridge electrostatic attraction between oppositely charged sidechains

salting out precipitation of proteins from solution by high salt concentrations

sarcomere the structural unit of a muscle fibre

SARS-CoV-2 virus responsible for COVID-19

Sasisekharan–Ramakrishnan–Ramachandran diagram plot of φ and ψ angles of protein mainchains, indicating sterically allowed regions

scale-free network a network that retains a similar connectivity distribution as the network grows by random addition of nodes and edges

SCOP Structural Classification of Proteins database

scrapie a prion disease of sheep

SDS-PAGE polyacrylamide gel electrophoresis of a sample of proteins treated with the denaturing detergent sodium dodecyl sulphate

secondary structure hydrogen-bonded mainchain conformations, including α-helices and β-sheets, common to many proteins

secondary structure prediction prediction of positions of α-helices and strands of β-sheet in proteins from the amino-acid sequence

senile plaques dense insoluble extracellular deposits associated with Alzheimer's disease

serpin a serine proteinase inhibitor or other protein related to α_1-antitrypsin

short tandem repeat polymorphism (STRP) variation in number of repeats in genomic regions comprising multiple successive copies of short (1–6 bp) sequences

shortest path minimal set of nodes from a graph, connected by edges between successive nodes, that links two selected nodes

sickle-cell disease a disease caused by a mutation in haemoglobin and resulting in polymerization of deoxyhaemoglobin within the red blood cell

sidechain set of atoms attached to a residue of a protein mainchain

signal transduction cascade a succession of control interactions

SILAC (stable isotope labelling with amino acids in cell culture) method for accurate determination of absolute abundances of proteins in a sample

similarity resemblance of some particular feature, independent of relationship

SINE short interspersed nuclear element, repeat element in human and other genomes

single-end read sequencing of region from only one end of a DNA fragment

single-nucleotide polymorphisms (SNPs) base changes at single sites

small-world network a type of network characterized by short path lengths between most pairs of nodes

sodium dodecyl sulphate (SDS) a detergent that denatures proteins and coats them with a uniform layer of negative charge

Southern blot detection, by hybridization probes, of DNA molecules that have been separated by electrophoresis

spliceosome a large macromolecular nucleoprotein complex responsible for removing introns from RNA

split decomposition presentation of the structure of a graph as a more general network than a tree

spongiform encephalopathies progressive degenerative neurological diseases characterized by cell death resulting from conformational change and aggregation of prion protein

spontaneous folding the adoption of the native state of proteins based exclusively on amino-acid sequence

SSearch program to align a shorter sequence within a longer one; that is, 'local alignment'

steady state non-equilibrium condition in which the amounts of certain substances remain constant

stimulatory causing an increase in an activity

strange attractor dynamical state of a chaotic system that gives some appearance of stability

structural alignment determination of residue–residue correspondences by spatial superposition

structural genomics goal of determining enough crystal structures of proteins to permit at least rough homology modelling of almost all proteins from their sequences

substitutional redundancy possessing multiple copies of a molecule providing a function

superfamily group of proteins with enough similarity to conjecture that they are related, but not enough to prove it

superposition bringing two sets of points into optimal spatial correspondence

supersecondary structure a local structure in proteins involving standard assemblies of α-helices and strands of β-sheet among regions adjacent in the sequence

Support Vector Machine supervised learning algorithm used for classification, based on finding boundaries maximizing separation of classes

SWISS-MODEL publicly available homology-modelling server

SWISS-PROT a database of protein sequences and related information

synaptic pruning feature of nervous system development in which synapses are eliminated

synchrotron a particle accelerator capable of providing a very bright, tunable, X-ray source

synonymous mutation a mutation that does not change the amino acid encoded

syntenic blocks regions of different chromosomes showing large-scale similarity in gene distribution

TATA box sequence upstream of the transcriptional start site of bacterial genes

tauopathies a group of diseases involving aggregates of tau protein

T-cell receptor (TCR) an immune-system protein that distinguishes self from non-self by binding to MHC–peptide complexes

T-COFFEE a program for multiple sequence alignment

tense (T) state a state of haemoglobin in which oxygen affinity is reduced

tertiary structure the assembly of helices and sheets in a protein domain

thalassaemias blood disorders arising from mutations producing reduced haemoglobin levels

thermophile an organism adapted to growth at high temperature

thrombosis formation of a blood clot in an undamaged blood vessel

tiling array a microarray containing overlapping fragments covering a long region or even an entire genome

transcriptome the set of RNA molecules transcribed from a genome

transition a mutation in which a purine (A or G) is replaced by another purine, or a pyrimidine (T or C) by another pyrimidine

translocation moving of a gene, or larger section of a chromosome, from one place in the genome to another

transversion a single-site mutation in which a purine (A or G) is replaced by a pyrimidine (T or C), or a pyrimidine is replaced by a purine

tree a graph in which there is only one path between any two nodes

TrEMBL a database of amino-acid sequence translations of gene sequences from EMBLBank

tumour suppressor a gene that protects against cancer; often involved in DNA repair. Mutations in tumour suppressor genes increase risk of cancer

twenty standard amino acids the set of amino acids specified in the standard genetic code that are translated into proteins

UniProt database integrating knowledge about proteins

Van der Waals forces weak, distance-dependent forces between all atoms

variable expressivity the range of severity of phenotypic manifestation of the same genotype

variable number of tandem repeats (VNTRs) length differences in regions of DNA containing multiple repetitions of a single short sequence

variable splicing formation of more than one mature messenger RNA from an initial transcript by including different sets of exons

variant Creutzfeld–Jacob disease (vCJD) type of Creutzfeld–Jacob disease appearing in people in their 20s in the absence of familial pattern

virulence factor extracellular products released by bacterial cells contributing to pathogenesis

whole-exome sequencing sequencing the regions of a genome that code for proteins

Worldwide Protein Data Bank (wwPDB) organization responsible for archiving, curating, and distributing structural information about biological macromolecules, and providing specialized information-retrieval tools for access to those data

X-ray crystallography a method of structure determination based on measurements of X-ray diffraction by a crystal

zinc finger a type of transcriptional regulator that contains zinc

INDEX

Tables, figures, and boxes are indicated by an italic *t*, *f*, and *b* following the page number.

N

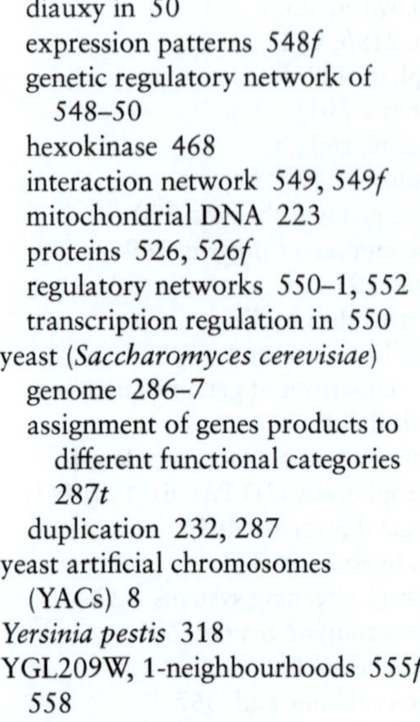